D1573880

Geophysical Monograph Series

Including

IUGG Volumes
Maurice Ewing Volumes
Mineral Physics Volumes

Geophysical Monograph Series

83 **Nonlinear Dynamics and Predictability of Geophysical Phenomena (IUGG Volume 18)** *William I. Newman, Andrei Gabrielov, and Donald L. Turcotte (Eds.)*

84 **Solar System Plasmas in Space and Time** *J. Burch, J. H. Waite, Jr. (Eds.)*

85 **The Polar Oceans and Their Role in Shaping the Global Environment** *O. M. Johannessen, R. D. Muench, and J. E. Overland (Eds.)*

86 **Space Plasmas: Coupling Between Small and Medium Scale Processes** *Maha Ashour-Abdalla, Tom Chang, and Paul Dusenbery (Eds.)*

87 **The Upper Mesosphere and Lower Thermosphere: A Review of Experiment and Theory** *R. M. Johnson and T. L. Killeen (Eds.)*

88 **Active Margins and Marginal Basins of the Western Pacific** *Brian Taylor and James Natland (Eds.)*

89 **Natural and Anthropogenic Influences in Fluvial Geomorphology** *John E. Costa, Andrew J. Miller, Kenneth W. Potter, and Peter R. Wilcock (Eds.)*

90 **Physics of the Magnetopause** *Paul Song, B.U.Ö. Sonnerup, and M.F. Thomsen (Eds.)*

91 **Seafloor Hydrothermal Systems: Physical, Chemical, Biological, and Geological Interactions** *Susan E. Humphris, Robert A. Zierenberg, Lauren S. Mullineaux, and Richard E. Thomson (Eds.)*

92 **Mauna Loa Revealed: Structure, Composition, History, and Hazards** *J. M. Rhodes and John P. Lockwood (Eds.)*

93 **Cross-Scale Coupling in Space Plasmas** *James L. Horwitz, Nagendra Singh, and James L. Burch (Eds.)*

94 **Double-Diffusive Convection** *Alan Brandt and H. J. S. Fernando (Eds.)*

95 **Earth Processes: Reading the Isotopic Code** *Asish Basu and Stan Hart (Eds.)*

96 **Subduction Top to Bottom** *Gray E. Bebout, David Scholl, Stephen Kirby, and John Platt (Eds.)*

97 **Radiation Belts: Models and Standards** *J. F. Lemaire, D. Heynderickx, and D. N. Baker (Eds.)*

98 **Magnetic Storms** *Bruce T. Tsurutani, Walter D. Gonzalez, Yohsuke Kamide, and John K. Arballo (Eds.)*

99 **Coronal Mass Ejections** *Nancy Crooker, Jo Ann Joselyn, and Joan Feynman (Eds.)*

100 **Large Igneous Provinces** *John J. Mahoney and Millard F. Coffin (Eds.)*

101 **Properties of Earth and Planetary Materials at High Pressure and Temperature** *Murli Manghnani and Takehiki Yagi (Eds.)*

102 **Measurement Techniques in Space Plasmas: Particles** *Robert F. Pfaff, Joseph E. Borovsky, and David T. Young (Eds.)*

103 **Measurement Techniques in Space Plasmas: Fields** *Robert F. Pfaff, Joseph E. Borovsky, and David T. Young (Eds.)*

104 **Geospace Mass and Energy Flow: Results From the International Solar-Terrestrial Physics Program** *James L. Horwitz, Dennis L. Gallagher, and William K. Peterson (Eds.)*

105 **New Perspectives on the Earth's Magnetotail** *A. Nishida, D. N. Baker, and S. W. H. Cowley (Eds.)*

106 **Faulting and Magmatism at Mid-Ocean Ridges** *W. Roger Buck, Paul T. Delaney, Jeffrey A. Karson, and Yves Lagabrielle (Eds.)*

107 **Rivers Over Rock: Fluvial Processes in Bedrock Channels** *Keith J. Tinkler and Ellen E. Wohl (Eds.)*

108 **Assessment of Non-Point Source Pollution in the Vadose Zone** *Dennis L. Corwin, Keith Loague, and Timothy R. Ellsworth (Eds.)*

109 **Sun-Earth Plasma Interactions** *J. L. Burch, R. L. Carovillano, and S. K. Antiochos (Eds.)*

110 **The Controlled Flood in Grand Canyon** *Robert H. Webb, John C. Schmidt, G. Richard Marzolf, and Richard A. Valdez (Eds.)*

111 **Magnetic Helicity in Space and Laboratory Plasmas** *Michael R. Brown, Richard C. Canfield, and Alexei A. Pevtsov (Eds.)*

112 **Mechanisms of Global Climate Change at Millennial Time Scales** *Peter U. Clark, Robert S. Webb, and Lloyd D. Keigwin (Eds.)*

113 **Faults and Subsurface Fluid Flow in the Shallow Crust** *William C. Haneberg, Peter S. Mozley, J. Casey Moore, and Laurel B. Goodwin (Eds.)*

114 **Inverse Methods in Global Biogeochemical Cycles** *Prasad Kasibhatla, Martin Heimann, Peter Rayner, Natalie Mahowald, Ronald G. Prinn, and Dana E. Hartley (Eds.)*

115 **Atlantic Rifts and Continental Margins** *Webster Mohriak and Manik Talwani (Eds.)*

116 **Remote Sensing of Active Volcanism** *Peter J. Mouginis-Mark, Joy A. Crisp, and Jonathan H. Fink (Eds.)*

117 **Earth's Deep Interior: Mineral Physics and Tomography From the Atomic to the Global Scale** *Shun-ichiro Karato, Alessandro Forte, Robert Liebermann, Guy Masters, Lars Stixrude (Eds.)*

118 **Magnetospheric Current Systems** *Shin-Ichi Ohtani, Ryoichi Fujii, Michael Hesse, and Robert L. Lysak (Eds.)*

Geophysical Monograph 119

Radio Astronomy at Long Wavelengths

Robert G. Stone
Kurt W. Weiler
Melvyn L. Goldstein
Jean-Louis Bougeret
Editors

American Geophysical Union
Washington, DC

Published under the aegis of the AGU Books Board

Roberta M. Johnson, Chair; John E. Costa, Jeffrey M. Forbes, W. Rockwell Geyer, Rebecca Lange, Douglas S. Luther, Walter H. F. Smith, Darrell Strobel, and R. Eugene Turner, members.

Library of Congress Cataloging-in-Publication Data
Radio astronomy at long wavelengths/Robert G. Stone...[et al.], editors.
      p.cm -- (Geophysical monograph ; 119)
   Includes bibliographical references
   ISBN 0-87590-977-9
   1. Radio astronomy. I. Stone, Robert G. (Robert Gilbert), 1928- II. Series
QB476.5.R33 2000
582'.682--dc21   00-038940

ISBN 0-87590-977-9
ISSN 0065-8448

Copyright 2000 by the American Geophysical Union
2000 Florida Avenue, N.W.
Washington, DC 20009

     Figures, tables, and short excerpts may be reprinted in scientific books and journals if the source is properly cited.

     Authorization to photocopy items for internal or personal use, or the internal or personal use of specific clients, is granted by the American Geophysical Union for libraries and other users registered with the Copyright Clearance Center (CCC) Transactional Reporting Service, provided that the base fee of $1.50 per copy plus $0.35 per page is paid directly to CCC, 222 Rosewood Dr., Danvers, MA 01923. 0065-8448/00/$01.50+0.35.

     This consent does not extend to other kinds of copying, such as copying for creating new collective works or for resale. The reproduction of multiple copies and the use of full articles or the use of extracts, including figures and tables, for commercial purposes requires permission from the American Geophysical Union.

Printed in the United States of America.

CONTENTS

Preface
Robert G. Stone, Kurt W. Weiler, Melvyn L. Goldstein, and Jean-Louis Bougeret ix

The Current Status of Low Frequency Radio Astronomy from Space
M. L. Kaiser and K. W. Weiler 1

GENERATION OF RADIO WAVES

Planetary Radio Emission Mechanisms: A Tutorial
Rudolf A. Treumann 13

Roles Played by Electrostatic Waves in Producing Radio Emissions
Iver H. Cairns and P. A. Robinson 27

Theory of Type III And Type II Solar Radio Emissions
P. A. Robinson and I. H. Cairns 37

On the Harmonic Component of Type III Solar Radio Bursts
C. S. Wu, Y. Li, and Peter H. Yoon 47

Synchrotron Maser: A "New" Emission Process
V. V. Zheleznyakov, G. Thejappa, S. A. Koryagin, and R. G. Stone 57

Plasma Thermal Noise: The Long Wavelength Radio Limit
Nicole Meyer-Vernet, Sang Hoang, Karine Issautier, Michel Moncuquet, and Gregory Marcos 67

PROPAGATION AND SCATTERING

Radio Wave Propagation in the Earth's Magnetosphere
J.-L. Steinberg, C. Lacombe, and S. Hoang 75

Propagation of Radio Waves in the Corona and Solar Wind
T. S. Bastian 85

Scattering in the Solar Wind at Long Wavelengths
B. J. Rickett and W. A. Coles 97

Interstellar Scattering: Radio Sensing of Deep Space Through the Turbulent Interstellar Medium
James M. Cordes 105

LONG WAVELENGTH RADIO EMISSION FROM THE SOLAR SYSTEM

The Sun

Type III Solar Radio Bursts at Long Wavelengths
George A. Dulk 115

CONTENTS

Type II Solar Radio Bursts
N. Gopalswamy 123

Interplanetary Type II Radio Emissions Associated With CMEs
Michael J. Reiner 137

ISEE-3 Observations of Radio Emission from Coronal and Interplanetary Shocks
H. V. Cane 147

Radar Studies of the Solar Corona: A Review of Experiments Using HF Wavelengths
Paul Rodriguez 155

The Planets

Radio Emissions from the Planets and Their Moons
Philippe Zarka 167

Planetary Radio Emission from Lightning: Discharge and Detectability
William M. Farrell 179

Terrestrial Continuum Radiation in the Magnetotail: Geotail Observations
H. Matsumoto, I. Nagano, and Y. Kasaba 187

Terrestrial LF Bursts: Escape Paths and Wave Intensification
Michael D. Desch and William M. Farrell 205

The Influence of the Galilean Satellites on Radio Emissions from the Jovian System
W. S. Kurth, D. A. Gurnett, and J. D. Menietti 213

SL-9: The Impact of Comet Shoemaker-Levy 9 at Jupiter
Yolande Leblanc 227

LONG WAVELENGTH RADIO EMISSION FROM GALACTIC AND EXTRAGALACTIC SOURCES

Long Wavelength Astrophysics
W. C. Erickson 237

The Promise of Long Wavelength Radio Astronomy
K. W. Weiler 243

What Would the Sky Look Like at Long Radio Wavelengths?
K. S. Dwarakanath 257

Capabilities and Limitations of Long Wavelength Observations from Space
Graham Woan 267

CONTENTS

Low-Frequency Radio Astronomy and the Origin of Cosmic Rays
Nebojsa Duric 277

Long Wavelength Observations of Supernova Remnants
Namir E. Kassim and Farhad Yusef-Zadeh 287

RADIO TELESCOPES FOR LONG WAVELENGTH OBSERVATIONS AND SOUNDING

Ground-Based

The Giant Metrewave Radio Telescope
G. Swarup 297

The VLA at 74 MHz and Plans for a Long Wavelength Array
William C. Erickson, Namir E. Kassim, and Richard A. Perley 303

Ukraine Decameter Wave Radio Astronomy Systems and Their Perspectives
A. A. Konovalenko 311

The Nançay Decameter Array: A Useful Step Towards Giant, New Generation Radio Telescopes for Long Wavelength Radio Astronomy
Alain Lecacheux 321

Space-Based

Instrumentation for Space-Based Low Frequency Radio Astronomy
Robert Manning 329

The Astronomical Low Frequency Array: A Proposed Explorer Mission for Radio Astronomy
D. Jones, R. Allen, J. Basart, T. Bastian, W. Blume, J.-L. Bougeret, B. Dennison, M. Desch, K. Dwarakanath, W. Erickson, D. Finley, N. Gopalswamy, R. Howard, M. Kaiser, N. Kassim, T. Kuiper, R. MacDowall, M. Mahoney, R. Perley, R. Preston, M. Reiner, P. Rodriguez, R. Stone, S. Unwin, K. Weiler, G. Woan, and R. Woo 339

Lunar Surface Arrays
T. B. H. Kuiper and D. Jones 351

Magnetospheric Sounding

Radio Sounding in the Earth's Magnetosphere
J. L. Green, R. F. Benson, S. F. Fung, W. W. L. Taylor, S. A. Boardsen, and B. W. Reinisch 359

PREFACE

The spectacular success of Earth-based radio astronomy is due to several factors. A broad atmospheric window of more than four orders-of-magnitude in frequency extends from ~30 MHz (wavelength 10 m), where ionospheric distortions and opacity begin to become a problem, to ~300 GHz (wavelength 1 mm), where atmospheric absorption becomes excessive, even from high mountain sites. This radio window reveals a rich variety of astrophysical phenomena. Also key to the success of radio astronomy has been the development of interferometry which provides high resolution, even at long wavelengths, without the construction of impossibly large and expensive single dish radio telescopes.

At still lower radio frequencies from ~30 kHz (wavelength 10 km), just above the local plasma frequency of the interplanetary medium, to ~30 MHz (wavelength 10 m), where observations through the ionosphere become possible most of the time, three orders of magnitude in frequency are essentially unexplored. This is the last unstudied region of the electromagnetic spectrum accessible from the vicinity of Earth. Many important astrophysical questions concerning the Sun, Earth, planets, interplanetary medium, Galaxy, interstellar medium, and the extragalactic universe require observations at these very long radio wavelengths.

In the solar system, long wavelength radio investigations include the emissions generated by suprathermal solar electrons, by interplanetary collisionless shock waves, and by the Earth's magnetosphere and the magnetospheres of the magnetized planets (particularly, Jupiter, Saturn, Uranus and Neptune). The boundary of the heliosphere appears to be a source of radio emission at long wavelengths and long wavelength radio emissions have proven useful in defining the spiral structure of the interplanetary magnetic field. Coronal mass ejections propagating from the Sun to the vicinity of Earth also emit long wavelength radio waves. Unfortunately, while space-based observations carried out for the past four decades have provided great insight into these problems, space-borne long wavelength radio antennas have been limited to electrically short dipoles with resolutions of only ~ 1 steradian or poorer. The need to go to space-based interferometric imaging arrays is pressing.

Beyond the solar system, the thermal interstellar medium, supernova remnants, pulsars, interstellar plasma refractive and diffractive scattering, cosmic rays, the galactic background, the galactic halo, radio galaxies, quasars, and old "fossil" electron populations are all copious emitters of low frequency radio waves and need to be investigated at frequencies of a few MHz and angular resolution near an arcminute. Additionally, this unexplored frontier undoubtedly hides many serendipitous discoveries. While high resolution ground-based observations at frequencies as low as ~300 MHz are routinely available, extending these down to ~30 MHz from the ground, and to even lower frequencies from space, is critical. Again, the exploitation of interferometric array technology at long wavelengths, both ground- and space-based is necessary.

To summarize the current state of long wavelength radio astrophysics and instrumentation and to look to the future in a coherent manner, an AGU Chapman Conference was convened in October 1998 in Paris, France. The conference brought together over 90 scientists and students interested in long wavelength radio astronomy and radio astrophysics from Australia, France, Germany, India, Japan, Poland, Russia, Sweden, Taiwan, Ukraine, United Kingdom, and the United States. This Monograph, which is composed of tutorials and reviews from that conference, summarizes our current knowledge and future prospects for this poorly explored, low frequency part of the electromagnetic spectrum.

We thank NASA Headquarters and the Goddard Space Flight Center for supporting our Chapman conference and this monograph, the National Science Foundation for supporting student travel to the conference, and the French Ministry of Research for welcoming the conference within the historic Le Carré des Sciences buildings. To Yolande Leblanc of the Paris Observatory we owe our appreciation for outstanding local arrangements. We also thank James L. Green and Jay S. Frielander of the Goddard Space Flight Center for assistance with the monograph cover illustrations. KWW thanks the Office of Naval Research (ONR) for the 6.1 funding supporting his work and research. This monograph would not have been possible without the generous assistance of over 40 volunteer reviewers for the manuscripts.

Robert G. Stone
Kurt W. Weiler
Melvyn L. Goldstein
Jean-Louis Bougeret
Editors

The Current Status of Low Frequency Radio Astronomy from Space

M. L. Kaiser

NASA Goddard Space Flight Center, Greenbelt, MD

K. W. Weiler

Naval Research Laboratory, Code 7213, Washington, DC

Ground-based radio astronomy is severely limited by the Earth's ionosphere. Below 15 – 20 MHz, space-based radio observations are superior or even mandatory. Three different areas of astronomical research manifest themselves at low radio frequencies: solar, planetary, and galactic-extragalactic. Space-based observations of solar phenomena at low frequencies are a natural extension of high-frequency ground-based observations that have been carried out since the beginnings of radio astronomy. Measurements of known solar phenomena such as Types II and III bursts have been extended from the few solar radii altitude range reachable by ground-based techniques out to 1 AU and beyond. These space-based solar measurements have become critical in our developing an understanding of "space weather." In contrast, non-thermal planetary radio emissions are almost exclusively a space radio astronomy phenomenon. With the exception of two components of Jupiter's complex radio spectrum, the magnetospheric and Auroral radio emissions of Earth, Jupiter, Saturn, Uranus, and Neptune have all been discovered by space radio astronomy techniques. For astrophysical applications, the lack of angular resolution from space at low frequencies has thwarted progress such that most areas still remain to be fully exploited. Results to date have only included overall cosmic background spectra and extremely crude ($\sim$1 steradian resolution) "maps." In this overview we will briefly summarize the current status of science in the three areas of research and outline some future concepts for low-frequency, space-based instruments for solar, planetary, and astrophysical problems.

1. SOLAR AND PLANETARY ASTROPHYSICS

1.1. Introduction

The history of low frequency radio astronomy from space has been largely in the area of solar and planetary studies. By the mid 1980s, *Kaiser and Desch,* [1984] reported that more than 25 spacecraft had made or were going to make radio observations of the planets and Sun. Since that time, several additional spacecraft have been added to this list.

The study of low frequency solar radio emissions differs significantly from the study of planetary emissions. To a very large extent, the low frequency solar emissions are a low frequency extension of the solar radio emissions that have been studied by ground-based observers for decades. On the other hand, with the sole exception of Jovian decametric and decimetric emissions, the planetary radio emissions were completely unknown prior to the space age. Thus, we have had extended or expanded studies of known phenomena in the case of solar radio astronomy and have had essentially pure discovery for planetary radio astronomy.

1.2. Major Accomplishments

Choosing highlights or important milestones in any field is difficult because invariably some paper is overlooked and its authors are offended. Nevertheless, with apologies afore hand to those whose work is not mentioned, I have picked five or six topics that I consider of crucial importance, not only in the history of solar and planetary research but also for future progress in these areas.

1.2.1. Solar. Certainly, the series of three papers by *Fainberg and Stone* [1970a,b, 1971] analyzing a low frequency Type III storm would be high on any list of major milestones. In that series of papers, they were able to determine the solar wind density from the Sun to 1 AU (the so-called RAE density scale), determine the solar wind speed and the speed of the exciter electrons, and give a useful upper limit to the size of density inhomogeneities.

With dipole antennas, the angular resolution available at low frequencies is extremely poor, literally half the sky in many cases. Thus, the analysis technique first used by *Fainberg, et al.* [1972], whereby the rotation of a spacecraft was used to deduce the direction of the radio source, is of fundamental importance to all branches of low frequency radio astronomy. Plate 1 shows the original direction finding figure [*Fainberg, et al.*, 1972] and a recent contribution [*Reiner, et al.*, 1995] using an antenna system which electrically combines spin plane and spin axis dipoles, a technique which is in common use now.

The detection of the in situ electrons responsible for radio emissions is impossible for most solar bursts. However, for low frequency solar bursts where the emission is generated in interplanetary space, the simultaneous detection of waves and particles is possible. Among the first instances reported for interplanetary Type III bursts were papers by *Lin, et al.,* [1973] and *Fitzenreiter, et al.* [1976], where they found a relationship between electron energies and radio emitted power. Simultaneous measurement of Type II radio emissions and associated particles has remained very elusive. Only recently have the first reported observations been made [*Bale, et al.,* 1999], where they show that the emission mechanism is similar to that which produces the $2f_p$ emission at Earth's bow shock.

The early work on interplanetary Type II bursts by Cane and others [*Cane and Stone*, 1984; *Cane, et al.,* 1982] has held up well over the years. Recent work [*Reiner and Kaiser,* 1999a; *Reiner, et al.,* 1998] has confirmed that interplanetary Type II bursts are essentially all associated with CME driven shocks as opposed to the flare driven 'blast wave' shocks thought to be the cause of many, or most, metric wavelength Type II bursts. These metric Type IIs usually die out in the low corona, presumably because they lack a 'piston' driving the shock [*Gopalswamy, et al.,* 1998].

Finally, in this short and incomplete list of solar radio accomplishments, I would add the discovery of so-called SA (shock accelerated) events [*Cane, et al.,* 1981; *MacDowall, et al.,* 1987]. These Type III-like bursts were originally thought to be caused by electrons accelerated by interplanetary shocks and were considered the low frequency extension of "herringbone" bursts observed by ground based telescopes. Recent new observations [*Reiner and Kaiser,* 1999b] by the WAVES instrument [*Bougeret, et al.,* 1995] on the WIND spacecraft has renewed interest in these events and will be a subject of intensive study in the future.

1.2.2. Planetary. For low frequency planetary radio astronomy, the discoveries of non-thermal radio emissions from Earth, Saturn, Uranus, and Neptune are the major highlights. (Some components of Jupiter's complex radio spectrum were also discovered by space-based radio telescopes, although the major Jovian components were already known from earlier ground-based observations.) Plate 2 shows a sample of the dynamic radio spectra of all the radio planets.

Earth was first recognized as a radio astronomical object by *Gurnett* [1974], and since that time there have been hundreds of observational and theoretical papers on Earth's radio spectrum. More is known about the details of Earth's radio emission than for any of the other planets. Many spacecraft have actually flown through the emission regions, which has not been possible at the other (outer) radio planets.

Most of the knowledge of the other radio planets

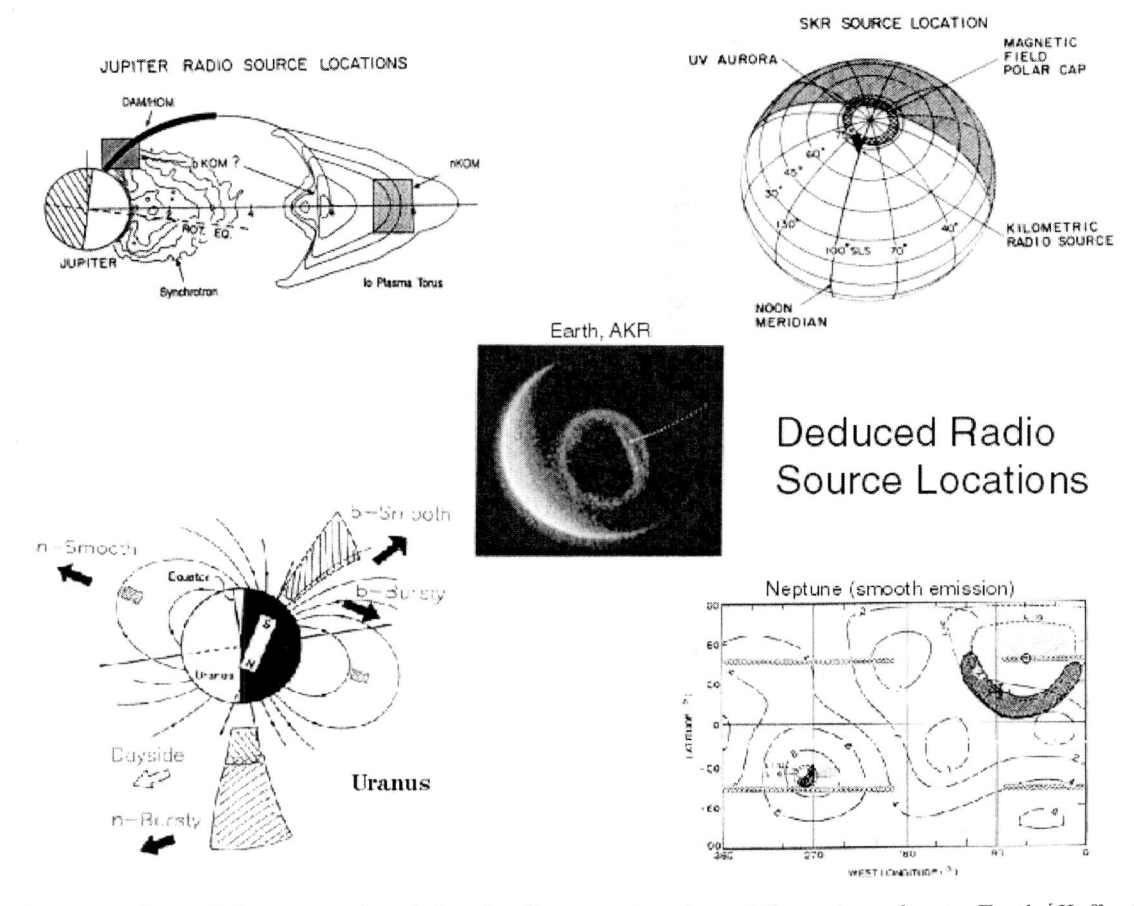

Figure 1. Some of the measured or deduced radio source locations at the various planets, Earth [*Huff, et al.*, 1988], Jupiter [*Kaiser*, 1993], Saturn [*Kaiser, et al.*, 1984], Uranus [*Desch, et al.*, 1991], and Neptune [*Zarka, et al.*, 1995].

comes from the observations made by the two Voyager spacecraft during their historic encounters with Jupiter (1979), Saturn (1980-81), Uranus (1986), and Neptune (1989). Ulysses (1992) and Galileo, presently in orbit at Jupiter, revealed additional details of Jupiter's radio spectrum.

One of the major by-products of the planetary radio emission work has been the determination of the intrinsic rotation period of all the outer planets, Saturn through Neptune [*Desch, et al.*, 1986; *Desch and Kaiser*, 1981; *Lecacheux, et al.*, 1993]. These determinations were based on the observed rotational modulation of the radio signals and the implicit assumption that the radio emissions are fixed relative to a planet's magnetic field.

Determination of the emission regions at the various planets has been quite difficult with the Voyager spacecraft because they had no inherent direction finding capability. Therefore, researchers had to resort to a variety of techniques including theoretical considerations, modeling, and some occultations as spacecraft went behind planets or moons. Figure 1 shows the best estimates to date for the source locations of most of the radio components at the various planets.

Finally, the theoretical paper by *Wu and Lee* [1979] has been crucial to low frequency planetary radio astronomy. They outlined the electron cyclotron maser mechanism that is, to this day, considered the most likely mechanism at work to produce the most intense emissions at Earth and all the radio planets. Since that initial paper, a very large number of papers refining the theory have been published and particle and field measurements in Earth's radio source regions have essentially verified the basic ideas.

4 LOW FREQUENCY RADIO ASTRONOMY FROM SPACE

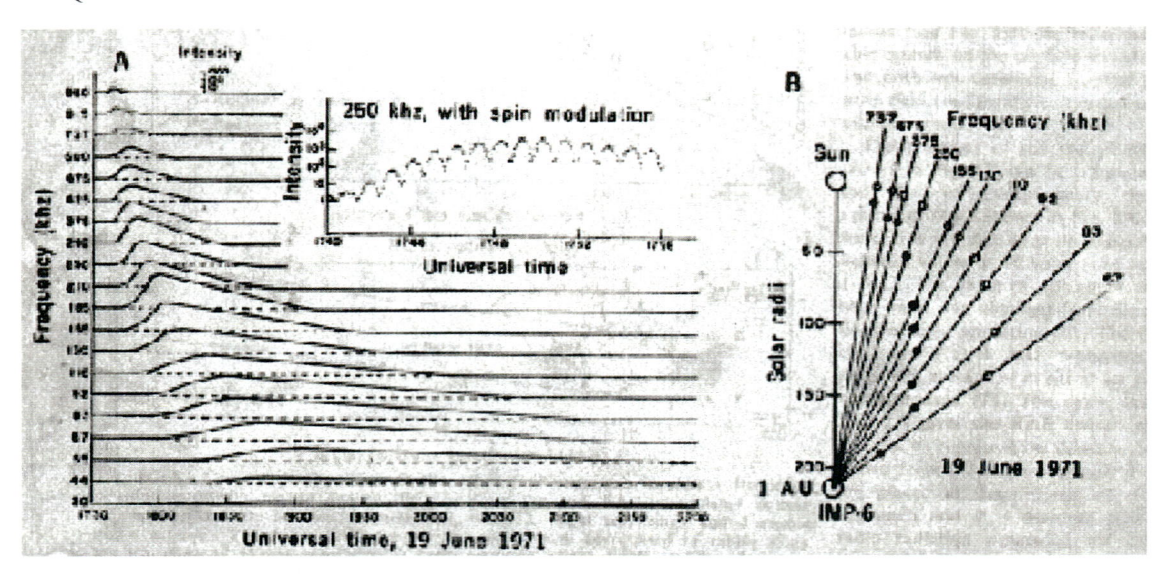

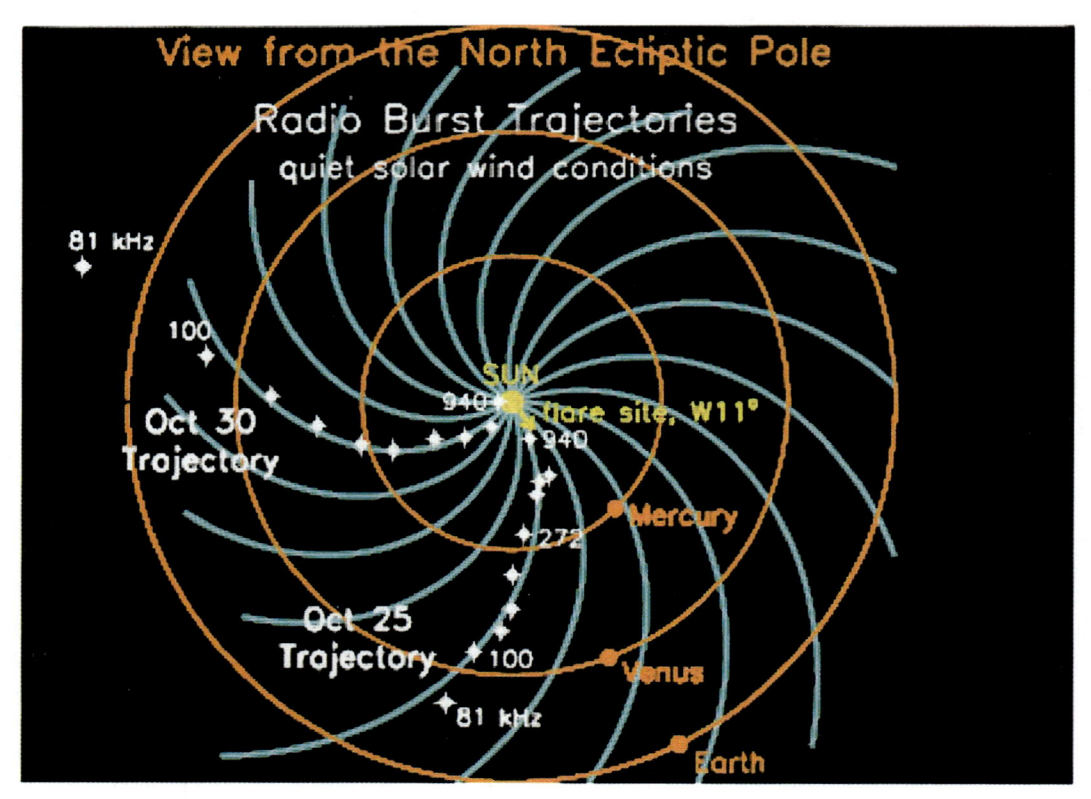

Plate 1. Top panel shows the first effort at determination of the direction of arrival of a low frequency Type III solar burst [*Fainberg, et al.,* 1972]. Bottom panel shows the complete trajectory of Type III solar bursts as observed from high above the ecliptic plane [*Reiner, et al.,* 1995].

KAISER AND WEILER 5

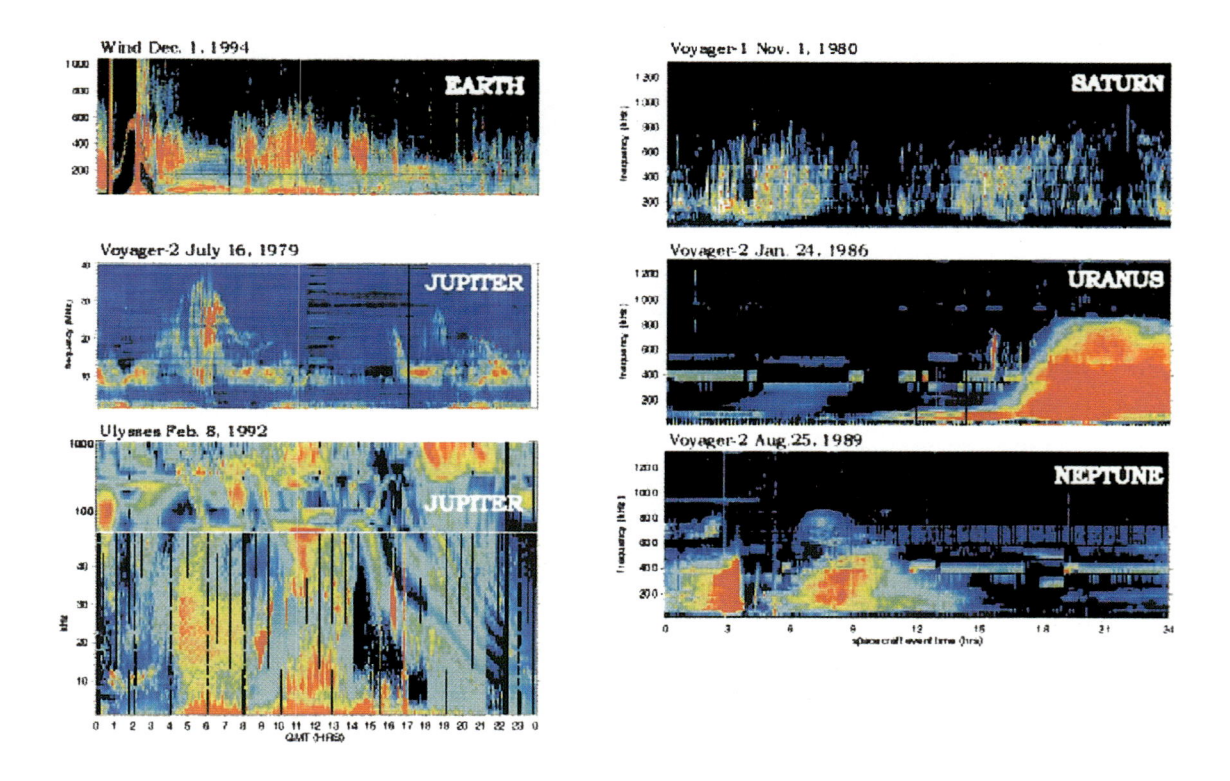

Plate 2. Dynamic spectra of emission from each of the radio planets as observed by the indicated spacecraft. Each panel shows color-coded (arbitrary scale) intensity as a function of observing frequency and time.

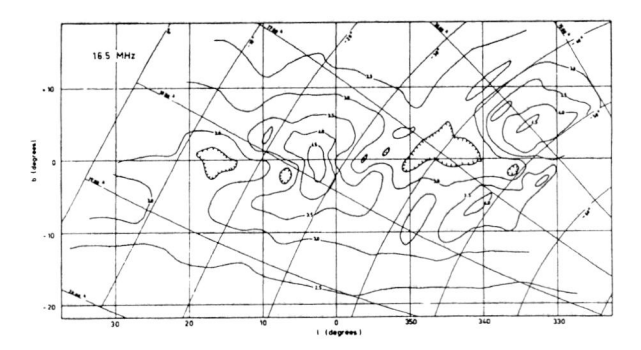

Figure 2. A 16.5 MHz map with a resolution of 1.6° of a large region near the galactic center taken with the Llanherne array in Tasmania, Australia. The contours indicated with tags enclose regions of depressed brightness resulting from absorption in HII regions. [*Cane and Whitham*, 1977].

1.3. Current Research and Future Prospects

Current research areas in low frequency solar astronomy include the study of:

- The relationship between different types of shocks in the solar corona and Type II bursts

- The detailed correlation between Type II and "SA" radio events and optical features as revealed by SOHO imaging

- "Space weather" - trying to predict the terrestrial effects of CMEs as measured by their radio emissions

- Detailed comparison between Type III emission mechanism theories and in situ particle and field observations

In planetary radio astronomy, the main efforts in progress today are the same as those in the past, namely:

- Determination of source locations or improvement on those deduced from earlier observations

- Determination of emission mechanisms in addition to the electron cyclotron maser and energy sources

The future of both solar and planetary radio astronomy from space is fairly bright. For solar radio astronomy the STEREO Mission, with launch date in 2004, will use radio direction finding techniques to triangulate the source locations of Type II and III bursts by viewing from two identical spacecraft separated by 1-2 AU. There is also the expectation that a low frequency aperture synthesis imaging instrument will be chosen as a mission in the future. Such an instrument will be capable of providing two-dimensional images of both the quiet sun and solar transient events and will have the added capability of imaging the galactic and extragalactic sky.

For planetary radio emissions, the Cassini spacecraft is on its way to an encounter with Saturn in 2004, where it will go into elongated orbit around the planet. Cassini carries a sophisticated radio and plasma wave instrument capable of measuring direction of arrival and all polarization parameters for Saturn's magnetospheric and atmospheric (lightning) emissions. Future possible planetary missions that might carry radio receivers include the Europa Orbiter, the Jupiter Polar Orbiter, and the French ORAJES dedicated radio astronomy missions.

These future planned and possible missions combined with the continuing present-day missions should keep the solar and planetary low frequency radio communities active for at least another decade. By that time, additional new missions, and possibly even a lunar-based observatory, should be in the planning stages.

2. GALACTIC AND EXTRAGALACTIC ASTROPHYSICS

2.1. Introduction

Galactic and extragalactic radio astronomy has been spectacularly successful from the Earth's surface due to the broad atmospheric window, more than 4 orders-of-magnitude wide, between ~30 MHz (wavelength 10 m), where ionospheric distortions and opacity begin to become a problem, and ~300 GHz (wavelength 1 mm) where atmospheric absorption becomes excessive at most times even from high mountain sites. This success for radio astronomy has been largely gained through the development of interferometry which has allowed high resolution, even at such long wavelengths, without the construction of impossibly large and expensive single dish radio telescopes.

The low frequency window, on the other hand, from ~30 kHz (wavelength 10 km), just above the local plasma frequency of the interplanetary medium (IPM), to ~30 MHz (wavelength 10 m), where high resolution observations from the ground become possible most of the time, spans three orders of magnitude in frequency, is almost as wide as the entire "traditional" radio astronomy window, and is wider than the infrared window opened by IRAS and ISO or the ultraviolet window opened by IUE and EUVE. This low frequency window

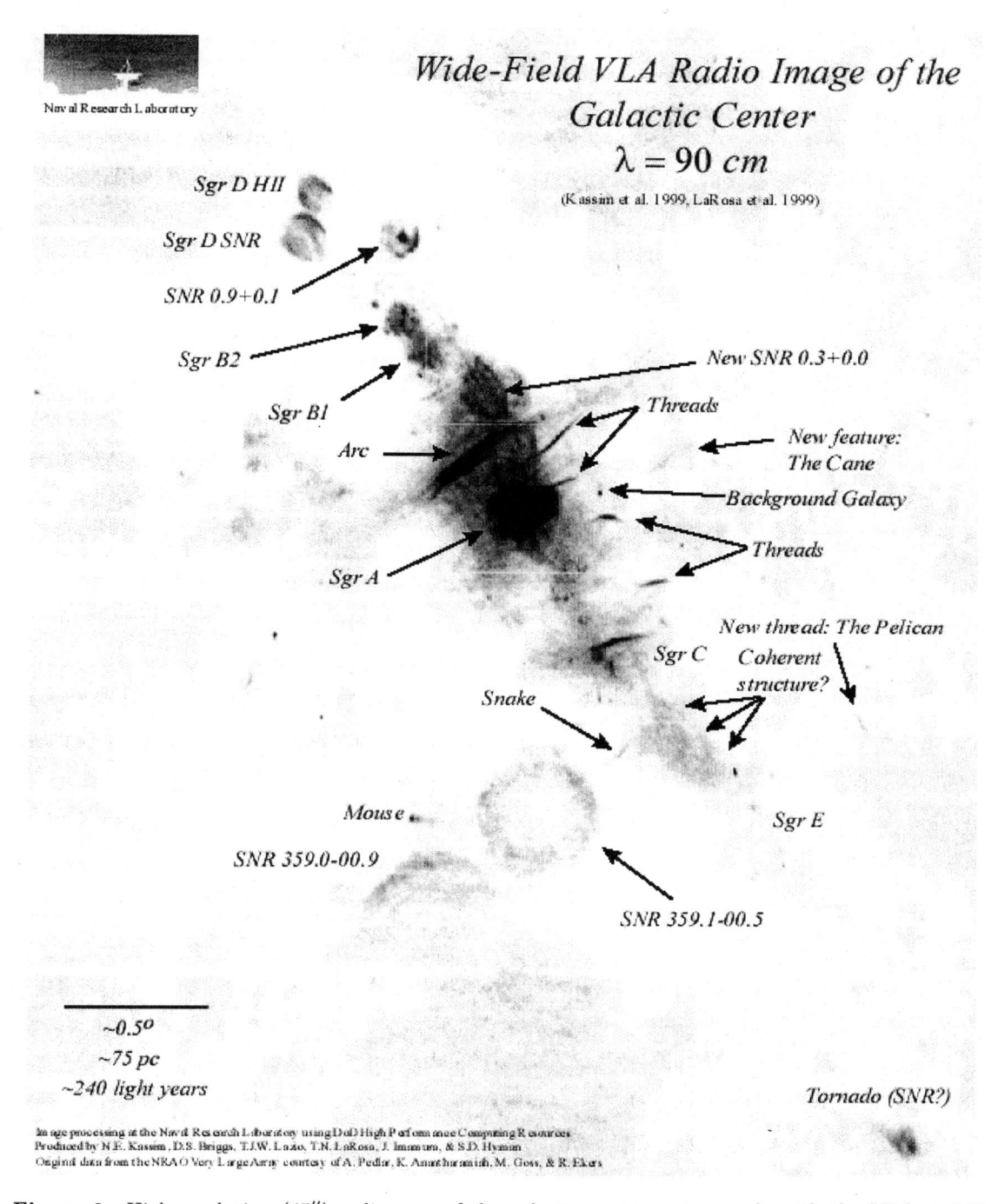

Figure 3. High resolution (45″) radio map of the galactic center region made with the VLA at 327 MHz. The image illustrates the great amount of astrophysical detail which is available with arcminute resolution. [*Kassim, et al., 2000*]

is essentially unexplored, and is the last region of the electromagnetic spectrum which is accessible from the vicinity of Earth yet unstudied. Many important astrophysical questions concerning the solar system, Galaxy, and distant universe, such as the lifetimes and evolution of extended radio sources, the origin of cosmic rays, and the distribution and turbulence properties of interstellar and interplanetary plasma can be answered with

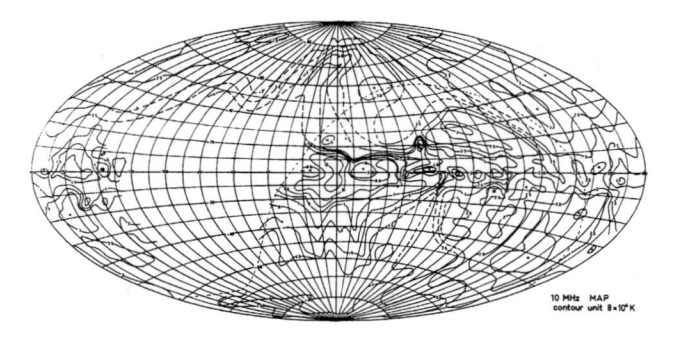

Figure 4. A 10 MHz map (resolution ∼5°) of the southern sky in galactic coordinates. [*Cane and Erickson*, 2000]

observations at frequencies of a few MHz and angular resolution near an arcminute. However, because of limits imposed by the ionosphere, such data can only be obtained with a radio interferometric array in space, a Low Frequency Space Array (LFSA).

2.2. Past Work

A few examples of the very limited low-frequency information which is available from ground-based and early space-based observations can be seen in the following figures. For example, Figure 2 shows a survey of the Galactic center region taken with the Llanherne array in Tasmania at the relatively high frequency of 16.5 MHz. It has a resolution of only 1.6° and shows relatively little of the extremely complex structure in the galactic center region. For comparison, one might contrast Figure 2 with a 327 MHz image of the galactic center shown in Figure 3 taken with the VLA[1].

As one goes to lower frequencies than the 16.5 MHz map shown in Figure 2, the situation becomes even poorer. A 10 MHz map of the southern sky is shown in Figure 4. It displays relatively little detail even of the galactic plane. The very complex structure of emitting and absorbing regions, which must be present, is lost in the low resolution imaging.

At still lower frequencies, such as the 2.1 MHz (resolution ∼8°) and 1.6 MHz (resolution ∼30°) maps shown in Figures 5 and Figure 6, respectively, almost no detail of the radio sky remains. It should also be kept in mind that even these very low resolution maps must be made at preferred locations on the surface of the Earth near the geomagnetic poles and at preferred times at night and near sunspot minimum because of the severe ionospheric effects.

---

[1]The VLA is operated by the National Radio Astronomy Observatory of the Associated Universities, Inc., under a cooperative agreement with the National Science Foundation.

It is clear from these pioneering and difficult, but ultimately limited, efforts that low frequency radio astronomy must move to space to escape the ionospheric barrier. Two attempts were made to do this in the late 1960's and early 1970's with the Radio Astronomy Explorer (RAE) satellites. However, the use of only a single satellite each time, and no capability for interferometric imaging, meant that the resolutions were very poor, only ∼1 steradian, and little detail of the low frequency radio sky could be discerned (Figure 7).

In the intervening 30 years since the RAE satellites were launched, interferometric imaging techniques and data processing algorithms have advanced rapidly, so that it is now possible to build and launch a radio interferometric array into space to image the low frequency sky to arcminute resolution. At such a resolution, as is illustrated in Figure 3, the details of the astrophysics stand out and yield new information in a frequency band several orders-of-magnitude lower than is possible from the ground. It is fully expected that such an unexplored frequency realm will yield new discoveries, as well as new insight into known astrophysical processes.

2.3. Low Frequency Astrophysics

Even though opening an entirely new region of parameter space will likely produce many surprises, there are significant areas which we can immediately identify for fruitful work at low frequencies.

2.3.1. The origin of cosmic rays - an 83 year old mystery. Perhaps the most fundamental question still remaining from the era of classical physics is the origin of cosmic rays. Cosmic rays (CR) represent the most energetic form of matter and trace the highest energy phenomena. At frequencies below 30 MHz there is a real possibility for probing the particle acceleration process

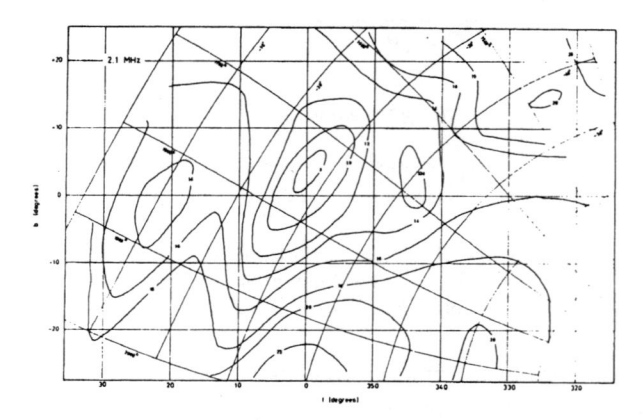

Figure 5. A 2.1 MHz map of the galactic center region with a resolution of 8°. [*Cane and Withham*, 1977]

in supernova remnants (SNRs) and addressing the issue of the origin of CRs.

2.3.2. Galactic nonthermal background. In studies of the distributed nonthermal background emission of the Milky Way, different frequencies emphasize different physical processes: the surveys of γ-ray emission are sensitive to the interaction of cosmic rays with the ambient interstellar gas; optical surveys emphasize stars and ionized hydrogen (HII) regions; and IR surveys enhance visibility of the relatively cold interstellar dust. Radio frequency studies are most sensitive to the relativistic CR electrons and interstellar magnetic fields. However, there are problems in explaining the observed break in the cosmic ray electron energy spectrum near 3 GeV, which is equivalent to the background radio spectrum break at $\sim$300 MHz, and low frequency observations of the galactic background will provide information on the relevant loss and injection mechanisms.

2.3.3. Galactic diffuse free-free absorption. By observing a large number of extragalactic radio sources and determining their low frequency spectra as a function of Galactic latitude and longitude, it will be possible to measure the distribution of absorption by the diffuse, interstellar gas in the Milky Way. Combining survey results at the several frequencies available to a LFSA and the low resolution, higher frequency maps from the literature, one can successfully separate the thermal absorption and nonthermal emission components of the Galaxy and finally establish their relative contributions to the energy balance in the ISM.

2.3.4. Interstellar scattering and refraction. It is generally accepted that small scale ($\sim 10^9$ cm) fluctuations in electron density in the interstellar medium can diffractively scatter radio waves from a background source. Less clear is the ability of somewhat larger irregularities ($\sim 10^{13}$ cm) to refractively focus and defocus radio waves. The questions which can be addressed at

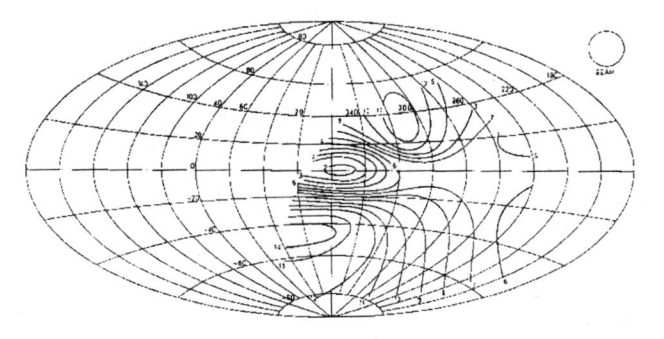

Figure 6. A 1.6 MHz map of the southern sky in galactic coordinates with resolution $\sim 30°$ [*Ellis and Mendillo*, 1987].

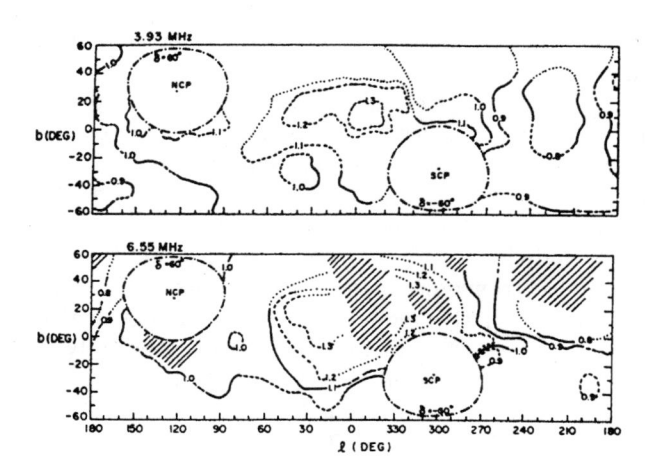

Figure 7. 3.93 MHz and 6.55 MHz maps of the sky from space by the Radio Astronomy Explorer 2 (RAE-2) satellite with $\sim$1 steradian resolution [*Alexander and Novaco*, 1974].

low frequencies are: What is the correct form of the irregularity power spectrum? Is the irregularity spectrum ever anisotropic? How common are refractive distortions and refractive scintillation? What is the origin of the turbulence and how is it distributed throughout the Galaxy?

2.3.5. Extragalactic sources. A LFSA will be able to detect thousands of discrete sources for greatly extended source counts [log(N)/Log(S)] and study the brighter ones for such properties as integrated spectrum, surface brightness, and spectral index distribution. This is especially important since, due to synchrotron radiative lifetimes, the relativistic electrons which a LFSA detects are far older than those normally studied by radio astronomy.

2.3.6. Source spectra. It has long been known that the radio spectral index (α) is a function of frequency $\alpha(\nu)$ and that the measurement of this frequency dependence is important for understanding the physics of the emission and absorption process in the sources. At present, very little is known about source spectra at frequencies as low as 20 MHz and practically nothing has been measured for $\nu < 10$ MHz.

2.3.7. Supernova remnants and pulsars. Low frequency radio observations provide a unique means for investigating SNRs, their interaction with the ISM, and the shock acceleration processes. A number of millisecond pulsars have spectra that are very steep with flux densities that continue to increase down to the lowest observed frequency of $\sim$10 MHz. These pulsars, although non-pulsing due to interstellar scattering at these frequencies, will be among the strongest sources in the sky at 10 MHz and low frequency observations can distinguish among the several particle acceleration mechanisms.

2.3.8. Coherent emission. A very exciting possibility at low frequencies is the detection of coherent radiation. Low frequency coherent emission is well known from solar system objects and there are valid physical reasons to anticipate that the smaller distance between individual radiating electrons measured in terms of the electromagnetic wavelength is likely to amplify collective radiative modes in galactic and extragalactic objects. Such observations will open an entirely new regime of astrophysical parameters for study.

2.4. Concept and Requirements for a LFSA

2.4.1. Hardware. A conceptual Low Frequency Space Array will consist of a large number (>12) of small spacecraft arranged in a spherical array with baselines up to ~100 km. To avoid terrestrial interference, the array should be placed at a large distance from the Earth or on the backside of the Moon. For a deep space location, a libration point, a solar orbit far from the Earth, or a distant retrograde orbit around the Earth-Moon system will fulfill this requirement with the choice being made on the basis of minimum launch energy and calculable array stability criteria. The operating frequencies for a LFSA should cover from ~30 kHz, which is just above the local plasma frequency in the Interplanetary Medium (IPM), to ~30 MHz, where observations can be taken satisfactorily from the ground. Because of the long wavelengths involved, it is not possible to construct individual array elements with antenna gain, such as parabolic dishes or array beds, so that the spacecraft antennas will likely be limited to simple, electrically short crossed dipoles. Fortunately, such space hardware presents no major technical challenges.

2.4.2. Software. Major new development will be required in software. The concept of mapping the full 3-D sky at all times in all directions is a new concept in interferometric imaging and data processing. This is even more complex due to the presence of moving, bursting, occulting, and refracting sources such as the Sun, Jupiter, the Moon, and the Earth with its large magnetosphere. While the concepts for handling such a complex data processing problem are currently available, the full algorithms and application packages employing such techniques for the data analysis of a LFSA have yet to be developed.

2.4.3. Other locations. Other possible locations for a LFSA, such as lunar orbit or the lunar surface (with the "holy grail" of low frequency radio astronomy being the shielded backside of the Moon) have not been addressed in this introductory talk. They will be discussed, however, later in this volume [*Weiler*, 2000].

2.5. Conclusions

The opening of a new spectral window for astronomical investigations has always resulted in major discoveries, significant new insights into astrophysical processes, and an enrichment of our understanding of the universe. Frequencies below ~30 MHz are a range which is totally inaccessible or extremely difficult to observe from the ground with high resolution due to ionospheric absorption and scattering. These frequencies represent a region of the spectrum that is essentially unexplored by astronomy but which, at ~10^6 Hz, is likely to display phenomena as different from those at centimeter radio wavelengths (~10^9 Hz) as centimeter radio phenomena are from infrared (~10^{12} Hz), or infrared are from the ultraviolet (~10^{15} Hz), or ultraviolet are from the X-ray (~10^{18} Hz). Because of this large gap in our knowledge, the likelihood of discovering new processes and objects is great, even though many worthwhile projects can already be defined for a new instrument.

Developing and launching a LFSA will open a new window on the universe, explore a new parameter space in sensitivity, frequency, and resolution, and study a new sky with very different absorption and emission parameters from centimeter wavelengths. It will certainly discover new astrophysics.

Acknowledgments. KWW wishes to thank the Office of Naval Research (ONR) for the 6.1 funding supporting this work.

REFERENCES

Alexander, J. K. and Novaco, J. C., Survey of the galactic background radiation at 3.93 and 6.55 MHz, *Astron. J.*, *79*, 777-785, 1974.

Bale, S. D., Reiner, M. J., Bougeret, J.-L., Kaiser, M. L., Krucker, S., Larson, D. E., and Lin, R. P., The source region of an interplanetary Type II radio burst, *Geophys. Res. Lett.*, *26*, 1573, 1999.

Bougeret, J.-L., et al., The WIND/WAVES instrument, *Spa. Sci. Rev.*, *71*, 231, 1995.

Cane, H. V. and Stone, R. G., Type II solar radio bursts, interplanetary shocks, and energetic particle events, *Astrophys. J.*, *282*, 339, 1984.

Cane, H. V., Stone, R. G., Fainberg, J., Steinberg, J.-L., and Hoang, S., Type II solar radio events observed in the interplanetary medium, *Solar Phys.*, *78*, 187, 1982.

Cane, H. V., Stone, R. G., Fainberg, J., Stewart, R. T., Steinberg, J.-L., and Hoang, S., Radio evidence for shock acceleration of electrons in the solar corona, *Geophys. Res. Lett.*, *8*, 1285, 1981.

Cane, H. V. and Whitham, P. S., Observations of the southern sky at five frequencies in the range 2-20 MHz, *Mon. Not. R. Astron. Soc., 179,* 21-29, 1977.

Cane, H. V. and Erickson, W. C., A 10 MHz map of the Galaxy, *Radio Sci.,* in press, 2000.

Desch, M. D., Connerney, J. E. P., and Kaiser, M. L., The Rotation period of Uranus, *Nature, 322,* 42, 1986.

Desch, M. D., and Kaiser, M. L., Saturnian kilometric radiation: Satellite modulation, *Nature, 292,* 739, 1981.

Desch, M. D., Kaiser, M. L., Lecacheux, A., Aubier, M., Zarka, P., and Leblanc, Y., Uranus as a radio source, in *Uranus,* edited by J. Bergstrahl and M. Mathews, U. of Arizona Press, Tucson, 1991.

Ellis, G. R. A. and Mendillo, M., A 1.6 MHz survey of the galactic background radio emission, *Australian J. Phys., 40,* 705-708, 1987.

Fainberg, J., Evans, L. G., and Stone, R. G., Radio tracking of solar energetic particles through interplanetary space, *Science, 178,* 743, 1972.

Fainberg, J. and Stone, R. G., Type II solar radio burst storms observed at low frequencies. II: Average exciter speed, *Solar Phys., 15,* 433, 1970a.

Fainberg, J. and Stone, R. G., Type III solar radio burst storms observed at low frequencies. I: Storm morphology, *Solar Phys., 15,* 222, 1970b.

Fainberg, J. and Stone, R. G., Type III solar radio burst storms observed at low frequencies. III: Streamer density, inhomogeneities, and solar wind speed, *Solar Phys., 17,* 392, 1971.

Fitzenreiter, R. J., Evans, L. G., and Lin, R. P., Quantitative comparisons of Type III radio burst intensity and fast electron flux at 1 AU, *Solar Phys., 46,* 437, 1976.

Gopalswamy, N., Kaiser, M. L., Kundu, M. R., Kahler, S. W., Kondo, T., Isobe, T., and Akioka, M., Origin of coronal and interplanetary shocks: A new look with WIND spacecraft data, *J. Geophys. Res., 103,* 307, 1998.

Gurnett, D. A., The Earth as a radio source: Terrestrial kilometric radiation, *J. Geophys. Res., 79,* 4227, 1974.

Huff, R. L., Calvert, W., Craven, J. D., Frank, L. A., and Gurnett, D. A., Mapping of auroral kilometric radiation sources to the aurora, *J. Geophys. Res., 93,* 11445-11454, 1988.

Kaiser, M.L., Time variable magnetospheric radio emissions from Jupiter, *J. Geophys. Res., 98,* 18757, 1993.

Kaiser, M. L. and Desch, M. D., Radio emission from the planets Earth, Jupiter and Saturn, *Rev. Geophys. Space Phys., 22,* 373, 1984.

Kaiser, M. L., Desch, M. D., Kurth, W. S., Lecacheux, A., Genova, F., and Pedersen, B. M., Saturn as a radio source, in *Saturn,* edited by T. Gehrels, Univ. of Arizona Press, Tucson, 1984.

Kassim, N. E., LaRosa, T. N., Lazio, T. J. W., and Hyman, S. D., Wide field radio imaging of the galactic center, *Proc. Conf., The Central Parsecs – Galactic Center Workshop,* edited by A. Cotera and H. Falcke, *Publ. A. S. P.,* in press, 2000.

Lecacheux, A., Zarka, P., Desch, M. D., and Evans, D. R., The sidereal rotation period of Neptune, *Geophys. Res. Lett., 20,* 2711, 1993.

Lin, R. P., Evans, L. G., and Fainberg, J., Simultaneous observations of fast solar electrons and Type III radio burst emission near 1 AU, *Astrophys. Lett. and Comm., 14,* 191-198, 1973.

MacDowall, R. J., Stone, R. G., and Kundu, M. R., Characteristics of shock-associated fast-drift kilometric radio bursts, *Solar Phys., 111,* 397, 1987.

Reiner, M. J., Fainberg, J., and Stone, R. G., Large scale interplanetary magnetic field configuration revealed by solar radio bursts, *Science, 270,* 461, 1995.

Reiner, M. J. and Kaiser, M. L., High-frequency Type II radio emissions associated with shocks driven by CMEs, *J. Geophys. Res., 104* 16979, 1999a.

Reiner, M. J. and Kaiser, M. L., Complex Type III-like radio emissions observed from 1 to 14 MHz, *Geophys. Res. Lett., 26,* 397, 1999b.

Reiner, M. J., Kaiser, M. L., Fainberg, J., and Stone, R. G., A new method for studying remote Type II radio emissions from coronal mass ejection-driven shocks, *J. Geophys. Res., 103,* 29651-29664, 1998.

Weiler, K. W., The promise of long wavelength radio astronomy, in *Space-Based Radio Observations at Long Wavelengths,* edited by R. Stone, K. Weiler, M. Goldstein, and J-L Bougeret, AGU, in press, 2000.

Wu, C. S., and Lee, L. C., A theory of terrestrial kilometric radiation, *Astrophys. J., 230,* 621, 1979.

Zarka, P., et al., Radio Emissions from Neptune, in *Neptune and Triton,* edited by D. P. Cruikshank, U. of Arizona Press, Tucson, 1995.

M. L. Kaiser, NASA Goddard Space Flight Center, Greenbelt, MD 20771, USA (e-mail: kaiser@lepmlk.gsfc.nasa.gov)

K. W. Weiler, Naval Research Laboratory, Code 7213, Washington, DC 20375-5320, USA
(e-mail: weiler@rsd.nrl.navy.mil)

Planetary Radio Emission Mechanisms: A Tutorial

Rudolf A. Treumann

Max-Planck-Institute for extraterrestrial Physics, Garching, Germany
International Space Science Institute, Bern, Switzerland

This tutorial reviews the basic modes and processes of radio wave emission from planetary plasmas without specification to a particular planet. The Earth's magnetosphere is taken as a planetary paradigm mostly because it is the best studied planetary radiator. The various source regions in the magnetosphere are identified. Of the various possible mechanisms for generating a radiation current the most important three are identified as the direct emission process via instability of one of the free-space modes (cyclotron maser), linear wave transformation, and nonlinear wave-wave/wave-particle interactions. In dense plasma cases the latter dominate over maser emission.

1. INTRODUCTION

The intense radio emission occasionally observed originating near strongly magnetized planets like Earth, Jupiter, and Saturn, points to the presence of non-thermal sources in the vicinity of magnetized planets. It requires violent, time-variable, sometimes even coherent processes. A number of conditions must be satisfied for a planet to become a natural radio emitter. The most important and crucial element is that the planet is strongly magnetized and develops a magnetosphere in the interaction with the stellar wind, in our solar system the solar wind. Because of this reason Mercury and possibly also Jupiter's moon Ganymede may be radio emitters because they possess their own magnetospheres. The Earth's magnetosphere is the canonical paradigm of a planetary magnetosphere as a natural radio source. Figure 1 presents a schematic (not-to-scale) view of the field and plasma content of the terrestrial magnetosphere.

A first summary of magnetospheric radio emissions has been given a quarter of a century ago in a famous paper by *Gurnett* [1974] entitled "The Earth as a Radio Source: Terrestrial Kilometric Radiation". Gurnett found that the Earth's magnetosphere contained at least two components of radio waves: a trapped continuum caused by the ring current-radiation belt particles, and a very intense radiation originating in the nightside auroral zone in correlation with the degree of auroral disturbance, the Auroral Kilometric Radiation (AKR). In addition there is a weak escaping continuum with high enough frequency to overcome the density barrier provided by the presence of the mantle/low latitude boundary layer/magnetosheath plasma.

More recent observations have identified a whole lot of structures in different radio emissions from the environment of the Earth. Starting from proximity to the Earth, the high frequency auroral roar radio emission is known at around the harmonics of the local electron cyclotron frequency which is of the order of a few MHz. Its presence has been know since the late fiftieth and in the sixtieth by the ISIS spacecraft. It is generated somewhere in the upper ionosphere during auroral disturbances and is restricted to the auroral region. The most interesting fact about this weak emission is that it is highly structured in time and frequency and consist of a number of components. This has been realized only very recently [*Weatherwax et al.*, 1995; *LaBelle et al.*, 1995; *Shepherd et al.*, 1998]. The origin of these structured emission has not yet been clarified. They are very weak in intensity.

Radio Astronomy at Long Wavelengths
Geophysical Monograph 119
Copyright 2000 by the American Geophysical Union

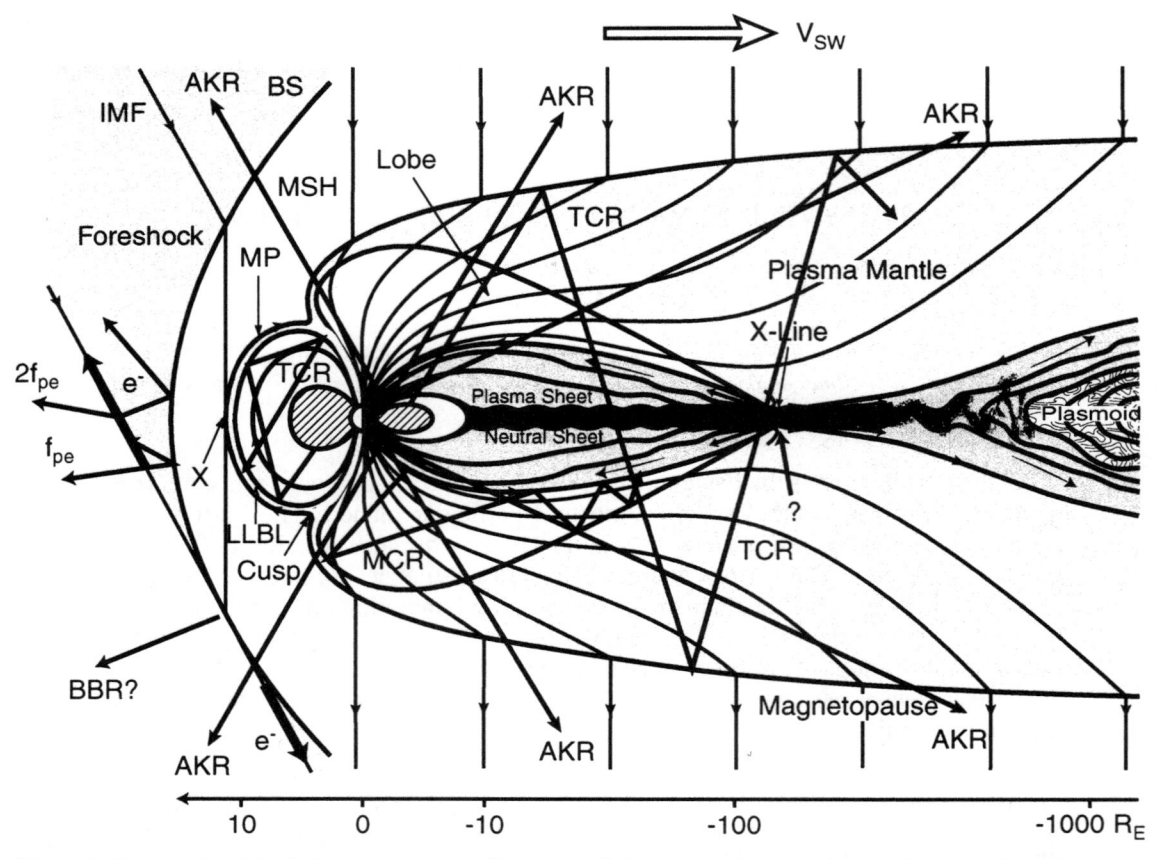

Figure 1. Schematic of the field and plasma configuration of the terrestrial magnetosphere (not to scale). Some of the known modes of radiation are indicated: TCR = trapped continuum radiation; MCR = Myriametric continuum; the auroral kilometric radiation source is marked as AKR, due to its high frequency it dan readily escape from the magnetosphere and at large distances is the main natural terrestrial radio signal. Fundamental and harmonic radiation from the foreshock of the bow shock are shown. These radiations are weak and are reflected at the shock; BBR is indicated with question mark for the undetected back-bone radiation; a question mark is also put at a possible undetected radiation source from the X-line or slow shocks in reconnection; escaping radiation is not shown. Note that similar radiations are observed at other magnetized planets with Jupiter an exception because there Io contributes most to the radio emission.

Some of them might even come from below the ionosphere from the region where electron cyclotron maser emission becomes possible in a transition region where the local plasma frequency drops below the cyclotron frequency while the collision frequency is still small enough to prevent instantaneous damping of the waves. Under favorable conditions, the waves may leak out along channels of ionospheric density perforations. Other unexplained observations have been reported by *Malingre et al.* [1997].

The trapped radiation is relatively quiet and unstructured. It forms a continuum whose variation is determined by the temporal variations of the ring current-radiation belt intensities. It is gyro-synchrotron in nature. Its trapped part is the band below roughly 50 kHz (wave length $\lambda > 10$ km), the typical plasma frequency of the magnetosheath adjacent to the magnetopause. (Higher frequencies escape from the magnetosphere.) It is reflected back and forth between magnetopause and plasmapause and is thus isotropic. The AKR, on the other hand, has a low frequency cut-off close to ~300 kHz, is highly variable in time, bandwidth and spatial extension, and is closely related to the AE index. Stereoscopic measurements and ray tracing techniques have shown that it originates somewhere within a few 1000 km altitude in density depletions [e.g., *Hilgers*, 1992] in the auroral zone identifying its emission frequency as the local auroral cyclotron frequency. Its integrated intensity corresponds to about 1% of the total substorm energy. AKR thus provides an astonishingly efficient energy release process during a substorm.

In the plasma gap between the plasma sheet boundary layer and the inner mantle boundary still longer wave length radiation may become trapped. Such radiation has been detected by Geotail [*Nagano et al.*, 1994; *Matsumoto et al.*, 1998] and has been called myriametric. Its generation is probably due to wave-wave interaction in the plasma sheet boundary layer. This region is in the high-β regime where energy release through particle heating, generation of macroscopic turbulence and plasma wave generation is more favorable than in low-β dilute plasmas. The plasma waves needed are generated by field aligned electron beams [*Schriver et al.*, 1991; *Matsumoto et al.*, 1998]. Similarly, the low latitude boundary layer should contain sources of radiation. However, the main process in this region, i.e. reconnection, has not left any marks in radiation. The same argument applies to the reconnection site in the high-β distant tail.

Returning to the high-β bow shock region we note that weak and moderately intense radiation is known to exist here. Its observation confirms the widely used claim that shock waves are efficient radio wave emitters. The most famous example is the solar type II radio burst. However, there is an important difference between type II (and presumably other) shock emissions and the radiation from the bow shock. Bow shock radiation is excited by bow shock reflected electrons in the electron foreshock as indicated in Figure 1. The bow shock does not exhibit any backbone emission which dominates type II shocks [*McLean and Labrum*, 1985]. Backbones are conventionally attributed to hypothetical emission from the shock ramp. But no measurement so far has ever seen any indication of such intense emission from the bow shock ramp or from the downstream magnetosheath.

2. PRINCIPLES OF RADIATION

Radiation is emitted whenever electrons are accelerated. All radiation mechanisms can be traced back to a specific kind of acceleration. Plasma sources include a very large number of electrons all of which, when accelerated, contribute to the radiation. Hence one must integrate over the collective of particles described by the particle distribution. This leads to a time varying current density $\mathbf{j}(\mathbf{r}, t)$. The physics is reduced to the determination of this current density.

At sufficiently large distance from the source in the far field only the lowest momenta of the acceleration contribute to surviving radiation power. The relation between the vector potential of the radiation field $\mathbf{A}(\mathbf{r}, t)$ and the source current density is given by the solution of the vector Poisson equation

$$\nabla^2 \mathbf{A} = -\mu_0 \mathbf{j} \quad (1)$$

and is obtained as

$$\mathbf{A} = \frac{\mu_0}{4\pi} \int_{V,t} d\mathbf{r}' dt' \delta\left(t' - t + \frac{1}{c}|\mathbf{r} - \mathbf{r}'|\right) \frac{\mathbf{j}(\mathbf{r}', t')}{|\mathbf{r} - \mathbf{r}'|} \quad (2)$$

When replacing the time variability with a Fourier integral assuming quasiperiodic variations, the integration can be formally performed yielding

$$\mathbf{A}(\omega, \mathbf{r}) = \frac{\mu_0}{4\pi} \int_V d\mathbf{r}' \frac{\mathbf{j}(\omega, \mathbf{r}')}{|\mathbf{r} - \mathbf{r}'|} \exp\left(ik|\mathbf{r} - \mathbf{r}'|\right) \quad (3)$$

where ω is the radiation frequency and $k = \omega/c$. The next approximation is that $|\mathbf{r} - \mathbf{r}'| \approx r - \mathbf{n} \cdot \mathbf{r}'$ which holds for free-space waves at large distance r from the source. This permits to write for the vector potential

$$\mathbf{A}(\omega, \mathbf{r}) \approx \frac{\mu_0 e^{ikr}}{4\pi r} \int_V d\mathbf{r}' \mathbf{j}(\omega, \mathbf{r}') \quad (4)$$

The radiation current, which is a pure electron current, can always be represented as a moment of the electron distribution function $f_e(\omega, \mathbf{r}, \mathbf{v})$

$$\mathbf{j}(\omega, \mathbf{r}) = -e n_e(\omega, \mathbf{r}) \mathbf{v}_e(\omega, \mathbf{r}) = -e n_0 \int d\mathbf{v}\, \mathbf{v} f_{e,\mathrm{rad}} \quad (5)$$

where the distribution function has been explicitly normalized to the total density n_0. Clearly, this current is a nonlinear quantity, the product of the partial electron density and the partial-average electron velocity. Both are moments only of that part of the electron distribution function that is actively engaged in radiation. Hence, the distribution itself is the product of some nonlinear process. In a kinetic treatment it must be calculated from the interaction Vlasov equation taking into account the relevant acceleration mechanisms of the group of radiating particles. This is, in most cases, circumvented by considering only simplified particular processes.

Equation (4) is the final result since from the vector potential one can calculate the magnetic and electric fields by standard methods from $\mathbf{B} = \nabla \times \mathbf{A}$ and $\mathbf{E} = -(\partial \mathbf{A}/\partial t)$. Knowing the vector potential, the total energy radiated per unit solid angle Ω is the time integral of the square modulus of the radiation vector potential [e.g., *Jackson*, 1962]

$$\frac{dW}{d\Omega} = \int_{-\infty}^{\infty} dt |\mathbf{A}(t)|^2 \quad (6)$$

When inserting the Fourier transform for $\mathbf{A}(t)$, it is then simple matter to obtain the relation between the radiated energy per solid angle and frequency

$$\frac{dW}{d\Omega d\omega} = 2|\mathbf{A}(\omega)|^2 \qquad (7)$$

which is the same as the radiated power per unit solid angle. This relation closes the system.

Before, however, presenting a brief classification of such models it is important to point in passing on one generally overlooked simplification that concerns the use of the far field approximation. Being sufficiently remote from the source at $r \gg L$, where L is the linear extension of the source, the far field approximation is an excellent approximation [e.g., *Landau and Lifshitz*, 1975; *Jackson*, 1962]. When the spacecraft passes close to the radiation source in dense plasma, the approximation breaks down for emission at the fundamental. Here the frequency matches the plasma $\omega \approx \omega_{pe}$ (or X-mode cut-off $\omega \approx \omega_{X,co}$) frequency. In this case, for instance from the dispersion relation for the O-mode, the wave length of the radiation

$$\lambda^2 = \frac{4\pi^2 c^2/\omega^2}{1 - \omega_{pe}^2/\omega^2} \to \infty \qquad (8)$$

approaches infinity and may become of the same size as the extension of the source L. The above approximation becomes invalid in this case and the full relativistic retardation of the radiation potential must be taken into account. So far all calculations of fundamental emission have ignored this fact though of course it is not known if it will make a measurable effect.

In the following sections we review and comment on the most relevant non-thermal plasma emission mechanisms. Because gyro-synchrotron and bremsstrahlung emissions are not of interest here we refer to the well-known literature on these subjects [e.g., *Bekefi*, 1966; *Rybicki and Lightman*, 1979].

3. LINEAR PLASMA EMISSIONS

Weak interaction theory is based on the existence of a small expansion parameter. If such a parameter exists the radiation leads to only a small perturbation of the system. This parameter is the ratio of the radiation intensity to the thermal or bulk plasma energy. When a perturbation treatment is possible, the lowest order approximation should cover the effect and one expects that any first-order radiation mechanism is the dominant one. Historically, two kinds of such *linear* mechanisms have been considered. The first mechanism is linear wave *transformation* in dense high-β plasmas. The other is *direct radiation* from the plasma via negative wave absorption known under the (somewhat misleading) name of *plasma* (or electron cyclotron) *maser* radiation. These processes are not necessarily mutually exclusive but may work simultaneously. In particular, wave transformation becomes most important when enabling a linearly excited radio wave to bridge the gap between the excitation frequency ω and the lower cut-off $\omega_{X,co}$ of the free space propagation frequency in order to escape into free space. This may happen at the plasma boundary where the plasma is inhomogeneous. Usually the maser mechanism is believed to be most effective in very low-β plasmas under the condition that $\omega_{pe} \ll \omega_{ce}$. But more recent investigation has also shown that it may work in higher density plasmas in particular when they are inhomogeneous. Moreover, another mechanism may have become even more important than the conventional cyclotron maser in loss-cone plasmas. This mechanism is based on the existence of a field-aligned electric field and causes electron ring distributions which are much more efficient in generating radiation [*Pritchett*, 1984; *Pritchett and Strangeway*, 1985; *Ergun et al.*, 1998] than the loss-cone distribution.

3.1. Linear Mode Transformation

Plasmas are active media. This implies that the linear modes which are allowed to propagate in a plasma are the solutions of an eigenvalue problem. The modes form a discrete (though not equally spaced) spectrum of waves that is calculated from the dispersion relation

$$\frac{k^2 c^2}{\omega^2} = N^2 \qquad (9)$$

$N(\omega, \mathbf{k})$ is the refraction index found from linear wave theory in the appropriate approximations.

Neither the two free space modes, the O and X modes, may penetrate the plasma. The O-mode is reflected from the plasma at the point where its frequency locally matches the plasma frequency, $\omega_O = \omega_{pe}$, the X-mode is cut off at an even higher frequency, $\omega_{X,co} = \frac{1}{2}[\omega_{ce} + (4\omega_{pe}^2 + \omega_{ce}^2)^{1/2}]$. The two cut-offs are closest to each other in dense plasmas with $\omega_{pe} \gg \omega_{ce}$. Free space modes *as such* cannot be generated inside the plasma because they do not propagate. They are either composed of other modes, adding up their frequencies until the frequency exceeds the plasma frequency and permits the free space wave to propagate.

The other possibility is that the dispersion curve of some plasma mode, electrostatic or electromagnetic, experiences the local inhomogeneity of the plasma when approaching the plasma boundary in such a way that it comes close enough to a free space dispersion curve. In this case an intense plasma mode may tunnel over to the free space mode dispersion curve and may leave the plasma. No doubt, plasmas use this possibility to get rid of some of the wave energy stored in plasma modes near the boundary and to radiate the energy away into space.

In the linear transformation mechanism the inhomogeity of the plasma plays an extraordinarily important role. Usually this case is treated in the eikonal approximation by calculating the dispersion of plasma modes for the homogeneous (though anisotropic) plasma and afterwards following its change across a weakly inhomogeneous plasma. For sufficiently high frequency and short wavelength plasma waves such an approach is justified since the typical scale of inhomogeneity is much longer than the wavelength. The phase and group velocities of the plasma mode are much less than those of the free space mode ($\omega/k \ll c$). At the tunneling (coupling) point the frequency of the wave stays constant. Hence the wavelength must increases drastically in order to become a free space mode. Only a small part of the plasma wave momentum is lost in an elementary transformation event to the free space mode while the amount of energy loss is determined by the efficiency of the tunneling.

Tunneling is most efficient in dense plasmas where high frequency plasma modes may approach the free space dispersion curves near the upper hybrid wave resonance and the gap to the cut off is narrow. It is also possible in low density plasmas for waves which tunnel over to the whistler or Z-mode dispersion. Since this process is linear it should not be ignored as an important generator of radiation. Its unpleasant feature is that it is a boring process not containing interesting physics but propagation effects in inhomogeneous plasma. However, nature doesn't care whether or not we are pleased by some mechanism. If it feels that some dirty way is convenient it will not dare to go even if it is not appealing to us.

An important recent development in mode conversion and transformation processes is the radiation from the mode conversion layer in electron cyclotron resonance when the plasma wave that cannot propagate goes into resonance, is absorbed, and the layer starts radiating at the higher electron gyroharmonics above cut-off [for review see, e.g., *Swanson*, 1995]. When expanding the general plasma wave dispersion relation $D(k,x) = 0$ in an inhomogeneous plasma with either varying magnetic field $B(x) = B_0(1 + x/L_B)$ or density $n(x) = n_0(1 + x/L_n)$, we can easily understand what happens at the resonance $k = k_c$:

$$D(k,x) = P(x) + Q(x)(k - k_c)^2 + \ldots = 0 \quad (10)$$

Defining the shifted wave number $k_s = k - k_c$, this may be rewritten as

$$k_s^2 = -P(x)/Q(x) \quad (11)$$

Clearly, $P(x_0) = 0$ is a reflection point or cut-off, and $Q(x_r) = 0$ is a resonance. But at resonance the next higher order term $\propto k_s^3$ must be taken into account which describes dispersion and hence coupling to another wave which can be an escaping mode. Clearly this mechanism is of most interest in the layer where $\omega_{pe} \sim \omega_{ce}$ in order to obtain radiation at not too high harmonics. But in such a case it may contribute to emission and may explain the observation of higher harmonics than the first.

3.2. Direct Radiation: Cyclotron Maser

In the last two decades the cyclotron maser has experienced an enormous increase in the price of its stocks. This is essentially due to a famous paper by *Wu and Lee* [1979] who realized that the inclusion of even a weak relativistic effect into the calculation of the direct instability of a free space wave in plasma makes an enormous difference in the efficiency of the excitation of the free space mode. Earlier calculations of the possibility of exciting a free space mode (X- or O-mode) by linear instability [e.g., *Melrose*, 1976] made use of a temperature anisotropy in a trapping magnetic field configuration. It turned out that extraordinarily high anisotropies would be required for direct amplification. This can be understood when asking what the mechanism of instability would be. For low energy electrons gyrating in a strong magnetic field the persistent acceleration of the electrons around the magnetic field causes them to radiate weakly at the harmonics of the electron cyclotron frequency, $\ell\omega_{ce}$. The higher the harmonic number the weaker is the radiation intensity. Most of the energy is fed into the fundamental mode $\ell = 1$ leaving little for higher harmonics. In a plasma part of the radiated energy at the fundamental is reabsorbed by other electrons. In order for the fundamental to be emitted, one needs $\omega_{ce} > \omega_{pe}$. This implies dilute plasmas and strong magnetic fields. The energy of the radiation is extracted from the perpendicular energy of the electrons, $m_e v_\perp^2/2$. It is thus clear that for sufficiently high radiation energies the transverse energy of the electrons must be high and the absorption negligible. The group of electrons that radiates and absorbs satisfies the resonance condition

$$\omega - \ell\omega_{ce} = k_\parallel v_\parallel \quad (12)$$

In a thermal plasma the resonance curves in $(v_\perp, v_\parallel)$-space are circles, and there are always plenty of low energy electrons close to resonance which absorb the wave energy. To overcome absorption, one needs more particles at higher than at lower perpendicular velocities. Expressed in terms of the derivative of the particle distribution function this implies that

$$\partial f_e/\partial v_\perp > 0 \quad (13)$$

Thus one needs very large anisotropies to overcome the absorption. However, taking into account even weak relativistic effects in trapped electrons, the electron cyclotron frequency in (12) becomes velocity dependent

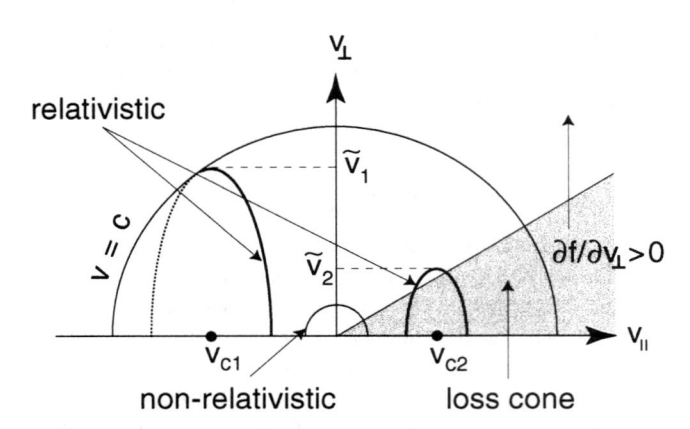

Figure 2. The resonance condition for non-relativistic (small origin centered circle) and relativistic cases. The large limiting circle is the condition $v = c$. Indicated are $v_\parallel = v_c$ and $v_\perp = \tilde{v}$ and a hypothetical loss cone. Since there are missing particles, one has $\partial f/\partial v_\perp > 0$ in the direction of the arrow parallel to the $v_\perp$ axis. Since one of the relativistic resonance ellipses falls into this region inversion of absorption and maser emission becomes possible.

$$\omega_{ce} = \frac{eB}{m_e \gamma} \quad \text{where} \quad \gamma^2 = \left[1 - \frac{v_\perp^2 + v_\parallel^2}{c^2}\right]^{-1} \quad (14)$$

This implies that the resonance curves become ellipses in phase space that are shifted away from the origin. Any resonance ellipse is centered on the $v_\parallel$ axis at parallel velocity $v_\parallel = v_c$, has excentricity e and semi-major axis $\tilde{v}$ perpendicular to the $v_\parallel$. Solving the relativistic resonance condition one finds for these quantities

$$\begin{aligned} v_c &= \omega c^2 k_\parallel / D, \quad D = (k_\parallel^2 c^2 + \ell^2 \omega_{ce,0}^2) \\ e^2 &= k_\parallel^2 c^2 / D \\ \tilde{v}^2 &= c^2 (k_\parallel^2 c^2 - \omega^2 + \ell^2 \omega_{ce,0}^2)/D \end{aligned} \quad (15)$$

Physically the cyclotron absorption of the plasma due to the excess of particles at large perpendicular velocities is turned to negative values, and the whole bulk of the plasma starts acting as a radiator instead of an absorber. This is similar to the Laser or Maser radiation from pumped levels. In our case the pumped level is populated by the excess particles at perpendicular speeds which then give up part of their energy to radiation since this is the most efficient way for the plasma of getting rid of the excess energy. The difference to a real maser it that this process is not coherent from the point of view of the particles.

Figure 2 shows examples of resonance curves. Clearly, no ellipse radius can exceed the light circle $v = c$. Hence all physically possible resonance ellipses lie inside it. One realizes, however, that it is now possible to have a favorable discrepancy between emitting and absorbing particles. The resonance ellipse needs only to fall into the domain where condition (13) holds. This can be most easily realized for loss cone distributions.

The theory of the cyclotron plasma maser is the theme of a number of more specialized contributions to these proceedings. We therefore only sketch it in passing. A more complete account can be found in *Wu* [1986], *Melrose* [1980], and *Melrose and McPhedran* [1991]. The former presentation makes use of the general formalism of growth rates, the latter two use (in a semi-quantum mechanical jargon) emission probabilities looking more complicated but containing essentially the same physics. We should note at this place that the above resonance condition (12) accounts only for the resonant part. For high radiation frequencies $[\omega > \max(\omega_{pe}, \omega_{ce})]$ this is reasonable because in the short time the wave interacts with the particles, non-resonant effects have no time to evolve and to contribute. (Note that this remark implies a criticism on any quasilinear treatment of cyclotron masers.) In the above jargon the "emission probability" is

$$w_{\ell,\mathbf{p},\mathbf{k}} = \frac{2\pi e^2 R(\mathbf{k})}{\epsilon_0 \hbar \omega} |\mathbf{e}^* \cdot \mathbf{V}|^2 \delta(\omega - \ell\omega_{ce} - k_\parallel v_\parallel) \quad (16)$$

where the Delta-function simply accounts for resonance (along the resonance ellipse), and the meaning of the symbols is as follows: e is the polarization vector of the radiation at frequency ω and wave number $\mathbf{k}$, $R(\mathbf{k}) = \gamma v_\perp / \omega_{ce,0}$ the relativistic gyro-radius, and

$$\mathbf{V} = \left(\frac{v_\perp \ell}{k_\perp R} J_\ell, i v_\perp J'_\ell, v_\parallel J_\ell\right) \quad (17)$$

is a vector containing Bessel functions of argument $k_\perp R$. In order to find the emissivity one must insert the polarization of the wanted free space mode (X- or O-mode). This gives an expression elaborated just for the mode of interest. In order to obtain the emitted power, one multiplies with the quantum of energy at the particular frequency, $\hbar \omega$, sums over all harmonic numbers ℓ and integrates over the entire $\mathbf{k}$-space with volume element $d\mathbf{k}/8\pi^3$. The resulting expressions are involved, but the physics contained is not more than has been told above.

The most important fact is that the radiation for being emitted has to have frequency above the X-mode cut-off at

$$\omega_{X,co} \approx \omega_{ce,0} + \omega_{pe}^2/\omega_{ce,0} \quad (18)$$

which only for the plasma frequency being considerably smaller than cyclotron frequency is close to ω_{ce}. This implies that emission at the fundamental $\ell = 1$ is possible only in a very dilute plasma. Also, emission at strictly perpendicular angles is excluded because there the resonance condition (12) requires $\omega < \omega_{ce} < \omega_{ce,0}$ which is below cut-off.

There is, however, a most important exception from this general rule. The above consideration holds if the plasma is non-relativistic. In presence of a field-aligned electric field $E_\parallel$ the situation changes fundamentally. If the field is strong enough, it may locally lift the entire plasma electron distribution function to sufficiently high energies such that no cold non-relativistic plasma remains (except for the non-relativistic protons). In such a case the cut-off condition must be modified. Because relativistically the plasma frequency is unchanged and only the cyclotron frequency changes according to $\omega_{ce} \to \omega_{ce}/\gamma$, tow effects can be observed. First, the emission frequency near the cyclotron frequency may drop below the nonrelativistic cyclotron frequency such that $\omega \sim \omega_{ce}/\gamma < \omega_{ce}$. Second, under conditions of a dilute plasma even the X-mode cutoff may decrease sufficiently to allow for escaping radiation at such low frequencies. This becomes obvious from

$$\omega_{X,co} \approx \omega_{ce}/\gamma + \omega_{pe}^2 \gamma/\omega_{ce} \qquad (19)$$

which replaces the ordinary cut-off. If the plasma is dilute enough and $\gamma > 1$ not too large, the cut-off is lowered by the relativistic effect. In such cases one will observe radiation coming from below the local cyclotron frequency as in the FAST measurements of *Ergun et al.* [1998] performed in the auroral magnetosphere.

With these restrictions and returning to the conventional case one can use the emission probability, follow the above prescription and, in order to find the growth rate of the X mode, sum and integrate over the probability, and integrate over the Delta-function. This implies integration over the resonance ellipse. If approximating this ellipse by a shifted circle with parameters

$$\begin{aligned} v_c &= k_\parallel c^2/\ell\omega_{ce,0} \\ \tilde{v} &= [v_c - 2(\omega - \ell\omega_{ce,0})c^2/\ell\omega_{ce,0}]^{1/2} \end{aligned} \qquad (20)$$

the integration reduces to an integration over the circle with coordinates v', ϕ' relative to the circle center. The Delta-function is replaced by

$$\delta(\omega - \ell\omega_{ce} - k_\parallel v_\parallel) \approx c^2 \delta(v' - \tilde{v})/\ell\omega_{ce,0}\tilde{v} \qquad (21)$$

and the simplified absorption rate of the cyclotron maser emission becomes

$$\gamma_{MX} = -\sum_{\ell=-\infty}^{\infty} \frac{16\pi^3 m_e^4 c^4 r_0 \tilde{v} R}{\ell\omega_{ce,0}\omega} \int_0^\pi d\phi' \times \qquad (22)$$
$$\times \left\{ |\mathbf{e}^* \cdot \mathbf{V}|^2 \left[\frac{\ell\omega_{ce}}{v_\perp} \frac{\partial}{\partial p_\perp} + k_\parallel \frac{\partial}{\partial p_\parallel} \right] f_e(\mathbf{p}) \right\}\bigg|_{\tilde{v}}$$

This equation has to be solved numerically for given distributions. In the spirit of our former discussion it is clear that the absorption must be negative in order to become a positive growth rate and produce emission. This requires that the distribution function is of the type of a loss-cone with perpendicular derivative large enough. This can be seen for instance for $\ell = 1$, the fundamental emission, where all coefficients are positive, and therefore the distribution must definitely have positive slope in the perpendicular momentum along the resonance ellipse (shifted circle). For this particular case a simple analytical estimate for X-mode emission is [*Hewitt et al.*, 1982]

$$\gamma_{MX} \approx -\frac{\pi\omega_{pe}^2 c^2}{\omega_{ce,0}\langle v\rangle^2} \frac{n_{lc}}{n_0} (\sin\alpha_0)^{-1} \qquad (23)$$

where n_{lc} is the number of missing particles in the loss cone, $\langle v \rangle$ is the average speed, and α_0 is the loss cone angle. This rate is small because of the very small frequency ratio but is enhanced by the smallness of the average speed to light velocity ratio. (Note that the smallness of the loss cone angle cancels against the smallness of the missing particle number inside the loss cone!) One thus obtains susceptible emission rates of the order of $|\gamma_{MX}/\omega_{ce,0}| \approx 10^{-5} - 10^{-3}$. A typical example of emission rates is shown in Figure 3. Usually the emission rate covers a broad spectral interval above ω_{ce} which provides difficulty to explain the sometimes observed [*Gurnett*, 1974] extremely narrow band emissions.

So far we discussed only the direct generation of X-mode radiation. The electron cyclotron maser is not restricted to X-mode excitation, however. Z-mode and O-mode generation is possible as well though less efficient. The advantage of O-mode generation is that it is less restricted to escape than the X-mode. But the weaker growth rate makes it less interesting as a loss mechanism of particle energy. Z-mode radiation is as efficient as X-mode radiation but has the problem that it must bridge the gap between the upper hybrid and X-mode cut-off in order to reach free space and radiate away. This can, for not too large gaps be achieved either by tunneling a narrow gap or by mode conversion. On the other hand, intense Z-mode waves can, close to the upper hybrid frequency, also participate in nonlinear wave-wave interaction described in the next section and lead to escaping waves. Observationally, all those waves have been reported in the literature. In particular, O-mode radiation has been seen in inhomogeneous plasmas.

It is interesting to note a number of more recent developments. One of them concerns the coherent amplification of a maser signal in a hollow ionospheric wave guide put forward by *Calvert*, [1982, 1995, 1998] (see Figure 4). The other is the explicit calculation of cyclotron maser emission in

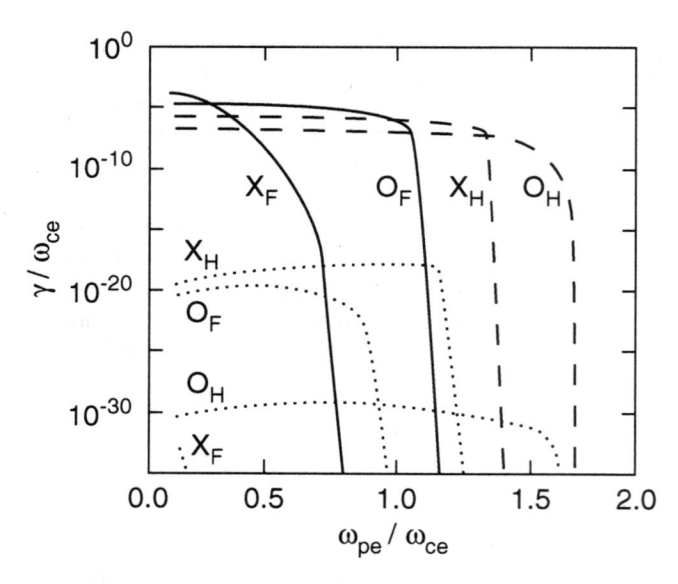

Figure 3. Linear cyclotron maser emission rates for fundamental (F, solid lines) and harmonic (H, dashed lines) as function of the plasma to cyclotron frequency ratio. Cyclotron damping (dotted) by cold electrons is also shown for a temperature ratio $T_c/T_h = 1\%$ and $T_h = 10\,\text{keV}$ [after *Aschwanden and Benz*, 1988]. Cyclotron damping is very inefficient. However, the resonance limits the emission at higher ω_{pe}/ω_{ce}. As discussed in the text, the emission rate is as well low and susceptible only for X_F and small ω_{pe}/ω_{ce} though very weak emission is still possible up to higher ratios in particular for O_F. Hence, trapping radiation in a cavity such that the waves cannot escape but may grow for longer time as in *Calvert's* [1982] model may cause enhanced emission efficiencies.

an inhomogeneous medium [*Louarn and Le Quéau*, 1996]. Another important development is the quasilinear theory of maser radiation [*Aschwanden and Benz*, 1988]. Since maser emission is at the expense of the loss cone properties of the particle distribution it may be suspected that it will readily deplete condition (13) driving the particles into the loss cone and switching the radiation off. For waves propagating at the speed of light this assumption requires that either the waves are confined to the radiation region by walls (as in Calvert's case) or that the quasilinear diffusion of particles proceeds in a time short with respect to wave amplification. Only the former condition is realistic, but whence it is satisfied, the self-quenching of the instability becomes violent causing drifting narrowband emissions and short time emission spikes similar to those observed. Thus combination of Calvert's proposal and quasilinear diffusion may provide a tool to generate narrow emission lines.

Yoon and Weatherwax [1998] recently argued that extreme gradients at the boundary of the loss cone may also produce narrow emissions. In agreement with our former discussion of the relativistic effect on the emission and propagation,

they found that the emission maxima shift to frequencies below the cyclotron frequency for the fundamental. Such radiation can escape, however, only under severe restrictions in their case of a relatively dense background in the maser mechanism, though for higher harmonic excitation this effect may lead to narrow radiation band structures.

A much more favorable mechanism can instead be based on the calculations of *Pritchett* [1984]. Pritchett used the model of a relativistic electron beam that spreads in phase space in the perpendicular direction due to the mirror force acting on it when entering a mirror magnetic field configuration. The resulting distribution is called a horseshoe distribution [*Ergun et al.*, 1998]. For dense enough beams, i. e. in the absence of a dense background plasma, the direct maser emission from this beam is intense, extends to frequencies below the local electron cyclotron frequency, and may even escape because of the lowering of the X-mode cut-off frequency in the absence of dense cold background electrons. The radiation then grows on the expense of the spread of the beam in phase space and does not depend on the loss cone. It requires only sufficiently many electrons at oblique velocities. Most easily such a situation is reached in a very dilute plasma with $\omega_{ce} \gg \omega_{pe}$ and in presence of field-aligned electric fields. These fields lift the electron distribution up to high parallel velocities while the mirror force acts at the same time spreading the beam into perpendicular direction. The problem then lies in the generation of the parallel electric field. This can be achieved by the mirror force or also by the existence of beam-excited solitary structures near the electron plasma frequency, so-called electron holes. Many such holes may produce a non-vanishing component in the parallel field. In such a case the beam is responsible for both, its spreading, and the generation of the field. As a consequence, then, radiation is produced in the same range as by the conventional maser.

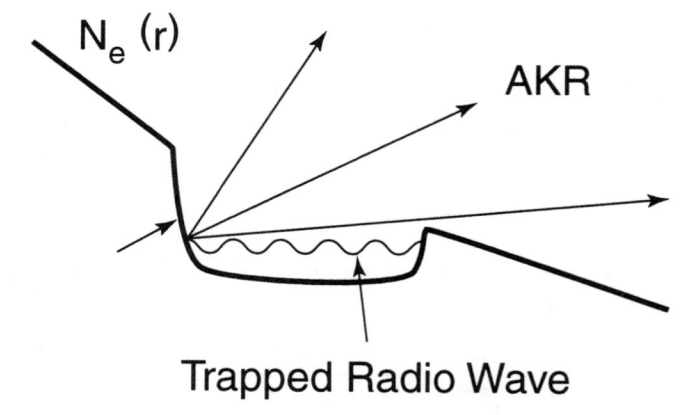

Figure 4. Trapping of AKR in auroral density depletions as proposed by *Calvert* [1982] can cause intensification of the AKR before escape is possible.

What concerns the emission of the radiation from the large density cavity it is usually speculated that reflection of the radiation from the density walls will finally guide the radiation out of the cavity where it is trapped in. As mentioned above, reduction in the X-mode cut-off by the relativistic effect helps in escaping. But in addition one even does not have to call for reflection at the walls. It is known that the density cavity is not as homogeneous as drawn in simplistic pictures. It may contain blobs of denser material or channels or filaments. When the ray of radiation is passing through one will observe scattering of the wave like in a stochastic medium. Such scattering leads to a wide range of propagation angles such that a considerable fraction of radiation can ultimately leave the cavity and escape into free space.

4. NON-LINEAR EMISSION PROCESSES

Cyclotron maser emission, because it is a direct linear amplification process, is the most important nonthermal radiation mechanism in underdense plasmas. It does, however, not apply to the overdense case. Overdense plasmas are in the majority in planetary magnetospheres. Which mechanisms will be at work there?

4.1. Fundamental and Harmonic Radiation

The simplest mechanism is one which has been proposed decades ago on the basis of the presence of Langmuir waves. Clearly, any hot plasma, because of the thermal motion of the electrons, spontaneously emits a thermal level of Langmuir waves. These waves are most intense at short wavelengths $k\lambda_D \sim 1$. But the integrated thermal level is $W/nk_BT \sim 1/n\lambda_D^3 \ll 1$ because of the large number of particles in the Debye sphere. Coalescence of such waves can be neglected. In order to achieve sufficient radiation intensity one needs to call for some violent non-thermal mechanism. All such mechanisms suffer from the fact that they are nonlinear in the sense of an expansion procedure with respect to the above small parameter. This implies that the emissivity will naturally be low.

Most of the mechanisms proposed are based on the assumption that one is dealing with head-on collisions of plasma waves of sufficiently high intensity. The problem of such an approach is always twofold: (1) there must be a mechanism that leads to excitation of very strong plasma waves (large amplitude, presumably high enough frequency, modes favorable for coalescence), (2) these modes must exist in the same region of spacetime and must satisfy the resonance conditions for coalescence. The simplest and thus most favorable process will be a three wave interaction. Remember that the escaping free space mode (phase speed $\omega/k = c$) has $k \ll k_{\rm pl}$, where $k_{\rm pl}$ is the wave number of a plasma

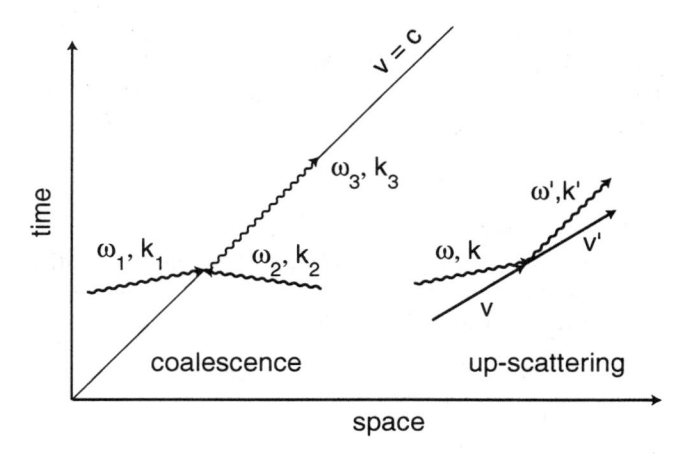

Figure 5. Two examples of Feynman diagrams for emission of radiation. *Left*: coalescence of two high-frequency plasmons of opposite momenta into a free-space mode that escapes at light speed. *Right*: scattering of a high frequency wave at a fast particle into a free-space mode with slight change of the particle momentum.

wave. The resonance condition for such a three-wave process is most easily written down in terms of the particle picture. It consists of the energy and momentum conservation equations

$$\begin{aligned} \hbar\omega_1 + \hbar\omega_2 &= \hbar\omega_3 \\ \hbar\mathbf{k}_1 + \hbar\mathbf{k}_2 &= \hbar\mathbf{k}_3 \end{aligned} \quad (24)$$

The three wave dispersion relations $\omega_\ell = \omega_\ell(\mathbf{k}_\ell)$, $\ell = 1, 2, 3$ relate the frequencies and wave numbers to each other. Only a restricted number of dispersion relations satisfy these conditions simultaneously if the escape conditions for free space modes are imposed in addition. In particular, when only one of the waves is the free space mode, then the two wave numbers of the two plasma modes must practically cancel. This implies that they be almost equal in value and of antiparallel direction, which requires very special conditions. The most probable and also most successful mechanism is still the one that involves a fast electron beam of speed $v_b > \sqrt{3}v_e$ (with v_e the electron thermal speed) which generates large amplitude Langmuir waves L via the gentle beam instability. It has been most extensively investigated in relation to solar type III radio bursts [for the best available discussion see *Robinson and Cairns*, 1998] but applies equally well to all cases where electron beams are present.

The idea is simple (cf. Figure 5): two L waves of opposite directions collide and merge into the (third) transverse T escaping free-space mode. Energy conservation then implies that the escape mode has harmonic frequency $\omega_H \approx 2\omega_{pe}$. It can be argued that the most probable mechanism includes an initial decay $L \to L' + S$ of the mother L wave into a daughter L' and a very low frequency $\omega_s \ll \omega_{pe}$ sound wave S.

This makes the process $L + L' \to T$ possible. $L'(-k_{pl})$ moves in opposite direction which is accounted for by the long wavelength $S(2k_{pl})$ wave. On the other hand, the decay reaction $L \to T + S$ just excites fundamental emission at

$$\omega_F \approx \omega_{pe}\left(1 + \frac{3}{2}\frac{v_e^2}{v_b^2} - \mu\frac{v_e}{v_b}\right) \quad (25)$$

where $\mu^2 = (1 + 3T_i/T_e)(m_i/m_e)$. Denoting the positive slope width of the gentle electron beam by Δv_b, these emissions have bandwidth of the order of

$$\frac{\Delta\omega}{\omega} \approx 3\frac{v_e^2}{v_b^2}\frac{\Delta v_b}{v_b} < \frac{\Delta v_b}{v_b} \quad (26)$$

The emission will thus fade away for very fast or very cold beams. In addition to these two merging processes, emission can also be produced by the following two scattering processes: by Raman scattering $T + L \to T'$, and by Brillouin scattering $T + S \to T'$. Both processes require the initial presence of a transverse wave. If $T = T_F$ is the fundamental frequency, Raman scattering generates harmonic emission. If $T = T_H$ is the harmonic, it may generate third harmonic emission at $\omega \approx 3\omega_{pe}$. Under the same conditions Brillouin scattering merely excites slightly shifted emissions at either the fundamental or the harmonic and is of minor interest.

The above-mentioned Raman scattering is a particularly simple example [see, e.g., *Kruer*, 1993] because two free space modes are involved. Assume the strong free space mode is an O-mode with $\omega_0 > 2\omega_{pe}$. The wave amplitude is $\mathbf{E}_O$. Electrons oscillating in this field have velocity $\mathbf{v}_O = e\mathbf{E}_O/m_e\omega_0$ parallel to the transverse field and generate a transverse current $\mathbf{j} = -e\mathbf{v}_O\delta n$. The density fluctuation here is the effect of the Langmuir wave at ω_{pe} present in the plasma. This transverse current generates another transverse wave with field $\delta\mathbf{E}$. Moreover, the O-wave and this field produce a wave pressure $\nabla P_O = \epsilon_0 \nabla(\mathbf{E}_O \cdot \delta\mathbf{E})$. This pressure modulates the density causing a contribution to the above density perturbation in the current. Hence this process is not only nonlinear but also self-modulating. The equations needed to describe this three-wave interaction are Maxwell's equations plus the equations for the electron motion and charge conservation in the wave fields. Retaining only small perturbations $\tilde{n}_e$ and $\tilde{\mathbf{A}}$ in the density vector potential variation for the scattered wave (assuming a weakly effective process) one obtains two coupled equations

$$\left(\frac{\partial}{\partial t^2} - c^2\nabla^2 + \omega_{pe}^2\right)\tilde{\mathbf{A}} = -\omega_{pe}^2\frac{\delta n_e}{n_0}\mathbf{A}_O \quad (27)$$

$$\left(\frac{\partial}{\partial t^2} + \omega_{pe}^2 - 3v_e^2\nabla^2\right)\delta n_e = \frac{\epsilon_0\omega_{pe}^2}{m_e c^2}\nabla^2\left(\mathbf{A}_O \cdot \tilde{\mathbf{A}}\right)$$

These equations describe the evolution of the amplitudes of the fields and the density fluctuations. To proceed one simply Fourier analyzes them and finds the dispersion relation

$$\omega^2 - \omega_e^2(k) = \frac{1}{4}\omega_{pe}^2 k^2 v_O^2\left(\frac{1}{\epsilon_-} + \frac{1}{\epsilon_+}\right) \quad (28)$$

where $\epsilon_\pm = \epsilon(\omega \pm \omega_0, \mathbf{k} - \mathbf{k}_0)$, $\epsilon(\omega, k) = \omega^2 - \omega_{pe}^2 - k^2c^2$, and $\omega_e^2(k) = \omega_{pe}^2 + 3k^2v_e^2$. From this one determines the condition for instability and the maximum growth rate, setting $\omega = \omega_e(k) + \delta\omega$ and considering only resonant interactions. With $\delta\omega = i\gamma$ one finds

$$\gamma \approx \frac{1}{4}\omega_{pe} k v_O[\omega_e(k)(\omega_0 - \omega_e(k))]^{-1/2} \quad (29)$$

For backscatter of the wave, k is given by $(k - k_0)^2 c^2/\omega_0^2 \approx 1 - 2\omega_{pe}/\omega_0$. This step merely fixes the excited wave mode and its emission rate.

In order to obtain the intensity of the emitted radiation one must calculate the nonlinear radiation current generated in the process of merging. This is a formidable task. The relevant expressions for wave-wave interaction can, for instance, be found in *Davidson* [1972]. Their physical content is that the fluctuation amplitudes in density and electron speed depend on the dielectric polarization properties of the Langmuir and sound waves which cause the electron oscillation. These depend on the initial electron distribution function, usually a Maxwellian. Combining these fluctuations yields the wanted radiation current density $\mathbf{j}_{rad}$ and finally the emitted power.

Similar considerations can be applied to other types of waves like electron acoustic waves, upper hybrid and lower hybrid waves, and to electron cyclotron harmonics. Electron acoustic waves are excited by electron beams in two-temperature electron plasmas and have been detected as nonlinear structure in the mid-altitude polar cusp [*Pottelette and Treumann*, 1998] and in the plasma sheet boundary layer [*Matsumoto et al.*, 1998]. Their frequency

$$\omega \approx \omega_{pe}(n_c/n_h)^{1/2} \quad (30)$$

(indices c, h are for cold and hot electrons) lies below the plasma frequency for lesser cold electrons. Since their wave numbers are parallel to the magnetic field they behave similar to Langmuir waves. This implies that merging of two electron acoustic waves can lead to a broad range of fundamental emission with frequency $\omega_{pe} < \omega < 2\omega_{pe}$. Candidates for such an emission are in the first place the shock ramps where two-temperature plasmas with denser hot than cold electrons are the rule (Figure 6). Possibly this is the mechanism for generating the back-bone emission of strong shocks. The other waves mentioned propagate either perpendicular or oblique to the magnetic field. This makes it more difficult to satisfy the kinematic merging conditions.

4.2. Collapse

In an extended plasma with waves propagating in all directions the probability of just catching up two oppositely directed sufficiently intense waves for coalescence is not high. In strongly magnetized plasmas, plasma waves that propagate strictly parallel to the magnetic field and are thus nearly one-dimensional have a greater chance to coalesce. We already mentioned that Langmuir waves excited by electron beams along the magnetic field must be scattered into opposite direction for efficient radiation. One particularly promising mechanism is collapse of plasma waves originally discovered by *Zakharov* [1972]. Collapse may turn the wave vector around by 180°, and at the same time cause high emission intensities.

Collapse is a non-resonant instability of high frequency plasma waves in more than one dimension when the local plasma pressure cannot maintain the equilibrium with the plasmon wave pressure. In this case the plasmon pushes the plasma out of the high field region causing a hole in configuration space. This hole corresponds to a broad spectrum of ion-acoustic waves and may trap other plasmons such that the plasmon field inside the hole strongly intensifies during deepening. Since the trapped plasmons are confined to a small spatial scale they become reflected from the walls of the hole such that in a very natural way waves of both propagation directions are generated. These waves may interact and give up their energy into escaping radiation at the harmonic or even at the fundamental.

The basics of the collapse [*Sagdeev*, 1979, *Robinson*, 1997, and references therein] can be easily understood from inspection of the Langmuir wave dispersion relation

$$\omega_e(\mathbf{k}) = \omega_{pe} + \frac{3k^2 v_e^2}{2\omega_{pe}} + \frac{\delta n}{2n_0}\omega_{pe} - i\gamma_L(\mathbf{k}) \quad (31)$$

where we included Landau damping γ_L. Interpreting the frequencies and wave numbers as operators, this equation leads to the famous Zakharov equation

$$\nabla \cdot \left(i\frac{\partial}{\partial t} + \frac{3v_e^2}{2\omega_{pe}}\nabla^2 + i\hat{\gamma}_L\right)\mathbf{E} = \nabla \cdot \left(\frac{\omega_{pe}\delta n}{2n_0}\mathbf{E}\right) \quad (32)$$

for the Langmuir wave amplitude. It is complemented by an equation which describes the low-frequency modulation of the density caused by the high frequency wave field. It can be obtained from considering the ion motion and averaging over the high frequency field

$$\left(\frac{\partial^2}{\partial t^2} + 2\hat{\gamma}_s\frac{\partial}{\partial t} - c_s^2\nabla^2\right)\delta n = \frac{\epsilon_0}{4m_i}\nabla^2|\mathbf{E}|^2 \quad (33)$$

The right-hand side is the force the HF-wave exerts on the plasma. c_s is the sound speed, and Landau damping of sound

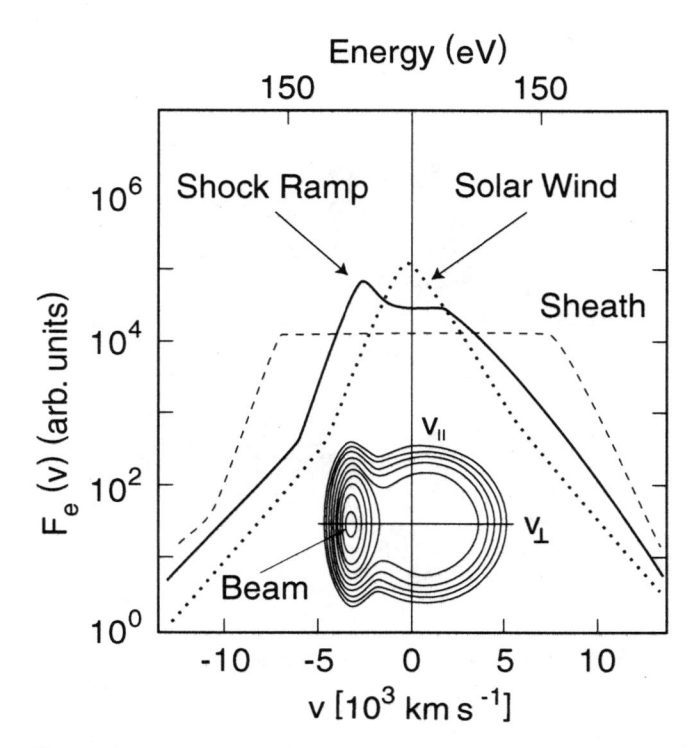

Figure 6. Schematic cuts through the electron distribution function near to and in the shock ramp of the bow shock. Behind the shock in the turbulent sheath the distribution is top-flat due to heating. In the ramp one has the mixture of a hot heated component and the remainders of the fast cold impacting solar wind electrons forming a beam. The upstream region is characterized by the solar wind distribution, consisting of the cold beam and a dilute halo. The inlay shows a velocity space section for the ramp distribution showing the incoming beam. [after *Feldman et al.*, 1982].

waves is taken into account. Figure 7 shows a numerical calculation of the formation of collapsons or cavities based on these equations. Interestingly, for more than one dimensional systems there is no stationary state on the basis of these equations. Hence the cavities shrink, and the trapped plasmons interact ever stronger.

The intensity of the radiation produced in this mechanism [Figure 8] is a strong function of time. This has been shown by *Hafizi and Goldman* [1981]. It starts at the very weak two plasmon coalescence level of the original wave and increases by orders of magnitude towards the end of the collapse at time t_c. Then the wavelength of the trapped waves becomes very short, of the order of the Debye length, and the emission process ceases. Hence, the resulting radiation should be very bursty.

Unfortunately, there is a problem with this beautiful mechanism. Observations near the bow shock where it is expected to work do not support its reality [*Cairns and Robinson*, 1999]. The cavitons do quickly burn out by transit time

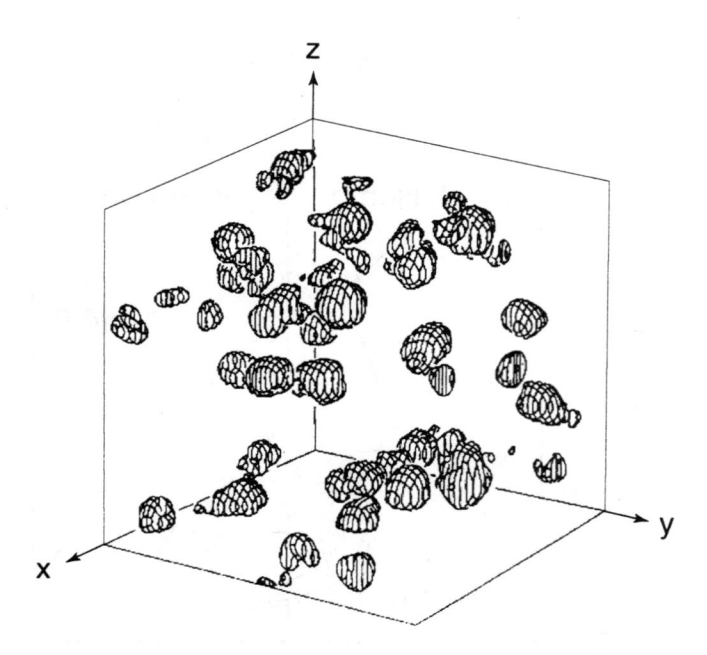

Figure 7. A numerical simulation result showing the threedimensional evolution of localized collapsing cavitons (collapsons) from a given initial state based on the two Zakharov equations. Note the irregular distribution of the localized wave packets that resembles a turbulent state. Outside these wave packets very low levels of irregular waves are found [from *Robinson*, 1997].

damping of the trapped plasmons which give up their energy to beam electrons. This helps maintaining the beam over long distances. Because it involves many particles passing for a short time across the caviton and being accelerated weakly, it does not lead to appreciable radiation. Such burnt-out cavities are slowly decaying ion acoustic waves which trap other Langmuir waves and cause radiation in an uncorrelated way, called stochastic radiation as investigated by *Robinson and Cairns* [1998].

Similarly, in the auroral zone collapse may not be very efficient as well. Rather one expects solitons of low frequency waves like perpendicular lower hybrid waves at $\omega \approx \omega_{lh}$ to be generated which when passing across the plasma can couple to perpendicular high frequency waves near the upper hybrid frequency ω_{uh}. This kind of 'vacuum' cleaner (or antenna) effect has been investigated by *Pottelette et al.* [1992, 1993] and should lead to narrow band emission. It will work in competition with the maser mechanism because the conditions for both mechanisms are similar.

5. SUMMARY

In summarizing our brief overview of planetary radio emission mechanisms we may conclude that almost all planetary radio wave emission is nonthermal and needs some nonthermal energy source. Though such sources are in excess in any of the planetary magnetospheres, nonthermal radiation can only be emitted if it is related to particle acceleration. The most important nonthermal sources are thus weakly relativistic loss cones in underdense plasmas, horseshoe distributions of electron beams in plasmas of low cold plasma density, and electron beams in dense plasmas. The most promising mechanisms are the electron cyclotron and horseshoe masers in underdense plasmas, as well as the stochastic model of Langmuir wave merging in dense plasmas containing pre-formed density cavities that may trap the Langmuir waves (or equivalently in burnt-out collapsons). The former models and mechanisms apply to the auroral magnetosphere, while the latter works in the vicinity of the bow shock or other shock waves in magnetospheres as well as in the dense regions of the magnetospheric magnetotails. We would like to stress that the present state of the art of our current knowledge favors in the first line the horseshoe radiation mechanism in the relatively low-density regions of the magnetosphere. This has been strongly suggested by

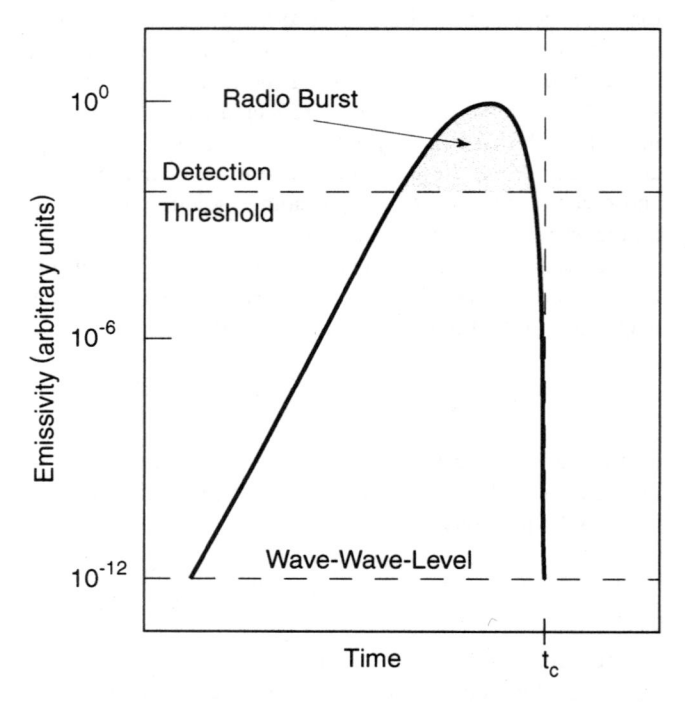

Figure 8. Radiation intensity emitted in harmonic emission from a collapsing Langmuir caviton as function of time. The radiation starts at a very low level corresponding to initial three-wave interaction. During collapse the intensity increases by orders of magnitude as consequence of the trapping of plasmons and increasingly intense interaction of the trapped waves until shortly before collapse time τ_c the emissivity drops to zero. Such behavior should cause intense bursts of harmonic radiation of typical length of a fraction of τ_c emitted just before end of the collapse [after *Hafizi and Goldman*, 1981].

the recent observations of the FAST spacecraft. These measurements showed that a precipitating electron beam in presence of a field-aligned electric field in a converging magnetic field tends to be scattered in pitch-angle into the perpendicular direction while the whole electron plasma component is lifted by the electric field to higher energies. In this case the plasma is void of a cold component that could damp the wave, and linear maser emission becomes ever more favorable than in any loss-cone cyclotron maser case. We may expect such a mechanism to work in all magnetospheres as well as in the solar coronal and upper chromospheric loops and cause radiation of the kind observed there in solar radio bursts.

A large number of unresolved problems still remain to be solved. Such problems include the narrow bandedness of the AKR and its observed very fast temporal changes. The current models will have to be developed further in order to account for such properties of the emission. It may be expected that the local inhomogeneity of the plasma, scattering of waves at density striations and trapping of waves in density cavities are responsible for producing these fine structures and variations in the radiation. On the other hand, the acceleration mechanism of the beam will be another source of such modulations of the radiation.

Other problems involve the observation of highly structured radio emissions from the deeper ionosphere. Also the lack of observation of radiation from the reconnection site poses a barely understood fact. Electron beams ejected from these regions may cause weak highly variable (type III like) emissions in the tail. But the presumed slow shocks emanating from the reconnection site do not radiate. This is probably a cause of their subcritical nature. They heat the plasma but do not reflect particles and therefore show no signature in the radio while their X-ray emission measure is too low for detection.

Finally, we should mention that though collapse is probably not as important as previously believed in generating radiation and though stochastic emission seems more probable, we have yet by far not investigated all possible mechanisms. In particular, what is the role of higher order processes in a non-perturbative treatment, what the role of d.c. fields in double layers, macroscopic or microscopic, what the role of electron holes, ion holes, BGK modes and other more exotic beasts one encounters in plasma turbulence, and finally what is the role of a plasma consisting of large numbers of those structures, i.e. what is the role of turbulence in affecting the radiation? Future research should concentrate on some of these items. A number of processes are detailed in the more specialized contributions to these proceedings.

Acknowledgments. The author thanks the editors and organizers for their kind invitation to the Chapman Conference. He also thanks Iver Cairns, Bob Ergun, Andrew Melatos, Raymond Pottelette, and Peter Robinson for valuable discussions. He also thanks the referee for his constructive remarks which led to considerable improvement of the presentation. Part of this work has been supported by the Procope program under contract number D/9822921.

The Editor would like to thank the reviewer of this manuscript.

REFERENCES

Aschwanden, M. J. and A. O. Benz, On the electron-cyclotron maser instability. I. Quasi-linear diffusion in the loss cone, *Astrophys. J.*, *332*, 447-465, 1988.

Bekefi, G., *Radiation Processes in Plasmas*, 377 pp., J. Wiley & Sons, New York, 1966.

Cairns, I. H. and P. A. Robinson, Strong evidence for stochastic growth of Langmuir-like waves in Earth's foreshock, *Phys. Rev. Lett.*, **82**, 3066-3069, 1999.

Calvert, W., A feedback model for the source of auroral kilometric radiation, *J. Geophys. Res.*, *87*, 8199-8214, 1982.

Calvert, W., An explanation for auroral structure and the triggering of auroral kilometric radiation, *J. Geophys. Res.*, *100*, 14887-14894, 1995.

Calvert, W., The gotcha-kata-kata or "domino" theory for substorm expansion, in *Substorms-4*, edited by S. Kokubun and Y. Kamide, pp. 259-263, Kluwer, Dordrecht, Holland, 1998.

Davidson, R. C., *Methods in Nonlinear Plasma Theory*, 356 pp., Academic Press., New York, 1972.

Ergun, R. E., C. W. Carlson, J. P. McFadden, F. S. Mozer, G. T. Delory, W. Peria, C. C. Chaston, M. Temerin, R. Elphic, R. Strangeway, R. Pfaff, C. A. Cattell, D. Klumpar, E. Shelly, W. Peterson, E. Moebius, and L. Kistler, FAST satellite wave observations in the AKR source region, *Geophys. Res. Lett.*, *25*, 2061, 1998.

Feldman, W. C., S. J. Bame, S. P. Gary, J. T. Gosling, D. McComas, M. F. Thomsen, G. Paschmann, N. Sckopke, M. M. Hoppe, and C. T. Russell, Electron heating within the earth's bow shock, *Phys. Rev. Lett.*, *49*, 199-202, 1982.

Gurnett, D. A., The Earth as a radio source: Terrestrial kilometric radiation, *J. Geophys. Res.*, *79*, 4277-4238, 1974.

Hafizi, B. and M. V. Goldman, Harmonic emission from adiabatically collapsing Langmuir solitons *Phys. Fluids*, *24*, 145, 1981.

Hewitt, R. G., D. B. Melrose, and K. G. Rönnmark, The loss-cone driven eloectron cyclotron maser, *Austr. J. Physics*, *35*, 447-471, 1982.

Hilgers, A., The auroral radiating plasma cavities, *Geophys. Res. Lett.*, *19*, 237-240, 1992.

Jackson, J. D., *Classical Electrodynamics*, 641 pp., John Wiley & Sons, New York, 1962.

Kruer, W., *The Physics of Laser-Plasma Interactions*, 182 pp., Addison-Wesley, Redwood City, 1993.

LaBelle, J., M. L. Trimpi, R. Brittain, and A. T. Weatherwax, Fine structure of auroral roar emissions, *J. Geophys. Res.*, *100*, 21935, 1995.

Landau, L. D., and E. M. Lifshitz, *The Classical Theory of Fields*, 392 pp., Addison-Wesley, Reading, Mass., 1951.

Louarn, P. and D. Le Quéau, Generation of auroral kilometric radiation in plasma cavities - II. The cyclotron maser instability in small size sources, *Planet. Space Sci.*, *44*, 211-224, 1996.

Malingre, M., R. Pottelette, R. A. Treumann, and M. Berthomier, Observation of broadband wave bursts with power law spectra above the plasma frequency in the underdense auroral plasma, *J. Geophys. Res.*, *102*, 19861-19871, 1997.

Matsumoto, H., H. Kojima, Y. Omura, and I. Nagano, Plasma waves in geospace: Geotail observations, in *New Perspectives on the Earth's Magnetotail*, Geophys. Monograph 105, edited by A. Nishida, D. N. Baker, and S. W. H. Cowley, pp. 259-319, AGU, Washington D.C., 1998.

McLean, D. J. and N. R. Labrum (Eds.), *Solar Radiophysics*, 300 pp., Cambridge University Press, Cambridge, 1985.

Melrose, D. B., The interpretation of Jupiter's decametric radiation and the terrestrial kilometric radiation as direct amplified gyroemisson, *Astrophys. J.*, 207, 651-662, 1976.

Melrose, D. B., Emission at cyclotron harmonics due to coalescence of Z-mode waves, *Astophys. J.*, 380, 256-267, 1991.

Melrose, D. B., Kinetic Plasma Physics, in *Plasma Astrophysics*, Saas-Fee Advanced Course 24, edited by J. G. Kirk, D. B. Melrose, and E. R. Priest, pp. 113-224, Springer-Verlag, Berlin, 1994.

Nagano, I., S. Yagitani, H. Kojima, Y. Kakehi, T. Shiozaki, H. Matsumoto, K. Hashimoto, T. Okada, S. Kokubun, and T. Yamamoto, Wave form analysis of the continuum radiation observed by Geotail, *Geophys. Res. Lett.*, 96, 5631-5634, 1994.

Pottelette, R., R. A. Treumann, G. Holmgren, N. Dubouloz, and M. Malingre, Acceleration and radiation from auroral cavities, in *Auroral Plasma Dynamics*, edited by R. L. Lysak, pp. 253-265, Geophys. Monograph 80, AGU, Washington, D. C., 1993.

Pottelette, R., R. A. Treumann, and N. Dubouloz, Generation of auroral kilometric radiation in upper-hybrid wave - lower-hybrid soliton interaction, *J. Geophys. Res.*, 97, 12029-1239, 1992.

Pottelette, R. and R. A. Treumann, Impulsive broadband electrostatic noise in the cleft: A signature of dayside reconnection, *J. Geophys. Res.*, 103, 9299-9307, 1998.

Pritchett, P. L., Relativistic dispersion and the generation of auroral kilometric radiation, *Geophys. Res. Lett.*, 11, 143, 1984.

Pritchett, P. L. and R. J. Strangeway, A simulation study of kilometric radiation along an auroral field line, *J. Geophys. Res.*, 90, 9650, 1985.

Robinson, P. A., Nonlinear wave collapse and strong turbulence, *Rev. Mod. Phys.*, 69, 507-574, 1997.

Robinson, P. A. and I. H. Cairns, Fundamental and harmonic emission in type III solar radio bursts. I-III, *Solar Phys.* 181, 363-437, 1998.

Rybicki, G. B. and A. P. Lightman, *Radiative Processes in Astrophysics*, 382 pp., J. Wiley & Sons, New York, 1979.

Sagdeev, R. Z., The 1976 Oppenheimer Lectures: Critical problems in plasma astrophysics. I. Turbulence and nonlinear waves, *Rev. Mod. Phys.*, 51, 1-9, 1979.

Schriver, D., M. Ashour-Abdalla, R. A. Treumann, M. Nakamura, and L. M. Kistler, The lobe to plasma sheet boundary layer transition: Theory and observations, *Geophys. Res. Lett.*, 17, 2027-2031, 1990.

Shepherd, S. G., J. LaBelle, and M. L. Trimpi, Further investigation of auroral roar fine structure, *J. Geophys. Res.*, 103, 2219-2227, 1998.

Swanson, D. G., Cyclotron absorption and emission in mode conversion layers - a new paradigm, *Rev. Mod. Phys.*, 67, 837-862, 1995.

Treumann, R. A., and R. Pottelette, Solar flare radio emission: An application of the horseshoe maser mechanism, *Astrophys. J.*, to be submitted, 2000.

Weatherwax, A. T., J. LaBelle, M. L. Trimpi, R. A. Treumann, J. Minow, and C. Deehr, Statistical case studies of radio emissions observed near $2f_{ce}$ and $3f_{ce}$ in the auroral zone, *J. Geophys. Res.*, 100, 7745-2257, 1995.

Wu, C. S., Kinetic cyclotron and synchrotron maser instabilities: Radio emission processes by direct amplification of radiation, *Space Sci. Rev.*, 41, 215-298, 1986.

Wu, C.S. and L. C. Lee, A theory of the terrestrial kilometric radiation, *Astrophys. J.*, 230, 621-626, 1979.

Zakharov, V. E., Collapse of Langmuir waves, *Sov. Phys.-JETP*, 35, 908-918, 1972.

R. A. Treumann, Max-Planck-Institute for extraterrestrial Physics, Postfach 1603, D-85740 Garching bei München, Germany, and International Space Science Institute, Hallerstr. 6, CH-3012 Bern, Switzerland. (e-mail: tre@mpe.mpg.de; treumann@issi.unibe.ch)

Roles Played by Electrostatic Waves in Producing Radio Emissions

Iver H. Cairns and P. A. Robinson

School of Physics, University of Sydney, Sydney, Australia

Processes in which electromagnetic radiation is produced directly or indirectly via intermediate waves are reviewed. It is shown that strict theoretical constraints exist for electrons to produce nonthermal levels of radiation directly by the Cerenkov or cyclotron resonances. In contrast, indirect emission processes in which intermediary plasma waves are converted into radiation are often favored on general and specific grounds. Four classes of mechanisms involving the conversion of electrostatic waves into radiation are linear mode conversion, hybrid linear/nonlinear mechanisms, nonlinear wave-wave and wave-particle processes, and radiation from localized wave packets. These processes are reviewed theoretically and observational evidence summarized for their occurrence. Strong evidence exists that specific nonlinear wave processes and mode conversion can explain quantitatively phenomena involving type III solar radio bursts and ionospheric emissions. On the other hand, no convincing evidence exists that magnetospheric continuum radiation is produced by mode conversion instead of nonlinear wave processes. Further research on these processes is needed.

1. INTRODUCTION

Plasmas directly produce thermal levels of radiation due to spontaneous emission. However, while thermal radio emissions are ubiquitous, thermal damping limits their brightness temperatures T_b to less than approximately the electron temperature T_e of the source plasma [e.g., *Melrose*, 1970]. Radiation with $T_b \gtrsim T_e$ is therefore necessarily produced by nonthermal processes, associated with particle distributions or radiation processes that are nonthermal. Since almost all known solar system and astrophysical plasmas have $T_e \lesssim 10^8$ K, a rule-of-thumb is that radio emissions with $T_b \gtrsim 10^8$ K are produced by nonthermal processes.

Direct emission processes are those in which a charged particle radiates waves or radiation directly by the Cerenkov or cyclotron resonances. Familiar examples include Langmuir waves driven by an electron beam and gyrosynchrotron or synchrotron radiation from gyrating electrons. A necessary condition for generating nonthermal levels of waves or radiation directly is then that the electron distribution be nonthermal, whether due to the presence of energetic electrons with an anisotropy in velocity space or a non-Maxwellian distribution function such as a power law.

Indirect emission processes are those in which one or more sources of free energy drive nonthermal levels of intermediate plasma waves which are then converted into escaping radiation by another process. Three basic indirect emission mechanisms exist for converting wave energy into free-space radio emission: (1) linear mode conversion in regions with gradients in plasma characteristics [e.g., *Budden*, 1961; *Jones*, 1976]; (2) nonlinear wave-wave and wave-particle processes [e.g., *Tsytovich*, 1972; *Melrose*, 1980b]; and (3) radiation from

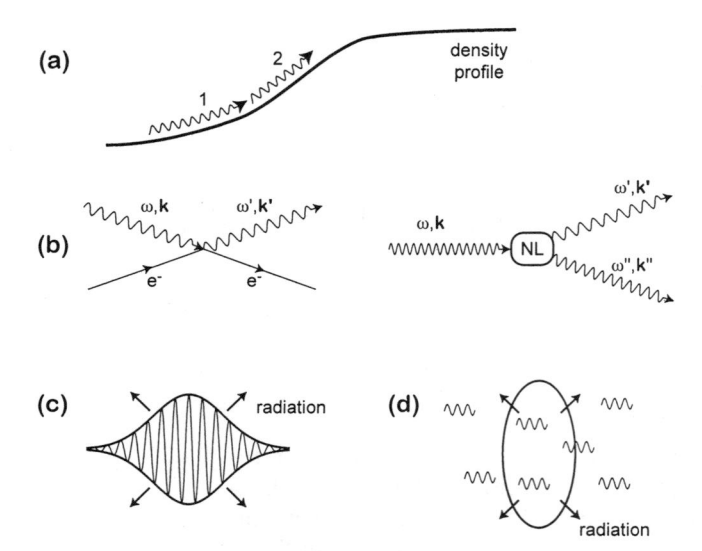

Figure 1. Schematics of (a) linear mode conversion of mode 1 into mode 2 at a density gradient, (b) nonlinear wave-particle and wave-wave processes involving multiple waves, (c) radiation from a collapsing Langmuir wave packet, and (d) a localized wave packet acting as an antenna for the conversion of other waves into radiation. More details are given in Sections 4 to 7.

localized wave packets [e.g., *Papadopoulos and Freund*, 1978; *Pottelette et al.*, 1992]. These processes are illustrated qualitatively in Figure 1. Not all proposed radiation processes are classifiable uniquely into one of the three basic mechanisms above because one overall process may involve several steps, each categorized under a different basic mechanism; e.g., a nonlinear process may produce an electrostatic wave which undergoes linear mode conversion into escaping radiation.

This review focuses on indirect emission processes which involve intermediate plasma waves that are electrostatic, in the process updating and extending previous reviews [e.g., *Budden*, 1961; *Tsytovich*, 1972; *Melrose*, 1980a,b, 1986; *Goldman*, 1983]. The emphasis is on analytic/numerical theory and observations of solar system phenomena, with few references to numerical simulations or laboratory experiments. Reasons for these emphases, in addition to this book's strong focus on space plasmas and to length limitations, include (1) the lack of multiple-dimensional simulations able to self-consistently follow the evolution of unstable particles, electrostatic waves, and electromagnetic radiation in inhomogeneous plasmas, (2) the inability of current simulations to adequately account for the almost universal burstyness, statistics, and typically low fields (values usually much less than the thresholds for relevant non-linear processes) of electrostatic waves in space [e.g., *Robinson et al.*, 1993; *Cairns and Robinson*, 1999, and references therein], meaning that little confidence can be attached to predictions for the typically weak electromagnetic radiation resulting therefrom, and (3) the lack of laboratory experiments for non-relativistic particle distributions and plasma conditions typical of solar system plasmas.

One aim of the paper is to present general arguments why indirect emission processes are often favored to dominate over direct emission processes in producing radiation in nonrelativistic plasmas. This is done by describing the basic constraints on, and associated difficulties for, direct emission processes to produce high levels of radiation (Section 2) and then presenting reasons why indirect emission processes, particularly those involving electrostatic intermediary waves, are often favored to produce radiation (Section 3). The second aim of the paper is to describe the theory and observational status of the three basic indirect emission mechanisms plus one hybrid linear/nonlinear mechanism (Sections 4 to 7). Each of these mechanisms is illustrated and discussed in terms of observational data, with a focus on continuum radiation in Earth's magnetosphere and type III solar radio bursts in the interplanetary medium.

2. CONSTRAINTS ON DIRECT GENERATION OF NONTHERMAL RADIO EMISSIONS

Direct generation of nonthermal radio emissions encounters strict theoretical constraints associated with the large phase speeds $v_\phi \approx c$ of electromagnetic radiation in the free-space o and x modes. Figure 2 shows the predictions of cold plasma theory for the refractive indexes $N = c/v_\phi$ of the o, x, z, and whistler modes as functions of the ratio ω/ω_p of the radiation's (angular) frequency ω to the electron plasma frequency ω_p. (Here the electron gyrofrequency $\Omega_e = 0.5\omega_p$ and $\theta = 30°$ is the angle between the wave vector **k** and the ambient magnetic field **B**.) Note that the o and x modes have $N \approx 1$ and $v_\phi \approx c$ except very close to their cutoffs at $N^2 = 0$, that cold plasma theory is usually justified when v_ϕ greatly exceeds relevant particle thermal speeds, and that the whistler and z modes cannot propagate where $\omega > \min(\omega_p, \Omega_e)$ and where ω exceeds the upper hybrid frequency ω_{uh}, respectively.

Using standard kinetic plasma theory the growth rate Γ of waves in a magnetized plasma can be written, ignoring ion effects, as

$$\Gamma \propto - \sum_{s=-\infty}^{+\infty} \int d^3\mathbf{p} \left(\frac{s\Omega_e}{\gamma v_\perp} \frac{\partial}{\partial p_\perp} + k_\| \frac{\partial}{\partial p_\|} \right) f(\mathbf{p}) \quad (1)$$
$$\times \delta(\omega - s\Omega_e/\gamma - k_\| v_\|) \,.$$

Here $f(\mathbf{p})$ is the distribution function of electrons with momentum $\mathbf{p}$, Lorentz factor γ, and speeds $v_\perp$ and $v_\|$ perpendicular and parallel to the magnetic field. The resonance condition in the delta function yields

$$v_\phi = v_\| \cos\theta + s\Omega/k\gamma \,. \quad (2)$$

Accordingly, large values of Γ require sizable $\|$ or $\perp$ gradients in $f(\mathbf{p})$ where the constraint (2) on v_ϕ can be satisfied.

Combining the properties of the o and x modes with Eqs (1) and (2), direct generation of radiation therefore requires (i) the presence of electrons with $v_\| \sim c$ and $\theta \sim 0$ when magnetization effects are unimportant, or (ii) the cyclotron term $s\Omega_e/k_\|\gamma$ to dominate in (2). Typically the cyclotron term dominates when γ is large and $\Omega_e \gtrsim 0.1\omega_p$, as found in studies of cyclotron maser radiation [*Wu and Lee*, 1979; *Melrose*, 1986; *Ergun et al.*, 1998]. In either case, direct generation of nonthermal radio emissions requires the presence of energetic electrons whose energies are at least semi-relativistic and which have suitable gradients in $f(\mathbf{p})$.

3. REASONS WHY INDIRECT EMISSION PROCESSES CAN DOMINATE

Three basic reasons exist why indirect emission processes can dominate direct emission processes in producing high levels of radiation. The first is that plasma waves with $v_\phi \ll c$ are typically much easier to drive in nonrelativistic plasmas than o- and x-mode radiation with $v_\phi \approx c$, for instance via resonant linear instabilities which efficiently and rapidly couple unstable features in the particle distribution to growing waves. The second reason is that such instabilities routinely produce large amplitude waves whose energy density $k_B T_w V_w$ can be comparable to the energy density in the driving particles and for which the effective wave temperature T_w greatly exceeds T_e (here k_B is Boltzmann's constant and V_w is the volume of wavenumber space occupied by the waves). The third reason is that multiple processes exist to convert such large amplitude waves into escaping radiation with

$$T_w \gtrsim T_b \gg T_e \,. \quad (3)$$

Even when the efficiency for this final conversion process is small, say $\sim 10^{-6}$ as in some of the examples be-

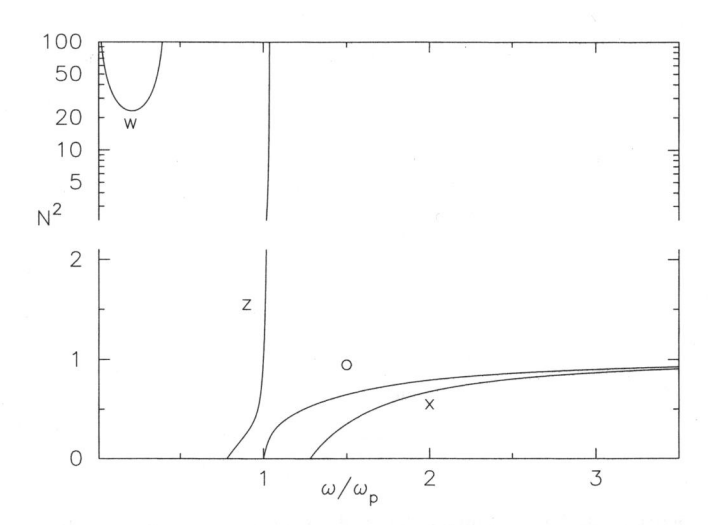

Figure 2. Dispersion diagram of N^2 versus ω/ω_p for the o, x, z, and whistler (w) modes for $\omega_p = 2\Omega_e$ and $\theta = 30°$, following *Melrose* [1980b].

low, the overall conversion efficiency from particle free energy to driven (intermediate) plasma waves and then to radiation can be sufficient to dominate any competing direct emission process. One example of this is in type III solar radio bursts [*Robinson and Cairns*, 1993, 1998a,b, 1999], as discussed below.

The three reasons above are general and do not require that the intermediate waves be electrostatic rather than electromagnetic. Instead, the basic requirement is that at least one of the intermediate wave modes participating in the overall process must have a frequency comparable to or greater than the low frequency cutoffs of the o and x modes, at ω_p and $\omega_x = \Omega_e/2 + (\omega_p^2 + 4\Omega_e^2)^{1/2}/2$, respectively. This requirement is imposed by energy (frequency) conservation for the conversion process. In electron-proton and electron-ion plasmas this frequency constraint can be satisfied only if at least one of the participating wave modes has a high frequency comparable with ω_p or Ω_e and is thus determined primarily by electron physics. Suitable modes are the Langmuir, z, and upper hybrid modes in plasmas with $\omega_p \gg \Omega_e$, and the Bernstein, upper hybrid, z, and whistler modes when $\omega_p \ll \Omega_e$. Note that the Langmuir and upper hybrid modes are special cases of the more general z mode which, like the whistler mode, is strongly electromagnetic under some plasma conditions and electrostatic under others.

This review concentrates on indirect emission processes with electrostatic intermediate waves. This situation is encountered frequently, due to the typically nonrelativistic nature of the driving electrons and due to

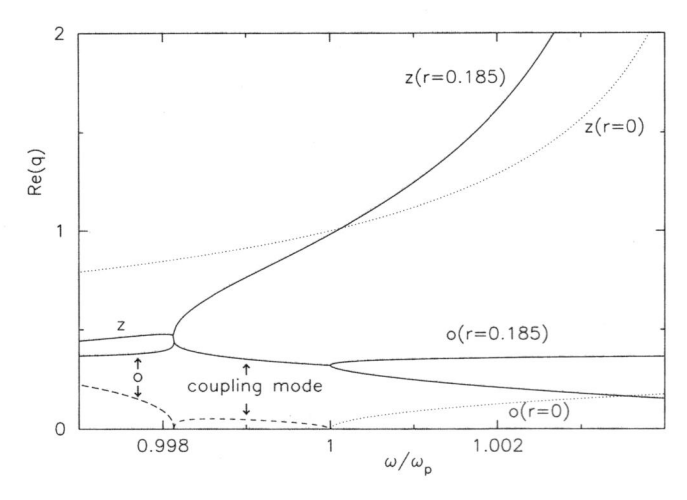

Figure 3. Dispersion diagram of q versus ω/ω_p, showing the mode coupling solution joining the o and z modes (solid and dashed curves show real and imaginary parts, respectively). Here $\Omega_e = 0.2\omega_p$, $\psi = \phi = 30°$, and the modes for $r = 0$ (no coupling) and $r = 0.185$ are shown. The o mode is evanescent for $\omega/\omega_p \lesssim 0.99982$.

magnetization effects often being unimportant: under these conditions the resonance condition in (1) and (2) implies that $N \cos\theta \approx c/v_\parallel$ and the condition $v_\parallel \ll c$ then implies

$$N \cos\theta = c/v_\phi \sim c/v_\parallel \gg 1 \ ; \qquad (4)$$

i.e., the waves must have large N and (from above) high frequencies $\gtrsim \min(\omega_p, \Omega_e)$. Reviewing the plasma modes predicted by kinetic theory and cold plasma theory, as well as the list above, it becomes apparent that modes obeying these constraints are usually (but not always) electrostatic; examples include Langmuir, upper hybrid, and (usually) Bernstein waves. Note for completeness, however, an indirect emission process involving electromagnetic z-mode waves [*Willes et al.*, 1998].

4. LINEAR MODE CONVERSION

An example of linear mode conversion (Figure 1) is the transformation into o-mode radiation of a Langmuir wave as it propagates with constant ω but decreasing k into a region with increasing density [*Budden*, 1961; *Melrose*, 1980b]. Linear mode conversion can also be a loss mechanism; e.g., o-mode radiation can be transformed into Langmuir waves as it propagates into a region where $\omega_p \gtrsim \omega$. The standard formalism for studying linear conversion of o, x, z, and whistler waves into one another is based on the cold plasma dispersion equation, which can be written as a quadratic in N^2 with coefficients that depend only on ω/ω_p, ω_p/Ω_e, and θ.

Spatial variations in plasma quantities, which permit waves in different modes to couple together, are often assumed to occur in parallel, stratified planes with a unit normal $\hat{n}$ relative to **B** [e.g., *Budden*, 1961]. In this case the cold plasma dispersion equation can be rearranged into the so-called Booker quartic in the variable $q = N\mathbf{k}.\hat{n}/k$, which depends only on ω and the wavenumber parallel to $\hat{n}$:

$$aq^4 + bq^3 + cq^2 + bq + e = 0, \qquad (5)$$

with $N^2 = q^2 + r^2$, $a = a(\omega/\omega_p, \omega/\Omega_e, r, \psi, \phi)$, and likewise for the other coefficients. Here r is the component of N perpendicular to $\hat{n}$, and the angles ψ and ϕ are the polar and azimuthal angles of $\hat{n}$ relative to **B**. Snell's Law implies that waves have constant r.

A wave propagating through time-invariant plasma moves with constant frequency but varying **k** and N. Wave energy may move from one mode to another via evanescent (complex) solutions for q, the so-called Ellis windows, which join the normal modes. For significant coupling to occur, the normal modes must be close together in $N - \omega/\omega_p$ space. Accordingly, from Figure 2 and its analog for $\Omega_e > \omega_p$, linear mode conversion only involves the production or loss of radiation in the o mode. Put another way, mode conversion produces radiation 100% polarized in the sense of the o mode. In the situation $\omega_p \gg \Omega_e$ the coupling is between the o and z modes, referred to as the first Ellis window. The second Ellis window occurs in the opposite regime, $\omega_p \lesssim \Omega_e$, and involves coupling between the whistler, o, and z modes.

Figure 3 illustrates linear mode conversion for the first Ellis window. The usual o and z mode solutions are apparent for $r = 0$, together with the modified modes for $r = 0.185$. A z-mode wave moving toward lower density travels rightward from the low ω/ω_p side of Figure 3, eventually travelling along the evanescent mode-coupling solution to the o mode or else continuing in the z mode. Alternatively, incident o-mode radiation moving into a higher density region can couple into the z mode as it travels leftward in Figure 3. Theoretically, the attenuation factor for the coupling waves is

$$A = \exp\left[\int (\mathrm{Im}\ \mathbf{q}).\mathbf{dl}\right] = \exp[B\ q_{i,max}\ \Delta\omega\ L_N] ,$$
(6)

thereby depending on the length scale L_N of the density gradient and the extent $\Delta\omega$ of the mode coupling region.

Many radio phenomena have been interpreted in terms of linear mode conversion, including ionospheric emissions from sounding [*Budden*, 1961] and the aurora

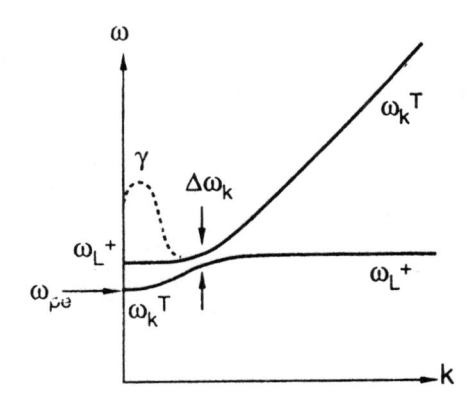

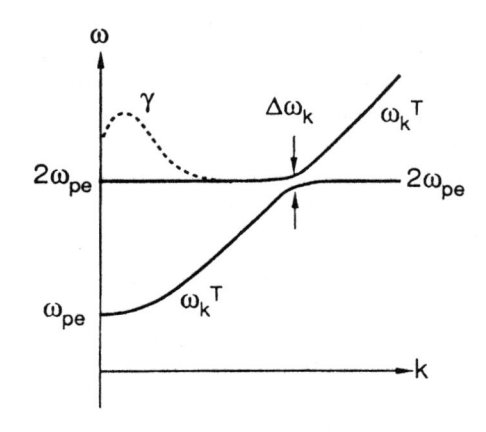

Figure 4. Modifications to the normal Langmuir and o/x modes due to the presence of pre-existing turbulence following *Yoon* [1995].

[*Willes et al.*, 1998; *Yoon et al.*, 1998], continuum radiation from planetary magnetospheres, and solar radio bursts. The last two examples are now discussed in more detail. *Jones* [1976] proposed that continuum radiation in Earth's magnetosphere [*Gurnett and Shaw*, 1973] is produced in the o mode by mode conversion $U \to o$ of intense upper hybrid waves U found near the magnetic equator. *Jones* [1976] argues that the mechanism can explain the observed levels of radiation, although *Rönnmark* [1989] disputes this. The principal predictions of the theory are that the radiation should be in the o-mode and that each source region should produce two beams of radiation directed at specific angles dependent on ω_p/Ω_e above and below the magnetic equator. *Jones et al.* [1987] published spacecraft data from one specific period which appear to show the predicted radiation pattern. However, *Morgan and Gurnett* [1991] published a detailed analysis of this radiation, concluding that little direct evidence exists for the mode conversion model. This may be due to the existence of multiple simultaneous, distributed, and time-variable sources which obscure the predicted radiation patterns. More work on mode conversion and on competing nonlinear processes [*Melrose*, 1981] is needed.

Linear mode conversion of Langmuir or z-mode waves into o-mode radiation near ω_p has been proposed as both a generation process and a loss process in the context of type III solar radio bursts. Using length scales L_N large compared with the wavelength λ of the radiation, but comparable to the length scales of MHD turbulence, it has been argued that mode conversion cannot explain the levels and dynamic spectra of type III bursts in the corona and interplanetary medium [*Ginzburg and Zheleznyakov*, 1959; *Melrose*, 1980a,b; *Robinson and Cairns*, 1998a]. More recently *Bale et al.* [1998] and *Yin et al.* [1998] have interpreted Langmuir wave data in terms of linear mode conversion, with *Yin et al.* arguing that the conversion efficiency is increased for $L_N \lesssim \lambda$. Further research on mode conversion should be pursued in this context, despite the existence of a quantitative and so-far successful theory for type III bursts based primarily on nonlinear processes [*Robinson and Cairns*, 1993, 1998a,b; *Robinson et al.*, 1993, 1994; *Cairns and Robinson*, 1995a]. Interestingly, *Robinson and Cairns* [1998a] found that mode conversion $o \to z$ is an effective loss process for type III bursts: in combination with large angle scattering this process explains the time scale for the exponential decay of type III radiation [*Evans et al.*, 1973] to within a factor of 2 over frequencies from 50 MHz to $\sim$ 10 kHz.

5. HYBRID LINEAR/NONLINEAR MECHANISMS

Two novel hybrid linear and nonlinear mechanisms have been proposed by P. H. Yoon and C. S. Wu for the production of radiation near ω_p and $2\omega_p$, as summarized and extended by *Yoon* [1995, 1997]. These mechanisms are derived using weak turbulence expansions of the Vlasov equation with (1) a persisting, unstable electron beam and (2) significant levels of pre-existing wave turbulence and their associated particle motions. The ω_p and $2\omega_p$ mechanisms assume the presence of ion acoustic and Langmuir waves propagating parallel and antiparallel to the beam, respectively, but do not include any pre-existing Langmuir waves driven by the beam or any thermal waves. This inclusion of pre-existing turbulence is novel and leads to nonlinear modifications of the normal Langmuir (or z), o, and x modes near ω_p and near $2\omega_p$, as illustrated schematically in Figure 4. Importantly, the modified modes have nonlinear growth rates that are essentially independent of the levels of pre-existing turbulence, reminiscent of a linear process, and the standard beam instability for Langmuir/z mode waves remains essentially unchanged. Another reason for labelling the processes "hybrid linear/nonlinear" is that the modified modes near ω_p and $2\omega_p$ are essentially electrostatic, so that linear mode conversion is necessary to produce freely propagating radiation (i.e., by moving into regions with lower density).

While the properties of these mechanisms are potentially very attractive, a number of reasons for concern are apparent. First, the calculations are not self-consistent, since (1) pre-existing beam-driven Langmuir waves are not included despite the presence of a persisting electron beam and inclusion of either backward-propagating Langmuir waves or ion acoustic waves, both of which are best explained in terms of nonlinear processes involving beam-generated Langmuir waves, and (2) thermal populations of all participating waves are neglected. If inclusion of one wave population in each mechanism leads to such dramatic changes, then all possible participating waves should be included for confidence to be attached to the results. Second, the mechanism predicts growth rates for the radiation which are commensurate with those for Langmuir waves driven directly by the beam, thereby plausibly leading to much higher levels of radiation than observed and to rapid beam relaxation. Third, confirmation of these ideas using numerical simulations is necessary.

6. NONLINEAR WAVE PROCESSES

Standard linear instabilities involve the emission of waves stimulated by the power input $\mathbf{j}_{ind}.\mathbf{E}$ of the current $\mathbf{j}_{ind} \sim \mathbf{E}$ induced in the plasma by a wave with electric field $\mathbf{E}$. Similarly, in nonlinear wave processes the induced currents for multiple waves are coupled together or coupled with the currents carried directly by charged particles, forming a nonlinear current $\mathbf{j}_{NL}$ that then drives emission or absorption of waves through $\mathbf{j}_{NL}.\mathbf{E}$ effects. Nonlinear wave-wave and wave-particle processes are best illustrated using Feynman diagrams of wave quanta which coalesce, decay, or scatter off particles (Figure 1). These processes are usually classified and derived by expanding the nonlinear current in powers of E during weak turbulence analyses of the Vlasov equation and then retaining terms up to a given order [*Tsytovich*, 1972; *Melrose*, 1980a,b, 1986]. The order is related to the number of participating waves. For instance, a second order nonlinear process retains terms in $\mathbf{j}_{NL}$ up to $O(E^2)$ and is either a wave-particle scattering process with two waves or a three-wave process, while a third-order process involves terms up to $O(E^3)$ in $\mathbf{j}_{NL}$ and represents either a four-wave process, two coupled three-wave processes, or wave-particle processes involving three waves. Figure 1 illustrates Thomson scattering and a nonlinear three-wave decay process. These nonlinear processes proceed only when high levels of at least one wave population are present and the nonlinear growth/loss rates for each wave species are proportional to the product of the energy densities of the other participating waves. Moreover, a loss process exists for every generation process.

Thomson scattering $L \to T(\omega_p)$ of Langmuir waves L into radiation T near ω_p, sometimes known as induced scattering or scattering off thermal ions, is a second order wave-particle scattering process. Originally proposed to explain fundamental radiation in type III solar radio bursts [*Ginzburg and Zheleznyakov*, 1959], the nonlinear rate is too low to explain observations of type III bursts and in Earth's foreshock [*Melrose*, 1980a,b; *Cairns*, 1988; *Robinson et al.*, 1994].

Three-wave processes have been proposed in multiple contexts. In particular, the modern theory for type III bursts involves Langmuir-like waves L being driven by the type III electron beam and then producing radiation and other product waves via nonlinear three-wave processes [*Robinson et al.*, 1993, 1994; *Cairns and Robinson*, 1995a; *Robinson and Cairns*, 1998a,b, 1999]: the electrostatic (ES) decay $L \to L' + S$ involves the decay of L waves into backscattered Langmuir waves L' and forward-going ion acoustic waves S, the electromagnetic (EM) decay $L \to T(\omega_p) + S'$ produces radiation near ω_p and ion acoustic waves S', and the coalescence $L + L' \to T(2\omega_p)$ produces radiation near $2\omega_p$ using the beam-driven and backscattered Langmuir waves. Observational tests of the theory address the existence, characteristics, and levels of product S waves and radiation. Discovered originally by *Lin et al* [1986], detailed studies show that low frequency waves exist in type III source regions with the correct timings, amplitudes, and frequencies (Figure 5) to be S waves produced by the ES decay [*Cairns*, 1995; *Cairns and Robinson*, 1995a]. Predictions for the volume emissivities and fluxes of specific type III bursts, as well as model dynamic spectra from the corona to 1 AU, have been calculated and found to be consistent with the available data [*Robinson and Cairns*, 1993, 1998a,b; *Robinson et al.*, 1994]. In particular, Figure 6 compares observational data (squares) and theory (line) for the normalized flux of type III bursts as a function of the observed beam speed and plasma parameters, finding excellent agreement over three orders of magnitude in the normalized flux [*Dulk et al.*, 1998].

The Langmuir wave processes just described for type III bursts form the basis of a detailed theory for f_p and $2f_p$ radiation from Earth's foreshock [*Cairns*, 1988]. Efforts to simulate the radiation processes for such waves exist [e.g., *Pritchett and Dawson*, 1983; *Kasaba et al.*, 1997; *Yin et al.*, 1998, and references therein], with mixed results, but confidence cannot be attached to the results until the evolution of the bursty Langmuir waves observed is adequately reproduced. Despite the successes described above for nonlinear processes in explaining type III bursts, further work on linear mode conversion should be performed for type III bursts and the foreshock radiation [cf. *Bale et al.*, 1998; *Yin et al.*, 1998a,b].

Magnetospheric continuum radiation has been interpreted in terms of nonlinear three-wave decays and coalescences of the intense upper hybrid waves U often observed in the source regions [*Melrose*, 1981]: $U \pm l \rightarrow T$ and $U + U' \rightarrow T$ for radiation near one and two times ω_{uh}, respectively, where l is a low frequency wave. However, the mode of the proposed l waves remains relatively unconstrained by theory and detailed observational testing of these ideas has not been performed. Similar processes have been proposed for Jovian decametric radiation [*Fung and Papadopoulos*, 1987].

Finally, 4-wave processes include the modulational instabilities and collapse of Langmuir waves [e.g., *Pa-*

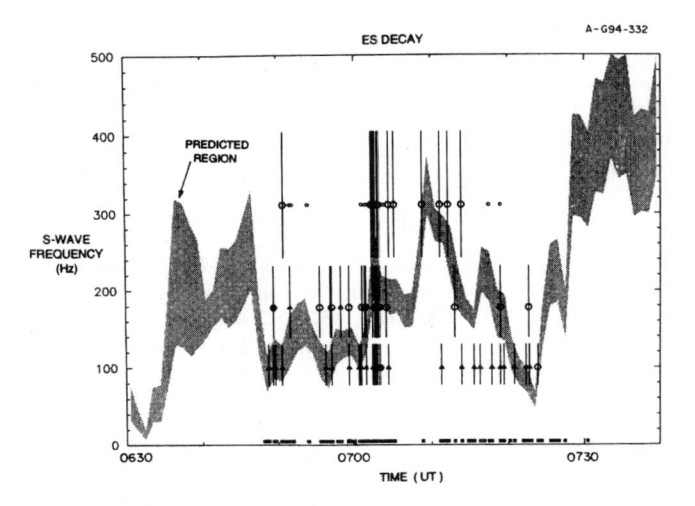

Figure 5. The frequency range for S waves produced by the ES decay is predicted using the observed electron beam and plasma data (gray region) and compared with the observed wave frequencies (points with error bars) for a type III bursts observed by ISEE 3 [*Cairns and Robinson*, 1995].

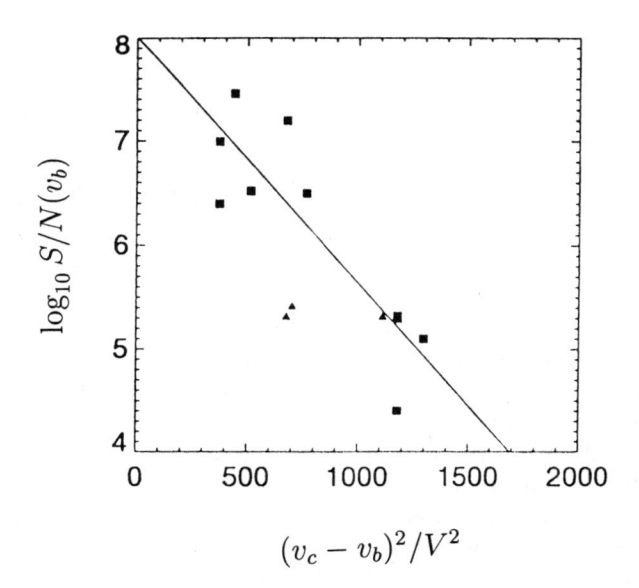

Figure 6. Comparison of the predicted relationship (solid line) between the normalized radiation flux and the characteristics of the beams and solar wind observed by Wind (square and triangle symbols) for a group of type III bursts [*Dulk et al.*, 1998]. Here S is the radiation flux, $N(v_b)$ is the electron flux at the beam speed v_b, $v_c = 3V^2/2V_s$, and V and V_s are the electron thermal and ion acoustic speeds, respectively.

padopoulos et al., 1974; *Goldman*, 1983; *Zakharov et al.*, 1985; *Robinson*, 1997], where wave collapse can be viewed as the modulational instability of a localized wave packet rather than an infinite wave train. Long considered in the contexts of type III solar bursts [*Papadopoulos et al.*, 1974; *Goldman*, 1983] and Earth's foreshock, modern analyses present strong arguments against either process being important in these contexts [*Cairns and Robinson*, 1992, 1995b, 1998; *Cairns et al.*, 1997]. These processes are important, however, in ionospheric heating experiments and (for lower hybrid waves) in Earth's auroral regions [*Robinson*, 1997].

7. RADIATION FROM LOCALIZED WAVE PACKETS

These mechanisms include radiation emitted by Langmuir wave packets as they undergo wave collapse [*Papadopoulos and Freund*, 1978; *Goldman et al.*, 1980] and novel ideas involving wave packets, solitons, or double layers acting as antennas for the production of radiation [*Borovsky*, 1988; *Pottelette et al.*, 1992]. Only the first of these examples is a special case of a nonlinear wave-wave process involving localized waves.

Theoretically the nonlinear currents $\mathbf{j}_{NL}$ associated with intense wave packets can produce radiation directly through $\mathbf{j}_{NL}.\mathbf{E}$ effects. This is the basis for the proposal [*Papadopoulos and Freund*, 1978; *Goldman et al.*, 1980] that Langmuir wave packets generate ω_p and $2\omega_p$ radiation while undergoing strong turbulence wave collapse. This mechanism has not yet been observed, due to the absence of evidence for wave collapse in Earth's foreshock and type III sources [*Cairns and Robinson*, 1992, 1995b].

Borovsky [1988] proposed a theory for AKR in which double layers act as antennas due to their production of beams of gyrophase-bunched electrons which emit radiation collectively. *Pottelette et al.*'s [1992] theory for AKR involves localized wave packets acting as antennas for the conversion of other plasma waves into radiation: radiation is generated directly by electrons moving through and being accelerated by the superposed electric fields of lower hybrid solitons and upper hybrid waves. The mechanism is therefore an example of Larmor radiation by accelerated charged particles. The analysis predicts generation of the radiation at the sum of the lower hybrid and upper hybrid frequencies, as expected for a nonlinear three-wave process. However, in contrast to a three-wave process, *Pottelette et al.* [1992] predict that the emission rate is proportional only to the energy density W_{uh} in upper hybrid waves and not that of the lower hybrid waves. Unfortunately, detailed examination of AKR source regions show no evidence for upper hybrid waves or double layers but instead strong evidence for a cyclotron-maser mechanism for AKR [*Ergun et al.*, 1998]. However, this class of "antenna" emission mechanisms should be pursued and applied in other contexts.

8. CONCLUSIONS

Nonthermal radio emissions are generated either directly by nonthermal distributions of electrons or indirectly by the conversion of intermediary plasma waves into radiation. Indirect processes thus involve collective plasma effects while direct processes need not. The high phase speeds $\approx c$ of the free-space o- and x-modes place significant constraints on the electron distribution for direct emission processes to be effective, requiring that the driving electrons be at least semi-relativistic and have a nonthermal distribution function. In contrast, indirect generation of radiation via the production and subsequent conversion of electrostatic waves is a natural and often favored means to transform plasma free energy into nonthermal escaping radiation for nonrelativistic plasmas. Reasons include the typically much greater ease of resonantly driving low phase speed, electrostatic waves to levels with $T_w \gg T_e$, and the existence of many conversion mechanisms with sufficient efficiency to convert electrostatic waves into radiation with $T_w \gg T_b \gg T_e$.

The conversion mechanisms proposed include linear mode conversion, hybrid linear/nonlinear processes, nonlinear wave-wave and wave-particle processes, radiation from collapsing Langmuir wave packets, and localized wave structures acting as antennas for the conversion of plasma energy or other waves into radiation. Well-developed theories exist for linear mode conversion and many nonlinear wave processes, but problems remain in current formulations of the hybrid linear/nonlinear processes. Strong quantitative evidence exists that specific nonlinear wave processes and linear mode conversion can account for the detailed properties of type III solar radio bursts and various ionospheric phenomena. In contrast, although linear mode conversion and nonlinear wave processes remain the only favored theories for magnetospheric continuum radiation, strong evidence exists for neither mechanism. The remaining mechanisms are all interesting but no convincing data or theoretical arguments exist that these processes actually proceed or dominate mode conversion, nonlinear wave processes, or direct cyclotron maser emission in the proposed applications. In conclusion, indirect emission processes involving electrostatic waves remain favored to explain many natural radio emissions but further research into all the mechanisms above is desirable.

Acknowledgments. The Australian Research Council, NASA grants NAG5-61271 and -6369, and the American Geophysical Union supported this research. The authors thank S.D. Bale, D.B. Melrose, and A.J. Willes for helpful conversations.

REFERENCES

Bale, S. D., P. J. Kellogg, K. Goetz, and S. J. Monson, Transverse z-mode waves in the terrestrial electron foreshock, *Geophys. Res. Lett.*, 25, 9, 1998.

Borovsky, J. E., Production of auroral kilometric radiation by gyrophase-bunched double-layer-emitted electrons: antennae in the magnetospheric current regions, *J. Geophys. Res.*, 93, 5727, 1988.

Budden, K. G., *Radio Waves in the Ionosphere*, Cambridge Univ. Press, Cambridge, 1961.

Cairns, I. H., A semi-quantitative theory for the $2f_p$ radiation observed upstream from the Earth's bow shock, *J. Geophys. Res.*, 93, 3958, 1988.

Cairns, I. H., Detectability of electrostatic decay products in Ulysses and Galileo observations of type III solar radio sources, *Astrophys. J. 449*, L95, 1995.

Cairns, I. H., and P. A. Robinson, Strong Langmuir turbulence at Jupiter?, *Geophys. Res. Lett.*, *19*, 1069, 1992.

Cairns, I. H., and P. A. Robinson, Ion acoustic wave frequencies and onset times during type III solar radio bursts, *Astrophys. J.*, *453*, 959, 1995a.

Cairns, I. H., and P. A. Robinson, Inconsistency of Ulysses millisecond Langmuir spikes with wave collapse in type III radio sources, *Geophys. Res. Lett.*, *22*, 3437, 1995b.

Cairns, I. H., and P. A. Robinson, Constraints on nonlinear and stochastic growth theories for type III solar radio bursts from the corona to 1 AU, *Astrophys. J.*, *509*, 471, 1998.

Cairns, I. H., and P. A. Robinson, Strong evidence for stochastic growth of Langmuir-like waves in Earth's foreshock, *Phys. Rev. Lett.*, *82*, 3066, 1999.

Cairns, I. H., P. A. Robinson, and N. I. Smith, Arguments against modulational instability of Langmuir waves in Earth's foreshock, *J. Geophys. Res.*, *103*, 287, 1998.

Dulk, G. A., Y. Leblanc, P. A. Robinson, J.-L. Bougeret, and R. P. Lin, Electron beams and radio waves of solar type III bursts, *J. Geophys. Res.*, *103*, 17,223, 1998.

Ergun, R. E., C. W. Carlson, J. P. McFadden et al., FAST satellite wave observations in the AKR source region, *Geophys. Res. Lett.*, *25*, 2061, 1998.

Evans, L. G., J. Fainberg, and R. G. Stone, Characteristics of type III exciters derived from low frequency radio observations, *Sol. Phys.*, *31*, 501-511, 1973.

Fung, S. F., and K. D. Papadopoulos, The emission of narrowband Jovian kilometric radiation, *J. Geophys. Res.*, *92*, 8579, 1987.

Ginzburg, V. L., and V. V. Zheleznyakov, On the possible mechanisms of sporadic radio emission (radiation in an isotropic plasma), *Sov. Astron., Engl. Trans.*, *2*, 653, 1959.

Goldman, M. V., Progress and problems in the theory of type III solar radio emission, *Sol. Phys.*, *89*, 403, 1983.

Goldman, M. V., F. G. Reiter, and D. R. Nicholson, Radiation from a strongly turbulent plasma: Application to electron beam-excited solar emissions, *Phys. Fluids*, *23*, 388, 1980.

Ginzburg, V. L., and V. V. Zheleznyakov, On the possible mechanisms of sporadic solar radio emission (radiation in an isotropic plasma) *Sov. Astron. AJ*, *2*, 653-668, 1958.

Gurnett, D. A., and R. R. Shaw, Electromagnetic radiation trapped in the magnetosphere above the plasma frequency, *J. Geophys. Res.*, *78*, 8136, 1973.

Jones, D., Source of terrestrial nonthermal radiation, *Nature*, *260*, 686, 1976.

Jones, D., W. Calvert, D. A. Gurnett, and R. L. Huff, Observed beaming of terrestrial myriametric radiation, *Nature*, *328*, 391, 1987.

Kasaba, Y., et al., Spacecraft observations and numerical simulations of $2f_p$ waves in the electron foreshock, in *Proc. of the Fifth Int. Sch./Symp. for Space Simulations*, Kyoto University, Japan, p. 29, 1997.

Lin, R. P., W. K. Levedahl, W. Lotko, D. A. Gurnett, and F. L. Scarf, Evidence for nonlinear wave-wave interactions in solar type III radio bursts, *Astrophys. J.*, *308*, 954-965, 1986.

Melrose, D. B., On the theory of type II and III solar radio bursts, 1, The impossibility of nonthermal emission due to combination scattering off thermal fluctuations, *Aust. J. Phys.*, *23*, 871, 1970.

Melrose, D. B., The emission mechanisms for solar radio bursts, *Space Sci. Rev.*, *26*, 3, 1980a.

Melrose, D. B., *Plasma Astrophysics, Vol. II*, Gordon and Breach, New York, 1980b.

Melrose, D. B., A theory for the nonthermal radio continuum in the terrestrial and Jovian magnetospheres, *J. Geophys. Res.*, *86*, 30, 1981.

Melrose, D. B. *Instabilities in Space and Laboratory Plasmas*, Cambridge, Cambridge, 1986.

Morgan, D. D., and D. A. Gurnett, The source location and beaming of terrestrial continuum radiation, *J. Geophys. Res.*, *96*, 9595, 1991.

Papadopoulos, K. D., and H. P. Freund, Solitons and second harmonic radiation in type III bursts, *Geophys. Res. Lett.*, *5*, 881, 1978.

Papadopoulos, K. D., M. L. Goldstein, and R. A. Smith, Stabilization of electron streams in type III solar radio bursts, *Astrophys. J.*, *190*, 175, 1974.

Pottelette, R., R. A. Treumann, and N. Dubouloz, Generation of auroral kilometric radiation in upper hybrid wave - lower hybrid soliton interaction, *J. Geophys. Res.*, *97*, 12,029, 1992.

Pritchett, P.L., and J.M. Dawson, Electromagnetic radiation from beam-plasma instabilities, *Phys. Fluids*, *26*, 1114, 1983.

Robinson, P. A., Nonlinear wave collapse and strong turbulence, *Rev. Mod. Phys.*, *69*, 507, 1997.

Robinson, P. A., and I. H. Cairns, Stochastic growth theory of type III solar radio emission, *Astrophys. J.*, *418*, 506-509, 1993.

Robinson, P. A., and I. H. Cairns, Fundamental and harmonic emission in type III solar bursts. I. Emission at a single location or frequency, *Sol. Phys.*, *181*, 363-394, 1998a.

Robinson, P. A., and I. H. Cairns, Fundamental and harmonic emission in type III solar radio bursts. II. Dominant modes and dynamic spectra, *Sol. Phys.*, *181*, 395-428, 1998b.

Robinson, P. A., and I. H. Cairns, Theory of type III and type II solar radio emissions, this issue, 1999.

Robinson, P. A., I. H. Cairns, and D. A. Gurnett, Clumpy Langmuir waves in type III radio sources: Comparison of stochastic-growth theory with observations, *Astrophys. J.*, *407*, 790-800, 1993.

Robinson, P. A., I. H. Cairns, and A. J. Willes, Dynamics and efficiency of type III solar radio emission, *Astrophys. J.*, *422*, 870-882, 1994.

Rönnmark, K., Myriametric radiation and the efficiency of linear mode conversion, *Geophys. Res. Lett.*, *16*, 731, 1989.

Tsytovich, V. N., *An Introduction to the Theory of Plasma Turbulence*, Pergamon, New York, 1972.

Willes, A. J., S. D. Bale, and Z. Kunzic, A z mode electron-cyclotron maser model for bottomside ionospheric harmonic radio emissions, *J. Geophys. Res.*, *103*, 7017, 1998.

Wu, C. S., and L. C. Lee, A theory of terrestrial kilometric radiation, *Astrophys. J.*, *230*, 621, 1979.

Yin, L., M. Ashour-Abdalla, M. El-Alaoui, J.M. Bosqued, and J.L. Bougeret, Generation of electromagnetic f_{pe} and

$2f_{pe}$ waves in the Earth's electron foreshock via linear mode conversion, *Geophys. Res. Lett.*, *25*, 2609, 1998a.

Yin, L., M. Ashour-Abdalla, M. El-Alaoui, J.M. Bosqued, and J.L. Bougeret, Plasma waves in the Earth's electron foreshock: 2. Simulations using time-of-flight electron distributions in a generalized Lorentzian plasma, *J. Geophys. Res.*, *103*, 29,619, 1998b.

Yoon, P. H., Plasma emission by a nonlinear beam instability, *Phys. Plasmas*, *2*, 537, 1995.

Yoon, P. H., Plasma emission by a nonlinear beam instability in a weakly magnetized plasma, *Phys. Plasmas*, *4*, 3863, 1997.

Yoon, P. H., A. T. Weatherwax, T. J. Rosenberg, J. LaBelle, and S.G. Shepherd, Propagation of medium frequency (1–4 MHz) auroral radio waves to the ground via the Z-mode radio window, *J. Geophys. Res.*, *103*, 29,267, 1998.

Zakharov, V. E., S. L. Musher, and A. M. Rubenchik, Hamiltonian approach to the description of non-linear plasma phenomena, *Phys. Rep.*, *129*, 285, 1985.

Iver H. Cairns and P. A. Robinson, School of Physics, University of Sydney, NSW 2006, Australia. (e-mail: cairns@physics.usyd.edu.au & p.robinson@physics.usyd.edu.au)

Theory of Type III and Type II Solar Radio Emissions

P. A. Robinson and I. H. Cairns

School of Physics, University of Sydney, Sydney, Australia

The main features of some current theories of type III and type II bursts are outlined. Among the most common solar radio bursts, type III bursts are produced at frequencies of 10 kHz to a few GHz when electron beams are ejected from solar active regions, entering the corona and solar wind at typical speeds of $0.1c$. These beams provide energy to generate Langmuir waves via a streaming instability. In the current stochastic-growth theory, Langmuir waves grow in clumps associated with random low-frequency density fluctuations, leading to the observed spiky waves. Nonlinear wave-wave interactions then lead to secondary emission of observable radio waves near the fundamental and harmonic of the plasma frequency. Subsequent scattering processes modify the dynamic radio spectra, while back-reaction of Langmuir waves on the beam causes it to fluctuate about a state of marginal stability. Theories based on these ideas can account for the observed properties of type III bursts, including the in situ waves and the dynamic spectra of the radiation. Type II bursts are associated with shock waves propagating through the corona and interplanetary space and radiating from roughly 30 kHz to 1 GHz. Their basic emission mechanisms are believed to be similar to those of type III events and radiation from Earth's foreshock. However, several sub-classes of type II bursts may exist with different source regions and detailed characteristics. Theoretical models for type II bursts are briefly reviewed, focusing on a model with emission from a foreshock region upstream of the shock for which observational evidence has just been reported.

1. INTRODUCTION

Solar radio bursts have been observed from several GHz down to a few kHz, with ground-based observations restricted to frequencies above the ionospheric propagation cutoff of about 5 – 20 MHz. Frequencies above about 10 MHz (metric or decimetric bursts) are associated with coronal sources, while lower frequencies (decametric and kilometric bursts) are produced in interplanetary space. Several major classes of radio bursts have been identified, mostly in the 1940s and 1950s; notable among them are type II and III bursts.

Type III bursts have dynamic spectra that are characterized by a frequency that falls rapidly with time, as shown in Fig. 1 [*Dulk* 1985]. In contrast, type II bursts drift slowly downward in frequency (Fig. 1). If the emission is assumed to be near the plasma frequency or its harmonic in both cases, then both type II and III bursts are interpreted in terms of sources moving outward from the Sun, but with far higher velocities for type III bursts.

The purpose of this tutorial review is to outline some of the current theories of these two types of solar radio

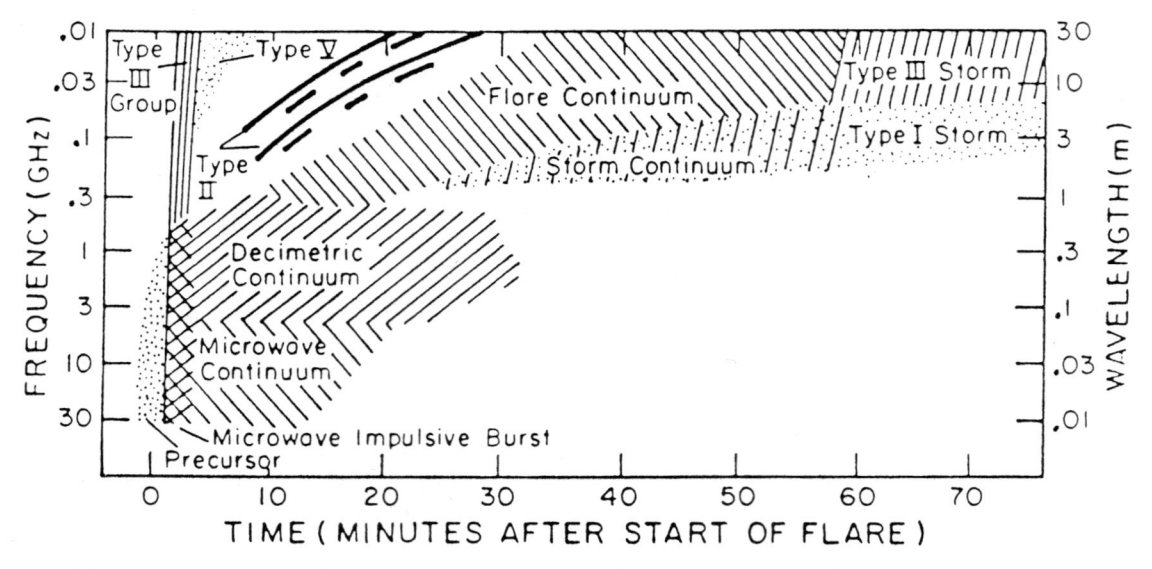

Figure 1. Schematic dynamic spectra of commonly observed solar radio bursts [*Dulk* 1985].

bursts, indicating where they are reasonably well confirmed by observations, and where substantially more work needs to be done. Space limitations prevent us from providing an exhaustive review; we thus emphasize those areas which appear to be most promising at present. Reviews of this area include those of *Dulk* [1985], *Suzuki and Dulk* [1985], *Nelson and Melrose* [1985], *Goldman and Smith* [1986], and *Melrose* [1980, 1997].

2. TYPE III BURSTS

Type III solar radio bursts have been observed for roughly 50 years [*Wild*, 1950]. They are associated with electrons ejected from solar active regions with typical velocities of 0.1c, but sometimes considerably higher or lower than this. The brightness temperatures of these bursts typically exceed 10^{10} K, and can reach 10^{15} K, requiring a coherent emission mechanism, since there are no source plasmas of this temperature [*Melrose*, 1986]. Type III dynamic spectra peak at a frequency that decreases with time (at least for bursts that commence below a few hundred MHz), and they sometimes show harmonic structure, with two components separated by a factor of nearly two in frequency [*Suzuki and Dulk*, 1985]. Often, multiple type III bursts occur in type III *storms* [*Suzuki and Dulk*, 1985]. At a single frequency the intensity of type III emission rises rapidly, then decays exponentially on a timescale much greater than that required for light or the underlying electron beam to cross the source [*Evans et al.*, 1973].

In the following subsections, we outline theoretical interpretations of the above features of type III bursts.

2.1. Beams and Langmuir Waves

Since the work of *Ginzburg and Zheleznyakov* [1958] it has been argued that type III emission is produced by processes that start with Langmuir waves being generated by an electron beam that is ejected by a solar flare. As they propagate outward from the Sun along the magnetic field, fast beam electrons tend to outrun slow ones and arrive first at a given location. This leads to formation of a high-velocity bump on the tail of the ambient plasma velocity distribution $f(v)$. Langmuir waves L near the electron plasma frequency ω_p then grow via the bump-on-tail (or "streaming") instability with growth rate

$$\Gamma \propto \frac{\partial f(v)}{\partial v}. \qquad (1)$$

Spacecraft have observed beam-like features and intense, extremely spiky, Langmuir waves simultaneously in type III sources, supporting this theoretical picture [*Gurnett and Anderson*, 1976, 1977; *Lin et al.*, 1981, 1986].

As Langmuir waves grow, they take energy from the beam in the region of maximal slope, causing it to relax toward a marginally stable plateau distribution [$\partial f(v)/\partial v = 0$]. This process competes with rebuilding of the beam due to the tendency of fast electrons to outrun slow ones. Numerical calculations of the competition between these processes indicate that most of the energy lost by electrons near the leading edge of a beam is reabsorbed by those further back, allowing the beam to propagate large distances before its energy dissipates [*Grognard*, 1985]. Observed distributions are very similar to those predicted numerically in the 1980s, but the extreme spikiness of the observed waves was not predicted by early theories.

2.2. Stochastic Growth Theory

In the solar corona and wind, buildup and relaxation of the beam compete amid ambient density fluctuations which are sustained by other processes and are always present in the solar wind. *Robinson* [1992] argued that this causes the beam to fluctuate about marginal stability, with the Langmuir growth rate fluctuating about a value near zero. In this case, the Langmuir waves would grow *stochastically*, undergoing a random walk in the gain G, with

$$G = \int \Gamma dt. \quad (2)$$

A gaussian probability distribution of $G \propto \ln E$ results from this random walk, satisfying

$$P(G) \propto \exp\left[\frac{-(G - \langle G \rangle)^2}{2\sigma^2}\right], \quad (3)$$

where $\langle G \rangle$ and σ characterize the center and width of the distribution, respectively. Figure 2 shows that this prediction is in good accord with ISEE 3 observations [*Robinson et al.*, 1993]. The detailed parameters of the distribution (3) and of the distribution of Langmuir clump sizes have been estimated theoretically from the underlying plasma, beam, and ambient fluctuation characteristics and found to be consistent with those actually observed [*Robinson*, 1992, 1993, 1995, 1996; *Robinson, Cairns, and Gurnett*, 1992, 1993].

Note that [*Smith and Sime*, 1979] suggested that plasma fluctuations may lead to markedly differing growth of waves following neighboring paths, a qualitative idea somewhat akin to the one used here. *Melrose and Goldman* [1986] attempted to account for bursty Langmuir waves via nonlinear wave collapse, discussed further below.

2.3. Electromagnetic Emission

Langmuir waves L, L' generate radio waves T, T' and ion-sound waves S via the nonlinear three-wave processes [e.g., *Melrose*, 1986] $L \to L' + S$,

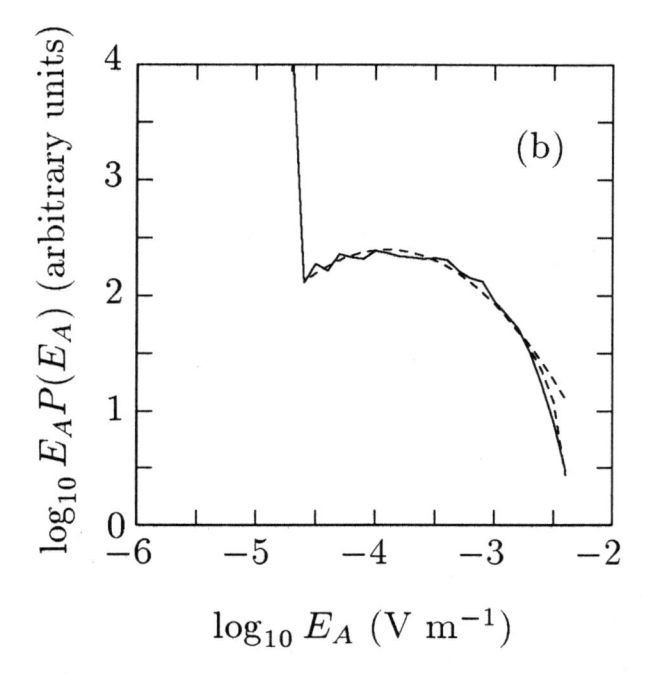

$$L \to T(\omega_p) + S, \quad (4)$$
$$L + L' \to T'(2\omega_p), \quad (5)$$

Figure 2. Comparison between theoretical (upper dashed) and observed (solid) electric field probability distributions, with $G \propto \log_{10} E_A$ and $P(G) \propto E_A \log_{10} P(E_A)$ [*Robinson et al.* 1993]. The lower dashed curve is an improved theoretical prediction including additional saturation effects, not discussed in this paper.

the latter two of which represent emission at the fundamental (F) and harmonic (H) of the plasma frequency, respectively. The emission frequencies fall as the beam moves away from the Sun into plasma of decreasing density, which accounts for the observed trend in type III frequencies with time and allows beam velocities to be inferred from dynamic spectra using the observed density profile of the solar corona and wind.

The theory of three-wave interactions has been used to estimate the conversion efficiencies ϕ_M of Langmuir energy into radio waves of type $M = F, H$ [*Melrose*, 1986; *Robinson and Cairns*, 1993; *Robinson et al.*, 1994]. The total emissivity j_M is then given by multiplying by the rate of build-up of beam energy, and dividing by the angular range $\Delta\Omega_M$ of the emission:

$$j_M \approx \frac{N_b m_e v_b^3}{3r} \frac{\Delta v_b}{v_b} \frac{\phi_M}{\Delta\Omega_M}, \quad (6)$$

where N_b, v_b and Δv_b are the beam number density, velocity, and width, and r is the heliocentric distance.

The conversion efficiencies depend strongly on v_b and Δv_b, with

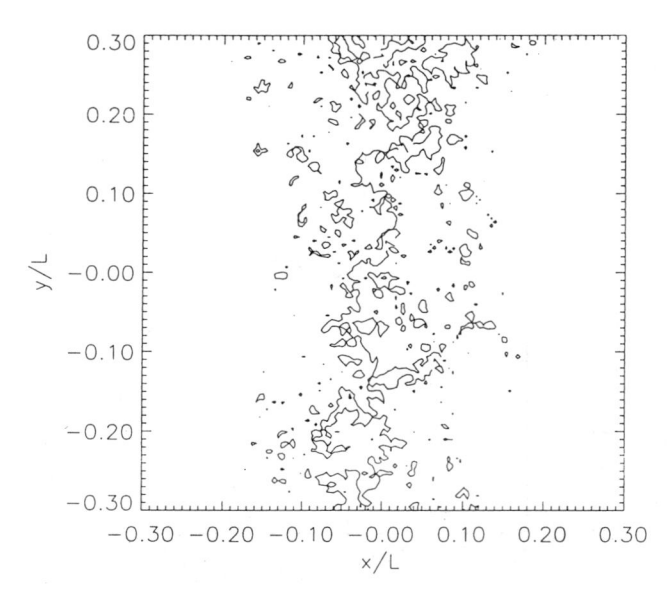

Figure 3. Contour of constant ω_p in a turbulent plasma with mean density decreasing to the right [*Robinson and Cairns* 1998a]. Scattering of fundamental radiation emitted with $\omega \approx \omega_p$ occurs off this surface. The Sun is located far to the left.

$$\phi_F \propto \exp[-2C(v_b/\Delta v_b)^2(v_b - v_c)^2], \quad (7)$$

$$\phi_H \propto \exp[-5C(v_b/\Delta v_b)^2 v_b^2], \quad (8)$$

where C is a constant that can be expressed in terms of the beam parameters [*Robinson and Cairns*, 1998b]. The steep decrease in ϕ_M as $\Delta v_b/v_b \to 0$ implies the existence of a low-frequency cutoff to type III emission, because fractional beam widths are predicted to decrease with heliocentric distance [*Robinson and Cairns*, 1998c]. This is consistent with the observed cutoff at 10 kHz, below which type III emission is not seen, even when beams are present [*Reiner et al.*, 1992; *Leblanc et al.*, 1995].

Three-wave interactions are invoked in type III applications because they have lower thresholds than competing processes such as nonlinear wave collapse, which was once widely invoked to explain type III emission (see *Robinson* [1997] for extensive references and discussion), and they explain the observed association between intense Langmuir waves and ion-sound waves in situ and the observed upper bound to Langmuir fields [*Robinson et al.*, 1993; *Cairns and Robinson*, 1995]. Wave collapse would also yield a power-law field-strength distribution, and imply a positive correlation between short spatial scales and strong fields, with significant clumping only at the highest field strengths — all these predictions contradict observations [*Robinson*,

1997]. In another context *Muschietti et al.* [1996] suggested an alternative wave clumping mechanism, based on kinetic wave-particle interactions, but this has not yet been investigated for type III bursts.

2.4. Propagation of Beam and Waves

To calculate dynamic spectra for type III bursts, we must allow for the propagation of the beam to the source, and of radiation from source to observer. The beam propagates along the Archimedean spiral of the interplanetary field lines, while harmonic radiation travels in almost a straight line between source and observer, since its frequency is well above the electromagnetic cutoff frequency near ω_p. Fundamental emission, however, is generated close to the cutoff and scatters strongly off ambient density fluctuations in the source, which distort this cutoff layer into numerous islands, as seen in Fig. 3 [*Robinson and Cairns*, 1998a]. At a given heliocentric distance the overall emission thus has a relatively large bandwidth due to fluctuations in ω_p, despite the narrow intrinsic bandwidth of the emission process at a single point. Fundamental radiation then random walks out of the source, during which time the part nearest ω_p is reabsorbed. This raises the mean F emission frequency, thereby accounting for the anomalous harmonic ratio of $\sim 1.8 : 1$ seen in H-F pairs [*Stewart*, 1974; *Suzuki and Dulk*, 1985]. Multiple scattering also delays the F radiation, producing an exponential tail to the emission, whose predicted decay time is within a factor of 2 – 3 of those observed over three orders of magnitude in frequency [*Robinson and Cairns*, 1998a]. This is in contrast with previous explanations in terms of collisional damping, which predicted decay times that erred by factors of 10 – 100, as was pointed out by *Evans et al.* [1973].

2.5. Dynamic Spectra

When one combines the radial variation of beam and source parameters, propagation of beam and radiation, and stochastic conversion efficiencies from beam energy to electromagnetic emission, it is possible to predict observable dynamic spectra from beam and background plasma parameters and their known typical variations with heliospheric distance [*Robinson and Cairns*, 1998b]. Figure 4 shows a typical spectrum predicted for an observer at 1 AU from the Sun [*Robinson and Cairns*, 1998b]. The predicted emissivity is consistent with observations, as is the range of brightness temperatures, 10^{10} K – 10^{17} K. At fixed frequency, F emission typically dominates before the peak, followed by H emission near the peak, then an exponential F tail [*Robinson and Cairns, 1998b]. The dominance of F early in the burst, and H near the peak, have previously been inferred from observations [Dulk et al., 1984]. It has also been shown that the radial variations of emissivity, brightness temperature, and other beam and source parameters, are all simultaneously consistent with a range of independent ground-based and in situ observations [Robinson and Cairns, 1998a,b,c], within the framework of the above theory.

3. TYPE II BURSTS

Type II solar radio bursts were also discovered roughly 50 years ago [Wild, 1950], as reviewed by Nelson and Melrose [1986]. Their dynamic spectra show emission drifting to lower frequencies with time, but with a source velocity of only ~ 1000 km s^{-1}, much slower than for type III bursts, and which has long been interpreted in terms of emission from a moving shock. The emission has two main parts — a broad *backbone*, including F and H components, which drift slowly in frequency and short-lived *herringbone* bursts that emerge from the backbone and drift rapidly up or down. These features are illustrated in Fig. 5. Often the backbone shows additional fine structure, sometimes having the F and H components each subdivided into two parts separated by $\sim 10\%$ in frequency (split-band type IIs) or else having multiple sets of F and H bands which mimic one another or else vary independently (multiple-lane and multiple-band events).

Like type III bursts, the observed type II brightness temperatures of 10^7 K to 10^{13} K require a coherent emission mechanism; these brightness temperatures and their common F/H character tend to imply strong similarities between these two classes of emission. However, observationally, it is not yet clear whether numerous sub-classes of type II bursts exist, corresponding to emissions from different types of shocks in different plasma conditions. In particular, type II bursts may be divided into distinct subtypes depending on whether (1) they are in the corona or interplanetary medium, (2) emission is from upstream or downstream of the shock, (3) they are associated with coronal mass ejections (CMEs), flares, or co-rotating interaction regions (CIRs, where fast solar wind streams impact slower ones further from the Sun), or (4) their underlying shocks are piston-driven, blast waves, or CIR shocks [Nelson and Melrose, 1985; Cairns and Robinson, 1987; Hoang et al., 1991; Gopalswamy et al., 1998; Reiner et al., 1997, 1998; Kaiser et al., 1998; Reiner and Kaiser, 1999a]. Overall, the observational characterization and theory of type II

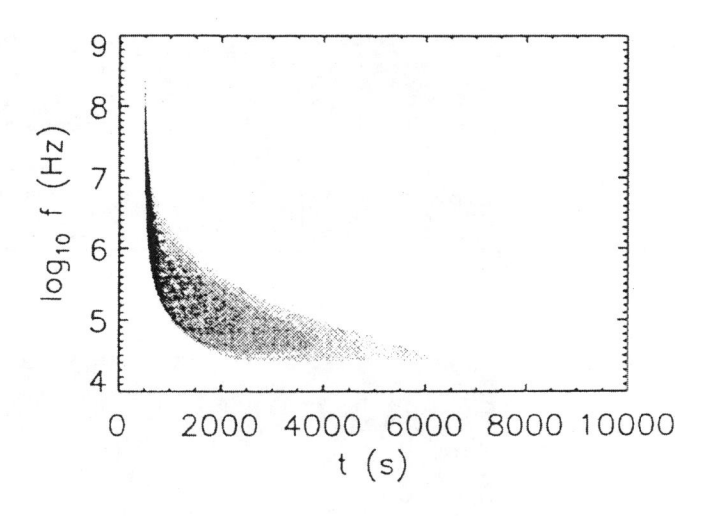

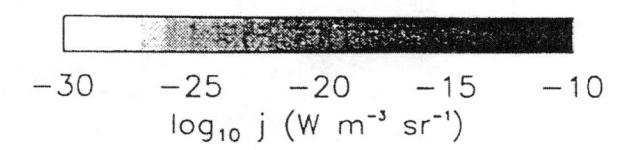

Figure 4. Typical type III dynamic spectrum predicted for an observer at 1 AU. The grayscale shows the contour levels [Robinson and Cairns 1998b].

bursts are far less developed and related to one another than for type IIIs. However, very recent ground- and space-based observations of type II bursts have started to clarify the situation and significant progress is expected over the next few years.

3.1. Backbone Emission

The low drift speeds of type II bursts and their very high associations with CMEs and interplanetary shocks imply that these sources are associated with MHD shocks in the corona and solar wind [Cane et al., 1984; Nelson and Melrose, 1985; Reiner et al., 1997, 1998; Gopalswamy et al., 1998; Kaiser et al., 1998]. These shocks have large Alfvén Mach numbers $M_A > 5$ near 1 AU for typical interplanetary conditions [e.g., Bale et al., 1999] but their Mach numbers are not well contrained in the corona and inner solar wind. There is evidence for three types of shocks being relevant to type II bursts, illustrated in Fig. 6: (i) piston-driven bow shocks formed ahead of CMEs [Cane et al., 1984; Reiner et al., 1997, 1998; Reiner and Kaiser, 1999a], (ii) blast waves [Gopalswamy et al., 1998; Kaiser et al., 1998; Reiner and Kaiser, 1999a], and (iii) CIR shocks where they interact with a CME-driven shock [Reiner et al., 1997, 1998].

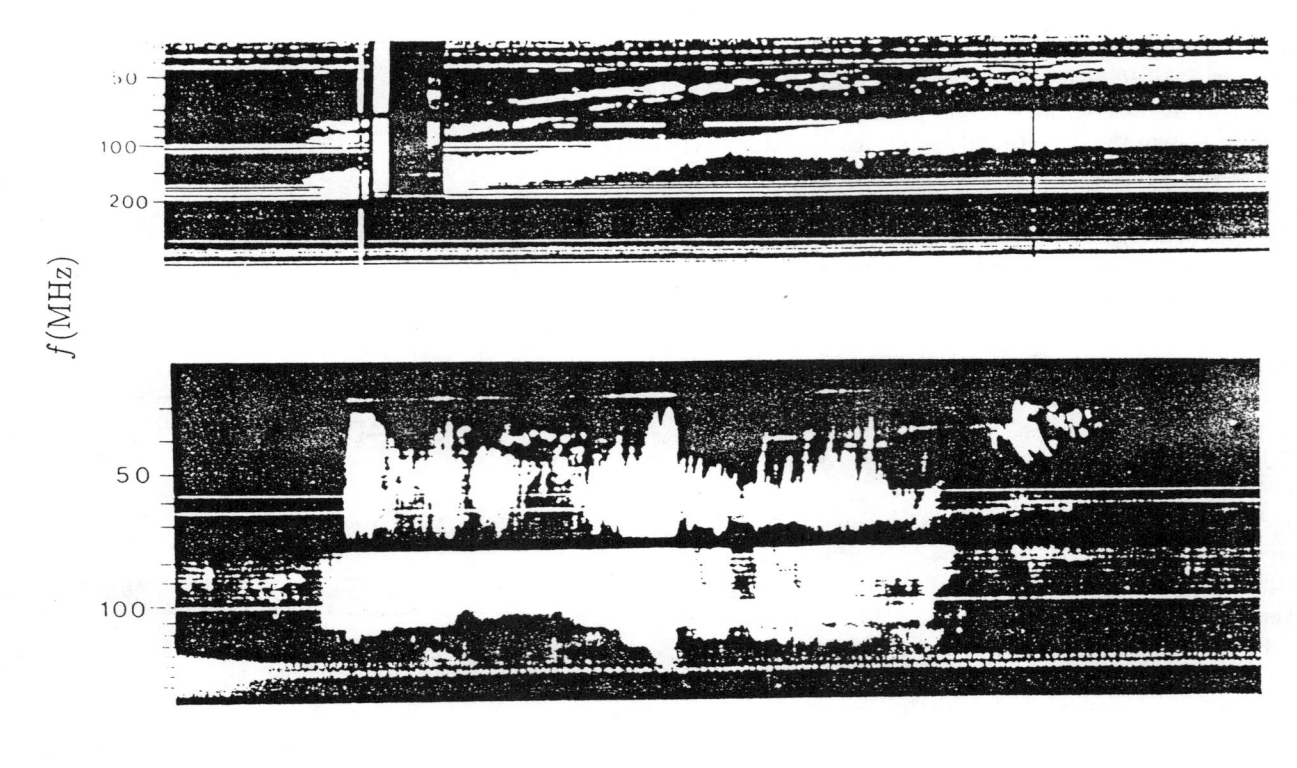

Figure 5. Type II dynamic spectra, showing the backbone (the two broad features in the upper frame with an approximately 2:1 frequency ratio), lane structure (in the fundamental backbone feature around the upper middle of the upper frame), and herringbone features (narrow, upward-pointing, thorn-shaped features in the lower frame) [*Cairns and Robinson* 1987]. The total duration of these spectra is 500 s for the upper frame and 360 s for the lower one.

As mentioned above, observations to date cannot definitively answer whether multiple subclasses of type II bursts exist which are differentiated by the source being upstream/downstream of the shock, different origins for the shock, and emission in the corona vs. the solar wind. Additional unanswered questions include: why do so few interplanetary shocks have observable type II emission in the corona and/or solar wind, and why is F/H structure rare? However, as described next, new observations from the Wind spacecraft support a specific theoretical model for at least a large fraction of interplanetary type II bursts.

The F/H structure of many coronal and some interplanetary type II bursts [*Wild*, 1950; *Nelson and Melrose*, 1985] supports models in which type II bursts are plasma radiation produced via the same processes involved in type III bursts, in this case associated with electron beams energized at a shock. Very recent Wind observations of the frequencies and drift rates of interplanetary type IIs [*Reiner et al.*, 1997, 1998; *Reiner and Kaiser*, 1999a], and of Langmuir waves and energized electrons in a foreshock upstream of an interplanetary type II shock [*Bale et al.*, 1999], confirm and strengthen this interpretation. Furthermore, these Wind observations strongly imply that many, if not all, interplanetary type II bursts are generated in foreshock regions upstream of CME-driven shocks. In the upstream foreshock model for type II bursts [*Cairns*, 1986], spatial variations in the velocity of electrons required to reach a given location in the foreshock naturally provide a continuous source of electron beams (and free energy for growth of Langmuir waves and radiation) in the foreshock. This model is based on the theory for electron beams in Earth's foreshock [*Filbert and Kellogg*, 1979], where analogous Langmuir waves and F/H radiation are also observed. These strong similarities suggest that existing foreshock theory, including the stochastic growth theory of Langmuir waves in Earth's foreshock [*Cairns and Robinson*, 1997, 1999], may be applicable to at least this class of type II bursts.

Other emission mechanisms, including electron cyclotron maser emission in front of or behind the shock (see Fig. 6), have also been investigated for type II bursts [e.g., *Kainer and MacDowall*, 1996]. Most face serious difficulties. For example, cyclotron maser models require semirelativistic beams not observed thus far and cannot account naturally for the emission occurring in harmonic bands near f_p and $2f_p$.

Whether emission occurs ahead of the shock for all type II bursts, and never behind the shock, is not clear observationally. The strongest evidence for occasional downstream emission involves Ulysses observations [*Hoang et al.*, 1991] of radiation apparently produced downstream of a CIR shock. However, these observations showed no Langmuir waves downstream of the CIR shock, similar to the usual situation at Earth's bow shock, and it is unclear theoretically how electron beams might be produced downstream. In principle, harmonic structure could be the result of simultaneous F emission from both sides of a shock with a 4:1 density ratio; however, the low Mach numbers inferred for coronal type II bursts make this unlikely [*Smerd et al.*, 1975] and *Bale et al.*'s [1999] observations show no downstream Langmuir waves.

Smerd et al. [1975] suggested that split-band events may result from F and H emission on both sides of a weak shock. Another plausible model, applying also to multiple-band and multiple-lane events, is that each band is emitted from a different localized region of the shock as it moves through an inhomogenous plasma whose plasma density and Alfven speed vary [*Nelson and Melrose*, 1985; *Cairns*, 1986; *Reiner et al.*, 1997, 1998]. Finally, based on observations in Earth's foreshock, the splitting in events with similar drift rates (split-band and multiple-band events) might be an intrinsic fine structure caused by magnetization effects on the plasma emission processes [*Cairns*, 1994].

3.2. Herringbone Emission

Herringbone emission (Fig. 5, lower panel) differs strongly from backbone emission by having much shorter timescales, faster drift rates, and higher polarizations [*Nelson and Melrose*, 1985; *Cairns and Robinson*, 1987]. The drift rates suggest an association with fast, shock-accelerated beams, analogous to type III beams. However, the dynamic spectra show a different thorn-shaped morphology [*Cairns and Robinson*, 1987], rather than the fan-like shape of type III bursts (cf., Fig. 4 with the lower panel of Fig. 5). Herringbone emission has also been associated with type III-like *shock-associated* (SA) emission [*Cane et al.*, 1981; *Bougeret*

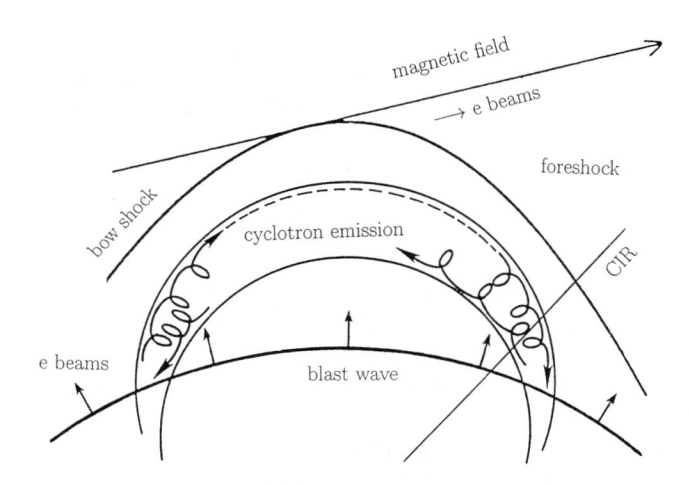

Figure 6. Schematic of possible type II source regions.

et al., 1998; *Reiner and Kaiser*, 1999b]. The SA emission starts at frequencies below the herringbone bursts, but one-to-one correpondences sometimes observed between individual herringbone bursts and enhancements in SA emission make an association via shared underlying beams likely.

The favored model for herringbone emission is plasma radiation produced by very fast beams accelerated at the shock [*Nelson and Melrose*, 1985; *Cairns and Robinson*, 1987], thereby being analogous but different in detail from both type III bursts and the backbone type II emission. The cause of the different dynamic spectra for herringbone and backbone emission is not known, but may result from the different geometry causing the efficiencies ϕ_F and ϕ_H to evolve differently with distance from the shock. The subsequent appearance of SA emission might then be the result of merging between the beams associated with individual herringbone fine structures, causing Δv_b to increase and thereby raising the conversion efficiencies (7) and (8). More work is required to resolve these questions.

4. CONCLUSIONS

A stochastic growth theory for type III bursts is well developed and its predictions are in good agreement with observations. In contrast, earlier theories based on modulational instabilities and wave collapse are inconsistent with observations. The SGT theory incorporates beam propagation near marginal stability, stochastic growth of Langmuir waves, wave-wave interaction efficiencies and dynamics, scattering and partial reabsorption of fundamental emission, and propagation of beams and radiation. It successfully predicts Langmuir field statistics, burst exponential decay times, anomalous harmonic ratios, dynamic spectra (including intensities and timings), modal dominance, and the low-frequency cutoff to type III emission. More work on predicting the polarization of emission and the properties of coronal type III bursts above a few hundred MHz is needed.

Theories of type II bursts are relatively poorly developed and connected to observations. However, it is well accepted that the radiation is produced by electron beams associated with a shock wave, and that the emission is probably produced by similar three-wave mechanisms to those involved in type III bursts. It is possible that several different subclasses of type II burst exist, corresponding to different source locations located relative to the shock, different drivers for the shock wave (CME, blast wave, and/or CIR), and possibly different emission mechanisms. However, recent Wind data are consistent with an existing model in which many, and perhaps all, interplanetary type II bursts are produced upstream of the shock in a foreshock region which contains Langmuir waves driven by electrons energized at the shock.

Acknowledgments. The authors thank G. A. Dulk for permission to reproduce Fig. 1. This work was supported by the Australian Research Council, NASA grants NAG5-6127 and -6369, and the American Geophysical Union.

REFERENCES

Bale, S. D., M. J. Reiner, et al., The source region of an interplanetary type II radio burst, *Geophys. Res. Lett.*, 26, 1573, 1999.

Bougeret, J.-L., P. Zarka, et al., A shock associated (SA) radio event and related phenomena observed from the base of the corona to 1 AU, *Geophys. Res. Lett.*, 25, 2513, 1998.

Cairns, I. H., The source of free energy for type II solar radio bursts, *Proc. Astron. Soc. Austral.*, 6, 444, 1986.

Cairns, I. H., Fine structure in plasma waves and radiation near the plasma frequency in Earth's foreshock, *J. Geophys. Res.*, 99, 23,505, 1994.

Cairns, I. H. and P. A. Robinson, Ion acoustic wave frequencies and onset times during type III solar radio bursts, *Astrophys. J.*, 453, 959, 1995.

Cairns, I. H., and P. A. Robinson, First test of stochastic growth theory for the Langmuir waves in Earth's foreshock, *Geophys. Res. Lett.*, 24, 369, 1997.

Cairns, I. H., and P. A. Robinson, Strong evidence for stochastic growth of Langmuir-like waves in Earth's foreshock, *Phys. Rev. Lett.*, 82, 3069, 1999.

Cairns, I. H., and R. D. Robinson, Herringbone bursts associated with type II solar radio emission, *Sol. Phys.*, 111, 365, 1987.

Cane, H.V., R. G. Stone, J. Fainberg, R. T. Stewart, and J.-L. Steinberg, Radio evidence for shock acceleration of electrons in the corona, *Geophys. Res. Lett.*, 8, 1285, 1981.

Cane, H. V., R. G. Stone, et al., Type II solar radio events observed in the interplanetary medium, *Sol. Phys.*, 78, 187, 1982.

Dulk, G. A., Radio emission from the Sun and stars, *Ann. Rev. Astron. Astrophys.*, 23, 169, 1985.

Dulk, G. A., J. L. Steinberg, and S. Hoang, Type III bursts in interplanetary space. Fundamental or harmonic?, *Astron. Astrophys.*, 141, 30, 1984.

Evans, L. G., J. Fainberg, and R. G. Stone, Characteristics of type III exciters derived from low frequency radio observations, *Sol. Phys.*, 31, 501, 1973.

Filbert, P. C., and P. J. Kellogg, Electrostatic noise at the plasma frequency beyond the Earth's bow shock, *J. Geophys. Res.*, 84, 1369, 1979.

Ginzburg, V. L., and V. V. Zheleznyakov, On the possible mechanisms of sporadic solar radio emission (radiation in an isotropic plasma) *Sov. Astron. AJ*, 2, 653, 1958.

Goldman, M. V., and D. F. Smith, Solar Radio Emission, in *Physics of the Sun*, Ed. P. A. Sturrock, Reidel, New York, 1986.

Gopalswamy, N., M. L. Kaiser, et al., Origin of coronal and interplanetary shocks: A new look with WIND spacecraft data, *J. Geophys. Res.*, 103, 307, 1998.

Grognard, R. J.-M., Numerical simulation of the weak turbulence excited by a beam of electrons in the interplanetary plasma, *Sol. Phys.*, 81, 173, 1982.

Gurnett, D. A., and R. R. Anderson, Electron plasma oscillations associated with type III radio bursts, *Science*, 194, 1159, 1976.

Gurnett, D. A., and R. R. Anderson, Plasma wave fields in the solar wind: Initial results from Helios 1 *J. Geophys. Res.*, 82, 632, 1977.

Hoang, S., F. Pantellini, et al., Interplanetary fast shock diagnosis with the radio receiver on Ulysses, in *Solar Wind Seven*, Eds E. Marsch and R. Schwenn, 465, 1991.

Kainer, S., and R. J. MacDowall, A ring bem mechanism for radio wave emission in the interplanetary medium, *J. Geophys. Res.*, 101, 495, 1996.

Kaiser, M. L., M. J. Reiner et al., Type II radio emissions from 1 - 14 MHz associated with the April 7, 1997 solar event, *Geophys. Res. Lett.*, 25, 2501, 1998.

Leblanc, Y., G. A. Dulk, and S. Hoang, The low radio frequency limit of solar type III bursts: Ulysses observations in and out of the ecliptic, *Geophys. Res. Lett.*, 22, 3429, 1995.

Lin, R. P., W. K. Levedahl, et al. Evidence for nonlinear wave-wave interactions in solar type III radio bursts, *Astrophys. J.*, 308, 954, 1986.

Lin, R. P., D. W. Potter, et al., Energetic electrons and plasma waves associated with a solar type III radio burst, *Astrophys. J.*, 251, 364, 1981.

Melrose, D. B., *Plasma Astrophysics*, Vol. 2, Gordon and Breach, New York, 1980.

Melrose, D. B., The brightness temperatures of solar type III bursts, *Sol. Phys.*, 120, 369, 1989.

Melrose, D. B. *Instabilities in Space and Laboratory Plasmas*, Cambridge, Cambridge, 1986.

Melrose, D. B., Particle acceleration and nonthermal radiation in space plasmas, *Astrophys. Space Sci.*, 242, 209, 1997.

Melrose, D. B., and Goldman, M. V., Microstructures in

type III events in the solar wind, *Sol. Phys.*, *107*, 329, 1986.

Muschietti, L., I. Roth, and R. E. Ergun, On the formation of wave packets in planetary foreshocks, *J. Geophys. Res.*, *101*, 15 605, 1996.

Nelson, G. S., and D. B. Melrose, Type II bursts, in *Solar Radiophysics*, Eds D. J. McLean and N. R. Labrum, Cambridge, 1985.

Reiner, M. J., and M. L. Kaiser, High-frequency type II radio emissions associated with shocks driven by coronal mass ejections, *J. Geophys. Res.*, *104*, 16,979, 1999a.

Reiner, M. J., and M. L. Kaiser, Complex type III-like radio emissions observed from 1 to 14 MHz, *Geophys. Res. Lett.*, *26*, 397, 1999b.

Reiner, M. J., R. G. Stone, and J. Fainberg, Detection of fundamental and harmonic type III radio emission and the associated Langmuir waves at the source region, *Astrophys. J.*, *394*, 340, 1992.

Reiner, M. J., M. L. Kaiser et al., Remote radio tracking of interplanetary CMEs, in *Correlated Phenomena at the Sun, in the Heliosphere and in Geospace*, ESA SP-415, 183, 1997.

Reiner, M. J., M. L. Kaiser et al., On the origin of radio emissions associated with the January 6-11, 1997, CME, *Geophys. Res. Lett.*, *25*, 2493, 1998.

Robinson, P. A., Clumpy Langmuir waves in type III solar radio sources, *Sol. Phys.*, *139*, 147, 1992.

Robinson, P. A., Stochastic-growth theory of Langmuir growth-rate fluctuations in type III solar radio sources, *Sol. Phys.*, *146*, 357, 1993.

Robinson, P. A., Stochastic wave growth, *Phys. Plasmas*, *2*, 1466, 1995.

Robinson, P. A., Stochastic wave growth, power balance, and beam evolution in type III solar radio sources, *Sol. Phys.*, *168*, 357, 1996.

Robinson, P. A., Nonlinear wave collapse and strong turbulence, *Rev. Mod. Phys.*, *69*, 507, 1997.

Robinson, P. A., and I. H. Cairns, Stochastic growth theory of type III solar radio emission, *Astrophys. J.*, *418*, 506, 1993.

Robinson, P. A., and I. H. Cairns, Fundamental and harmonic emission in type III solar bursts. I. Emission at a single location or frequency, *Sol. Phys.*, *181*, 363, 1998a.

Robinson, P. A., and I. H. Cairns, Fundamental and harmonic emission in type III solar radio bursts. II. Dominant modes and dynamic spectra, *Sol. Phys.*, *181*, 395, 1998b.

Robinson, P. A., and I. H. Cairns, Fundamental and harmonic emission in type III solar radio bursts. III. Heliocentric variation of interplanetary beam and source parameters, *Sol. Phys.*, *181*, 429, 1998c.

Robinson, P. A., I. H. Cairns, and D. A. Gurnett, Connection between ambient density fluctuations and clumpy Langmuir waves in type III radio sources, *Astrophys. J.*, *387*, L101, 1992.

Robinson, P. A., I. H. Cairns, and D. A. Gurnett, Clumpy Langmuir waves in type III radio sources: Comparison of stochastic-growth theory with observations, *Astrophys. J.*, *407*, 790, 1993.

Robinson, P. A., I. H. Cairns, and A. J. Willes, Dynamics and efficiency of type III solar radio emission, *Astrophys. J.*, *422*, 870, 1994.

Smith, D. F., and D. Sime, Origin of plasma-wave clumping in type III solar radio burst sources, *Astrophys. J.*, *233*, 998, 1979.

Stewart, R. T., Harmonic ratios of inverted-U type III bursts, *Sol. Phys.*, *39*, 451, 1974.

Smerd, S. F., K. V. Sheridan, and R. T. Stewart, Split-band structure in type II radio bursts from the Sun, *Astrophys. Lett.*, *16*, 23, 1975.

Suzuki, S. and G. A. Dulk, Bursts of type III and type V, in *Solar Radiophysics*, edited by D. J. McLean and N. R. Labrum, p. 289, Cambridge, Cambridge, 1985.

Wild, J. P., Observations of the spectrum of high-intensity solar radiation at metre wavelengths, *Aust. J. Sci. Res.*, *A3*, 541, 1950.

P. A. Robinson and I. H. Cairns, School of Physics, University of Sydney, New South Wales, 2006, Australia (e-mail: p.robinson@physics.usyd.edu.au i.cairns@physics.usyd.edu.au)

On the Harmonic Component of Type III Solar Radio Bursts

C. S. Wu[1] and Y. Li

Institute of Space Science, National Central University, Chungli, Taiwan, ROC

Peter H. Yoon

Institute for Physical Science and Technology, University of Maryland, College Park, MD 20742

A model for type III solar radio bursts is proposed, which provides an alternative explanation of the harmonic component. The present theory relies on an instability which amplifies extraordinary-Bernstein waves. Among the excited electron gyroharmonic modes, particular emphasis is placed on those waves with frequencies close to the local plasma frequency, ω_{pe}. It is suggested that Bernstein modes with frequencies just above ω_{pe} can convert to inward propagating electromagnetic waves, while those modes just below ω_{pe} may convert to outward-propagating radiation. The inward waves eventually turn around and propagate outward when they encounter the local cutoff frequency. The observed "fundamental-harmonic" band structure results from the time delay between the arrival times of the two waves. The reflected waves appear as a separate "harmonic" component while the outward waves are detected as the "fundamental," although at the source location the two waves possess very small frequency separation (of only a couple of electron gyrofrequencies apart). It is concluded that the observed frequency ratio of the two components at a given time should strongly depend on the geometrical configuration of the source region and physical parameters. A preliminary calculation shows that under certain conditions this ratio ranges from ~ 1.5 to ~ 2.

1. INTRODUCTION

One of the outstanding features of solar type III radio bursts is the fundamental (F) and harmonic (H) emission band structure that characterizes some observations. The F-H pair structure in the dynamic spectra of solar radio emissions was first observed in type II bursts and later in type III bursts [*Wild et al.*, 1954]. The two bands are so named because their spectra have similar characteristics, and also because at a given time the ratio of the central frequencies appears to be roughly equal to two. Although only a small fraction of the observed events exhibit the pair feature, this phenomenon has attracted a considerable amount of interests, particularly from a theoretical viewpoint.

In the standard interpretation of type III bursts, it is widely accepted that Langmuir waves play a primary role in the generation of the radiation. It is believed

[1]Institute for Physical Science and Technology, University of Maryland, College Park, MD 20742

Radio Astronomy at Long Wavelengths
Geophysical Monograph 119
Copyright 2000 by the American Geophysical Union

that the streaming energetic electrons produced during large solar flare events first excite Langmuir waves, which are partly converted to electromagnetic waves. This scenario is supported by the observed frequency drift which is attributed to the passage of the moving source. This is known as the "plasma emission hypothesis" which has been the conceptual foundation for almost all existing theories of type III radio emission. The principal issue of interest to theorists has been to search for the most likely conversion mechanism of electrostatic Langmuir wave to electromagnetic radiation.

The theory proposed by *Ginzburg and Zheleznyakov* [1958] suggests that the emission at the F band results from Langmuir waves nonlinearly interacting with thermal ions, while the emission at the H band results from the coalescence of two oppositely traveling Langmuir waves. This early work has profoundly influenced the subsequent researchers. Although the original theory by *Ginzburg and Zheleznyakov* [1958] has been modified and even some alternatives have been suggested [*Kaplan and Tsytovich*, 1968; *Zheleznyakov and Zaitsev*, 1970; *Melrose*, 1970, 1980, 1982a,b, 1985, 1987; *Smith and Fung*, 1971; *Melrose and Sy*, 1972a,b; *Smith*, 1974; *Papadopoulos et al.*, 1974; *Melrose et al.*, 1978; *Smith et al.*, 1979; *Goldstein et al.*, 1979; *Goldman*, 1983; *Cairns*, 1987; *Robinson and Cairns*, 1994, 1998a, 1998b; *Robinson et al.*, 1993, 1994; *Willes et al.*, 1996; *Yoon and Wu*, 1994; *Wu et al.*, 1994; *Yoon*, 1995, 1997, 1998; *Cairns and Robinson*, 1998], all these subsequent works involve Langmuir waves in one way or another.

It is important to note that the plasma emission theories of type III bursts implicitly assume that the F and H bands are generated within the same source region. However, the exact relation between the fundamental and harmonic components is clouded by intriguing issues raised by observations. For instance, *Mercier and Rosenberg* [1974], *Rosenberg* [1975], *Daigne* [1975a,b] and *Raoult and Pick* [1980] doubt the existence of the fundamental component. These authors discussed cases involving high frequency emissions (> 150 MHz) and argued that the observed pairs are both harmonic waves. Later, *Dulk and Suzuki* [1980] also pointed out that at high frequencies, indeed the fundamental components are rarely seen. However, the physical reason for the suppression of F band at high frequencies remains unknown.

Another unresolved but related issue first discussed by *Smerd et al.* [1962] is concerned with the relative heights of the two components. These authors point out some serious inconsistencies between observations and the accepted notion implied in the F and H components of type III bursts: The details will be explained in the next section. The same difficulty was later confirmed by others [*Bougeret et al.*, 1970; *Labrum*, 1971; *McLean*, 1971; *Stewart*, 1972, 1974; *Dulk and Suzuki*, 1980]. Since all existing theories of type III radio bursts do not resolve this issue satisfactorily, it seems timely to reassess the conventional models and search for a possible new scenario. It is appropriate to remark that although the latest discussions by *Robinson and Cairns* [1998] address many unresolved issues and ambiguities in the conventional theories of type III bursts in light of the stochastic growth theory [*Robinson*, 1992, 1993; *Cairns and Robinson*, 1999] with some success, yet they do not directly address the issue of source positions.

Motivated by these thoughts we are currently carrying out a series of theoretical investigations. In this paper we present a preliminary discussion of a possible new theory for type III bursts. The main objective of our analysis is to explain the origin of the F-H pair in the context of the new paradigm, which also resolves the inconsistencies between the observed source positions and the traditional F-H hypothesis.

The organization of the paper is as follows. In Section 2, we briefly review the issue related to the relative positions and heights of F and H sources. Then, Section 3 presents our model in a simple and intuitive manner. Finally discussion and conclusions are presented in Sections 4 and 5.

2. POSITIONS FOR F AND H SOURCES

Smerd et al. [1962] investigated F-H pair emissions (for both type III and type II emissions) by studying their relative positions. To avoid the effects of ionospheric refraction, they conducted observations of F-H pair emissions at a given frequency. They found that some of the source positions for the harmonic component coincide with that of the fundamental. Such a finding is incompatible with the standard interpretation, since the H component at a fixed frequency is supposed to be generated at a greater height than the F component where the local plasma frequency is half the observed emission frequency. At that time the finding was considered to be inconclusive.

Subsequent observations by others also led to the similar conclusion [*McLean*, 1971; *Stewart*, 1972, 1974; *Dulk and Suzuki*, 1980]. It was also found that the observed heights of F sources at a given frequency f, are always greater than those of H sources observed simultaneously at frequency $2f$. This result is again unexpected because one would anticipate the two components to be generated at the same location. In particular, *Stewart* [1972] discussed the radial distances of F and H components at 80 MHz, and on the basis of his data set, he concludes that, on the average, $R_{F80} = (1.0 \pm 0.06) R_{H80}$, where R_{F80} and R_{H80} denote the radial distances for the F and H components at a fixed frequency, 80 MHz. He also finds that the average radial distances of the F component at 80 MHz and that of the H components at 158 MHz measured about the same time, are related by $R_{H160} = (0.7 \pm 0.1) R_{F80}$. Although a small number of cases in which the F and H components have close but separate source regions are also reported [e.g., *Labrum*, 1971; *Kundu et al.*, 1970], the issue concerning the conventional "fundamental-harmonic" component hypothesis is still outstanding.

Since the observed coincidence of F/H source positions appears to be occurring nearly in all cases and definitely not fortuitous, we find that any explanation involving statistical and random processes are not convincing. On the other hand, we feel that it is also desirable to explain the causality of the observed physical phenomenon by a scenario inherent to the emission mechanism responsible for the type III radio bursts. In the following section we present a model that is derived from this notion.

3. AN ALTERNATIVE SCENARIO

As already stated, the starting point of the existing theories of type III bursts is the excitation of Langmuir waves by a stream of energetic electrons traversing through the corona. However, *in situ* observations of energetic electrons in the solar wind often show very little or no postitive slope $[\partial F(v_\parallel)/\partial v_\parallel > 0]$ in the reduced distribution function, $[F(v_\parallel)$ being the velocity space distribution function integrated over $v_\perp$, $F(v_\parallel) = 2\pi \int_0^\infty dv_\perp\, v_\perp\, f(v_\perp, v_\parallel)]$, which is required for the excitation of Langmuir waves [*Lin et al.*, 1981, 1986; *Ergun et al.*, 1998]. This has prompted researchers to investigate the role of small-scale density irregularities on the electron beam propagation, the nature of clumpy Langmuir waves, and ultimately culminated in the stochastic growth theory [*Melrose and Goldman*, 1987; *Robinson*, 1992, 1993; *Cairns and Robinson*, 1999]. The basic idea is that small-scale density irregularities scattered along the beam propagation path in a random way cause the Langmuir waves to grow and get reabsorbed in a stochastic manner, leading to the clumpiness of Langmuir waves and to overall marginally stable beam distributions. (There are competing explanations on the beam stabilization which are based upon strong turbulence and coherent nonlinear processes [see e.g., *Papadopoulos et al.*, 1974; *Smith et al.*, 1979; *Goldstein et al.*, 1979].)

For a magnetized plasma, on the other hand, the energetic electrons can excite a different class of instabilities as discussed in the paper by *Yoon et al.* [1999], who make note of the fact that the energetic electrons streaming along the ambient magnetic field can be affected by intrinsic hydromagnetic turbulence and undergo pitch-angle scattering. The resulting electron distribution possesses positive slope in the perpendicular momentum space, $\partial f(v_\perp, v_\parallel)/\partial v_\perp > 0$. The instability of interest is driven by this perpendicular positive slope in the electron distribution function, and excites waves propagating nearly perpendicular to the ambient magnetic field, such that these waves can be excited even for those electron distributions that do not meet the requirement for Langmuir wave growth. These waves belong to the extraordinary mode but with wavelengths comparable to the thermal electron gyroradius, and are basically the classical Bernstein waves [*Bernstein*, 1958; *Stix*, 1962, 1992; *Puri et al.*, 1973, 1975].

Instabilities of this type have been investigated for magnetospheric applications [*Crawford et al.*, 1967; *Fredericks*, 1971; *Young et al.*, 1973; *Karpman et al.*, 1975; *Ashour-Abdalla and Kennell*, 1978] and for solar microwave bursts [*Zheleznyakov and Zlotnik*, 1975; *Winglee and Dulk*, 1986a,b; *Robinson*, 1991; *Fleishman and Yastrebov*, 1994; *Willes and Robinson*, 1996, 1997; *Ledenev*, 1998], and therefore, are not entirely new. However, they have never been applied to the solar type III situation.

We pay attention to unstable Bernstein gyroharmonic modes with frequencies around ω_{pe}. A salient point, as discussed in the recent paper by *Yoon et al.* [1999], is that in the presence of magnetic field and density gradients, modes with frequencies above the X-mode cutoff frequency can only propagate toward regions with higher field and/or higher density, while the mode which possesses frequency just below ω_{pe} can propagate toward regions with lower field and/or density.

The mode which propagates outward from the source region into decreasing density and magnetic field may convert to forward propagating O mode wave through the so-called Ellis radio window [*Ellis*, 1956]. The partial conversion of Z mode to O mode wave is a well-established concept in the ionospheric radio wave propagation problem. If such a conversion should take place in the present situation, then the outward propagating mode which originally has wave vector perpendicular to the ambient magnetic field, will turn into forward prop-

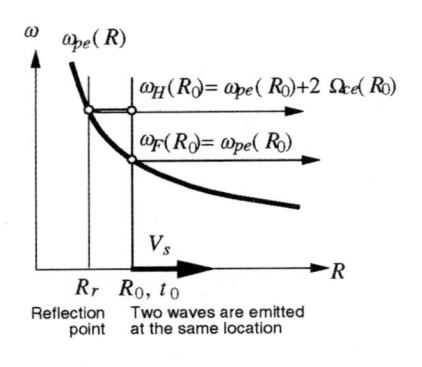

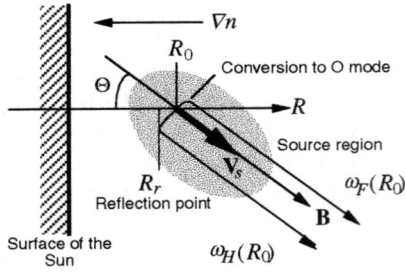

Figure 1. Two waves are emitted at R_0. The backward wave is reflected at R_r. This reflected wave and the forward wave emitted at R_0 are separated by a time-delay. In a general situation under consideration the local magnetic field vector $\mathbf{B}(R)$ is tilted with respect to the density gradient $\nabla n(R)$ at an angle Θ, as depicted on the bottom panel.

agating electromagnetic signal with $\mathbf{k}$ vector directed predominantly along B-field. We note, however, that the detailed demonstration of the conversion of quasi-electrostatic Bernstein mode to either of the fast waves is rather complicated and is beyond the scope of the present discussion.

The conversion of the mode that propagates inward, toward the Sun, into X (and possibly O) mode is conceptually more straightforward. When such a mode approaches $k \to 0$ limit (X or O mode cutoff), as the local electron density and magnetic field strength increases, the mode is expected to be reflected at that point and begin to propagate outward.

Henceforth, to facilitate the discussion of the essential scenario, we assume that the conversions of backward propagating Bernstein gyroharmonic mode and the forward propagating mode to O-mode are almost instantaneous. Hereafter, we shall employ the unmagnetized transverse wave dispersion relation for the converted backward and forward waves, as an approximation,

$$\omega^2 = \omega_{pe}^2 + c^2 k^2. \qquad (1)$$

The above implies that we are considering the situation where $\omega_{pe}^2 \gg \Omega_{ce}^2$, where $\omega_{pe} = (4\pi n_e e^2/m_e)^{1/2}$ and $\Omega_{ce} = eB/m_e c$ are electron plasma frequency and gyrofrequency, respectively. The forward propagating O-mode, which is described by (1), is expected to have frequency very close to the local plasma frequency, $\omega \sim \omega_{pe}$. Similarly, the backward X-mode, which also follows the dispersion relation (1), and is expected to possess frequency higher than that of the forward wave by approximately $2\Omega_{ce}$.

Let us denote the frequencies associated with forward and backward waves by ω_H and ω_F, respectively. Subscript H is meant to denote the fact that the backward-propagating mode (toward the photosphere) will eventually be associated with what is traditionally interpreted as H component. Similarly, subscript F denotes the outward-propagating mode, which will be associated with the F component. Thus,

$$\omega_H(R) \approx \omega_{pe}(R) + 2\Omega_{ce}(R),$$
$$\omega_F(R) \approx \omega_{pe}(R), \qquad (2)$$

where $\Omega_{ce}(R)$ denotes the local electron gyrofrequency, $\omega_{pe}(R)$ is the local plasma frequency, R is the height measured from the center of the Sun.

Let us consider that at moment t_0, the source situated at R_0 emits two waves with frequencies $\omega_H(R_0) = \omega_{pe}(R_0) + 2\Omega_{ce}(R_0)$ and $\omega_F(R_0) = \omega_{pe}(R_0)$. This physical situation is graphically depicted in Figure 1, in which the top panel shows the profile of the plasma frequency as a function of solar altitude R, and simultaneous forward and backward waves being emitted at R_0, at time t_0. The higher frequency wave $\omega_H(R_0)$ propagates inward until it reaches a point, R_r, where

$$\omega_{pe}(R_r) = \omega_H(R_0).$$

At this point, this wave is reflected and propagates outward with a group speed $v_g = c\,[1-\omega_{pe}^2(R)/\omega_H^2(R_0)]^{1/2}$. The lower frequency wave $\omega_F(R_0)$ simply propagates outward with group speed $v_g = c\,[1-\omega_{pe}^2(R)/\omega_F^2(R_0)]^{1/2}$.

We are interested in a geometrical configuration in which the magnetic field vector $\mathbf{B}(R)$ is tilted at an angle Θ with respect to the density gradient $\nabla n(R)$. The source is moving parallel to the B-field with velocity $\mathbf{V}_s$. The simplest configuration would be the one in which the beam velocity and the magnetic field vector are parallel to the density gradient. However, this situation does not appear to be very interesting, since for those waves propagating in directions nearly perpendicular to the density gradient, the reflection would not be very effective. For this reason, we assume that $\mathbf{B}(R)$ is oblique to $\nabla n(R)$.

As we have already discussed, the forward wave (F) is

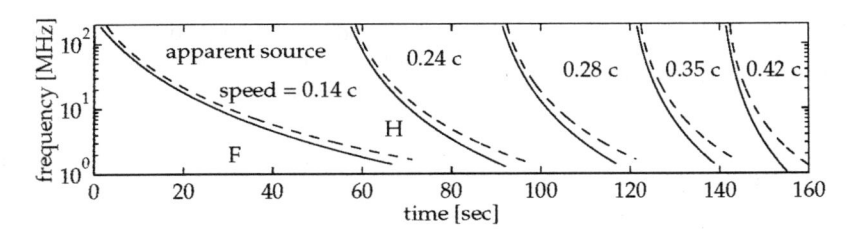

Figure 2. Plots of F and H component frequencies, $\omega_F(R)$ (solid lines) and $\omega_H(R)$ (dashes), versus the arrival times, t_F and t_H, computed on the basis of (6), for tilt angle $\Theta = 45°$, and for different source speeds, $V_s = 0.2c, c/3, 0.4c$, and $0.5c$. Apparent source speeds (projection along R) are computed by $V_s \cos \Theta$.

assumed to have rapidly undergone quasi-perpendicular Bernstein mode to quasi-parallel O-mode conversion within the source region before it escapes. This situation is depicted in the bottom panel of Figure 1. After the reflection, the backward propagating wave (H) also propagates primarily along the ambient magnetic field, as shown in Figure 1. As a result of the present geometrical configuration, the effective source velocity is $V_s \cos \Theta$, while the wave group speed projected along R is $v_g \cos \Theta$ for the forward wave, and $v_g \sin \Theta$ for the backward wave, where v_g is the group speed considered previously. For numerical calculation we examine the case $\Theta = \pi/4$.

To calculate the reflection point, we adopt the following density and magnetic field model:

$$n_e(R) = 8 \times 10^8 \, (1.55 \, r^{-6} + 2.99 \, r^{-16}) \, [\text{cm}^{-3}],$$

$$B(R) = 0.5 \, (r-1)^{-1.5} \, [\text{gauss}], \quad r = R/R_\odot. \quad (3)$$

where $R_\odot$ is the solar radius. The density model is basically the Baumbach-Allen model described in the review paper by *Wild* [1985], except that the numerical factor 8 is imposed to model the streamer region [*Newkirk*, 1967]. The magnetic field model is the same as that given by *Dulk and McLean* [1978]. To simplify the analysis, we ignore the second term in the density model in (3). This leads to

$$f_{pe}(R) = \omega_{pe}(R)/2\pi = 3.16 \times 10^2 (R/R_\odot)^{-3} \, [\text{MHz}],$$

$$f_{ce}(R) = \Omega_{ce}(R)/2\pi = 1.4 \, (R/R_\odot - 1)^{-1.5} \, [\text{MHz}]. \quad (4)$$

From (4), the reflection point R_r for the backward wave ω_H originated from R is obtained as

$$R_r(R) = R \left(1 + \frac{8.86 \times 10^{-3} (R/R_\odot)^3}{(R/R_\odot - 1)^{1.5}} \right)^{-1/3}. \quad (5)$$

The arrival times of the two signals to an observer situated at R_E (we use $R_E = 1$ A.U.) are given, respectively, by

$$t_H(R_0) = t_0 + \frac{1}{c \sin \Theta} \left(\int_{R_r(R_0)}^{R_0} + \int_{R_r(R_0)}^{R_E} \right)$$

$$\times \frac{dR}{\{1 - \omega_{pe}^2(R)/\omega_{pe}^2[R_r(R_0)]\}^{1/2}}, \quad (6)$$

$$t_F(R_0) = t_0 + \frac{1}{c \cos \Theta} \int_{R_0}^{R_E} \frac{dR}{[1 - \omega_{pe}^2(R)/\omega_{pe}^2(R_0)]^{1/2}}.$$

Suppose that, after an interval Δt, the source moves to a higher altitude $R = R_0 + V_s \cos \Theta \Delta t$ with speed V_s. The arrival times of the two waves emitted at the new location can be easily calculated from (6), provided we iteratively update t_0 and R_0 by

$$t_0 \rightarrow t_0 + \Delta t, \quad R_0 \rightarrow R_0 + V_s \cos \Theta \Delta t.$$

By continuously varying the time step Δt we can generate a theoretical dynamic spectrum. For a given beam speed, V_s, we calculate the frequency of the two wave bands emitted from each point along the passage of the moving source, $\omega_F(R)$ and $\omega_H(R)$, and plot the result against the respective arrival times, $t_F(R)$ and $t_H(R)$.

Figure 2 plots the result of the numerical calculation in which we have considered $V_s = 0.2c$, $c/3$, $0.4c$, $c/2$, and $0.6c$. The apparent source speeds are $V_s \cos \Theta = 0.14c, 0.24c, 0.28c, 0.35c$, and $0.42c$, respectively, as indicated in Figure 2. On the basis of the model plasma- and gyro-frequencies, given by (4), we find that $\omega_{pe}(R) \gg \Omega_{ce}(R)$. The ratio of these two frequencies ranges from ~ 10 (near the chromosphere) to ~ 28 (in the lower corona), which implies that the two bands $\omega_H(R) \approx \omega_F(R) + 2\Omega_{ce}(R)$ and $\omega_F(R) \approx \omega_{pe}(R)$ are fairly close in frequency. Nevertheless, Figure 2 shows that the nominal F and H bands are separated by a significant time delay of several seconds.

In Figure 3 we plot the frequency ratios of the H to F bands, $\omega_H(R)/\omega_F(R)$ versus the arrival time, corresponding to the five source speeds considered in Figure 2. According to Figure 3, the ratio $\omega_H(R)/\omega_F(R)$

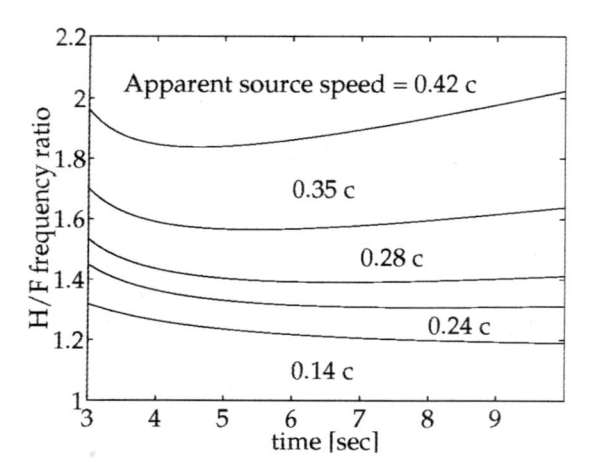

Figure 3. H to F component frequency ratio versus time for the five cases of source speed considered in Figure 2.

ranges from ~1.3 to ~2. The inferred source speed based on observations is roughly $V_s \cos\Theta$. For $\Theta = \pi/4$, one finds that the observed source speed ranges from $0.35c$ to $0.42c$ for $V_s \approx 0.5c \sim 0.6c$. For these cases, the frequency ratio falls within the range 1.5 to 2.

The choice of $\Theta = \pi/4$ in the above discussion was made because it appears to be the most representative case. As can be easily envisioned, the reflected waves may not be observable if Θ is very small. In this case, the apparent source region would be greatly displaced along the longitudinal direction, while the ray paths are expected to be nearly parallel to the surface of the Sun. On the other hand, magnetic field configuration which renders $\Theta \sim \pi/2$ is unlikely to be realized.

Admittedly, the result depends upon the specific density model, the tilt angle Θ, and other parameters under consideration. The important point is that the initially backward propagating waves in the source region can lead to a delayed component which appears as a "harmonic" component, while in fact the frequency of the outward propagating waves differs only slightly from that of the inward propagating waves at the same location in the source region. However, at each location the combined effects of the source speed and group delay near the reflection point of the inward waves can result in a significant separation of the two types of waves in time.

According to the present model an observer situated far away would see that the wave ω_H is emitted at the reflection point rather than the true source point, and the arrival time of wave ω_H is delayed after the wave ω_F has been observed. Thus, one expects to see that at each spatial position there are two observable waves "emitted" with the same frequency equal to the local plasma frequency (but at different times), and that the observed "harmonic" wave at a given time is emitted at a lower altitude than the "fundamental."

4. DISCUSSION

Smerd et al. [1962] first suggested that wave reflection may explain the issue of the observed harmonic component source height. They reasoned that the generation of $2\omega_{pe}$ electromagnetic waves by a nonlinear process [*Ginzburg and Zheleznyakov*, 1958] can initially propagate backward, in anti-beam direction, and subsequently get reflected when they encounter the cutoff frequency. In this sense, the basic notion has some basic similarity with our model, but the difficulty is that if the harmonic wave is produced at a site where the local plasma frequency is ω_{pe}, and is later reflected at lower altitude where the local plasma frequency is $2\omega_{pe}$, then the combined effect of the source propagation and group delay should make the ratio ω_H/ω_F significantly greater than 2.

To demonstrate this point we have repeated the arrival time calculations except that the mode reflection point, according to (4), is given by

$$R_r(R) = R/2^{1/3},$$

instead of (5), and that relations,

$$\omega_H(R) = 2\omega_{pe}(R) \text{ and } \omega_F(R) = \omega_{pe}(R),$$

are used instead of (2). The result of the calculation is displayed in Figure 4. As the reader can appreciate, the model suggsted by *Smerd et al.* [1962] predicts very

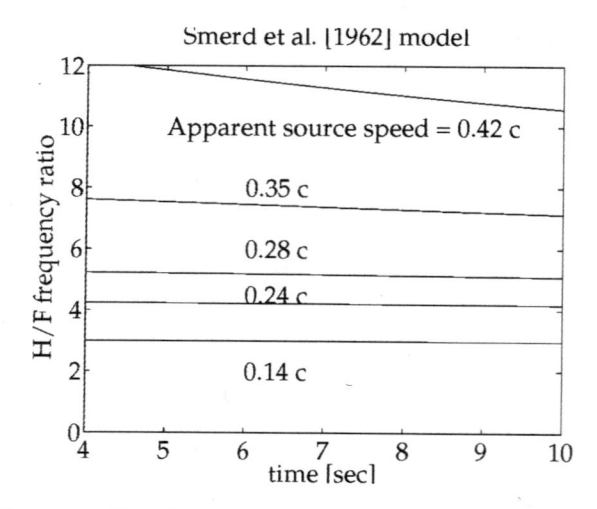

Figure 4. H to F component frequency ratio versus time for a model discussed by *Smerd et al.* [1962].

high H/F frequency ratio. Regardless of the specific values of V_s, the model does not work.

A comment which pertains to the absence of nominal F component in the high-frequency regime [*Dulk and Suzuki*, 1980] is in order. In the lower corona (immediately above the chromosphere), say at altitudes around 1.2 $R_\odot$, the magnetic field falls off more rapidly than the density so that the ratio of plasma frequency to gyrofrequency increases considerably over a small radial distance. For such a high frequency region, the outward Bernstein wave may not be able to convert to O-mode since the ratio ω_{pe}/Ω_{ce} increases too rapidly, as the mode propagates outward. As a result, in this case, such a mode may not escape. At higher altitudes, on the other hand, where the ratio ω_{pe}/Ω_{ce} varies slowly as altitude increases and eventually remains nearly constant, Bernstein to O mode conversion is anticipated to occur without any problem. Thus, in the low-altitude or equivalently, the high-frequency regime, the F band might not be able to escape while the H band can escape via reflection. This may explain why at high-frequency regime, the F component is often absent, as reported by *Dulk and Suzuki* [1980], for instance.

We note that the primary interest of this paper is to study type III emissions in the corona. In the interplanetary space where the magnetic field strength is weak and the ratio ω_p/Ω is very high, the conventional models which rely on the excitation of Langmuir waves followed by nonlinear conversion processes into transverse waves may be operative. On the other hand, if the reduced electron distribution function possesses little or no positive slope in velocity space, then the condition for the Langmuir instability may not be satisfied, and the present Bernstein mode instability may be more viable.

Finally, we should remark that a wave ducting model was suggested by *Duncan* [1979] in order to explain the issue of the source position. The basic idea is that waves are generated within an underdense duct, with the electron density outside the wall higher by a factor of ~4. As a consequence, while the H component can leave the source, the F component is confined to the duct until it arrives to a region where the density becomes much lower such that the local $2\omega_{pe}$ is equal to the wave frequency. As a result, apparent source locations for both the F and H component would nearly coincide. This model was later modified by *Robinson* [1983], who attributes the duct to fibrous structures in the corona and emphasizes the effect of anisotropic scattering. The dicussions by both Duncan (1979) and Robinson (1983) do not address the emission mechanism. In contrast, the present theory suggests that several types of pair emission shown in the classic paper by Wild et al. (1954) may be explained within the context of the new emission scenario. However, we concede that our theory does not explain every case, including the U burst exmaple reported by Stewart (1975). Of course, for this type of pair emission, the observational result that the source positions of F-H bands at a fixed frequency coincide does not apply anyway, because the two components have entirely different frequency intervals. We also believe that anisotropic scattering can complement our model.

5. CONCLUSIONS

We have presented a new scenario of type III solar radio emission which does not rely on enhanced Langmuir waves and subsequent nonlinear processes. Rather, the theory emphasizes amplification of generalized Bernstein electron gyroharmonic modes, their mode conversion, and propagation effects. In the proposed scenario, the streaming energetic electrons amplify generalized Bernstein waves, that propagate in directions nearly perpendicular to the ambient magnetic field. Of these waves, a group of gyroharmonic waves with frequencies close to the local plasma frequency, ω_{pe}, are of most interest.

Owing to their dispersion characteristics, those waves with frequencies slightly higher than ω_{pe} can only propagate toward regions with increasing B-field and density in an inhomogeneous plasma, and convert to fast X mode (and to a less extent, O mode). When these waves propagate to a region for which the wave frequency equals the X-mode cutoff frequency, they will get reflected and propagate in the opposite direction until they escape to free space. On the other hand, those waves with frequencies slightly below the local ω_{pe} can propagate toward regions with lower B-field and/or density. When these waves approach the region where the frequencies match the local O-mode cutoff frequency, mode-conversion to O mode may take place, and the converted O mode will escape to free space.

In short, at a given site in the source region, the forward propagating waves and the reflected waves are emitted at different times. As a result of the separation time between the two signals, which is on the order of a few seconds, the emission is observed to consist of two components. The detailed frequency ratio of the two bands at a fixed time depends upon the density and magnetic field models, beam velocity, and the direction of the beam with respect to the observer. Under certain assumed conditions, however, we find that the values are reasonable and consistent with observations.

Here we reiterate that the results reported in this paper are preliminary. Many questions must be addressed in subsequent discussions. These include the directivity and polarization of the radiation, to name a couple. To study the directivity of the radiation, observations based on both ground and spacecraft data [*Poquerusse and Bougeret*, 1981; *Poquerusse et al.*, 1988] can be valuable. Along this line, the issues raised by *Mercier and Rosenberg* [1974], *Rosenberg* [1975], *Daigne* [1975a,b] may also be discussed. In our opinion, the model discussed in this paper is qualitatively compatible with the observations reported by these authors. However, a quantitative discussion of the observable results may not be easy. The difficulty stems from uncertainties with the density variation and magnetic field configuration in the source region which are crucial to our analysis.

To recap the main conclusions, our model suggests that type III emissions at the so-called "harmonic" bands may result from physical processes very different from the conventional theories. The proposed scenario explains the dilemma that for a fixed frequency the source regions of the F-H pairs nearly coincide, and at a fixed time the source height of the H band is lower than that of the F band. Because of the occurrence of the F-H pair emission depends on many conditions such as the direction of the beam velocity or local magnetic field, wave refraction, group delay, and other propagation effects, it is not surprising that only a small fraction of the type III events exhibit the pair emissions.

Acknowledgments. The present research at the University of Maryland was supported by the National Science Foundation grant ATM-9802498, while efforts at the National Central University, Chungli, Taiwan, were supported by the National Science Council in Taiwan under grant NSC 89-2111-M-008-019. C.S.W. is very much indebted to the Foundation for Development of Outstanding Scholarship in Taiwan for an award granted to him during his periodic visits to the National Central University in 1989-1999. He is grateful to M. Poquerusse, G. Dulk and Y. LeBlanc for very helpful discussions from observational point of view during his visit to Observatory of Paris-Meudon. Finally, Y.L. gratefully acknowledges the financial support received from the National Science Council in Taiwan during his visit to the National Central University.

REFERENCES

Ashour-Abdalla, M., and C. F. Kennell, Nonconvective and convective electron cyclotron harmonic instabilities, *J. Geophys. Res., 83,* 1531, 1978.

Bernstein, I. B., Waves in a plasma in a magnetic field, *Phys. Rev., 109,* 10, 1958.

Bougeret, J. L., C. Caroubalos, C. Mercier, and M. Pick, Sources of type III solar bursts observed at 169 MHz with the Nançay radioheliograph, *Astron. Astrophys., 6,* 406, 1970.

Cairns, I. H., Fundamental plasma emission involving ion sound waves, *J. Plasma Phys., 38,* 169; Second harmonic plasma emission involving ion sound waves, *ibid.,* 179; Third and higher harmonic plasma emission due to Raman scattering, *ibid.,* 199, 1987.

Cairns, I. H., and P. A. Robinson, Roles played by electrostatic waves in producing radio emissions, this proceeding, 1998; Strong evidence for stochastic growth of Langmuir-like waves in Earth's foreshock, *Phys. Rev. Lett., 82,* 3066, 1999.

Caroubalos, C., M. Poquerusse, and J. L. Steinberg, The directivity of type III bursts, *Astron. Astrophys., 32,* 255, 1974.

Crawford, F. W., R. S. Harp, and T. D. Mantei, On the interpretation of ionospheric resonances stimulated by Alouette 1, *J. Geophys. Res., 72,* 57, 1967.

Daigne, G., On the 'harmonic structure' in type III solar radio bursts, *Astron. Astrophys., 38,* 141, 1975a; Characteristic pairs of type III bursts as the electromagnetic response to a dispersive stream exciter, *ibid, 42,* 71, 1975b.

Dulk, G. A., in *Solar Radiophysics*, edited by D. J. McLean and N. R. Labrum, Cambridge University Press, New York, p. 19, 1985.

Dulk, G. A., and D. J., McLean, Coronal magnetic fields, *Solar Phys., 57,* 279, 1978.

Dulk, G. A., and S. Suzuki, The position and polarization of type III solar bursts, *Astron. Astrophys., 88,* 203, 1980.

Duncan, R. A., Wave ducting of solar metre-wave radio emission as an explanation of funcdamental/harmonic source coincidence and other anomalies, *Solar Phys., 63,* 389, 1979.

Ellis, G. R., The Z propagation hole in the ionosphere, *J. Atmos. Terr. Phys., 8,* 43, 1956.

Ergun, R. E., et al., Wind spacecraft observations of solar impulsive electron events associated with solar type III radio bursts, *Astrophys. J., 503,* 435, 1998.

Fleishman, G. D., and G. Yastrebov, On the harmonic structure of solar radio spikes, *Solar Phys., 154,* 361, 1994.

Fredericks, R. W., Plasma instability at $(n+1/2) f_c$ and its relationship to some satellite observations, *J. Geophys. Res., 76,* 5344, 1971.

Ginzburg, V. L., and V. V. Zheleznyakov, On the possible mechanisms of sporadic solar radio emissions (radiation in an isotropic plasma), *Sov. Astron. - AJ., 2,* 653, 1958.

Goldman, M. V., Progress and problems in the theory of type III solar radio emission, *Solar Phys., 89,* 403, 1983.

Goldstein, M. L., R. A., Smith, and K., Papadopoulos, K., Nonlinear stability of solar type III radio bursts II. Application to observations near 1 AU, *Astrophys. J., 234,* 683, 1979.

Kaplan, S. A., and V. N. Tsytovich, Radio emission from beams of fast particles under cosmic conditions, *Sov. Phys. - Astron., 11,* 956, 1968.

Karpman, V. I., Ju. K. Alekhin, N. D. Borisov, and N. A. Rjabova, Electrostatic electron cyclotron waves in plasma with a loss-cone distribution, *Plasma Phys., 17,* 361, 1975.

Kundu, M. R., W. C. Erickson, P. D. Jackson, and J. Fainberg, Positions and motions of solar bursts at decimeter wavelengths, *Solar Phys., 14,* 394, 1970.

Labrum, N. R., Observations of harmonic structure in type III solar radio bursts, *Aust. J. Phys., 24,* 193, 1971.

Ledenev, V. G., Generation of electromagnetic radiation by an electron beam with a bump on the tail distribution function, *Solar Phys., 179,* 405, 1998.

Lin, R. P., D. W. Potter, D. A. Gurnett, and F. L. Scarf, Energetic electrons and plasma waves associated with a solar type III radio burst, *Astrophys. J., 251,* 364, 1981.

Lin, R. P., W. K. Levedahl, W. Lotko, D. A. Gurnett, and F. L. Scarf, Evidence for nonlinear wave-wave interactions in solar type III radio bursts, *Astrophys. J., 308,* 954, 1986.

McLean, D. J., Radioheliograph observations of harmonic type III solar bursts, *Aust. J. Phys., 24,* 201, 1971.

Melrose, D. B., On the theory of type II and type III solar radio bursts, *Aust. J. Phys., 23,* 871, 1970; The emission mechanisms for solar radio bursts, *Space Sci. Rev., 26,* 3, 1980; Fundamental emission for type III bursts in the interplanetary medium: the role of ion-sound turbulence, *Solar Phys., 79,* 173, 1982a; Plasma emission without Langmuir waves, *Aust. J. Phys., 35,* 67, 1982b; in *Solar Radiophysics,* edited by D. J. McLean and N. R. Labrum, Cambridge University Press, New York, p.177, 1985; Plasma emission: a review, *Solar Phys., 111,* 89, 1987.

Melrose, D. B., and W. N. Sy, Scattering of waves in a magnetoactive plasma, *Astrophys. Space Sci., 17,* 343, 1972a; Plasma emission processes in a magnetoactive plasma, *Aust. J. Phys., 25,* 387, 1972b.

Melrose, D. B., G. A. Dulk, and S. F. Smerd, The polarization of second harmonic plasma emission, *Astron. Astrophys., 66,* 315, 1978.

Melrose, D. B., and M. V. Goldman, Microstructures in type III events in the solar wind, *Solar Phys., 107,* 329, 1987.

Mercier, C., and H. Rosenberg, Type III solar radio bursts observed at 169 MHz: height and relative positions in pairs, *Solar Phys., 39,* 193, 1974.

Newkirk, G., Structure of the solar corona, *Ann. Rev. Astron. Astrophys., 5,* 213, 1967.

Papadopoulos, K., M. L. Goldstein, and R. A. Smith, Stabilization of electron streams in type III solar radio bursts, *Astrophys. J., 190,* 175, 1974.

Poquerusse, M., and J. L. Bougeret, Radiation mode and coronal propagation of solar type III radio bursts observed on 14 November 1971 during stereo-1 experiment, *Astron. Astrophys., 97,* 36, 1981.

Poquerusse, M., J. L. Steinberg, C. Caroubalos, G. A. Dulk, and R. M. MacQueen, Measurement of the 3-dimensional position of type III bursts in the solar corona, *Astron. Astrophys., 192,* 323, 1988.

Puri, S., F. Leuterer, and M. Tutter, The totality of waves in a homogeneous Vlasov plasma, *J. Plasma Phys., 9,* 89, 1973; Dispersion curves for the generalized Bernstein modes, *ibid., 14,* 169, 1975.

Raoult, A., and M. Pick, Space-time evolution of type III burst sources observed with the Nançay radioheliograph – Implications for the size of the emitting source, *Astron. Astrophys., 87,* 63, 1980.

Robinson, P. A., Electron-cyclotron maser emission in solar microwave spike bursts, *Solar Phys., 134,* 299, 1991; Clumpy Langmuir waves in type III radio sources, *Solar Phys., 139,* 147, 1992; Stochastic-growth theory of Langmuir growth-rate fluctuations in type III solar radio sources, *ibid, 146,* 357, 1993.

Robinson, P. A., and I. H. Cairns, Fundamental and harmonic radiation in type III solar radio bursts, *Solar Phys., 154,* 335, 1994; Fundamental and harmonic emission in type III solar radio bursts, *ibid, 181,* 363, 1998a; Theory of type III & type II solar radio emissions, this proceeding, 1998b.

Robinson, P. A., A. J. Willes, and I. H. Cairns, Dynamics of Langmuir and ion-sound waves in type III solar radio sources, *Astrophys. J., 408,* 720, 1993.

Robinson, P. A., I. H. Cairns, and A. J. Willes, Dynamics and efficiency of type III solar radio emission, *Astrophys. J., 422,* 870, 1994.

Robinson, R. D., Scattering of radio waves in the solar corona, *Proc. Astron. Soc. Aust., 5,* 208, 1983.

Rosenberg, H., Type III solar radio bursts and the fundamental - harmonic hypothesis, *Solar Phys., 42,* 274, 1975.

Smerd, S. F., J. P. Wild, and K. V. Sheridan, On the relative position and origin of harmonics in the spectra of solar radio bursts of spectral type II and III, *Austr. J. Phys., 15,* 180, 1962.

Smith, D. F., Type-III radio bursts and their interpretation, *Space Sci. Rev., 16,* 91, 1974.

Smith, D. F., and P. C. W. Fung, A weak turbulence analysis of the two-stream instability, *J. Plasma Phys., 5,* 1, 1971.

Smith, R. A., M. L. Goldstein, and K. Papadopoulos, Nonlinear stability of solar type III radio bursts I. Theory, *Astrophys. J., 234,* 348, 1979.

Stewart, R. T., Relative positions of fundamental and second harmonic type III bursts, *Proc. Astron. Soc. Australia, 2,* 100, 1972; in *Coronal Disturbances,* Ed. G. A. Newkirk, IAU Symp. No. 57, Dordrecht: Reidel, p. 161, 1974.

Stewart, R. T., An example of a fundamental type IIIb radio burst, *Solar Phys., 40,* 417, 1975.

Stix, T. H., *The Theory of Plasma Waves,* McGraw-Hill, New York, 1962; *Waves in Plasmas,* American Institute of Physics, New York, 1992.

Wild, J. P., in *Solar Radiophysics,* edited by D. J. McLean and N. R. Labrum, Cambridge University Press, New York, p. 3, 1985.

Wild, J. P., J. D. Murray, and W. C. Rowe, Harmonics in the spectra of solar radio disturbances, *Aust. J. Phys., 7,* 439, 1954.

Willes, A. J., and P. A. Robinson, Electron-cyclotron maser theory for noninteger ratio emission frequencies in solar microwave spike bursts, *Astrophys. J., 467,* 465, 1996; The electron-cyclotron maser theory for extraordinary Bernstein waves, *J. Plasma Phys., 58,* 171, 1997.

Willes, A. J., P. A. Robinson, and D. B. Melrose, Second harmonic electromagnetic emission via Langmuir wave coalescence, *Phys. Plasmas, 3,* 149, 1996.

Winglee, R. M., and G. A. Dulk, The electron-cyclotron maser instability as the source of plasma radiation, *Astrophys. J., 307,* 808, 1986a; The electron-cyclotron maser instability as the source of solar type V continuum, *ibid., 310,* 432, 1986b.

Wu, C. S., P. H. Yoon, and G. C. Zhou, Generation of radiation in solar corona and interplanetary space by energetic electrons, *Astrophys. J.*, *429*, 406, 1994.

Yoon, P. H., Plasma emission by a nonlinear beam instability, *Phys. Plasmas*, *2*, 537, 1995; Plasma emission by a nonlinear beam instability in a weakly magnetized plasma, *ibid*, *4*, 3863, 1997; On the higher-order nonlinear corrections to the theory of plasma emission by a nonlinear beam instability, *ibid*, *5*, 2590, 1998.

Yoon, P. H., and C. S. Wu, Plasma emission via a beam instability with density modulation, *Phys. Plasmas*, *1*, 76, 1994.

Yoon, P. H., C. S. Wu, and Y. Li, Excitation of extraordinary-Bernstein waves by a beam of energetic electrons, *J. Geophys. Res.*, , *104*, 19,801, 1999.

Young, T. S. T., J. D. Callen, and J. E. McCune, High-frequency electrostatic waves in the magnetosphere, *J. Geophys. Res.*, *78*, 1082, 1973.

Zheleznyakov, V. V., and V. V. Zaitsev, The theory of type III solar radio bursts, II, *Sov. Astron. - AJ*, *14*, 250, 1970.

Zheleznyakov, V. V., and E. Ya. Zlotnik, Cyclotron wave instability in the corona and origin of solar radio emission with fine structure, *Solar Phys.*, *43*, 431, 1975.

Y. Li and C. S. Wu, Institute of Space Science, National Central University, Chungli, Taiwan, ROC

P. H. Yoon, Institute for Physical Science and Technology, University of Maryland, College Park, Maryland 20742 (email: yoonp@ipst.umd.edu)

Synchrotron Maser: A "New" Emission Process

V. V. Zheleznyakov,[1] G. Thejappa,[2] S. A. Koryagin,[1] and R. G. Stone[3]

Incoherent synchrotron mechanism is usually invoked to interpret the diffuse radio emission of our Galaxy, the radio emissions of the supernovae remnants and other galactic and extragalactic discrete sources. Under certain conditions, the realization of synchrotron mechanism in its coherent form (synchrotron maser) becomes possible in the low-frequency band of the radio spectrum. The coherent synchrotron radiation resulting from negative reabsorption can account for very intense, highly variable and strongly polarized radio emissions from quasars and other cosmic sources with a two-maximum ("camel" type) frequency spectra. The conditions for the existence of the "synchrotron masers" in astrophysical situations are analyzed. The possible role of synchrotron masers in the radio afterglow of gamma-ray burst of GRB 970508, radio emission from the discrete source GR 0538-49 (quasar 3C147, red shift $Z=0.54$), and intense radio emission from the solar wind as observed by Ulysses is discussed.

1. INTRODUCTION

Zheleznyakov [1966] and *McCray* [1966] were the first to propose the idea of the synchrotron maser. This is the coherent variant of the synchrotron emission mechanism, which involves a system of relativistic or ultra-relativistic electrons spiralling around the magnetic field in a plasma. In thise case, the amplification of synchrotron radiation occurs. This coherent process can yield very intense, circularly polarized, and highly variable emissions in comparison with usually invoked incoherent synchrotron mechanism. Even though *Zheleznyakov* [1967] had attempted to interpret some of the radio observations of quasars, supernova remnants and other radio sources in terms of coherent synchrotron mechanism, there is no observational evidence for synchrotron masers in astrophysical situations. This may be due to very stringent conditions required for its realization.

The most essential requirement is that the reabsorption coefficient μ of synchrotron radiation should be negative. This can be possible only in the low frequency band of the radio spectrum, where the background plasma can have a significant influence on the synchrotron emission. In order to detect the synchrotron masers in cosmic conditions, it is highly desirable to have space-based low frequency radio facilities with high frequency and angular resolutions

The purpose of this paper is to provide a comprehensive description of the synchrotron maser mechanism, and to identify the main observable features which will enable distinguishing it from other emission processes. One of such specific features is the camel-like spectrum with two maxima at low frequencies, produced by a realistic distribution of relativistic electrons. Other features include the strong circular polarization and high intensity variability. These features can be

---

[1]Astrophysics and Cosmic Plasma Physics Department, Institute of Applied Physics, Nizhny Novgorod
[2]Department of Astronomy, University of Maryland, College Park, Maryland
[3]NASA, Goddard Space Flight Center, Greenbelt, Maryland

used for conducting a systematic search for the synchrotron maser signatures in the existing observational data, and also for planning the future observational facilities at low frequencies, especially, space-based radio telescopes for detection purposes. We discuss the observations of the radio afterglow of a cosmic gamma-ray burst GRB 970508 [*Frail et al.*, 1997], the radio emission from quasar 3C147 [*Braude et al.*, 1995], and the intense radio emission from the solar wind [*Reiner et al.*, 1992], detected by the Ulysses Unified Radio and Plasma Wave Experiment (URAP) [*Stone et al.*, 1992].

2. REABSORPTION COEFFICIENT OF SYNCHROTRON RADIATION

The reabsorption coefficient of synchrotron radiation obtained by Einstein coefficient method is [*Wild et al.*, 1963]:

$$\mu = \frac{8\pi^3 c^2}{\omega^2} \int_0^\infty \frac{N(E)}{E^2} \frac{\mathrm{d}\left[E^2 p_\omega(\omega, E)\right]}{\mathrm{d}E} \mathrm{d}E, \quad (1)$$

where ω is the circular frequency, c is the speed of light, $N(E)$ is the energy spectrum of the relativistic electrons, and $p_\omega(\omega, E)$ is the specific power of the synchrotron radiation of one electron with energy E in a unit frequency interval and in a single normal mode. Even though equation (1) is written for a system of relativistic electrons in a vacuum (the refractive index n is equal to unity), its form remains intact, even if the background plasma with $n \neq 1$, and $(1 - n) \ll 1$ is taken into account, since the inclusion of the background plasma affects only the expression for the spectral power $p_\omega(\omega, E)$.

From equation (1), it is clear that the reabsorption coefficient μ is always positive if

$$\frac{\mathrm{d}}{\mathrm{d}E}\left[E^2 p_\omega(\omega, E)\right] > 0 \quad (2)$$

across the entire energy and frequency interval. The inequality (2) is always satisfied, in the case of relativistic electrons spiralling around the magnetic field in vacuum, i. e., synchrotron masers cannot be realized in vacuum. The reabsorption coefficient μ can be negative, if

$$\frac{\mathrm{d}}{\mathrm{d}E}\left[E^2 p_\omega(\omega, E)\right] < 0 \quad (3)$$

in the energy interval, where $N(E)$ is defined. This inequality is easily satisfied, if influence of the ambient "cold" (non-relativistic) plasma on $p_\omega(\omega, E)$ is taken into account, i. e., the synchrotron masers can be realized for a specific distribution of relativistic electrons spiralling around the magnetic field in a cold plasma.

3. SYNCHROTRON RADIATION BY A SINGLE ELECTRON

In an isotropic and sufficiently dilute plasma, for which $\omega_B/\omega \ll 1$, $\omega_L^2/2\omega^2 \ll 1$ and $n^2 \sim 1 - \omega_L^2/\omega^2$, the normal modes (ordinary and extraordinary) are circularly polarized. Here, $\omega_B = \frac{eB}{mc}$ is the electron cyclotron frequency, and $\omega_L = \left(4\pi N_e e^2/m\right)^{1/2}$ is the electron plasma frequency, N_e is the electron density of the ambient plasma, and B is the ambient magnetic field, where e and m are the electron charge and rest mass, respectively. The power radiated in one of these modes by an electron of energy E, with a pitch angle θ with respect to B in a "cold" plasma is [*Zheleznyakov*, 1996]:

$$p_\omega(\omega, E) = \frac{\sqrt{3} e^2 \omega_B \sin\theta}{4\pi c \sqrt{1 + \frac{\omega_L^2}{\omega^2}\left(\frac{E}{mc^2}\right)^2}} \frac{\omega}{\omega_c'} \int_{\frac{\omega}{\omega_c'}}^\infty K_{5/3}(x)\,\mathrm{d}x, \quad (4)$$

where "critical frequency" ω_c' is defined as

$$\omega_c' = \frac{3}{2}\omega_B \sin\theta \left(\frac{E}{mc^2}\right)^2 \left[1 + \frac{\omega_L^2}{\omega^2}\left(\frac{E}{mc^2}\right)^2\right]^{-3/2}, \quad (5)$$

and $K_{5/3}(x)$ is the McDonald function. Due to high directivity of synchrotron radiation, the angle α between magnetic field and the direction of radiation coincides with the pitch-angle θ. The term $\frac{\omega_L^2}{\omega^2}\left(\frac{E}{mc^2}\right)^2$ in equations (4) and (5) reflects the influence of the ambient plasma on synchrotron radiation. If $\frac{\omega_L^2}{\omega^2}\left(\frac{E}{mc^2}\right)^2 \ll 1$, the influence of the ambient plasma on synchrotron radiation is negligible, and the power emitted by an electron becomes:

$$p_\omega(\omega, E) = \frac{\sqrt{3}}{4\pi} \frac{e^2 \omega_B \sin\theta}{c} \frac{\omega}{\omega_c} \int_{\frac{\omega}{\omega_c}}^\infty K_{5/3}(x)\,\mathrm{d}x, \quad (6)$$

where, the critical frequency ω_c is defined as

$$\omega_c = \frac{3}{2}\omega_B \sin\theta \left(\frac{E}{mc^2}\right)^2. \quad (7)$$

The emitted spectrum $p_\omega(\omega, E)$ of the electron of energy E in vacuum is rather sharply peaked at

$$\omega_{\max} \simeq 0.3\,\omega_c \simeq \frac{\omega_B}{2}\sin\theta\left(\frac{E}{mc^2}\right)^2, \quad (8)$$

(solid line in Fig. 1), with the peak value of [*Zheleznyakov*, 1996]:

$$p_\omega^{\max} \equiv p_\omega(\omega_{\max}, E) = \frac{1.6\,e^2 \omega_B \sin\theta}{4\pi c}. \quad (9)$$

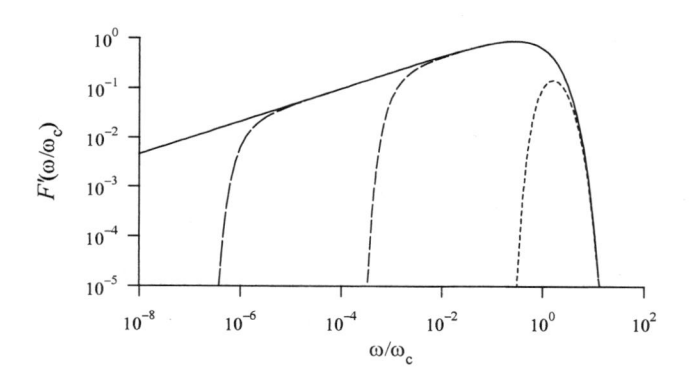

Figure 1. Frequency spectrum of synchrotron radiation of a single electron in a vacuum (solid line) and in a cold plasma (dashed lines). The function $F'(\omega/\omega_c) = \frac{4\pi c}{\sqrt{3}e^2 \omega_B \sin\theta} p_\omega(\omega, E)$. The curves with increasing length of dashes correspond to the parameters $\frac{2}{3}\frac{mc^2}{E}\frac{\omega_L}{\omega_B \sin\theta} = 1$, 10^{-2}, and 10^{-4}.

From equations (8) and (9), it is clear that $p_\omega^{\max}$ is determined solely by the ambient magnetic field B, whereas, $\omega_{\max}$ and the spectral width $\Delta\omega$ depend not only on B, but also on the energy of the electron, E. At high frequencies, the McDonald function $K_{5/3}(\omega/\omega_c)$ asymptotically tends to an exponential function, and therefore the spectral power decreases exponentially $p_\omega \propto (\omega/\omega_c)^{1/2} \exp(-\omega/\omega_c)$ for $\omega \gg \omega_{\max}$. At low frequencies, $p_\omega \propto (\omega/\omega_c)^{1/3}$ for $\omega \ll \omega_{\max}$.

In the case when the influence of the ambient plasma on synchrotron radiation is significant, two energy intervals can be distinguished: (1) $E/mc^2 \ll \omega_L/(\omega_B \sin\theta)$, and (2) $E/mc^2 \gg \omega_L/(\omega_B \sin\theta)$. In the low energy regime, $E/mc^2 \ll \omega_L/(\omega_B \sin\theta)$, the synchrotron radiation is effectively damped at all frequencies. This is because: (1) the background plasma causes the suppression of the synchrotron radiation at low frequencies

$$\omega < \sqrt{2}\omega_L \frac{E}{mc^2} = \frac{2\sqrt{2}}{3}\frac{\omega_L}{\omega_B \sin\theta}\frac{mc^2}{E}\omega_c,$$

and (2) even though an electron emits as if it is in a vacuum at higher frequencies,

$$\omega > \sqrt{2}\omega_L \frac{E}{mc^2} \gg \omega_c,$$

the synchrotron emission is exponentially small. This causes a shift in the frequency, corresponding to the peak of $p_\omega(\omega, E)$ from $\omega_{\max} \simeq 0.3\omega_c$ (vacuum) to $\omega'_{\max} \simeq 0.9\omega_c(mc^2/E)(\omega_L/\omega_B \sin\theta)$ at the exponential slope of $p_\omega(\omega, E)$ in a vacuum (see dashed line in Fig. 1 for the parameter $\frac{2}{3}\frac{mc^2}{E}\frac{\omega_L}{\omega_B \sin\theta} = 1$). In the high energy interval

$E/mc^2 \gg \omega_L/(\omega_B \sin\theta)$, the ambient plasma damps the synchrotron radiation in the frequency interval:

$$\omega < \omega_c \left(\frac{\omega_L}{\omega_B \sin\theta}\frac{mc^2}{E}\right)^{3/2} \ll \omega_c,$$

i. e., the electron emits via synchrotron mechanism, as if it is in a vacuum in a narrow frequency interval

$$\omega_c \left(\frac{\omega_L}{\omega_B \sin\theta}\frac{mc^2}{E}\right)^{3/2} < \omega < \omega_c.$$

The frequency $\omega_{\max} \simeq 0.3\omega_c$, corresponding to the spectral peak, also lies in this frequency interval (see dashed lines in Fig. 1 for the parameters $\frac{2}{3}\frac{mc^2}{E}\frac{\omega_L}{\omega_B \sin\theta} = 10^{-4}$ and 10^{-2}).

The absorption coefficient μ (see, equation (1)) is proportional to

$$E^{-2}\frac{d}{dE}\left[E^2 p_\omega(\omega, E)\right] \propto z^{-2}\Phi(z),$$

where

$$\Phi(z) = 2z \int_z^\infty K_{5/3}(x)\,dx - z^2 K_{5/3}(z),$$

and

$$z = \frac{\omega}{\omega'_c} \sim \frac{2}{3}\frac{E}{mc^2}\frac{\omega_L^3}{\omega^2 \omega_B \sin\theta}$$

[Zheleznyakov, 1967]. The function $\Phi(z)$ is positive in the low energy region $z < 1.35$ (see, Fig. 2). In this case, the inequality $\frac{d}{dE}\left[E^2 p_\omega(\omega, E)\right] > 0$ is satisfied. Therefore, in this energy range, the reabsorption coefficient μ is always positive. On the other hand, $\Phi(z)$ is negative at high energies $z > 1.35$, i. e., $\frac{d}{dE}\left[E^2 p_\omega(\omega, E)\right] < 0$ (see, Fig. 2). In this case, the reabsorption coefficient μ can be negative.

4. SYNCHROTRON EMISSION BY A SYSTEM OF RELATIVISTIC ELECTRONS WITH A POSITIVE SPECTRAL INDEX

The specific intensity I_ω of electromagnetic radiation emitted by a homogeneous source of linear size L containing a system of relativistic electrons is

$$I_\omega = \frac{a_\omega}{\mu}\left[1 - e^{-\tau}\right], \qquad (10)$$

where a_ω is the emissivity and $\tau = \mu L$ is the optical depth. For an optically thick source $\tau \gg 1$, equation (10) can be approximated as $I_\omega \simeq a_\omega/\mu$, whereas for an optically thin source ($\tau \ll 1$), $I_\omega \simeq a_\omega L$. The emissivity a_ω of synchrotron radiation for a system of energetic electrons is given by

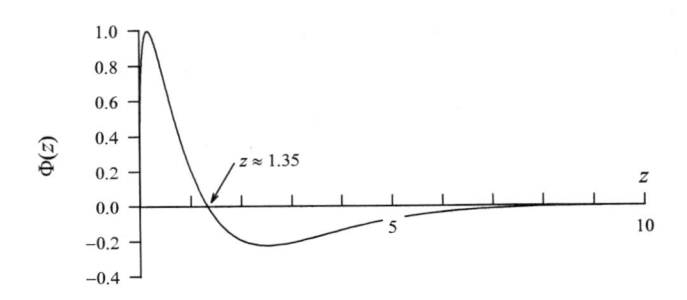

Figure 2. The variation of the function $\Phi(\omega/\omega_c)$.

$$a_\omega = \int_0^\infty p_\omega(\omega, E) N(E) \, dE. \qquad (11)$$

For example, in the case of a power law distribution of relativistic electrons,

$$N(E) = AE^{-\gamma} \qquad (12)$$

(where A is a constant), $a_\omega \propto \omega^{-\frac{\gamma-1}{2}}$. This implies that at high frequencies corresponding to the optically thin ($\tau \ll 1$) spectral regions,

$$I_\omega \propto \omega^{-\delta}, \qquad (13)$$

where the spectral index $\delta = \frac{\gamma-1}{2}$. At low frequencies, corresponding to optically thick spectral regions ($\tau > 1$),

$$I_\omega \propto \omega^{-5/2},$$

since $I_\omega \simeq a_\omega/\mu$, where $a_\omega \propto \omega^{-\frac{\gamma-1}{2}}$ and $\mu \propto \omega^{-\frac{\gamma+4}{2}}$. This suggests that in an optically thick spectral region (see, Fig. 3), the spectral index δ is independent of power law index γ of relativistic electrons. The frequency corresponding to the spectral peak, which occurs at $\tau \simeq 1$ is denoted as ω_r.

At high frequencies, where $\tau < 1$ and $\omega > \omega_L^2/\omega_B \sin\theta$, equation (13) is valid for vacuum as well as for an ionized medium. The spectral peak of the synchrotron radiation occurs at $\omega \sim \omega_L^2/(\omega_B \sin\theta)$ in the plasma (see dashed line in Fig. 4 for the parameter $\frac{2}{3}\frac{\omega_L^2}{\omega_r \omega_B \sin\theta} = 10^2$). At low frequencies, $\omega < \omega_L^2/(\omega_B \sin\theta)$, or $f < 30 N_e/(B \sin\theta)$, the plasma considerably suppresses the synchrotron emission. As a result, the intensity I_ω falls as the frequency decreases according to the formula:

$$I_\omega \propto \omega^{1-\gamma} e^{-\sqrt{3}\omega_L^2/(\omega\omega_B \sin\theta)}. \qquad (14)$$

5. POWER-LAW ELECTRON DISTRIBUTION WITH A NEGATIVE SPECTRAL INDEX

In the case of a power-law electron distribution with a negative spectral index γ,

$$N(E) = \begin{cases} AE^{-\gamma} & \text{if } E_1 < E < E_2 \\ 0 & \text{if } E < E_1 \text{ or } E > E_2 \end{cases}, \qquad (15)$$

in a wide interval $[E_1, E_2]$, where $E_2/E_1 \gg 1$, the values E_1 and E_2 are not essential to calculate the reabsorption coefficient μ at frequencies

$$\frac{2}{3}\frac{E_1}{mc^2}\frac{\omega_L^3}{\omega_B \sin\theta} \ll \omega^2 \ll \frac{2}{3}\frac{E_2}{mc^2}\frac{\omega_L^3}{\omega_B \sin\theta}. \qquad (16)$$

For this type of power law distribution and for $\gamma < -2/3$ [*Zheleznyakov*, 1996]

$$\mu \propto -A\left(\frac{2}{\gamma}+1\right)\int_0^\infty x^{-\gamma} K_{5/3}(x)\,dx, \qquad (17)$$

which is negative for $(2/\gamma+1) > 0$, or $\gamma < -2$. This implies that the relativistic electrons having a power-law distribution with $\gamma < -2$ can excite synchrotron radiation in its coherent form. In this case, the optical depth $|\mu|L$ can exceed unity, and therefore the specific intensity I_ω can be approximated as:

$$I_\omega \simeq \frac{a_\omega}{|\mu|}\exp(|\mu|L). \qquad (18)$$

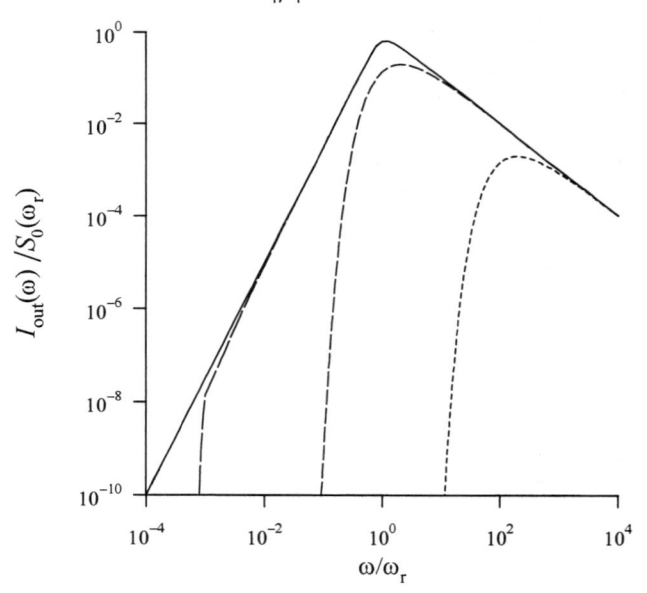

Figure 3. Frequency spectrum of synchrotron radiation by a system of relativistic electrons with a power-law distribution $N(E) \propto E^{-\gamma}$ in a vacuum (reabsorption is taken into account). The solid line marks the spectrum for the power index $\gamma = 2$. The dashed lines with decreasing length of dashes correspond to the indexes γ equal to 3, 4.3, and 11 respectively. At the frequency ω_r the optical depth τ equals unity. The outgoing intensities I_ω are normalized to the source functions $S_0 \equiv a_\omega/\mu$ in a vacuum for the given parameters γ at the frequency ω_r.

Here we note that, even though the power-law distributions of relativistic electrons are capable of generating synchrotron emission in a coherent way, they are not observed in nature.

6. ELECTRON DISTRIBUTION WITH A MAXIMUM AT $E=E_0$

The relativistic electron distribution is usually observed to have a peak at a particular energy, $E = E_0$. One such simple distribution is

$$N(E) = \begin{cases} A(E/E_0)^{-\gamma_1} & (\gamma_1 < 0) \text{ if } E < E_0 \\ A(E/E_0)^{-\gamma_2} & (\gamma_2 > 2) \text{ if } E > E_0 \end{cases}, \quad (19)$$

where

$$A = \frac{N_s}{4\pi E_0} \frac{(\gamma_2 - 1)(1 - \gamma_1)}{\gamma_2 - \gamma_1},$$

N_s is the number density of the electrons, and $\int_0^\infty dE \times EN(E)$ is a finite value. This distribution function, which peaks at $E = E_0$ (see Fig. 5) is very convenient for numerical calculations. The reabsorption coefficient μ is computed numerically using formula (1).

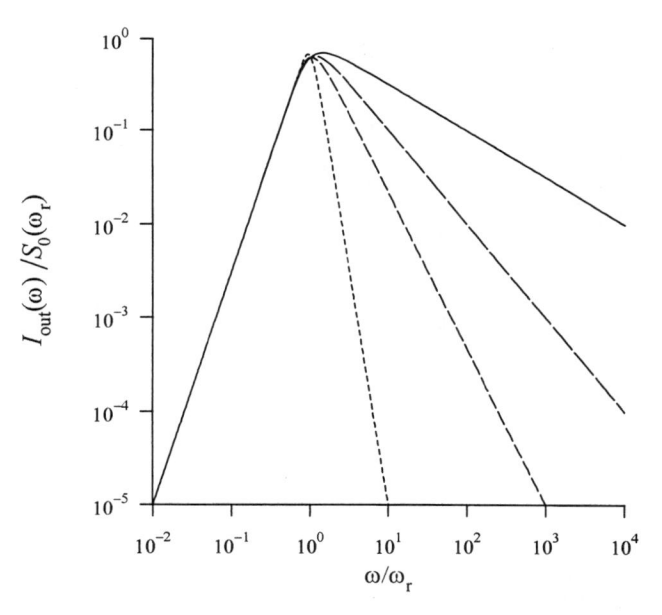

Figure 4. Frequency spectrum of synchrotron radiation by a system of relativistic electrons with a power-law distribution $N(E) \propto E^{-3}$ in a cold plasma (reabsorption is taken into account). The dashed lines with increasing length of dashes mark the spectra for the parameters $\frac{2}{3}\frac{\omega_L^2}{\omega_r \omega_B \sin\alpha} = 10^2$, 10^0, and 10^{-2}. The solid line is identical to one in Fig. 3 for $\gamma=3$. The frequency ω_r, corresponding to the spectral peak where the optical depth $\tau \simeq 1$ and the factor $S_0(\omega_r)$ are defined for the vacuum.

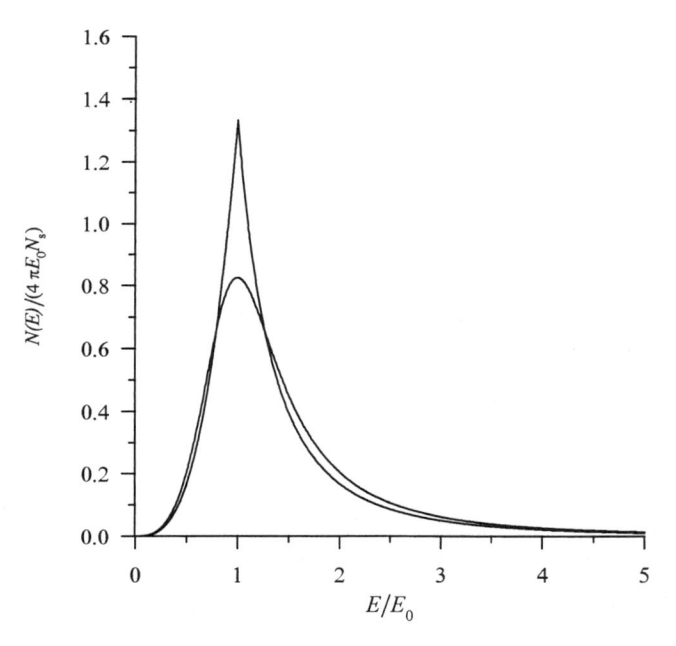

Figure 5. Energy distribution of relativistic electrons: (I) with a sharp maximum $N(E) = A(E/E_0)^{-\gamma_1}$ for $E < E_0$, and $N(E) = A(E/E_0)^{-\gamma_2}$ for $E > E_0$, with $\gamma_1 = -3$; and $\gamma_2 = 3$, and (II) with a smooth maximum.

It is clear from Fig. 6 that reabsorption coefficient $\mu < 0$ for $\omega/\omega_c < (\Gamma_0 \omega_L/\omega_c)^{3/2}$ and $\mu > 0$ for $\omega/\omega_c > (\Gamma_0 \omega_L/\omega_c)^{3/2}$. The maximum of the function $(-\mu) \sim eN_s/(B\sin\alpha \Gamma_0^5) \times (\Gamma_0 \omega_L/\omega_c)^{-5/2}$ occurs very close to the frequency where μ changes its sign. The results obtained for a smooth distribution of relativistic electrons agree very well with those obtained for a distribution with a sharp peak (as given by equation (19)).

In Fig. 7, the normalized outgoing specific intensity $I_\omega \times 32\pi^3 mc^2/(9e^2 B^2 \Gamma_0^5 \sin^2\alpha)$ is plotted as a function of $\omega/\omega_c[\Gamma_0]$. It is clear from this figure that in the low frequency range $\omega/\omega_c < (\Gamma_0 \omega_L/\omega_c)^{3/2}$, the specific intensity increases exponentially, by reaching a strong peak at $\omega/\omega_c \sim (\Gamma_0 \omega_L/\omega_c)^{3/2}$. This part of the spectrum is due to coherent synchrotron mechanism. At high frequencies where $\tau \ll 1$, the intensity I_ω varies as a power-law. In this case, μ is positive and therefore the spectrum is due to incoherent synchrotron emission. The second peak at $\omega \sim \omega_r$ corresponds to the condition $\tau \sim 1$ for the parameters $\omega_r/\omega_c[\Gamma_0] > 1$. In the opposite case $\omega_r/\omega_c[\Gamma_0] < 1$, this second peak coincides with the peak of the emissivity a_ω at $\omega \sim \omega_c[\Gamma_0]$. This suggests that any astrophysical object with a relativistic electron distribution, peaking at a particular energy $E = E_0$ produces synchrotron radiation spectra with two maxima:

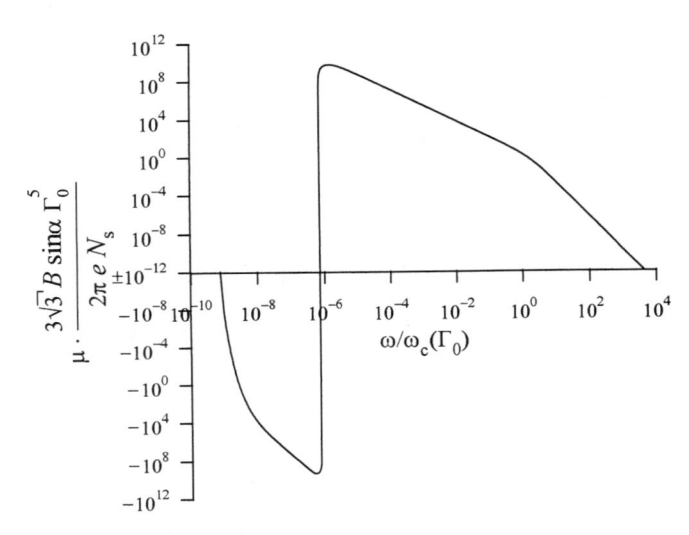

Figure 6. The reabsorption coefficient of synchrotron radiation by a system of relativistic electrons with a sharp maximum $N(E) = A(E/E_0)^{-\gamma_1}$ for $E < E_0$, and $N(E) = A(E/E_0)^{-\gamma_2}$ for $E > E_0$, with $\gamma_1 = -3$; and $\gamma_2 = 3$, for $\frac{\omega_L}{\omega_c[\Gamma_0]}\Gamma_0 = 10^{-4}$.

the first one due to coherent and the second one due to incoherent mechanisms.

The back influence of the strong synchrotron radiation on electron distribution (quasi-linear relaxation process in plasma physics terminology) can lead to the variation of the distribution function to $N(E) \to E^2$, i. e., $\gamma_1 \to -2$,. This is equivalent to a plateau formation in plasma physics, because $N(E) \sim E^2$ corresponds to the distribution function of relativistic electrons over momentum $f(\vec{p}) = \text{const}$. This leads to the increase of the reabsorption coefficient to $\mu \to 0$. The frequency spectrum in this case has two maxima (see, Fig. 8). The numerical calculations show that the low frequency maximum disappears for γ_1 slightly bigger than -2 (for example, for $\gamma_1 \geq -1.95$). Here one should note that $N(E)$ can also differ from E^2 due to acceleration of relativistic particles, collisions etc.; this quasi-linear relaxation process is not yet investigated for synchrotron masers.

7. DISCUSSION AND CONCLUSIONS

As shown in the previous section, the specific intensity I_ω of the synchrotron maser emission is characterized by a two-maximum frequency spectrum ("camel" type), where the low-frequency spectral peak corresponds to the maser emission. The high temporal variations of radio flux due to the strong influence of source parameters on the level of radiation, $I_\omega \propto e^{|\mu L|}$ is another signature of the synchrotron maser; it can also produce high circularly polarized radiation resulting from the difference of the reabsorption coefficients for the ordinary (o) and extraordinary (e) modes in the ambient plasma with a quasi-homogeneous, (non-chaotic) magnetic field.

7.1. Polarization of Radio Emission

In a rarefied plasma ($\omega_L^2 \ll \omega^2$) with a relatively weak magnetic field ($\omega_B \ll \omega$), the ordinary (o) and extraordinary (e) modes are circularly polarized for a wide range of angles between $\vec{B}$ and the ray path. In the first approximation, the absorption coefficient of e-mode (μ_e) is approximately equal to that of o-mode μ_o, i. e., $\mu_e \simeq \mu_o$. If higher order effects are considered, $\mu_e \neq \mu_o$, and the difference between the absorption coefficients of these two modes is

$$\mu_e - \mu_o \sim \sqrt{(mc^2/E)^2 + \omega_L^2/\omega^2}. \quad (20)$$

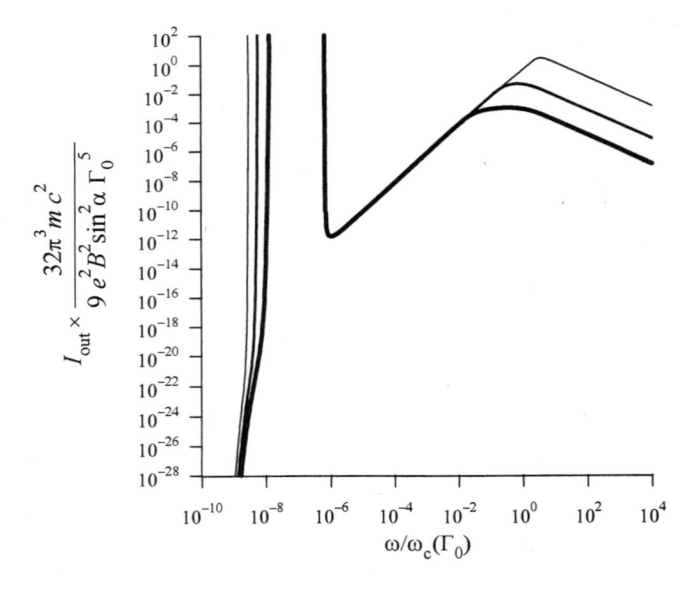

Figure 7. Frequency spectrum of the synchrotron radiation emitted by energetic electrons with a sharp maximum $N(E) = A(E/E_0)^{-\gamma_1}$ for $E < E_0$, and $N(E) = A(E/E_0)^{-\gamma_2}$ for $E > E_0$, with $\gamma_1 = -3$; and $\gamma_2 = 3$, for $\frac{\omega_L}{\omega_c[\Gamma_0]}\Gamma_0 = 10^{-4}$. The thicker lines correspond to smaller path lengths L and the ratios $\omega_r(L)/\omega_{\max a_\omega} = 10$, 1, and 0.1, respectively. The optical depth τ of a source equals unity at the frequency ω_r for a given length L. At the frequency $\omega_{\max a_\omega}$ the emissivity a_ω reaches its peak value. For low frequencies $\omega/\omega_c[\Gamma_0] < 10^{-6}$, the intensity of radiation sharply increases since the reabsorption coefficient μ is negative at these frequencies.

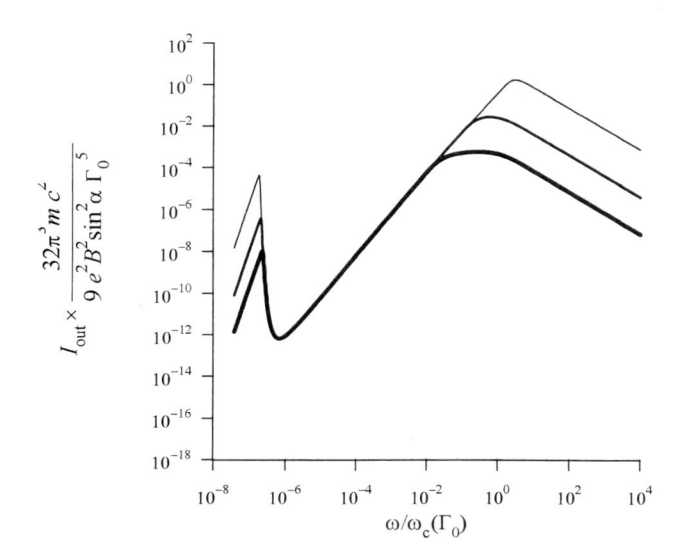

Figure 8. The same as Fig. 7 for $\gamma_1 = -2$ and $\gamma_2 = 3$.

Eventhough, this can be very small [*Zheleznyakov*, 1996], the Faraday rotation can be very large, i.e., $|\mu_e - \mu_o| L \gg 1$ due to large path lengths, L, in the synchrotron maser source regions. This can lead to a highly circularly polarized radiation. At present, there are no observations of such highly circularly polarized sources in astrophysical situations. Therefore, detection of circularly polarized radio emissions from the galactic or extragalactic radio sources will provide a good evidence for the synchrotron masers. One should note that the conditions required for the polarized radiation are more severe than the maser criterion alone.

7.2. Where to Search

The realization of a synchrotron maser is possible only if highly energetic relativistic electrons $E_0/mc^2 \gg \omega_L/\omega_B$ are present in the source. In the opposite case the synchrotron increment is exponentially small. The synchrotron increment peaks at

$$f_{\max}^{LF} \sim \left(\frac{E_0}{mc^2}\right)^{1/2} \frac{\omega_L^{3/2}}{2\pi\omega_B^{1/2}} = 250 \frac{N_e^{3/4}}{B^{1/2}} \left(\frac{E_0}{mc^2}\right)^{1/2} \text{ Hz.} \quad (21)$$

For an isotropic refraction index n, where $(1-n) \ll 1$,

$$f_{\max}^{LF} \gg \max\{\omega_L, \omega_B\}/2\pi,$$

which is valid for $E_0/mc^2 \gg \omega_L/\omega_B$ when $\omega_L > \omega_B$ and for $E_0/mc^2 \gg (\omega_B/\omega_L)^3$ when $\omega_L < \omega_B$. Thus, the relativistic electrons with energies

$$\frac{E_0}{mc^2} \gg \max\left\{\frac{\omega_L}{\omega_B}, \left(\frac{\omega_B}{\omega_L}\right)^3\right\} \quad (22)$$

easily excite synchrotron maser emission at frequencies given by equation (21).

The peak frequency of the maser emission (equation (21)) depends on an unknown energy E_0 of relativistic electrons. This can be eliminated by estimating the lower and upper limits of $f_{\max}^{LF}$ that depend only on the ambient plasma density, N_e, and magnetic field B. Since the energy E_0 satisfies the inequality (22), one can write an equality for the frequency as:

$$f_{\max}^{LF} \gg f_{LL} \equiv \max\left\{\frac{\omega_L^2}{2\pi\omega_B}, \frac{\omega_B}{2\pi}\right\} \simeq \max\left\{\frac{30 N_e}{B}, 3 \cdot 10^6 B\right\}. \quad (23)$$

On the other hand, the low frequency maximum is located well below the second maximum $f_{\max}^{HF} \sim \omega_r/2\pi$, caused by the reabsorption. The outgoing radiation at $f_{\max}^{HF}$ is determined by the higher energy electrons $E \geq E_0$ with the "critical" frequency $\omega_c \sim 2\pi f_{\max}^{HF}$. This can be used to estimate the energy $E_0/mc^2 \leq (2\pi f_{\max}^{HF}/\omega_B)^{1/2}$ and the upper limit on the frequency

$$f_{\max}^{LF} \leq f_{UL} \equiv \left(f_{\max}^{HF}\right)^{1/4} \left(\frac{\omega_L^2}{2\pi\omega_B}\right)^{3/4} \leq f_{LL} \left(\frac{f_{\max}^{HF}}{f_{LL}}\right)^{1/4}. \quad (24)$$

It should be noted that the upper limit f_{UL} is higher than the lower limit f_{LL} by a factor of $(f_{\max}^{HF}/f_{LL})^{1/4}$. The power index $1/4$ of this factor is rather small (for example, for the ratio $f_{\max}^{HF}/f_{LL} = 10^4$ the upper limit $f_{UL} < 10 f_{LL}$).

In Table 1, we give the estimated frequency limits for the possible detection of synchrotron masers in some astrophysical situations. The electron plasma frequency is higher than the electron cyclotron frequency, i.e., $\omega_L/\omega_B > 1$ for all the objects presented in Table 1, except for solar corona. Therefore, the lower limit on the energy $E_0 > \omega_L/\omega_B$ (see, equation (22)) is appropriate, and accordingly, the frequency $f_{LL}^{(1)} = 30 N_e/B$ Hz determines the lower limit of the frequency band for the synchrotron masers. If one considers the energy $E_0 = 10^2 \omega_L/\omega_B$, the frequency corresponding to the maser radiation peak is $f_{\max}^{LF} = 10 f_{LL}^{(1)}$. Even though, the dispersion in the frequency limits $f_{LL}^{(1)}$ is very large, the tendency is obvious; if the synchrotron maser is operative, it should be observable in the low frequency radio band.

7.3. Examples

The unified Radio and Plasma Wave (URAP) experiment [*Stone et al.*, 1992] on Ulysses spacecraft has

Table 1. Parameters of Possible Synchrotron Maser Source Regions

Source	N_e (cm^{-3})	B (Gauss)	ω_L/ω_B	$30N_e/B$
HI Regions	10^{-2}	10^{-6}	$3 \cdot 10^2$	300 kHz
HII Regions	10^2	10^{-6}	$3 \cdot 10^4$	3 GHz
Crab Nebula	10^3	10^{-3}	$1 \cdot 10^2$	30 MHz
Solar corona	10^8	10^2	$3 \cdot 10^{-1}$	30 MHz
Solar wind (1 A. U)	10	10^{-4}	$1 \cdot 10^2$	3 MHz
Solar wind (4 A. U)	1	10^{-4}	$3 \cdot 10^1$	300 kHz

observed non-drifting type III-like emissions in the frequency band 10 to 100 kHz on March 25–March 27, 1991 [Reiner et al., 1992]. In Plate 1, we present the dynamic spectrum corresponding to this event. The observed flux was reported to be 3 orders of magnitude too high to be accounted for by incoherent synchrotron mechanism. During this event, relativistic electrons with energies of $E \sim 10$ MeV were observed. However, the energy distribution of these relativistic electrons is not measured. In this case, the high intensities of these type III-like emissions from the solar wind can be interpreted in terms of synchrotron maser emission.

The radio spectrum of the discrete source GR 0538-49 (quasar 3C147, red shift $Z = 0.54$) is presented in Fig. 9. Here, the low frequency points are from Braude et al. [1995], and the high frequency points are from other radio surveys. These data points appear to lie on a "camel" like spectrum, as shown in the figure. Low-frequency maximum corresponds either to the synchrotron maser, or due to a double incoherent radio source.

A good example of a very distant object having extremely small linear size is the radio afterglow of the γ-ray burst of 1997 May 8 (GRB 970508; $0.86 < Z < 2.3$). One of the models of the origin of this radio source is the fireball originated as a result of merging of two neutron stars. The energy released during this process is $E_m > 10^{52}$ ergs. This fireball generates a relativistic expanding shell, which in turn generates a shock wave propagating in the surrounding plasma. The radio observations of the afterglow (one month after the γ-ray burst) were obtained at three frequencies, 1.43 GHz, 4.86 GHz, and 8.46 GHz using VLA and VLBI by Frail et al. [1997]. One can extrapolate these points in the form of a probable frequency spectrum. Brainerd [1998] has discussed the afterglow in terms of incoherent synchrotron mechanism and has interpreted the drop at 1.43 GHz due to strong reabsorption in the shell.

For this γ-ray burst, we have reexamined the incoherent synchrotron radiation process. According to our analysis, the observed radio emission can be explained by incoherent processes, if $B > 10^{-2}$ G for real values of parameters, $\tau(W_e/W_B)(d/R) < 1$, $\Gamma_s = (1 - v^2/c^2)^{-1/2} > 5$, and $W_e > 10^{50}$ ergs. Here, τ is the optical depth of the source Over synchrotron reabsorption, W_e is the total energy of radiating electrons, W_B is the energy of the magnetic field in the shell; Γ_s is the Lorenz factor and v is the velocity of expansion of the shell. If $B < 1$ G, synchrotron losses are small and the radiating electrons have the primary origin; acceleration in the shell is not needed. These constraints on parameters of the radiating shell permit to link the type of synchrotron mechanism with the type of the fireball expansion.

In the case of adiabatic expansion, the total emission in γ-rays, x-rays and other wavelengths is $\sim 10^{52}$ ergs, which is much less than the primary energy release. This energy is sufficient to store the relativistic electrons in the shell with energy $W_e \geq 10^{50}$ ergs, needed for incoherent synchrotron mechanism in radio band. In the case of a fast radiating expansion, almost all the energy is lost on γ- and x-ray emissions in the first stage of radiation (< 1 day) itself. By this time, the energy of the fireball drops down to 10^{49} ergs, which is not sufficient to support the relativistic shell with energy $W_e \geq 10^{50}$ ergs. Therefore, in the case of highly radiating (non-adiabatic) expansion, the incoherent synchrotron mechanism cannot explain the observed radio emission, whereas, the maser variant of synchrotron mechanism, which requires considerably fewer relativistic electrons can easily account for the observed radio flux of the afterglow.

Therefore, the search of extraterrestrial synchrotron masers based on radio observations from the Earth has not yet yielded any definite results. The successful detection of these masers requires space-based radio observations at long wavelengths ($f < 10$ MHz). The synchrotron masers probably occur in at least two astrophysical situations. These include: (1) discrete cosmic

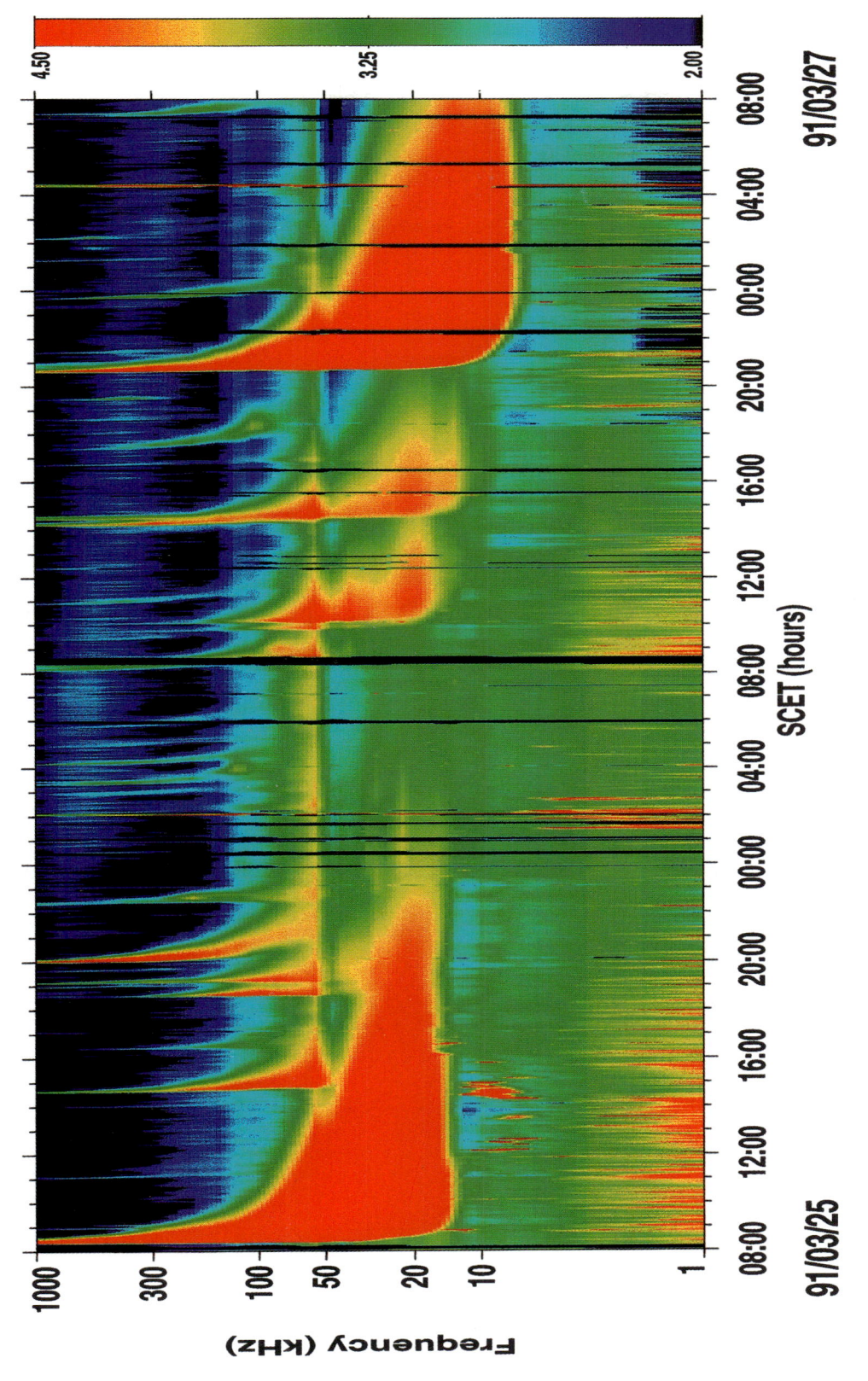

Plate 1. Dynamic spectral representation of the high- and low-band radio receiver data in the frequency range from 3 to 940 kHz showing three radio bursts separated by 4.5 hours. The bar chart on the right shows the dynamic range of the display. (adapted from *Reiner et al.* [1992])

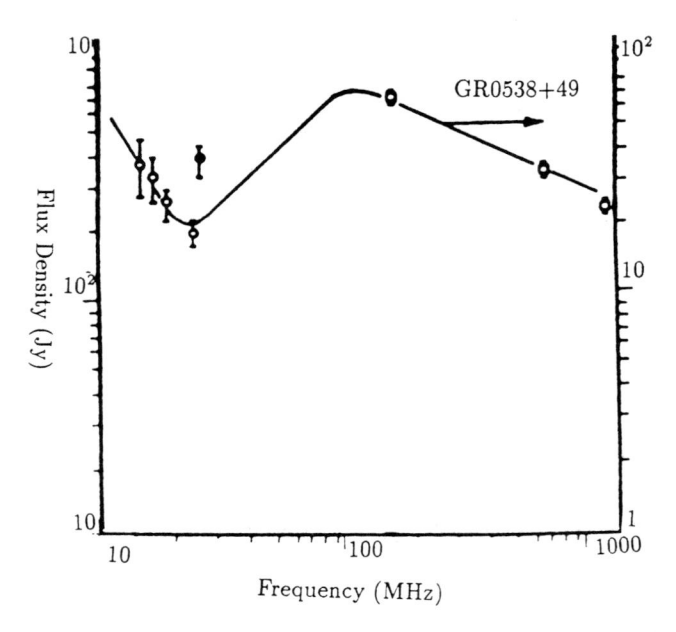

Figure 9. Spectral plot of GR 0538-49 (quasar 3C147, red shift $Z = 0.54$) from the UTR-2 sky survey for declinations 41 to 52 degrees (adapted from *Braude et al.* [1995]).

radio sources with two-maximum frequency spectra, in association with strong circular polarization, and (2) synchrotron sources in the interplanetary medium with very intense low frequency radio flux which cannot be accounted for by the incoherent processes. In the latter case, the knowledge of the *in situ* energy distribution of relativistic electrons is essential.

Acknowledgments. The research of GT is supported by the NASA grants NAG-57145, NCC5-372 and NAG-56059. The research of VZ and SK has been supported by Russian Foundation for Basic Research (the project 99-0218244) and by the Council for State Support of Leading Scientific Schools (the project 96-15-96739).

REFERENCES

Brainerd, J. J, Physical constraints on GRB 970508 from its radio afterglow, *Astrophys. J. Lett.*, *496*, 67–70, 1998.

Braude, S. Ya., K. P. Sokolov, N. K. Sharykin, and S. M. Zakharenko, Decametric Survey of Discrete Sources in the Northern Sky, *Astrophys. Space Sci.*, *226*, 245–271, 1995.

McCray, R., *Science*, *154*, 1320–1323, 1966.

Frail, D. A., S. R. Kulkarni, L. Nicastro, M. Feroci, and G. B. Taylor, The radio afterglow from the γ-ray burst of 8 May 1997, *Nature*, *389*, 261–263, 1997.

Reiner, M. J., R. G. Stone, and J. Fainberg, Observation of non-drifting radio emissions associated with the intense solar activity in March 1991, *Geophys. Res. Lett.*, *19*, 1275–1278, 1992.

Stone, R. G., J. L. Bougeret, J. Caldwell, P. Canu, Y. de Conchy, N. Cornilleau-Wehrlin, M. D. Desch, J. Fainberg, K. Goetz, and M. L. Goldstein, The unified radio and plasma wave investigation, *Astron. Astrophys. (Supp.)*, *92*, 291–316, 1992.

Wild, J. P., S. F. Smerd, and A. A. Weiss, *Ann. Rev. Astron. Astrophys.*, *1*, 291, 1963.

Zheleznyakov, V. V., On the negative reabsorption of synchrotron radiation (in Russian), *Soviet Phys. JETP*, *51*, 570–578, 1966.

Zheleznyakov, V. V., On a coherent synchrotron mechanism of radio emission of some cosmic sources, *Soviet Astron.*, Engl. Transl., *11*, 33, 1967.

Zheleznyakov, V. V. *Radiation in Astrophysical Plasmas*, 472 pp., Kluwer Academic Publishers, Dordrecht, 1996.

V. V. Zheleznyakov and S. A. Koryagin, Astrophysics and Cosmic Plasma Physics Department, Institute of Applied Physics, Ulyanov St., 46, Nizhny Novgorod 603600 Russia

G. Thejappa, Department of Astronomy, University of Maryland, College Park, MD 20742 USA

R. G. Stone, NASA, Goddard Space Flight Center, Greenbelt, MD 20771 USA

Plasma Thermal Noise: The Long Wavelength Radio Limit

Nicole Meyer-Vernet, Sang Hoang, Karine Issautier, Michel Moncuquet, and Gregory Marcos

Dt. de Recherche Spatiale, CNRS UMR-8632, Observatoire de Paris, Meudon, France

The thermal noise in a plasma is due to the electrostatic fluctuations produced by the thermal motion of the ambient particles - a concept which can be easily generalized to the quasi-thermal conditions arising in space. This noise is detected at the ports of any electric antenna immersed in a plasma, and this ubiquity makes it the ultimate limit to detect weak radio signals. Since this noise reveals the local plasma properties, an analysis of its spectrum provides accurate *in situ* plasma measurements. This paper explains the basic underlying physics in various space conditions, and gives estimates of the corresponding detection threshold for radio signals. This threshold can be lowered by calculating the plasma thermal noise and subtracting it from the data. As a by-product, we obtain the antenna impedance and its reception pattern for electrostatic waves. It turns out that antennas in plasmas often mock the radio astronomer's intuition, which might result in erroneous interpretation of space data.

1. INTRODUCTION

What is hiding behind the name: "plasma thermal noise"? How can it be estimated? Why bothering about it?

The underlying concepts are simple, although the calculations are not. Any plasma is made up of charged particles moving around, thereby producing a fluctuating electric field which can be detected at the ports of an electric antenna. Just as an antenna immersed in blackbody radiation detects the corresponding thermal noise and measures the blackbody temperature, so an antenna immersed in a plasma detects the plasma thermal noise and measures the plasma temperature.

There are, however, two basic differences. First, in contrast to blackbody radiation - which is electromagnetic - the plasma thermal noise mainly consists of electrostatic waves or fluctuations, produced by the motion of the ambient charged particles. Except for very special circumstances or frequencies, these waves are heavily damped in the medium, so that the noise reveals the local plasma. Second, contrary to blackbody radiation, which is completely defined by its temperature, an equilibrium plasma needs one more parameter - the particle number density - to be completely defined. Hence the plasma thermal noise reveals not only the plasma temperature but also the density, which is very convenient for diagnostic purposes.

What happens if the plasma is not in equilibrium, which is the usual situation in space? In that case, one needs still more parameters to define the medium, but if the plasma is stable, the electrostatic fluctuations are completely determined by the particle velocity distributions [Rostoker, 1961], so that the corresponding noise, thus called "quasi-thermal noise" can be calculated from them. Hence, in that case, measuring the noise can yield more plasma parameters than just the particle number density and temperature.

How does this noise look like? There is no unique

Radio Astronomy at Long Wavelengths
Geophysical Monograph 119
Copyright 2000 by the American Geophysical Union

answer, because the phenomenon has many different disguises. Indeed, one needs an antenna to measure it, so that the signal depends on the antenna length and shape, and also on the ambient magnetic field. As a consequence, this noise has been sometimes erroneously interpreted as more fashionable "new" emissions or instabilities, although it was first studied a long time ago [Andronov, 1966; Fejer and Kan, 1969], and was calculated and shown to explain quantitatively the typical spectrum observed in interplanetary space below 100 kHz [Meyer-Vernet, 1979], as also the diffuse magnetospheric "emissions" [Sentman, 1982].

This noise sets the ultimate limit of detection of weak signals by antennas immersed in plasmas. However, it is not always a hindrance. Indeed, the spectroscopy of this noise has turned out to be an efficient tool for *in situ* plasma measurements [Meyer-Vernet and Perche, 1989], which is now widely applied in space [Meyer-Vernet et al., 1998 and references therein] and can serve to calibrate other measurements [Maksimovic et al., 1998]. Indeed, this technique has often proved to be more accurate than the classical particle analysers to measure the electron density and thermal temperature [e.g., Issautier et al. 1999] because it is relatively insensitive to space potential and photoelectron emission, since it is based on wave measurements.

2. FROM BLACKBODY RADIATION TO PLASMA THERMAL NOISE

2.1. Blackbody Radiation

It may be worth recalling some basics. Let us put an electric antenna of length L in blackbody radiation at temperature T. The mean electromagnetic energy in the enclosure is $k_B T$ per radiation mode, where k_B is Boltzmann's constant. The mean energy dW per unit volume in the frequency interval df is thus $k_B T$ times the number of corresponding radiation modes. This number is $2 \times V_k/(2\pi)^3$, where the factor 2 takes into account the two polarizations and V_k is the corresponding volume in wave vector space; because of isotropy, this volume is that of a spherical shell of radius k and thickness dk, i.e., $V_k = 4\pi k^2 dk$ with $k = 2\pi f/c$, c being the velocity of light, so that $V_k = 4\pi \times (2\pi)^3 f^2 df/c^3$.

We deduce the spectral density of the electric field, E_f^2, by noting that the mean energy per unit volume per unit frequency, $dW/df = k_B T \times 2 \times V_k/(2\pi)^3 /df$, is twice the electric energy, $\epsilon_0 E_f^2/2$ (because half of the energy is magnetic). This yields

$$E_f^2 = \frac{8\pi k_B T f^2}{\epsilon_0 c^3} \qquad (1)$$

which corresponds to the well-known Rayleigh-Jeans limit of the Planck spectrum.

2.2. Plasma Thermal Noise

Now, let us put the antenna in a plasma, made up of n electrons and n positive ions per unit volume, in equilibrium at temperature T. As is well-known, the plasma affects the propagation of electromagnetic waves through its refractive index, which changes accordingly the above estimate. However, something more important happens: the thermal motion of the charged particles yields a fluctuating space charge, which produces a fluctuating electric field. How can we estimate simply the field spectral density?

We know that a *bona fide* plasma tries to preserve its electrical neutrality, but cannot do so at scales smaller than the Debye length, L_D. Hence the largest volume in which the particle motion can produce an electric field is a Debye sphere of radius L_D. This volume contains a mean number $N \approx n \times 4\pi L_D^3/3$ of electrons, which, at equilibrium, fluctuates by $\Delta N \approx \sqrt{N}$. This produces electric field fluctuations whose mean square has the order of magnitude

$$E^2 \sim \left(\frac{\Delta N \times e}{4\pi \epsilon_0 L_D^2}\right)^2 \qquad (2)$$

To estimate the power spectral density, E_f^2, we divide E^2 by the bandwidth of the fluctuations, which is of the order of the plasma frequency f_p. Substituting the electron thermal velocity $v_{th} = \sqrt{k_B T/m_e}$ where m_e is the electron mass, and $L_D = v_{th}/(2\pi f_p)$, we find

$$E_f^2 \sim \frac{2\pi^2 k_B T f_p^2}{3\epsilon_0 v_{th}^3} \qquad (3)$$

The ratio of the plasma thermal electrostatic noise to that of blackbody radiation at the same temperature is thus approximately

$$\frac{E_f^2 \text{ plasma}}{E_f^2 \text{ black-body}} \sim \frac{f_p^2}{f^2} \times \frac{c^3}{v_{th}^3} \qquad (4)$$

[see Bekefi, 1966]. For example, with a temperature of 10^5 K, this ratio amounts to $(4 \times 10^3 \times f_p/f)^2$, so that the thermal electrostatic noise plays generally a major role at long wavelengths in a plasma.

One must keep in mind that this order of magnitude estimate considers only the fluctuations at the scale of the Debye length and does not take into account the thermal Langmuir waves, which propagate above the plasma frequency with a wavelength greater than L_D. The above estimate, therefore, is only relevant below f_p

and if the noise is detected with an antenna of length of order L_D [Meyer-Vernet, 1993]. The Langmuir waves excited by the particle thermal motion produce a peak in the noise spectrum just above f_p (Figures 4 and 5). Because the Langmuir wavelength is greater than L_D, the detection of this peak requires an antenna adapted to measure these waves, i.e., of typical length $L > L_D$. Note, finally, that we have ignored the ions because of their large mass: they only play a role at frequencies of order of or smaller than their characteristic frequencies, i.e., much below f_p, except when there is a significant Doppler-shift due to their bulk motion in the antenna frame [Issautier et al., 1996].

3. COMPLICATIONS

In practice, one has to consider some further effects which complicate the problem.

3.1. Space Plasmas Are not in Local Thermal Equilibrium

Because most natural plasmas are dilute and warm, collisions cannot generally ensure local thermal equilibrium, given the non-equilibrium processes at work. Furthermore, plasmas share an important property which distinguishes them from neutral gases: because the particle encounters are governed by the Coulomb potential, the particle cross-sections for collisions vary as the inverse square of their energy; hence the energetic particles are not in equilibrium, even when the bulk of the particles is roughly so. Therefore, even though the particle velocity distribution remains close to a Maxwellian of temperature T at thermal energies, it exhibits a non-Maxwellian high-energy tail; that is, the plasma becomes "quasi-thermal".

The velocity distributions are often approximated by a superposition of a Maxwellian of temperature T which represents the bulk of the particles and of a second Maxwellian with a smaller density and a greater temperature [Feldman et al., 1975], so that the resulting mean temperature is still of the order of T. A better approximation for the particle distribution might be a Kappa function - i.e. a generalized Lorenzian - since the suprathermal tail is often observed to have a velocity distribution decreasing as an inverse power law rather than as a Maxwellian [e.g., Maksimovic et al., 1997].

How is the plasma thermal noise modified if the plasma is quasi-thermal?

The plasma particles interact with the waves having a phase velocity close to their own velocity. Except in the vicinity of f_p, the phase velocity of Langmuir waves is of the order of magnitude of the particle mean square velocity, so that the noise is determined by the thermal particles; thus it is not very different from the thermal noise at temperature T. As the frequency approaches f_p, however, the phase velocity increases, and when it matches the typical speed of the suprathermal tail, the noise is ultimately determined by those suprathermal particles. If the suprathermal tail velocity distribution is a Maxwellian of temperature $T_H > T$, then the quasi-thermal noise at the plasma frequency peak is greater than the thermal noise by the factor T_H/T.

More rigorous calculations [Meyer-Vernet and Perche, 1989; Chateau and Meyer-Vernet, 1991] yield results of the same order of magnitude as estimated above, provided that the length of the measuring antenna is greater than L_D, in order to be adapted to the waves. For example, in a plasma of thermal temperature $T \approx 10^5$ K with a suprathermal tail 10^2 times hotter and a plasma frequency $f_p \approx 20$ kHz, the quasi-thermal electric field spectral density at the plasma frequency peak is roughly given by (3) with the temperature T_H, which yields approximately 0.2 μV/m/$\sqrt{\text{Hz}}$.

3.2. Space Plasmas Are Generally Magnetised

In that case the relevant frequency scale is the electron gyro frequency $f_g = \omega_g/2\pi$, where $\omega_g = eB/m_e$, B being the modulus of the magnetic field. The electron gyration in the magnetic field plays a major role in determining the noise at frequencies not significantly greater than f_g. In that case, the electron thermal motion excites Bernstein waves, which propagate roughly normal to B, at frequencies between the gyro harmonics, with a wave number $k_\perp$ of the order of the inverse of the electron gyro radius $\rho = v_{th}/\omega_g$. The presence of suprathermal electrons yields noise maxima between the gyro harmonics, which are produced by the waves being negligibly damped by the thermal electrons. These waves should have a sufficiently small parallel wave number $k_\parallel$ for the thermal electrons not to meet the Cerenkov condition; this requires $k_\parallel v_{th}$ to be sufficiently small compared to $\omega - n\omega_g$, n being the order of the harmonic band considered, i.e., typically $|k_\parallel| \leq 0.1/\rho$.

The mean electrostatic energy W of quasi-thermal fluctuations per unit volume at the mid-band maxima is $k_B T_H$ times the number of modes per unit volume. This number is $V_k/(2\pi)^3$ where V_k is the volume in k space occupied by the relevant modes (which has a cylindrical symmetry around **B**). The volume V_k corresponding to a gyro harmonic band can be approximated by a cylinder of radius $k_\perp$ and half-height the maximum value of $k_\parallel$ estimated above, which gives $V_k \approx 0.2\pi/\rho^3$. We deduce the energy per unit volume and frequency, dW/df, by dividing the energy W by the width f_g of a gyro harmonic band, and finally estimate the electric field spectral density E_f^2 from $dW/df \sim \epsilon_0 E_f^2/2$. This yields approximately

$$E_f^2 \sim \frac{\pi k_B T_H f_g^2}{2\epsilon_0 v_{th}^3} \qquad (5)$$

A more rigorous calculation [Sentman, 1982] yields roughly the same result. With the quasi-thermal plasma temperatures considered above and a gyro frequency $f_g \approx 20$ kHz, (5) gives a spectral density around $1\ \mu$V/m/$\sqrt{\text{Hz}}$. This is the level measured with the URAP radio receiver on Ulysses [Stone et al., 1992] in the Io plasma torus for these values of the plasma parameters [Meyer-Vernet et al., 1993].

3.3. A Short Antenna May Be Long

When is the antenna short? The question arises because it is only under this condition that the voltage power spectrum at the antenna ports is obtained by just multiplying the mean electrostatic field power spectrum projected on the antenna by the square of the antenna length, L.

For blackbody radiation, the answer is well-known. If the refractive index is sufficiently close to one, the wave number is $k = 2\pi f/c$, so that the antenna can be considered as short under the condition

$$2\pi f L/c < 1 \qquad (6)$$

Now, consider the plasma thermal noise. As already noted, the typical scale of the electrostatic fluctuations below f_p is the Debye length so that the antenna is short if $L < L_D$. By contrast, above f_p, the noise is determined by the Langmuir waves, whose wave number is of the order of $k \approx 2\pi f/v_{th}$ (except very close to f_p). Hence the antenna is short under the stronger condition

$$2\pi f L/v_{th} < 1 \qquad (7)$$

Consider for example an antenna of length $L = 50$ m at the frequency $f = 100$ kHz. Condition (6) holds since $2\pi f L/c \approx 0.1$. But in order to meet condition (7), the plasma temperature would have to satisfy the inequality $T > 3 \times 10^7$ K, which generally does not hold in space. Hence, although the antenna is conveniently short for blackbody radiation, it is not generally so for the plasma thermal noise.

One has, of course, to take these results with a pinch of salt. First, as the frequency approaches f_p, the Langmuir wavelength increases without limit, making ultimately all antennas short in the close vicinity of the plasma frequency. Second, when the frequency is not much greater than the electron gyro frequency, the magnetic field cannot be neglected. As we already noted, Bernstein waves play a major role in determining the noise at frequencies between the gyro harmonics, so that the relevant scale for evaluating the shortness of the antenna becomes the electron gyro radius. One should remember, however, a special property of Bernstein waves: their dispersion curve is made of a series of gyro harmonic bands with $k \to \infty$ (or 0, respectively) at the bottom (or top, respectively) of each band. Therefore, as the frequency approaches a gyro harmonic from above, the wave number increase makes ultimately all antennas long; by contrast, when approaching a gyro harmonic from below, the wave number decrease makes ultimately all antennas short.

4. BEWARE: ANTENNAS IN PLASMAS ARE TRICKY

What happens when the antenna turns out not to be short? How does the angular pattern look like? What is the value of the impedance?

If we stop to think about these questions, we will find that antennas in plasmas behave in a rather tricky way: their angular pattern shifts, and their impedance has resonances, even when they are short from the electromagnetic point of view. The lack of addressing properly these questions, as well as inadequate use of the intuition acquired in radio astronomy, may result in erroneous measurements of the signal amplitude and incorrect wave identification. A little reflection, however, shows that this seemingly surprising antenna behaviour results from very simple physics.

4.1. Angular Pattern for Electrostatic Waves

Consider an antenna made up of two thin wires, each of length L, and an electrostatic wave of wave vector **k**, making an angle θ with the antenna axis z (Figure 1). Because the electrostatic field lies along **k**, the antenna response, i.e., the ratio of the squared voltage to the squared electric field, is proportional to $L^2 \cos^2 \theta$ when the antenna is short. This is the pattern shown on the left hand side of Figure 2. Now, what happens as kL increases? Because the electric potential of each arm is a mean over its length, the antenna is adapted to measure the waves whose wavelength along it is twice the length of one arm, i.e., $k_z = k\cos\theta \approx \pi/L$. If $kL \gg 1$, this matching requires $\cos\theta \ll 1$. In other words, a long antenna is adapted for waves whose wave vector is roughly normal to its direction. The antenna angular pattern, therefore, depends strongly on kL as soon as $kL \geq \pi$; the direction for which the response is maximum shifts by 90 degrees when the antenna becomes very long (Figure 2).

This behaviour is sometimes overlooked, thereby producing a 90 degrees error in wave direction measurement. Yet, a proper analysis allows one to deduce k from the observed pattern of a spinning antenna, if this pattern varies strongly with k, i.e., if the antenna length is of the order of half the wavelength or longer. Measuring k in this way for a set of frequencies yields the wave dispersion curve, thereby giving access to basic plasma properties. This has been applied in Jupiter's magnetosphere with the URAP radio receiver on the Ulysses spacecraft; the measured spin modulation of the quasi-thermal noise yielded the dispersion curve of Bernstein waves, thereby providing an accurate measurement of the electron thermal temperature [Meyer-Vernet et al., 1993; Moncuquet et al., 1995].

It may be worth noting that an antenna made of two small spheres has a much more complicated pattern with several lobes when it is long, so that it is less adapted to this kind of measurement because the lobes cannot be precisely calculated (see Section 5.1) nor measured.

4.2. Antenna Impedance

Why bothering about the antenna impedance? This is because, for all practical purposes, the signal is measured at the ports of a radio receiver whose impedance, Z_R, is of the same order of magnitude as that of the antenna itself, since it is mainly due to the antenna base capacitance. Hence the voltage power spectrum at the ports of the electric antenna, V_f^2 - which is the

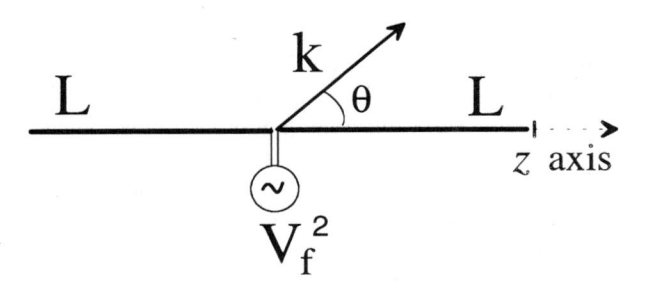

Figure 1. Wave vector making an angle θ with a thin wire dipole antenna. For electrostatic waves, the electric field lies along **k**.

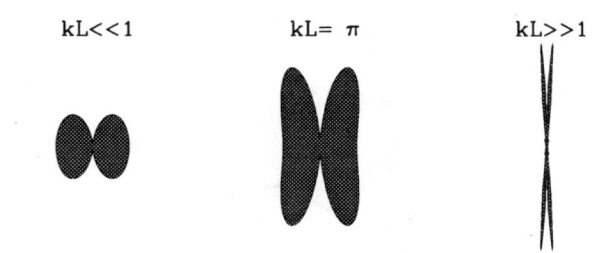

Figure 2. Angular pattern for electrostatic waves for a short, a half-wave, and a long wire antenna lying in the horizontal direction, as shown in Figure 1. For a short antenna, the pattern varies as $\cos^2\theta$ and does not depend on k. As kL increases, the pattern becomes strongly dependent on k, and for a very long antenna the maximum response is ultimately at 90° from the antenna direction (adapted from [Meyer-Vernet, 1994]).

interesting quantity - is related to that measured by the receiver, V_R^2, by the relation

$$V_R^2 = V_f^2 \times \left|\frac{Z_R}{Z_R + Z_A}\right|^2 \quad (8)$$

The antenna impedance Z_A, therefore, has to be precisely known; this is not a trivial matter, especially near the plasma characteristic frequencies.

Nyquist's theorem tells us that the antenna resistance R in an equilibrium plasma at temperature T is related to the thermal noise by $R = V_f^2/4k_B T$. Basically because of Eq.(4), this resistance turns out to be several orders of magnitude greater than the electromagnetic value given in the textbooks, even far from the plasma resonances. Still worse, the capacitance is very different from the free space value below the plasma frequency, and becomes generally negative just below f_p, thereby producing resonances in the circuit. The longer the antenna (compared to the Debye length), the spikier the response curve (Figure 3). As a consequence, the signal actually measured is not related to the signal to be measured in the manner ordinary intuition would tell us.

We show in the next section how to solve this problem.

5. CALCULATING THE QUASI-THERMAL NOISE AND USING IT

5.1. Basics

Since the response of the antenna depends on the wave number, the plasma quasi-thermal noise and the antenna impedance are given by an integration over the wave numbers. The voltage power spectrum of the

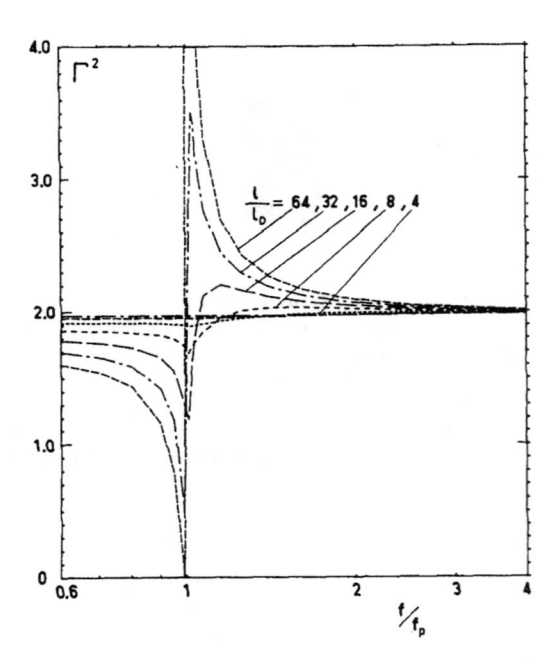

Figure 3. Antenna/receiver transfer function V_f^2/V_R^2 as a function of the frequency normalised to the plasma frequency, for different ratios of the antenna length to the Debye length (adapted from [Couturier et al., 1981]).

quasi-thermal noise at the terminals of an antenna in a plasma drifting with velocity **V** is

$$V_f^2 = \frac{2}{(2\pi)^3} \int d^3k \left|\frac{\mathbf{k}\cdot\mathbf{J}}{k}\right|^2 E^2(\mathbf{k}, \omega - \mathbf{k}\cdot\mathbf{V}) \quad (9)$$

where $\omega = 2\pi f$ is the angular frequency. The first term in the integral involves the antenna response to electrostatic waves discussed in section 4. This response can be expressed in terms of the Fourier transform $\mathbf{J}(\mathbf{k})$ of the current distribution along the antenna. For a wire dipole antenna made of two thin filaments each of length L along the **z** axis [Kuehl, 1966]:

$$\mathbf{k}\cdot\mathbf{J} = 4\,\frac{\sin^2(k_z L/2)}{k_z L} \quad (10)$$

The patterns shown in Figure 2 represent the square of this quantity in polar co-ordinates.

The second term is the auto correlation function of the electrostatic field fluctuations in the antenna frame, which is a function of the particle velocity distributions [Sitenko, 1967]. The antenna impedance is given by an expression similar to (9), just replacing E^2 by $i/2\omega\epsilon_0\epsilon_L$, where ϵ_L is the longitudinal dielectric function.

In general, the integration requires numerical aids, but simple analytical approximations are available in a number of cases [Meyer-Vernet and Perche, 1989; Meyer-Vernet et al., 1993; Issautier et al., 1999]. For example, with a wire dipole antenna being long and thin at the scale of the Debye length, the quasi-thermal noise is $V_f^2 \approx (2\pi)^{-1/2} k_B T/(2\epsilon_0 f_p L)$ at frequencies below f_p, whereas the low frequency antenna impedance is mainly given by the capacitance $C_A \approx \pi\epsilon_0 L/\ln(L_D/a)$, a being the wire radius. This result, which can be easily deduced from the low frequency limit of the theoretical impedance given above, can be understood as the contribution of the plasma Debye shielding around the wire antenna.

The integral (9) (with the response (10)) represents very accurately the quasi-thermal noise on a thin wire dipole antenna [e.g., Issautier et al., 1999]. However, with an antenna made of two spheres distant by L, the noise is more difficult to calculate accurately, even though the theoretical response is simpler since in that case, $\mathbf{k}\cdot\mathbf{J} = 2\sin(k_z L/2)$. This is because, for all practical purposes, the spheres must be mounted at the end of booms, which are difficult to model, so that the actual response is very different from the calculated one, even with a short antenna [Manning, 1998]. Moreover, the shot noise due to particle impacts on (and/or emission from) the antenna is very large with a sphere antenna, although it is negligible for a sufficiently thin wire antenna because the latter has a much larger capacitance.

5.2. Applications

As we already mentioned, an accurate diagnostic of the ambient electrons is obtained from the quasi-thermal noise measured around f_p. As explained above, the position of the peak reveals the total density, the overall shape and level reveals the thermal temperature, whereas the very fine structure of the peak reveals the suprathermal electrons. In practice, one assumes a model for the distribution functions, computes the quasi-thermal noise spectrum from (9), (10) and (8), and deduces the parameters of the model by fitting the theory to the observations [Maksimovic et al., 1995; Issautier et al., 1998, 1999]. This requires a radio receiver sufficiently sensitive and well-calibrated, at the ports of an electric antenna made of two wires being long and thin compared to the Debye length. A badly calibrated receiver, however, would still allow measuring the plasma density since this parameter is revealed by the frequency of the spectral peak. By contrast, the antenna length is a critical parameter because, as the length decreases and approaches the Debye length, the antenna becomes less adapted to measure the Langmuir waves and the peak broadens so that both the density and temperature become difficult to measure accurately. Because the technique relies on a wave measurement, it senses a large plasma volume and is relatively immune to the spacecraft potential and photoelectron perturbations which generally pollute the particle analysers and the Langmuir probes. When the electron gyro frequency f_g is not much smaller than f_p, the minima at the gyro harmonics also allow a simple measurement of the modulus of the magnetic field [Meyer-Vernet et al., 1993].

Radio astronomers have often to calculate the thermal noise above f_p - where low-frequency radio emission can be observed - in order to be able to subtract the thermal background from the data. A simple analytical expression holds for the voltage power spectral density received at frequencies $f \gg f_p$ by a thin wire dipole antenna of length $L \gg L_D$:

$$V_f^2 \approx 5 \times 10^{-13} \times \frac{T f_p^2}{L f^3} \quad (11)$$

where L is the length of one arm and T the mean electron temperature of the quasi-thermal plasma (in SI units); the transfer function is given by (8) with the antenna impedance approximated in that case by the free-space capacitance $C_A \approx \pi \epsilon_0 L / [\ln(L/a) - 1]$, a being the wire radius.

Figures 4 and 5 show two examples of low-frequency

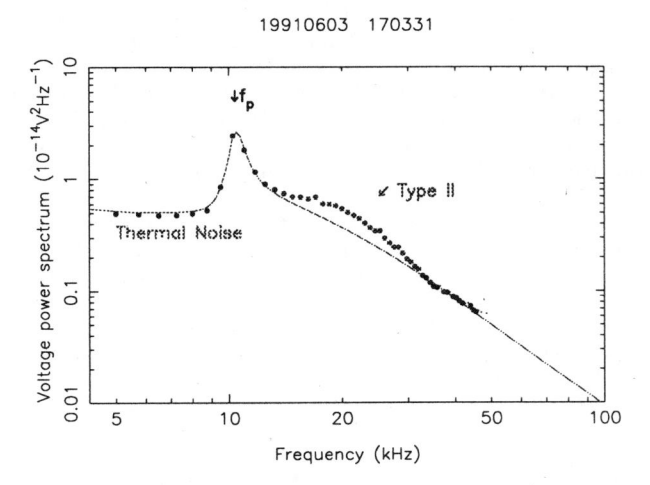

Figure 4. Voltage power spectrum measured with the URAP radio receiver on Ulysses in the interplanetary medium (dots) when a so-called "type II-like radio burst" is superimposed on the plasma quasi-thermal noise. The continuous line is the theoretical spectrum which fits best the measured thermal noise; deducing the density and temperature from this fitting allows one to calculate the thermal background to be subtracted for determining the actual spectrum of the radio burst.

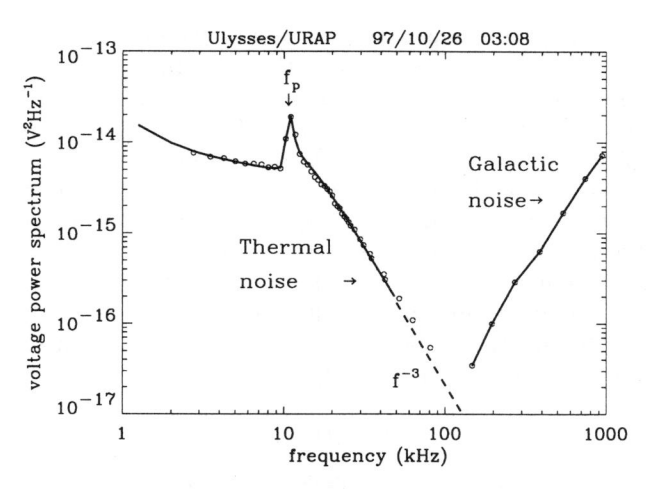

Figure 5. Typical voltage power spectrum measured with the URAP radio receiver on Ulysses in the interplanetary medium (dots). The plasma quasi-thermal noise is observed below 100 kHz; the continuous line is the theoretical spectrum which fits best the data; this fitting yields an accurate measurement of the electron density and temperature and of the plasma drift speed. The dashed line is the analytical approximation (11) with these plasma parameters. Determination of the galactic radio noise at low frequencies requires careful substraction of this thermal background.

radio emissions observed on Ulysses to be superimposed on the plasma quasi-thermal noise. In Figure 4, we can see the spectrum of a so-called interplanetary "type II"-like slow drift radio emission whose intensity is comparable to the thermal noise continuum. (This radio emission is associated with interplanetary shocks [e.g., Hoang et al., 1992].) In Figure 5, the ubiquitous galactic radio background [e.g., Brown, 1973] shows up clearly above the thermal noise; determination of the galactic noise at low frequencies requires careful substraction of the thermal background.

6. FINAL REMARKS

In the stone-age of space radioastronomy, when measuring the galactic emission on the IMP-6 spacecraft, Brown [1973] detected a new "low-frequency source radiating continuously but highly variable", with a "spectral index of -2.8 from 30 to 110 kHz", which he tried to remove empirically from his data to deduce the galactic noise. This emission was most probably the high-frequency thermal noise given in Eq.(11) and shown in Figure 5, which he observed near the Earth. A few years later, a paper was submitted to the Journal of Geophysical Research, providing the expression (11) of the plasma thermal noise at high frequencies, and suggesting possible space applications [Meyer-Vernet, 1979].

Ironically, this paper was initially in the process of being rejected, on the grounds that the theory was too simplified to be applied in space and lacked a detailed comparison with data. It was finally accepted because of a lucky conjuncture: in the meantime, the ISEE 3 spacecraft had just been launched and it carried the most sensitive and well-calibrated radio receiver ever flown [Knoll et al., 1978]; this instrument turned out to measure exactly the plasma noise level predicted in the paper submitted. One lesson from this story may be Dyson's third rule: "Don't be afraid of the scorn of theoreticians" [Dyson, 1970].

REFERENCES

Andronov, A. A., Antenna impedance and noise in a space plasma, *Kosmich. Issled., 4*, 558, 1966.

Bame, S. J., et al., The Ulysses solar wind plasma experiment, *Astron. Astrophys. Suppl. Ser., 92*, 237, 1992.

Bekefi, G., *Radiation processes in Plasmas* (Wiley, London), 1966.

Brown, L. W., The galactic spectrum between 130 and 2600 kHz, *ApJ., 180*, 359, 1973.

Chateau, Y. F., and N. Meyer-Vernet, Electrostatic noise in non-Maxwellian plasmas: Generic properties and "kappa" distributions, *J. Geophys. Res., 96*, 5825, 1991.

Dyson, F. J., The future of physics, *Physics Today, 23*, 23, 1970.

Fejer, J. A. and J. R. Kan, Noise spectrum received by an antenna in a plasma, *J. Geophys. Res., 4*, 721, 1969.

Feldman, W. C., J. R. Asbridge, S. J. Bame, M. D. Montgomery, and S. P. Gary, Solar wind electrons, *J. Geophys. Res., 80*, 4181, 1975.

Hoang, S. et al., Interplanetary fast shock diagnosis with the radio receiver on Ulysses, *Solar Wind Seven, proceed. of the 3rd COSPAR Coll.*, ed. E. Marsch and R. Schwenn, 465, 1992.

Issautier, K., N. Meyer-Vernet, M. Moncuquet, and S. Hoang, A novel method to measure the solar wind speed, *Geophys. Res. Lett., 23*, 1649, 1996.

Issautier, K., N. Meyer-Vernet, M. Moncuquet, and S. Hoang, Solar wind radial and latitudinal structure: Electron density and core temperature from Ulysses thermal noise spectroscopy, *J. Geophys. Res., 103*, 1969, 1998.

Issautier, K., N. Meyer-Vernet, M. Moncuquet, S. Hoang, and D. J. McComas, Quasi-thermal noise in a drifting plasma: Theory and application to solar wind diagnostic on Ulysses, *J. Geophys. Res., 104*, 6691, 1999.

Knoll, R., G. Epstein, S. Hoang, G. Huntzinger, J. L. Steinberg, J. Fainberg, F. Grena, S. R. Mozier, and R. G. Stone, The 3-Dimensional radio mapping experiment (SBH) on ISEE-C, *IEEE Trans. on Geosci. Electron., GE-16*, 199, 1978.

Kuehl, H. H., Resistance of a short antenna in a warm plasma, *Radio Sci., 1*, 971-976, 1966.

Maksimovic, M., S. Hoang, N. Meyer-Vernet, M. Moncuquet, J.-L. Bougeret, J. L. Phillips, and P. Canu, The solar wind electron parameters from quasi-thermal noise spectroscopy and comparison with other measurements on Ulysses, *J. Geophys. Res., 100*, 19881, 1995.

Maksimovic, M., V. Pierrard, and P. Riley, Ulysses electron distributions fitted with Kappa functions, *Geophys. Res. Lett., 24*, 1151, 1997.

Maksimovic, M., et al., Solar wind density intercomparisons on the wind spacecraft using WAVES and SWE experiments, *Geophys. Res. Lett., 25*, 1265, 1998.

Manning, R., A simulation of the behavior of a spherical probe antenna in an AC field, in *Measurement Techniques in Space Plasmas: Fields, Geophys. Monogr. Ser.*, Vol.103, edited by Robert F. Pfaff, J. Borovsky, and David T. Young, pp.205-210, AGU, Washington, D.C., 1998.

Meyer-Vernet, N., On natural noises detected by antennas in plasmas, *J. Geophys. Res., 84*, 5373, 1979.

Meyer-Vernet, N., Aspects of Debye Shielding, *Am. J. Phys., 61*, 249, 1993.

Meyer-Vernet, N., On the thermal noise "temperature" in an anisotropic plasma, *Geophys. Res. Lett., 21*, 397, 1994.

Meyer-Vernet, N., and C. Perche, Tool kit for antennae and thermal noise near the plasma frequency, *J. Geophys. Res., 94*, 2405, 1989.

Meyer-Vernet, N., S. Hoang, and M. Moncuquet, Bernstein waves in the Io torus plasma: A novel kind of electron temperature sensor, *J. Geophys. Res., 98*, 21163, 1993.

Meyer-Vernet, N., S. Hoang, K. Issautier, M. Maksimovic, R. Manning, M. Moncuquet, and R. G. Stone, Measuring plasma parameters with thermal noise spectroscopy, in *Measurement Techniques in Space Plasmas: Fields, Geophys. Monogr. Ser.*, Vol.103, edited by Robert F. Pfaff, J. Borovsky, and David T. Young, pp.205-210, AGU, Washington, D.C., 1998.

Moncuquet, M., N. Meyer-Vernet, and S. Hoang, Dispersion of electrostatic waves in the Io plasma torus and derived electron temperature, *J. Geophys. Res., 100*, 21697, 1995.

Rostoker, N., Fluctuations of a plasma, *Nucl. Fusion, 1*, 101, 1961.

Sentman, D. D., Thermal fluctuations and the diffuse electrostatic emissions, *J. Geophys. Res., 87*, 1455, 1982.

Sitenko, A. G., *Electromagnetic Fluctuations in Plasma*, Academic, San Diego, Calif., 1967.

Stone, R. G., et al., The unified radio and plasma wave investigation on Ulysses, *Astron. Astrophys. Suppl. Ser., 92*, 291, 1992.

S. Hoang, K. Issautier, N. Meyer-Vernet, M. Moncuquet, DESPA, Observatoire de Paris, 92195 Meudon Cedex, France. (e-mail: nicole.meyer@obspm.fr)

Radio Wave Propagation in the Earth's Magnetosphere

J.-L. Steinberg, C. Lacombe and S. Hoang

Observatoire de Paris, DESPA, UMR 8632 CNRS, 92195 Meudon, France

When low frequency terrestrial radio waves are observed far from their source, in the remote magnetospheric tail or in the solar wind, strong frequency-dependent variations of their apparent source direction and/or angular size are observed. For instance, far in the tail, the Non Thermal Continuum (NTC) source appears nearly isotropic at the lowest frequencies. An observer located in the solar wind, at $\simeq$ 03:00 LT, 200 R_E from Earth sees the Auroral Kilometric Radiation (AKR) source away from the Earth's direction, the farther tailward, the lower the frequency. A similar phenomenon is observed from the Lagrange point: the AKR source appears to move to large angular distances away from the Earth's direction as the observing frequency decreases. In all these cases, large scale propagation effects (refraction and scattering) are most probably at work. These effects will be described and discussed. They are controlled by density structures. Their deconvolution is not easy but it can provide a way to remotely sound the electron density in the magnetosheath and the bow shock.

INTRODUCTION

There are two main types of terrestrial radio emissions: the Auroral Kilometric Radiation [AKR, Gurnett, 1974] and the Non Thermal Continuum [NTC, Gurnett and Shaw, 1973; Gurnett, 1975]. Their sources have been observed at close enough distances for propagation effects between them and the observer to be negligible; we thus know the position and size of these primary sources and how they eventually vary with the radiated frequency. The Low Frequency (LF) bursts [Steinberg et al., 1988a, 1990a; Kaiser et al., 1996] appear to be an extension of some AKR bursts to low frequencies.

When these emissions are observed from a remote location, the direction Φ in the XOY (GSE) plane and the angular radius γ of their apparent source differ widely from the Φ_o and γ_o expected if their primary source were seen through a vacuum. These differences increase as the observing frequency f decreases; they are due to the propagation of the radio waves through various media such as the bow shock and/or the magnetosheath or to their reflection on plasma boundaries such as the plasmapause or the magnetopause.

When these effects can be interpreted quantitatively, they can be used to remotely probe some regions of the Earth's plasma environment. In what follows we shall describe and discuss some observations of the Earth's radio emissions made in the period 1980-1983 on board ISEE 3 with the radioastronomy receiver [Knoll et al., 1978]. We are not yet able to fully interpret all of them. But they raise interesting problems which are the subject of further studies using multispacecraft data.

The direction and size of the (apparent) sources can be measured using a spacecraft spinning around an axis Z (for instance the Z (GSE) axis) and fitted with two dipoles, one along Z (the Z dipole), the other one, the

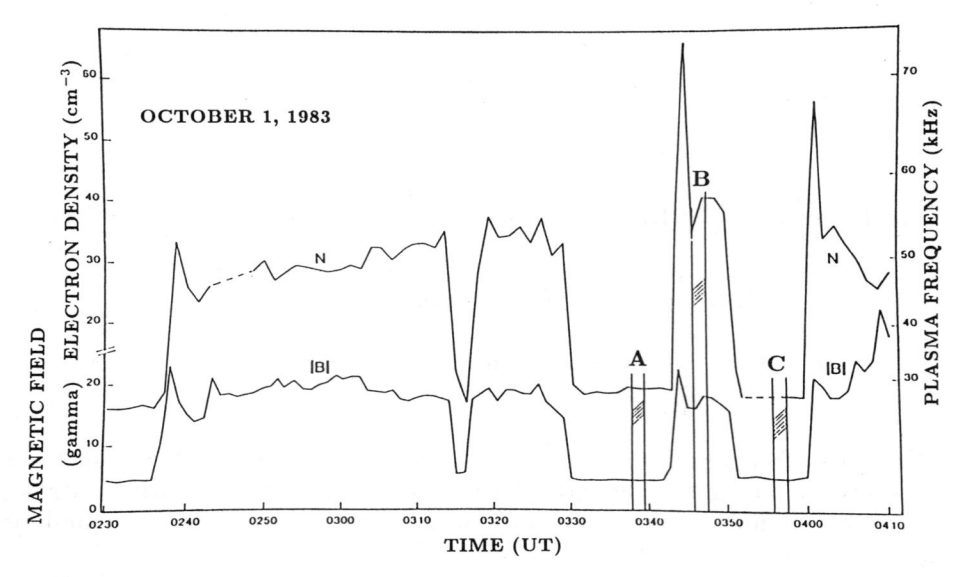

Figure 1. The variations of the magnetic field magnitude $|B|$ (1 point every 64 s) and of the electron density N (1 point every 84 s) showing that ISEE 3 crossed the moving bow shock several times. For instance at C, the spacecraft is again in the solar wind as it was at A because the shock moved between these two observations (From Steinberg et al. 1988b).

S dipole, spinning in the XOY (GSE) plane. Then, the signal from the S dipole is spin modulated and, from the analysis of that modulation, one can get the direction Φ of the source in the spin plane and its elevation Θ with respect to that plane from a comparison of the signals from the S and Z dipoles. When the signal-to-noise ratio is too low on the Z dipole, Θ cannot be measured and is assumed equal to zero. The source angular radius γ can be derived from the modulation factor α of the signal received by a spinning dipole:

$$\alpha = (M - m)/(M + m) \qquad (1)$$

, where M and m are respectively the maximum and the minimum signal received over half a spin (a full period of the modulation). From α the angular radius γ of the source is obtained [Fainberg, 1980] from:

$$\alpha = \cos^2 \Theta / (K - \cos^2 \Theta)$$

$$K = [8 + 2\cos\gamma(1 + \cos\gamma)]/[3\cos\gamma(1 + \cos\gamma)]$$

If $M \to m$, $\alpha \to 0$, $\gamma \to 90°$ and the source becomes isotropic. If $m \to 0$, $\alpha \to 1$ and $\gamma \to 0$, we are dealing with a point source. In general, the angles Φ, Θ and γ are determined under two assumptions: the source is circular and uniformly bright and it is unpolarized or circularly polarized.

1. PROPAGATION OF THE NON THERMAL CONTINUUM

The NTC is a smoothly varying emission of relatively low intensity. Its source is at the plasmapause, about 2 to 5 R_E from Earth. Its spectrum is smooth and its spectral index negative: as frequency decreases, its intensity increases so that its propagation can be studied down to low frequencies.

1.1. Blockage of the NTC Emission by the Bow Shock

On October 1, 1983 between 02:30 and 04:00 UT ISEE 3 crossed the bow shock near its nose [Steinberg et al., 1988b]. Over that time period, the solar wind dynamic pressure varied several times so that the bow shock and the magnetosheath moved back and forth. In Figure 1, the periods when ISEE 3 was in the solar wind correspond to relatively low values of the density N and the magnetic field magnitude B. When the spacecraft was in the magnetosheath N and B were higher. N and B thus went from high to low values several times when the shock moved across ISEE 3. At about 03:43 UT the spacecraft flew across a narrow dense region: an overshoot of the shock where the density reached 65 cm^{-3} and the plasma frequency 73 kHz; only one data point was acquired during that short crossing. The NTC spectra acquired during interval B (in the magnetosheath) and A and C (in the solar wind) are shown on Figure 2. To plot the cut-off of these spectra down to the lowest frequencies where the signal was weak, it was necessary to subtract from it the Quasi Thermal (QT) noise spectrum which was calculated from the plasma data [Meyer-Vernet and Perche, 1989]. During interval B, ISEE 3 was in the magnetosheath where the plasma frequency was about 60 kHz, so that the NTC (-3 dB) cut-off was about 63 kHz (Figure 2). During intervals A and C, in the solar wind, the spectral (-3 dB) cut-off was at about 74 kHz; it was not due to the local plasma frequency of about 30 kHz: the NTC emitted at the plasmapause [Morgan and Gurnett, 1991] close to Earth was occulted by the shock and its density overshoot, the plasma frequency of which was at least 75 kHz. The cut-off frequency of the NTC measured just upstream of the bow shock thus allows a remote sounding of the shock density.

1.2. Propagation of the NTC down the Magnetotail

Coroniti et al. [1984] and Steinberg et al. [1990b] studied the NTC radiation far down the tail to $\simeq -200$ R_E.

Coroniti et al. [1984] analyzed 2 crossings of the magnetopause, Steinberg et al. [1990b] 60 of them. Both teams observed that as the spacecraft went from the tail lobes into the magnetosheath where the local f_p is much higher, the intensity of the apparent NTC source decreased (observation 1) and its angular radius γ increased (observation 2) at a given f. (And, in all these media, γ decreased with increasing f). Furthermore, γ measured in the tail lobes remained larger than expected from the reflection conditions of a wave at frequency f on the magnetopause where the plasma frequency is $f_{p.msh}$ and the refractive index μ_{msh}. Such a wave will only be reflected if its angle of incidence measured from the local normal is larger than i_{max}: $\sin i_{max} = \mu_{msh}$. If the magnetotail was a perfect waveguide, one should never observe inside it an angular radius larger than $90° - i_{max}$. But this is commonly observed (observation 3). Observations 1, 2 and 3 are illustrated in Figures 1, 2 and 4 in [Steinberg et al., 1990b]

About observation 1, it should be noted that when analyzing data acquired when the s/c crossed the magnetopause and went from a low density medium to a much denser one, it is necessary to take into account the fact that the antenna is immersed in a plasma [Steinberg et al., 1990b]: this is important when deriving the voltage spectral density of the signal at the antenna terminals V_a^2 from the measured one at the receiver terminals V_r^2. For instance, the radiation resistance of a

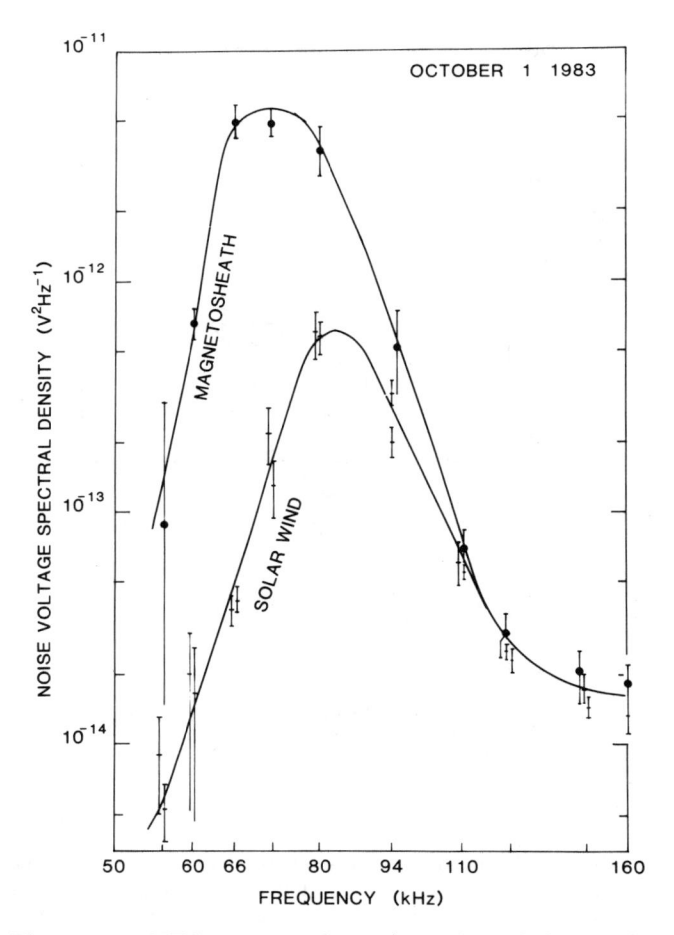

Figure 2. NTC spectra after subtraction of the quasi-thermal (QT) noise background. The NTC low frequency cut-off observed while ISEE 3 was in the solar wind is due to the large overshoot in the bow shock which is clearly seen in Figure 1. Only one solar wind spectrum is plotted because no significant difference is found between those acquired during intervals A and C (Figure 1) (Steinberg et al., 1988b).

dipole in a plasma is μ times what it is in a vacuum and the antenna capacity is proportional to μ^2 (μ is the refractive index) (see for instance Lacombe et al., [1988]). Thus a decrease of μ might explain a decrease of the measured intensity of the source. Steinberg et al. [1990b] also subtracted from the low level NTC signal the calculated QT noise.

About observation 3: the polar plot of the NTC signal received over half a spin of the dipole can be approximated by an ellipse whose major axis (direction of maximum received signal M) is nearly parallel to the GSE axis OX, i.e., to the Earth's direction. A decrease of the measured α implies that the semi-major axis M of that ellipse and its semi-minor axis m tend to become

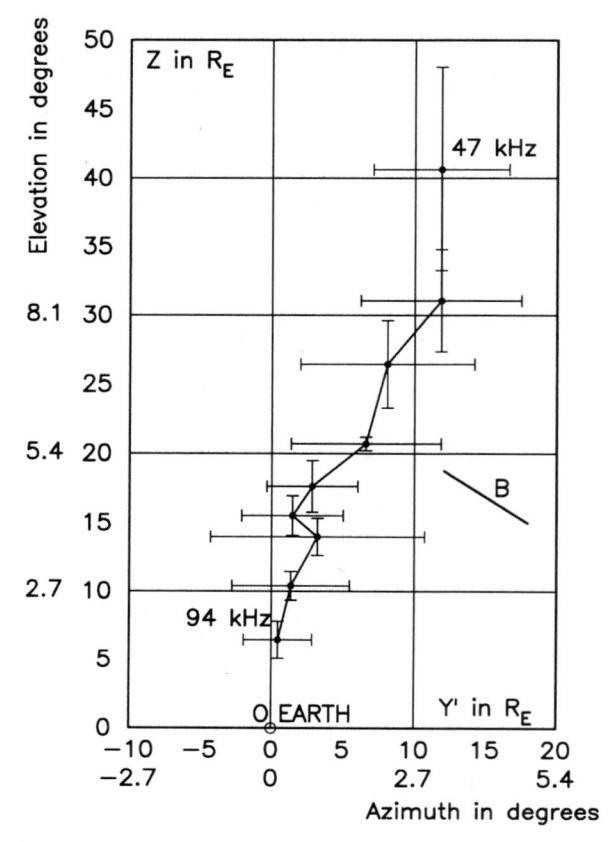

Figure 3. Variation with frequency of the direction of the AKR source on November 5, 1982. Each point represents the intersection of the direction measured on board ISEE 3 with a plane P perpendicular to the direction of the Earth's center O as seen from ISEE 3 and containing O. **B** is the projection on P of the interplanetary magnetic field.

equal (eqn 1). To increase m, and thus to increase the angular radius γ above $90° - i_{max}$ in the lobes, it can be suggested that some radiation, besides that received directly from the NTC source, is received in a direction perpendicular to OX.

The two teams suggested different mechanisms: 1) Coroniti et al. [1984] invoked some new kind of radiation produced in the magnetosheath where they had observed strong spiky signals. This could explain observation 3 as well as observation 2. 2) Steinberg et al. [1990b] suggested that the tail magnetopause "walls" were somewhat rough and the magnetosheath itself inhomogeneous enough to scatter the waves propagating through it. The roughness of the "walls" can allow some rays to enter the magnetosheath: those NTC rays would not have crossed a smooth magnetopause because of too large incidence angles. After crossing it, those rays might be refracted and scattered back towards the lobes

and exit into the lobes in a direction perpendicular to OX, thus increasing m. Following that line of thought, there is no need of a magnetosheath emission and observations 2 and 3 can be interpreted. The invoked effects were simulated with some success by tracing rays through the magnetopause and the magnetosheath and in the tail lobes [Steinberg and Hoang, 1995]. Observation 1 was simply interpreted in terms of refractive effects.

2. PROPAGATION OF THE AURORAL KILOMETRIC RADIATION

The source of AKR is on the line of force of invariant latitude 70° at about 23:00 LT and a few R_E from Earth. This radiation is much more intense than NTC and that should make its use for sounding studies much easier. However AKR is strongly and quickly variable: direction measurements must be made quickly so that the signal level does not change too much while its modulation factor α is measured over half a spacecraft spin period; the spin period must therefore be as short as possible and, over that period, as many data points as possible should be acquired. From that point of view the data acquired with the ISEE 3 receivers are the best available at the present time: 11 data points are acquired over 1.5 s. Propagation studies of AKR are impossible below about twice the solar wind plasma frequency $f_{p.sw}$ because this is the minimum frequency at which AKR is usually received.

2.1. Apparent Position of the AKR Source seen from the Lagrange Point

The primary source of the AKR is known to be located within a few R_E's from Earth where the gyrofrequency is equal to our observing frequencies of a few tens of kHz. But seen from the ISEE 3 orbit around the Lagrange point L1, that is 200 to 250 R_E away from Earth in the Sun's direction, the AKR source appears farther and farther from the Earth as frequency decreases. In Figure 3 we show the apparent motion of the trace S of the source direction on the plane P perpendicular to the direction of ISEE 3 which contains the Earth's center O. On 82 11 05, as the observing frequency f decreases, S moves along a nearly straight line to 40 R_E away from O. Indeed, at 47 kHz, the source is seen out of any plausible model of the bow shock. The trajectory of S in P is nearly perpendicular to the projection on P of the interplanetary magnetic field **B**. This trajectory might be the projection on the plane P of the locus W on the shock surface of the points where the interplanetary (IP) magnetic field **B** is tangent to that shock. Along W, the magnetic field **B** is perpendicular to the shock normal, there is no proton foreshock, the density fluctuations are low as compared to what they are in other parts of the foreshock; and there the propagation of AKR from the magnetosheath to the solar wind should be easier. This would provide an interpretation of Figure 3. But there are cases where S appears to move parallel to XOY, not perpendicular to B and many other geometries are observed.

The fact that S appears as far from the Earth's center as 40 R_E implies that there are propagation effects. But we have not yet tried to simulate them because we do not understand how refraction and scattering can bring the apparent source sometimes near the Z axis and sometimes in the XOY plane. The shock density structure might not be cylindrically symmetric around OX and the source directivity might have to be taken into account. Moreover we only have a dozen such observations with ISEE 3. The problem will be tackled again using WIND data.

2.2. The AKR Source as seen from inside the Solar Wind on the Dawn Side of the Shock

As early as 1976 [Kaiser and Alexander, 1976; Alexander and Kaiser, 1976] the moon orbiting spacecraft RAE 2 yielded 2-D positions of the AKR source. Even at frequencies as high as 200 kHz, several of these sources were sometimes found as far as 25 R_E from Earth. There was a minimum frequency f_o below which the source was not occulted because its center appeared too far from Earth or its angular size was too large. f_o was well correlated to the solar wind plasma frequency $f_{p.sw}$ with $f_o \simeq 4 f_{p.sw}$. This clearly pointed to propagation effects. Alexander et al. [1979] also noted that some AKR sources were observed at very large distances from Earth when large density enhancements were present in the IP medium.

In 1989, directions of the AKR source were measured on board ISEE 3 when that spacecraft travelled through various regions of the Earth's environment [Steinberg et al., 1989]. Figure 4 shows some results obtained at 56 kHz. Direction measurements with ISEE 3 are more accurate and more frequently available than those obtained from lunar occultations. When the spacecraft is in the IP medium in the 03:00-04:00 LT sector, the source is seen tailward from Earth, sometimes as far as 50° away (Figure 5). The source angular distance Φ to Earth increases as f decreases. Its angular radius also increases with decreasing f. These two effects are functions of $f_{p.sw}$.

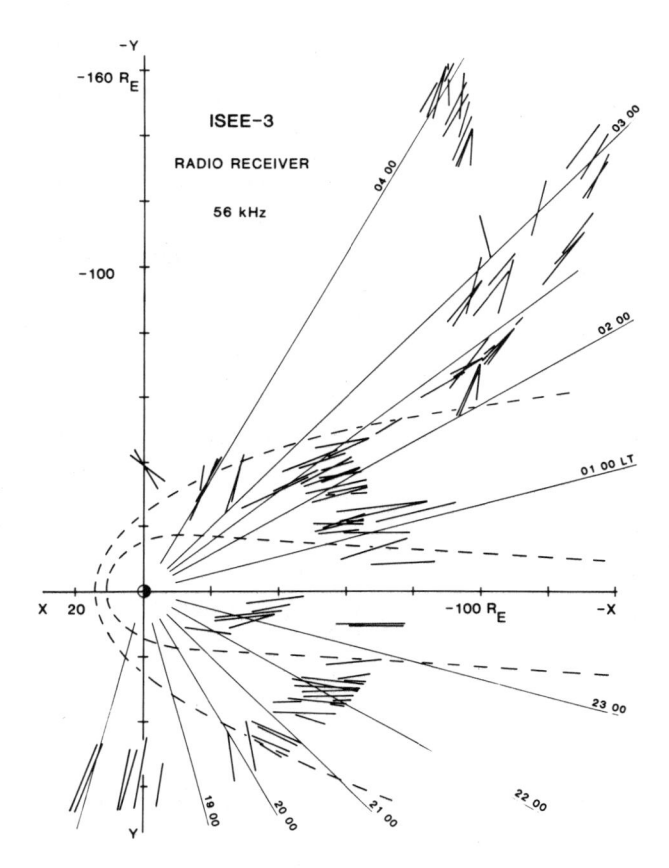

Figure 4. Directions of the apparent AKR sources measured from ISEE 3 at 56 kHz. The dashed lines indicate the probable positions of the bow shock and of the magnetopause. The radial lines which give the local time (LT) help to see that the source angular distance to the Earth changes sign when ISEE 3 crosses the bow shock (From Steinberg et al, 1989).

To explain these observations Steinberg and Hoang [1993] simulated the propagation of AKR at low frequencies through the magnetopause, the magnetosheath and the bow shock. To do that one needs a numerical model of the plasma boundaries and of the density distribution in the magnetosheath. The electron density model is cylindrically symmetric about the aberrated Sun-Earth line; it was derived from time-averaged in-situ measurements on board several spacecraft [Steinberg and Lacombe, 1992]. The density jump $(f_{max.sh}/f_{p.sw})^2$ across the shock decreased with X but remained constant and equal to 1.8 for $X < -130$ R_E (Model M2). In that model, all densities were expressed in units of the IP medium density. To fit the data, the scattering power in the magnetosheath $b(2f_p)$ at twice the local plasma frequency was set at $4\ 10^{-9}$ rad^2 m^{-1} which is about 10 times that of the solar

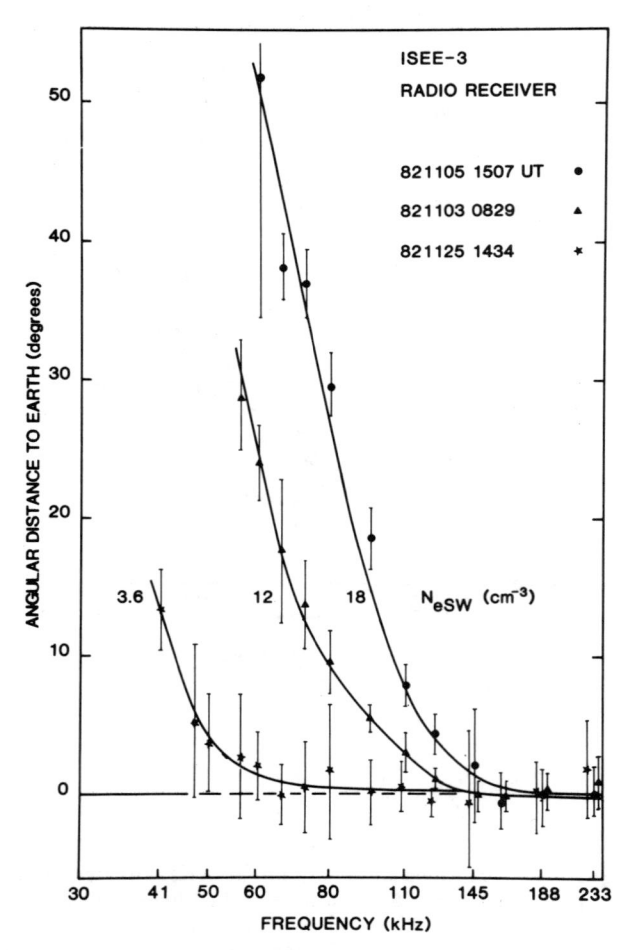

Figure 5. Angular distance Φ to Earth of three AKR sources observed when ISEE 3 was in the solar wind. Φ increases as the observing frequency decreases and when the solar wind density N_{esw} increases (From Steinberg et al., 1989).

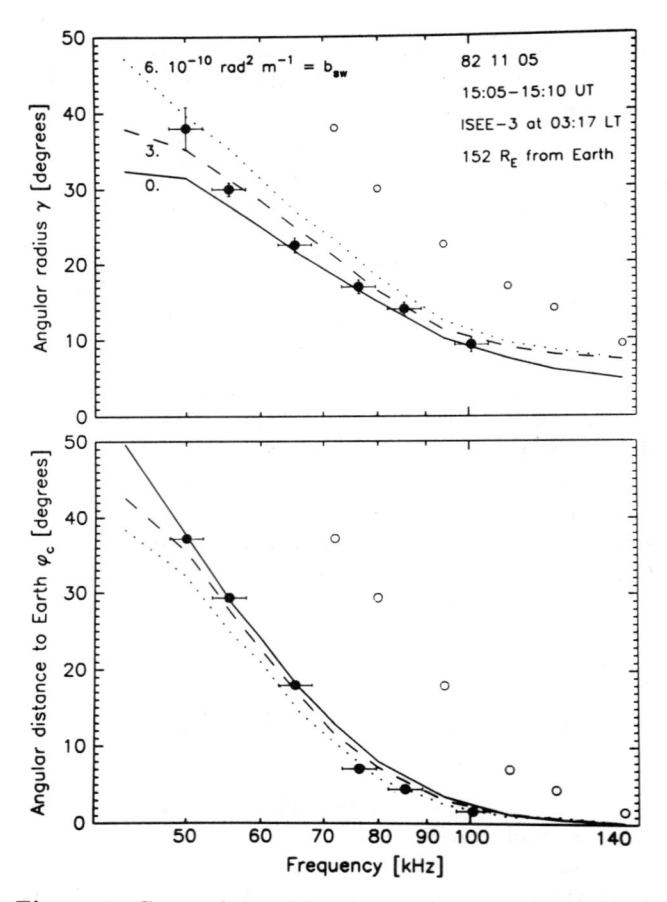

Figure 6. Comparison of the observed angular radius (top) and the angular distance to Earth (bottom) of the AKR source scattered image on November 5, 1982 to their calculated values (model M2). The open circles represent the data acquired at the observing frequencies; the black dots with error bars represent the same data after a frequency shift $25/f_{p.sw}$ where $f_{p.sw}$ is the solar wind plasma frequency measured at the time of the event (From Steinberg and Hoang, 1993).

wind [Lacombe et al., 1997]. $b(2f_p)$ was assumed to be only a function of the local f_p. Scattering in the solar wind between the shock surface and the observer was taken into account. The calculations were carried out with $f_{p.sw}=25$ kHz. In Figure 6 the calculations fit the observations pretty well after the frequency scale had been adjusted in such a way as to bring the $f_{p.sw}$ during the observations to 25 kHz. In the same paper [Steinberg and Hoang, 1993], three such good fits were shown. We have now reduced 50 such observations of the AKR source direction and size; some of them can be fitted with model M2. But for many of them the shock strength (the density jump across it) has to be larger or smaller than that of model M2. This model is certainly not adequate in all interplanetary situations.

2.3. Propagation of the LF Burst Spike through the Bow Shock

Seen from within the solar wind, LF bursts are made of a high frequency non-drifting component which spans the frequency range from about 500 kHz down to $\simeq 2f_{p.sw}$ and a much longer lasting, smooth component which can be sometimes observed down to $f_{p.sw}$. This smooth component appears isotropic and was named Isotropic Terrestrial Kilometric Radiation (ITKR) by Steinberg et al. [1988a, 1990a]. On the rising front of the ITKR, there are a few spikes of radiation whose angular radius is small enough for the corresponding signal to be spin modulated and its source direction measured (Figure 7). They can sometimes be observed down to $1.3 \times f_{p.sw}$. Among those spikes, we select the one which is most intense as compared to the isotropic component intensity.

At the lowest frequencies where it can be measured, the direction **D** of the spike source is found very far away tailward of the Earth's direction, much farther than the simultaneously observed AKR source (Figure 8); as seen on the shock surface, its source should be farther tailward from Earth than -200 R_E. Sometimes, **D** does not even intersect any plausible aberrated shock model. When there is an intersection, one may assume that the source really is farther tailward than -200 R_E in the magnetosheath before its radiation has crossed the shock. Indeed, for two out of our nine observations [Steinberg et al., 1998], the spacecraft was close enough to the aberrated shock to remove any doubt that the source on the shock was that far. But that interpretation is not valid when the measured direction misses the shock surface.

Scattering in the IP medium can help explaining the source direction. We simulated the propagation from an extended source on the shock surface at a variable position X(GSE) to the spacecraft in the IP medium. In those simulations no assumptions are made about the shock plasma structure; a slice of the geometric shock surface, centered at G, is assumed to emit radio waves (Figure 9). These ray tracing simulations show that, using the known IP medium scattering paramet-

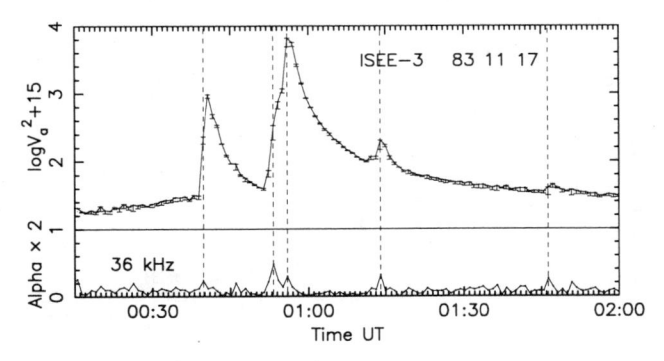

Figure 7. (Top) Time variation of the maximum M over half a spacecraft spin of the 36 kHz signal from three LF bursts. The lower end of each "error bar" is the minimum signal m received over the same half spin. (bottom) Time variation of twice the modulation factor $\alpha = (M-m)/(M+m)$ of the signal. The value of α is a rough indication of the source angular radius γ: γ decreases as α increases. When $\alpha \simeq 0$, the source is isotropic. Note that α peaks on the rising part of each LF burst or near its peak (dashed vertical lines): this is the signature of the "spike" (From Steinberg et al., 1998).

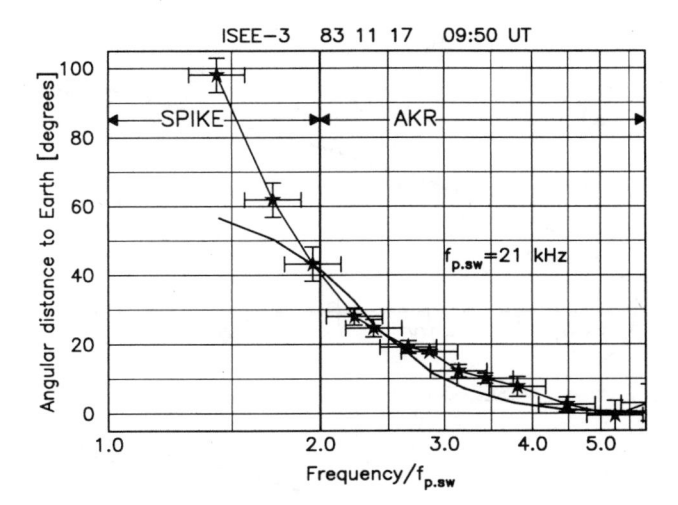

Figure 8. Variation with $f/f_{p.sw}$ of the angular distance Φ to Earth of the kilometric source: AKR for $f \gtrsim 2f_{p.sw}$ or the LF burst spike below $2f_{p.sw}$. Thin lines and stars denote the observed direction of the apparent source S. Thick line denotes the apparent source direction of a terrestrial source computed by ray tracing in the magnetosheath density model M2 [Steinberg and Hoang, 1993] taking interplanetary scattering into account (From Steinberg et al., 1998).

ers, the source centroid S cannot be seen as far as -200 R_E unless it is actually very remote from Earth on the shock surface. This is simply due to the fact that the path length from the source S to the s/c must be long enough for the total angular scattering over it to be large enough (Figure 9). The angular distance to Earth of the source cannot be increased by scattering by more than $20°$; but this is enough to explain why some source directions do not intersect any shock model.

LF bursts are associated with geomagnetic storms and AKR; so that their source is probably close to Earth. One might assume that S is seen as far as -200 R_E because of propagation effects inside the magnetosheath. Simple ray tracing calculations with model M2 (see section 2.2.) cannot, in general, bring the scattered source centroid S that far on the shock surface. In addition to scattering, one has to invoke blockage of the spike radiation at the lowest frequencies by the shock where the density increases (sunward) with increasing X.

Indeed Steinberg et al. [1998] showed that in 7 cases out of 9, the lowest frequency $f_{min.sp}$ at which the spike source was intense enough to be located is about equal to the maximum plasma frequency across the shock $f_{max.sh}$. The frequency $f_{max.sh}$ was calculated from the plasma parameters using the Rankine-Hugoniot equations. $f_{max.sh}$ is always smaller than $2f_{p.sw}$ because the shock densities generally larger than those of model M2.

New shock models were built where the variation of $f_{max.sh}$ with X was shaped to produce the required blockage phenomenon. There are, in general, two observing frequencies smaller than $2f_{p.sw}$. For each of them, we determine a GSE abscissa $X_{occ}(f)$ sunward of which the radiation at f cannot cross the shock. By a trial-and-error process $X_{occ}(f)$ is adjusted in such a way that the source at f is seen from ISEE 3 in the observed direction, after taking IP scattering into account. At large distances from Earth, $f_{max.sh}$ must tend towards the value calculated from Rankine-Hugoniot equations. It thus became possible to interpret three sets of observations of the spike source at frequencies smaller than $2f_{p.sw}$.

In Figure 10 $f_{max.sh}/f_{p.sw}$ was 1.7 at the source S and tailward of it, close to what it was in model M2 for X <-130 R_E. Then it was possible to account for the variation of the angular distance to Earth with frequency for both AKR and the LF burst spike. On the contrary, on Figure 11, we had to take $f_{max.sh}/f_{p.sw}$ =2.7 at S as given by the Rankine-Hugoniot equations for the new model to fit the spike source data. But then the calculated AKR source direction is too far away from Earth as compared to the observations.

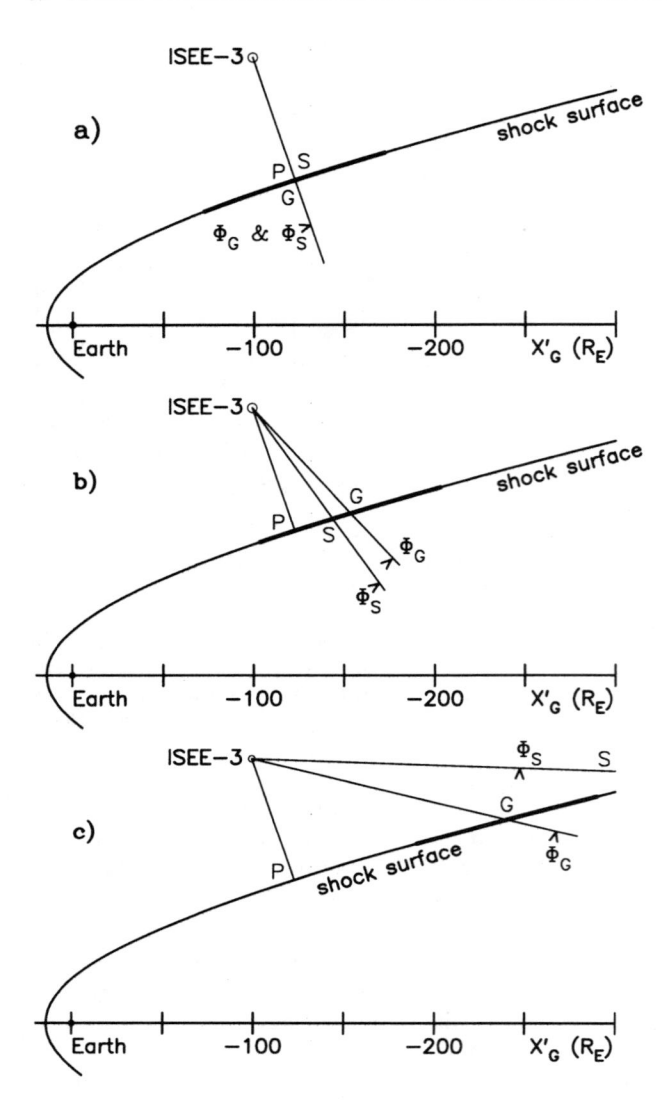

Figure 9. Variation, in the presence of IP scattering, of the apparent direction Φ_S of a source radiating on the shock as a function of the position of its geometric center G (abscissa X'_G) seen in the direction Φ_G. The radio source is represented by the thickened part of the shock. (a) the source is about centered at the foot P of the normal to the shock through ISEE 3: $\Phi_S = \Phi_G$. (b) G is shifted tailward, but P remains inside the source; then radiation from the radiating elements close to P and therefore to ISEE 3 contribute much more to building up the scattered image than source elements located farther tailward: $\Phi_S < \Phi_G$. (c) The radiation from all elements of the source is heavily scattered on its way to ISEE 3: $\Phi_S > \Phi_G$ (From Steinberg et al., 1998).

density jump across the shock cannot exceed a factor 4, especially on the shock flanks. So that measuring the direction of the spike source at $f < 2f_{p.sw}$ provides a means of sounding the shock at frequencies lower than when using AKR as a source. When doing so, we find

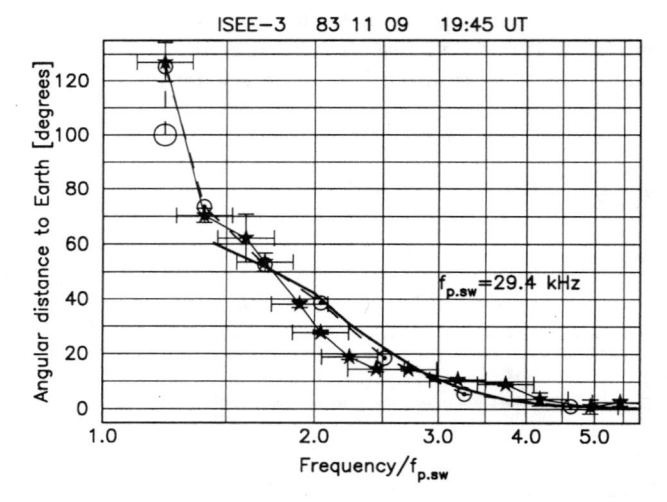

Figure 10. Variation with $f/f_{p.sw}$ of the angular distance to Earth Φ of the AKR source ($f \gtrsim 2f_{p.sw}$) or the LF burst spike for $f \lesssim 2f_{p.sw}$. Thin line and stars denote observations, thick line denotes results of ray tracing simulations in model M2. The empty circle shows where the observed spike source would have been seen at 36 kHz in the absence of IP scattering. The circled dots and the dashed line represent the $\Phi(f/f_{p.sw})$ variation calculated with a new model including an X'-dependent blockage by the shock and taking IP scattering into account (From Steinberg et al., 1998).

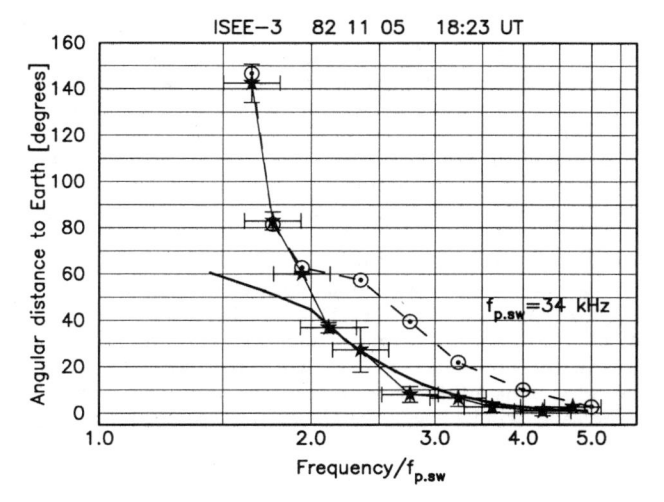

Figure 11. Same format as Figure 10. But the calculated density jump at the source S on the shock surface at the lowest observed frequency was 2.7 (From Steinberg et al., 1998).

We can only conclude that a unique density model cannot, in general, account for the simultaneous observations of the spike and those of AKR. Most probably AKR and the spike radiation do not cross the shock in the same region. That might be due to the fact that AKR is beamed towards the Earth's poles [Green and Gallagher, 1985] while the spike radiation is more strongly scattered in the magnetosheath at frequencies closer to $f_{p.sw}$ and therefore appears to cross the shock close to the XOY plane. Models which are symmetric around OX should probably be given up.

3. CONCLUSION

Large propagation effects are seen when Earth's radio sources are observed at low frequencies through the bow shock or the magnetosheath or reflected at boundaries such as the magnetopause. These effects are functions of the structure of these regions and therefore of the level of geomagnetic activity and of the solar wind parameters.

To derive information on these regions is not easy because we are dealing with global effects. However it can sometimes be done using average plasma and magnetic field parameters. In any case, at the present time, the derivation may be uncertain. It is most probable that using data acquired simultaneously from several spacecraft will improve the situation. For instance, Hashimoto et al. [1998] provide evidence of the blockage of AKR by the plasmasphere using GEOTAIL and WIND data. But, as far as direction finding is concerned, ISEE 3 data remain very valuable.

Acknowledgments. We are grateful to our referee for several useful comments and suggestions. The ISEE 3 radio experiment was a joint project of Observatoire de Paris, DESPA and Laboratory for Extraterrestrial Physics in NASA-GSFC.

REFERENCES

Alexander, J.K. and M.L. Kaiser Terrestrial Kilometric Radiation, 1. Spatial Structure Studies, *J. Geophys. Res.*, *81*, 5948-5956, 1976.

Alexander, J.K., M.L. Kaiser and P. Rodriguez, Scattering of terrestrial kilometric radiation at very high altitudes, *J. Geophys. Res.*, *84*, 2619-2629, 1979.

Coroniti, F.V., F.L. Scarf, C.F. Kennel and D.A. Gurnett, Continuum radiation and electron plasma oscillations in the distant geomagnetic tail, *Geophys. Res. Lett.*, *11*, 661-664, 1984

Fainberg, J., Technique to determine location of radio sources from measurements taken on spinning spacecraft, *NASA Techn. Memo 80598*, 1980

Green, J.L. and D.L. Gallagher. The detailed intensity distribution of the AKR emission cone, *J. Geophys. Res.*, *90*, 9641-9649, 1985

Gurnett, D.A. and R.R. Shaw, Electromagnetic radiation trapped in the magnetosheath above the plasma frequency, *J. Geophys. Res.*, *78*, 8136-8149, 1973

Gurnett, D.A., The Earth as a radio source: terrestrial kilometric radiation, *J. Geophys. Res.*, *79*, 4227-4238, 1974

Gurnett, D.A., The Earth as a radio source: the non thermal continuum, *J. Geophys. Res.*, *80*, 2751-2763, 1975

Hashimoto, K., H. Matsumoto, T. Murata, M.L. Kaiser and J.-L. Bougeret, Comparison of AKR simultaneously observed by GEOTAIL and WIND spacecraft, *Geophys. Res. Lett.*, *25*, 853-856, 1998

Hollweg, J.V.. Angular broadening of radio sources by solar wind turbulence. *J. Geophys. Res.*, *75*, 3715-3727, 1970

Kaiser, M.L. and J.K. Alexander, Source location measurements of terrestrial kilometric radiation obtained from lunar orbit, *Geophys. Res. Lett.*, *3*, 37-40, 1976

Kaiser, M.L., M.D. Desch, W.M. Farrell, J.-L. Steinberg and M.J. Reiner, LF band terrestrial radio bursts observed by Wind/WAVES, *Geophys. Res. Lett.*, *23*, 1283-1286, 1996

Knoll, R., G. Epstein, S. Hoang, G. Huntzinger, J.-L. Steinberg, J. Fainberg, F. Grena, S.R. Mosier and R.G. Stone, The 3-Dimensional Radio Mapping Experiment (SBH) on ISEE C, *IEEE Trans. Geosci. Electron.*, *GE-16*, 199-204, 1978

Lacombe, C., C.C. Harvey, S. Hoang, A. Mangeney, J.-L. Steinberg and D. Burgess, ISEE observations of radiation at twice the solar wind plasma frequency, *Ann. Geophysicae*, *6*, 113-128, 1988

Lacombe, C., J.-L. Steinberg, C.C. Harvey, D. Hubert, A. Mangeney and M. Moncuquet, Density fluctuations measured by ISEE 1-2 in the Earth's magnetosheath and the resultant scattering of radio waves, *Ann. Geophysicae*, *15*, 387-396, 1997

Meyer-Vernet, N. and C. Perche, Tool kit for antennae and thermal noise near the plasma frequency, *J. Geophys. Res., 94*, 2405-2415, 1989

Morgan, D.D., and D.A. Gurnett, The source location and beaming of the terrestrial continuum radiation, *J. Geophys. Res., 96*, 9595-9613, 1991

Steinberg, J.-L., C. Lacombe and S. Hoang, A new component of terrestrial radio emission observed from ISEE 3 and ISEE 1 in the solar wind, *Geophys. Res. Lett., 15*, 176-179, 1988a

Steinberg, J.-L., S. Hoang, C. Lacombe and R.D. Zwickl, ISEE 3 observations of the Earth's radio continuum through the bow shock and magnetosheath and in the magnetosphere, *Ann. Geophysicae, 6*, 309-318, 1988b

Steinberg, J.-L., S. Hoang and C. Lacombe, Propagation of terrestrial kilometric radiation through the magnetosheath: ISEE 3 observations, *Ann. Geophysicae, 7*, 151-160, 1989

Steinberg, J.-L., S. Hoang and J.-M. Bosqued, Isotropic Terrestrial Kilometric Radiation: a new component of the Earth's radio emission, *Ann. Geophysicae, 8*, 671-686, 1990a

Steinberg, J.-L., S. Hoang and M.F. Thomsen, Observations of the Earth's continuum radiation in the magnetotail with ISEE 3, *J. Geophys. Res., 95*, 20781-20791, 1990b

Steinberg, J.-L. and C. Lacombe, An empirical model of the plasma density distribution in the distant magnetosheath, *Geophys. Res. Lett., 19*, 2285- 2288, 1992

Steinberg, J.-L. and S. Hoang, Magnetosheath density fluctuations from a simulation of auroral kilometric radiation radio propagation, *J. Geophys. Res., 98*, 13.467-13475, 1993

Steinberg, J.-L. and S. Hoang, The magnetotail: a rough and leaky plasma waveguide for terrestrial radio waves, *J. Geophys. Res., 100*, 17.241-17.251, 1995

Steinberg, J.-L., C, Lacombe and S. Hoang, Sounding the flanks of the Earths's bow shock to −230 R_E: ISEE 3 observations of terrestrial radio sources down to 1.3× the solar wind plasma frequency, *J. Geophys. Res., 103*, 23.565-23.579, 1998

S. Hoang, C. Lacombe and J.-L. Steinberg, Observatoire de Paris, DESPA, UMR 8632 CNRS, 92195 Meudon, France (e-mail: SAng.Hoang@obspm.fr; Catherine.Lacombe@obspm.fr; Steinberg@obspm.fr)

Propagation of Radio Waves in the Corona and Solar Wind

T. S. Bastian[1]

National Radio Astronomy Observatory, Socorro, New Mexico

The solar corona and solar wind are plasmas characterized by large scale MHD structures, waves, and turbulence. These introduce both systematic and random variations in the refractive index which affect the propagation of radio waves. A variety of propagation phenomena occur – regular refraction; angular, temporal, and spectral broadening; scintillations in amplitude and phase – widely referred to as *scattering* phenomena. In this tutorial I review the physical basis of these phenomena and describe a variety of techniques designed to exploit observations of scattering phenomena to deduce properties of the corona and solar wind plasma.

1. INTRODUCTION

Propagation phenomena in the solar wind and corona, phenomena caused by the refraction and diffraction of radiation, have been studied at radio wavelengths for more than forty years. They represent a two-edged sword. On the one hand, propagation phenomena distort or degrade observable properties of the radiation and therefore impose limitations on the precision with which one can deduce the intrinsic radiating properties of the source. For example, angular broadening, in analogy to optical "seeing", limits the angular resolution with which radio sources can be studied. On the other hand, propagation phenomena offer a variety of tools for studying the medium responsible for the scattering and have been widely used for studies of the interstellar medium, planetary atmospheres, the ionosphere, and – of particular interest here – the solar corona and the solar wind.

This tutorial provides a brief introduction to propagation phenomena, their underlying physical basis, and their uses as diagnostic probes of physical conditions in the solar corona and the interplanetary medium (IPM). The organization of the tutorial is as follows: a phenomenological overview of propagation phenomena is presented in §2. A theoretical framework based in statistical optics is outlined in §3. Relations connecting observables with physical quantities of interest are emphasized. Experimental techniques that exploit scattering phenomena as probes of the corona and solar wind are discussed in §4. I discuss the closely related problem of solar and interplanetary (IP) radio bursts in §5.

2. PHENOMENOLGICAL OVERVIEW

Scattering phenomena in the solar wind and corona result from the interaction of radio waves with inhomogeneities in the plasma they traverse. The interaction between radio waves and plasma is determined by the refractive index μ, or equivalently, by the dielectric constant $\epsilon = \mu^2$. In many practical cases of interest, the magnetic field is sufficiently weak and the plasma is sufficiently tenuous that $\omega \gg \omega_{pe} \gg \omega_{Be}$,

---

[1]Currently at DESPA, Observatoire de Paris, Section de Meudon, 92195 Meudon Cedex, France

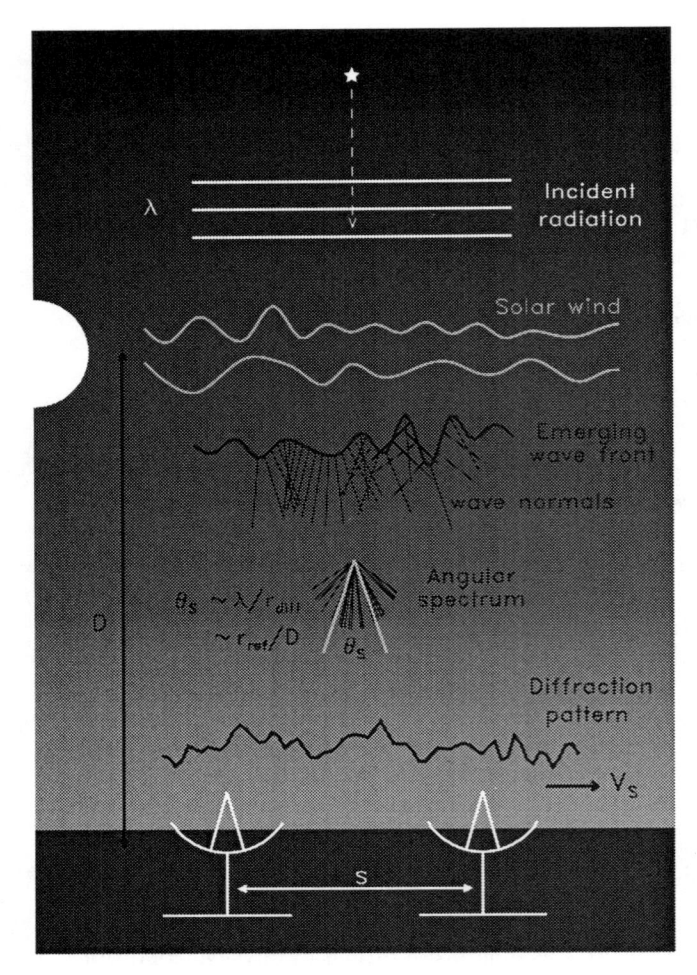

Figure 1. Cartoon outlining the basic scattering phenomena. A plane wave is incident on a thin, phase-changing screen. The emerging wave form is corrugated, equivalent to the radiation being scattered into a cone of angular width $\theta_S \sim \lambda/r_{diff}$ (angular broadening). In strong scattering, the scattered radiation interferes with itself as it propagates to an observer at a distance D from the screen, producing a diffraction pattern on the ground. If the screen drifts over the observer, the intensity varies in time with 100% modulation; i.e., strong scintillations are observed.

where ω is the angular frequency of the propagating wave, $\omega_{pe} = (4\pi n_e e^2/m_e)^{1/2}$ is the electron plasma frequency and $\omega_{Be} = eB/m_e c$ is the electron gyrofrequency. Absorption by the plasma is taken to be negligible. Cases where these conditions are not necessarily valid are touched upon in §5.

The dielectric constant is $\epsilon = 1 - \omega_{pe}^2/\omega^2 = 1 - \lambda^2 r_e n_e/\pi$ where λ is the wavelength of the radio waves and $r_e = e^2/m_e c^2$ is the classical electron radius. Under the conditions assumed above, the basic interaction between a radio wave and a plasma inhomogeneity characterized by a fluctuation in the refractive index $\delta\mu \approx r_e \lambda^2 \delta n_e/2\pi$, with a characteristic size scale Δz, is a change in the phase $\delta\phi$ of the incident radio wave given by $\delta\phi = k\Delta z \delta\mu = \lambda r_e \Delta z \delta n_e$. It follows that a gradient in the electron number density yields a gradient in the phase of the incident wave. Taking the direction of the wave propagation to be normal to the wave front, it is easily seen that a density gradient – due to a wedge of plasma, for example – changes the direction of the incident wave; i.e., refracts the wave.

A purely refractive description of propagation phenomena is sometimes adequate. For this to be so the plasma must 1) vary smoothly, with the spatial scale of variations in ϵ such that $l_\epsilon \gg \lambda$, and 2) $r_F = \sqrt{\lambda D/2\pi} \ll l_\epsilon$, where D is the distance the wave propagates and r_F is the Fresnel scale. The origin of the second condition, under which diffraction can be neglected, is as follows: consider an inhomogeneity in ϵ with a size l_ϵ. When illuminated by a plane wave it casts a shadow of size l_ϵ on a screen at a distance D from the inhomogeneity. The inhomogeneity also diffractively scatters the incident wave into a cone with an angular size $\lambda/2\pi l_\epsilon$ which illuminates an area of size $l_D \sim \lambda D/2\pi l_\epsilon$ on the screen at distance D. Diffraction may be neglected so long as $l_D \ll l_\epsilon$, from which condition (2) follows. In such cases, the *geometrical optics approximation* is exploited, allowing the propagation of radiation to be described in terms of rays. Powerful *ray tracing* techniques can then be brought to bear on problems of interest (see §5), techniques which allow the wave character of the radiation to be ignored. However, for some phenomena – e.g., intensity scintillations – the wave character of the radiation is of fundamental importance. A wave description of propagation phenomena is therefore more general and will be emphasized here.

To illustrate the various propagation phenomena observed in the corona and IPM, consider a right-handed coordinate system such that an observer is in the (x, y)-plane at a distance D along the z-axis (Figure 1). A plane wave of unit amplitude and wavelength λ propagates from $z < 0$ to $z = 0$. The scattering medium is idealized as a thin, two-dimensional, screen located at $z = 0$. This idealization is well justified for many problems of interest and greatly simplifies the analysis. The thin screen embodies the action of the three-dimensional turbulent medium traversed by the wave. Its role is to change the phase of the incident wave randomly as a function of position. The statistics of the scattering medium can be described by a spatial correlation function, a structure function, or a spatial power spectrum (see §4). In practice, the spatial spectrum of the fluctuating plasma is broadband, with power across a wide range of wave numbers.

After passing through the thin, phase-changing, screen at $z = 0$ the wave front is corrugated and can be analyzed into a spectrum of plane waves with wave normals in range of directions. To an observer at distance D from the scattering screen, radiation is received from a range of angles: the source has suffered *angular broadening* with a characteristic width θ_S. Because the source is no longer point-like as viewed in the observer's plane, there is a geometrical time delay $t_D \sim D\theta_S^2/2c$. An impulsive source of radio waves therefore experiences *temporal broadening* of this order. Suppose the screen moves in a direction transverse to the observer with a velocity v_S. In addition to angular and/or temporal broadening, a monochromatic source will experience *spectral broadening* of order $\Delta\nu/\nu \sim 2v_\parallel/c = 2v_S\theta_S/c$ as a result of the Doppler shifts caused by the radial projection of the screen velocity over the characteristic angular width θ_s.

The action of the screen is to introduce introduce a position-dependent phase change $\phi_S(\mathbf{r})$ to the wave at $z = 0$. The complex wave amplitude u measured on the observer's plane at a position $\mathbf{b}$ is given by the Fresnel-Kirchoff integral of scalar diffraction theory [e.g., *Goodman and Narayan* 1989]:

$$u(\mathbf{b}, D) = \frac{e^{ikD}}{2\pi r_F^2} \int d\mathbf{r} \, \exp\left[i\frac{(\mathbf{r}-\mathbf{b})^2}{2r_F^2} + i\phi_S(\mathbf{r})\right] \quad (1)$$

Ignoring the random phase term $\phi_S(\mathbf{r})$ due to the scattering screen for the moment, the complex wave amplitude is determined by the integral over a phase term which, beyond the radius of the Fresnel scale, r_F, oscillates and therefore does not contribute; that is, r_F is the scale over which the phase sums coherently. We now include the random phase term and define the *coherence scale* r_{diff} as the transverse scale over which the mean square phase varies by 1 rad^2; i.e., $\langle|\phi(0, D) - \phi(r_{diff}, D)|^2\rangle = 1$. The coherence scale r_{diff} is often referred to as the *diffractive scale* for reasons which will become clear. A comparison of r_F and r_{diff} defines two scattering regimes. The regime in which $r_{diff} \gg r_F$ is referred to as the *weak scattering* regime while the opposite inequality corresponds to the *strong scattering* regime.

In weak scattering, the phase variations $\phi_S(\mathbf{r})$ introduced by the thin screen are only minor perturbations on the propagation of the radio radiation and the Fresnel scale retains its role in Eqn. 1 as the scale of a low pass (spatial) filter. The action of the small phase perturbations introduced by the screen are to slightly increase or decrease the effective size of the Fresnel patch as a function of position $\mathbf{b}$, resulting in an increase or decrease of the flux $F(\mathbf{b}) = u(\mathbf{b})u^*(\mathbf{b})$. Letting σ_F represent the rms flux variation, the modulation index of the flux scintillation is $m = \sigma_F/F$. In the weak scattering regime, $m \ll 1$. Another way to think of the flux variation as a function of position in the weak scattering regime is in terms of weak focusing or defocusing by spatial inhomogeneities in the thin screen with a characteristic size r_F (Figure 2). If the thin scattering screen moves transverse to the observer's plane with a velocity v_S, weak flux scintillations are seen at a fixed point. The time scale on which the flux scintillations occur in the weak scattering regime is $t_F \sim r_F/v_S$.

In strong scattering, $r_{diff} \ll r_F$; i.e., the phase is coherent over r_{diff} but varies by many radians over r_F. The Fresnel scale is therefore no longer relevant in the strong scattering regime. Consider a phase coherent patch on the scattering screen of size r_{diff}; it is separated from neighboring patches by a similar scale. It diffractively scatters radiation into a cone with an angular width $\theta_S \sim 1/kr_{diff}$. At a distance D, the cone illuminates an area of scale $r_{ref} \sim D\theta_S = D/kr_{diff}$. With all such patches illuminating similar areas on the observer's plane, a given point on the observer's plane receives radiation from many phase-independent patches, the number of such patches being roughly $(r_{ref}/r_{diff})^2$. The radiation emitted into each diffractive cone interferes with that of the other overlapping cones, yielding a diffraction pattern in the observer's plane, the correlation length of which is $1/k\theta_S \sim r_{diff} \ll r_F$. The scintillation time is therefore $t_{diff} \sim r_{diff}/v_S \ll t_F$. In contrast to the weak scattering regime, the flux modulation due to diffractive scintillation satisfies Rayleigh statistics [e.g., *Narayan*, 1992] with $\langle F \rangle = 1$ and $\langle (F - \langle F \rangle)^2 \rangle = 1$; i.e., the fluctuations are saturated with a modulation index $m \sim 1$.

An interesting feature of the strong scattering regime is the appearance of a second spatial scale, the so-called refractive scale r_{ref}. Note that $r_{ref}r_{diff} = r_F^2$. In strong scattering, the observer is illuminated by many phase-independent diffractive cones. A mild focusing or defocusing occurs on the scale r_{ref} in analogy to that which occurs on the Fresnel scale in weak scattering. Greater numbers, or fewer numbers, of diffractive cones are directed towards the observer and a weak scintillation is observed. As is the case for scintillations in weak scattering, the modulation index of *refractive scintillation* is $m_{ref} < 1$. The spatial correlation scale

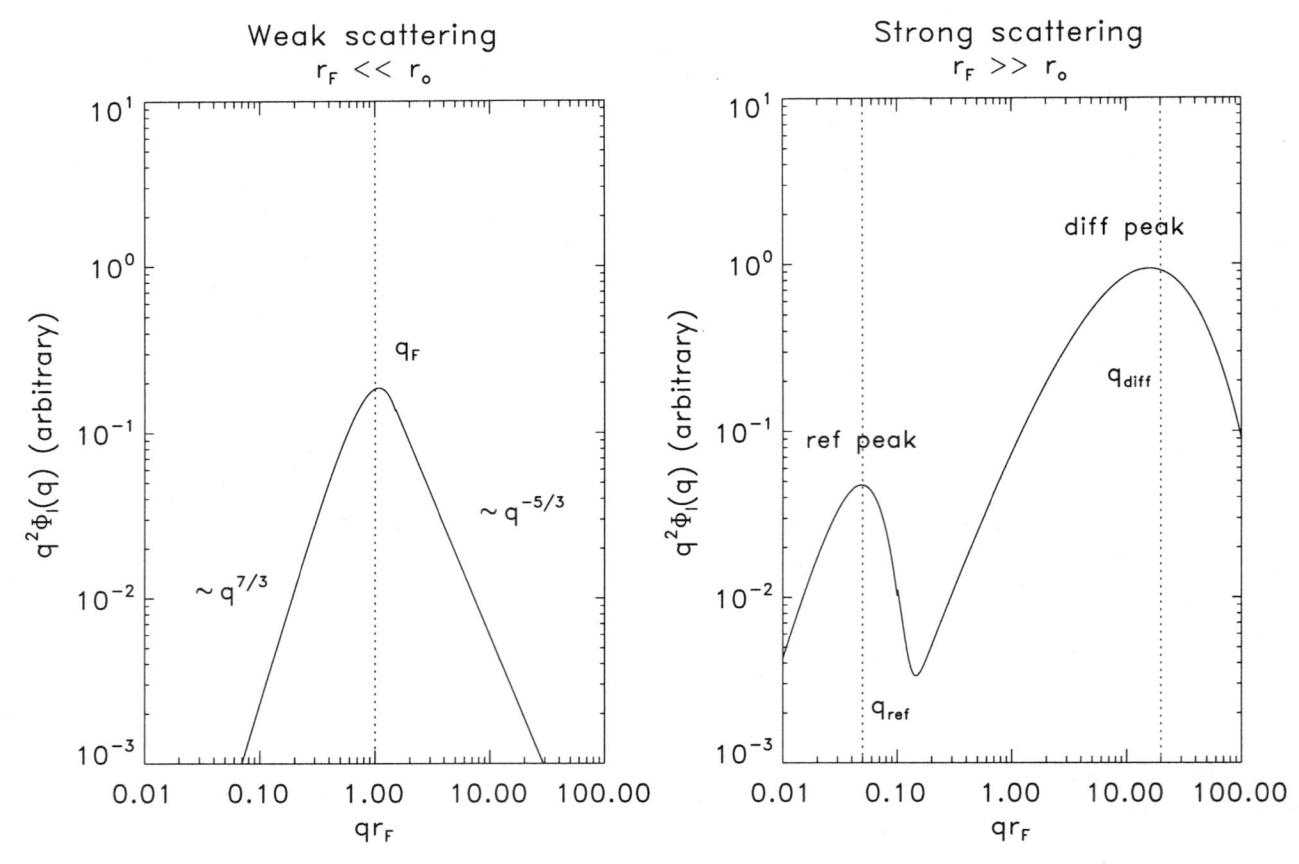

Figure 2. Plots of the power spectrum of the intensity scintillations observed in the weak (left panel) and strong (right panel) scattering regimes. Note that $q^2\Phi_I(q)$ is plotted, the power per logarithmic interval in q. There is a single maximum in the power spectrum in the weak scattering regime at a wave number corresponding to the Fresnel scale r_F. In contrast, there are two maxima in the spectrum for the strong scattering regime at wave numbers corresponding to the diffractive and refractive spatial scales, r_{diff} and r_{ref}, respectively.

is $r_{ref} \gg r_F$ (Figure 2) and hence the scintillation time is $t_{ref} \sim r_{ref}/v_S \gg t_F \gg t_{diff}$.

These simple considerations have revealed the presence, then, of several observable propagation phenomena: angular broadening, temporal broadening, spectral broadening, and scintillations. In the case of scintillations, it is useful to distinguish between the weak and strong scattering regimes. The terms have nothing to do with the strength of the plasma turbulence responsible for the scattering, instead referring to size of the modulation index m. The distinction between strong and weak scattering is an important one because the theoretical description of scintillations in the weak scattering regime is invalid as the scintillations saturate. Both scattering regimes are encountered in the solar corona and IPM and hence care must be taken when interpreting observational results.

3. THEORY

A theoretical treatment of radio propagation in an inhomogeneous plasmas is well beyond the scope of this tutorial. The reader is referred to the reviews of *Rickett* [1971, 1990] for an introduction, or to the proceedings edited by *Tatarskii, et al.* [1993]. Detailed treatments may be found in, e.g., *Ishimaru* [1978], *Goodman* [1985], or *Rytov, et al.* [1989]. Here, the basic theoretical framework is outlined and relevant terminology is introduced.

A powerful theory for understanding propagation phenomena, and for leveraging constraints on physical quantities of interest, is based on a statistical description of the scattering medium and of the resulting scattered radiation field. The statistics of the scattering medium are described by spatial correlation functions B_ϵ, B_μ, or B_{n_e} in $\delta\epsilon$, $\delta\mu$, or δn_e, respectively, with

$$\begin{aligned}B_\epsilon(\mathbf{r_1}, \mathbf{r_2}) &= \langle \delta\epsilon(\mathbf{r_1})\delta\epsilon(\mathbf{r_2})\rangle \\ &= 4B_\mu(\mathbf{r_1},\mathbf{r_2}) \\ &= (r_e^2\lambda^4/\pi)B_{n_e}(\mathbf{r_1},\mathbf{r_2}).\end{aligned}$$

For a statistically homogeneous, isotropic medium, which we assume to be the case here, the spatial correlation functions depend only on the magnitude of the difference in spatial coordinates, not the coordinates themselves; i.e. $r = |\mathbf{r_1} - \mathbf{r_2}|$. The three-dimensional Fourier transform of the spatial correlation functions $B_\epsilon(r)$, $B_\mu(r)$, or $B_{n_e}(r)$ yield the spatial power spectra $\Phi_\epsilon(q)$, $\Phi_\mu(q)$, or $\Phi_{n_e}(q)$, respectively, of fluctuations in the medium, where $q = 2\pi/r$ is the wave number. $\Phi_{n_e}(q)$ is most often the function of interest, for it is the spatial spectrum of the electron density fluctuations that embodies the physics of energy input, its transfer over a wide band of q via a turbulent cascade, and its ultimate dissipation in the corona and IPM (see §4).

Turning to the radiation field, it is straightforward [e.g., *Lee and Jokipii*, 1975a; *Ishimaru*, 1978] to reduce the wave equation for (monochromatic) electromagnetic waves $fe^{-i\omega t}$ in a plasma characterized by a spectrum of random inhomogeneities to a scalar Helmholtz equation in which polarization of the radiation is ignored:

$$\nabla^2 f + k^2(1 + \epsilon^*)^2 f = 0. \qquad (2)$$

Here $\epsilon^* = (\epsilon - \langle\epsilon\rangle)/\langle\epsilon\rangle$ represents the (normalized) fluctuating part of the dielectric constant and $k^2 = \omega^2\langle\epsilon\rangle/c^2$. The scalar Helmholtz equation is a good approximation so long as $\lambda \ll l_\epsilon$, where l_ϵ is the minimum spatial scale on which ϵ varies. Taking $f = ue^{ikz}$ and substituting, we obtain

$$2ik\frac{\partial u}{\partial z} + \nabla^2 u + k^2\epsilon^* u = 0.$$

If small-angle scattering is assumed (i.e., large-angle scattering, and reflected or back-scattered waves are ignored), the term $\partial^2 u/\partial z^2$ may be neglected, an approximation commonly referred to as the "parabolic approximation", which results in the *parabolic wave equation*:

$$2ik\frac{\partial u}{\partial u} + \nabla_\perp^2 u + k^2\epsilon^* u = 0 \qquad (3)$$

where $\nabla_\perp^2 = \partial^2/\partial x^2 + \partial^2/\partial y^2$.

The radiation field is fully characterized by the following (infinite) set of correlation functions:

$$\Gamma_{m,n}(z, \rho_1, ...\rho_m, \rho_1', ..., \rho_n') = \\ \langle u(z,\rho_1)...u(z,\rho_m)u^*(z,\rho_1')...u(z,\rho_n')\rangle$$

where $\rho_i = (x_i, y_i)$ is a transverse coordinate. We first consider the second moment of the electric field, $\Gamma_{1,1}(z, \rho_1, \rho_2) = \langle u(z,\rho_1)u^*(z,\rho_2)\rangle$. *Lee and Jokipii* [1975a] show that under the Markov approximation[1], and using the parabolic wave equation, one obtains a parabolic equation for $\Gamma_{1,1}$:

$$\frac{\partial\Gamma_{1,1}(z,\rho_1,\rho_2)}{\partial z} = -\frac{k^2}{4}[A(0) - A(\rho_1-\rho_2)]\Gamma_{1,1}(z,\rho_1,\rho_2). \qquad (4)$$

where

$$\delta(z-z')A(\rho_1-\rho_2) = \langle\epsilon(z,\rho_1)\epsilon(z',\rho_2)\rangle.$$

Note that $(\nabla_{\perp 1}^2 - \nabla_{\perp 1}^2)\Gamma_{1,1}$ has been dropped because ϵ^* is statistically homogeneous in the transverse direction (by assumption) and $\Gamma_{1,1}$ therefore depends only on differences ρ between the transverse coordinates ρ_i. It should be noted that parabolic equations can be obtained for all higher moments of the electric field. However, an exact solution is only available for $\Gamma_{1,1}$. Eqn. (4) can be simply integrated, yielding the important result

$$\Gamma_{1,1}(z,\rho) = \exp\{-\frac{zk^2}{4}[A(0) - A(\rho)]\}. \qquad (5)$$

It can be shown [e.g., *Ishimaru* 1978] that

$$A(\rho) = 2\pi \int_0^\infty \Phi_\epsilon(\mathbf{q})e^{i\mathbf{q}\cdot\rho}d\mathbf{q}.$$

which, for an isotropic spectrum becomes

$$A(\rho) = (2\pi)^2 \int_0^\infty J_\circ(q\rho)\Phi_\epsilon(q)q\,dq. \qquad (6)$$

Substituting Eqn. 6 in Eqn. 5, one obtains $\Gamma_{1,1}(z,\rho) = \exp[-D(\rho)/2]$ where

$$\begin{aligned}D(\rho) &= 2\pi^2 k^2 z\int_0^\infty [1 - J_\circ(q\rho)]\Phi_\epsilon(q)\,q\,dq \\ &= 8\pi^2\lambda^2 r_e^2 z\int_0^\infty [1 - J_\circ(q\rho)]\Phi_{n_e}(q)\,q\,dq \quad (7)\end{aligned}$$

---

[1] The Markov approximation states that u varies by only a small amount in the z direction over a correlation length of ϵ, or that u is δ-correlated in z.

It should be emphasized that, within the assumptions made, the above solution for $\Gamma_{1,1}$ is exact and is valid regardless of whether the scattering is in the weak or strong scattering regimes. In practical terms, $D(\rho)$ is relatively straightforward to measure using a variety of techniques outlined in §4. These measurements can then be related to Φ_ϵ and, hence, Φ_{n_e} in the medium of interest through Eqn. (7).

For a spatially coherent source of radio waves (a point source) $\Gamma_{1,1}$, also called the mutual coherence function, only differs from the visibility function of radio astronomy by a normalization factor. Using the Van Cittert-Zernicke theorem, the visibility function is related to the radio brightness distribution on the sky through a Fourier transform relationship. For a distant point source, we have identically

$$\Gamma_{1,1}(\rho) = \frac{V(\rho)}{V(0)} = e^{-D(\rho)/2} \qquad (8)$$

In the absence of the intervening scattering medium, $D(\rho) = 0$ and $V(\rho) = $ const for all ρ and the inverse Fourier transform yields a point source. The presence of a scattering medium causes a loss of coherence with increasing ρ. The inverse Fourier transform is therefore broader than a point source. We see that the second moment of the electric field describes angular broadening. The angular width of a point source is conveniently characterized by $\theta_S = 1/kr_{diff}$, where we recognize r_{diff} from our earlier discussion to be the coherence scale, on which the mean square phase difference is 1 rad^2; i.e., $D(r_{diff}) = 1$. Temporal broadening can also be described by a more general treatment of $\Gamma_{1,1}$, wherein $k_1 \neq k_2$ [*Lee and Jokipii*, 1975b].

The statistics of intensity scintillations are described by the fourth moment of the electric field

$$\Gamma_{2,2} = \langle u(r,t) u^*(r,t) u(r+s,t) u^*(r+s,t) \rangle,$$

which we recognize as the intensity covariance. As mentioned above, no exact solutions for $\Gamma_{m,n}$ exist except for $m,n = 1$. Nevertheless, approximate methods exist which allow progress to made under certain limiting assumptions.

In the weak scattering regime ($m \ll 1$) the Born approximation and the Rytov approximation (also called the "method of smooth perturbations") have both been employed to solve for the correlation functions and variances of the amplitude and phase of a propagating wave in a randomly fluctuating medium. The Born approximation involves expanding the transverse electric field in a series and solving the scalar Helmholtz equation (Eqn. 2). The Rytov or MSP approximation involves an expansion of the logarithm of the transverse field. Here I simply quote the result for the wave-number spectrum of the intensity scintillations in the weak scattering regime for a slab of plasma of thickness Δz valid in the Born approximation [*Rickett*, 1990]:

$$\Phi_I(q_x, q_y) = 8\pi r_e^2 \lambda^2 \sin^2\left(\frac{q^2 \lambda z}{4\pi}\right)$$
$$\times \Phi_{n_e}(q_x, q_y, q_z = 0) \Delta z, \qquad (9)$$

which is simply the spatial spectrum of the electron density fluctuations multiplied by a "Fresnel filter" term which acts as a high-pass filter, suppressing wave numbers corresponding to spatial scales greater than the Fresnel scale.

As the modulation index saturates, the linear relationship between Φ_I and Φ_{n_e} in the Born approximation breaks down. In strong scattering, scintillations are characterized by two widely separated scales r_{diff} and r_{ref} as discussed in §3. Two approximate solutions are relevant. At high wave numbers the spectrum Φ_I is approximated as the Fourier transform of the *square* of the electric field covariance $\Gamma_{1,1}$; i.e.,

$$\Phi_I(q) = \frac{1}{4\pi^2} \int e^{-D(r)} e^{i\mathbf{q}\cdot\mathbf{r}} d^2 r,$$

which describes the spectrum of diffractive scintillations in strong scattering. At the opposite end of the wave number spectrum it is possible to perform a low-wave-number expansion in order to obtain an approximate expression for Φ_I which is analogous to that obtained under the Born for weak scattering (Eqn. 10) except that it is multiplied by a cutoff which suppresses the contribution of wave numbers above the reciprocal of r_{ref}. Unfortunately, no simple expressions exist for the regime between the limiting cases $r_F \ll r_{diff}$ (weak scattering) and $r_F \gg r_{diff}$ (strong scattering). Figure 2 shows the normalized power per logarithmic interval of wave number q in Φ_I in the two scattering regimes [*Narayan*, 1992].

4. DIAGNOSTIC USES OF SCATTERING PHENOMENA

Figure 3 summarizes various experimental configurations of natural and man-made radio sources that have been used to probe the corona and solar wind. Most employ a variant of detecting spatially or temporally coherent radio waves which have passed through the corona or solar wind. While *in situ* observations have

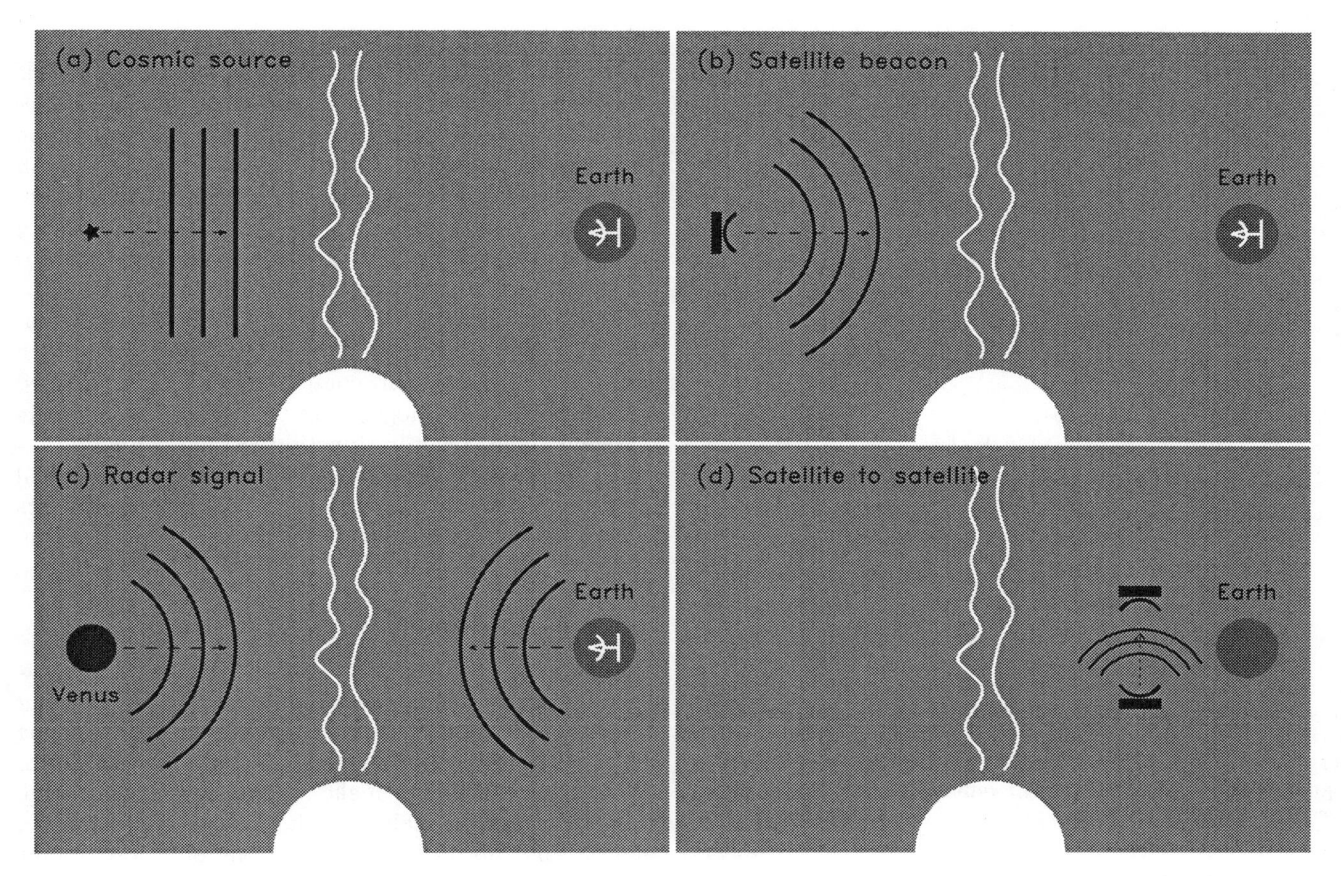

Figure 3. An example of a source which has undergone angular broadening by the solar wind. An unresolved background source was observed by the VLA at a wavelength of 6.1 cm at an apparent position of 3.1 $R_\odot$ and a postion angle $\phi = 81°$, measured east from solar north. The scattering is clearly anisotropic as a result of inhomogeneities in the solar wind being themselves spatially anisotropic. The magnetic field is assumed to be aligned with the long dimension of the inhomogeneities. The direction of the center of the solar disk is as indicated. Note that the source centroid has been shifted by Δr as a result of general refraction.

been available in the solar wind for some time, they sample conditions in specific locations; e.g., near 1 AU for the ISEE missions, along a trajectory between 0.3-1 AU for the *Helios* mission, or on a trajectory between 1-5 AU for the *Ulysses* mission. Propagation studies have been one of the few means available to probe the inner heliosphere. They also provide the means to characterize global, rather than local, aspects of the solar wind.

Eqn. 7 gives the basic relationship between the wave structure function and the spatial power spectrum of the electron number density $\Phi_{n_e}(q)$. For a spatial spectrum that is a power law, $\Phi_{n_e}(q) \sim q^{-\alpha}$, the corresponding structure function depends on the wave number as $D(s) \propto s^{\alpha-2}$ (for $2 < \alpha < 4$). For a spectrum with a dissipation scale (inner scale) q_i, the structure function breaks from a $a^{\alpha-2}$ dependence to a square law dependence $D(s) \sim s^2$ at a spatial scale $\sim 3/q_i$. Experiments associated with measuring the temporal and spatial electric field covariance have played an important role in establishing the basic character of Φ_{n_e} in the outer corona and solar wind [e.g., *Coles and Harmon*, 1989]. Scintillation experiments have provided unique measurements of the solar wind speed in the inner heliosphere as well as maps of large scale coherent structures in the heliosphere. The most common experimental techniques are now briefly described.

4.1. Phase Scintillations

Phase scintillations are simply the phase fluctuations observed in a wave that has traversed a randomly inhomogeneous medium (see discussion in §2). Phase scintillations can be measured using a single antenna ob-

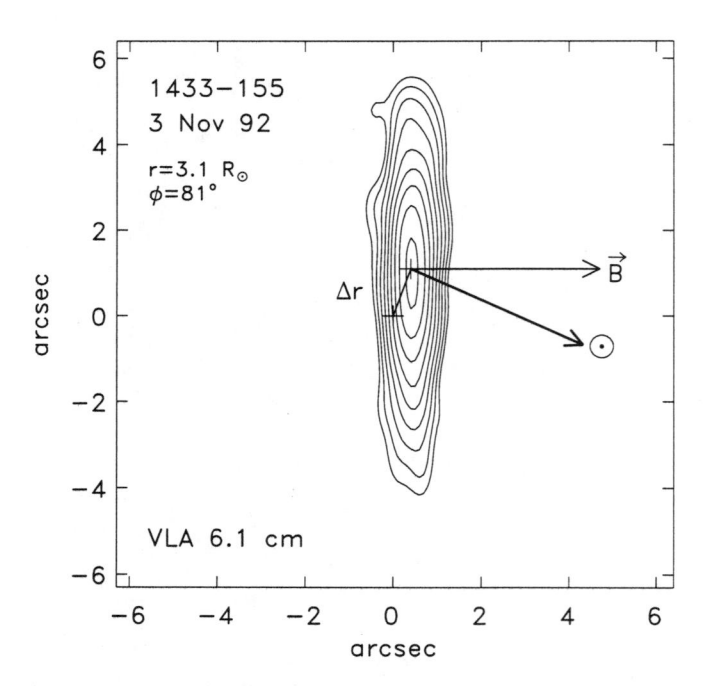

Figure 4. A cartoon summarizing experimental configurations that have been employed to probe the corona and solar wind: a) observations of distant background sources; b) observations of spacecraft beacons; c) observations of a radar signal transmitted from Earth, reflected from Venus, and received on Earth; d) spacecraft to spacecraft observations.

serving a temporally coherent (quasi-monochromatic) signal emitted by a spacecraft beacon, or by making interferometric observations of a spatially coherent source (point-like) with a pair of antennas.

In the case of spacecraft beacons, a two-frequency scheme is typically employed, using the standard telemetry bands at 2.3 (S band) and 8.4 GHz (X band). Using the known dispersion properties of the plasma, the S and X band phases can be differenced to separate the plasma contribution to the scintillations from that due to free space propagation. While it is possible to construct the wave structure function from the time series of phase scintillation measurements, it is more usual to compute a power spectrum of the time series $P_\phi(f)$. If the spatial spectrum of electron number density is a power law $\Phi_{n_e}(q) \propto q^{-alpha}$ then $P_\phi(f) \propto f^{\alpha-1}$. Examples of phase scintillation measurements are described by *Woo and Armstrong* [1979] and, more recently, by *Pätzold, et al.* [1996].

Spacecraft-to-spacecraft measurements of phase scintillations have also been made. *Celnikier, et al.* [1983, 1987] used a dual-frequency system to measure phase fluctuations in a 683 kHz signal referenced to a VHF signal transmitted between the ISEE 1 and ISEE 2 spacecraft. The spacecraft were typically separated by only a few hundred kilometers. Since the spacecraft speeds were much less than that of the solar wind, and because the propagation time of the signal between the two spacecraft was so short compared to the time in which the solar wind parameters changed, the time series of phase fluctuations could again be converted to a spatial power spectrum of the electron density fluctuations.

Using an interferometer it is straightforward to measure the differential phase $\Delta\phi(\mathbf{s}) = \phi(\mathbf{r}) - \phi(\mathbf{r}+\mathbf{s})$ on a time scale T that is short compared to the time required for solar wind inhomogeneities to drift the scale of the baseline. The structure function is then estimated directly as $D(s) \approx \langle[\Delta\phi(\mathbf{s})]^2\rangle$, as has been done in several very long baseline interferometry (VLBI) experiments (see *Coles and Harmon*, 1989; *Spangler and Sakurai*, 1995).

4.2. Angular Broadening

A particularly straightforward experiment is to measure the electric field spatial covariance directly using radio interferometers. Fourier synthesis radio telescope such as the Very Large Array have interferometers in abundance, which can make measurements on a large number of spatial scales simultaneously. As shown in §3, suitably normalized observations of distant point sources through the solar wind and corona are related to the wave structure function through an exponential (Eqn. 8), a relation which holds regardless of scattering strength. Rather than fitting the brightness distribution of the scatter-broadened source, it is more typical to fit functional forms for the structure function directly to the visibility data [e.g., *Armstrong, et al.*, 1990; *Narayan, et al.*, 1989; *Anantharamaiah, et al.*, 1994]. One of the important contributions of angular broadening observations has been to establish that solar wind turbulence is highly anisotropic in the inner heliosphere. An example of a scatter-broadened image of a radio source is shown in Figure 4.

As an aside, it is interesting to point out the work of *Goodman and Narayan*, [1989] on mapping radio sources in the strong scattering regime on short, intermediate, and long time scales. On time scales $t < t_{diff}$, the solar wind motion is effectively frozen and a map is the inverse Fourier transform of the fixed diffraction pattern [see *Cornwell, et al.*, 1990] and the resulting image is analogous to a "speckle" image in the optical regime. On time scales $t_{diff} < t < t_{ref}$, diffractive effects are suppressed by averaging, but a weak granularity remains due to refractive scintillation, a granularity that manifests itself as an enhanced variance in visibility amplitudes measured on large spatial scales. It is not until one averages the data over a time scale $t \gg t_{ref}$ that accurate estimates of the wave structure function may be obtained. In the solar wind, t_{diff} is some 10s of msec, while t_{ref} is several seconds to several 10s of seconds. All three imaging regimes are accessible to study by a Fourier synthesis telescopes like the VLA, so the solar wind is an ideal test bed for theories of certain propagation phenomena.

4.3. Spectral Broadening

In addition to experiments which measure the angular broadening of a spatially coherent source, the *spectral broadening* of a temporally coherent radio sources can also be used to probe the corona and IPM. Coherent beacons on spacecraft have been employed for this purpose, as well as ground based radar signals. As shown heuristically in §2, a monochromatic signal undergoes spectral broadening as it interacts with a scattering screen in motion relative to the observer. The spectrum $P(f)$ of the signal can be measured with receiving equipment on the ground. The Fourier transform of $P(f)$ is the temporal correlation function $\Gamma(\tau)$ which can be related to the spatial correlation function (or mutual coherence function) by a simple Galilean transformation [*Coles and Harmon*, 1989]:

$$\Gamma(\tau) = \frac{\langle E(r,t)E^*(r,t+\tau)\rangle}{\langle |E|^2 \rangle} = \Gamma(s = V_{sw}\tau)$$

where V_{sw} is the solar wind speed. As before, the wave structure function may be estimated and constraints placed on the spatial power spectrum of the electron number density. Examples of observations of spectral broadening of coherent radio beacons can be found in *Woo* [1978], *Woo and Armstrong* [1979], *Yakovlev, et al.* [1980], *Armstrong and Woo* [1981].

Radar experiments have used Venus as a passive reflector [*Harmon and Coles*, 1983; *Coles and Harmon*, 1989]. As Venus moves into opposition behind the Sun, a radio signal can be transmitted from Earth to Venus. There, the signal is reflected and returns to Earth, where it is detected and spectrum-analyzed. Both spacecraft signals and reflections from Venus are sufficiently local that the wavefronts involved are no longer planar. Furthermore, in the case of radar experiments, the wave makes a double pass through the solar wind. These complications aside, radar and spacecraft signals have been successfully employed for a variety of studies.

4.4. Intensity Scintillations

In view of the difficulties surrounding theoretical treatments of strong scintillations it is not surprising that most experimental work has been confined to the weak scattering regime where a convenient, linear relationship exists between the spatial spectrum of electron density inhomogeneities and the spatial spectrum of intensity scintillations.

Observations of interplanetary scintillations have been commonly exploited in at least three ways. First, scintillation is caused by the movement of the diffraction pattern of the incident radiation across the ground as the solar wind travels outward from the Sun. The intensity fluctuations can be measured by a single antenna and the autocorrelation function of a time series of such measurements can be formed. The Fourier transform of the autocorrelation function of the intensity fluctuations is the power spectrum Φ_I of the intensity fluctuations at the location of the antenna [*Coles and Harmon*, 1978]. Since $P(f) \propto \Phi_{N_e}(q)$ in weak scattering (Eqn. 9), single station measurements of scintillations of background radio sources can be used to constrain $\Phi_{N_e}(q)$ at various elongations from the Sun, so long as the elongation and wavelength employed are such that $m \ll 1$. Recent examples include the work of *Manoharan, et al.* [1994] and *Yamauchi, et al.* [1998]

A second way in which scintillations are used is as a passive tracer of solar wind velocity. In a procedure similar to that outlined above for one antenna, the temporal cross-spectrum can be measured by each pair of antennas and the diffraction pattern velocity vector can be extracted from the cross-correlation function (the Fourier transform of the cross-spectrum), or the cross-correlation function can be measured directly. As the background source changes in solar elongation, V_{sw} can be measured as a function of elongation and its acceleration profile can be inferred. While conceptually simple, complications enter into such measurements in practice. One of the main complications is the fact that the solar wind is composed of both slow speed and high speed streams. In addition, the electron density inhomogeneities are anisotropic. See *Grall* [1995], *Grall et al.* [1996].

A third use of scintillation in the weak scattering regime is to observe a grid of background cosmic sources and to compute the scintillation index m and its mean value $\langle m \rangle$ in each case. The Cambridge disturbance factor is then computed as $G = m/\langle m \rangle$. Typically, hundreds of measurements are made each day to provide a dense sampling of the scintillation properties of the IPM in the form of a "G map". Lines of sight along which enhanced regions of electron density occur yield enhanced values of G. Coherent density structures in the IPM can be mapped out using G maps and their evolution in time can be tracked by forming daily G maps. For example, heliospheric structures that co-rotate with the Sun have been observed using this technique. Recently, tomographic techniques have been exploited to infer the three-dimensional structure of such disturbances [*Jackson, et al.*, 1997, 1998; *Asai, et al.*, 1998; *Kojima, et al.*, 1998].

4.5. Faraday Rotation

The polarization properties of electromagnetic waves propagating through the corona and IPM have been ignored in this tutorial. A final class of experiments is worth mentioning, however, which makes explicit use of the polarization of the propagating waves, specifically, the Faraday rotation experienced by waves as they propagate through the corona and IPM, or fluctuations therein. Faraday rotation allows constraints to be placed on both the electron number density and the line-of-sight component of the magnetic field while fluctuations in the Faraday rotation allow, in principle, constraints to be placed on fluctuations in the density and/or magnetic field in the plasma as it is advected past the line of sight.

Two observational approaches have been exploited. One uses observations of the Faraday rotation of a linearly polarized spacecraft beacon [*Bird*, 1982; *Pätzold, et al.*, 1987]. The other approach uses naturally occurring linearly polarized signals such as those associated with pulsars [e.g., *Bird, et al.*, 1980] or with the incoherent synchrotron emission emitted by radio galaxies [e.g., *Soboleva and Timofeeva*, 1983; *Sakurai and Spangler*, 1994].

5. SOLAR AND INTERPLANETARY RADIO BURSTS

I conclude this brief tutorial with some remarks on studies of low frequency solar and interplanetary radio bursts. Although statistical wave treatments have been brought to bear on aspects of solar [*Bastian*, 1994; *Uralov*, 1998] and/or IP bursts [*Cairns*, 1998], studies of these phenomena have largely followed a different interpretive tradition from that described in previous sections. This is because the problem posed by low frequency solar and IP bursts differs from those described above in two key respects. First, unlike cosmic background sources, spacecraft beacons, or radar signals, whose intrinsic properties are known, the intrinsic properties of low frequency solar and IP radio bursts are unknown *a priori*. It is, in fact, an overriding goal of propagation studies, in this context, to strip away the influence of the scattering medium in order to reveal these properties. Second, in the case of low frequency solar and IP radio bursts it is not necessarily the case that $\omega_{pe} \ll \omega$ throughout the plasma traversed by the burst radiation – it can vary from a small value to unity between the source and the observer! Furthermore, low frequency solar and IP radio bursts typically occur in sources where large scale density gradients and hence, large scale gradients in the refractive index, occur. Refraction is of critical importance.

The techniques described in previous sections, based in statistical optics, are of little practical use for the study of low frequency solar and IP radio bursts in their present formulation. As pointed out by *Lacombe, et al.* [1997], there are presently no interferometric radio instruments available in space with which to measure the coherence properties of these radio sources (although see Jones, et al., this volume, which describes an instrument concept for a low-frequency array in space). More importantly, however, the theoretical framework sketched out in previous sections makes no provision for large scale refraction.

Ray tracing techniques offer a convenient and powerful means of understanding the influence of scattering on the emitted radio radiation in order to better understand the intrinsic properties of the source. These techniques were initially developed to study the effects of scattering on solar radio bursts [*Fokker* 1965] and to study the scattering properties of the solar corona [*Hollweg*, 1968]. In a classic paper by *Steinberg, et al.* [1971], the ray tracing technique was refined and extended to explore angular broadening, temporal broadening, and absorption of type III radio bursts in a spherically symmetric corona, including the effects of both regular refraction and scattering on random inhomogeneities. Subsequent work included analyses of non-symmetric large scale inhomogeneities such as coronal streamers [e.g., *Riddle* 1974], and took into account the "fibrous" structure of the coronal medium [*Bougeret and Steinberg*, 1977; *Duncan*, 1979, *Robinson*, 1983]. With the advent of radio receivers on spacecraft, ray tracing techniques have been used as a tool for understanding the propagation of radiation from interplanetary type III radio bursts [*Steinberg, et al.*, 1984; 1985], and the propagation of low-frequency radio waves in the near-Earth environment; for example, in the magnetosheath [*Steinberg and Hoang*, 1993; *Lacombe, et al.*, 1997].

Ray tracing techniques offer at least two significant advantages over the statistical optics approach for the study of scattering phenomena: 1) The technique, based in the geometrical optics approximation, ignores diffractive phenomena. It is is therefore a purely refractive theory and is in many ways more intuitive than alternative treatments. 2) Ray tracing techniques offer a convenient and flexible means of introducing both systematic and random variations in the refractive index. Consequently it is far easier to model complex propagation scenarios, including those in which the refractive index departs significantly from unity, a situation encountered in modeling the propagation of fundamental type III radio burst emission near the source, for example.

In physical regimes where both the geometrical optics and wave optics treatments are valid, both approaches should yield the same result for angular broadening. This is indeed the case, as demonstrated by *Strohbehn* [1967] for the case $\mu \approx 1$, and by *Cairns* [1998] for the more general case in which the refractive index is no longer unity. However, since both approaches rely on the assumption of small (integral) angle scattering (discounting general refraction), it is important to gain a quantitative understanding of propagation phenomena involving large-angle scattering [*Dulk et al.*, 1985] or backscatter, problems for which existing tools appear to be inadequate.

REFERENCES

Anantharamaiah, K. R., Gothoskar, P., Cornwell, T. J., Radio synthesis imaging of anisotropic angular broadening in the solar wind, *J. Astronom. Astrophys.*, 15, 387-414, 1994.

Armstrong, J. W., Woo, R., Solar wind motion within 30 $R_\odot$: spacecraft radio scintillation observations, *Astron. Astrophys.*, 103, 415, 1981.

Armstrong, J. W., Coles, W. A., Kojima, M., Rickett, B. J., Observations of field-aligned density fluctuations in the inner solar wind, *Astrophys. J.*, 358, 685-692, 1990.

Asai, K., Kojima, M., Tokumaru, M., Yokobe, A., Jackson, B. V., Hick, P. L., Manoharan, P. K., Heliospheric tomography using interplanetary scintillation observations: III. Correlation between speed and electron density fluctuations in the solar wind, *J. Geophys. Res.*, 103, 1991, 1998.

Bird, M. K., Coronal investigations with occulted spacecraft signals, *Sp. Sci. Rev.*, 33, 99, 1982.

Bird, M. K., Schruefer, E., Volland, H., Sieber, W., Coronal Faraday rotation during solar occultation of PSR0525+21, *Nature*, 283, 459, 1980.

Bastian, T. S., Angular scattering of solar radio emission by coronal turbulence, *Astrophys. J.*, 426, 774, 1994.

Bougeret, J.-L., Steinberg, J.-L., A new scattering process above solar active regions: propagation in a fibrous medium, *Astron. Astrophys.*, 61, 777-783, 1977.

Cairns, I., Angular broadening: effects of nonzero, spatially varying plasma frequency between the source and observer, *Astrophys. J.*, 506, 456-463, 1998.

Celnikier, L. M., Harvey, C. C., Jegou, R., Kemp, M., Moricet, P., A determination of the electron density fluctuation spectrum in the solar wind using the ISEE propagation experiment, *Astron. Astrophys.*, 126, 293-298, 1983.

Celnikier, L. K., Muschietti, L., Goldman, M. V., Aspects of interplanetary plasma turbulence, *Astron. Astrophys.*, 181, 138-154, 1987.

Coles, W. A., Harmon, J. K., Propagation observations of the solar wind near the Sun, *Astrophys. J.*, , 337, 1023-1034, 1989.

Cornwell, T. J., Anantharamaiah, K. R., Narayan, R., Propagation of coherence in scattering: an experiment using interplanetary scintillation, *J. Opt. Soc. Am. A*, 6, 977-985, 1990.

Dulk, G., Steinberg, J.-L., Lecacheux, A., Hoang, S., MacDowell, R. J., The visibility of type III radio bursts originating behind the Sun, *Astron. Astrophys.*, 150, 28, 1985.

Duncan, R. A., Wave ducting of solar metre-wave radio emission as an explanation of fundamental/harmonic source coincidence and other anomalies, *Solar Phys.*, 63, 389-398, 1979.

Fokker, A. D., Scattering of type I sources in the solar corona, *Bull. Astron. Inst. Netherlands*, 11, 118, 1965.

Goodman, J. W., *Statistical Optics* 550 pp., John Wiley and Sons, New York, 1985.

Goodman, J., Narayan, R., The shape of a scatter-broadened image – II. Interferometric visibilities, *Mon. Not. R. Astron. Soc.*, , 238, 995-1028, 1989.

Grall, R. R., Remote Sensing Observations of the Solar Wind Near the Sun, Ph.D Thesis, 141 pp., Univ. of California, San Diego, 1995.

Grall, R. R., Coles, W. A., Klinglesmith, M. T., Breen, A. R., Williams, P. J., Markkanen, J., Esser, R., Measurements of the solar wind speed in the south polar stream near the Sun, *Nature*, 379, 429, 1996.

Harmon, J. K., Coles, W. A., Spectral broadening of planetary radar signals by the solar wind, *Astrophys. J.*, , 270, 748-757, 1983.

Hollweg, J. V., A statistical ray analysis of the scattering of radio waves by the solar corona, *Astronom. J.*, 73, 972, 1968.

Ishimaru, A. *Wave propagation and scattering in random media*, 2 vol., Academic Press, New York, 1978.

Jackson, B. V., Hick, P. L., Kojima, M., Yokobe, A., Heliospheric tomography using interplanetary scintillation observations, *Adv. Sp. Res.*, 20, 23, 1997.

Jackson, B. V., Hick, P. L., Kojima, M., Yokobe, A., Heliospheric tomography using interplanetary scintillation observations: I. Combined Nagoya and Cambridge data, *J. Geophys. Res.*, , 103, 12049, 1998.

Kojima, M., Tokumaru, M., Watanabe, H., Yokobe, A., Asai, K., Jackson, B. V., Hick, P. L., Heliospheric tomography using interplanetary scintillation observations: II. Latitude and heliocentric distance dependence of solar wind structure at 0.1-1 AU, *J. Geophys. Res.*, , *103*, 1981, 1998.

Lacombe, C., Steinberg, J.-L., Harvey, C. C., Hubert, D., Mangeney, A., Moncuquet, M., Density fluctuations measured by ISEE 1-2 in the Earth's magnetosheath and the resultant Scattering of radio waves, *Ann. Geophysicae*, *15*, 387-396, 1997.

Lee, L. C., Jokipii, J. R. 1975, Strong scintillations in astrophysics: I - The Markov approximation, its validity and application to angular broadening, *Astrophys. J.*, *196*, 695, 1975a.

Lee, L. C., Jokipii, J. R. 1975, Strong scintillations in astrophysics: II - A theory of temporal broadening of pulses, *Astrophys. J.*, *201*, 532, 1975b.

Manoharan, P. K., Kojima, M., Misawa, H., The spectrum of electron density fluctuations in the solar wind and its variation with solar wind speed, *J. Geophys. Res.*, , *99*, 23411-23420, 1994.

Narayan, R., The physics of pulsar scintillation, *Phil. Trans. R. Soc. Lond. A.*, *341*, 151-165, 1992.

Narayan, R., Anantharamaiah, K.R., and Cornwell, T.J., Refractive radio scintillation in the solar wind, *Mon. Not. R. Astron. Soc.*, , *241*, 403-413, 1989.

Pätzold, M., Bird, M. K., Volland, H., Levy, G. S., Seidel, B. L., Stelzried, C. T., The mean coronal magnetic field determined from Helios Faraday rotation measurements, *Solar Phys.*, *109*, 91-105, 1987.

Pätzold, M., Karl, J., Bird, M. K., Coronal radio sounding with Ulysses: dual-frequency phase scintillation spectra in coronal holes and streamers, *Astron. Astrophys.*, *316*, 316, 1996.

Rickett, B. J., Radio propagation through the turbulent interstellar plasma, *Ann. Rev. Astron. Astrophys.*, *28*, 561-605, 1990.

Riddle, A. C., On the observation of scattered radio emission from sources in the solar corona, *Solar Phys.*, *35*, 153-169, 1974.

Robinson, R. D., Scattering of radio waves in the solar corona, *Proc. ASA*, *5*, 208-211, 1983.

Rytov, S. M., Kravtsov, Yu. A., Tartarskii, V., *Principles of Statistical Radiophysics*, 4 vol., Springer-Verlag, New York, 1989.

Sakurai, T., Spangler, S. R., The study of coronal plasma structures and fluctuations with Faraday rotation measurements, *Astrophys. J.*, *434*, 773-785, 1994.

Soboleva, N. S., Timofeeva, G. M., The Faraday effect in the solar supercorona during its 1977–1982 radio occultations of the Crab nebular, *Soviet Astron. Lett.*, *9*, 216, 1983.

Spangler, S. R., Sakurai, T., Radio interferometer observations of solar wind turbulence from the orbit of Helios to the solar corona, *Astrophys. J.*, *445*, 999-1016, 1995.

Steinberg, J.-L., Aubier-Giraud, M., Leblanc, Y., Boischot, A., Coronal scattering, absorption, and refraction of solar radio bursts, *Astron. Astrophys.*, *10*, 362, 1971.

Steinberg, J.-L., Hoang, S., Dulk, G., Evidence of scattering effects on the size of interplanetary type III radio bursts, *Astron. Astrophys.*, *150*, 205, 1985.

Steinberg, J.-L., Hoang, S., Lecacheux, A., Aubier, M., Dulk, G., Type III radio bursts in the interplanetary medium - The role of propagation, *Astron. Astrophys.*, *140*, 39, 1984.

Steinberg, J.-L., Hoang, S., Magnetosheath density fluctuations from a simulation of auroral kilometric radiation propagation, *J. Geophys. Res.*, , *98*, 13467, 1993.

Steinberg, J.-L., Lacombe, C., Hoang, S., Sounding the flanks of the Earth's bow shock to -230 R_E: ISEE 3 observations of terrestrial radio sources down to 1.3 times the solar wind plasma frequency, *J. Geophys. Res.*, *103*, 23565, 1998.

Strohbehn, J. W., Polarization and angle-of-arrival fluctuations for a plane wave propagating through a turbulent medium, *IEEE Trans. AP-15*, 416-420, 1967.

Tatarskii, V. I., Ishimaru, A., Zovorotny, V. U. (eds.), *Wave propagation in random media (scintillations)*, SPIE Press, Bellingham, 1993.

Uralov, A. M., Scintillation of solar radio sources: Implications for spikes, *Solar. Phys.*, *183*, 133, 1998.

Woo, R., Radial dependence of solar wind properties deduced from Helios 1/2 and Pioneer 10/11 radio scattering observations, *Astrophys. J.*, *219*, 727, 1978.

Woo, R., Armstrong, J., Spacecraft radio scattering observations of the power spectrum of electron density fluctuations in the solar wind, *J. Geophys. Res.*, , *84*, 7288 , 1979.

Yakovlev, O. I., Efimov, A. I., Razmanov, V. M.m and Shtrykov, V. K., inhomogeneous structure and velocity of the circumsolar plasma based on data of the Venera-10 stationm *Sov. Astron.*, *24*, 454-459, 1980.

Yamauchi, Y., Tokumaru, M., Kojima, M., Manoharan, P., Esser, R., A study of density fluctuations in the solar wind acceleration region, *J. Geophys. Res.*, *103*, 6571, 1998.

T. S. Bastian, National Radio Astronomy Observatory, Socorro, New Mexico 87801 (e-mail: tbastian@nrao.edu)

Scattering in the Solar Wind at Long Wavelengths

B. J. Rickett and W. A. Coles

Electrical and Computer Engineering, University of California, San Diego, California, USA

The performance of a space-based low frequency array will be limited by angular scattering in both the interstellar medium (ISM) and the solar wind. As it happens the two effects are roughly equal when the source is in the antisolar direction. When the source is closer to the Sun the solar wind scattering will dominate. Image correction methods developed to remove ionospheric and atmospheric effects at higher frequencies work well in weak scintillation conditions. However, below about 25 MHz interplanetary scintillation becomes strong in all directions, and the presence of interstellar scattering then makes it impractical to correct the images for the solar wind scattering. As a rough guide serious loss of visibility will occur on baselines longer than about $20 f_{\rm MHz}$ km.

1. RADIO PROPAGATION IN THE SOLAR WIND

Electron density inhomogeneities in the interplanetary medium (IPM) cause angular broadening and intensity scintillation of distant radio sources. Though angular broadening has been used to study the solar wind, for low-frequency astronomy it provides a serious seeing limit on angular resolution. Ground-based observations of the scattering phenomena have built a statistical picture for the microstructure in the electron density of the solar wind. In this paper we apply this picture to predict the low frequency propagation environment in the solar wind.

Pioneered in the 60's, angular broadening involves interferometry observations of a compact radio source, as its line of sight passes close to the Sun (e.g. Hewish 1958; Erickson 1964). The observed angle of scattering varies with observing frequency, solar elongation, solar latitude, and with phase of the solar cycle. The observed rate of decrease of scattering with solar distance implies that the small scale density fluctuations responsible for scattering decrease with mean density, in such a way that the ratio of rms to mean remains roughly constant.

Close to the Sun the scattering is highly anisotropic, indicating that the density irregularities are highly field-aligned (e. g. Armstrong et al. 1990, Grall et al. 1997). The anisotropy decreases rapidly with distance and the scattering is essentially isotropic at elongations greater than a few degrees.

Intensity scintillations, also pioneered in the 60's, have been used both for studying the spatial spectrum of the electron density and the distribution of flow velocity. Observations over three decades have successfully delineated how the solar wind speed varies with solar distance, solar latitude, and phase of the solar cycle. The density spectrum estimated from these observations are consistent with those derived from angular scattering and also with those derived from analysis of spacecraft telemetry signals (Woo & Armstrong 1979; Coles et al. 1991). Near the Sun the spectrum is a Kolmogorov power law at low wavenumbers (large scales) but at higher wavenumbers (scales of the order of 300 km) it flattens to a spectral exponent of about -3. At still higher wavenumbers (near the ion inertial scale) the spectrum drops steeply, apparently due to turbulent dissipation (Coles & Harmon 1989). As the solar distance increases the flattened region of the spectrum becomes less distinct and the inner scale, at which dissipation becomes important, increases linearly, reaching 250 km at 1 AU. In situ plasma measurements have filled out this picture, showing power law behavior for the spectra of velocity and magnetic field. Spacecraft observations generally cannot reach small enough scales to observe the inner scale directly, due to data rate limitations; however, they do confirm the flattened region of the spectrum determined by propagation measurements. Direct observations of velocity and magnetic field confirm that magneto-hydromagnetic turbulence is an important part of the underlying physics (Tu and Marsch, 1995). There is also a latitude dependent modulation of the level of turbulence over the solar cycle, as described by Coles et al. (1995).

However, in the context of low-frequency observations, which must be made away from the Sun, the density spectrum near the Earth determines the scattering. This can be modelled as $P_N(\vec{\kappa}, R) \propto R^{-4}\kappa^{-11/3}$ with an inner scale cut-off at $(250\text{km})^{-1}$. Here κ is the magnitude of the three dimensional wavenumber vector and R is the distance form the Sun's center. An immediate consequence of the inner scale is that, for baselines shorter than 250 km, the associated angular broadening at low frequencies is given by a Gaussian function.

2. INTERFEROMETRY THROUGH A RANDOM PHSE SCREEN

As a prelude to discussing the seeing limitations imposed by the IPM on low-frequency arrays, we review the theoretical aspects of interferometry through a random phase-changing screen. The ionosphere and atmosphere are well modelled as phase screens, but an extended scattering medium is required to model both the interplanetary and insterstellar plasmas. Nevertheless we discuss the screen model at length since the limits can be stated more simply, and provide a lower limit to the interplanetary and interstellar scattering.

Consider a pair of antennas separated by a baseline b at a distance L from such a phase screen. The basic angular resolution, $\sim \lambda/b$, is only achievable if the screen perturbations can be corrected. When the intensity scintillations are weak this can be achieved using various image correction methods, which have been successfully developed by radio astronomers for ground-based interferometry. When the intensity scintillations become strong the restoration problem becomes substantially more difficult.

2.1. Weak Scintillation Conditions

If the screen is very close to the observing plane, it imposes phase fluctuations with no accompanying amplitude fluctuations. In this case one can model the effect of propagation as introducing a phase error at each antenna. The phase errors can be estimated and corrected if the array has enough redundancy that phase closure can be used. As the distance increases, amplitude fluctuations build up. They are classified as weak if their rms amplitude is a small fraction of the mean. In weak scintillation the differential phase fluctuations across the Fresnel scale $r_F = \sqrt{L\lambda/2\pi}$ are much less than 1 radian rms. One can also model such amplitude fluctuations as a propagation induced multiplier at each antenna, which can also be corrected with sufficient baseline redundancy.

When the phase screen is further from the array, the antenna-based phase and amplitude errors depend on the position of the source. This makes estimation and correction of phase errors more complex. The effective range of angles over which the propagation errors remain sensibly constant is called the "isoplanatic patch" (radius θ_{iso}). Normally sources are smaller than θ_{iso} and various model fitting techniques have been developed to determine and correct for the antenna-based multipliers; these include self-calibration, hybrid mapping and triple-product methods, each with its own advantages (Pearson and Readhead, 1984; Cornwell, 1987). As the observing frequency is reduced the typical field of view increases, and in observations using the VLA at 74 MHz, Erickson reports (elsewhere in this volume) that separate antenna-based corrections had to be found for each 2°x2° patch in a 12°x12° field. Here the ionosphere caused the main perturbations. In more typical radio-astronomical image synthesis observations at higher frequencies the *amplitude* variations (from the ionosphere or atmosphere) are small enough to be ignored, but in principle the technique can be used up to conditions that the rms amplitude fluctuation approaches the mean. The conditions needed for successful corrections are:

(a) the integration time must be short enough that the diffraction pattern remains effectively constant; if the pattern crosses the observing array with speed V, we require $T_{\text{int}} \lesssim \min(r_F/V, b/V)$, and

(b) the angular extent of the mapped source-region

must be small compared with the "isoplanatic patch" of radius $\theta_{\rm iso}$, which in weak scintillation is r_F/L, and

(c) each antenna must have a diameter small compared to r_F

If any of these conditions fails the diffraction pattern cannot be observed and the visibility on each baseline is reduced by the factor $\exp[-0.5 D_\phi(b)]$, where $D_\phi(b) \sim (b/s_o)^{5/3}$ for a screen described by a Kolmogorov phase spectrum. Here s_o is the field coherence length and weak scintillations require $s_o > r_F$. The quantity $D_\phi(b)$ is the structure function of the phase perturbations caused by the screen. In extended medium scattering the overall phase perturbations integrated along a straight path through the medium provides an equivalent function (see reviews by Rickett, 1990 and Narayan, 1992).

There is another scale which enters the problem when the baseline becomes comparable or shorter than the inner scale. This is the physical scale at which the supposed turbulence described by the Kolmogorov spectrum is dissipated. Expressions for $D_\phi(b)$ are also given by Coles et al. (1987), covering cases with b both greater and smaller than the inner scale. In the limit of small b $D_\phi \propto b^2$.

2.2. Strong Scintillation Conditions

When the screen is sufficiently distant that $s_o < r_F$, the rms amplitude fluctuations exceed the mean and the scintillations are strong. The intensity diffraction pattern due to a point source then takes on a two-scale character. The total normalized variance m^2 can exceed unity, of which $(m^2 - 1)/2$ is in a refractive process with a scale of $s_r = L\lambda/2\pi s_o$ and $(m^2+1)/2$ is in a diffractive process with a scale of s_o. In correcting for antenna-based errors the diffractive process dominates because it has larger variance and smaller scale so it varies more rapidly. Consequently the appropriate $\theta_{\rm iso}$ becomes s_o/L, which becomes rapidly smaller than r_F/L as the frequency decreases. In addition the pattern becomes narrow band in nature, with a decorrelation bandwidth decreasing as the inverse fourth power of frequency (or faster). The conditions for resolving these small scale narrow-band fluctuations are just those needed to estimate the antenna-based multipliers and correct for the scattering:

(a) the integration time must be short enough $T_{\rm int} \lesssim \min(s_o/V, b/V)$, and

(b) the angular extent of the source-region must be small compared with the "isoplanatic patch" of angular radius s_o/L, and

(c) the receiver bandwidth must be narrow compared to the decorrelation bandwidth, and

(d) antennas must be small compared to s_o

If any of these conditions fails the diffraction pattern cannot be observed and the visibility on each baseline is reduced by the factor $\exp[-0.5 D_\phi(b)]$. Hence for baselines greater than s_o the angular resolution will be limited by scattering to an angular radius $\sim \lambda/(2\pi s_o)$, which is exactly the rms scattering angle.

Cornwell & Narayan (1993) have described a method to circumvent the isoplanatic restriction (b). It relies on the fact that for a screen the diffraction pattern due to a point source is shifted laterally as the point source is moved in angle. The technique requires pairs of parallel and equal baselines (as for example with three antennas at equal spacings in a straight line). The method provides an estimate of the squared magnitude of the source visibility function, so the brightness distribution itself cannot be recovered, only its auto-covariance.

3. LIMITS DUE TO INTERPLANETARY AND INTERSTELLAR SCATTERING

Though a space-based low frequency array will avoid atmospheric and ionospheric scattering, it must still contend with interplanetary scattering and, in most cases, with interstellar scattering. In both media the scintillations will be strong in all directions at the low frequencies of concern to this meeting. The conditions for correcting the strong scattering in the two regimes are given in Table 1. The analysis assumes an equivalent scattering screen. A full calculation for scattering extended along the line of sight cannot be summarized so simply, but it is clear that the conditions for correction become somewhat more stringent than these screen-based limits.

3.1. Interstellar Scattering

Scattering in the ionized ISM is now a well-studied phenomenon. For extragalactic sources viewed normal to the Galactic plane strong scattering applies at all frequencies below 2-5 GHz (see Cordes, this volume, Narayan 1992, Rickett 1990). The angular broadening is extremely patchy, but has a minimum angle of scattering normal to the Galactic plane given approximately as an angular diameter in Table 1. The associated field coherence scale is $s_{o,ISS} \sim 21 f_{\rm MHz}^{1.2}$ km. We note, however, that baselines as short as a few hundred kilometers

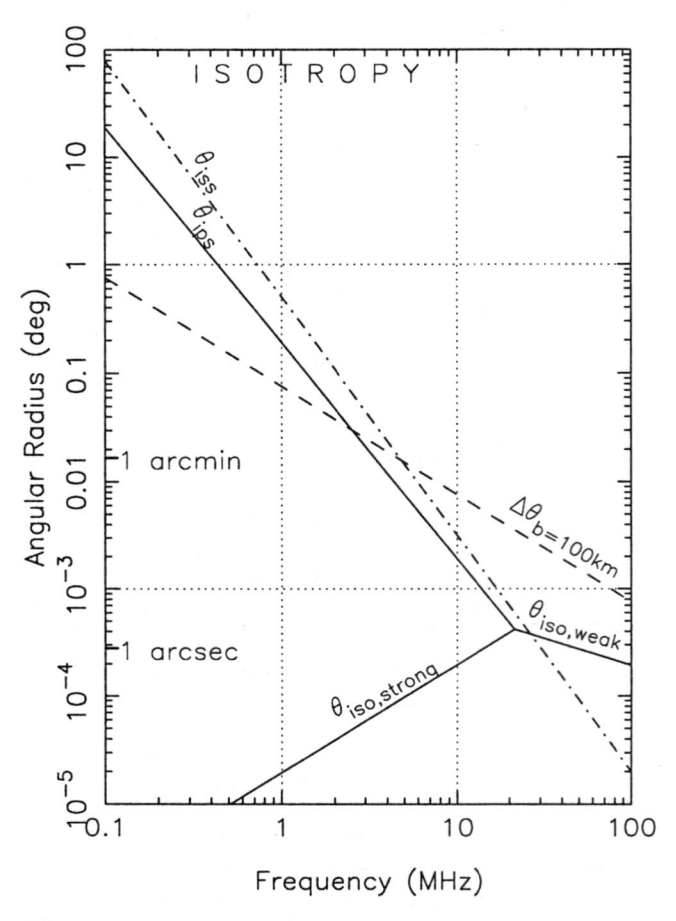

Figure 1. Angular radius versus observing frequency. The diffraction limited resolution of a 100 km baseline is indicated by the dashed line. The solid line shows the angular broadening in the solar wind (ips) of a point source observing at 90 degrees from the Sun. The Condition (b), to remove the effects of such broadening by determining the antenna based phase and amplitude, requires the angular radius of the waves incident on the IPM to be smaller than the lower solid curve which is labelled $\theta_{\text{iso,strong}}$. For Galactic and extra-galactic sources, broadening in the ISM will already have increased the effective radius well above this limit indicated by the dash-dot line, for scattering normal to the Galactic plane.

warm ionized medium will exceed unity for long Galactic paths. This will limit the effective scattered path lengths. At low frequencies the decorrelation bandwidth for the diffractive scintillations will be *extremely narrow* (0.2 Hz at 10 MHz). Thus receivers narrow enough to resolve and correct for the interstellar scintillations will have insufficient signal to noise ratio, except for any radio sources within a few parsecs of the Sun. Furthermore, the conditions on the source diameter to avoid smearing the diffractive scintillation pattern require sources of unrealisably high intrinsic brightness. While the technique proposed by Cornwell & Narayan (1993) may be used to overcome condition (b), it too requires condition (c), which as noted leads to an extreme signal to noise problem. Effectively no image correction technique will be possible to overcome the angular broadening in the ISM.

3.2. Interplanetary Scattering

Using the wavenumber spectrum of the solar wind plasma, discussed above, we estimate that there will be strong scintillations in the solar wind at all frequencies below about 25 MHz in all directions relative to the Sun. For observations at $\gtrsim 90°$ from the Sun, the coherence scale is smaller than the 250 km inner scale (at 1AU) for all frequencies below about 70 MHz. Hence we use $D_\phi(b) \sim (b/s_o)^2$ in our modelling. At 90 degree elongation, the corresponding expressions for the spatial scales applicable to diffractive and refractive scintillation are given by

$$s_o \sim 15 \text{ km } f_{\text{MHz}}$$
$$r_F \sim 1500 \text{ km } f_{\text{MHz}}^{-0.5}$$
$$s_r \sim 150,000 \text{ km } f_{\text{MHz}}^{-2}$$

The associated angular scattering radius is given by

$$\theta_{\text{IPS}} \sim 0.2 f_{\text{MHz}}^{-2} \text{ degrees.}$$

Note that for an arbitrary elongation ϵ, the angle of scattering should be multiplied by

$$(\sin \epsilon)^{-1.5} \sqrt{2(\pi - \epsilon)/\pi + \sin 2\epsilon/\pi}.$$

At 180° it would be about 0.6 times smaller and at 30° 5 times bigger.

Figure 1 shows angular radii plotted versus frequency. θ_{IPS} is shown for 90° elongation. It meets the isoplanatic angle limits in weak and strong scintillation at 25 MHz. Also shown are the angular resolution for a 100 km baseline typical of proposals for the ALFA array (Jones, this volume) and the LFSA (Weiler et al. 1988)

will probably be less than the inner scale in the ISM. This would somewhat increase $s_{o,ISS}$ and make it linearly proportional to frequency and somewhat lower the angular scattering plotted in Figure 1. Low frequency interferometry would provide constraints on this inner scale, if the interplanetary scattering could be removed.

At low Galactic latitude θ_{ISS} can be 10-1000 times greater; thus the point spread function can vary greatly across the sky. One factor that will reduce the effect at very low frequencies is that the optical depth of the

Table 1. Conditions for Image Correction in Strong Scintillation

Condition	IPM	ISM
a - Integration time	$\lesssim 0.04\, f_{\text{MHz}}$ secs	$\lesssim 0.7\, f_{\text{MHz}}^{1.2}$ secs
b - Source Diameter	$\lesssim 4.6 \times 10^{-5}\, f_{\text{MHz}}$ deg.	$\lesssim 3.0 \times 10^{-9}\, f_{\text{MHz}}^{1.2}$ arcsec
c - Bandwidth	$\lesssim 2.0 \times 10^{-4}\, f_{\text{MHz}}^{4}$ MHz	$\lesssim 6.0 \times 10^{-6}\, f_{\text{MHz}}^{4.4}$ Hz
Scattering angle	$\theta_{\text{IPS}} \sim 0.2 f_{\text{MHz}}^{-2}$ deg	$\theta_{\text{ISS}} \sim 0.5 f_{\text{MHz}}^{-2.2}$ deg

and also the (irreducible) scattering angle imposed by the ISM. The plot shows that the full resolution of a 100 km baseline can be obtained for frequencies above about 4 MHz, when observing in the anti-solar hemisphere. In contrast, observations of galactic or extra-galactic sources at lower frequencies will be limited by interplanetary angular scattering, and also by interstellar scattering, to lower angular resolution. This will be impractical to correct, since the angular broadening in the ISM will ensure that condition (b) for the correction of interplanetary scattering will not be valid (since $\theta_{\text{iss}} \gg \theta_{\text{iso,strong}}$).

Evidently, sources nearer than a few parsecs from the Sun (planets for example) will not suffer interstellar scattering and then correction of the interplanetary scattering can be considered. This can, in principle, give diffraction limited resolution below 1 MHz. However the low signal to noise ratio caused by a short integration times to meet conditon (a) will make it difficult except for extremely bright sources. The integration time could be increased by perhaps a factor 5 for observations within 10 degrees of the anti-solar direction, where the transverse solar wind velocity is low. To study extended sources the redundant baseline technique proposed by Cornwell & Narayan (1993) could, in principle, be used but would still be subject to short integration times and the associated low signal-to noise.

A related angular broadening must occur for the 3 kHz radiation detected by Voyagers 1 and 2 at 20-60 AU from the Sun (e. g. Gurnett, Allendorf and Kurth, 1998). The signals apparently originate at or near the termination shock of the solar wind, and as they travel through the inhomogeneous solar wind, they will suffer angular broadening. We have extrapolated our scattering model from 1 AU to the outer heliosphere, including the influence of both the inner scale and a reduced turbulence level away from the solar wind current sheet; we find that the 3 kHz radiation should have been greatly broadened - to cover as much as 1-2 radians. The observed depth of modulation, due to the rotation of the dipole antenna used, is consistent with that expected from such a heavily scattered source (Armstrong, Coles and Rickett, 2000). This illustrates the importance of propagation in low-frequency observations.

The solar wind is also highly variable and far from spherically symmetric. There are often transient increases in the scattering that will exceed these estimates by factors ten or more, making it impossible to obtain the full resolution of 100 km baselines even at 20 MHz.

Observations of the Sun are a special case, evidently free from interstellar scattering, but emission from active regions occurs deep inside the strong scattering of the corona. Bastian (1994) has reviewed such effects in the context of the smooth appearance of solar features when mapped at high resolution at cm wavelengths. Such coronal broadening also scales as the square of the wavelength, masking the shape of the emitting regions.

One could imagine observing intensity scintillations (IPS) at low frequencies to estimate angular diameters, in the same way that it was used to discover arcsecond sized sources at 80 MHz (Readhead and Hewish, 1974). A source represented by a point in Figure 1 (angular radius θ observed at some frequency f) which lies below the line marked $\theta_{\text{iso,strong}}$ should show rapid (faster than 1 second) diffractive IPS. Such scintillations would probe sources much smaller than the baseline resolution limit. In a similar fashion sources below the line marked θ_{ips} should show refractive scintillations in the solar wind on time scales of 10 seconds to minutes (increasing as λ^2). Unfortunately, except for nearby sources, the interstellar scattering will increase the angular radius to θ_{iss} which will quench the diffractive IPS. A measurement of refractive IPS may well be possible (e. g. Narayan et al. , 1989), but will give an effective angular resolution of about θ_{ips}, which is the limit on interferometric resolution below about 3 MHz, anyway. One can also see that above about 25 MHz weak IPS is not quenched by interstellar broadening.

We can also consider whether space-based interferometry has the potential to improve our knowledge of the density structures in the interstellar plasma and the solar wind. From the reduction in visibility one obtains information on the scattering medium at scales comparable to the baseline; it would be interesting to

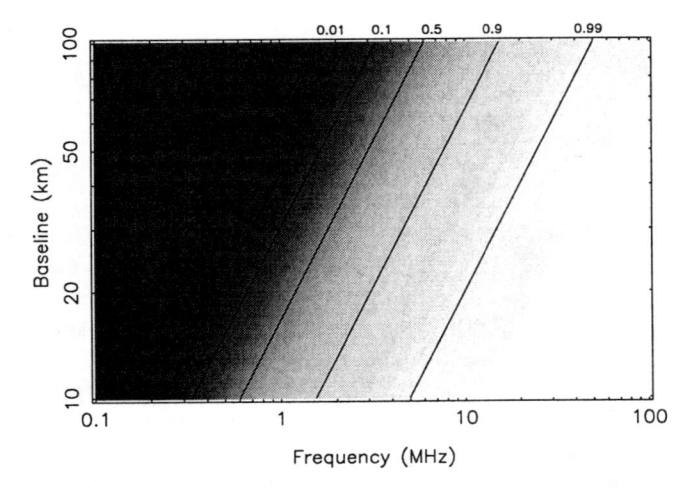

Figure 2. A graphical representation of the loss of visibility due to scattering in the IPM for observations at about 90° from the Sun. Contours of visibility amplitude are shown at 0.99, 0.9, 0.5, 0.1 and 0.01

confirm that the inner scale is indeed larger than 100 km as presently believed. However, for waves travelling through both media it will be difficult to distinguish their respective effects, making interpretation difficult. The measurement of apparent flux variations due to refractive interplanetary scintillations, on times scales of seconds to minutes, will probe scales larger by one or two orders of magnitude and so could add to the rather sparse new information from coherent interferometry.

An array of antennas in the solar wind will clearly require precise information on their relative locations. If this is done using a microwave telemetry signal along each baseline, it would be relatively easy to add a phase coherent low frequency transmission between the antennas, to measure the electron column density on each path. The spatial and temporal variations of such density measurements would provide a very valuable way to probe the small scale structure in the solar wind plasma, which was not subject to averaging over the much longer paths for incoming waves from astronomical sources. Such density measurements could be made at a much higher rate than from those made using plasma analyzers as flown on recent spacecraft, and so could probe the inner scale in the density spectrum.

4. SUMMARY

We conclude that low frequency observations will be in the regime of strong scintillations for both the IPM and the ISM. This means that image correction techniques which rely on determining antenna-based phase and amplitude errors will be impractical. In the absence of such corrections scattering in the solar wind will reduce visibility estimates on a baseline b by a factor:

$$V(b) = \exp[-0.5(b/s_o)^2] \sim \exp[-(b_{km}/20 f_{MHz})^2]$$

This is a very approximate expression for observing at 90 degrees from the Sun and is displayed in Figure 2. Only a slight improvement would come from observing in the anti-solar direction; the reduction worsens quite rapidly toward the Sun and at times of disturbances in the solar wind. Scattering in the ISM for sources beyond 100 pc will cause a reduction by a similar factor, and is worse for longer path lengths through the ionized ISM.

It is clear that, while interferometry from space with baselines in the range of 100 km will provide important new information on the low frequency sky, there is little to be gained from extending the baselines to 1000 km. Though measurements of the scattering itself will be hard to interpret in terms of the solar wind plasma, it appears that the addition of electron column density measurements between each antenna pair could provide a valuable new tool for studying the structure of the plasma on very small scales.

Acknowledgment. Our solar wind research is supported by the NSF under grant ATM-9627228.

REFERENCES

Armstrong, J. W., Coles, W. A., Kojima, M. and Rickett, B. J., Observations of field-algned density fluctuations in the solar wind, *Astrophys. J.*, , *358*, 685-692, 1990.

Armstrong, J. W., Coles, W. A. and Rickett, B. J., Radiowave scattering in the outer heliosphere, *J. Geophys. Res.*, , in press 2000.

Bastian, T. S., Angular scattering of solar radio emission by coronal turbulence, *Astrophys. J.*, , *426*, 774-781, 1994.

Coles, W. A., Frehlich, R. G., Rickett, B. J. and Codona, J. L., Refractive Scintillation in the interstellar Medium, *Astrophys. J.*, , *315*, 666-674, 1987.

Coles, W. A. and Harmon, J. K., Propagation observations of the solar wind near the sun, *Astrophys. J.*, , *337*, 1023, 1989.

Coles , W. A., Liu, W., Harmon, J. K. and Martin, C. L., The solar wind density spectrum near the sun: results from voyager radio measurements, *J. Geophys. Res.*, , *96*, 1745-1755, 1991.

Coles , W. A., Grall, R. R., Klinglesmith, M. T. and Bourgois, G., Solar cycle changes in the level of compressive microturbulence near the sun, *J. Geophys. Res.*, , *100*, 17069-17079, 1995.

Cornwell, T. J., *Astron. Astrophys.*, *180*, 269-274, 1987.

Cornwell, T. J. and Narayan, R., Imaging with ultra resolution in the presence of strong scattering, *Astrophys. J.*, , *408*, L69-72, 1993.

Erickson, W. C., The radio-wave scattering properties of the solar corona, *Astrophys. J.*, , *139*, 1290-1311, 1964.

Grall, R. R., Coles, W. A., Spangler, S. R., Sakurai, T. and Harmon, J. K., Observations of filed-aligned density microstructure near the sun, *J. Geophys. Res.*, , *102*, 263-273, 1997.

Gurnett, D. A., Allendorf, S. C. and Kurth, W. S., Direction-finding measurements of heliospheric 2-3 kHz radio emissions, *Geophys. Res. Lett.*, , *25*, 4433-4436, 1998.

Hewish, A., The scattering of radio waves in the solar corona, *Mon. Not. R. Astron. Soc.*, , *6*, 534-546, 1958.

Narayan, R., The physics od pulsar scintillation, *Phil. Trans. Roy. Soc. A., 341*, 151-165, 1992.

Narayan, R., Anantharamaiah, K. R. and Cornwell, T. J. Refractive radio scintillation in the solar wind, *Mon. Not. R. Astron. Soc.*, , *241*, 403-413, 1989.

Pearson, T. J. and Readhead, A. C. S., Image formation by self-calibration in radio astronomy, *Annu. Rev. Astron. Astrophys., 22*, 97-130, 1984.

Rickett, B. J., Radio propagation through the turbulent interstellar plasma, *Annu. Rev. Astron. Astrophys., 28*, 561-605, 1990.

Readhead, A. C. S., Fine structure in radio sources at 81.5 MHz III: the survey, *Mon. Not. R. Astron. Soc.*, , *78*, 1-49, 1974.

Tu, C.-Y. and Marsch, E., MHD structures, waves and turbulence in the solar wind: observations and theory, *Space Science Reviews, 73*, 1-210, 1995.

Weiler, K. W., Dennison, B. K., Johnston, K. J., Simon, R. S., Erickson, W. C., Kaiser, M. L., Cane, H. V., Desch, M. D. and Hammarstrom, L. M., A low frequency radio array for space, *Astron. Astrophys., 195*, 372-379, 1988.

Woo, R. and Armstrong, J. W., Spacecraft radio scattering observations of the power spectrum of electron density fluctuations in the solar wind, *J. Geophys. Res.*, , *84*, 7288-7296, 1979.

B. J. Rickett and W. A. Coles, Department of Electrical and Computer Engineering, University of California at San Diego, La Jolla, CA 92093-0407. (e-mail: bjrickett@ucsd.edu; bcoles@ucsd.edu)

Interstellar Scattering: Radio Sensing of Deep Space Through the Turbulent Interstellar Medium

James M. Cordes

Department of Astronomy, Cornell University, Ithaca, New York

Diffraction and refraction of radio waves by irregularities in the interstellar electron density produce a wide range of phenomena that allow inferences of the wavenumber spectrum of the irregularities, anisotropies of the fluctuations, the strength of the variations as a function of location in the Galaxy, and characteristic velocities. I summarize the empirical constraints on these aspects of density microstructure on length scales from about 100 km up to ~ 1 pc. As an astronomical "seeing" phenomenon, interstellar scattering is a hindrence to discerning spatial and temporal structure in radio sources. However, analogous to optical and infrared speckle techniques, appropriate analyses allow much to be inferred about radio sources themselves, including pulsars, gamma-ray burst counterparts, and active galactic nuclei. I review the importance of seeing effects that are relevant to high-angular resolution radio observations, such as those that might be made from low-frequency ($\lesssim 100$ MHz) space interferometers.

INTRODUCTION

In this paper I discuss turbulence in the diffuse, ionized component (DIG) of the interstellar medium (ISM). The DIG has been probed on scales that are phenomenally small, by astronomical standards ($\gtrsim 100$ km), using innovative radio astronomical techniques. I discuss the propagation effects underlying those techniques and summarize the current state of knowledge about turbulence in the DIG. The information gleaned includes the galactic distribution of the free-electron density, n_e, and its fluctuations; the wavenumber spectrum of δn_e, including the level, shape and extent in wavenumber; and the conclusion that, in addition to the Kolmogorov-like fluctuations that pervade the ISM, there are additional structures on ~ 1 AU scales which appear to be independent of the Kolmogorov-like fluctuations.

The turbulent ISM distorts images, pulses and spectra of celestial sources viewed through it. Distortions scale strongly with radio frequency, so design of high-resolution telescopes can be optimized with respect to frequency.

RADIO-WAVE PROPAGATION EFFECTS

Radio waves are strongly influenced by the free electron density along the line of sight. Dispersive and scattering effects are determined by the cold-plasma refractive index, $n_r(\nu) = \left(1 - \nu_p^2/\nu^2\right)^{1/2}$. For the interstellar medium (ISM), the plasma frequency $\nu_p \approx 2\,\mathrm{kHz}\,(n_e/0.03\,\mathrm{cm}^{-3})^{1/2}$. Magnetic fields introduce birefringence that is most easily detected as Faraday rotation. Electron-ion scattering is responsible for free-free absorption. Overall, many effects may be used to diagnose n_e and its fluctuations, δn_e, with various scaling laws in frequency $\nu = c\lambda^{-1}$:

- Dispersive arrival times of pulsar pulses: arrival times $\Delta t(\nu) \propto \nu^{-2}\mathrm{DM}$, where the *dispersion measure* is the integral $\mathrm{DM} = \int_0^D ds\, n_e(s)$.

- Faraday rotation: the plane of polarization rotates as $\lambda^2 \mathrm{RM}$ where the *rotation measure* is $\mathrm{RM} = \int_0^D ds\, n_e(s) B_\parallel(s)$.

- Scattering: underlies radio-wave 'seeing' (angular broadening), pulse broadening, DM variations, and intensity scintillations of various types. All such observations yield estimates of the *scattering measure* $\mathrm{SM} = \int_0^D ds\, C_n^2(s)$, where C_n^2 is the spectral coefficient of the wavenumber spectrum of δn_e:

$$P_{\delta n_e}(q) = C_n^2 q^{-\alpha}, \quad \frac{2\pi}{\ell_0} \leq q \leq \frac{2\pi}{\ell_1}, \quad (1)$$

where ℓ_1, ℓ_0 are the inner and outer scales of the turbulence and the spectrum vanishes outside the specified interval. Eq. 1 explicitly assumes isotropic irregularities. Evidence for anisotropies exists but it is not clear on what length scale these are present (see below).

- Free-free absorption is described by the *emission measure* $\mathrm{EM} = \int_0^D ds\, n_e^2$ and the temperature of ionized gas.

If we assume that scattering, dispersing and absorbing gas is distributed in clouds with filling factor f, internal density, $\overline{n}_e$, and with an outer scale, ℓ_0, we have

$$\mathrm{EM} = \zeta \overline{n}_e \mathrm{DM} + C_{\mathrm{SM}}^{-1} \ell_0^{2/3} \mathrm{SM} \quad (2)$$

$$\frac{\mathrm{SM}}{\mathrm{DM}} = C_{\mathrm{SM}} \langle n_e \rangle_{\mathrm{l.o.s.}} \left[\frac{\zeta \epsilon^2}{f \ell_0^{2/3}}\right], \quad (3)$$

where $\zeta \sim 1$, $\epsilon \equiv \sigma_{n_e}/n_e$, the line of sight average electron density $\langle n_e \rangle_{\mathrm{l.o.s.}} \sim 0.025$ cm^{-3}, $C_{\mathrm{SM}} \equiv (\alpha-3)[2(2\pi)^{4-\alpha}]^{-1}$, and we assume a Kolmogorov spectrum with $\alpha = 11/3$. Eq. 2 states that EM has a contribution from both the mean and fluctuating n_e inside clouds and that the "SM" term from fluctuations is no larger than the value of the first, DM, term. Eq. 3 implies that, if all clouds are the same, and clouds are uniformly distributed, then SM/DM = constant (cf. Cordes *et al.* 1991). As discussed later, we find that the marked variations in SM/DM can be attributed to strong variations in the bracketed quantity in Eq. 3 between different regions in the Galaxy.

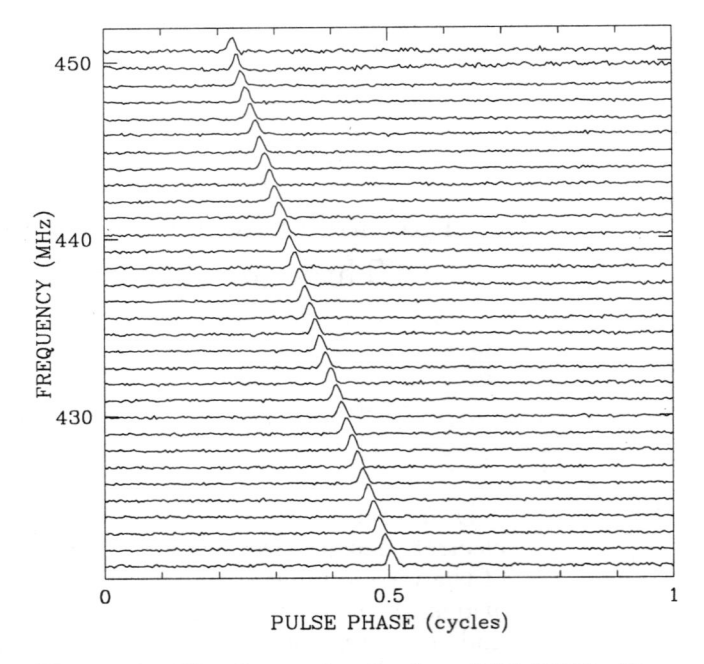

Figure 1. The dispersed pulse from PSR B1749−28 as seen in 32 separate frequency channels, illustrating dispersive propagation. The locus of arrival times follows a curve, $\Delta t(\nu) \propto \nu^{-2}\mathrm{DM}$, where $\mathrm{DM} = 50.9$ pc cm^{-3}.

OBSERVED PHENOMENA

The phase perturbation, $\phi = -\lambda r_e \int_0^D ds\, \delta n_e(s)$, underlies propagation phenomena, where r_e is the classical electron radius and we integrate along an unperturbed ray path, a procedure justified by the small scattering and refraction angles.

Pulsar Dispersion Measures and their Variability

Pulsar dispersion measures are a routine byproduct of radio pulsar surveys and are known for about 1000 objects, a number that will double in just a few years in ongoing surveys. Figure 1 shows the dispersed pulse from a pulsar. Independent distance estimates of ~ 70 pulsars allow models for the galactic distribution of n_e to be calibrated (e.g. Taylor & Cordes 1993; hereafter TC93). Dispersion measures of pulsars show time variations that sample irregularities in the ISM on length scales $V_{p\perp} T \sim 23\, \mathrm{AU}\, V_{400} T_{100}$, in dual-frequency pulse-timing programs on time scales of 100 days and for typical transverse speeds ~ 400 km s^{-1}. The structure function of temporal DM variations and the phase structure function $D_\phi(b)$ are related (Rickett 1990; Cordes *et al.* 1990; Kaspi *et al.* 1994),

$$D_{\rm DM}(\tau) \equiv \langle [{\rm DM}(t+\tau) - {\rm DM}(t)]^2 \rangle$$
$$= (\lambda r_e)^{-2} D_\phi(V_{p_\perp}\tau), \qquad (4)$$

where the phase structure function is the mean-square difference of the phase at two locations separated by transverse distance **b**:

$$D_\phi({\bf b}) \equiv \left\langle [\phi({\bf x}+{\bf b}) - \phi({\bf x})]^2 \right\rangle. \qquad (5)$$

Use of the structure function to constrain the wavenumber spectrum is discussed below. An alternative interpretation is given by Backer et al. (1993).

Refraction Effects (RISS)

Refraction effects are associated with larger irregularities in δn_e than those responsible for diffraction. They include slow intensity variations from focusing and defocusing of wavefronts; arrival-time variations of pulsars; occasional caustic and multiple-imaging events; wander of source images. Their mere existence indicates that a wide range of length scales exists—those that cause diffraction as well as refraction—and, quantita-

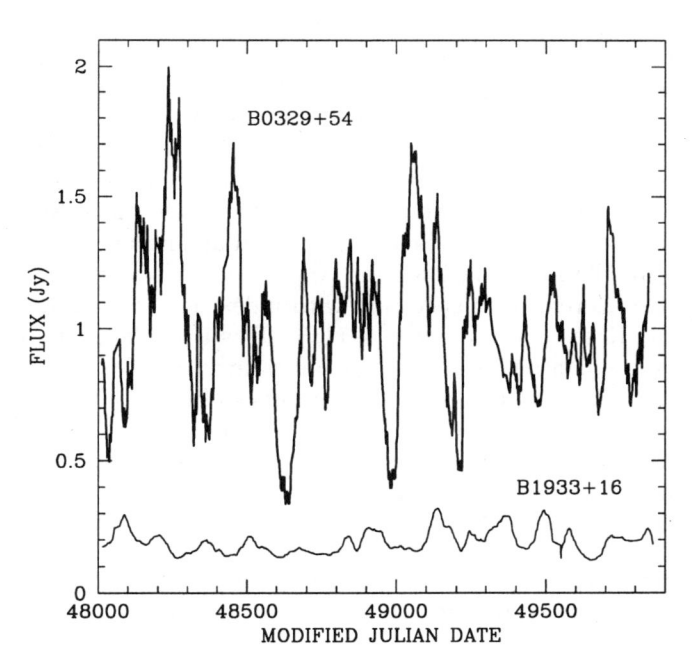

Figure 2. Refractive interstellar scintillation of two pulsars at 610 MHz (Stinebring et al. 1998). The intrinsic flux density of these pulsars is essentially constant (when averaged over a pulse period). The fractional variation for B0329+54 is larger than it is for B1933+16 because the refraction time scale ℓ_r/V is smaller than the plotted time interval for the former but much larger for the latter. Therefore, the refractive modulation for B1933+16 is essentially constant over the plot.

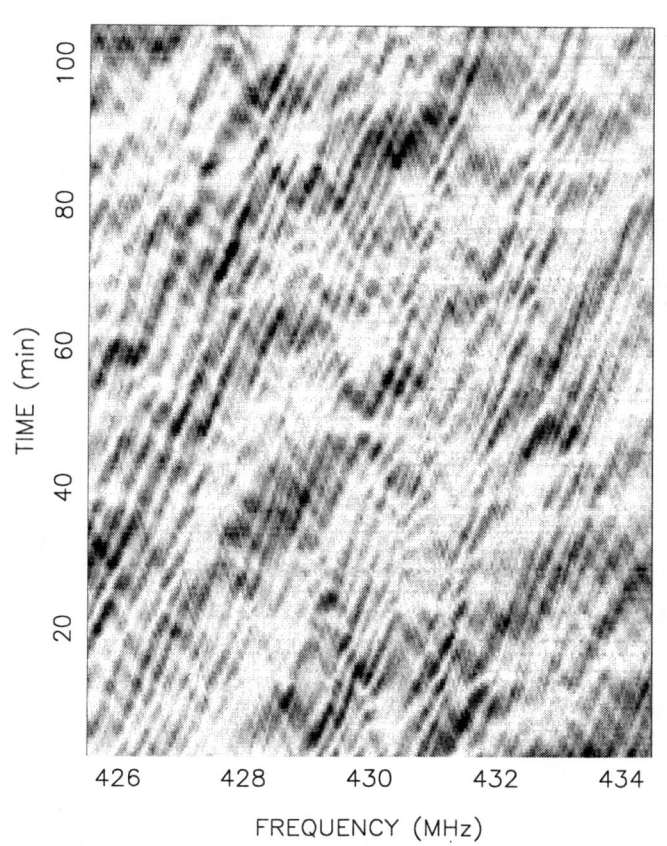

Figure 3. Dynamic spectrum $I(\nu,t)$ for PSR B1133+16 that shows constructive and destructive interference from multipath propagation.

tively, one may estimate the slope of the wavenumber spectrum for δn_e as well as its level in narrow wavenumber bands. Figure 2 shows RISS from two pulsars.

Diffractive Scattering Effects (DISS)

Fluctuations δn_e typically induce wavefront phase variations $\phi_{\rm rms} \gg 1$ rad on a transverse scale equal to the Fresnel scale $\ell_F \sim \sqrt{\lambda D} \sim 10^{11}$ cm for radio frequencies $\nu \lesssim 1$ GHz. The resultant scattering or diffraction causes angular broadening (radio "seeing"), pulse broadening, intensity scintillations vs. t, ν, and other effects. Figure 3 shows intensity scintillations of a radio pulsar plotted as a grey-scale against time and frequency. Figure 4 shows pulse broadening for a pulsar that is about 8 kpc away. The broadening at 1.4 GHz is the largest known to date. Figure 5 shows the visibility functions for two OH masers in the direc-

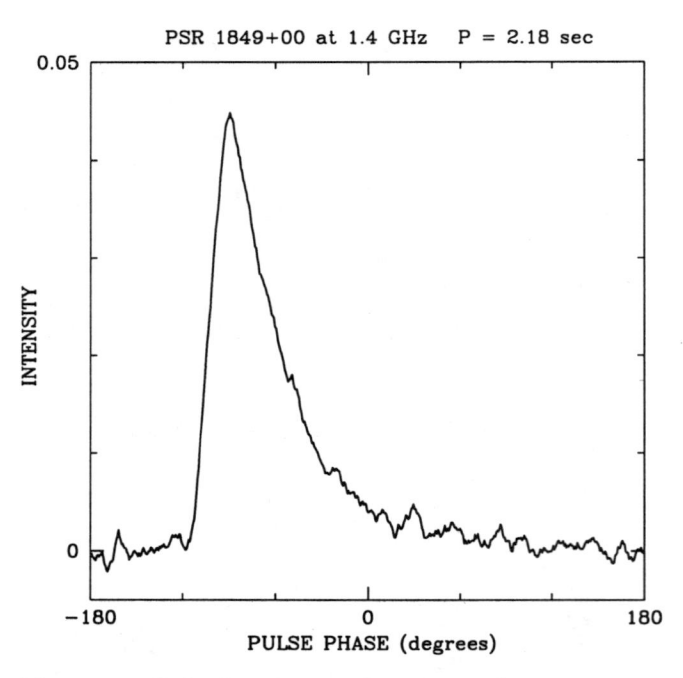

Figure 4. Pulse broadening of a pulsar pulse.

tion of the Galactic center, one of which is unscattered, the other heavily scattered, producing a Gaussian-like image and corresponding Gaussian-like visibility function. Figure 6 summarizes the pulse broadening (referenced to 1 GHz) of a large number of pulsars plotted against DM. The points rise more quickly with DM than expected for a medium with homogeneous turbulence statistics (e.g. Cordes et al. 1991), signifying that lines of sight toward the inner Galaxy intersect denser regions that also must have a higher degree of density fluctuations.

Diffraction effects yield estimates of the scattering measure and, less directly, C_n^2, on length scales $\ell_d \approx \lambda/\theta_d$, where θ_d is the scattering disk size. Thus ℓ_d ranges from 10^2 to 10^5 km for sources observed to date. For extragalactic sources, a seeing disk size $\theta_{\rm FWHM}$ yields SM = $(\theta_{\rm FWHM}/128 \text{ mas})^{5/3} \nu^{11/3}$ (for ν in GHz). Galactic sources require a larger SM to produce a given $\theta_{\rm FWHM}$ owing to wavefront sphericity. For pulsars, the pulse broadening time, $\tau_d \propto D\theta_d^2/2c$, yields SM = $292 \left(\tau_d/D\right)^{5/6} \nu^{11/3}$. The ISS bandwidth, $\Delta\nu_d$, is the characteristic width of diffraction features in dynamic spectra (cf. Figure 3). The 'uncertainty' relation $2\pi\tau_d\Delta\nu_d \approx C_1$ holds, where $C_1 \sim 1$ depends on both the wavenumber spectrum and the large-scale spatial distribution of C_n^2 along the line of sight (Cordes & Rickett 1998).

WAVENUMBER SPECTRUM FOR δn_e

Figure 7 shows the wavenumber spectrum for δn_e, with estimates for C_n^2 from many different kinds of measurements. Wavenumbers $q \gtrsim 10^{-7}$ m^{-1} are sampled by diffraction effects, while wavenumbers $10^{-12} \lesssim q \lesssim 10^{-8}$ are sampled by refractive scintillations and variations in DM. Variations in rotation measure, RM, are measured on scales as small as ~ 0.01 pc ($q \sim 10^{-16}$ m^{-1}). Some measurements yield only an estimate of $P_{\delta n_e}(q)$ in a relatively narrow band (e.g. ~ 1 octave) of wavenumbers. Others, through scaling laws in wavelength or interferometry baseline yield nearly direct determinations of the spectral index, α. Fluctuations of measureables sometimes yield important constraints on α, because steep spectra can yield greater variations than shallow spectra.

The totality of measurements suggests that *portions* of $P_{\delta n_e}$ are *approximately* Kolmogorov in form, $P_{\delta n_e}(q) \propto q^{-11/3}$, at least for *some* lines of sight.[1] However, *strong departures* from Kolmogorov are present along some lines of sight. The evidence is detailed below. Suffice it to say here, however, that the composite spectrum in Figure 7 (and in Armstrong et al. 1995) is deceptive in showing apparent consistency with a power law of *any* slope. First, C_n^2 itself varies by nearly 10 orders of magnitude between different locations. Second, large, factor-of-ten bumps in the spectrum (for the same line of sight) are masked by the large dynamic range plotted. Third, there are gaps in the spectrum where we have no knowledge about the spectral amplitude.

In the following I will use the phrase 'Kolmogorov-like' to refer to a spectrum with index $\alpha <$, including the Kolmogorov value $\alpha = 11/3$. This is for linguistic ease and also to acknowledge that estimates of the index are not exact and vary between lines of sight. It also attaches significance to the fact that the index is often not found to be precisely $\alpha = 4$, a value that could signify dominance of fluctuations by shocks rather than by any cascade process (Rickett, private communication).

Evidence for a Kolmogorov-like Spectrum and its Wavenumber Extent

The best evidence for a Kolmogorov-like spectrum along some lines of sight includes:

---

[1] Of course the original Kolmogorov spectrum applied to an incompressible, neutral fluid. Density fluctuations could be a passive tracer of such turbulence: on the small scales where we actually have measurements, e.g. $\lesssim 1$ AU, the fractional fluctuation satisfies $\delta n_e/n_e \ll 1$.

- *Scaling of ISS bandwidth and pulse broadening time with ν*: For $\alpha < 4$, $\Delta\nu_d \propto \nu^{2\alpha/(\alpha-2)}$. Results on five pulsars yield $\langle\alpha\rangle = 3.63 \pm 0.2$. Note that such objects show only moderate scattering, owing to the selection effect that $\Delta\nu_d$ must be in a measureable range with typical radio astronomy bandwidths and its reciprocal, τ_d, must be small enough that pulsations from a pulsar are not quenched.

- *Angular broadening vs. ν*: $\theta_d \propto \nu^{-\alpha/(\alpha-2)}$ for $\alpha < 4$ and if all interferometer baselines used to map a source satisfy $b \gg \ell_1$; for $b \lesssim \ell_1$ (for all baselines used), $\theta_d \propto \nu^{-2}$ (Cordes & Lazio 1991). Achieving large dynamic range in ν is difficult using available baselines of terrestrial interferometers. Most heavily scattered sources are consistent with $\alpha < 4$ but tend toward a value nearer to 4 than to the Kolmogorov value, 11/3. This may signify that the measurements on heavily scattered sources are influenced by an inner scale. For Cyg X-3 (Molnar *et al.* 1995; Wilkinson *et al.* 1994), a fit to the visibility function's shape is consistent with $\alpha = 11/3$.

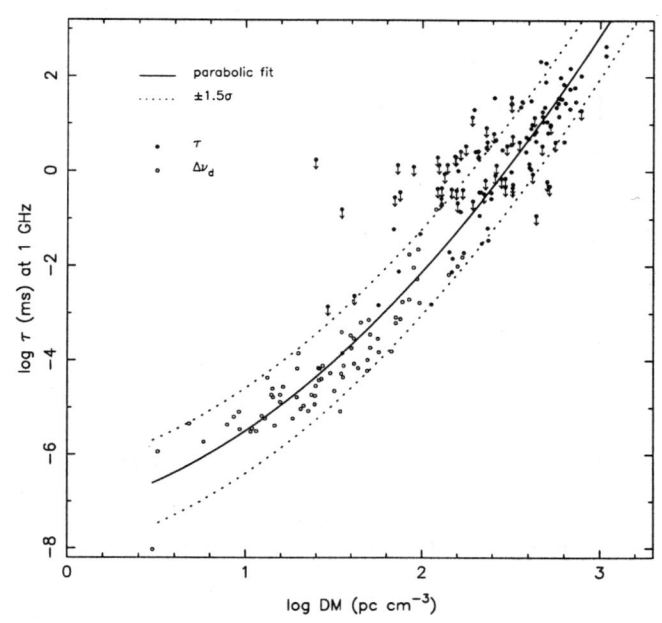

Figure 6. Pulse broadening time vs. DM for 206 pulsars. The plotted lines are the best fit parabola to log τ-log DM and $\pm 1.5\sigma$.

- *Variations of pulsar DM*: For the line of sight to pulsar B1937+21, the DM structure function indicates $\alpha = 3.87 \pm 0.01$ (Kaspi *et al.* 1994). While formally larger than 11/3, the arrival-time analysis ignores other contributions that can bias the deduced α.

- *Refractive ISS from Pulsars*: The fractional modulation due to RISS depends on α and ℓ_1 even when DISS is saturated and independent of these quantities.

The ranges in wavenumber that pertain to these spectral-index determinations are determinated by the 'optics' of the radio propagation and the data time spans. The diffraction scale, ℓ_d (defined earlier), is the smallest scale while the physical size of the scattering disk (at its largest point along the line of sight) is called the 'refraction' or multipath scale, $\ell_r \sim D\theta_d$. Roughly, $\ell_d \ell_r \sim \ell_F^2$. Measurements represent an integral combination of wavenumbers that encompass these scales. So, consistency with a Kolmogorov spectrum suggests that the spectrum extends over the corresponding range of wavenumbers. Observations of the most heavily scattered objects (e.g. the most distant or at the lowest ν) yield the largest ratio, ℓ_r/ℓ_d. These suggest that the relevant scales are 100 km $\leq \ell \leq 10^{10}$ km, or 10^{-12} m$^{-1} \leq q \leq 10^{-6}$ m^{-1}. More direct con-

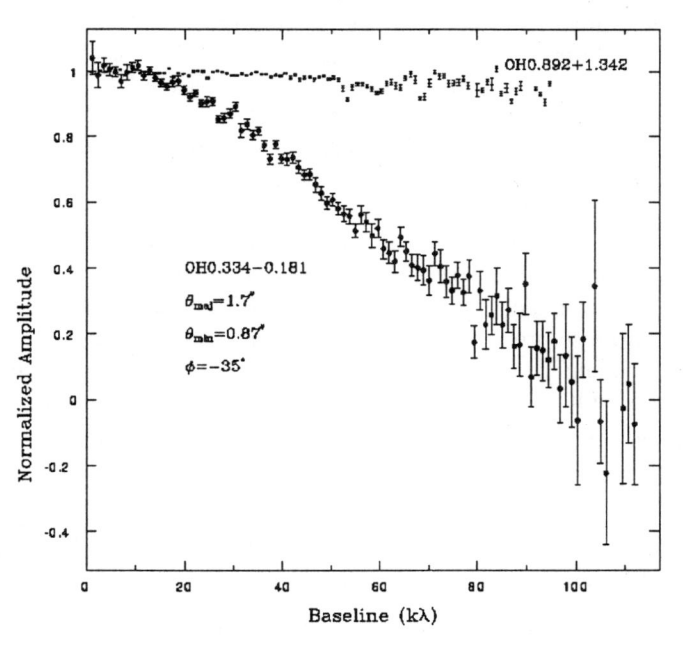

Figure 5. Angular broadening as shown in the visibility function $\Gamma(b)$ of an OH maser near the galactic center (declining curve). The flat curve is an unscattered maser that looks like a point source on the baselines shown (after Frail *et al.* 1993).

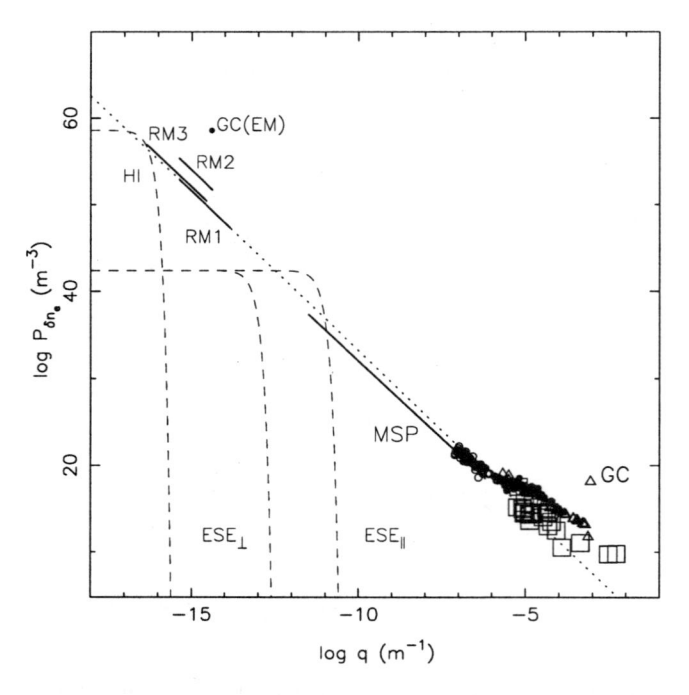

Figure 7. Wavenumber spectrum for δn_e estimated from a variety of observations. The dotted line $\propto q^{-11/3}$, the Kolmogorov spectrum. Plotted symbols for wavenumbers $q \gtrsim 10^{-7}$ m^{-1} are from diffraction measurements. The point labelled 'GC' takes into account localized, intense scattering to Sgr A*; 'GC(EM)' is an estimate from free-free absorption toward Sgr A*. The line labelled 'MSP' is the fit for DM variations from Cordes et al. (1990), also consistent with Kaspi et al. (1994). ESE$_{\parallel,\perp}$ are estimates for refracting structures that cause 'extreme scattering events' (Fiedler et al. 1987; Romani et al. 1987). 'RM1,2,3' inferences from RM variations (Simonetti & Cordes 1988; Lazio et al. 1990; Minter & Spangler 1996). 'HI' is an estimate for ionized extremities of HI clouds.

straints on the inner scale, ℓ_1, derive from interferometry of heavily scattered sources for which $\ell_d \lesssim \ell_1$. Results indicate 100 km $\lesssim \ell_1 \lesssim 10^5$ km (Spangler & Gwinn 1990; Wilkinson et al. 1994). Once again, it should be noted that the heavily scattered sources on which the interferometer results rest may not be representative probes of the bulk of the ISM.

Departures from a Kolmogorov-like Spectrum

The Kolmogorov spectrum, combined with inferred levels of C_n^2, distances, and wavenumber cutoffs, implies particular relative levels of refraction and diffraction effects. For example, diffractive angular scattering is predicted to be much larger than refractive 'wandering' of radio sources. Also, episodes where the pulsar is multiply imaged with image separations larger than the diffractive seeing disk should be very rare. Similarly, refractive scintillation is expected to be small: $\sigma_{I,RISS}/I \sim 0.1\text{-}0.2$ compared to the 100% modulation of DISS (in the strong scattering regime). Some pulsars indeed are consistent with this picture, at least at some epochs. However, there are very clear episodes of multiple imaging, with image separations much larger than the seeing disk size (Cordes & Wolszczan 1986; Gupta, Rickett & Lyne 1994). These require much more power in $P_{\delta n_e}(q)$ on scales $\gtrsim \ell_r$ than is contained in the Kolmogorov spectrum, at least when normalized by the power needed on scales $\ll \ell_F$.

Possible sources of this 'enhanced' power are (1) a spectrum steeper than $\alpha > 4$; (2) an inner scale that is relatively large, $\ell_1 \sim \ell_F$, thus reducing the relative amount of 'diffractive' power in the spectrum; and (3) 'bumps' in the spectrum on scales $\ell \sim$ 1-10 AU. These possibilities are testable. Case (1) appears to be ruled out because steep spectra cause much larger volatility in observable quantities (flux, angular size and shape, pulse shape) than is seen. Case (2) is inconsistent with detailed studies of RISS compared with DISS (D. R. Stinebring & B. J. Rickett, private communication). Case (3), bumps in the spectrum, seems to be the favored interpretation.

Case (3) receives support from the fact that multiple imaging is episodic, suggesting the presence of discrete structures along the line of sight. Also, discrete events in the light curves of extragalactic sources ('extreme scattering events' or ESEs; Fiedler et al. 1987) and in light curves and $\delta DM(t)$ of a pulsar (Cognard et al. 1993) suggest the existence of discrete plasma structures that may be generically related to those that cause pulsar multiple imaging (Romani et al. 1987). The evidence very strongly suggests that the observables result from a combination of Kolmogorov-like fluctuations and discrete structures that perhaps are a totally independent process. This superposition picture makes it difficult to deconstruct the extent in wavenumber space and the line-of-sight distributions of the individual components.

Anisotropies in Density Microstructure

Anisotropies in δn_e are suggested by the axial ratios $\sim$ 1.2 : 1 to 4:1 of scattered source images, as measured through aperture synthesis of Cyg X-3 (Wilkinson et al. 1994; Sgr A* (Lo et al. 1993); OH/IR masers in the galactic center (Frail et al. 1993); OH masers in W49N (Desai et al. 1994); and various AGNs (Spangler

Table 1. The Low-frequency Sky

Effect	λ scaling	Typical Values at 10 MHz
ANGULAR BROADENING	λ^2	30 arc sec
PULSE BROADENING	λ^4	10^4 sec
SPECTRAL BROADENING	λ^1	10 Hz
SCINTILLATION TIME SCALE	λ^{-1}	1 sec
SCINTILLATION BANDWIDTH	λ^{-4}	10^{-4} Hz
FREE-FREE OPTICAL DEPTH	λ^2	1
FARADAY ROTATION	λ^2	10^5 rad

[a]Scaling laws of the first 5 entries are approximate and apply when the phase structure function is square-law in form. This case holds when the diffraction length scale is smaller than the inner (or smallest) scale of a distribution of scales.

[b]Actual values can deviate by orders of magnitude from the quoted typical values, depending on direction and path length through the Galactic plane.

& Cordes 1988; Trotter et al. 1998). The simplest interpretation is that irregularities are themselves elongated by the same amount on scales $\sim \ell_d = \lambda/\theta_d$ (though in directions orthogonal to the image elongations). Alternatively, media with striations comparable to the refractive scale, $\ell_r \sim D\theta_d$, could also produce the observed elongations, a suggestion supported by the fact that the axial ratio is wavelength dependent for Cyg X-3 (Wilkinson et al. 1994).

The Outer Scale

Estimates for the outer scale can be made by comparing EM and SM on the same lines of sight while assuming that both terms in Eq. 2 are comparable. For heavily scattered sources (Cyg X-3, NGC6334B and a few pulsars), $\ell_0 \lesssim 0.04$ pc, while nearby pulsars and high-latitude AGNs suggest, for the local interstellar medium, $\ell_0 \sim 200$ pc. Variations in DM between pulsars in globular clusters, due to foreground material in the galactic disk, imply $\ell_0 \gtrsim 0.2$ pc.

CONSTRAINTS ON TURBULENT VELOCITIES

There are not many constraints on *temporal* variations of the microturbulence in δn_e. Improving such constraints is important in order to assess whether the observed δn_e is due to *active* turbulence driven by a velocity field and influenced by large scale magnetic fields; alternatively, the bulk of δn_e might be fossil remnants of dynamic processes or it might have nothing to do with turbulence whatsoever.. What *can* be stated derives from pulsar DISS. The time scale for DISS, Δt_d, is determined by $V_{\rm ISS}$, the effective velocity by which the diffraction pattern convects across the line of sight and/or re-organizes. A frozen pattern exists only if there are no turbulent velocities. For pulsars (the only objects for which DISS has been observed in the strong scattering regime), $V_{\rm ISS}$ is dominated by the large pulsar speeds. The ISS speed can be estimated as $V_{\rm ISS} = \ell_d/\Delta t_d$, but the diffraction scale, ℓ_d, is difficult to measure directly; it may be estimated (model dependent!) from the diffraction bandwidth, $\Delta\nu_d$, and other quantities. With a Kolmogorov medium having uniform statistics along the line of sight, $V_{\rm ISS} = 10^{4.4}$ km s^{-1} $\sqrt{D\Delta\nu_d}/\nu\Delta t_d$, for ν in GHz, Δt_d in s, $\Delta\nu_d$ in MHz and D in kpc (Cordes & Rickett 1998). Nonuniformities in C_n^2 introduce strong leverage effects that weight differently the contributions to $V_{\rm ISS}$ from the pulsar, observer and ISM's velocities. For a uniform medium, negligible observer motion, and only pulsar and random ISM motions, $V_{\rm ISS} \approx \left(V_{p\perp}^2 + 3\sigma_{V_{\rm ism}}^2\right)^{1/2}$. Pulsar speeds, $V_{p\perp}$, (from interferometry or pulse timing) $\sim$ 10-1600 km s^{-1}. For pulsars for which we have both proper-motion velocities ($V_{\rm PM}$) and ISS-speed estimates ($V_{\rm ISS}$), there is a correlation between $V_{\rm ISS}$ and $V_{\rm PM}$, suggesting $\sigma_{V_{\rm ism}} \ll V_{p\perp}$. For the slowest pulsars, in fact, $V_{\rm ISS}$ is found to be only 10 km s^{-1}, indicating that $\sigma_{V_{\rm ism}} \lesssim 6$ km s^{-1}. It is clearly important to estimate turbulent velocities using more detailed and innovative techniques in order to diagnose the energy budget of the microturbulence in δn_e as well as establishing that δn_e involves *bona fide* turbulence.

CONTOURS OF LOG ANGULAR BROADENING
0.10 GHz

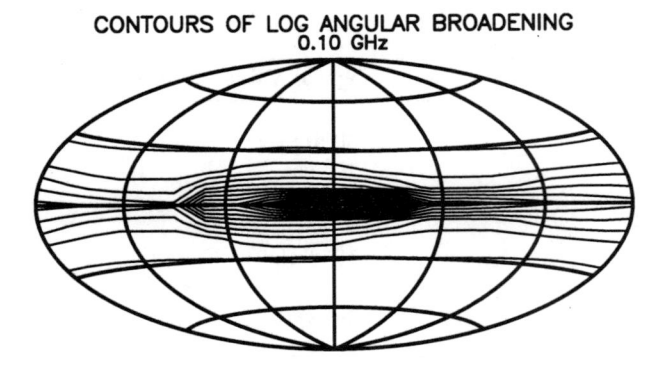

Figure 8. Aitoff projection showing angular broadening at 0.1 GHz as contours in galactic coordinates. (The galactic plane is the equator of this figure, while the galactic poles are at the top and bottom.) The 15 contours range from 0.1 arc sec at about $\pm 30°$ galactic latitude, to 54 arc sec in the galactic plane. The values scale as λ^2.

GALACTIC DISTRIBUTION OF THE ELECTRON DENSITY & ITS FLUCTUATIONS

The galactic distribution of n_e and C_n^2 is determined largely from DMs, independent distance estimates, and ISS observations of pulsars, AGNs, masers, Cyg X-3 and galactic center sources. The most extensive model includes spiral arms, thin and thick disk components and galactocentric radial dependence (TC93). The salient features are:

1. A tenuous component of C_n^2 extends to high z's (scale height $\sim 1/2$ kpc) and to large galactocentric radius (~ 20 kpc; Lazio & Cordes 1998ab).

2. The stronger, thin-disk component has a Population I distribution, implying that C_n^2 is driven by the 'usual suspects' in such populations (stellar winds, supernovae).

3. Especially intense C_n^2 occurs in regions that are rare, including directions toward Cyg X-3 (Wilkinson *et al.* 1994; Molnar *et al.* 1995), NGC6334B (Moran *et al.* 1990; Trotter *et al.* 1998); and especially Sgr A* (Lazio & Cordes 1998cd) and OH/IR stars toward the galactic center (Frail *et al.* 1993).

4. The nature of clouds is quite different in the inner Galaxy compared to the local ISM: the composite quantity $\zeta\epsilon^2/f\ell_0^{2/3}$ (Eq. 3) ~ 0.43 near the Sun but is ~ 22 for pulsars in the inner Galaxy. This suggests that the filling factor (f) or outer scale (ℓ_0) is much smaller there or that the fractional modulation (ϵ) is much larger.

The TC93 model is deficient in how it handles the GC and outer galaxy components and in estimates of the z scale heights. Work by J. Lazio (1997) and the author will lead to improvements in the model.

THE LOW FREQUENCY SKY

The appearance of the sky at frequencies $\lesssim 100$ MHz is strongly affected by ionized plasma between us and sources in the Galaxy and beyond. In Table 1, the scaling laws for various effects are given along with typical values at 10 MHz for a source in the plane of the Galaxy at a few kpc from us. From these values, it is clear that any point source will be angularly broadened by an amount that must be considered in designing any space interferometer. Also, other effects, such as absorption, angular, temporal and spectral broadening, will make sources difficult to even detect by conventional means.

Figure 8 shows contours of angular broadening at 0.1 GHz plotted against Galactic coordinates in an Aitoff projection. The contours are based on the TC93 model. It should be emphasized that the model does not include the effects of discrete HII regions, which might scatter radiation by very large angles. Figure 9 shows contours

CONTOURS OF ANGULAR BROADENING
0.10 GHz

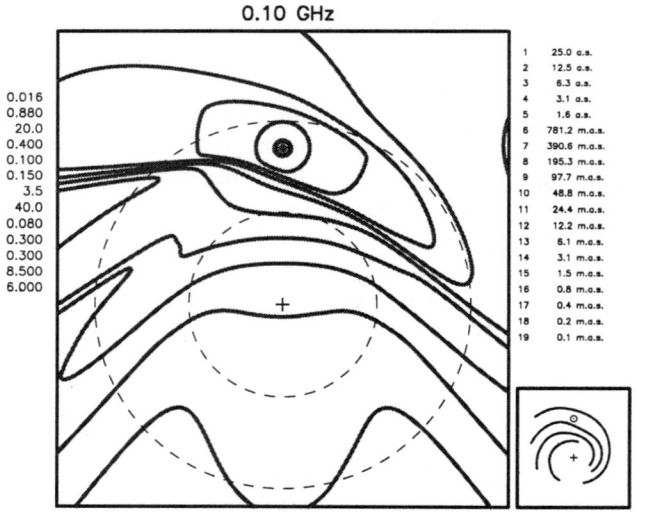

1	25.0 a.s.
2	12.5 a.s.
3	6.3 a.s.
4	3.1 a.s.
5	1.6 a.s.
6	781.2 m.a.s.
7	390.6 m.a.s.
8	195.3 m.a.s.
9	97.7 m.a.s.
10	48.8 m.a.s.
11	24.4 m.a.s.
12	12.2 m.a.s.
13	6.1 m.a.s.
14	3.1 m.a.s.
15	1.5 m.a.s.
16	0.8 m.a.s.
17	0.4 m.a.s.
18	0.2 m.a.s.
19	0.1 m.a.s.

Figure 9. Contours of angular broadening at 0.1 GHz for sources in the Galactic plane. The dashed lines represent distances of 5 and 10 kpc from the Galactic center, which is denoted by the '+' sign. The filled circle is the Sun. The side plot shows the locations of spiral arms used in the TC93 model. The contour furthest from the Sun is 25 arc sec while the circular contour centered on the Sun that just touches the 10 kpc dashed circle is 0.1 arc sec.

Table 2. Low-frequency Interferometry

Frequency	Scattering Diameter	Isoplanatic Scale
Galactic Plane: $b = 0°$, $\ell = 0°$		
10 MHz	3 deg	0.6 km
30	0.3 deg	2.0 km
300	11 arc sec	19 km
Galactic Pole: $b = 90°$		
10 MHz	8 arc sec	800 km
30	0.9 arc sec	2300 km
300	0.009 arc sec	23 Mm

[a] Quoted values are based on the model of Taylor & Cordes 1993.

of angular broadening at 0.1 GHz for sources in the plane of the Galaxy.

Image asymmetries of scattered sources should be very obvious at low frequencies owing to the following curious effect. At low-enough frequencies, scattering angles may exceed the angular size of the scattering region (e.g. an HII region, a spiral arm, or even the entire Galaxy). The image of a background source will then be strongly influenced by the shape of the scattering region as well as by the shapes of the small-scale, diffracting irregularities.

Source Detection Issues

Pulsed emission from ordinary pulsars is rendered undetectable by pulse broadening that exceeds many pulse periods. Sources typically will be free-free absorbed and Faraday rotation likely will depolarize radiation. Free-free absorption toward extragalactic sources exceeds optical depth unity for: (a) all directions at 1 MHz; (b) $|b| \lesssim 5°$, $|\ell| \lesssim 30°$ at 10 MHz; and (c) $|b| \lesssim 1°$, $|\ell| \lesssim 10°$ at 30 MHz.

Incoherent continuum sources (active galactic nuclei and gamma-ray burst sources) will show refractive scintillation, though the time scale will be very long (years). No diffractive scintillation is expected from these sources or even from pulsars because intrinsic source sizes are too large (in the case of AGNs) and also because the characteristic time and frequency scales of DISS will be much smaller than receiver resolutions.

Interferometry Issues

Scattering in the ISM places a lower bound on apparent source sizes that imposes practical limits on baseline lengths. Table 2 gives values for scattering diameters (or "seeing disk" sizes) when looking at extragalactic sources in and perpendicular to the Galactic plane. Also given are the baselines (expressed as isoplanatic scales) needed to resolve scattered sources in these various directions. Scattering at low frequencies defines a clear upper bound on the baselines needed to study the radio sky.

Acknowledgments. I thank Joe Lazio, Barney Rickett, Steve Spangler and Dan Stinebring for useful discussions about most of the topics of this paper and Bill Coles for comments on the manuscript. This work was supported by NSF Grant AST 95-28394 to Cornell University and also by the National Astronomy and Ionosphere Center, which is managed by Cornell University under a cooperative agreement with the NSF.

REFERENCES

Armstrong, J. W., Rickett, B. J. & Spangler, S. R. 1995, Electron Density Power Spectrum in the Local Interstellar Medium, *ApJ*, 443, 209.

Backer, D. C., Hama, S., van Hook, S., Foster, R. S. 1993, Temporal Variations of Pulsar Dispersion Measures, *ApJ*, 404, 636.

Cognard, I. et al. 1993, An Extreme Scattering Event in the Direction of the Millisecond Pulsar 1937+21, *Nature*, 366, 320.

Cordes, J. M. 1990, Timing and Scintillations of the Millisecond Pulsar 1937+214, *ApJ*, 349, 245

Cordes, J.M., Weisberg, J.M., Frail, D. A., Spangler, S. R. & Ryan, M. 1991, The Galactic Distribution of Free Electrons, *Nature*, 354, 121.

Cordes, J. M., Lazio, T. J. W. 1991, Interstellar Scattering Effects on the Detection of Narrow-band Signals, *ApJ*, 475, 557

Cordes, J. M. & Wolszczan, A. 1986, Multiple Imaging of Pulsars by Refraction in the Interstellar Medium, *ApJLetts*, 307, L27.

Cordes, J.M. & Rickett, B. J. 1998 Diffractive Interstellar Scintillation Time Scales & Velocities, *ApJ*, 507, 846.

Desai, K. M., Gwinn, C. R. & Diamond, P. J. 1994, Evidence for Magnetic Field Induced Anisotropy of the Interstellar Medium, *Nature*, 372, 754

Dewey, R. J., Cordes, J. M., Wolszczan, A. & Weisberg, J. M. 1988, Interstellar Scintillations of Binary Pulsars, in Radio Wave Scattering in the Interstellar Medium, AIP Conf. Proc. 174, eds. J. M. Cordes, B. J. Rickett, & D. C. Backer, New York: AIP, 217.

Fiedler, R. L., Dennison, B., Johnston, K. J. & Hewish, A. 1987, Extreme Scattering Events Caused by Compact Structures in the Interstellar Medium, *Nature*, 326, 675.

Frail, D. A. *et al.* 1993, Anisotropic Scattering of OH/IR Stars Toward the Galactic Center, *ApJLetts*, 427, 43.

Gupta, Y., Rickett, B.J. & Lyne, A.G. 1994, Refractive Interstellar Scintillation in Pulsar Dynamic Spectra, *MNRAS*, 269, 1035.

Kaspi, V. M., Taylor, J. H. & Ryba, M. F. 1994, High-Precision Timing of Millisecond Pulsars, *ApJ*, 428, 713.

Lazio, T. J. W. 1997, Genetic Algorithms, Pulsar Planets and Ionized Interstellar Microturbulence, *PhD Thesis, Cornell University*

Lazio, T.J.W.L., Spangler, S. R. & Cordes, J. M. 1990, Faraday Rotation Measure Variations in the Cygnus Region and the Spectrum of Interstellar Plasma Turbulence, *ApJ*, 363, 515.

Lazio, T. J. W. & Cordes, J. M. 1998ab, The Radial Extent and Warp of the Ionized Galactic Disk. I.,II., it ApJSuppl, 115, 225 and *ApJ*, 497, 238.

Lazio, T. J. W. & Cordes, J. M. 1998cd, Hyperstrong Radio-Wave Scattering in the Galactic Center. I., II. it ApJSuppl, 118, 201 and *ApJ*, 505, 715.

Lee, L.C. & Jokipii, J. R. 1975, Strong Scintillations in Astrophysics. I., *ApJ*, 196, 695.

Lo, K. Y. *et al.* 1993 High-resolution VLBA Imaging of the Radio Source Sgr A* at the Galactic Center, *Nature*, 362, 38.

Minter, A. H. & Spangler, S. R 1996, Observation of Turbulent Flucutations in the Interstellar Plasma Density and Magnetic Field on Spatial Scales of 0.01 to 100 Parsecs, *ApJ*, 458, 194.

Molnar, L. A., Mutel, R. L., Reid, M. J. & Johnston, K. 1995, Interstellar Scattering Toward Cygnus X-3: Measurements of Anisotropy and of the Inner Scale, *ApJ*, 438, 708

Moran, J. M., Greene, B., Rodriquez, L. F. & Backer, D. C. 1990, The Large Scattering Disk of NGC6334B, *ApJ*, 348, 147

Rickett, B. J. 1990, Radio Propagation Through the Turbulent Interstellar Plasma, *ARAA*, 28, 561

Rickett, B. J., Coles, W. A. & Bourgois, G. 1984, Slow Scintillation in the Interstellar Medium, *AstAp*, 134, 390

Romani R. W., Blandford, R. D. & Cordes, J. M. 1987, Radio Caustics from Localized Interstellar Medium Plasma Structures, *Nature*, 328, 324

Simonetti, J. H. & Cordes, J. M. 1988, Interstellar Electron Density and Magnetic Field Irregularities on 0.001 Parsec to 100 Parsec Scales, in Radio Wave Scattering in the Interstellar Medium, AIP Conf. Proc. 174, eds. J. M. Cordes, B. J. Rickett, & D. C. Backer, New York: AIP, 134.

Spangler, S. R. & Cordes, J. M. 1988, Interstellar Scattering of the Radio Source 2013+370, *ApJ*, 332, 346.

Spangler, S. R. & Gwinn, C. R. 1990, Evidence for an Inner Scale to the Density Turbulence in the Interstellar Medium, *ApJLetts*, 353, L29

Stinebring, D. R. *et al.* 1998, in preparation.

Taylor, J. H. & Cordes, J. M. 1993, Pulsar Distances and the Galactic Distribution of Free Electrons, *ApJ*, 411, 674.

Trotter, A. S., Moran, J. M. & Rodriguez, L. F., 1998, Anisotropic Radio Scattering of NGC6334B, *ApJ*, 493, 666.

Wilkinson, P.N., Narayan, R. & Spencer, R. E. 1994, The Scatter-Broadened Image of Cygnus X-3, *MNRAS*, 269, 67

J. M. Cordes, Astronomy Department, Cornell University, Ithaca, NY 14853 (email: cordes@spacenet.tn.cornell.edu)

Type III Solar Radio Bursts at Long Wavelengths

George A. Dulk

Departement de Recherche Spatiale, Observatoire de Paris, 92195 Meudon, France

Properties of solar type III bursts discovered or confirmed from space observations at long wavelengths include: 1) The speed of the exciting electrons averages 0.14 c and varies by a factor of about two. 2) Statistically, type III electron streams travel in coronal/solar wind regions of normal density, NOT, in general, in regions of ~ 10 times enhanced density as is often assumed. 3) When a spacecraft is immersed in the beam of fast electrons, the initial radiation recorded is almost always F. In late phases it may be F or H. 4) Two spacecraft, viewing a given type III from different locations, observe different apparent source positions, and do not necessarily see the same F or H component. 5) The low-frequency limit is usually much higher than the plasma frequency at the spacecraft f_{pS}. Sometimes it is just at f_{pS} when $f_{pS} \gtrsim 20$ kHz, and it is almost never lower than about 10 kHz even when $f_{pS} \lesssim 3$ kHz. 6) Observed sources are scatter images of the true sources, lie at considerably greater radial distances than true sources, and are considerably larger, particularly at kilometric wavelengths. 7) Intense, kilometric type III bursts are visible irrespective of their heliographic positions relative to the spacecraft, even when the burst location is directly behind the sun (then it has a flux density ~ 100 lower). 8) Many bursts appear only at decametric and shorter wavelengths, and many appear only at hectometric and longer wavelengths. 9) Type III bursts are the most intense of all radio emissions. Some bursts that continue from decametric to kilometric wavelengths have flux densities up to $S_{\max} \approx 10^{-14}$ W m^{-2} Hz$^{-1} = 10^{12}$ Jy near frequency $f_{\max} \sim 1$ MHz. Values of $f_{\max}$ vary by a factor $\gtrsim 3$. 10) The radio flux density S is related to the number N and energy E of the causative fast electrons approximately as $S \propto N \times E^4$. 11) Flare-initiated, type II-emitting shock waves travelling through the corona often accelerate electrons to ~ 0.1 c, and these "shock-accelerated type III bursts" continue outward to $\gtrsim 1$ AU.

1. INTRODUCTION

In his discovery paper, *Wild* [1950] proposed the general picture for solar type III bursts that remains valid today. The bursts are generated by beams of energetic electrons accelerated near the Sun's surface and travel outward along open magnetic field lines through the corona and the interplanetary space. Along their path they generate radio emission at the plasma frequency $f = f_p$, the second harmonic, $f = 2f_p$, or both. Since their discovery, type III bursts have been observed intensively with ground-based and space-based instruments, and in situ electrons and Langmuir waves have been directly associated with type III emission [*Lin*, 1970; *Gurnett and Anderson*, 1976]. In particular, *Lin* et al. [1981] and *Ergun et al.* [1998] showed that the presence of Langmuir waves observed in situ coincided with the bump on tail in the electron parallel velocity distribution (e.g. Fig. 4 of *Ergun et al.*). Then *Dulk et al.* [1998] showed how the flux of radio emission is related to the number and energy of fast electrons in the beam.

A theoretical picture of type III emission was first formulated by *Ginzburg and Zheleznyakov* [1958], and the core of their picture is still current: Langmuir waves are generated by a bump-on-tail instability of the electron beam. Growth of Langmuir waves then tends to relax the beam toward marginal stability by flattening its velocity distribution. Nonlinear wave-wave interactions involving the Langmuir waves then lead to fundamental and harmonic radio emission. However, early versions of the theory could not account for propagation of the electron stream over long distances, nor for the effects of inhomogeneities in the solar wind. The recently developed stochastic-growth theory (SGT) [e.g. *Robinson and Cairns*, 1993; *Cairns and Robinson*, 1995] has shown that propagation of the beam with fluctuations about marginal stability can account for beam propagation to and beyond 1 AU, the clumpiness of the observed Langmuir waves, the production of associated ion-sound waves, and the observed radio emissivities.

The purpose of this paper is to review what has been learned from observations of type III radio bursts at long wavelengths. In general, attention is concentrated on bursts occurring at decameter to kilometer wavelengths, i.e., those that are generated in the corona above about 2 R_o, and in the solar wind to 1 AU and beyond. Few remarks will be made on the historical development because that has been covered in the tutorial lecture of *Bougeret* at this conference. Similarly, few remarks will be made about the theoretical interpretation because that has been reviewed by *Robinson* at this conference.

2. OBSERVATIONS ON EARTH AND IN SPACE

Observations at long wavelengths from the ground are hindered or prevented by terrestrial interference and by absorption and reflection by the ionosphere. Only in special circumstances is it possible to observe below about 20 MHz. One special circumstance is at moderately high magnetic latitudes where the ionospheric density is typically lower than elsewhere. Then, using techniques to search for narrow windows among the interference, it is possible to observe down to 5-10 MHz in daytime, as is done with the Bruny Island Radio Spectrometer in Tasmania, Australia [*Erickson*, 1997].

Observations from space (e.g. ISEE-3, Ulysses, Geotail) have usually been confined to very low frequencies, $\lesssim 1 - 2$ MHz. An exception is the WAVES experiment [*Bougeret et al.*, 1995] on the Wind spacecraft that observes from 13.8 MHz to a few kHz. This experiment consists of long dipole antennas in the spin plane (2×50 m and 2×7.5 m) and a short (2×5.7 m) antenna along the spin axis, feeding into three receivers covering the range from 4 kHz to 13.8 MHz. Wind is always in the ecliptic plane, near 1 AU.

Complemented by ground based data at decametric wavelengths (sometimes down to 5-10 MHz) the WAVES experiment allows a measure of the flux density spectrum of emissions occurring from about 1.4 R_o to 1 AU. Time profiles over this large range of frequency give accurate information on how groups of bursts at coronal heights merge together to form one interplanetary burst, and on the disappearance and/or emergence of burst components at intermediate or low frequencies. The composite then gives rise to a very broad low frequency burst that sometimes descends just to the plasma frequency.

2.1. Type III Bursts and Solar Flares

A general review of radio emission from solar flares has recently been given by *Bastian, Benz and Gary* [1998]. Regarding low frequencies, type III bursts that are recorded at less than a few MHz are almost always associated with flares observed in Hα and/or X-rays. Sometimes the flares are of minor importance, and sometimes major. In small flares type IIIs may be the only radio bursts produced. In major flares the type IIIs usually are intense, occur at the beginning of the impulsive phase, and are accompanied by other phenomena such as type II bursts from shock waves and type IV bursts from electrons trapped in coronal loops. The flares usually produce a group of a few to several tens of distinct type IIIs at frequencies $\gtrsim 10$ MHz, and these merge to form one burst at frequencies $\lesssim 1$ MHz.

2.2. Fundamental (F) and Harmonic (H) Components

At decametric and metric wavelengths it is often possible to distinguish two components, one, the harmonic (H) at roughly twice the frequency of the other, the fundamental (F). Statistically, *Stewart* [1974] finds an average frequency ratio of 1.8, with a range from 1.6 to 2.0. The difference from the expected value of 2.0 is usually attributed to the blockage by overlying plasma of much of the emitted fundamental band, the part very close to the plasma frequency.

When both components are observed, the flux density of the F component is usually more intense than that of the H component. At frequencies $\gtrsim 20$ MHz the degree of circular polarization of the F component is 2-3 times higher than that of the H component: typically 35% and 15% respectively [*Dulk and Suzuki*, 1980]. The sense of polarization of both F and H components of type IIIs is always that of the o-mode. However, in late phases of long lasting type IIIs and the type V extensions, the polarization switches to the opposite sense, i.e., x-mode [*Dulk, Suzuki, and Gary*, 1980].

Below 10-20 MHz it is difficult to distinguish the two components, and no measurements of the polarization have been published. However, at $\lesssim 10$ MHz, it is possible to use the method originated by *Kellogg* [1980], to determine which component is present, F or H, when the electron streams intersect the spacecraft where the radio emission is observed. In those events, one can measure the onset times of the type III burst at frequencies from $\gtrsim 10$ MHz, emitted in the corona, to frequencies near the plasma frequency at the spacecraft. The trace of these onset times almost always terminates quite close to the time of onset of Langmuir waves at the spacecraft (e.g. Fig. 1 of *Dulk et al.* [1987]). This correspondence demonstrates forcibly that the initial radiation of these type III bursts is almost always at the fundamental, i.e., Langmuir waves at plasma frequency f_p emit radio waves at that frequency $f = f_p$.

Consider radiation generated by Langmuir waves at the local plasma frequency at the spacecraft, f_{pS}. H radiation at $f = 2 \times f_{pS}$ can only begin when the Langmuir waves commence. But radiation at $f = 2 \times f_{pS}$ has begun some time earlier, usually about 30 minutes, and typically is reaching maximum intensity when the Langmuir waves commence. During those ≈ 30 min the radiation at $f = 2 \times f_{pS}$ must be the F component, emitted at about 0.5 AU.

A controversy as to whether this initial radiation is F or H has arisen because of the absence of spin modulation of the signal at $f \gtrsim 2 \times f_{pS}$. This absence implies that the source of radiation occupies $\gtrsim 2\pi$ steradians, i.e., that the source essentially surrounds the spacecraft, and has been interpreted as evidence that the radiation is locally generated, even though no Langmuir waves are present. Confusion about this point has raised questions in the past about Langmuir waves being the generator of radio waves [e.g. *Lin et al.*, 1981]. In Sec. 2.6 we discuss the modern view, that the radiation at $f \approx 2 \times f_{pS}$ actually is emitted at about 0.5 AU and appears to surround the spacecraft because of scattering by irregularities in the solar wind.

These remarks do not necessarily imply that the radiation at and after the time of maximum intensity is F radiation. Usually there is no known way to be sure. However, by comparing the theoretical vs. observational relationship between radio flux and the number and energy of fast electrons, *Dulk et al.* [1998], showed that the radiation at the time of maximum intensity is compatible only with F, and not H radiation. Later, during the declining phase of the burst, there is no evidence as to whether F or H radiation dominates. It may be H radiation generated some distance downstream of the plasma level, or F radiation that has been strongly scattered and delayed near the plasma level.

Reiner, Stone and Fainberg [1992] published three examples of type IIIs observed by the radio experiment on the Ulysses spacecraft [*Stone et al.*, 1992] that perfectly illustrate the relationship between the radio waves and the Langmuir waves. In one example, the event of 22 Feb. 1991, shown in Fig. 5 of *Reiner et al.*, the onset times vs. frequency of the type III led directly to the onset of Langmuir waves at $f_{pS} \approx 10$ kHz, implying that this initial radiation at 10 kHz was at the fundamental. Radiation at 20 kHz had begun about 30 min earlier. Then, when Langmuir waves commenced at 10 kHz, and for the 1.5 hour duration of the Langmuir waves, an enhancement of intensity was superimposed on the ongoing radiation at 20 kHz. The enhancement at 20 kHz is attributed to locally-produced radiation at the harmonic, whereas the preceding and ongoing 20 kHz radiation is attributed to remotely-produced radiation at the fundamental from the 20 kHz plasma level located nearer to the Sun.

As mentioned, the trace of onset times almost always leads to the approximate time of onset of Langmuir waves at the spacecraft. This was the case in all but four of the approximately 60 bursts that have been analyzed [*Kellogg*, 1980; *Hoang et al.*, 1994]. In three of four exceptions to the rule, the trace of onset times initially implied F radiation, but then the trace changed slope, indicating that H radiation was dominant when the burst exciter reached 1 AU. In the other exception, the initial H radiation changed to F near 1 AU.

When two spacecraft observe a type III burst from different directions, and when the electron stream intersects one of the spacecraft, *Hoang et al.* [1998] were able to demonstrate that the initial radiation observed by the spacecraft within the electron stream is at the fundamental, but that the radiation observed by the second spacecraft is at the harmonic. *Hoang et al.* found this to be true for two events, one when Ulysses was within the electron beam and Wind was not, and the other when Wind was within the electron beam and Ulysses was not. The probable explanation for this dichotomy is that the radiation pattern of F radiation is more directive than that of H radiation, and so the second spacecraft is out of the directivity pattern of F radiation.

To summarize, the initial radiation is almost always F when the spacecraft is located within the beam of fast electrons. In late phases it may be F or H.

2.3. Speeds of Type III-producing Electron Streams

The average speed of the type III-producing electrons from the Sun to the spacecraft can be derived from traces of onset time vs. frequency in events when the fast electrons and/or Langmuir waves are observed [e.g. *Dulk et al.*, 1987; *Hoang et al.*, 1994]. The time of electron acceleration at the Sun is known from the high frequency ($\gtrsim 10$ MHz) radio waves, the time of arrival at the spacecraft of type III-producing electrons is known from the onset of Langmuir waves, and the length of the Archimedean spiral travelled by the electrons is known from a measurement of the solar wind speed. The average speed thus derived is 0.14 c, and the range is about a factor of two from this value. While 0.14 c is smaller than expected from many published papers, it is compatible with the speeds of the electrons that are unstable to the production of Langmuir waves in events where the electron distribution functions are measured [e.g *Lin et al.*, 1981; *Ergun et al.*, 1998].

We emphasize that 0.14 c is the *average speed of the type III-producing electrons*. In the electron stream there are electrons travelling several times faster that arrive earlier than those of average speed, but there are not enough of them to produce the unstable bump-on-tail in competition with the background electrons of the solar wind. Similarly there are electrons travelling several times slower that arrive later than those of average speed, and there may or may not be enough of them to produce the bump-on-tail and continue to emit Langmuir and radio waves [e.g Figs. 3, 5 of *Lin et al.* 1981].

2.4. Do Type III Electron Streams Travel in Overdense Regions?

While there are often-quoted reports of occasional correspondences between radio positions and coronal bright features, there is no statistical evidence that type IIIs consistently propagate along dense structures. The statistical studies of *Leblanc et al.* [1974], *Leblanc and de la Noë* [1977], *Poquérusse and Bougeret* [1981], *Steinberg et al.* [1984], *Poquérusse et al.* [1988], and *Smerd, Sheridan and MacQueen* [personal communication] show that the sources of type III bursts rarely lie on dense structures in the corona. Instead they propagate along open field lines in the normal corona; that is, occasionally in overdense regions, occasionally in underdense regions, and frequently in average regions.

A model of the corona and solar wind from 1.7 R_o to 1 AU has been derived from observations of the trajectories of type III bursts by *Leblanc, Dulk and Bougeret* [1998]. This model agrees well with densities of the average corona/solar wind derived from coronagraph observations below about 2.5 R_o [*Saito, Poland and Munro*, 1977], and with densities measured in situ by the Helios spacecraft from 0.3 to 1 AU [*Bougeret et al.*, 1984].

2.5. Flux Densities, Brightness Temperatures, Source Sizes and Decay Times of Type III Bursts

Spectra of type III bursts are extremely variable, both in flux density and in frequency range. Many bursts occur in a limited band of frequencies, e.g. some occur only at decametric or shorter wavelengths, and some only at hectometric or longer wavelengths.

The following remarks concern the subset of type IIIs that continue from decametric to kilometric wavelengths. Some of them are the most intense of all solar radio emissions, particularly near 1 MHz. The remarks are based on the studies of *Evans, Fainberg and Stone* [1973], *Weber* [1978], *Steinberg et al.* [1984; 1985] and *Dulk et al.* [1984; 1998; 1999].

- The flux density S increases from metric to decametric to hectometric wavelenths, then decreases at the longer kilometric wavelengths.
- Maximum flux density occurs at $f_{peak} \sim 1$ MHz, but for some bursts f_{peak} is as low as 0.1 MHz and for others it is as high as 5 MHz [e.g. Fig. 2 of *Weber*, 1978].
- The flux density at f_{peak} varies over at least 6 orders of magnitude. The maximum so far reported was at ≈ 1 MHz: 10^{-14} W m^{-2} Hz^{-1}, or 10^8 SFU, or 10^{12} Jy.
- The source size varies approximately with wavelength. At 1 MHz, the half-width to e^{-1} brightness averages about 6° with a variation of about a factor of two. An empirical relationship between source size and frequency, valid from $\lesssim 0.1$ to $\gtrsim 1500$ MHz, is

$$\log(size) = 0.76 - 0.86 \times \log(f)$$

where f is in MHz and $size$ is in degrees.

- The brightness temperature T_B of the sources varies over 6 orders of magnitude. At 1 MHz it is sometimes as high as 10^{16} K. While T_B is inversely correlated with source size, the flux density is more important than source size in determining its value.
- The radio flux density is related to the flux and energy of the electrons that generate the Langmuir waves, and hence the radio waves. Letting E_{elec} be the energy of the electrons which are unstable [i.e., have a positive slope in the distribution $f(V_\parallel)$ vs. $V_\parallel$] and N_{elec} be the number flux of the electrons with $E = E_{elec}$, then the radio flux density S goes as:

$$S \propto N_{elec} \times E_{elec}^4$$

- Following the time of maximum intensity, the flux density decreases exponentially over several decades of flux. The *e*-folding decay time given by *Evans et al.* [1973] is

$$t_D = (2.0 \pm 1.2) \times 10^8 \, f^{-(1.09 \pm 0.05)}$$

where f is in Hz and t_D is in seconds.

2.6. High and Low Frequency Cutoffs

While some type III bursts commence at metric or shorter wavelengths and continue to kilometric wavelenths, the majority are observable only in a limited range of wavelengths. To our knowledge no statistical studies have been published of starting frequencies, i.e., high frequency cutoffs, in any wavelength band.

In an unpublished study, we have found that 70% of a set of 269 bursts recorded by WAVES had starting frequencies below 13.8 MHz, and about 35% below 3 MHz.

There is no evidence that the electron streams producing type III bursts lose their energy or stop: they are observed out to many AU. However, low frequency cutoffs of the radiation are the rule rather than the exception.

The following properties of low frequency cutoffs are summarized from papers by *Hoang et al.* [1994], *Leblanc et al.* [1995; 1996] and *Dulk et al.* [1996].

• The cutoffs are independent of the plasma frequency f_{pS} at the spacecraft except, of course, they cannot descend below f_{pS}.

• At the location of Ulysses, f_{pS} was as low as 3 kHz for several years, yet type III radiation seldom descended below about 20 kHz, and attained 10 MHz fewer than 10 times [e.g Fig. 2 of *Leblanc et al.* 1995]. Only once did it descend to 5 kHz [*S. Hoang*, personal communication].

• When f_{pS} was about 20, 10 and 5 kHz, only 15%, 5% and $\ll 1\%$ respectively of type III bursts approached within a factor of two of f_{pS}.

• The distribution of cutoffs is the same at Ulysses as at Wind. It is independent of distance of spacecraft from the Sun, and of ecliptic latitude.

• The cutoff frequency depends strongly on burst intensity; it is about 4 times lower for strong bursts than for weak ones.

• Intense bursts observed simultaneously at Ulysses and Wind usually have similar cutoff frequencies, but cutoffs of weak bursts differ substantially, often by more than a factor of three.

2.7. Propagation Effects

During the propagation of type III bursts from the source to the observer, the radiation suffers refraction in density gradients and scattering by inhomogeneities. The corona and solar wind are strongly inhomogeneous on all scales from $\lesssim 100$ km to $\gtrsim 0.3$ AU. In the solar wind, the radial distance of a given plasma density varies from one ecliptic longitude to another by a factor $\gtrsim 2$ from its average distance. For example, at some ecliptic longitudes the density of 44 cm^{-3} ($f_p = 60$ kHz) may occur at about 0.2 AU, and at other longitudes at about 1 AU (e.g. Fig. 9 of *Lecacheux et al.* [1989]).

Some consequences of these propagation effects are: i) the apparent source of radiation is shifted from its actual position, ii) the apparent source size is larger than its true size, iii) the apparent brightness is lower than its true brightness, and iv) its measured degree of polarization is lower than its true polarization.

The position shift was evident in the first observations of type III source locations with interferometers [*Wild, Sheridan and Neylan*, 1959]: the measured radial distances of sources were much larger than expected for radiation at the plasma frequency, i.e., the coronal density required to place the sources at their observed heights was $\gtrsim 10$ times the normal coronal density.

For some time it was accepted that type III electrons travel in overdense structures, but then further problems arose. Most important are the measurements of source positions and sizes of F and H components of given bursts: i) At a given frequency, the radial distance of H emission must be greater than that of the F emission observed a few seconds earlier. But the observed radial distance and source size of the H component are essentially the same as those of the F component. ii) At a given time, because the F and H radiation is emitted from a given location, the source locations should be identical. But in fact, the F source is observed to be at a greater radial distance and of larger size than the H source.

Clearly, propagation effects have to be considered, and calculations were undertaken by several groups, including *Fokker* [1965], *Steinberg et al.* [1971], *Leblanc* [1973] and *Duncan* [1979]. While progress has been made, the calculations have not yet fully explained the observations.

The following properties of propagation effects at long wavelengths are summarized from papers by *Fainberg, Evans and Stone* [1972], *Stewart* [1972], *Lin et al.* [1981], *Steinberg et al.* [1984], *Steinberg, Hoang and Dulk* [1985], *Dulk et al.* [1985], *Sawyer and Warwick* [1987], *Reiner and Stone* [1988; 1989; 1990], *Lecacheux et al.* [1989], *Bastian* [1995a; 1995b], *Hoang, Poquérusse and Bougeret* [1997] and *Hoang et al.* [1998].

• Radial distances of sources are too large to be compatible with plasma radiation in regions of normal coronal/solar wind density. Radiation observed to come from a given radial distance is at a frequency that averages 2 to 5 times the plasma frequency at that height. Thus the displacement from true to apparent source location is one or more scale heights, e.g. 0.5 AU for sources at 0.5 AU.

- As mentioned above, the radial distances of F and H components are statistically the same when measured within a few seconds of each other at a given frequency, even though the H component is emitted at a greater height where the plasma density is 4 times lower. Similarly, the radial distances and source sizes of F and H components are different when both are measured at a given time (the F component coming from a larger apparent source at a greater radial distance), even though the two components are emitted from the same source.

- Observed source sizes are impossibly large. For example, sources located at 0.5 AU appear to surround the spacecraft at 1 AU. As discussed in Sec. 2.2, this property has confused identification of F and H components, and has even led to questioning whether Langmuir waves are necessary for generation of radio emission.

- Source sizes increase with angular distance from Sun center.

- Tracing the trajectory of the type III electrons by measuring the azimuths of apparent sources leads to apparent trajectories that intersect the observer, rather than being along the expected interplanetary magnetic field lines.

- When two spacecraft observe the same burst from different directions, there is an anomalous delay of up to 500 sec in the ray path to one of the spacecraft.

- If one of the two spacecraft is located within the electron beam, the initial radiation it records is the F component, while the other spacecraft records only H radiation if it is out of the beam of the F radiation. The directions of the sources as observed by the two spacecraft do not coincide.

- The effects of large-scale density structures may mask the presence of an F component in burst profiles.

- Radiation is most intense in the direction of the magnetic field along which the electron beam is travelling, thus along the Archimedean spiral. The F component is more tightly beamed than the H component. But a fraction of the radiation, 1-10%, is scattered and isotropized.

- Because of this scattering and isotropization, intense, kilometer type III bursts are visible irrespective of the location of the source, even if the source lies directly behind the Sun. Compared with sources between the observer and the Sun, flux densities of sources above the limb of the Sun are typically 10 times weaker, and flux densities of sources behind the Sun are typically 100 times weaker.

- The apparent source of a burst emitted directly behind the Sun appears as a halo surrounding the Sun.

3. CONCLUSIONS

A summary of the properties of solar type III bursts at long wavelengths is given in the abstract. It is evident that much has been learned about the radio emissions, the electron streams and Langmuir waves that cause them, and the theory to explain them. Yet the understanding is incomplete. In particular, propagation effects obscure some of the burst properties. What we observe is only a phantom of what is real, a view through a frosted glass, a scatter image of the true source. Further progress depends largely on a better understanding of the scattering process, both from theory and from observations.

Today we do not have proper images of the radio sources; we infer their properties from measurements with spinning dipole antennas. These give the direction of the source centroid and a measure of the source size, supplemented in a few cases by stereoscopic measurements by two spacecraft. In the future there will be imaging arrays in space and more extensive stereo observations. These will be essential to a full understanding of how the radio source properties are affected by propagation in the very inhomogeneous medium that is the solar wind.

A major objective of arrays in space will be the observations of stellar, galactic and extragalactic sources. Depending on wavelength, the interpretation of these observations will depend on the understanding of the propagation effects, and solar bursts such as type IIIs will be considered as interference. Indeed, type III bursts which have flux densities near 1 MHz of up to 10^{-14} W m^{-2} Hz^{-1} or 10^{12} Jy, will provide the most intense interference. Fortunately for cosmic observations, the bursts occur only a small fraction of the time, particularly near the minimum of the sunspot cycle.

REFERENCES

Bastian, T.S., Angular scattering of solar radio emission by coronal turbulence, *Solar Phys., 439,* 494, 1995a.

Bastian, T.S., Angular scattering of radio waves: implications for mode coupling in the solar corona, *Astrophys. J., 439,* 494, 1995b.

Bastian, T.S., A.O. Benz and D.E. Gary, Radio emission from solar flares, *Ann. Rev. Astron. Astrophys., 36, 131,* 1998.

Bougeret, J.-L., J.H. King, and R. Schwenn, Solar radio burst and in situ determination of interplanetary electron density, *Solar Phys., 90,* 401, 1984.

Bougeret, J.-L., et al., Waves: The radio and plasma wave investigation on the Wind spacecraft, *Space Sci. Rev., 71,* 231, 1995.

Cairns, I. H., and P.A. Robinson, Ion acoustic wave frequencies and onset times during solar type III radio burts, *Astrophys. J., 453,* 959, 1995.

Dulk, G.A., and S. Suzuki, The positions and polarization

of type III solar bursts, *Astron. Astrophys.*, *88*, 203, 1980.

Dulk, G.A., S. Suzuki, and D.E. Gary, The positions and polarization of type V solar bursts, *Astron. Astrophys.*, *88*, 218, 1980.

Dulk, G.A., J.-L. Steinberg, and S. Hoang, Type III bursts in interplanetary space - fundamental or harmonic?, *Astron. Astrophys.*, *141*, 30, 1984.

Dulk, G.A., J.-L. Steinberg, A. Lecacheux, S. Hoang, and R.J. McDowall, The visibility of type III radio bursts originating behind the Sun, *Astron. Astrophys.*, *150*, L28, 1985.

Dulk, G.A., J.-L. Steinberg, S. Hoang, and M.V. Goldman, The speeds of electrons that excite solar radio bursts of type III, *Astron. Astrophys.*, *173*, 366, 1987.

Dulk, G.A., Y. Leblanc, J.-L. Bougeret, and S. Hoang, Type III bursts observed simultaneously by Wind and Ulysses, *Geophys. Res. Lett.*, *23*, 1203, 1996.

Dulk, G.A., Y. Leblanc, P.A. Robinson, J.-L. Bougeret, and R.P. Lin, Electron beams and radio waves of solar type III bursts, *J. Geophys. Res.*, *103*, 17223, 1998.

Dulk, G.A., W.C. Erickson, R. Manning, and J.-L. Bougeret, Spectra of solar type III radio bursts from 20 kHz to 47 MHz, with flux density calibration by the galactic background radiation, *Astron. Astrophys.*, to be submitted, 1999.

Duncan, R.A., Wave ducting of solar metre-wave radio emission as an explanation of fundamental/harmonic source coincidence and other anomalies, *Solar Phys.*, *63*, 389, 1979.

Ergun, R.E., et al., Wind spacecraft observations of solar impulsive electron events associated with solar type III radio bursts, *Astrophys. J.*, *503*, 435, 1998.

Erickson, W.C., The Bruny Island Radio Spectrometer, *Proc. Astron. Soc. Aust.*, *14*, 278, 1997.

Evans, L.G., J. Fainberg, and R.G. Stone, Characteristics of type III exciters derived from low frequency radio observations, *Solar Phys.*, *31*, 501, 1973.

Fainberg, J., L.G. Evans, and R.G. Stone, Radio tracking of solar energetic particles through interplanetary space, *Science*, *178*, 743, 1972.

Fokker, A.D., Coronal scattering of radiation from solar radio sources, *Bull. Astron. Inst. Neth.*, *18*, 111, 1965.

Ginzburg, V.L. and V.V. Zheleznyakov, On the possible mechanisms of sporadic radio emission (Radiation in an isotropic plasma), *Astron. Zh.*, *35*, 694, 1958; transl. *Sov. Astron.-AJ*, *2*, 653, 1958.

Gurnett, D.A., and R.R. Anderson, Electron plasma oscillations associated with type III radio bursts, *Science*, *194*, 1159, 1976.

Hoang, S., G.A. Dulk, and Y. Leblanc, Interplanetary type III radio bursts that approach the plasma frequency: Ulysses observations, *Astron. Astrophys.*, *289*, 957, 1994.

Hoang, S., M. Poquérusse, and J.-L. Bougeret, The directivity of solar kilometric type III bursts: Ulysses-Artemis observations in and out of the ecliptic plane, *Solar Phys.*, *172*, 307, 1997.

Hoang, S., G.A. Dulk, J.-L. Bougeret, and Y. Leblanc, Solar type III kilometric radio bursts associated with Langmuir waves: Ulysses-Wind stereoscopic observations, this conference, 1998.

Kellogg, P.J., Fundamental emission in three type III radio bursts, *Astrophys. J.*, *236*, 696, 1980.

Leblanc, Y., Scattering effects on the relative positions and intensities of fundamental and harmonic emission of solar radio bursts, *Astrophys. Lett.*, *14*, 41, 1973.

Leblanc, Y., T.B.H. Kuiper, and S.F. Hansen, Coronal density structures in regions of type III activity, *Solar Phys.*, *37*, 215, 1974.

Leblanc, Y. and J. de la Noë, Solar radio type III bursts and coronal density structures, *Solar Phys.*, *52*, 133, 1977.

Leblanc, Y., G.A. Dulk, and S. Hoang, The low radio frequency limit of solar type III bursts: Ulysses observations in and out of the ecliptic, *Geophys. Res. Lett.*, *22*, 3429, 1995.

Leblanc, Y., G.A. Dulk, S. Hoang, J.-L. Bougeret, and P.A. Robinson, Type III radio bursts observed by Ulysses pole to pole, and simultaneously by Wind, *Astron. Astrophys.*, *316*, 406, 1996.

Leblanc, Y., G.A. Dulk, and J.-L. Bougeret, Tracing the electron density from the corona to 1 AU, *Solar Phys.*, *183*, 165, 1998.

Lecacheux, A., J.-L. Steinberg, S. Hoang, and G.A. Dulk, Characteristics of type III bursts in the solar wind from simultaneous observations from ISEE-3 and Voyager, *Astron. Astrophys.*, *217*, 237, 1989.

Lin, R.P., The emission and propagation of ~ 40 keV solar flare electrons, *Solar Phys.*, *12*, 266, 1970.

Lin, R.P., D.W. Potter, D.A. Gurnett, and F.L. Scarf, Energetic electrons and plasma waves associated with a solar type III radio burst, *Astrophys. J.*, *251*, 364, 1981.

Poquérusse, M., and J.-L. Bougeret, Radiation mode and coronal propagation of solar type III radio bursts observed on 14 November 1971 during Stereo-1 experiment, *Astron. Astrophys.*, *97*, 36, 1981.

Poquérusse, M., J.-L. Steinberg, C. Caroubalos, G.A. Dulk, and R.M. MacQueen, Measurement of the 3-dimensional positions of type III bursts in the solar corona, *Astron. Astrophys.*, *192*, 323, 1988.

Reiner, M.J., and R.G. Stone, Model interpretation of type III radio burst characteristics I. Spatial aspects, *Astron. Astrophys.*, *206*, 316, 1988.

Reiner, M.J., and R.G. Stone, Model interpretation of type III radio burst characteristics II. Temporal aspects, *Astron. Astrophys.*, *217*, 251, 1989.

Reiner, M.J., and R.G. Stone, Evidence for halo-like radio sources from kilometric type III burst observations, *Solar Phys.*, *125*, 371, 1990.

Reiner, M.J., R.G. Stone, and J. Fainberg, Detection of fundamental and harmonic type III radio emission and the associated Langmuir waves at the source region, *Astrophys. J.*, *394*, 349, 1992.

Robinson, P.A., and I.H. Cairns, Stochastic growth theory of type III solar radio emission, *Astrophys. J.*, *418*, 506, 1993.

Saito, K., A.I. Poland, and R.H. Munro, A study of the background corona near solar minimum, *Solar Phys.*, *55*, 121, 1977.

Sawyer, C., and J.W. Warwick, Wide visibility of kilometric type III bursts, *Astron. Astrophys.*, *177*, 277, 1987.

Steinberg, J.-L., M. Aubier-Giraud, Y. Leblanc, and A. Boischot, Coronal scattering, absorption and refraction of solar radio bursts, *Astron. Astrophys.*, *10*, 362, 1971.

Steinberg, J.-L., G.A. Dulk, S. Hoang, A. Lecacheux, and M.G. Aubier, Type III radio bursts in the interplanetary medium - The role of propagation, *Astron. Astrophys.*, *140*, 39, 1984.

Steinberg, J.-L., S. Hoang, and G.A. Dulk, Evidence of scattering effects on the sizes of interplanetary type III radio bursts, *Astron. Astrophys.*, *150*, 205, 1985.

Stewart, R.T., Relative positions of fundamental and second harmonic type III bursts, *Proc. Astron. Soc. Aust.*, *2*, 100, 1972.

Stewart, R.T., Harmonic ratios of inverted-U type III bursts, *Solar Phys.*, *39*, 451, 1974.

Stone, R.G., et al., The unified radio and plasma wave investigation, *Astron. Astrophys. Suppl.*, *92*, 291, 1992.

Weber, R.R., Low frequency spectra of type III solar radio bursts, *Solar Phys.*, *59*, 377, 1978.

Wild, J.P., Observations of the spectrum of high-intensity solar radiation at metre wavelengths - III. Isolated bursts, *Aust. J. Sci. Res. Ser. A*, *3*, 541, 1950.

Wild, J.P., K.V. Sheridan, and A.A. Neylan, An investigation of the speed of the solar disturbances responsible for type III radio bursts, *Aust. J. Phys.*, *12*, 369, 1959.

G.A. Dulk, DESPA, Observatoire de Paris, 92195 Meudon, France. (e-mail: george.dulk@obspm.fr)

Type II Solar Radio Bursts

N. Gopalswamy[1]

NASA Goddard Space Flight Center, Greenbelt, Maryland

Solar radio bursts of type II are thought to be caused by MHD shock waves propagating through the corona and interplanetary medium. They are identified as slowly drifting features in the dynamic spectra recorded by ground based and spaceborne radio instruments. The radio emission itself occurs as a final step in a series of physical processes: initiation of the shock, particle acceleration, generation of plasma waves and finally, conversion of the plasma waves into electromagnetic waves. Type II bursts play an important role in understanding the Sun-Earth connection, because of their association with flares and coronal mass ejections (CMEs). Images of type II bursts made by radioheliographs are crucial to understand the physical relationship between shocks, CMEs and solar flares. Observations of a new type of coronal waves by the EIT instrument on board the SOHO mission and long-decametric and hectometric type II bursts by the WIND/WAVES experiment have added new dimensions to the study of type II radio bursts. In this review, I summarize the basic properties and associated solar activities of type II bursts and discuss some of the current issues.

1. INTRODUCTION

More than half a century ago, Payne-Scott, Yabsley and Bolton observed the 'Large Outburst' of March 08, 1947 from the Sun at three frequencies (200, 100 and 60 MHz) and found that the lower frequency emission was delayed with respect to the higher frequencies [Payne-Scott et al., 1947]. They immediately recognized that some physical agency must be passing from higher to lower frequency plasma levels in the corona and estimated the speed of the agency to be $\sim$ 500-750 km s^{-1}. The physical agency was later identified with an MHD shock [Uchida, 1960]. Wild and McCready [1950] classified these outbursts as 'type II' radio bursts, the slowly drifting features in the dynamic spectrum, in contrast to the Type I storms and the fast-drifting type III bursts. [Type IV bursts from moving and stationary coronal structures and type V bursts, a variant of type III bursts, were subsequently added to the types of radio bursts]. All these radio bursts are due to non-thermal electrons and represent energy releases in the corona. The type II bursts are signatures of violent eruptions from the Sun that result in shock waves propagating through the corona and the interplanetary (IP) medium. The study of type II bursts is thus important for the understanding of the large scale structure and dynamics of the inner heliosphere.

The literature on type II bursts has been reviewed extensively [Roberts, 1959; Maxwell and Thompson, 1962; Weiss, 1965; Zheleznyakov, 1969; Kundu, 1965; Dodge, 1975; Sveztka, 1976; Kruger, 1979; Nelson and Melrose, 1985; Bougeret, 1985; Aurass, 1997]. In this tuto-

---

[1] NAS/NRC Senior Research Associate on leave from the Catholic University of America, Washington, DC.

rial, I summarize: (i) the main characteristics of type II bursts, (ii) associated phenomena and (iii) current issues and future prospects.

2. BASIC PROPERTIES

Early observations of type II bursts were made at single frequencies which showed the burst as a sudden increase in radio intensity. In fact, the type II bursts were originally discovered using single frequency observations with radio intensity recorded on paper charts [Payne-Scott, et al., 1947]. Later on, the intensities were recorded on films or stored digitally to construct what are known as dynamic spectra. A dynamic spectrum is a plot of the observed radio intensity as a function of frequency and time. Radio interferometric measurements give positional information of the bursts. Two dimensional interferometric arrays (known as radioheliographs) provide detailed information on the spatial structure and position of the radio bursts.

2.1. Frequency Regime

The type II bursts are typically observed in the meterwave regime at frequencies < 150 MHz. However, bursts have been observed to start at frequencies as high as 500 MHz [Nakajima et al., 1990; Vrsnak et al., 1995]. Ground based instruments observe type II bursts down to the ionospheric cut off (~ 15 MHz, sometimes down to ~ 5 MHz). Spaceborne radio instruments observe type II bursts at frequencies below the ionospheric cut off [Malitson et al., 1973] in the IP medium. Recently, the Wind/WAVES experiment [Bougeret et al., 1995] has detected type II bursts in the previously unexplored frequency regime of 1-14 MHz [Kaiser et al., 1998; Gopalswamy et al., 1999a,d; Reiner and Kaiser, 1999]. The Ulysses spacecraft has detected type II bursts at a distance of several AU from the Sun [Lengyel-Frey et al., 1997].

2.2. Dynamic Spectrum and Drift Rate

The dynamic spectrum of a typical type II burst is shown in Fig. 1 as recorded by the Culgoora Radiospectrograph. It is a plot of the radio intensity at several frequencies between 180 MHz an 18 MHz as a function of time. The slanted pair of features (marked F, H in Fig. 1) is the type II burst with the two components related harmonically (the emission frequencies are in the approximate ratio 1:2). The H component starts around 150 MHz at ~ 01:12 UT and ends at ~ 01:22 UT near 40 MHz. The F component roughly runs parallel to the H component. The rate at which the emission frequency changes is known as the drift rate and corresponds to the motion of the exciting agency

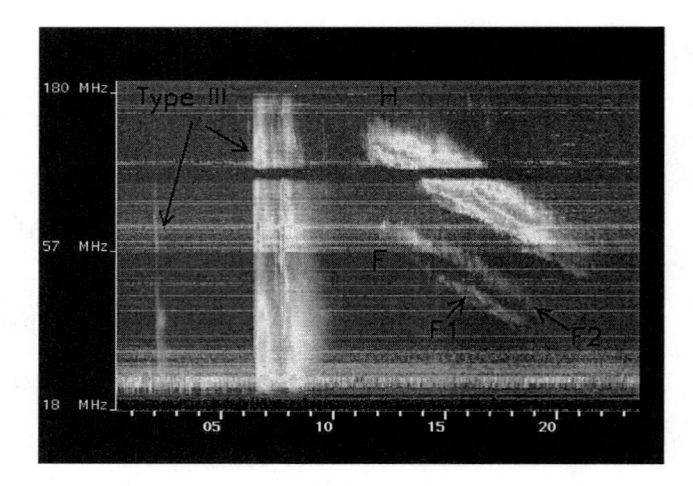

Figure 1. A typical dynamic spectrum of a type II radio burst recorded by the Culgoora radiospectrograph. The X-axis is time in minutes from 0100 UT on 1996 June 02 and Y-axis is frequency. Type II burst with fundamental (F) and harmonic (H) structure and a group of type III bursts can be seen. F1 and F2 are the lower and upper side bands of the fundamental. H component also shows band-splitting (H1, H2 not marked).

(an MHD shock) through the corona [Wild et al., 1959]. For the burst in Fig. 1, the F-component drifts from 75 MHz to 20 MHz in about 10 minutes, giving a drift rate of ~ 0.1 MHz s^{-1}. This value is rather typical of most of the type II bursts: Using statistical analysis of type II bursts at frequencies > 40 MHz, Mann et al., [1996] found that the type II bursts have a drift rate of -0.16 ± 0.11 Hz s^{-1}. The drift is negative because the burst drifts from higher to lower frequencies. For the H-component, the drift rate is ~ 0.2 MHz s^{-1}, but when normalized to the central frequency of emission, the rates are the same. Bursts starting at very high frequencies seem to have high drift rates. Mann et al., [1996] have found a linear relationship between drift rate (df/dt) and starting frequency (f_s) of the type II burst at frequencies > 40 MHz: $df/dt = 0.046$-$0.002 f_s$. The vertical streaks in Fig. 1 are type III radio bursts caused by electron beams streaming away from the Sun along open magnetic field lines. The drift rate of type III bursts is greater than that of type II bursts by more than an order of magnitude. Note that the type II burst at a given frequency lasts for at least a few minutes while the type III bursts last only for a few seconds or less. The overall life time of the type II burst in Fig. 1 is about 10 min. Statistical studies have shown that the life time of metric type II bursts is in the range 5 to 15 min. However, there are exceptional events which last for more than a day and cover a large frequency range from metric to kilometric regimes.

2.3. Emission Mechanism

The radio emission is thought to be due to a plasma emission process [see e.g. Mann, 1995] which consists of the following steps: (i) an instability condition is set up due to the propagation of an exciting agency through the corona resulting in the generation of high frequency plasma waves at the local plasma frequency (f_p), (ii) these plasma waves (also known as Langmuir waves) scatter on the background ions resulting in electromagnetic waves of roughly the same frequency which propagate towards the observer and are detected as fundamental (F) emission. Two plasma waves also can coalesce resulting in an electromagnetic wave at a frequency $2f_p$ which is observed as the harmonic (H) component. The condition for instability is maintained for the duration the agency passes through a given plasma layer. Once the agency leaves the layer, there is no more free energy available so the plasma waves decay to the thermal level and the generation of electromagnetic radiation ends. The agency is a shock wave in the case of type II bursts and an electron beam in the case of type III bursts.

2.4. Drift Rate and Shock Speed

As one moves away from the Sun, the coronal density (n) decreases and so does the local plasma frequency because $f_p = 9000\sqrt{n}$ Hz. If an agency moves away from the Sun exciting radio emission on its path at the local plasma frequency, one would expect lower frequencies at later times, as the dynamic spectrum indicates. Thus the frequency drift rate (df/dt) of the type II bursts can be converted into the speed of the agency if we know how the density varies in the corona. For fundamental emission, the observing frequency ($f \simeq f_p$) is given by the local electron density as,

$$f^2 = \frac{n(\mathbf{r})e^2}{\pi m}, \quad (1)$$

where e and m are the electronic charge and mass respectively and $\mathbf{r}$ is the position vector of the plasma layer in question. Differentiating this expression,

$$2f\frac{df}{dt} = \frac{e^2}{\pi m}\frac{dn}{d\mathbf{r}}\frac{d\mathbf{r}}{dt}. \quad (2)$$

If the angle between the density gradient ($dn/d\mathbf{r}$) and the shock velocity ($\mathbf{V} = d\mathbf{r}/dt$) is θ, the above expression can be reduced to,

$$V\cos\theta = 2L_n \frac{1}{f}\frac{df}{dt}, \quad (3)$$

where,

$$L_n = [\frac{1}{n}\frac{dn}{dr}]^{-1} \quad (4)$$

is the scale height of electron density variation. Note that the derived shock speed depends on the density scale height and the direction of propagation of the shock relative to the density gradient. Both of these parameters are difficult to determine. The density scale height is usually determined assuming a density model. As an example, let us use the Newkirk's density model,

$$n(r) = 4.2 \times 10^{4+4.32/r}, \quad (5)$$

which gives a scale height of $L_n = 0.23r^2$ ($R_\odot$). Here, r is the radial distance from the sun center in units of solar radii ($R_\odot$). For $r = 1.5$, the scale height is $0.5R_\odot = 3.6 \times 10^5$ km. For a shock propagating along the density gradient ($\theta = 0$) and emitting at a frequency $f = 50$ MHz with a drift rate of 0.1 MHz s^{-1}, we get a speed of ~ 1440 km s^{-1}. The shock is clearly a fast mode shock because the typical coronal Alfven speed at these heights is about 500 km s^{-1}. Note, however, that the speed derived is dependent on the density model used. Eclipse pictures, coronagraph images and X-ray images reveal that the corona is highly structured and time variable so we must consider radial, azimuthal and latitudinal variations of density. Simple density models such as the one used here will not give realistic speeds.

2.5. Imaging Observations

From images obtained by a radioheliograph, one can directly get the speed of the type II burst because we know the height of the type II burst at different frequencies. Fig. 2 shows a type II burst in contour representation observed by the Clark Lake multifrequency radioheliograph at 73.8 and 50 MHz [Gopalswamy and Kundu, 1990; Kundu and Gopalswamy, 1992]. The contours are in units of percentage of the peak brightness temperature. The centroids of 73.8 and 50 MHz sources were located at 1.38 and 1.47 $R_\odot$, respectively from the disk center. The burst starts at 20:24:18 UT at 73.8 MHz and at 20:25:1.5 UT at 50 MHz. The associated active region (AR 4713) was at S01 W32 so the deprojected distance between the two sources is about 0.17$R_\odot$. The shock travels between the two coronal layers in about 42 s so we can determine the speed to be $\sim$ 2800 km s^{-1}. If the shock originated from the flare site, we can also determine the average shock speed from the flare onset time (20:18:30 UT) and the time of arrival at the 73.8 MHz source as 3200 km s^{-1}, which is close to the above value. Note that we did not have to use any density model. In fact we can get the density distribution from the radio data because the emission frequency is the same as the local plasma frequency.

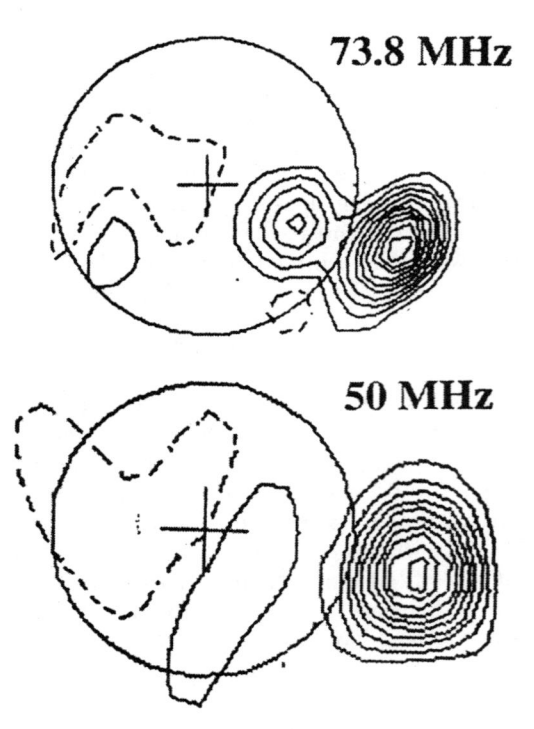

Figure 2. Two dimensional (contour) images of the 1996 February 10 type II burst observed by the Clark Lake Multifrequency radioheliograph at 73.8 (top) and 50 MHz (bottom). The circle represents the optical disk and the 'plus' sign is the disk center. The contour levels are at 5 to 95% of the peak brightness temperature in steps of 10%. The peak brightness temperature is 3×10^8 K at 73.8 MHz and 2.5×10^8 K at 50 MHz. The weak source on the disk at 73.8 MHz is a long lived continuum.

2.6. Directivity

According to plasma hypothesis, the fundamental component originates from the local plasma level where the refractive index deviates significantly from unity. Therefore, the F-component is best observed near the disk center. When the type II burst occurs at large central meridian distances, propagation effects become severe and the F component is severely attenuated. For this reason, backside type II bursts are mostly harmonic.

2.7. Fine Structures in Type II Bursts

2.7.1. Band splitting.
An interesting feature in Fig. 1 is that the F and H components are composed of two parallel lanes (F1, F2 in the fundamental) of emission. The frequency separation is about 10% of the central frequency. The phenomenon is known as band splitting and the individual features are called the split bands (upper and lower split bands for each harmonic). In two dimensional images, the lower split bands originate higher in the corona than the upper side band [Nelson and Robinson, 1975]. The band splitting seems to be related to the inhomogeneity in the medium.

McLean [1967] suggested that the quasi-spherical nature of the shock may be responsible for the band splitting. When the shock propagates outwards, it encounters different densities at different sections of the shock resulting in multiple emission lanes. However, it is not clear if this mechanism will produce well defined band splitting of 10%. Smerd et al., [1974] argued that the density jump across the shock front corresponds to the frequency difference between the two lanes. The density jump corresponding to the 10% band splitting is $\sim$ 20%. Treumann and LaBelle [1992] came up with an alternative explanation based on the inhomogeneity in the foreshock region caused by strong turbulence effects. Plasma waves generated in the foreshock region evolve into solitons ('Langmuir solitons') which result in low density regions known as cavitons. The frequency difference between plasma waves from inside and outside the cavitons can account for the band splitting.

2.7.2. Third harmonic.
The type II bursts also exhibit a third harmonic ($f = 3f_p$) component [see e.g., Zlotnik et al., 1998], which is usually weak and hence rarely observed. Recently, Zlotnik et al., [1998] studied three type II events which exhibited third harmonic. For two of these, they had positional information from the Nancay radioheliograph. They established that all the three harmonic components were found to originate from the same volume confirming the plasma emission hypothesis. Coalescence of three plasma waves ($f_p+f_p+f_p = 3f_p$) or a plasma wave and the second harmonic wave ($f_p+2f_p=3f_p$) are the two possible ways in which the third harmonic emission could occur. Since the conversion of plasma waves into electromagnetic waves involves higher order nonlinear processes, the third harmonic is rather weak. Zlotnik et al's calculations also show that the radiation pattern of the third harmonic is very narrow which is another reason why they are seldom observed.

2.7.3. Herringbone features.
When viewed with high time resolution, short duration bursts seem to emanate from the F-H stripes of some type II bursts. These bursts have properties similar to the normal type III bursts [Haddock, 1958]. Based on their appearance these fine structures are referred to as 'herringbones' [Roberts, 1959] connected to the F and H components ('backbones'). The herringbones are also referred to as type III fine structures. During the course of a type II, the backbones might fade out, but the herringbones would continue appearing from the expected location of the backbone. Even U type bursts (a variant of type III bursts) are observed as fine structures of type II

bursts. About 20% of all type II bursts exhibit herringbone fine structures. Herringbones drifting towards lower frequencies (away from the Sun) are more common. The herringbone bursts are thought to be produced when a curved shock front propagates across open magnetic field lines [Stewart and Magun, 1980]. Unusually small drift rates have been observed in type II bursts with herringbone structures [Weiss, 1963], consistent with shock propagation roughly parallel to the solar surface. Some herringbones associated with metric type II bursts extend into the outer corona and become IP type III bursts. Since the electrons responsible for these type III bursts were thought to be accelerated in the type II shocks, they were named as 'Shock Accelerated' (SA) events [Cane, et al., 1981]. Most of the herringbones that become IP type III bursts were found to originate at frequencies less than 50 MHz. Recently, Wind/WAVES experiment has observed many SA events with the type III bursts starting at a few MHz. Fig. 3 shows one such Wind/WAVES type II event with herringbone type III bursts starting at ~ 3 MHz. The type II burst was associated with a CME which occurred from the west limb on April 20, 1998. Note that the type II burst contains only one component, most likely the harmonic because the fundamental is not well observed in limb events. The type III bursts seem to originate below frequencies where the fundamental component of type II is expected, but not observed.

There is an alternative interpretation of the SA events: Electrons are accelerated low in the corona similar to the case of normal type III bursts and propagate past the shock front [Kundu and Stone, 1984]. This interpretation seems to be supported by the presence of microwave bursts at the time of the IP type III events [see e.g., Reiner et al., 1999]. MacDowall et al., [1987] reported evidence for shock accelerated electrons at least in some of the SA events. Thus, it is not clear if (i) the electrons are accelerated near the flare site, produce microwaves and also pass through the shock producing IP type III bursts, (ii) the shock accelerates electrons and those propagating away from the sun produce IP type III bursts and those propagating towards the Sun produce the microwave bursts, or (iii) a combination of these two processes takes place. Recently, Bougeret et al., [1998] proposed a model in which a shock induces particle acceleration via reconnection in helmet streamers. Kahler et al., [1986] found that metric type II bursts associated with well connected flares and SA events were well correlated with solar energetic particle (SEP) events. Thus, the electrons responsible for SA events and SEPs may be accelerated by the same shock.

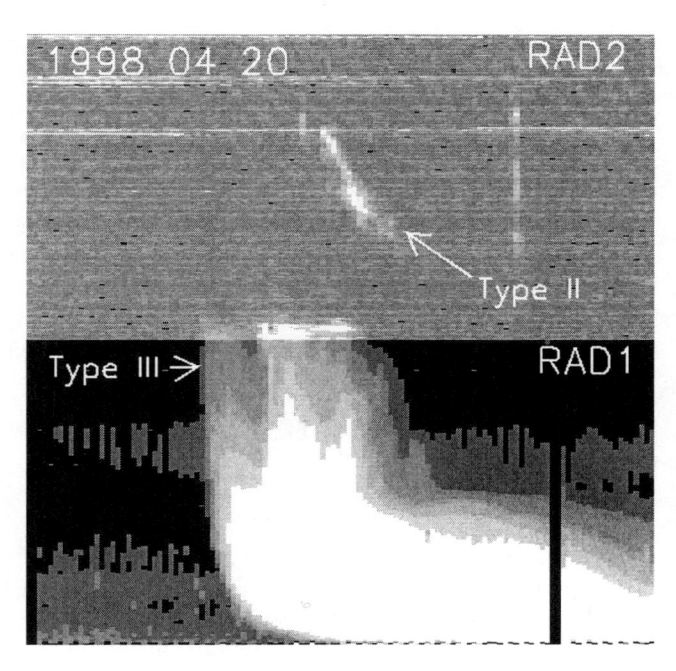

Figure 3. Herringbone bursts observed by the Wind/WAVES experiment on 1998 April 20. The X-axis is time from 9:30 to 11:30 UT. Herringbone type III bursts can be seen to start from the RAD2 (14 MHz - 1 MHz) range and becomes very intense in the RAD1 (20 kHz - 1 MHz) range. The type II lasts for only about 13 min. The herringbone type III is seen just under the span of the type II.

2.8. Source Sizes

In Fig. 2 we see that the burst size is typically less than a solar radius at 73 and 50 MHz. In this particular case, the north-south size is slightly larger than the east-west size. At 73.8 and 50 MHz we can measure the source diameters as $10' \times 7'$ and $13' \times 9'$, respectively. However, much larger sizes have also been reported [Smerd et al., 1974]. The spatial resolution of the Clark Lake radioheliograph was $5' \times 6'$ at 73.8 MHz and $7' \times 8'$ at 50 Hz (north-south × east-west directions). The trend of larger source size at lower frequencies is always observed.

2.9. Brightness Temperature

The brightness temperature of radio emission is the temperature of an equivalent black body that would produce the same amount of radio flux: For a given source size, higher brightness temperature source would emit more flux. The observed radio flux of type II bursts varies by several orders of magnitude from event to event (1 to 300,000 solar flux units (sfu); 1 sfu = 10^{-26} W m^{-2}Hz^{-1}sr^{-1}). The corresponding brightness temperature lies in the range from 10^7 K to 10^{13} K. In the solar atmosphere, one can easily find plasmas at a temperature of 10^7 K during flares, but never at 10^{13} K. This confirms that the type II radio emission is a not a thermal emission; it must be a coherent nonthermal emission such as the plasma emission. For the type II burst in Fig. 2, the brightness temperature is moderate: 3×10^8 K at 73.8 MHz and 2.5×10^8 K at 50 MHz.

When the type II burst occurs near the disk center, the fundamental usually dominates in brightness temperature [Nelson and Robinson, 1975]. Herringbones are the brightest features in type II bursts with a brightness temperature exceeding 10^{13} K. In the case of split-band sources, the lower split band seems to dominate most of the time.

2.10. Polarization of Type II Bursts

The type II radio emission is only weakly polarized (a few percent). However, the herringbones could have high degree of polarization (up to 70%), similar to the case of normal type III bursts. This is additional support for the idea that the herringbone emission is caused by electron beams streaming out of the shock front. The herringbones are polarized in the same sense as the backbones [Suzuki et al, 1980].

3. ASSOCIATED PHENOMENA

3.1. Flares, CMEs and Type II Bursts

Type II radio bursts are relatively rare compared to the number of eruptive events taking place on the Sun within any given time interval. Almost all of the type II bursts are associated with flares and coronal mass ejections (CMEs). The triangular relationship involving type II bursts, flares and CMEs is extremely complex and is highly controversial [Gopalswamy et al., 1998; Cliver et al, 1999]. Only a small number of flares are accompanied by type II bursts, but almost all type II bursts are associated with flares. Similarly, almost all type II bursts seem to be associated with CMEs [Cliver et al, 1999], although it was thought earlier that a third of the type II bursts are not associated with CMEs. Statistical studies have shown that 85% of all type II bursts are associated with Hα flares [Dodge, 1975]. The remaining 15% are probably associated with backside flares. Optical flares of all importance are associated with type II bursts (40% – subflares; 40% – importance 1 and 20% – importance 2 and 3). However, the fraction of type II bursts in each importance class increases with increasing importance. Vrsnak et al., [1995] compared the peak microwave flux of flares with and without type II bursts and found that the type II flares in each class are an order of magnitude more powerful than those without type II. Pearson et al., [1989] found little difference in flare size, impulsiveness and energetics between the hard X-ray flares with and without type II bursts. Cliver et al., [1999] found that the fractional number of type II bursts increased with > 25 keV hard X-ray count rate. When the hard X-ray flares were grouped based on their count rates as 10^2–10^3, 10^3 – 10^4, 10^4 – 10^5, and $\geq 10^5$ c s^{-1} Cliver et al., [1999] found that 2%, 17%, 41% and 100% of hard X-ray flares in each of these groups were associated with type II bursts.

Roberts [1959] was the first to recognize that some rare condition must be satisfied before a flare is accompanied by a type II burst. Material ejection was already considered [Dodson et al., 1953], but since such ejection is not uncommon during flares, a fast ejection was proposed to be the special condition [Giovanelli and Roberts, 1959]. Flare sprays attain very high speeds (> 700 km s^{-1}) and so they can drive shocks in the corona [see e.g., Garczynska, 1991]. The *Yohkoh* mission discovered fast X-ray ejecta with speeds sufficiently high to produce shock waves and hence type II bursts [Gopalswamy et al., 1997; 1999c; Klein et al, 1999]. A close relationship between fast CMEs and type II bursts was noted soon after the discovery of CMEs [Gosling et al, 1976; Munro et al., 1979]. Cliver et al., [1999] argue that presence of fast CMEs is essential for type II bursts. There are some problems, though: (i) not all fast CMEs are associated with type II bursts [Sheeley et al., 1984] so we may have to look for another special condition, (ii) a significant number of slow CMEs are associated with metric type II bursts; possible projection effect may be important for these CMEs, but many limb CMEs are known to be slow.

Statistical studies on flares, metric type II bursts and CMEs have shown a population of type II bursts without associated CMEs [Kahler et al., 1984; Sheeley et al., 1984]. Cliver et al., [1999] reexamined the same data set and concluded that the CMEs might have escaped detection in these cases.

How exactly the shocks responsible for the type II bursts originate in eruptive events is thus not clear. Since the shocks are definitely driven by CMEs in the IP medium [e.g. Sheeley et al., 1985], it is simplest to think that the same thing happens close to the Sun. The problem is the coronal medium near the Sun is so different from the IP medium in terms of magnetic field topology and local characteristic speeds. We point out that the following additional possibilities exist for metric type II shocks: (i) Flare blast waves: Blast waves are thought to be due to sudden heating of the coronal loops during flares. This seems to be consistent with the short life times of metric type II bursts [Cane and White, 1987; Bougeret, 1985; Vrsnak et al., 1995;

Gopalswamy et al., 1998]. A major problem with the blast wave scenario is the lack of an observable signature other than the metric the type II burst. (ii) Evaporation shocks: Electron precipitation at the feet of flaring loops causes heating and evaporation of the chromospheric plasma which could act as a piston to drive a shock. Shock formation times calculated by Smith and Brecht [1985] for typical flare conditions are less than 10 s so it is not clear if one can distinguish between evaporation shock and blast wave shocks. Klassen et al., [1999b] have shown evidence for two shocks from a flare site based on the observation of two spatially distinct type II bursts associated with the same flare event.

3.2. Moreton Waves and EIT Waves

Moreton waves were discovered in 1960 as a propagating feature in Hα pictures centered around the flare site and traveling with speeds comparable to those of coronal shocks inferred from type II bursts [Moreton, 1961; Smith and Harvey, 1971]. The Moreton waves have been interpreted as due to the up and down movement of the neighboring chromosphere appearing bright in Hα line center and dark in both wings. Sakurai et al., [1992] observed a Moreton wave propagating at 2500 km s^{-1} in the beginning and accelerating to 4000 km s^{-1} when it reached the north polar region. If the Moreton wave starts in the corona, should there not be a coronal signature? The Extreme-ultraviolet Imaging Telescope (EIT) on board the Solar and Heliospheric Observatory (SOHO) detected a new wave phenomena similar to the Moreton waves known as EIT waves [Thompson et al., 1998; 1999a]. A typical EIT wave is a slow wave that propagates symmetrically about the region of eruption. A preliminary examination of the EIT data corresponding to metric type II bursts suggests that almost all the type II bursts have a corresponding EIT wave signature [Thompson, 1999, private communication].

The relation between EIT waves and Hα Moreton waves was recently confirmed when Barry Reynolds, an amateur astronomer, photographed an Hα Moreton wave during the September 24, 1997 event [Thompson, et al., 1999b]. Figure 4 shows the Hα Moreton wave superposed on the EIT wave at two instances. A metric type II burst was observed by the Hiraiso Radiospectrograph at 02:49 UT, corresponding to the time of the EIT wave. Note that EIT wave has a 'brow' shape, morphologically similar to the Hα Moreton wave. The brow type EIT wave is rather rare, like the Hα Moreton wave. It is not clear if all the EIT waves have the brow feature early on; we need higher cadence images to clarify this. Maia et al., [1997] found that the April 07, 1997 type II burst was located to the south of an active region in the southeast quadrant, although the

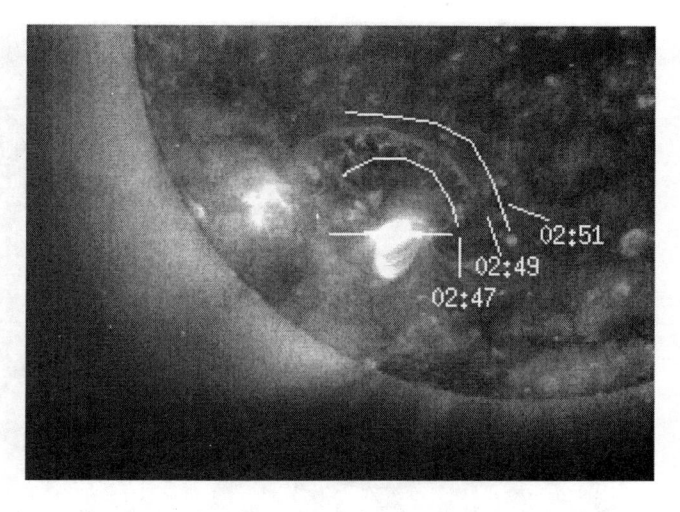

Figure 4. Hα Moreton wave (white lines at 02:47 and 02:51 UT) on either side of the EIT brow wave at 02:49 UT observed on September 24, 1997. Hiraiso radiospectrograph observed a type II starting at 02:49 UT. The horizontal streak in the active region is due to saturation. North is to the top and east is to the left. [Courtesy: B. J. Thompson]

EIT wave appeared to be symmetric around the active region. In another case, the metric type II burst was located precisely on the EIT brow wave. Fig. 5 shows a superposition of the type II burst contours at 164 MHz on an EIT difference image showing the close spatial relationship [Gopalswamy et al., 2000]. Note that the EIT brow and the type II burst are confined to the equatorial streamer region to the north of the flaring active region. This may be due to the refraction of the shock wave into the equatorial streamer region as suggested by Uchida [1974].

3.3. Association with Coronal Holes and Filaments

Shelke and Pande [1985] studied 84 type II bursts during April 1977 through June 1978 from 33 active regions. The type II-rich active regions were all found to be located near coronal holes. These authors suggested that the coronal shocks strengthen in coronal holes and propagate along the open field lines. This is contradictory to the work of Uchida [1968; 1974] that found shocks strengthening in the streamer region. Grib et al., [1996] studied the interaction of fast mode shocks with coronal structures such as streamers and coronal holes. They found that fast mode shocks refract into the streamers and coronal holes as fast mode and slow mode shocks, respectively. Nevertheless, the observation of Shelke and Pande [1985] is very interesting in that it points to a specific magnetic configuration in the vicinity of the eruption. Aurass and Rendtel [1989]

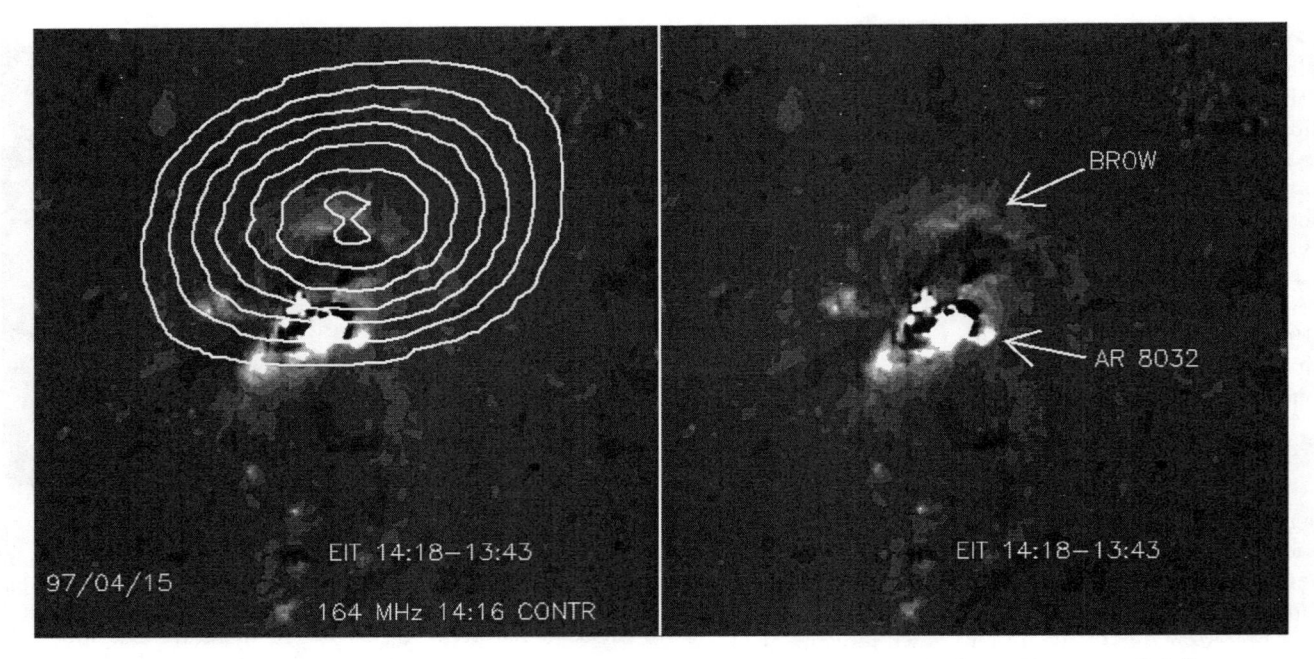

Figure 5. Contour map of the April 15, 1997 metric type II burst at 164 MHz (from Nancay radioheliograph) superposed on an EIT difference image showing the brow feature to the north of AR 8032. Note that the east west extent of the type II and the brow are similar and that the type II burst is confined to the northern side of the AR. The active region was located in the southeast quadrant. North is to the top and east is to the left.

found a similar prolific production of type II bursts from an active region accompanied by a low latitude coronal hole. One would expect the availability of shock drivers in these configurations because filaments located near coronal holes are more likely to erupt [Webb et al., 1978; Bravo and Blanco-Cano, 1998]. In another statistical study, Stewart [1984] found close spatial relationship between H-alpha filaments and type II bursts, suggesting that the type II radio emission most probably originates from within a coronal helmet streamer overlying the filament channel. Therefore, the type II burst must end when the shock reaches a coronal hole, unless type II bursts are also produced by slow mode shocks as suggested by Kundu et al., [1989].

4. CURRENT ISSUES AND FUTURE PROSPECTS

Type II bursts is a rare phenomenon that occurs during the general class of eruptive solar events consisting of CMEs and their associated flares. In spite of their long history, many questions remain. We shall touch upon some of these.

4.1. Deceleration of Type II Bursts

A significant number of metric type II bursts (15%) shows deceleration within their typical life time, identified as a change in slope in the dynamic spectra [Robinson, 1985]. For a sample of type II bursts with speeds exceeding 1100 km s^{-1}, the fraction of decelerating cases climbed to 40%. Most of the white light CMEs do not decelerate in the coronagraph field of view (out to $\sim 30\ R_\odot$), although they do decelerate to solar wind speed in the IP medium. If coronal shocks are driven by CMEs, it is hard to see why they would decelerate close to the Sun while their drivers do not. There are three possible explanations for the type II burst deceleration: (i) Shocks with speeds exceeding a critical value of the Mach number would result in the abrupt decrease in speed causing the shock to become turbulent and become highly dissipative [Karlicky et al., 1984]. (ii) Nonradial propagation of the shock at later stages may show a decrease in drift rate [Markeev et al., 1983]. (iii) When there is a change in the density gradient along the shock path, it is possible for the shock speed to decrease. The refraction of a fast mode shock into a coronal hole as a slow mode shock is one such situation. Propagation of brow type EIT waves towards the streamer region and location type II bursts to the side of hot ejecta [Gopalswamy et al, 1997; Klein et al., 1999] seem to suggest nonradial propagation of shocks.

4.2. Implications of Type II Burst Locations

Robinson and Stewart [1985] found a class of type metric II bursts with good the position angle and temporal correspondence with white light CMEs, but the type II sources were located radially below the CME leading edge. This is contradictory to the idea of CME driven shocks unless the type II emission originates only from the flanks of the shock. One of these events is shown in Fig. 6. Note that the CME was at a height of 3.3 $R_\odot$ (at 00:27 UT) even before the type II burst started. The type II burst appeared at a height of 1.9 $R_\odot$ at 00:46 UT. The CME height extrapolated to the type II burst time would be at 4.5 $R_\odot$ because the CME speed was measured to be 750 km s^{-1}. Even if the shock were draped around the CME, it is hard to explain a CME-driven shock at a distance of 2.5 $R_\odot$ behind the CME leading edge.

Sometimes, type II bursts have two sources located not only behind the CME leading edge but at the "legs" of the CME [Gary et al., 1984, Wagner and MacQueen, 1983]. Based on a number of similar observations, Wagner and MacQueen [1983] proposed that there are two independent shocks, one from the flare and the other driven by the CME [see also Cane, 1984]. Since CMEs precede flares, the flare shock would propagate through the CME and hence the type II burst can have any positional relationship with the CME depending upon the relative speeds and directions of propagation. This scenario will also explain the huge discrepancies between shock speeds and CME speeds found by Gergely et al., [1984] [see also Gopalswamy and Kundu, 1992]. Another support to such a two-shock model comes from the observation that two type II bursts are seen in ~ 20% of all type II events [Robinson and Sheridan 1982]. Holman and Pesses [1983] suggested that the shock-drift acceleration mechanism produces energetic electrons responsible for type II bursts only when the shock normal is nearly perpendicular to the upstream magnetic field. This condition seems to be met only at the flanks of the CME-driven shock and hence the type II source would appear to come from behind the CME. While this is certainly a plausible scenario, there are efficient mechanisms to produce particle acceleration in quasi-parallel shocks also [see e.g. Mann, 1995; Lembege, 1995]. Considering the number of type II bursts observed in the IP medium where the upstream magnetic field is along the shock normal, one definitely expects particle acceleration from quasiparallel shocks. We need to explore other possibilities such as relative shock strength in the nose and flank regions of the shock.

4.3. A New Spectral Feature

Klassen et al., [1999a] have identified an arc pattern at the beginning of type II bursts which they believe are signatures of the early disturbances getting ready to produce type II bursts. The arc pattern is seen at

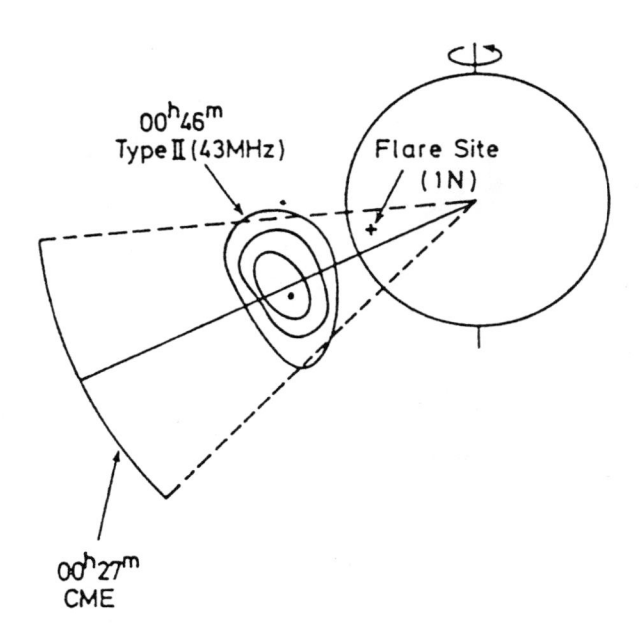

Figure 6. A 43 MHz type II burst (contours) located far behind the leading edge of the associated CME. The type II burst was observed by the Culgoora radioheliograph and the CME was observed by the *Solwind* coronagraph on board the P78-1 spacecraft. [After Robinson and Stewart, 1985].

the beginning of both fundamental and harmonic components and may indicate the propagation of the type II disturbance out of the flaring structure. An ideal set of observations would be to find the spatial location of the type II burst and the arc pattern coupled with imaging observations in X-rays and EUV so that we can pin down the origin of coronal shocks.

4.4. Coronal and IP Shocks

Relation between coronal and IP shocks has remained controversial. Gopalswamy et al [1998] compared metric type II bursts with IP shocks over an 18 month period and found minimal overlap. They concluded that the metric and coronal shocks are of independent origin. Reiner and Kaiser [1999] studied 4 type II bursts in the 1-14 MHz range and compared them with metric type II bursts. They concluded that both metric and longer wavelength type II bursts could not have been produced by a single shock. Only in some simple cases like the one in Fig. 7, we do see a nice continuation of the metric type II burst into the IP medium. In most other cases the situation calls for more than one shock or a highly inhomogeneous shock structure. Type II bursts in the D-H domain seem to be closely related to

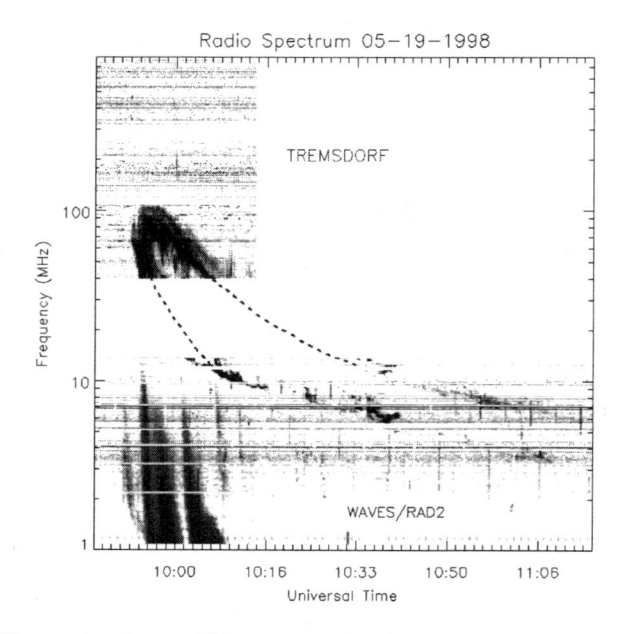

Figure 7. A type II burst associated with a filament eruption and CME on 1998 May 19. The type II burst can be seen to continue from the metric domain (Tremsdorf) into the WAVES/RAD2 spectral domain. Note the intense type III bursts (herringbones?) at and after the onset of type II burst. We have drawn dashed lines to fill the gap between Tremsdorf and WAVES data.

CMEs as in the case of kilometric type II bursts [see e.g., Cane, et al., 1987]. Gopalswamy et al., [1999b] searched for the distinguishing characteristics of CMEs associated with D-H type II bursts and found them to be faster (average speed = 854 km s^{-1}) and wider (average width = 160°) than normal CMEs (average speed = 450 km s^{-1}; average width = 45°). Figure 8 shows a comparison between speeds of CMEs associated with metric and D-H type II bursts. It appears that shocks associated with wider CMEs have a better chance to produce radio emission in the IP medium.

Simultaneous imaging observations of metric and IP type II bursts is needed to settle a number issues discussed above. The Culgoora and the Clark Lake radioheliographs imaged metric type II bursts in the seventies and eighties and are now closed. The only functional metric radioheliograph at Nancay operates at frequencies above 164 MHz. Since the type II bursts are more frequent at lower frequencies, the Nancay radioheliograph may miss many type II bursts. The Giant Meterwave Radio Telescope (GMRT) in India is ideal for imaging type II bursts but it is not a dedicated solar instrument. Imaging observations of IP type II bursts, of course, has to be done in space and is a logical step for low frequency radio astronomy [Jones et al., 1999].

It is easy to visualize a variety of mass motions near the Sun that could serve as shock sources. In the IP medium, CMEs are the main source of shocks. At large distances from the Sun, corotating interaction regions (CIRs) also result in shocks. The onset of these CIR regions and their role in particle acceleration is an important problem to investigate. CMEs propagating into these regions evolve, and so do shocks driven by them. The large variation in flow speed from the equatorial to the polar regions can have consequences for the evolution of large scale CMEs and the shocks driven by them [Odstracil and Pizzo, 1999]. Numerical simulations of these processes seem to indicate the possibilities of multiple shocks associated with a single CME. Gosling et al., [1994] have identified a new class of forward-reverse shock pairs indicating over expansion of CMEs in the solar wind. Possibilities of particle acceleration and radio emission from these shocks need to be explored.

Acknowledgments. The author is indebted to E. Cliver, A. Lara and D. Webb for comments on the paper and to the AGU for travel support. Work supported by NASA and NSF.

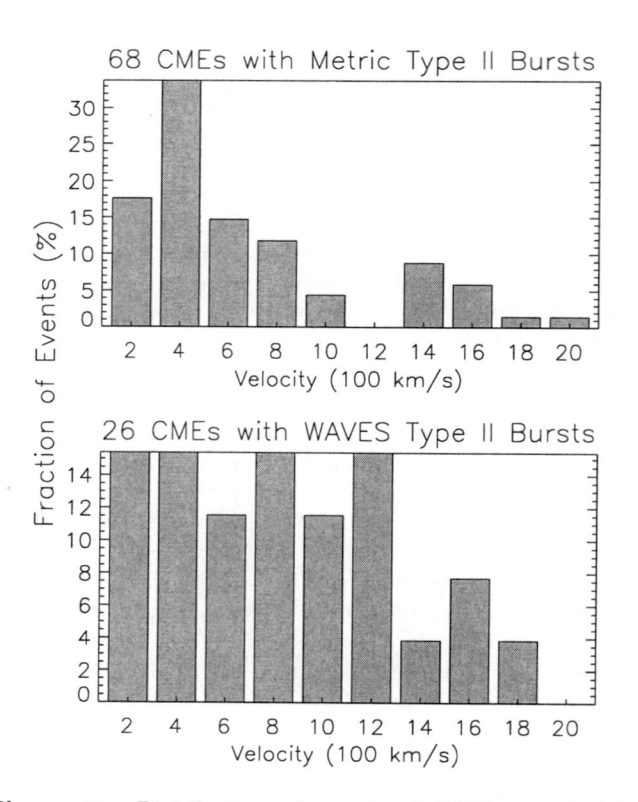

Figure 8. Distribution of speeds of CMEs associated with metric type II bursts (upper) and with decameter-hectometric (D-H) type II bursts (lower). Note that the speed distribution is much broader in the D-H case.

REFERENCES

Aurass, H. and J. Rendtel, Flare - Type II Burst Association in the February 1986 Activity Complex, *Solar Phys.*, 122, 381, 1989.

Aurass, H., Coronal mass ejections and type II radio bursts, in *Coronal Physics from Radio and Space Observations*, edited by G. Trottet, Springer-Verlag, New York, 135, 1997.

Boischot, A., A. C. Riddle, J. B. Pearce, J. W. and Warwick, Shock waves and type II radio bursts in the interplanetary medium, *Solar Phys.*, 65, 379, 1980.

Bougeret, J.-L., Observations of shock formation and evolution in the solar atmosphere, in *Collisionless Shocks in the Heliosphere: Reviews of Current Research*, edited by B. T. Tsurutani and R. G. Stone, AGU, Washington, D. C. p. 13, 1985.

Bougeret, J.-L. et al., Waves: The Radio and Plasma Wave Investigation on the Wind Spacecraft, *Space Sci. Rev.*, 71, p. 231, 1995.

Bougeret, J. L. et al., A shock-associated (SA) radio event and related phenomena observed from the base of the solar corona to 1 AU, *Geophys. Res. Lett.*, 25, 2513, 1998.

Bravo, S. and X. Blanco-Cano, Signatures of interplanetary transients and their associated near-surface solar activity, *Ann. Geophysicae*, 16, 359, 1998.

Cane, H. V., The relationship between coronal transients, type II bursts and interplanetary shocks, *Astron. Astrophys.*, 140, 205, 1984.

Cane, H. V., R. G. Stone, J. Fainberg, R. T. Stewart, J. L. Steinberg, and S. Hoang, Radio evidence for shock accelerated electrons in the Solar corona, *Geophys. Res. Lett.*, 8, 1285, 1981.

Cane, H. V., N. R. Sheeley, Jr., and R. A. Howard, Energetic interplanetary shocks, radio emission, and coronal mass ejections, *J. Geophys. Res.*, 92, 9869, 1987.

Cane, H. V. and White, S. M., On the source conditions for herringbone structure in type II solar radio bursts, *Solar Phys.*, 120, 137, 1989.

Cliver, E. W., D. F. Webb, and R. A. Howard, On the origin of solar metric type II bursts, *Solar Phys.*, 187, 89, 1999.

Dodson, H. W., E. R. Hedeman and J. Chamberlain, Ejection of hydrogen and ionized calcium atoms with high velocity at the time of solar flares, *Astrophys. J.*, 117, 66, 1953.

Dodge, J. C., Source regions for type II radio bursts, *Solar Phys.*, 42, 445, 1975.

Garczynska, I. N., Propagation of sprays with deceleration, *Solar Phys.*, 131, 129, 1991.

Gary, D. E., G. A., Dulk, L. House, R. Illing, C. Sawyer, W. J. Wagner, D. J. McLean, and E. Hildner, *Astron. Astrophys.*, 134, 222, 1984.

Gergely, T. E., M. R. Kundu, F. Erskine, C. Sawyer, W. J. Wagner, R. Illing, L. L. House, M. K. McCabe, R. T. Stewart, and G. J. Nelson, Radio and visible-light observations of a coronal arcade transient, *Solar Phys.*, 90, 161, 1984.

Giovanelli, R. G. and J. A. Roberts, in *Paris Symposium on Radio Astronomy*, edited by R. N. Bracewell, Stanford University Press, Stanford, p. 201, 1959.

Gopalswamy, N. and M. R. Kundu, The observation of an unusually fast type IV plasmoid, *Astrophys. J.*, 365, L31, 1990.

Gopalswamy, N. and M. R. Kundu, Are Coronal Shocks Piston Driven? in *AIP Conference Proceedings # 264: Particle Acceleration in Cosmic Plasmas*, edited by G. P. Zank and T. K. Gaisser, American Institute of Physics, New York, p. 257, 1992.

Gopalswamy, N., M. R. Kundu, P. K. Manoharan, A. Raoult, N. Nitta and P. Zarka, X-ray and Radio studies of a coronal eruption: Shock wave, plasmoid and coronal mass ejection, *Astrophys. J.*, 486, 1086, 1997.

Gopalswamy, N., M. L. Kaiser, R. P. Lepping, S. W. Kahler, K. Ogilvie, D. Berdichevsky, T. Kondo, T. Isobe and M. Akioka, Origin of coronal and interplanetary shocks - A new look with WIND spacecraft data, *J. Geophys. Res.*, 103, 307, 1998.

Gopalswamy, N., M. L. Kaiser, R. J. MacDowall, M. J. Reiner, B. J. Thompson, O. C. St. Cyr, Dynamical phenomena associated with a coronal mass ejection, in *Solar Wind Nine, AIP Conference Proceedings 471*, edited by S. Habbal et al., p. 641 1999a.

Gopalswamy, N., M. L. Kaiser, B. J. Thompson, L. Burlaga, A. Szabo, A. Lara, A. Vourlidas, S. Yashiro, and J.-L. Bougeret, Radio-rich solar eruptive events, *Geophys. Res. Lett.*, in press, 1999b.

Gopalswamy, N., N. Nitta, P. K. Manoharan, A. Raoult and M. Pick, X-ray and radio manifestations of a solar eruptive event, *Astron. Astrophys.*, 347, 684, 1999c.

Gopalswamy, N., S. Yashiro, M. L. Kaiser and B. J. Thompson, and S. Plunkett, Multi-wavelength signatures of a coronal mass ejection, in *Solar Physics with Radio Observations*, ed. T. Bastian, N. Gopalswamy and K. Shibasaki, in press, 1999d.

Gopalswamy, N., M. L. Kaiser, J. Sato and M. Pick, Shock waves and electron beams during a hard X-ray flare, in *Proc. HESSI Workshop*, edited by R. Ramaty et al., in press, 2000.

Gosling, J. T., E. Hildner, R. M. MacQueen, R. H. Munro, A. I. Poland, and C. L. Ross, The speeds of coronal mass ejection events, *Solar Phys.*, 48, 389, 1976.

Gosling, J. T., D. J. McComas, J. L. Phillips, L. A. Weiss, V. J. Pizzo, B. E. Goldstein, and R. J. Forsyth, A new class of forward-reverse shock pairs in the solar wind, *Geophys. Res. Lett.*, 21, 2271, 1994.

Grib, S. A., Koutchmy, S. and Sazonova, V. N., MHD shock interactions in coronal structures, *Solar Phys.*, 169, 151, 1996.

Haddock, F., Introduction to radio astronomy, *Proc. I. R. E.*, 46, 1, 1958.

Holman, G. D. and M. E. Pesses, Solar type II emission and the shock-drift acceleration of electrons, *Astrophys. J.*, 267, 837, 1983.

Jones, D. L. et al., Astronomical Low Frequency Array and the Solar Imaging Array: ALFA/SIRA, this volume.

Kahler, S. W., N. R. Sheeley, Jr., R. H. Howard, D. J. Michels, M. J. Koomen, Characteristics of flares producing metric type II bursts and coronal mass ejections, *Solar Phys.*, 93, 133, 1984.

Kahler, S. W., E. W. Cliver, and H. V. Cane, The relationship of shock-associated kilometric radio emission with metric type II bursts and energetic particles, *Astrophys. J., 302*, 504, 1986.

Kaiser, M. L., M. J. Reiner, R. A. Howard, N. Gopalswamy, O. C. St. Cyr, B. J. Thompson, and J.-L. Bougeret, Type II radio emissions in the frequency range from 1-14 MHz associated with the April 7, 1997 solar event, *Geophys. Res. Lett., 25*, 2501, 1998.

Karlicky, M. Narrow band dm-spikes as indication of flare mass ejection, *Solar Phys., 92*, 329, 1984.

Klassen, A., H. Aurass, K. L. Klein, A. Hofmann, and G. Mann, Radio evidence on shock wave formation in the solar corona, *Astron. Astrophys., 343*, 287, 1999a.

Klassen, A., M. Karlicky, H. Aurass, and K. Jiricka, On discriminating different flare shocks during the flare of 9 July, 1996, *Solar Phys.*, in press, 1999b.

Klein, K.-L., J. I. Khan, N. Vilmer, J.-M. DeLouis, and H. Aurass, X-ray and radio evidence on the origin of a coronal shock wave, *Astron. Astrophys., 346*, L53, 1999.

Kruger, A., Introduction to solar Radio astronomy and Radio Physics, Reidel, Dodrecht, 1979.

Kundu, M. R., Solar Radio Astronomy, Interscience Publishers, N. Y., 1965.

Kundu, M. R. and R. G. Stone, Observations of solar radio bursts from meter to kilometer wavelengths, *Adv. Space Res., 4*, 261, 1984.

Kundu, M. R., N. Gopalswamy, N., S. M. White, P. Cargill, E. J Schmahl, and E. Hildner, The radio signatures of a slow coronal mass ejection - Electron acceleration at slow-mode shocks? *Astrophys. J., 347*, 505, 1989.

Kundu, M. R. and N. Gopalswamy, Meter-decameter radio emissions associated with a CME and a disappearing filament, in *Eruptive Solar Flares*, Z. Svestka, B. V. Jackson and M. E. Machado (eds), Lecture Notes in Physics Series 399, Springer Verlag, New York, p. 268, 1992.

Lembege, B., Numerical simulations of shock electron acceleration in solar physics, in *Coronal Magnetic Energy Releases*, Edited by A. O. Benz, and A. Kruger, (Berlin: Springer), p. 201, 1995.

Lengyel-Frey, D., G. Thejappa, R.J. MacDowall, R.G. Stone, and J. L. Phillips, Ulysses Observations of Wave Activity at Interplanetary Shocks and Implications for Type II Radio Bursts, *J. Geophys. Res., 102*, 2611, 1997.

MacDowall, R. J., R. G. Stone, and M. R. Kundu, 1987, Characteristics of shock-associated fast-drift kilometric radio bursts, *Solar Phys., 111*, 397, 1987.

Maia, D., M. Pick, R. Howard, G. E. Brueckner, and P. Lamy, The April 7, 1997 Event: LASCO and Nancay Radioheliograph Joint Observations, Proc. of the Fifth SOHO Workshop, 'The Corona and Solar Wind Near Minimum Activity,' *ESA-SP 404*, p. 539, 1997.

Malitson, H. H., J. Fainberg, and R. G. Stone, Observations of a type II radio burst to 37 Ro, *Astrophys. Phys. Lett., 14*, 111, 1973.

Mann, G., Theory and observations of coronal shock waves, 1995; in *Coronal magnetic Energy Releases*, Edited by A. O. Benz, and A. Kruger, Springer, Berlin, p. 183, 1995.

Mann, G. et al., Catalogue of solar type II radio bursts observed from September 1990 to December 1993 and their statistical analysis, *Astron. Astrophys. Suppl. Ser., 119*, 489, 1996.

Markeev, A. K., V. V. Formichev, I. M. Chertok, A. Bhatnagar, R. M. Jain, R. N. Shelke, and R. V. Bhonsle, U-shaped type II solar radio bursts associated with the 1980 March 28 flare, *BASI, 11*, 318, 1983.

Maxwell, A. and A. R. Thompson, Spectral observations of solar radio bursts II: Slow drift bursts and coronal streamers, *Astrophys. J., 135*, 138, 1962.

McLean, D. J., *Proc. Astron. Soc. Australia, 1*, 47, 1967.

Moreton, G. F., Fast-moving disturbances on the Sun, *Sky and Telescope, 21*, 145, 1961.

Munro, R. H., J. T. Gosling, E. Hildner, R. M. MacQueen, A. I. Poland and C. L. Ross, The association of coronal mass ejection transients with other forms of solar activity, *Solar Phys., 61*, 201, 1979.

Nakajima, H., Kawashima, S., Shinohara, N., Shiomi, Y., Enome, S. and Rieger, E., A high-speed shock wave in the impulsive phase of 1984 April 24 flare, *Astrophys. J., 73*, 177, 1990.

Nelson, G. J, and R. D. Robinson, Multi-frequency heliograph observations of type II bursts, *PASA, 2*, 370, 1975.

Nelson, G. J., and D. B. Melrose, in *Solar Radio Physics*, edited by D. J. McLean and N. R. Labrum, Cambridge, New York, P. 333, 1985.

Odstrcil, D. and Pizzo, V., Three dimensional propagation of coronal mass ejections (CMEs) in a structured solar wind flow 1. CME launched within the streamer belt, *J. Geophys. Res., 104*, 483, 1999.

Payne-Scott, R., D. E. Yabsley and J. G. Bolton, Relative times of arrival of bursts of solar noise on different radio frequencies, *Nature, 160*, 256, 1947.

Pearson, D. H., R. Nelson, G. Kojoian, and J. Seal, Impulsiveness and energetics in solar flares with and without type II radio bursts - A comparison of hard X-ray characteristics for over 2500 solar flares *Astrophys. J., 336*, 1050, 1989.

Reiner, M. J. and M. L. Kaiser, High-frequency type II radio emissions associated with shocks driven by coronal mass ejections, *J. Geophys. Res., 104*, 16979, 1999.

Reiner, M. J., M. Karlicky, K. Jiricka, H. Auras, G. Mann, and . L. Kaiser, On the solar origin of complex type III-like bursts at and below 1 MHz, *J. Geophys. Res.*, (submitted), 1999.

Roberts, J. A., Solar radio bursts of spectral type II, *Aust. J. Phys., 12*, 327, 1959.

Robinson, R. D., Velocities of type II solar radio events, *Solar Phys., 95*, 343, 1985.

Robinson, R. D. and K. V. Sheridan, A study of multiple type II solar radio events, *Proc. Astron. Soc. Australia, 4*, 392, 1982.

Robinson, R. D. and R. T. Stewart, A positional comparison between coronal mass ejection events and solar type II bursts, *Solar phys., 97*, 145, 1985.

Sakurai, T., K. Ichimoto, K. Strong, E. Hiei, M. Irie, K. Kumagai, M. Miyashita, Y. Nishino, K. Yamaguchi, G. Fang, M. Kambry, Z. Zhao, and K. Shinoda, White-light flares of 1991 June in the NOAA region 6659, *PASJ, 44*, L7, 1992.

Sheeley, N. R. Jr., R. T. Stewart, R. D. Robinson, R. A. Howard, M. J. Koomen, and D. J. Michels, Associations between coronal mass ejections and metric type II bursts, *Astrophys. J., 279*, 839, 1984.

Sheeley, N. R. Jr., R. A. Howard, D. J. Michels, M. J. Koomen, R. Schwenn, K. H. Muehlhaeuser, and H. Rosenbauser, Coronal mass ejections and interplanetary shocks, *J. Geophys. Res., 90*, 163, 1985.

Shelke, R. N. and M. C. Pande, Propagation of type II burst shock waves along the radial magnetic field in the solar corona, *BASI, 13*, 62, 1985.

Smerd, S. F., K. V. Sheridan, and R. T. Stewart, On split-band structure in type II radio bursts from the Sun, in Proceedings from IAU Symposium no. 57, edited by G. A. Newkirk, Reidel, Boston, p. 389, 1974.

Smith, S. F. and K. L. Harvey, Observational effects of flare associated waves, in *Physics of the Solar Corona*, edited by C. J. Macris, D. Reidel, Boston, p.156, 1971.

Smith, D. F. and S. H. Brecht, Shock formation time and the variability of prompt Mev proton acceleration in solar flares, *J. Geophys. Res., 90*, 205, 1985.

Stewart, R. T., Association of type II solar radio bursts with coronal structures above H-alpha filament channels, *Solar Phys., 94*, 379, 1984.

Stewart, R. T. and A. Magun, Radio evidence for electron acceleration by transverse shock waves in herringbone type II solar radio bursts, *PASA, 4*, 53, 1980.

Suzuki, S., R. T. Stewart and A. Magun, Polarization of herringbone structure in Type II bursts, in *Radio Physics of the Sun*, edited by M. R. Kundu and T. E. Gergely, D. Reidel, Boston, p. 241, 1990.

Sveztka, Z., *Solar Flares*, D. Reidel, Dordrecht, 1976.

Thompson, B. J., S. P. Plunkett, J. B. Gurman, J. S. Newmark, O. C. St. Cyr, and D. J. Michels, SOHO/EIT observations of an Earth-directed coronal mass ejection on May 12, 1997, *Geophys. Res. Lett., 25*, 2465, 1998.

Thompson, B. J. et al., SOHO/EIT Observations of the 1997 April 7 Coronal Transient: Possible Evidence of Coronal Moreton Waves, *Astrophys. J., 517*, L151, 1999a.

Thompson, B. J., et al., Observations of the September 24, 1997 coronal flare waves, *Solar Phys.*, in press, 1999b.

Treumann, R. A. and J. LaBelle, Band splitting in solar type II radio bursts *Astrophys. J., 399*, L167, 1992.

Uchida, Y., On the exciters of type II and type III solar radio burst, *PASJ, 12*, 376, 1960.

Uchida, Y., Propagation of hydromagnetic disturbances in the solar corona and Moreton's wave phenomenon, *Solar Phys., 4*, 30, 1968.

Uchida, Y., Behavior of the flare produced coronal MHD wave front and the occurrence of type II radio bursts, *Solar Phys., 39*, 431, 1974.

Vrsnak, B., V. Ruzdjak, P. Zlobec and H. Aurass, Ignition of MHD shocks associated with solar flares, *Solar Phys., 158*, 331, 1995.

Wagner, W. J. and R. M. MacQueen, The excitation of type II radio bursts in the corona, *Astron. Astrophys., 120*, 136, 1983.

Webb, D. F., J. T. Nolte, C. V. Solodyna and P. S. McIntosh, Evidence linking coronal transients to the evolution of coronal holes, *Solar Phys., 58*, 389, 1978.

Weiss, A. A., The positions and movements of the sources of solar radio bursts of type II, *Australian J. Phys., 16*, 240, 1963.

Weiss, A. A., The nature and velocity of the source of type II solar radio bursts, *Australian J. Phys., 18*, 167, 1965.

Wild, J. P. and McCready, L. L., Observations of the spectrum of high intensity solar radiation at meter wavelengths, *Austral. J. Sci. Res., A 3*, 387, 1950.

Wild, J. P., Sheridan, K. V. and Trent, G. H., The transverse motions of the sources of solar radio bursts, in *Paris Symposium on Radio Astronomy*, edited by R. N. Bracewell, Stanford Univ. Press, p. 176, 1959.

Zheleznyakov, V. V., *Radio emission of the Sun and planets*, Pergamon Press, New York, 1969.

Zlotnik, E. Ya., A. Klassen, K.-L. Klein, H. Aurass, and G. Mann, Third harmonic plasma emission in solar type II radio bursts, *Astron. Astrophys., 311*, 1087, 1998.

N. Gopalswamy Center for Solar Physics and Space Weather, Institute for Astrophysics and Computational Sciences, Department of Physics, The Catholic University of America, Washington DC 20064, USA. (e-mail:gopals@fugee.gsfc.nasa.gov)

Interplanetary Type II Radio Emissions Associated With CMEs

Michael J. Reiner

Raytheon ITSS, Lanham, Maryland, USA

Some interplanetary shocks associated with coronal mass ejections (CMEs) generate type II radio emissions at the local plasma frequency and/or its harmonic. These type II radio emissions, at decameter-kilometer wavelengths, provide a means of remotely studying and tracking CMEs from the solar corona out to 1 AU and beyond. We review what was learned about the nature and evolution of interplanetary type II radio bursts during the ISEE-3 era and more recently from observations made by the Wind and Ulysses spacecraft.

1. INTRODUCTION

During the times of major solar flares, coronal mass ejections (CMEs) often lift off from the sun and propagate through the interplanetary medium–sometimes eventually encountering Earth. This paper reviews low frequency radio emissions that are associated with this process. The radio emissions associated with the interplanetary propagation of a CME are called *interplanetary type II radio bursts*.

Interplanetary type II radio bursts, i.e., remotely observed radio emissions in the frequency range from ~30 kHz to several megahertz, were first observed by *Malitson et al.* [1973a,b,1976]. These radio emissions are believed to be generated at the fundamental and/or harmonic of the plasma frequency by electrons accelerated at a (CME-driven) interplanetary shock [*Cane et al.*, 1981; *Cane and Stone*, 1984; *Kahler et al.*, 1989] The radio emissions drift slowly to lower frequencies with increasing time. This slow frequency drift results from the speed of propagation of the shock through the decreasing density of the interplanetary plasma with increasing heliocentric distance. Previous reviews that have discussed interplanetary type II radio bursts have been given by *Bougeret* [1985], *Nelson and Melrose* [1985], and *Simnett* [1986].

2. THE ISEE-3 ERA (1978 – 1984)

The ISEE-3 spacecraft was launched into a halo orbit at the Lagrange point (L1) from August of 1978 to ~1984. It contained very sensitive radio receivers covering the frequency range from 30 kHz to 2 MHz that were connected to long (90m tip to tip) wire antennas in the spacecraft spin plane [*Knoll et al.*, 1978]. The modulation of the signals produced by the 3 second spin of the spacecraft permitted the direction of the radio source to be determined [*Fainberg et al.*, 1972].

Most analyses of interplanetary type II radio emissions during the ISEE-3 era (1978 - ~1984) have been statistical studies based on large numbers of observed events to elucidate the characteristics of interplanetary type II radio bursts and their correlation with related phenomena such as CMEs, sudden commencement geomagnetic storms, and metric type II bursts.

Although some information on the characteristics of interplanetary type II radio bursts was obtained from an analysis of radio data from the Voyager spacecraft [*Boischot et al.*, 1980], the first comprehensive investigation of the general characteristics of 15 kilometric type II radio bursts, observed with the more sensitive radio receivers on the ISEE-3 spacecraft, was carried out by *Cane et al.* [1982].

Interplanetary type II radio emission is characterized by a diffuse backbone, with superposed comparatively short-duration sporadic intensifications, which slowly drift to lower radio frequencies from ~1 MHz to ~40 kHz [*Cane et al.*, 1982]. (The interplanetary type II burst could not be observed at lower frequencies because remote radio emissions cannot be observed at frequencies below the local plasma frequency, which at Earth averages at about 30 kHz, corresponding to an average plasma density of ~10 cm^{-3}). Some typical frequency profiles of interplanetary type II radio bursts are shown in Figure 1. Figure 1c shows the diffuse backbone, while Figures 1a and 1b show prominent intensifications. The intensity of the diffuse backbone emission is ~200 sfu (1 sfu = 10^{-22} W m^{-2} Hz^{-1}) at ~100 kHz, while the sporadic intensifications can reach intensities of ~1000 sfu or more [*Lengyel-Frey and Stone*, 1989]. The bandwidth of the diffuse backbone, $\Delta f/f$, is ~0.7 (full width at half maximum) [*Lengyel-Frey and Stone*, 1989]. The average measured frequency drift rate, ~1 kHz/min near 300 kHz, is consistent with an average shock (transit) speed of ~900 km/s [*Cane et al.*, 1982]. The interplanetary type II burst intensity appears to be correlated with the shock speed v, varying roughly as v^3 [*Lengyel-Frey and Stone*, 1989]. The occurrence rate of interplanetary type II bursts during the 1980-81 solar maximum was ~1/month [*Cane et al.*, 1982].

Fundamental and harmonic components, as well as band splitting, were noted for the interplanetary type II radio bursts observed at hectometer and kilometer wavelengths [*Malitson et al.*, 1973a; *Cane et al.*, 1982]. A more comprehensive study of the fundamental and harmonic components for three type II radio bursts obseved by ISEE-3 was carried out by *Lengyel-Frey et al.* [1985]. They found that the fundamental components had a greater average intensity than the harmonic component due primarily to comparatively short-duration narrowband brightenings observed for the fundamental. The fundamental spectral profile was found to be narrower than that of the harmonic and the fundamental source size was found to be significantly larger than that of the harmonic.

There has been a long-standing controversy regarding the location of the interplanetary type II source region relative to the shock. *Malitson et al.* [1973] assumed that the type II radio source was located in the upstream region of the interplanetary shock. *Chertok and Fomichev* [1976] and *Davis and Feynman* [1977] compared the type II frequencies with those deduced

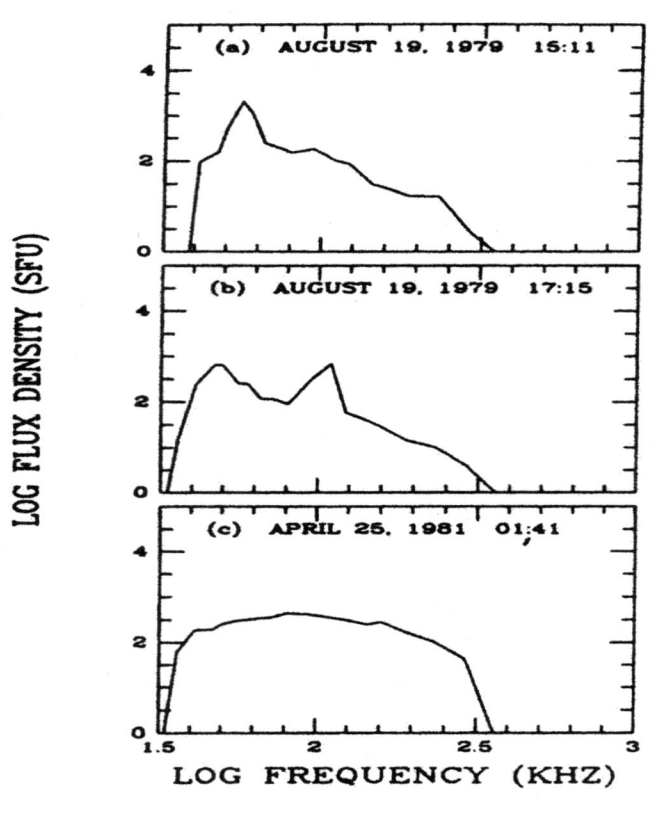

Figure 1. Spectra showing the major categories of type II profies in solar flux units is plotted versus log of observing frequency. (a) Prominent low-frequency component. (b) Spectrum showing evidence of two components. (c) Broad, fairly symmetrical profile. (after Lengyel-Frey et al. [1989]).

from density models of the interplanetary plasma for a given shock dynamics and concluded that the type II radio source was in the downstream region of the shock. *Lengyel-Frey* [1992] compared the type II emission frequencies with the densities in the upstream region of the shock as deduced by extrapolating from the actual measure densities at 1 AU. For the 20 interplanetary shocks that they analyzed, they found that the type II source occurred in regions of enhanced plasma density. They therefore also concluded that the type II radio source was most likely within or behind the shock.

Cane et al. [1985] established the clear relationship of interplanetary type II radio bursts with interplanetary shocks. Then *Sheeley et al.* [1985] established the association of interplanetary shocks with CMEs. Finally, although it was first noted by *Cane et al.* [1984] that interplanetary type II bursts were related to CMEs, it was *Cane et al.* [1987] who clearly demonstrated that all interplanetary shocks that generate interplanetary type II radio bursts were associated with CMEs and that CMEs associated with interplanetary type II events were the most massive and energetic, with shock speeds in excess of 500 km/s. Figure 2 displays histograms of occurances of CMEs versus plane-of-the-sky speeds, showing that CMEs with interplanetary type II radio bursts have plane-of-the-sky speeds greater than 500 km/s.

Interplanetary type II bursts were also found to be associated with solar flares, long-duration soft X-ray events (LDEs), and metric type II and type IV bursts [*Cane et al.*, 1982; *Cane et al.*, 1986]. The direct association with flares and with type II and type IV bursts have recently been called into question [*Gosling*, 1993; *Gopalswamy*, 1998; *Reiner et al.*, 1999c].

3. THE ULYSSES-WIND ERA (1990 –)

The Ulysses spacecraft, which was launched in October, 1990, is in an elliptical orbit about the sun, out of the ecliptic plane, with aphelion (perihelion) of about 5.3 (1.34) AU. The Unified Radio and Plasma (URAP) wave investigation [*Stone et al.*, 1992] consists of several radio receivers with a frequency range from 1.25 to 940 kHz. These receivers are connected to a dipole antenna (35 m elements) in the spin plane and a monopole antenna (7.5 m element) along the spacecraft spin axis. Since at 5.3 AU the average local density is about 0.2 cm^{-3}, Ulysses can observe interplanetary type II bursts to very low frequencies (~1 kHz).

The Wind spacecraft, which was launched in November of 1994, executes complex orbits that include excursions to the Lagrange point (L1) and series of near-Earth passes. The WAVES instrument on Wind includes several sensitive radio receivers that cover the frequency range from 4.0876 kHz to 13.825 MHz [*Bougeret et al.*, 1995]. The receivers are connected to dipole antennas (50m and 7.5m elements) in the spacecraft spin plane and a dipole antenna (5.28m elements) along the spacecraft spin axis. Analysis of the synthesized signals from the Ulysses and Wind receivers permits measurment of the two direction angles of the radio source as well as the four Stokes parameters of the source [*Manning and Fainberg*, 1980].

3.1. Kilometric Type II Radio Emissions

Since Wind was launched during solar minimum, the first interplanetary type II radio burst was not observed until January of 1997. In the first paper on interplanetary type II emission in the Wind era, *Reiner et al.* [1997] introduced a new way analyzing interplanetary type II radio bursts that directly reveals the dynam-

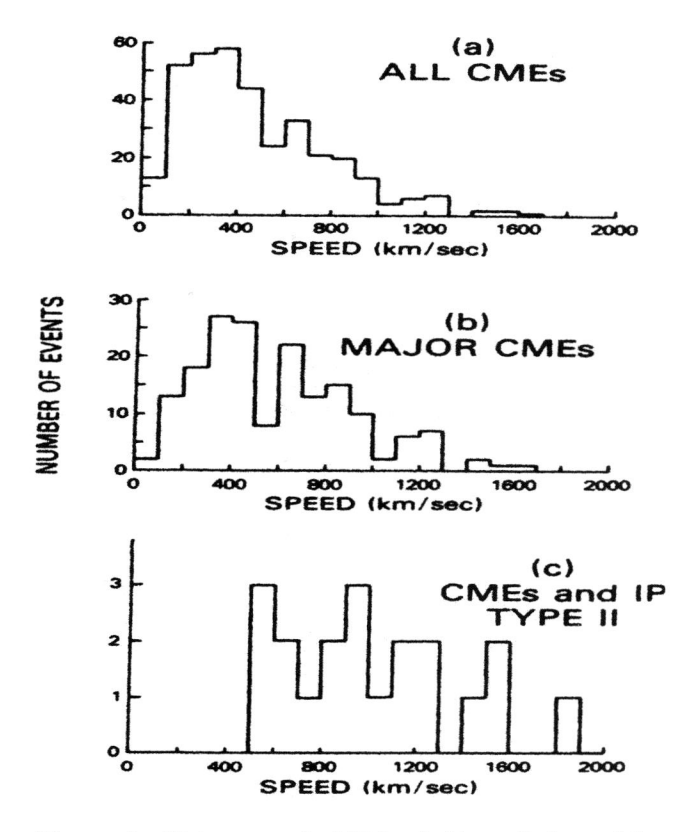

Figure 2. Histograms, in 100 km/s bins, of plane-of-the-sky speeds: (a) all CMEs; (b) all major CMEs; (c) CMEs associated with interplanetary type II events (after Cane et al. [1987]).

ics and evolution of CME induced interplanetary type II radio emissions. Instead of presenting the remotely observed radio data in the usual dynamic spectrum format, which plots the observed radio intensity as a function of observed frequency and time, they plotted the observed radio intensity as a function of inverse frequency, $1/f$, and time. This has the effect of organizing the data such that the relevant dynamic properties of the radio source are more clearly indicated. The logic behind the method is very simple.

The interplanetary plasma density is known, on average, to fall off approximately as $1/R^2$, where R is the heliocentric distance. Assuming that the type II radio emission is generated at the plasma frequency (or its harmonic) and noting that the plasma frequency is directly proportional to the square root of the plasma density, $f(\text{kHz}) = 9\sqrt{n(\text{cm}^{-3})}$, the plasma frequency must on average scale as $1/R$. Thus plotting the radio data as $1/f$ versus time is essentially equivalent to plotting it as R versus time [*Malitson et al.*, 1973a]. On this plot, the radio emission from a given source region along the shock front is expected to be organized along a straight line, since $R \approx v(t - t_o)$, where t is the time, t_o is the CME liftoff time, and v is the shock speed, assuming that the speed of the shock through the interplanetary medium is approximately constant. The slope of this line varies directly as the shock speed and inversely as the square root of the plasma density at 1 AU, slope = $v/(a\sqrt{n_o}R_o)$, where n_o is the density normalized to 1 AU, $R_o = 1.5 \times 10^8$ km, and $a = 9(18)$ depending on whether the emission is at the fundamental (harmonic) of the plasma frequency. Radio emissions generated from different regions of the large CME driven shock front, which has a different local plasma density, will be organized along another straight line with a different slope. By extrapolation, these lines of different local density on the $1/f$ versus time plot should all converge, at high frequency, to the time of liftoff of the CME from the sun. On the other hand, extrapolating these lines forward in time indicates the expected time of arrival of the shock at Earth (1 AU), provided that the local plasma density is known. Such observations have important space weather applicatons. Deviations of the radio emissions from these straight lines indicate either that the plasma density in the source region does not fall off exactly as $1/R^2$ or that the speed of the shock front is not constant. Thus, such deviations can reveal additional information about the physical and dynamical nature of the radio sources such as deceleration of the radio emitting shocks.

Using this method, Reiner et al. [1997] showed that both fundamental and harmonic emissions were observed and easily identified and that the radio emissions appear to be generated in the upstream region of the CME-driven shock. Analyses of wave phenomena in the upstream region of interplanetary shocks have also suggested an upstream radio source [Lengyel-Frey et al., 1997; Thejappa et al., 1997].

Plate 1 shows a radio dynamic spectrum, measured by the WAVES experiment onboard the Wind spacecraft, in August of 1998. This is a plot of the observed radio intensity (shown in color – red being the most intense) plotted as a function of inverse frequency along the vertical axis and time along the horizontal axis. The frequency ranges from 20 kHz to 13.8 MHz (right-hand scale) and the time period is from 22:00 UT on August 24 to 08:00 UT on August 26, 1998. Since the radio emission features shown in Plate 1 are generated by the plasma emission process, radio emissions at high frequencies (high plasma densities) occur very near the sun (~ 2.1 $R_\odot$ for 13.8 MHz), while those at low frequencies (low plasma densities) occur far from the sun (~ 1 AU for 20 kHz).

The very intense rapid frequency-drifting radio emission near 22:00 UT is a complex type III-like radio burst (sometimes called an SA event [Cane et al., 1981]) associated with the X1.0, 3B solar flare, near central meridian, that peaked at $\sim 22:00$ UT on August 24. This SA event [Reiner et al., 1999a] was produced by suprathermal electrons, with speeds of 0.1–0.3c, streaming along magnetic field lines that were open to the interplanetary medium. The weaker more slowly frequency-drifting radio emissions in Plate 1, starting shortly after the onset time of the flare, are the interplanetary type II radio emissions associated with the propagation of the CME from the sun to Earth.

Between $\sim 01:00$ and 06:00 UT on August 25, very intense type II radiation was observed drifting from ~ 500 to 150 kHz. The type II radio emissions produced later were weaker and more diffuse. For example, the diffuse type II radio emissions drifted in frequency from ~ 70 to 40 kHz from $\sim 22:00$ UT on August 25 to 06:40 UT on August 26. This emission continued right up until the arrival of the CME and shock at Wind at 06:40 UT on August 26, 1998. About 24 hours later, Earth's magnetosphere responded by a sudden onset of intense escaping continuum emissions (not shown). An important point to be made here is that these radio data provide us with a global view of the entire solar-terrestrial event, from the flare and CME liftoff on the sun, the CME propagation through the interplanetary medium, to the arrival at Earth, and to the response of Earth's magnetosphere.

Plate 1 shows that most of the type II radio emissions for the August 24–26 event lie along a straight line, labelled H, with a slope of 2.0×10^{-7} kHz^{-1} s^{-1}. A line of twice that slope, labelled F, is also shown. Note that this latter line extends to the local plasma frequency upstream of the shock at 06:40 UT on August 26. (The local plasma frequency corresponds to the lower edge of the quasi-thermal noise emissions [Meyer-Vernet et al., 1979], as labelled on the plot). Clearly any radio emissions organized along the line F would correspond to radio emissions generated at the fundamental of the plasma frequency upstream of the interplanetary shock. This then suggests that the radio emissions organized along the line labelled H must corresponds to emission at the *harmonic* of the plasma frequency in the *upstream* region of the interplanetary shock.

The difference between the time of the flare and the *in-situ* shock suggests an average CME/shock transit speed of 1275 km/s. If we use this speed, together with the measured slope of the line on the $1/f$ versus time dynamic spectrum, we can solve for the density in the source region (normalized to 1 AU). We find for the density, $n_o = 5.6$ cm^{-3}. This value is consistent with the plasma density from $5.4 - 9.0$ cm^{-3} measured at 1 AU upstream of the shock, again confirming that the harmonic radiation shown in Plate 1 was indeed generated in the upstream region of the shock.

This event was unusual in that the interplanetary type II radio emissions were observed until the time of the shock, suggesting that Wind was very close to the radio source region at that time. (After the passage of the shock, due to the very high plasma density region behind it, the harmonic radiation could no longer propagate back to the Wind spacecraft.) For this event both the electron beam and intense plasma (Langmuir) waves were detected just before the passage of the shock. Thus after many years of observational work, the electron beam in the radio source region of an interplanetary type II burst has finally been observed in-situ [*Bale et al.*, 1999], directly confirming the generation of the type II radiation in the upstream region of the CME-driven shock.

Reiner et al. [1998a,b] tried to identify solar wind structures that may be involved in the generation of the interplanetary type II radio emissions. By extrapolating the measured solar wind speed and density back along the Archimedean spiral, they constructed iso-frequency contours in the interplanetary medium that correspond to the radio observations. They then used the direction-finding capabilities on Wind to show that, in general, the type II radio emission sites originate from regions of differernt density along the shock front. However, *Reiner et al.* [1998b] showed for an interplanetary type II event on January 6-11, 1997 at least some of the radio emissions originated in the upstream region between the CME-driven shock and the highest density region of a corotating interaction region (CIR). The situation is shown Figure 3, which shows the location of the 196 kHz iso-frequency contour, the CME and driven shock at ~14:00 UT on January 8, 1997. The inset figures show the solar wind plasma density and speed as measured by Wind at 1 AU from which the iso-frequency contours were derived. The observed type II radio emissions at 196 kHz must originate where the shock intersects the iso-frequency contour, i.e., in the highest density region of the CIR. The direction-finding results from Wind at 196 kHz point in this direction, confirming this location as the source of the type II radio emissions. This was the first time that interplanetary type II emissions were traced to a specific interplanetary structure. *Hoang et al.* [1998] tried to confirm this source location by performing triangulation of the radio source between the Wind and Ulysses spacecraft. However, due to the great distance of Ulysses from the sun at that time (4.7 AU), propagation effects for the type II radio emissions rendered the triangulation intractable.

Reiner et al. [1998a] used the $1/f$ method of tracking type II radio emissions to determine the "true" speed of a CME-driven shock that was observed near the solar limb. These results suggested that the shock driven by the CME was not spherical beyond 50° from the longitude of the flare site. Finally, *Reiner et al.* [1999b] demonstrated that for one CME-driven shock for which they were able to track the radio emissions for three consecutive days, there was no evidence for deceleration of the CME-driven shock in the interplanetary medium, at least for a relatively slow shock, $v \approx 550$ km/s.

3.2. Decametric-Hectometric Type II Radio Emissions

The high-frequency receiver on Wind, with a frequency range from 1.075 to 13.825 MHz, covers the frequency gap which had previously existed between ground-based (metric) and space-based observations. This allows, for the first time, a detailed exploration of the relationship between metric and kilometric type II radio bursts.

Gopalswamy et al. [1998] studied 34 metric type II radio bursts observed from ground-based observatories and found that none of these bursts extended into the frequency range below 14 MHz observed by the Wind/WAVES radio receivers. While these results were consistent with an independent origin for metric and interplanetary type II radio bursts, in fact, no interplanetary type II radio emissions were observed by WAVES during this interval of study [*Cliver*, 1999]. More recently, *Kaiser et al.* [1997] studied an event on April 7, 1997 that did generate radio emissions in the frequency band between 1 and 14 MHz. They argued that some of these radio emissions were generated by a blast-wave coronal shock, which also produced a metric type II burst, and some were produced by a CME-driven shock. This analysis provided the first evidence that blast wave shocks could propagate beyond about $4R_\odot$, i.e., to frequencies less than 3 MHz. On the other hand, the radio emission associated with the CME-driven shock was observed at 4.5 MHz, the highest frequency reported for CME associated type II emission. However, this April 7 event also produced no interplanetary type II radio emissions at frequencies below 1 MHz that could be

Plate 1. A dynamic spectrum of the Wind/WAVES radio data from 22:00 UT on August 24 to 08:00 UT on August 26, 1998 showing interplanetary type II radio emissions generated by the propagation of a CME through the interplanetary medium (after Reiner et al. [1999b]).

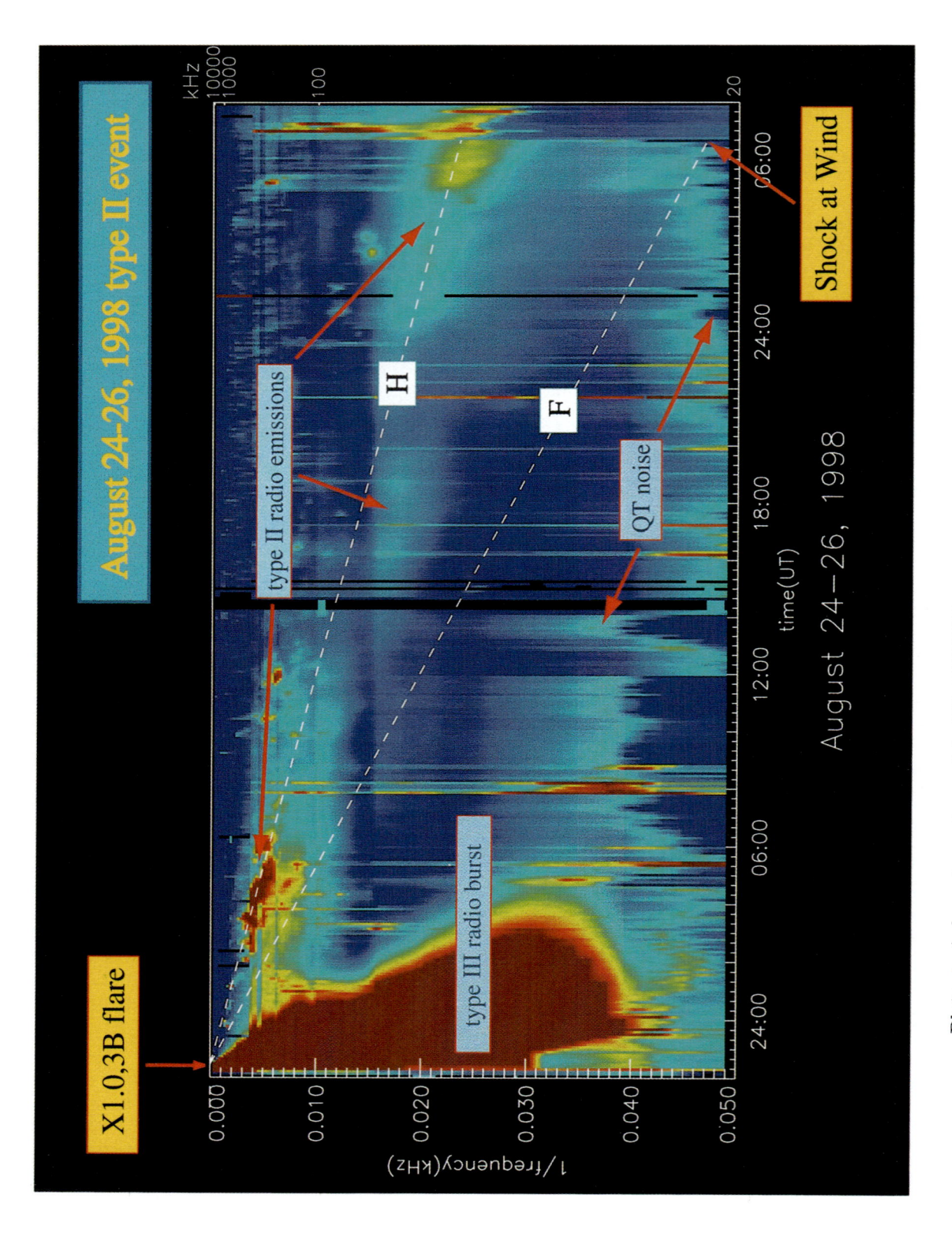

Plate 2. Dynamic spectra from Culgoora and from Wind/WAVES showing metric and interplanetary type II radio bursts. The pink lines represent a fit (obtained using the Saito coronal density model) to the drift rate of the fundamental and harmonic components of the metric type II burst, extrapolated into the frequency range measured by Wind/WAVES. Note that the extrapolation does not simultaneously fit the interplanetary type II radio emissions observed by Wind/WAVES.

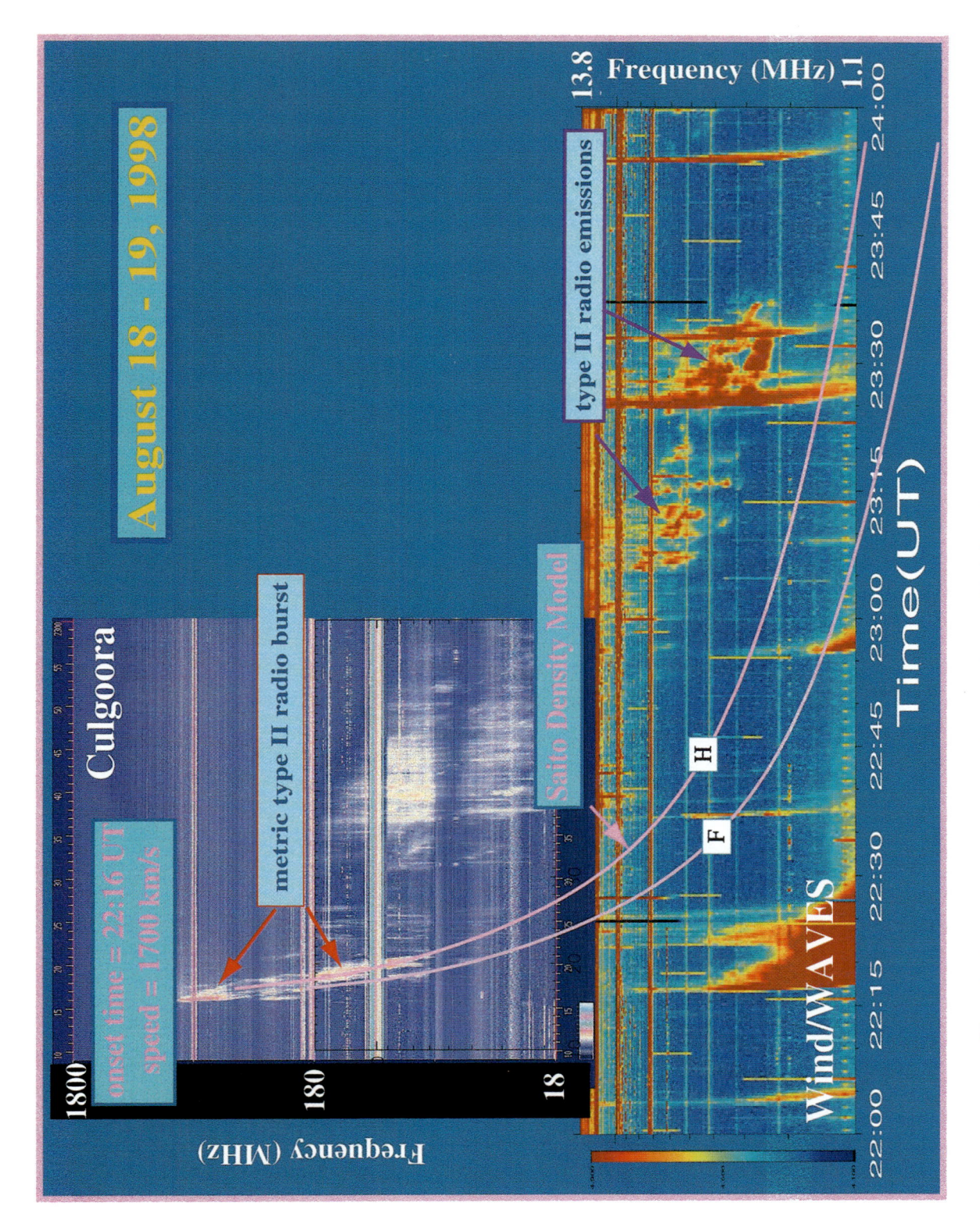

View from above the ecliptic plane

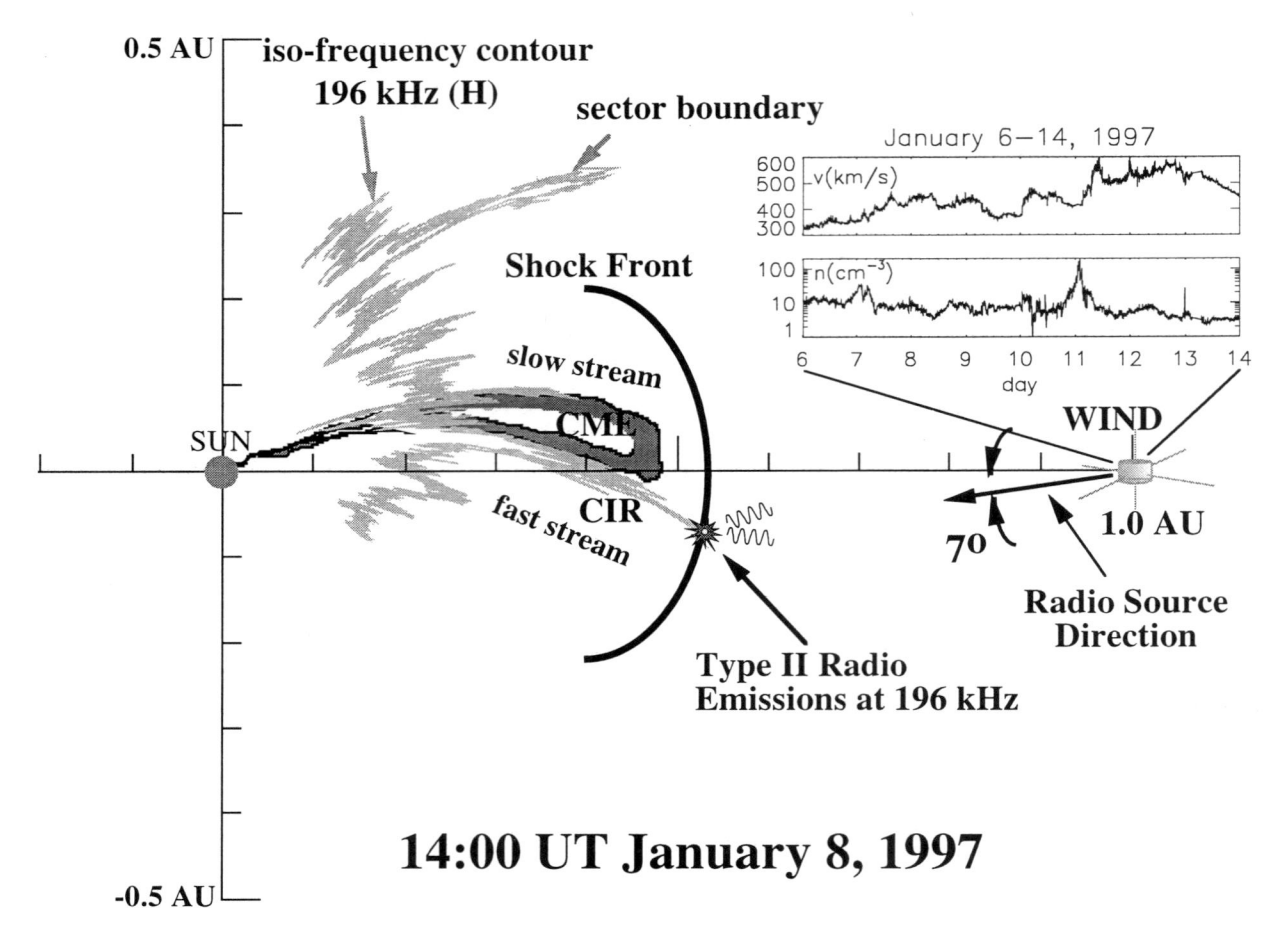

Figure 3. Illustration of the locations of the CME, the CME-driven shock and the 196 kHz iso-frequency contour (constructed from the measured solar wind plasma density and speed) at ~14:00 UT on January 8, 1997. The azimuthal direction to the radio source, as measured from Wind, is also indicated (after Reiner et al. [1998b]).

definitely associated with a CME-driven interplanetary shock.

Reiner et al. [1999c] studied one case where there was type II radio emissions observed all the way from the metric wavelength regime to the kilometric regime. Dynamic spectra of the radio emissions, observed from 1.1 MHz to 1800 MHz, are shown in Plate 2. The top dynamic spectrum from 18 to 1800 MHz is from Culgoora and shows the fundamental and harmonic components of a metric type II radio burst with an onset time of 22:16 UT and speed of 1700 km/s, based on a fit to the (unmodified) Saito model. The Saito model fit to the drift rates is shown by the pink lines for fundamental F and harmonic H emissions. The lower dynamic spectrum shows the Wind/WAVES radio data from 1.1 to 13.8 MHz. The logarithmic frequency scale is exactly matched to the Culgoora scale, as is the time scale. The Wind/WAVES dynamic spectrum shows type II radio emissions in the frequency range from 2 to 10 MHz between 23:07 UT to 23:40 UT. It is clear from Plate 2 that the extrapolation of the Saito model fit to the drift rate of the metric type II does not simultaneously fit the drift rate of the type II radio emissions observed by Wind/WAVES. Thus these observations required the existence of two temporally separated shocks in the solar coronal. The origin time and speed of the shock generating the metric type II burst has to be very different from the onset time and speed of the shock producing the decameter–kilometer radio emissions. *Reiner et al.* [1999c] also showed that the likely origin of the radio emissions observed in the decameter–hectometer regime was the CME-driven shock. They did this by demonstrating continuity between the decameter–hectometer radio emissions and the interplanetary radio emissions below 1 MHz. These observations indicate that the CME-driven shock must have formed deep in the corona, $< 4R_\odot$. They also studied three other decameter–hectometer type II radio emissions that required a origin time significantly different from the origin time of the corresponding metric type II bursts. Their results suggest that (1) there are two different shocks in the corona and it is natural to identify one as a blast-wave shock which is temporally associated with the flare and the second associated with the CME-driven shock, (2) CME–driven shocks can form very low in the corona, at distances, $< 1.1R_\odot$.

4. CONCLUSION

The study of interplanetary type II radio emissions is important because they provide the only means of remotely tracking CMEs through the interplanetary medium from the sun to earth—beyond the range of white-light coronagraph images. The observations of these radio emissions can therefore provide a more accurate prediction of the arrival of a CME at earth. It is therefore important to study the physical characteristics of and dynamics implied by the interplanetary type II radio emissions. The main contribution made during the ISEE-3 era was to identify and describe the physical characteristics of interplanetary type II radio emissions. During the Ulysses–Wind era, due to the increased importance of space weather, emphasis has shifted to describing the dynamic properties implied by the interplanetary type II radio emissions.

Acknowledgments. This work was supported, in part, by the NSF grant ATM-9713422. The Wind/WAVES experiment is a collaboration of NASA/Goddard Space Flight Center, the Observatoire of Paris-Meudon and the University of Minnesota.

REFERENCES

Bale, S. D., M. J. Reiner, J.-L. Bougeret, M. L. Kaiser, S. Krucker, D. E. Larson, and R. P. Lin, The source region of an interplanetary type II radio burst, *Geophys. Res. Lett., 26,* 1573-1576, 1999.

Boischot, A., A. C. Riddle, J. B. Pearce, and J. W. Warwick, Shock waves and type II radiobursts in the interplanetary medium, *Solar Phys., 65,* 397-404, 1980.

Bougeret, J.-L., Observations of shock formation and evolution in the solar atmosphere, in *Collisionless shocks in the heliosphere: Reviews of current research,* edited by B. T. Tsurutani and R. G. Stone, pp. 13-32, American Geophysical Union: Geophysical Monograph 35, Washington, D.C., 1985.

Bougeret, J-L., M. L. Kaiser, P. J. Kellogg, R. Manning, K. Goetz, S. J. Monson, N. Monge, L. Friel, C. A. Meetre, C. Perche, L. Sitruk, and S. Hoang, WAVES: The radio and plasma wave investigation on the Wind spacecraft, *Space Sci. Rev., 71,* 231-263, 1995.

Cane, H. V., R. G. Stone, J. Fainberg, R. T. Stewart, J. L. Steinberg, and S. Hoang, Radio evidence for shock acceleration of electrons in the solar corona, *Geophys. Res. Lett., 8,* 1285-1288, 1981.

Cane, H. V., R. G. Stone, J. Fainberg, J. L. Steinberg, and S. Hoang, Type II solar radio events observed in the interplanetary medium, *Solar Phys., 78,* 187-198, 1982.

Cane, H. V., and R. G. Stone, Type II solar radio bursts, interplanetary shocks, and energetic particle events, *Astrophys. J., 282,* 339-344, 1984.

Cane, H. V., The evolution of interplanetary shocks, *J. Geophys. Res., 90,* 191-197, 1985.

Cane, H. V., N. R. Sheeley, and R. A. Howard, Energetic interplanetary shocks, radio emission, and coronal mass ejections, *J. Geophys. Res., 92,* 9869-9874, 1987.

Cliver, E. W., Comment on "Origin of coronal and interplanetary shock: A new look with Wind spacecraft data" by N. Gopalswamy et al., *J. Geophys. Res., 104,* 4743-4747, 1999.

Fainberg J., L. G. Evans, and R. G. Stone, Radio tracking of solar energetic particles through interplanetary space, *Science, 178,* 743-745, 1972.

Gopalswamy, N., M. L. Kaiser, R. P. Lepping, S. W. Kahler, K. Ogilive, D. Berdichevsky, T. Kondo, T. Isobe, and M. Akioka, Origin of coronal and interplanetary shocks: A new look with WIND spacecraft data, *J. Geophys. Res., 103,* 307-316, 1998.

Gosling, J. T., The solar flare myth, *J. Geophys. Res., 98,* 18937-18949, 1993.

Hoang, S., M. Maksimovic, J.-L. Bougeret, M. J. Reiner, and M. L. Kaiser, Wind-Ulysses source location of radio emissions associated with the January 1997 coronal mass ejection, *Geophys. Res. Lett., 25,* 2497-2500, 1998.

Kahler, S. W., E. W. Cliver, and H. V. Cane, Shock-associated kilometric radio emission and solar metric type II bursts, *Solar Phys., 120,* 393-405, 1989.

Knoll, R., G. Epstein, S. Hoang, G. Huntzinger, J. L. Steinberg, J. Fainberg, F. Grena, S. R. Mosier, and R. G. Stone, The 3-dimensiona radio mapping experiment (SBH) on ISEE-C, *IEEE Transactions on Geoscience Electronics, GE-16,* 199-204, 1978.

Lengyel-Frey, D., R. G. Stone, and J. L. Bougeret, Fundamental and harmonic emission in interplanetary type II radio bursts, *Astron. Astrophys., 151,* 215-221, 1985.

Lengyel-Frey, D., and R. G. Stone, Characteristics of interplanetary type II radio emission and the relationship to shock and plasma properties, *J. Geophys. Res., 94,* 159-167, 1989.

Lengyel-Frey, D., Location of the radio emitting regions of interplanetary shocks, *J. Geophys. Res., 97,* 1609-1617, 1992.

Lengyel-Frey, D., G. Thejappa, R. J. MacDowall, R. G. Stone, and J. L. Phillips, Ulysses observations of wave activity at interplanetary shocks and implications for type II radio bursts, *J. Geophys. Res., 102,* 2611-2621, 1997.

Malitson, H. H., J. Fainberg, and R. G. Stone, Observation of a type II solar radio burst to 37 $R_\odot$, *Astrophys. Lett., 14,* 111-114, 1973.

Malitson, H. H., J. Fainberg, and R. G. Stone, A density scale for the interplanetary medium from observations of a type II solar radio burst out to 1 astronomical unit, *Astrophys. J., 183,* L35-L38, 1973.

Malitson, H. H., J. Fainberg, and R. G. Stone, Hectometric and kilometric solar radio emission observed from satellites in August 1972, *Space Sci. Rev., 19,* 511-531, 1976.

Manning, R., and J. Fainberg, A new method of measuring radio source parameters of a partially polarized distributed source from spacecraft observations, *Space Sci. Instrum., 5,* 161-181, 1980.

Meyer-Vernet, H., On natural noises detected by antennas in plasmas, *J. Geophys. Res., 84,* 5373-5377, 1979.

Nelson, G. J., and D. B. Melrose, Type II bursts, in *Solar Radiophysics*, edited by D. J. McLean and N. R. Labrum, pp. 333-359, Cambridge University Press, Cambridge, 1985.

Reiner, M. J., M. L. Kaiser, J. Fainberg, J.-L. Bougeret, and R. G. Stone, Remote radio tracking of interplanetary CMEs, *Proc. 31st ESLAB Sym., ESTEC, Noordwijk, The Netherlands, 22-25 September 1997*, ESA SP-415, 183-188, 1997.

Reiner, M. J., M. L. Kaiser, J. Fainberg, and R. G. Stone, A new method for studying remote type II radio emissions from coronal mass ejection-driven shocks, *J. Geophys. Res., 103,* 29651-29664, 1998a.

Reiner, M. J., M. L. Kaiser, J. Fainberg, J.-L. Bougeret, and R. G. Stone, On the origin of radio emissions associated with the January 6-11, 1997, CME, *Geophys. Res. Lett., 25,* 2493-2496, 1998b.

Reiner, M. J., and M. L. Kaiser, Complex type III-like radio emissions observed from 1 to 14 MHz, CME, *Geophys. Res. Lett., 26,* 397-400, 1999a.

Reiner, M. J., M. L. Kaiser, J. Fainberg, and R. G. Stone, Remote radio tracking of interplanetary CMEs in the solar corona and interplanetary medium, *Proc. Ninth International Solar Wind Conference, 5-9 October 1998, AIP Conference Proceedings 471*, eds., S. R. Habbal, R. Esser, J. V. Hollweg, and P. A. Isenberg, pp. 653-656, 1999b.

Reiner, M. J., M. L. Kaiser, High-frequency type II radio emissions associated with shocks driven by coronal mass ejections, *J. Geophys. Res., 104,* 16979-16991, 1999c.

Sheeley, N. R., Jr., R. A. Howard, M. J. Koomen, D. J. Michels, R. Schwenn, K. H. Muhlhauser, and H. Rosenbauer, Coronal mass ejections and interplanetary shocks, *J. Geophys. Res., 90,* 163-175, 1985.

Simnett, G. M., Interplanetary phenomena and solar radio bursts, *Solar Phys., 104,* 67-91, 1986.

Stone, R. G. et al., The Unified Radio and Plasma (URAP) wave investigation, *Astron. Astrophys., 92,* 291-316, 1992.

Thejappa, G., R. J. MacDowall, and A. F. Vinas, In situ wave phenomena in the upstream and downstream regions of interplanetary shocks: implications for type II burst theories, *Proc. 31st ESLAB Sym., ESTEC, Noordwijk, The Netherlands, 22-25 September 1997*, ESA SP-415, 189-194, 1997.

Michael J. Reiner, Raytheon ITSS, NASA Goddard Space Flight Center, Code 690.2, Greenbelt, MD 20771 (e-mail: reiner@urap.gsfc.nasa.gov)

ISEE-3 Observations of Radio Emission From Coronal and Interplanetary Shocks

H. V. Cane

School of Mathematics and Physics, University of Tasmania, Hobart, Australia.

The radio astronomy experiment on ISEE-3 was the first to observe a sufficient number of radio events associated with interplanetary shocks to enable their general characteristics to be derived. In addition to the 48 "IP type II events" there were also other slow drifting features with slightly different characteristics which were probably related to interplanetary phenomena, but these events have not been examined in any detail. ISEE-3 also observed for the first time a class of low frequency (less than 5 MHz) fast drift bursts which appeared to be caused by electron streams accelerated at shocks within a few solar radii of the Sun (i.e. to emanate from coronal type II bursts). However the relationships between ISEE-3 radio events (both slow and fast drift) and coronal type II bursts could not be fully investigated because of the large gap in frequency coverage between the ISEE-3 experiment and ground-based observations. The ISEE-3 observations are summarised and compared with recent observations from the Wind radio astronomy experiment.

1. INTRODUCTION

ISEE-3 was launched in August 1978 which was about 1 year after the onset of activity in solar cycle 21. Based on the recent evolution of sunspot number an equivalent time in the present solar cycle should have been late 1998 although the level of activity has been less than in cycle 21. The radio astronomy experiment on ISEE-3 [*Knoll et al.*, 1978] was the first one to detect long-lived slow drifting features clearly associated with interplanetary (IP) shocks. Except for one event from an extremely fast shock in August 1972 seen by an experiment on IMP 6 [*Malitson et al.*, 1973], previous identifications of shock-associated radio emissions at low frequencies (i.e. less than 5 MHz) were based on bursts lasting less than a few hours.

ISEE-3 was placed at L1, which is where SOHO and ACE are now located, and remained there until late 1982. During these $\sim$ 50 months, 47 "IP type II events" were identified, with one further event occurring in February 1983. The total list of events can be found in *Cane* [1985]. The radio phenomena were called 'events' for two reasons. First to distinguish them from coronal type II bursts, since our studies indicated that the IP phenomenom might not be a continuation of the bursts seen at meter wavelengths, (i.e. not just one shock propagating from the low corona into IP space), and second, because the slow-drift emissions were continuous in frequency and time over many hours. In order to select a uniform set of events, bursty or short-lived emissions were not included. Note that there were also many short-duration and long-duration, narrow-banded features seen in the ISEE-3 data which were

Radio Astronomy at Long Wavelengths
Geophysical Monograph 119
Copyright 2000 by the American Geophysical Union

informally called STs (Stripey Things!). Undoubtedly many of them will, in retrospect, be associated with IP shocks. Others are likely to result from interplanetary interactions.

This paper does not discuss or describe meter wavelength type II bursts. Good summaries of the results as of 1985 may be found in *Bougeret* [1985] and *Nelson and Melrose* [1985]. For a summery of more recent results see *Aurass* [1992].

2. GENERAL CHARACTERISTICS

The ISEE-3 experiment operated at 24 frequencies in the range 1.98 MHz to 30 kHz with bandwidths of 3 or 10 kHz. There were 23 independent frequencies since a 1 MHz channel had both 10 and 3 kHz bandwidths. The higher frequencies were sampled more often than the lower frequencies. Data sequences were usually represented by forming 108-s averages for all frequencies. Dynamic spectra were formed by a pseudo-grey shading system with various symbols representing intensity. Rows for each frequency were plotted one above the other. Examples are shown in *Cane et al.* [1982b]. Figure 2 in that paper shows the first IP type II event identified, commencing on September 23, 1978, and Figure 1 shows an event commencing August 18, 1979. Figure 1 of the present paper shows the 1979 event in a format more compatible with that presently used for the Wind data (kindly supplied by R. J. MacDowall). Data for each channel are shown using a logarithmic scale and a continuum is presented by interpolating between neighbouring frequencies. Note that such interpolation is not significant below 290 kHz but there are effectively only 4 channels (360, ~500, 1000 and 1980 kHz) above this frequency. The IP type event is the continuous band from about 1600 UT on DOY 230 (August 18) and ending about 0600 UT on DOY 232 (August 20). The brightest part occurs between about 300 and 70 kHz, ending at about 1200 UT on DOY 231. Note that, whereas there is structure in the IP type II emission with fluctuations in the intensity level, there is still continuous emission clearly seen in Figure 1, at least for the period from the onset until the middle of DOY 231, i.e. over at least 20 hours. The bright arc near 60 kHz around the beginning of DOY 231 is quasi-thermal noise [*Meyer-Vernet*, 1979] which has maximum intensity at the local plasma frequency. Late on DOY 231 the plasma frequency at ISEE-3 was unusually high. Another period when the thermal noise is apparent (at the lowest frequencies) commences at about 0600 UT on DOY 232. At this time the plasma frequency increased because the shock causing the IP type II event arrived at ISEE-3.

Note the intense, fast-drift (type III-like) burst preceding the IP type II emission. This burst occurred at the time of a meter wavelength type II burst. The type III emission seen at this time by ground-based observers was very weak and not reported by all stations to extend to meter wavelengths. Similar intense fast-drift bursts preceded all other IP type II events and, based on their characteristics, have been differentiated from type III bursts. They were called SA (shock associated) events because of their temporal association with herringbone structure in meter wavelength type II bursts; they are described in more detail below. Actually there is a second IP type II event in the period covered by Figure 1. This is the green band commencing around the middle of DOY 232. The SA event commenced at about 0900 UT. Thus the two bright, fast drift features in Figure 1 are both SA events.

Those readers familiar with meter wavelength type II bursts will be immediately struck by the absence of two bands in the IP type II event shown in Figure 1. About 60% of all type II *bursts* show two bands separated by a frequency ratio of about 2:1 [*Kundu*, 1965] consistent with the bands being fundamental and harmonic plasma emission. *Lengyel-Frey et al.* [1985] investigated bands in IP type II events and only found three cases (out of the 48 events) with prominent two-band structure. The bands had an average frequency ratio of slightly less than two but showed variation within each event. In all three events the fundamental band was brighter than the harmonic band. The fundamental band showed more intensity fluctuations. A more detailed analysis of one of the three events may be found in *Bougeret* [1986]. *Lengyel-Frey et al.* [1985] found no correlation between intensity fluctuations and source sizes, bandwidths or positions.

The ISEE-3 radio experiment had two antennas (dipoles) with one spinning in the ecliptic plane and the other aligned along the spin axis of the spacecraft. Using information from both antennas, source sizes and

Figure 1. Dynamic spectrum of a period of three days in August 1979. The intense fast-drift burst at ~1400 UT on DOY 230 is the SA event which preceded the IP type II event generated by a shock which originated on the east limb of the Sun on August 18 and reached ISEE-3 at 0552 UT August 20 (DOY 232). Later on this day there is another SA event and IP type II event commencing around 0900 UT. This IP type II event is considerably weaker than the first one. The bright emission at low frequencies, particularly near the start of DOY 231, is quasi-thermal noise.

ISEE-3 RADIO DATA - 79/08/18-79/08/20

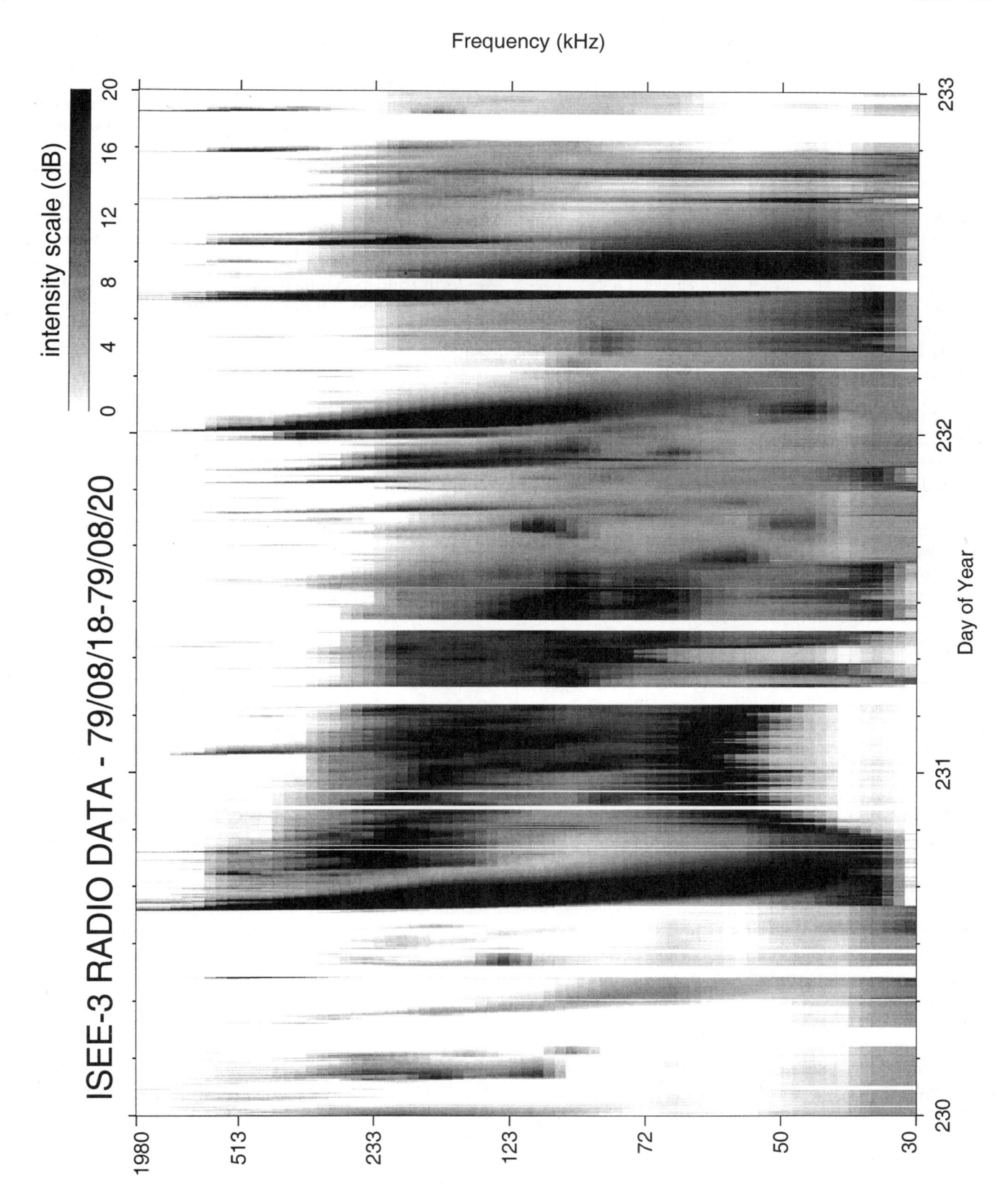

positions, east or west of the Sun-Earth line, could be determined. For type II studies source positions were determined just using the spin plane antenna and these positions agreed approximately with the expected location based on the flare longitude.

3. SOLAR ASSOCIATIONS

Most IP type II events were preceded by type II and/or type IV bursts seen at meter wavelengths. More than half had both [*Cane and Stone*, 1984]. This was expected based on the earlier work showing a good correlation between the combined type II/IV burst with geomagnetic storms, since the latter are well-associated with IP shocks. Many events had no preceding group of type III bursts. *Cane and Stone* [1984] noted that the starting frequencies of the associated meter wavelength type II bursts were rather low (see also *Robinson et al.*, [1983]). *Cane and Reames* [1988] interpreted this result to be related to the fact that the class of flare events associated with interplanetary shocks tend to occur higher in the corona.

The question of the relationship between flares and meter wavelength radio bursts to interplanetary shocks is still open. What we do know is that the solar phenomenom necessary for the generation of an interplanetary shock is a coronal mass ejection which may or may not have an associated flare. *Cane et al.* [1987] found that all IP type II events were preceded by a CME and furthermore that these CMEs were the most energetic observed by the Solwind coronagraph. This was consistent with the earlier work of *Cane* [1985] which suggested that shock speed was the important parameter which determined whether radio emission was detected or not and that the intensity of the emission was positively correlated with shock speed (see also *Lengyel-Frey and Stone*, 1989).

4. SHOCK CHARACTERISTICS

Since the IP type II events were associated with a group of shocks with similar features and possibly comprising all major shocks over a certain time period, it is possible to make some deductions about the characteristics of such shocks. The most important factor is that the radio emission allows one to make essentially unambiguous identifications of shock launch times and approximate source regions at the Sun. Results on shock shapes were presented by *Cane and Stone* [1984] and *Cane* [1985]. However with only 48 events in the latter study this is still a relatively small sample. Thus using the IP type II events as a starting point *Cane* [1988]

sought to find the commonality between all these events, apart from the radio emission. In retrospect the answer is obvious. Shocks accelerate other particle species besides the electrons responsible for the radio emission. Very energetic shocks sometimes accelerate protons to above a GeV. Thus all IP type II events have associated energetic particle events as pointed out by *Cane and Stone* [1984]. Starting with a list of all energetic particle events (> 20 MeV) detected over an ∼ 19 year period *Cane* [1988] determined which events were associated with shocks detected at Earth (by virtue of a storm sudden commencement) and derived an average shock shape. Of particular interest is whether the CME producing a shock is intercepted at Earth. Interplanetary CMEs are of great interest because they are responsible for many major geomagnetic storms. They also cause Forbush decreases in the cosmic ray intensity. Using the list of IP type IIs as a group of shocks with well-identified source regions on the Sun, *Richardson and Cane* [1993] showed that interplanetary CMEs extend at most 50° in longitude from the associated solar event. *Cliver et al.* [1990] used the IP type II event list to obtain a list of well-associated shocks to investigate shock speeds in the interplanetary medium while *McComas et al.* [1989] used the list to investigate draping of interplanetary magnetic field lines around CMEs.

Average shock speeds can be deduced from the time interval between the solar event and shock arrival at Earth. For the IP type II events the average shock speeds varied from 460 to 1220 km/sec [*Cane*, 1985]. Only one event had a speed below 610 km/sec and this was from a low intensity event from near the west limb. The shock was extremely weak at 1 AU. Based on the absence of speeds above 1220 km/sec *Cane* [1988] questioned the association of the famous event of August 1972 in which a transit speed of 2770 km/sec was deduced. However taking into consideration that the event was preceded by another fast CME which would have swept out the interplanetary medium so that the second August CME was moving through an unusually low density region, the association is accepted (see *Cliver et al.*, [1990]). Note, however, that there has not been another shock this fast. Using particle observations to relate shocks and solar events the fastest events in the years since the ISEE-3 study period had speeds less than 1800 km/sec [*Cane et al.*, 1993].

Of course one can investigate how some individual shocks propagate by analysing the frequency time structure of associated radio emission. At meter wavelengths this is fairly straight forward in terms of the radio structure since it is narrow banded, but a coronal density model needs to be assumed. For IP type II events the emission is broad banded and lasts for many hours at each frequency. Nevertheless the differences in drift rates from event to event were consistent with the variation in shock transit speeds. An analysis of a few events performed by *Cane* [1983] obtained the result that the fastest shocks decelerated more than the slower ones such that the range of shock speeds at 1 AU was smaller than the range closer to the Sun. Such a result was also obtained from a statistical analysis of in situ observations of shocks from the Helios spacecraft. It had been established quite early that shocks decelerate in transit from the Sun to the Earth [*Gosling et al.*, 1968]. The analysis of one event (that of Figure 1) also included information from radio scattering of spacecraft signals [*Cane et al.*, 1982a].

The *Cane* [1983] analysis obtained the result that the drift rates at 1 MHz and above did not seem consistent with those at lower frequencies. Thus it was suggested that the 1 and 2 MHz emissions were from the extension of the meter wavelength burst producer but that the interplanetary shock, which produced the lower frequencies, was a different phenomenom. (It is important to note that some IP type II events were not detected at 1 and 2 MHz. This may have been caused by the high background level of the Galaxy, which has a maximum intensity near 1 MHz) This decoupling of the type II burst shock and the interplanetary shock could account for some of the discrepancies which arose when comparing CMEs and type II bursts. These problems are listed in *Cane* [1997]. At present the existence of possibly two types of shock is still under investigation.

5. RADIO SOURCE LOCATION

The most detailed analysis attempting to determine the source location was made by *Lengyel-Frey* (1992). *Lengyel-Frey* used the densities observed at 1 AU to calculate expected densities that the shock might have encountered. This analysis assumes that the structure and density of the solar wind stays constant over several days and that interactions are insignificant. Given that the CME rams into the upstream plasma and is not part of the ambient solar wind it seems doubtful, unless the CME originates in the western hemisphere of the Sun, that useful information can be obtained about densities radially ahead of the CME or slightly to its east. This is exactly the region where the shock is the strongest and where the source region might be expected to be located. This method has also been used by *Reiner et al.* [1998a] in analysing recent data. *Lengyel-Frey's* analysis also assumed that the relevant plasma frequency at a particular time was the frequency at which there was the highest intensity. This might not be correct since there are intermittent brightenings.

6. COMPARISON WITH WIND OBSERVATIONS

Since ISEE-3 there has been no comparable space radio astronomy experiment until the launch of Wind in late 1994. The experiment on Ulysses was superior to that on ISEE-3 but the Ulysses orbit makes it difficult to compare observations. (For a comparison of these three experiments see *MacDowall et al.*, [1997]). As solar maximum is approached and given concurrent CME observations from SOHO there are great hopes of solving some of the outstanding problems. In particular the Waves experiment on Wind has a receiver covering the gap (almost) between conventual ground based observatories and the IP medium.

As of the end of 1998 Wind had not seen any broad band event like those seen on ISEE-3. The events have been very patchy and relatively narrow banded (see *Reiner et al.*, [1998a]). It has been suggested by *MacDowall et al.* [1997] that this might be because Wind was launched at solar minimum. However as stated in the introduction, the end of 1998 should be comparable to the end of 1978 when ISEE-3 detected three IP type II events. Assuming that there are no special conditions required, like a pre-existing population of low energy electrons in the interplanetary medium, the event of August 1998 with a shock transit speed of almost 1300 km/sec should have produced an IP type II event. Although there was shock emission again it was narrow banded and patchy. So it remains to be seen how the Wind observations relate to the ISEE-3 ones but the question should be settled within the next few years.

Note that ISEE-3 definitely saw phenomena like the Wind "type IIs". In the two initial papers on IP type II events it was stressed that ISEE-3 had seen "numerous slow drift features in the dynamic spectra" (see also *Stone et al.*, [1984]). The *Cane et al.* [1982b] paper discussed one event and showed the dynamic spectrum. This event occurred in November 1978. Subsequently, in discussing the interplanetary conditions which caused a major geomagnetic storm at this time, *Cane and Richardson* [1997] suggested that the stripey radio emission could result from the interaction of a CME and a high speed stream. Such an explanation had been proposed for the very intense narrow banded event seen by Wind in January 1997 [*Reiner et al.*, 1998b]. Thus it is likely that there will be different classes of slow drift radio features as the Wind events are analysed and catalogued. In listing the IP type II events an attempt was

7. SHOCK ASSOCIATED EVENTS

As mentioned earlier, many of the IP type II events were not preceded by type III bursts in meter wavelength observations. On the other hand all events were preceded by an intense fast drift burst in the ISEE-3 data. A detailed comparison of the ISEE-3 data and film records from the Culgoora observatory revealed that the low frequency fast drift bursts occurred at the times of meter wavelength type II bursts and also did not last much longer than the type II burst end times. Furthermore in many events there was a correspondence between intensity fluctuations in the ISEE-3 bursts and with herringbone structure on the type II bursts. (Herringbone structure is the phenomenom of fast drift bursts emanating from the type II 'backbone' both to lower and higher frequencies.) This lead *Cane et al.* [1981] to propose the existence of a new class of radio burst called an SA event. In the original paper it was proposed that since herringbones were presumed to be caused by shock accelerated electrons then the kilometer emission was Shock Accelerated. However in a later paper by *Cane and Stone* [1984] the acronym was changed to Shock Associated in order to include the possibility that the electrons were accelerated elsewhere but their radio emitting properties were influenced by the shock.

Several authors have challenged the concept of SA events and suggested that the electrons are accelerated low in the corona where microwave burst originate (e.g. *Klein and Trottet*, 1994). However these authors have not explained why there is a correspondence between the kilometer wavelength activity and herringbone structure nor why in many cases there is no associated microwave activity or no correspondence in profiles when there is. These properties were documented by *Kahler et al.* (1989) who concluded that SA events are the long wavelength extension of herringbone structure. Of course the difficulty in conclusively showing this with the ISEE-3 experiment was the gap between its highest frequency at ~2 MHz and the normal ground-based starting frequency of about 20 MHz. However a recent paper by *Bougeret et al.* [1998] using Wind and ground-base data has shown conclusively that a fast-drift burst seen down to 30 kHz started at the backbone of a type II burst at about the 50 MHz level. Another difficulty with the SA events is that it has been impossible to define a unique set of characteristics to unambiguously select events. *MacDowall et al.* [1987] developed a computer algorithm which was reasonably successful in selecting kilometer wavelength fast drift bursts associated with meter wavelength type II and/or type IV bursts, however it inherently could not select less intense or short duration candidate events.

A recent paper by *Reiner and Kaiser* [1999] discusses Wind observations of non-type III fast drift bursts. However their conclusions are somewhat less conclusive than for the event discussed by *Bougeret et al.* [1998] because they do not show the contemporaneous observations at higher frequencies. One should not simply rely on reported activity; the original records must be consulted. One of the arguments *Reiner and Kaiser* [1999] propose against the association with herringbone structure is that the Wind activity at 1 MHz is claimed to last well beyond the meter wavelength type II burst duration. However this is not shown and contradicts what was found with the original ISEE-3 SA events. Nevertheless their results are very interesting and the matter will surely be resolved as the Sun becomes more active.

Bougeret et al. [1998] suggest that instead of the SA event electrons being accelerated at the shock they are accelerated by reconnection stimulated by shock passage. This suggestion and also the discussion in *Reiner and Kaiser* [1999], about the presence of open and closed field lines, must be put in the context of the presence of CMEs since we know they are present. Since SA events precede IP type II events and are usually associated with proton acceleration [*Cane and Stone*, 1984; *Kahler et al.*, 1986], the question of the SA event origin cannot be treated in isolation from all of the CME phenomena.

Acknowledgments. I wish to thank R. J. MacDowall, W. C. Erickson and J.-L. Bougeret for useful discussions. I also thank R. J. MacDowall for the preparation of Figure 1.

REFERENCES

Aurass, H., Radio observations of coronal and interplanetary type-II bursts, *Ann. Geophysicae*, 10, 359, 1992.

Bougeret, J.-L., Observations of shock formation and evolution in the solar atmosphere, in *Collisionless shocks in the heliosphere: Reviews of current research*, eds. B. T. Tsurutani and R. G. Stone, p.13-32, AGU GM 35, Washington, 1985.

Bougeret, J.-L., Some results of STIP interval XI (October-November 1980), in Les Diablerets STIP Symposium on Retrospective Analyses, ed. M. A. Shea and D. F. Smart, 1986.

Bougeret, J.-L, P. Zarka, C. Caroubalos, M. Karlický, Y. Leblanc, D. Maroulis, A. Hillaris, X. Moussas, C. E. Alissandrakis, G. Dumas, and C. Perche, A shock associated (SA) radio event and related phenomena observed from the base of the corona to 1 AU, *Geophys. Res. Lett., 25*, 2513, 1998.

Cane, H. V., Velocity profiles of interplanetary shocks, *Solar Wind Five*, NASA Conf. Publ., CP-2280, 703, 1983.

Cane, H. V., The evolution of interplanetary shocks, *J. Geophys. Res., 90*, 191, 1985.

Cane, H. V., The large-scale structure of flare-associated interplanetary shocks, *J. Geophys. Res. 93*, 1,1988.

Cane, H. V.,The current status in our understanding of energetic particles, coronal mass ejections and flares", in *Coronal Mass Ejections:Causes and Consequences*, edited by N. Crooker, J. Joselyn, and J. Feynman, p.205 , Washington, DC, AGU, 1997.

Cane, H. V., and R. G. Stone, Type II radio bursts, interplanetary shocks and energetic particle events, *Astrophys. J., 282*, 339, 1984.

Cane, H. V., and D. V. Reames, Soft X-ray emissions, meter wavelength radio bursts and particle acceleration in solar flares, *Astrophys. J. 325*, 895, 1988

Cane, H. V., and I. G. Richardson, What caused the large geomagnetic storm of November 1978?, *J. Geophys. Res., 102*, 17,445, 1997.

Cane, H. V. , R. G. Stone, and R. Woo,Velocity of the shock generated by the large east limb flare on August 18, 1979, *Geophys. Res. Lett. 9*, 897, 1982a.

Cane, H. V., N. R. Sheeley, Jr., and R.A. Howard, Energetic interplanetary shocks, radio emission and coronal mass ejections, *J. Geophys. Res., 92*, 9869-9874, 1987.

Cane, H. V., I.G. Richardson, and T.T. von Rosenvinge, Cosmic ray decreases:1964-1994, *J. Geophys. Res., 101*, 21561, 1996.

Cane, H. V., R. G. Stone, J. Fainberg, J-L. Steinberg, and S. Hoang, Type II solar radio events observed in the interplanetary medium: Part I - general characteristics, *Solar Phys. 78*, 187, 1982b.

Cane, H. V., R.G. Stone, J. Fainberg, R. T. Stewart, J.-L. Steinberg, and S. Hoang, Radio evidence for shock acceleration of electrons in the solar corona, *Geophys. Res. Lett., 8*, 1285-1288, 1981.

Cliver, E.W., J. Feynman and H. B. Garrett, An estimate of the maximum speed of the solar wind, 1938-1989, *J. Geophys. Res., 95*, 17,103, 1990.

Gosling, J. T., J. R. Ashbridge, S. J. Bame, A. J. Hundhausen, and I. B.Strong, Satellite observations of interplanetary shock waves, , *J. Geophys. Res., 73*, 43, 1968.

Kahler, S. W., E. W. Cliver, and H. V. Cane, The relationship of shock-associated kilometric radio emission with metric type II bursts and energetic particles, *Adv. Space Res., 6(6)*, 319, 1986.

Kahler, S. W., E. W. Cliver, and H. V. Cane, Shock-Associated kilometric radio bursts and solar metric type II bursts, *Solar Phys. 120*, 393, 1989.

Klein, K-L, and G. Trottet, Energetic electron injection into the high corona during the gradual phase of flares: Evidence against acceleration by a large scale shock, in *High Energy Solar Phenomena, AIP Conf. Proc. 294*, 187-192, 1994.

Knoll, R., G. Epstein, S. Hoang, G. Huntzinger, J.L. Steinberg, J. Fainberg, F. Grena, S. R. Mosier and R. G. Stone, The 3-dimensional radio mapping experiment (SBH) on ISEE-3, *IEEE Trans. Geosci. Electron., GE-16*, 199, 1978.

Kundu, M. R., *Solar Radio Astronomy*, Wiley Interscience, New York, 1965.

Lengyel-Frey, D., Location of the radio emitting regions of interplanetary shocks, *J. Geophys. Res., 97*, 1609, 1992.

Lengyel-Frey, D., R. G. Stone, and J. L. Bougeret, Fundamental and harmonic emission in interplanetary Type II radio bursts, *Astron. Astrophys., 151*, 215, 1985.

Lengyel-Frey, D., and R. G. Stone, Characteristics of interplanetary type II radio emission and the relationship to shock and plasma properties, *J. Geophys. Res., 94*, 159, 1989.

MacDowall, R. J., R. G. Stone, and M. R. Kundu, Characteristics of shock-associated fast-drift kilometric radio bursts, *Sol. Phys, 111*, 397, 1987.

MacDowall, R. J., A. J. Klimas, D. Lengyel-Frey, R. G. Stone, and G. Thejappa, Comparison of interplanetary type II radio burst observations by ISEE-3, Ulysses, and wind with applications to space weather prediction, *ESA SP-415*, p. 533-538, 1997.

McComas, D. J., J. T. Gosling, S. J. Bame, E. J. Smith and H. V. Cane, A test of magnetic field draping induced Bz perturbations ahead of fast coronal mass ejecta, *J. Geophys. Res. 94*, 1465, 1989.

Meyer-Vernet, N., On natural noises detected by antennas in plasmas, *J. Geophys. Res. 84*, 5373, 1979.

Malitson, H. H., J. Fainberg, and R. G. Stone, A density scale for the interplanetary medium from observations of a type II solar radio burst out to 1 AU, *Astrophys. J., 183*, L35, 1973.

Nelson, G. S., and D. B. Melrose, Type II bursts, in *Solar Radiophysics*. Eds. D. J. McLean and N. R. Labrum, Cambridge Univ. Press, Cambridge, p. 333-359, 1985.

Reiner, M. J., and M. L. Kaiser, Complex type III-like radio emissions observed from 1 to 14 MHz, *Geophys. Res. Lett. 26*, 29,651, 1999.

Reiner, M. J., M. L. Kaiser, J. Fainberg, and R. G. Stone, A new method for studying remote type II radio emissions from coronal mass ejection-driven shocks, *J. Geophys. Res., 103*, 29,651, 1998a.

Reiner, M. J., M. L. Kaiser, J. Fainberg, J.-L. Bougeret, and R. G. Stone, On the origin of radio emissions associated with the January 6-11, 1997, coronal mass ejection, *Geophys. Res. Lett., 25*, 2493, 1998b.

Richardson, I. G., and H. V. Cane, Signatures of shock drivers in the solar wind and their dependence on the solar source location, *J. Geophys. Res., 98*, 15,295, 1993.

Robinson, R.D., R. T. Stewart, and H. V. Cane, Properties of meter-wavelength solar bursts associated with interplanetary type II emission, *Solar Phys. 91*, 159, 1983.

Stone, R. G. , H. V. Cane and J.-L. Bougeret, ISEE-3 radio observations of interplanetary solar shocks, in *Proc. of STIP symposium on solar/interplanetary intervals, Maynooth*, pp 371-382, eds. M. A. Shea, D. F. Smart and S. M. P. McKenna-Lawlor, Maynooth, 1984.

H. V. Cane, School of Mathe-matics and Physics, University of Tasmania, Hobart, Australia (email:hilary.cane@utas.edu.au)

Radar Studies of the Solar Corona: A Review of Experiments Using HF Wavelengths

Paul Rodriguez

Information Technology Division, Naval Research Laboratory, Washington, DC

The use of high frequency (9 to 40 MHz), high power radars to study the solar corona has a remarkable history. Solar radar experiments were proposed and started at the beginning of the modern era of space physics research. Early in the 1960s, the El Campo solar radar facility began routine operations. The published results from these pioneering experiments remain our largest resource of information on active probing of the solar corona. In 1969, solar radar experiments ceased even though experimental results suggested that significant diagnostics of the solar corona could be obtained with this technique. After the El Campo facility was decommissioned, a hiatus of about 25 years followed for further solar radar experiments. Recently, high frequency radar facilities in Russia and Ukraine have been used to conduct new solar radar experiments. The information that solar radars may provide is particularly relevant today, as we now recognize the important role of coronal mass ejections in geomagnetic disturbances. Solar radars offer the possibility of direct detection of earthward-moving coronal mass ejections, providing several days of advance warning to possible geomagnetic storms. In addition, investigations of wave scattering in the solar corona may provide information on coronal densities and irregularities. The techniques derived for studies of the earth's ionosphere using incoherent scatter radars are potentially applicable to the solar corona as well. We will review the early and current solar radar experiments and also discuss future directions for radar studies of the solar corona.

INTRODUCTION

Studies of the solar corona with high frequency radars were first done in the early 1960s by groups from Stanford University [*Eshelman et al.*, 1960] and the Massachusetts Institute of Technology's (MIT) Lincoln Laboratory [*Abel et al.*, 1961, 1963]. These experiments were begun to test theoretical ideas suggested by *Kerr* [1952] and by *Bass and Braude* [1957]. In subsequent years, from 1961 to 1969, the MIT group developed a sustained program of experiments using a radar facility specially designed for the solar studies and constructed in El Campo, Texas. The El Campo solar radar operated as both a transmitting and a receiving array for high frequency (HF) radio waves at 38.25 MHz. The results of the El Campo experiments have been summarized and reviewed by *James* [1968] and show that return signals were detected that suggested a radar cross section for the sun's corona of approximately the same size as the optical disk, with occasional cross-section enhancements by an order of magnitude or more. Consistently, Doppler shifts in the return signals of about 4 kHz to 15 kHz were detected, and these were attributed to mass motions in the solar corona. Such motions were identified with the outward flow of the solar wind from the base of the corona. Theoretical studies of the results of the El Campo solar radar tests have been published by *Gordon* [1968, 1969, 1973], *Gerasimova* [1975, 1979], *Wentzel* [1981], *Owocki et al.*, [1982], *Chashei and Shishov* [1994], and *Mel'nik* [1998]. These studies have suggested that radar return signals can provide important new diagnostic information on the structure and dynamics of the solar corona. Unfortunately, after 1969 further solar radar experiments were not possible because the El Campo solar radar was deactivated and no longer exists.

In the early experimental and theoretical studies, the possibility of detecting coronal mass ejections (CMEs) with solar radars was not specifically identified, probably because it was only later (in the 1970s) that CMEs were recognized as distinct large-scale coronal perturbations. It is possible that in some of the El Campo measurements, the signature of CMEs was recorded but remained unrecognized. The now-known fact that CMEs are capable of causing major geomagnetic storms at earth [monograph edited by *Crooker et al.*, 1997] has motivated new interest in solar radars to detect earthward-moving CMEs [*Rodriguez*, 1996]. For an earthward-moving CME, it is expected that a return signal might have a distinct Doppler shift associated with the earthward-directed velocity. Based on the known distribution of CME velocities, a wide range of Doppler shifts, up to 100 kHz, would be expected on HF return signals. The large-scale structure of a CME may also mean that the radar cross section would be larger than that of the quiescient corona. The early solar radar investigations also did not have the benefit of modern solar monitoring stations to provide correlative imaging of the corona, such as the SOHO satellite, the Mauna Loa coronameter, and various national observatories that provide daily reports of solar activity.

Beginning in the summer of 1996, an international consortium of investigators (see Appendix A) has conducted several new solar radar tests using facilities in Russia and Ukraine. The high power HF transmitting array SURA located near Nizhny Novgorod, Russia, has been used in a bistatic configuration with the radio astronomical array UTR-2 located near Kharkov, Ukraine, to conduct several experiments. In overall operational characteristics, the SURA/UTR-2 bistatic configuration approximates the El Campo facility except in operating frequency; El Campo used 38.25 MHz, whereas SURA/UTR-2 is presently restricted to a maximum of about 9.3 MHz. The principal motivations of new solar radar experiments have been to begin development of new techniques to identify earthward-moving CMEs and to use these techniques to study the solar corona. A greater understanding of coronal dynamical processes would be important to models of the development of CMEs and their subsequent effects at earth. The potential for providing several days' warning that large geomagnetic storms may occur is an attractive practical benefit. In the following sections of this report, we will discuss experimental results from both El Campo and SURA/UTR-2.

In the years since the El Campo experiments, the study of the earth's ionosphere with HF radars has provided well-established procedures and techniques that may carry over to studies of the solar corona. Such modern diagnostic techniques were not available to the early solar radar investigators. Also, the first solar radar experiment at microwave frequencies was reported by *Benz and Fitze* [1979] and *Fitze and Benz* [1981]. In addition, recent HF radar experiments with the near-earth space plasma have become possible. These experiments include studies of the solar wind [*Genkin and Erukhimov*, 1990; *Genkin et al.*, 1991] and the magnetosphere [*Gurevich et al.*, 1995]; experiments to detect the earth's magnetopause are also planned. New possibilities for the use of combined HF radars and satellite experiments have also been demonstrated in experiments with the NASA/WIND spacecraft [*van't Klooster et al.*, 1997; *Rodriguez et al.*, 1998, 1999].

REVIEW OF EL CAMPO OBSERVATIONS

The principal results of early solar radar observations can be found in the review by *J. C. James* [1968], several succeeding papers [*James*, 1970a; 1970b], and references therein. The fundamental basis of solar radar experiments is that coronal electron densities are typically in the range corresponding to electron plasma frequencies from about 5 to 100 MHz. The radial dependence of the average quiescent coronal electron density is generally an inverse power law for which several models have been derived. The range of plasma frequencies listed above corresponds to coronal radial distances of about 10 R_o to 1 R_o (where R_o is in solar radii) for the electron density models used by James (Pottasch and Baumbach-Allen models), [*Pottasch*, 1960; *Allen*, 1973]. An electromagnetic wave at a given frequency propagating into a plasma of increasing electron density encounters an index of refraction decreasing from the free-space value of 1 to 0 when the wave frequency equals the plasma frequency. This decrease of refractive index generally causes a turning in the direction of propagation of the wave and can result in reflection of the wave. Thus, an earth-launched electromagnetic wave entering the corona can be reflected from a coronal height where the wave frequency and the local plasma frequency are equal. However, for a turbulent and structured corona, the wave-reflection mechanism described above is probably too simple; a return signal is expected to be the result of more complex wave-plasma interactions. Such interactions, if properly diagnosed, may provide new information on coronal plasma dynamics. Fortunately, coronal electron densities are mostly greater than electron densities in the earth's ionosphere, so that the required frequencies for a solar radar can pass through the earth's ionosphere both on transmission and return.

In Figure 1, we show a photograph of the El Campo solar radar array. The array was formed from 1024 half-wave dipoles used as both transmitting and receiving antennas. A cross-polarized receive-only array was later added, consisting of 512 dipoles. The facililty operated at a total transmitted power of about 500 kW. At a half-power beamwidth of about 1° north-south by 6° east-west, the antenna gain factor was 32 to 36 dB, resulting in an effective radiated power of about 1300 MW. The array was phased manually to allow repositioning of the main lobe according to the change of the sun's apparent elevation throughout the year. During a given transit by the sun at local noon, the east-west beamwidth was wide enough to allow 16 minutes of transmission followed by 16 minutes of reception, where 16 minutes is the travel time of electromagnetic waves to the sun and back. The normal mode of operation was to transmit at two frequencies, offset from 38.25 MHz by several tens of kHz. The transmission was switched between the two frequencies according to a pseudo-random code that provided a known ON-OFF sequence of each frequency. The most common pulse widths used were 0.5 to 2 seconds, corresponding to a range resolution of about 0.1 to 0.4 R_o. The received signal was correlated with the known sequence of ON-OFF pulses and integrated in 20 range "boxes." Each range box corresponded to a time delay in the return signal and was associated with a radial distance interval in the solar corona. The recorded signal bandwidth was adjusted for various ranges of frequency shifts in the return signal, up to about 60 kHz, so that the final output consisted of a spectrum of total integrated power in about 20 by 10 range-frequency shift boxes. These spectra of the return signal, referred to as the solar echo spectra, were the basis of analysis and interpretation.

In Figure 2, we show a series of plots extracted from El Campo publications that illustrate the measurements. When examining these measurements, it is important to understand that all of the El Campo results were derived with integration techniques. The return signal is weak and always below the minimum background noise level, which is set by the solar HF background. Only by integration can the signal to noise ratio be made equal to or greater than 1. In Figure 2, the four data panels show a series of what were probably exploratory measurements taken during the first years of operation in which the pulse width of the transmitted waves was reduced from 8 sec to 1 sec. Reducing the pulse width increased the range resolution and simultaneously reduced the total range covered by the 20 integration boxes. Thus, with 8-sec pulse width, the measurement shows the maximum return power, relative to background, occurring in integration box number 16, whose time delay corresponds to a position centered on about 1.5 R_o. However, because of the long pulse width, there is little resolution of the region of maximum power return. The sequence of data plots as the pulse width is reduced illustrates the general increase of resolution of the region near 1.5 R_o, and the corresponding decrease of total range to about ± 2 R_o about the sun. The data plots of Figure 2 consist only of the total relative power versus range. In Figure 3, El Campo data are displayed in spectral

**Solar Radar Antenna
El Campo, Texas**

Transmitting λ/2 dipoles:	128 x 8 EW
Receiving λ/2 dipoles:	128 x 8 EW, 128 x 4 NS
Total Area: 18,000 m²	Power: 500 kW
Antenna Gain: 32-36 dB	ERP: 1300 MW
Frequency: 38.25 MHz	BW: 1° x 6° (NSxEW)

Figure 1. Aerial view of El Campo solar radar array. The long axis is in the north-south direction, thus providing a fan beam in the east-west direction. Characteristics of the array are listed.

format, in which pulse widths of 0.5 sec were used and the frequency shift for each range box is also resolved, in frequency intervals of about 5 kHz. Most of the published El Campo data are in the format shown in Figure 3. Total relative power in each spectral cell is indicated by the height of darkening. In this figure, the data indicate the strongest signal return occurring from a broad region centered at about 1.2 R_o with frequency shifts from about -5 kHz to about +25 kHz. The interpretation given to such data is that the solar echo comes from a range of coronal longitudes, including the eastern and western sides of the corona. Because the minimum beamwidth used is about 1°, the disk of the sun itself (about 0.5° in angle) cannot be resolved; thus, return power is integrated over a cross field range of about 2 R_o in dimension. Frequency shifts were interpreted as Doppler shifts resulting from moving coronal plasma that becomes the solar wind. During the years of El Campo operations, several categories of spectral plots were derived, which were based on certain recurring characteristics. Among these categories were the "high coronal echoes," corresponding to relative maximum

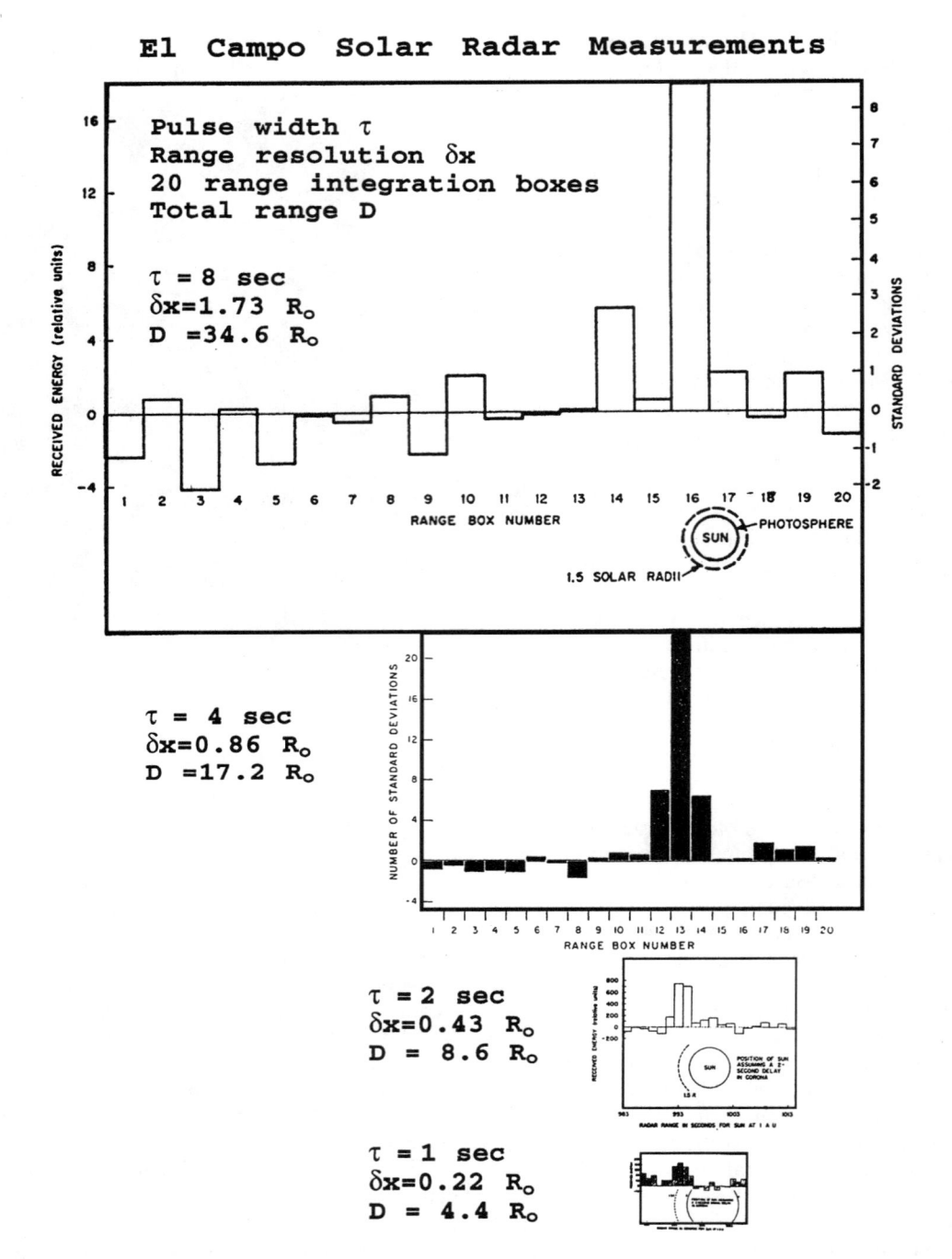

Figure 2. Composite of range-amplitude measurements from several publications of the El Campo results. As the transmitted pulse width was reduced from 8 s to 1 s, the corresponding range resolution increased, and the total range decreased. The plots are scaled to keep the solar diameter approximately the same.

signals located at about 2 R_o. These type spectra showed both "narrow" and "broad" Doppler shifts, suggesting various levels of ordering to coronal plasma motion.

In light of our current understanding of coronal dynamics, it is possible that some of these high coronal echoes may have been associated with coronal mass ejections. Unfortunately, the only correlative observations available at that time seems to have been the sunspot number. The El Campo results show that the radar cross section of the solar corona showed a positive correlation with sunspot number and was taken to indicate that increasing solar activity increased the power of return signals. Various reasonable interpretations for this correlation were given, and some models were suggested, which came close to what might be conceived of as coronal mass ejections, but the phenomenon of CMEs as we know it today (a large scale coronal disturbance) was not considered. Nevertheless, the published El Campo data are impressive, and our contemporary knowledge suggests that similar measurements made today could significantly increase our understanding of the solar corona, especially if these measurements were done in correlation with other coronal observations, such as those of coronagraph images. It is very fortuitous that the same radial range in the corona is covered by white-light coronagraph images and HF radar observations. Thus, the same electron densities that scatter white light for coronagraph images are responsible for the HF solar echo. There are few places in the space environment where such a coincidence of measurement parameters occurs.

NEW SOLAR RADAR EXPERIMENTS

The possibility that CMEs can be detected with HF solar radars is suggested by the El Campo observations. In order to test this concept, it would be desireable to use radar facilities that are close equivalents to El Campo. Such facilities may be found in some of the high power HF radars that are presently in operation. Among these are ionospheric "heating" facilities, which began to be developed at about the same time as the El Campo facility, although their objective was to study modifications and wave interactions in the earth's ionosphere. These facilities were designed to transmit at frequencies below about 10 MHz, so that the waves would be absorbed and scattered at ionospheric altitudes. In addition to the ionospheric radars, another type of high power radar available was developed for planetary studies. Planetary radars are designed to detect returns from planetary hard surfaces, such as those of Venus and the moon. For these experiments, frequencies in the several hundred MHz range are needed (for example, the Arecibo radar operates at 435 MHz). These higher frequencies are in fact too high for the solar corona investigation because they propagate far into the corona and reach photospheric heights where absorption dominates and no return signal occurs. The optimum frequency window for solar radar experiments is in the tens of MHz range, so El Campo was designed for 38.25 MHz. As of today, the only high power radars operating in this range are over-the-horizon radars designed by the military for long-range surveillance. These radars, however, have been constructed to form their transmit and receive beams at low elevation angles in specific azimuth directions so that the ionosphere can be used to reflect the waves around the curve of the earth's surface. This configuration is

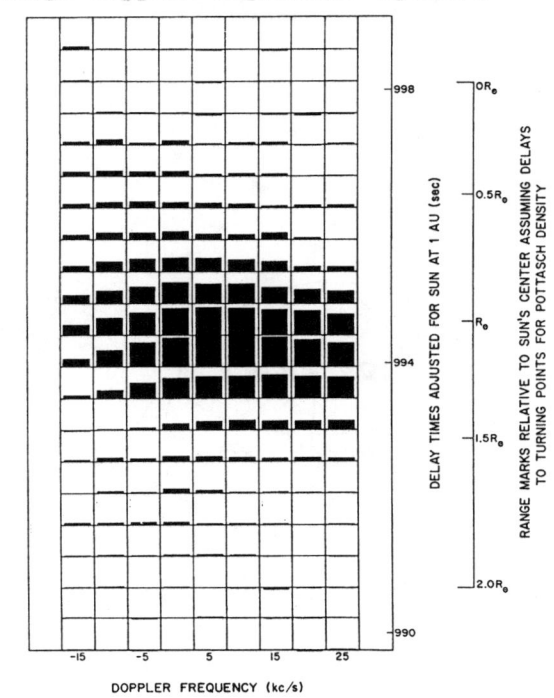

Figure 3. The typical El Campo range-doppler-amplitude spectra for pulse widths ≤ 1 s. The size of the black shading in each range-doppler cell is proportional to the integrated signal/noise ratio of the solar echo. The range scale is adjusted according to the coronal density model used. For the spectrum shown the maximum echo signal is observed between 1 and 1.5 R_o.

disadvantageous for a solar radar because only sunrise or sunset conditions can be used, and only if sunrise or sunset occurs within the azimuth direction of the main beam.

In terms of overall performance, the combination of the Russian SURA ionospheric heating facility and the Ukrainian UTR-2 radio astronomical array comes closest to the El Campo solar radar. Appendix B compares the system performance of SURA and UTR-2 facilities with El Campo. Except for the frequency of operation, the SURA/UTR-2 combined facility is similar to El Campo. A schematic drawing of the bistatic configuration used and the experiment concept is shown in Figure 4. The drawing of Figure 4 represents an earthward-directed CME, with the return signal coming from the 1 R_o to 5 R_o range. In the lower right corner is a schematic of the frequency spectrum of the solar echo that might occur. The transmitted signal is shown as a single frequency at high power; the return signal is broadened by both random and directed motion in the corona. If a CME is present, it is expected to be a generally large-scale structure moving approximately as a single entity; thus, the power return may be greater and the

Solar Radar Detection of Earthward-Directed Coronal Mass Ejections

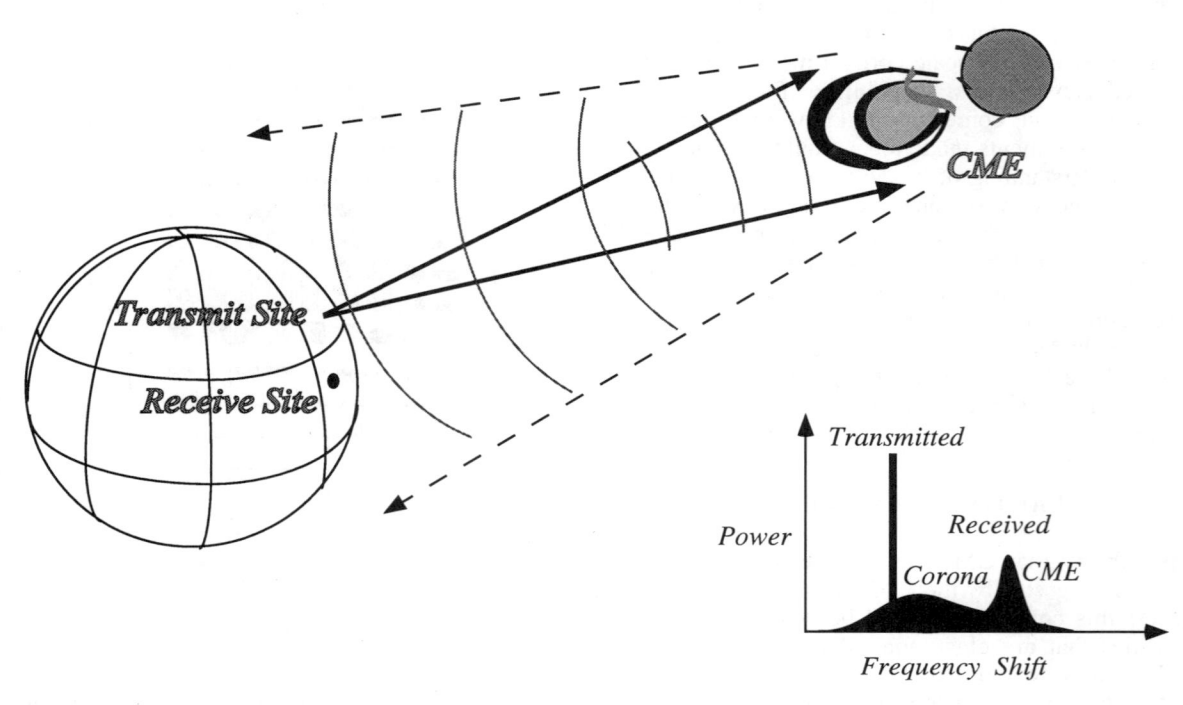

Figure 4. Schematic of the concept for detecting coronal mass ejections with an HF solar radar. A bistatic arrangement is shown, with the transmitting site separated from the receiving site. In the lower right-hand corner, a schematic of the echo signal frequency spectrum is shown. The transmitted signal at a given frequency is shown along with the expected echo spectra. A "normal quiescent" corona may produce a broad range of frequency shifts, while a CME event may produce a more distinct doppler shift.

frequency shift narrower than the background coronal plasma. Such a signature is reminiscent of some high coronal echoes of the El Campo data.

In addition to the ground-based facilities, we have made extensive use of the NASA/WIND spacecraft to provide a space-based receiver for the SURA-transmitted signals. Such use of a space experiments was not available to the early solar radar experiments and provides measurements of the influence of the earth's ionosphere on the high power signal launched toward the sun. An understanding and knowledge of possible ionospheric effects are important for subsequent analysis of return signals, especially at the lower frequency we must currently use.

EXPERIMENTS CONDUCTED

Beginning in the summer of 1996, we have conducted three limited series of experiments with the SURA/UTR-2 bistatic configuration. Initial tests involved transmission at full power with 5- to 10-sec pulses alternating between two frequencies in a periodic waveform. The transmission frequencies were determined by surveys of interference backgrounds observed at both transmit and receive sites. Generally, the two most common frequencies used for transmission from SURA were 8.920 and 8.880 MHz. At the UTR-2 site, reception was done with receivers with about 40 kHz bandwidth. Digital waveform measurements were made of the baseband frequency with 2x40 kHz sampling rate, and the analog signal was also recorded on magnetic tape. Several other types of received signal measurements were made using high-speed correlometers and total power detectors. On the selected summer days, SURA transmission occurred for the 16-minute interval *UT* 0911 to 0927, corresponding to near local noon, when the visible sun was within the main lobe of the SURA array. A periodic square wave modulation with pulse period of 20 sec switched the transmission frequency between 8.920 and 8.880 MHz, with ON pulse width of 10 sec. Thus, the full power of 750 kW was ON at all times, either in one frequency or the other. Other pulse periods and widths were also used; however, we discuss only this one mode in this report. Signals were recorded at UTR-2 in three phases: during transmission from SURA (with receiver front end attenuation of 30 dB), during the time interval of the expected return signal (receiver front end attenuation set to 0 dB), and after the return signal time period to obtain background measurements (receiver front end attenuation also at 0 dB). Data obtained during the return signal time interval (*UT* 0927 to 0943) were processed by removing interference signals (a standard process used for radio astronomy observations), Fourier analyzing the waveform data, and integrating the resulting spectrum over the basic pulse period of the transmissions. The analysis approach is basically similar to that used by the El Campo investigators.

ANALYSIS OF SELECTED DATA

In Plate 1, we show the results of analysis of the experiment of 21 July 1996. On that day the interference level was relatively low, and the probability is high that the coronal return signal was detected. In the upper part of the plate, the transmitted pulse waveforms are shown, consisting of 20-sec period with 10-sec pulse width at full power. The waveform at 8.920 MHz is in opposite phase to the waveform at 8.880 MHz, so that full power was being transmitted at all times. In the 16-min interval, 48 transmission cycles occur. For the recorded signal from *UT* 0927 to 0943, the Fourier-analyzed data were integrated in 48 cycles to produce the integrated spectrum shown in the lower part of Plate 1. The vertical axes of the spectra correspond to the bandwidth of the receivers, with the transmitted frequency offset to the bottom side of each spectrum; thus, the measured frequency shifts are positive and correspond to Doppler shifts that would be caused by motion toward the earth. The horizontal axes correspond to the transmitted pulse periods. The color shading represents relative integrated power level; the dynamic range is about 1 dB with yellow being the high end and blue the low end of the scale. Thus, the color variation is indicative of variations in the signal-to-noise ratio. The red regions prominent at the tops of both spectra should be ignored, as these correspond to the roll-off the receiver filters. The regions of color-coding at the green and yellow level correspond to detected signals, and a comparison of the two spectra shows that, while the signal is weak, it has the expected phase relationship of the transmitted waveform. Thus, at a given frequency, the received signal-to-noise ratio is higher during the expected ON phase compared with the OFF phase. Also, comparing the two spectra shows that the ON phases are in opposition, as was the transmitted signal. Thus, the averaged spectra are consistent with a positive return from the solar corona. Because the dynamic range of the signal-to-noise ratio is low, on the order of 1 dB, we cannot attribute much signficance to apparent spectral details; it is possible that some features are remnants of interference signals or other noise sources. Therefore, we refer only to the total power comparisons and phase relationships of the spectra as an indication of positive detection. In Figure 5, we show the same data integrated in frequency to obtain a total relative power comparison. The traces here show the relative phases of the integrated power levels appear to be consistent with the transmitted waveform. Thus, we conclude from these observations that the return signal from the solar corona at about 9 MHz has been detected.

Other experiments that we have conducted generally produced less reliable results, due primarily to higher

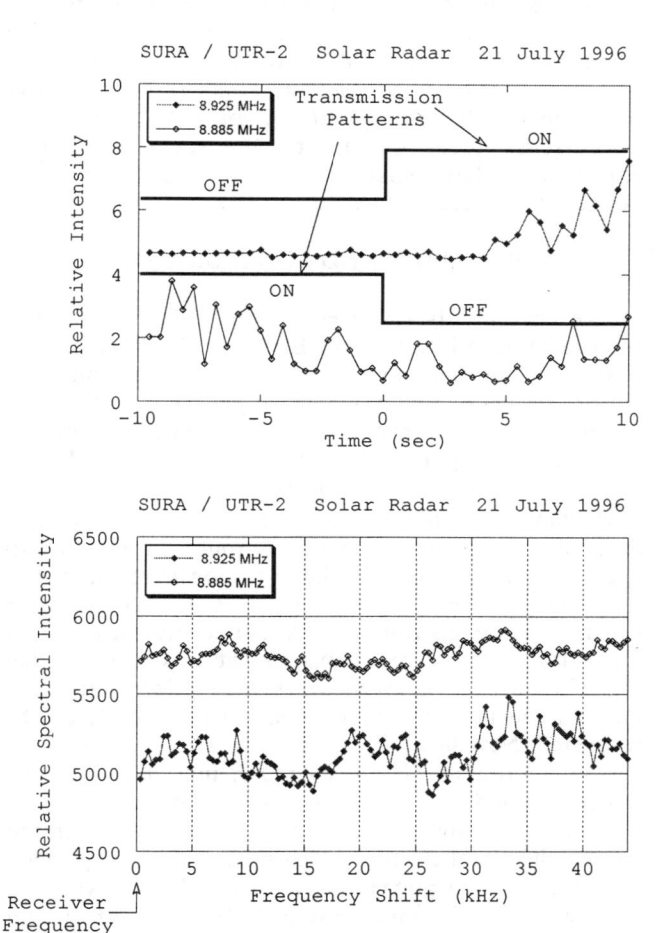

Figure 5. The data of Plate 1 displayed in two-dimensional integrated form. The upper panel is the time profile of the 16-min integrated echo signal at the two receiver frequencies. The pulse patttern is shown for comparison and shows that the integrated echo signal shows a similar phase relationship. The lower panel shows the 16-min integrated frequency spectra of the 44-kHz receiver bandwidth. No well-defined frequency "peaks" correpording to a doppler shift from a CME are evident; thus, only a quiescent coronal echo is suggested.

interference levels. Of the 8 experiments conducted in 1996, the data of 21 July were the best. Additional experiments were conducted in several weeks of the summers of 1997 and 1998, in which we investigated the use of several types of random coding to help improve discrimination against background noise. Although we do not discuss them in this report, several of these experiments have also produced apparent positive detections of a return signal.

SUMMARY AND CONCLUSIONS

We have briefly reviewed the highlights of the early solar radar experiments at the El Campo facility and have described the characterisitics of the experiments and general results of analyses of radar echoes from the solar corona. The early solar radar experiments are a remarkable example of research that appears to have been ahead of its time. It is unfortunate that no comparable facility exists today that can benefit from the several correlative and detailed observations that are fairly routine in the modern era of space research. We have demonstrated that it is possible to use existing facilities for solar radar experiments by reporting a weak, but positive detection of the return signal in a recent experiment. It is also clear that with upgrades to current facilities, mainly for operation at higher frequency, significant solar radar capability can be recovered. New facilities, such as the HF Active Auroral Research Program (HAARP) radar, may allow several frequencies and diagnostics approaches to be used to detect CMEs. For example, the HAARP radar is steerable through the use of electronic phasing, and the final planned effective radiated power will be about 3 GW, which is greater than El Campo provided. In terms of system performance, the completed HAARP facility, if combined with a receiving array similar in collecting area to UTR-2, could achieve a 5-10 dB signal-to-noise ratio in the solar echo with only a few seconds of integration. Thus, with the more advanced state of understanding and experimentation of coronal physics at present, solar radar experimental research may be poised for a promising new beginning.

Acknowledgements. The work at the Naval Research Laboratory and at the SURA and UTR-2 facilities is supported by the Office of Naval Research and the Navy International Cooperative Opportunities in Science and Technology Program. We also acknowledge with gratitude the participation of Dr. Lev Erukhimov, and express our regret at his passing in 1997.

APPENDIX A: SOLAR RADAR INVESTIGATORS

P. Rodriguez
(Information Technology Division, Naval Research Laboratory, Washington DC)

A. Konovalenko, O. M. Ulyanov, S. Stepkin, G. Inyutin, M. Sidorchuk, and E. Abranin
(Radio Astronomy Department, Institute of Radio Astronomy, Kharkov, Ukraine)

L. Erukhimov, Yu. I. Belov, Yu. V. Tokarev, A. Karashtin, and G. Komrakov
(Solar Terrestrial Physics Department, Radiophysical Research Institute, Nizhny Novgorod, Russia)

C. G .M. van't Klooster
(ESTEC XEA, Noordwijk, The Netherlands)

APPENDIX B: EL CAMPO - SURA/UTR-2 SYSTEM COMPARISON

The radar equation is

$$P_r = \frac{P_t G_t}{4\pi R^2} \frac{A_r \sigma p}{4\pi R^2}$$

where:

P_r - received power (watts)
P_t - transmitted power (watts)
G_t - transmitting antenna gain
A_r - receiving antenna area (m^2)

Plate 1. The data from the Sura/UTR-2 experiment of 21 July 1996. The upper panel shows the pulse pattern transmitted by Sura in the 16-min interval between *UT* 0911 and 0927 at two frequencies separated by 40 kHz. The lower panels show the dynamic spectra of the echo signals received during the 16-min interval between *UT* 0927 and 0943. The spectra are integrated over 20-s sub-intervals corresponding to the transmitted 20-s pulse period.The receiver frequency is shown on the frequency axis of each spectrum. The receiver bandwidth (44 kHz) determines the frequency range of each spectrum. The color coding covers a range of about 1 dB in signal to noise ratio. Above about 35 kHz the receiver filter roll-off determines the spectra color coding and should be ignored. The yellow and green colors correspond to the echo during the ON and OFF phases of the transmitted signal. A careful comparison of the total powers in the ON phases with the OFF phases illustrates that the echo intensities have the same phase relationship as the transmitted signals, providing high probability that a true echo signal is detected. However, no clear evidence of an echo from a CME is evident; the spectra appear to correspond to an echo from a normal corona.

Solar Radar Experiment 21 July 1996

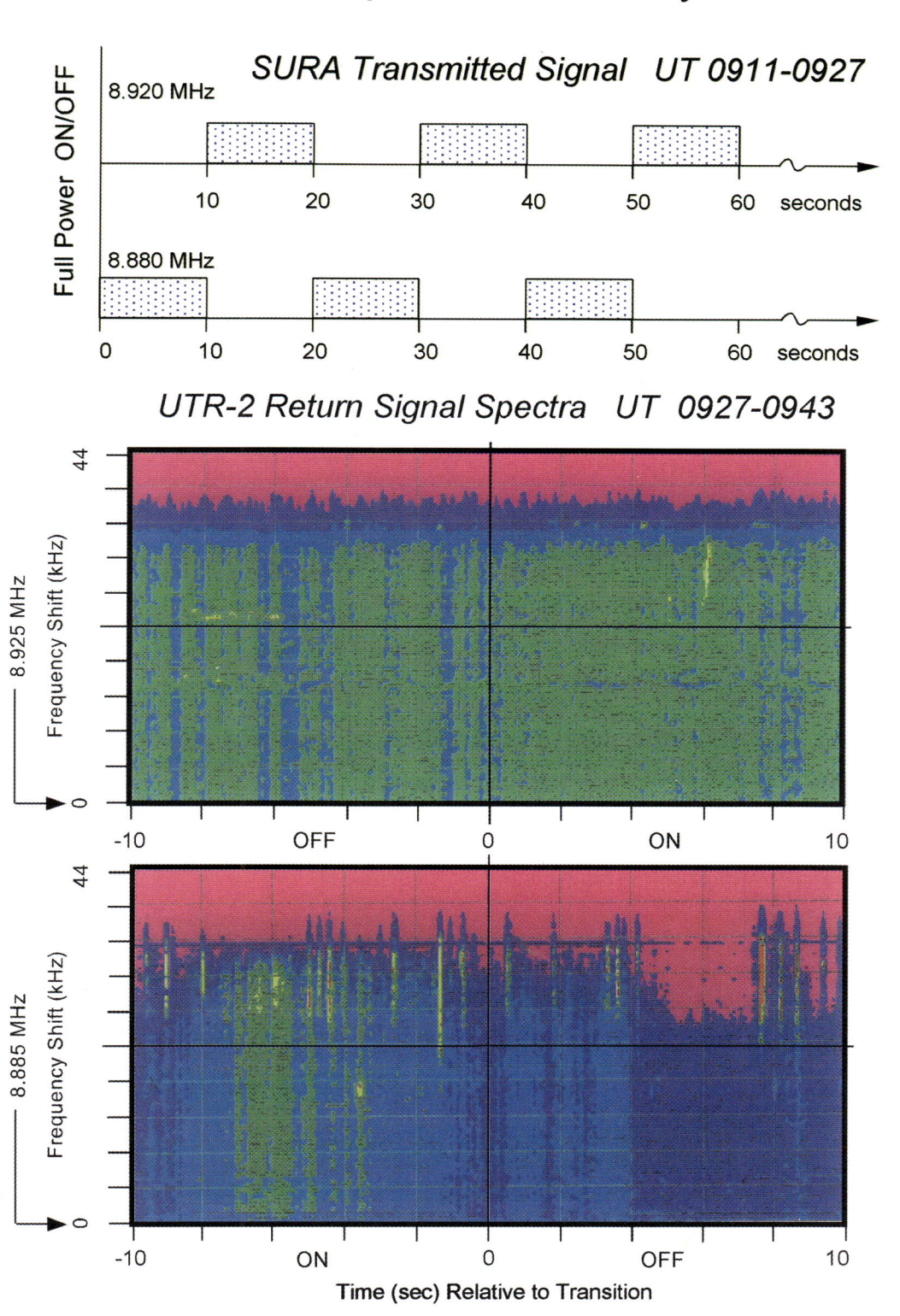

SURA Transmitted Signal UT 0911-0927

UTR-2 Return Signal Spectra UT 0927-0943

Time (sec) Relative to Transition

σ - scattering cross-section (m^2)
p - polarization fraction
R - range to scattering cross-section (m)
R$_o$ - solar radius (m)

Numerical values are:

R = 1.5 x 10^{11} m (1 AU)
R$_o$ = 7 x 10^8 m
$\sigma \sim 1\, \pi\, R_o^2$ = 1.54 x 10^{18} m^2
(typical value obtained by El Campo)
p = 0.5 (one of two polarizations)
N = k$_B$ T(f) B (noise power of sun)
k$_B$ = 1.38 x 10^{-23} watt-s °K^{-1} (Boltzman's constant)
T(f) = noise equivalent temperature of sun (°K)
B = receiver bandwidth (Hz)

System Comparison:

El Campo: SURA/UTR-2:

P$_t$ = 500 kW   P$_t$ = 750 kW
G$_t$ = 32 dB   G$_t$ = 19 dB
A$_{r1}$ = 18,000 m^2 A$_r$ = 150,000 m^2
A$_{r2}$ = 9,000 m^2 (cross polarization)
$\sigma \sim 1\, \pi\, R_o^2$ $\sigma \sim 1\, \pi\, R_o^2$
P$_r$ = 2.06 x 10^{-16} watts   P$_r$ = 8.23 x 10^{-17} watts

B = 50,000 Hz   B = 40,000 Hz
T(38 MHz) = 8 x 10^5 °K   T(9 MHz) = 3 x 10^5 °K
N = 5.5 x 10^{-13} watts   N = 1.7 x 10^{-13} watts

Signal to noise ratio, where S = P$_r$:

S/N = -34.0 dB   S/N = -33.0 dB

Signal processing (integration) gain: G $\sim \sqrt{Bt}$,
where t = 16 min = 960 s

G $\sim$ 38 dB   G $\sim$ 38 dB
Ionospheric Loss $\sim$ 1 dB   Ionospheric Loss $\sim$ 4 dB

Net S/N $\sim$ 3 dB   Net S/N $\sim$ 1 dB

REFERENCES

Abel, W. G., J. H. Chisholm, P. L. Fleck, and J. C. James, Radar reflections from the sun at very high frequencies, *J. Geophys. Res., 66*, 4303, 1961.

Abel, W. G., J. H. Chisholm, and J. C. James, Radar reflections from the sun at VHF, in *Space Res. III*, edited by W. Priester, North Holland Publishers, Amsterdam, p 635-643, 1963.

Allen, C. W., *Astrophysical Quantities*, 3rd ed., the Athlone Press, Univ. of London, 1973.

Bass, F. G., and S. Ya. Braude, On the question of reflecting radar signals from the sun, *Ukr. J. Phys., 2*, 149, 1957.

Benz, A. O., and H. R. Fitze, Microwave radar observations of the sun, *Astron. Astrophys., 76*, 354, 1979.

Braude, S. Ya., A. V. Megn, B. P. Ryabov, N. K. Sharykin and I. N. Zhouck, Decametric survey of discrete sources in the northern sky, *Astrophys. Space Sci., 54*, 3, 1978.

Chashei, I. V., and V. I. Shishov, Volume scattering model for interpretation of solar radar experiments, *Solar Phys., 149*, 413-416, 1994.

Crooker, N., J. A. Joselyn, and J. Feynman, editors, *Coronal Mass Ejections*, Geophysical Monograph 99, American Geophysical Union, 1997.

Eshleman, V. R., R. C. Barthle, and P. B. Gallagher, Radar echoes from the sun, *Science, 131*, 329-332, 1960.

Fitze, H. R., and A. O. Benz, The microwave solar radar experiment. I. Observations, *Astrophys J., 250*, 782-790, 1981.

Genkin, L. G., and L. M. Erukhimov, Interplanetary plasma irregularities and ion acoustic turbulence, *Phys. Repts., 186*, 97, 1990.

Genkin, L. G., L. M. Erukhimov, and Yu. V. Tokarev, Diagnostics of ion-acoustic turbulence in the solar wind by the method of radio wave scattering, *Proc. of the III URSI Symp. on Modification of the Ionosphere by Powerful Radio Waves(ISIM-3)*, Suzdal, USSR, Sept. 9-13, 171, 1991.

Gerasimova, N. N., Comparison of results of radar studies of the corona with solar activity, *Sov. Astron., 18*, 482, 1975.

Gerasimova, N. N., Efficiency of four-plasmon interactions in the reflection of a radar signal from the sun, *Sov. Astron., 23*, 738, 1979.

Gordon, I. M., Interpretation of radio echos from the sun, *Astrophys. Lett., 2*, 49, 1968.

Gordon, I. M., Radar exploration of the sun and physical processes of the solar corona, *Astrophys. Lett., 3*, 181, 1969.

Gordon, I, M., Plasma theory of radio echoes from the sun and its implications for the problem of the solar wind, *Space Sci. Rev., 15*, 157, 1973.

Gurevich A. V., A. N. Karashtin, A. M. Babichenko, I. N. Kazarov, and G. P. Komrakov, HF magnetosphere sounding at SURA, *Proc. 21st Remote Sensing Society Ann. Conf.*, Southampton, UK, p. 1333-1340, 1995.

James, J. C., Radar studies of the sun, in *Radar Astronomy*, edited by J. V. Evans and T. Hagfors, McGraw Hill Book Company, New York, 1968.

James, J. C., Some observed characteristics of solar radar echoes and their implications, *Solar Phys., 12*, 143-162, 1970a.

James, J. C., Some characteristics of the solar atmosphere that may be investigated by a VHF antenna system, *JPL Technical Report 32-1475*, 1970b.

Kerr, F. J., On the possibility of obtaining radar echoes from the sun and planets, *Proc. IRE, 40*, 660, 1952.

Mel'nik, V. N., Plasma theory of solar radar echoes, poster at this Chapman Conference, 1998.

Owocki, S. P., G. A. Newkirk, and D. G. Sime, Radar studies of the non-spherically symmetric solar corona, *Solar Phys., 78*, 317-331, 1982.

Pottasch, S. R., Use of the equation of hydrostatic equilibrium in determining the temperature distribution in the outer solar atmosphere, *Astrophys. J., 131*, 68-74, 1960.

Rodriguez, P., High frequency radar detection of coronal mass ejections, in *Solar Drivers of Interplanetary and Terrestrial Disturbances*, edited by K. S.Balasubramaniam, S. L. Keil, and R. N. Smartt, ASP Conference Series, Vol. 95, 180-188, 1996.

Rodriguez, P., E. J. Kennedy, M. J. Keskinen, C. L. Siefring, Sa. Basu, M. McCarrick, J. Preston, M. Engebretson, M. L. Kaiser, M. D. Desch, K. Goetz, J.-L. Bougeret, and R. Manning, The WIND-HAARP experiment: initial results of high power radiowave interactions with space plasmas, *Geophys. Res. Lett., 25*, 257, 1998.

Rodriguez, P., E. J. Kennedy, M. J. Keskinen, Sa. Basu, M. McCarrick, J. Preston, H. Zwi, M. Engebretson, A. Wong, R. Wuerker, M. L. Kaiser, M. D. Desch, K. Goetz, J.-L. Bougeret, and R. Manning, A wave interference experiment with HAARP, HIPAS, and WIND, *Geophys. Res. Lett., 26*, 2351, 1999.

van't Klooster, C. G. M., Yu. Tokarev, Yu. Belov, A. A. Konovalenko, and P. Rodriguez, Transmission of high power 9 MHz signals near ionospheric cut-off from 'Sura' towards the moon and reception of the echo with the radio telescope 'UTR2', in *10th International Conference on Antennas and Propagation*, Conference Publication No. 436, 2.35, 1997.

Wentzel, D. G., A new interpretation of James's solar radar echoes involving lower-hybrid waves, *Astrophys. J., 248*, 1132-1143, 1981.

P. Rodriguez, Information Technology Division, Naval Research Laboratory, Washington, DC 20375 (email: paul.rodriguez@nrl.navy.mil)

Radio Emissions from the Planets and their Moons

Philippe Zarka

Observatoire de Paris / CNRS, DESPA, Meudon, France

We review the present observational knowledge of planetary non-thermal radio emissions (spectrum, emitted power, source location, beaming, modulations and frequency-time structures), emphasizing high-latitude, auroral and satellite-induced components at the outer planets. Important results have been obtained in the past few years from the observations of Ulysses and Galileo at Jupiter, of Wind and other spacecraft in Earth orbit, from the reanalysis of Voyager data about Saturn, Uranus, and Neptune, from ground-based high frequency-time resolution and full polarization measurements, and from pioneering multi-spectral observations of the Jovian and Saturnian aurorae (radio/UV/IR). We try to organize those in a coherent frame, discuss similarities and differences at the five radio planets, and outline open questions, future observations, as well as the broad astrophysical interest of studying planetary magnetospheric, non-thermal radio emissions.

1. INTRODUCTION

Planetary radio emissions are electromagnetic radio waves propagating on the so-called left-handed polarized, ordinary (LH-O) magneto-ionic mode, or on the right-handed extraordinary (RH-X) mode. Those are the only modes detectable by a radioastronomy instrument in the free space. In a weakly magnetized plasma such as the solar wind in the interplanetary medium, the low-frequency (LF) cutoffs of these modes are both equal to the local electron plasma frequency

$$f_X \simeq f_O = f_{pe} = \tfrac{1}{2\pi}(N\ e^2/\epsilon_o\ m_e)^{1/2}$$

N being the electron density and e and m_e the electron charge and mass. As N is typically about $5 - 10\ cm^{-3}$ at the Earth orbit, and varies as the inverse square of the distance to the Sun, one obtains in Figure 1 the minimum frequency above which a radio emission is able to propagate in the solar wind, as a function of the distance to the Sun. The intersection of each planetary orbit with the shaded region of Figure 1 gives the minimum frequency above which a radio emission from the corresponding planet can be detected remotely. This minimum frequency is a few kHz for the outer planets, ~ 20 kHz for the Earth, but as high as $70 - 100$ kHz for Mercury. As the maximum surface magnetic field at that planet is expected to be $< 10^{-2}$ Gauss [*Ness*, 1979], the maximum surface electron gyrofrequency $f_{ce} = eB/2\pi m_e$ is thus < 30 kHz. As a consequence, Mercury's low-frequency auroral radio emissions, if they exist, cannot escape its magnetosphere to propagate in the solar wind.

2. PLANETARY RADIO COMPONENTS

We focus here on the five strongly magnetized planets (the Earth, Jupiter, Saturn, Uranus and Neptune). At least nine types of radio components are known to originate from these magnetospheres. None of the five radio planets emits all components, but Jupiter with

Radio Astronomy at Long Wavelengths
Geophysical Monograph 119
Copyright 2000 by the American Geophysical Union

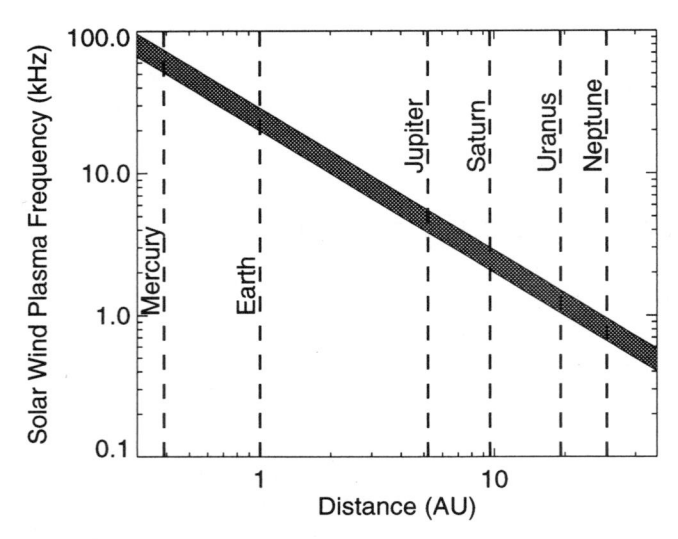

Figure 1. Solar wind plasma frequency versus distance to the Sun. Its value at a given distance fluctuates within the shaded area. No radio emission can propagate in the solar wind below this limit. The orbits of the strongly magnetized planets are indicated.

its intense magnetic field ($M = 4.2\ G.R_J^3$) and huge and complex magnetosphere, produces at least eight of them. Table 1 lists all these radio components together with the planets at which they are observed, their wavelength range and radiation process. Figure 2 displays the spectra of all Jovian radio components as well as those of the other planetary auroral radio emissions. Figure 3 sketches the locations of known radio sources in the Jovian magnetosphere.

About radio emissions from moons, let us add that the Galileo spacecraft recently revealed a mini-magnetosphere around the Jovian satellite Ganymede, producing LF (≤ 50 kHz) radio emissions similar to the so-called non-thermal continuum radiation [*Kurth et al.*, 1997]. *Menietti et al.* [1998a] have also attributed to Io, although on fragile grounds, a radio emission observed in the range 0.6 − 1.2 MHz by Galileo near Io.

As a detailed review of all these radio components is beyond the scope of this paper, we will focus hereafter on the planetary radio emissions originating from high-latitude magnetospheric regions, that is the auroral radio emissions and those from the Io-Jupiter (and possibly the Ganymede-Jupiter) electrodynamic circuit.

3. HIGH-LATITUDE RADIO EMISSIONS

These radio emissions all occur in strongly magnetized regions of the planets possessing a developed magnetosphere, where the local plasma density is low ($f_{pe}/f_{ce} \ll 1$) and where energetic (keV) electron populations are present [e.g., *Gurnett*, this book]. They are attributed to the cyclotron-maser mechanism [*Ladreiter*, this book]. At Earth, Saturn, Uranus, and Neptune, whose surface magnetic field is ≤ 1 Gauss, auroral radio emissions occur in the kilometer range and are thus named "XKR", where "X" refers to the planet name and "KR" stands for Kilometric Radiation. In the case of Earth, the name used by geophysicists is AKR (Auroral...), while astronomers rather use TKR (Terrestrial...). At Jupiter, the intense magnetic field (up to 14 Gauss at high northern latitudes) results in cyclotron emissions in the kilometer to decameter range. Jovian auroral components are thus named "bKOM" (for broadband kilometer — to distinguish from the narrowband kilometer emission from Io's plasma torus, or nKOM), "HOM" (for hectometer) and "Io-independent" or "non-Io" DAM (decameter). The Io-independent character of auroral radio emissions is explicitely stated only for the DAM component (although bKOM and HOM are also entirely Io-independent) in order to distinguish it from the "Io-controlled" DAM emission (or

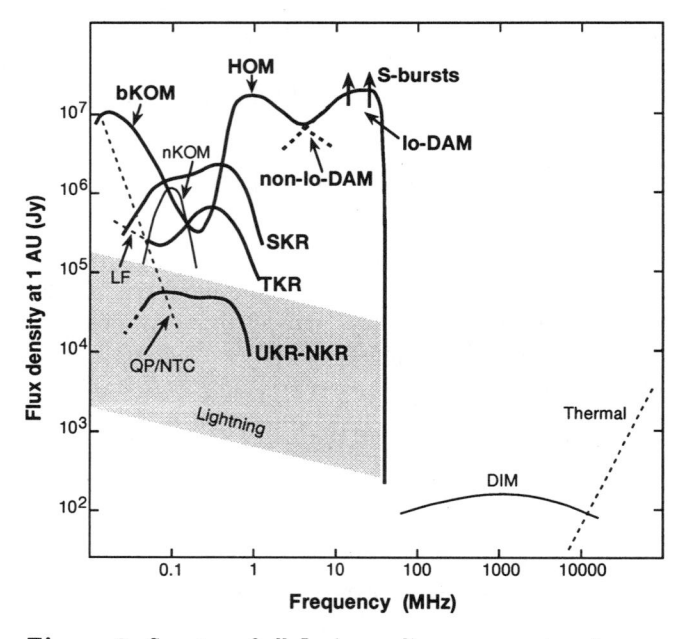

Figure 2. Spectra of all Jovian radio components, plus auroral radio emissions from the other radio planets ($1\ Jy = 10^{-26}\ Wm^{-2}Hz^{-1}$). Boldface lines emphasize high-latitude emission spectra. Part of Io-DAM consists of impulsive bursts ("S"=short). The range displayed for lightning corresponds to intensities detected at Saturn and Uranus; Terrestrial lightning can be orders of magnitude weaker. "DIM"=decimeter synchrotron emission from Jovian radiation belts. For "LF", "QP" and "NTC", see Table 1.

Table 1. Planetary Radio Components Characteristics

Radio Component	Planet[a]	Wavelength, m	Radiation Process	Refs.
Thermal	E J S U N	$10^{-2} - 10^{-1}$	Thermal !	1
from Van Allen belts	J (E)	10^{-1}	Incoherent synchrotron radiation from MeV electrons	1, 2
Lightning	E (J) S U (N)	$10 - 10^4$	Antenna radiation / Current discharge	3, 4
Auroral	E J S U N	$10 - 10^3$	Cyclotron-maser from unstable keV electron populations	5, 6
Satellite-induced (Io)	J	$10 - 10^2$	Cyclotron-maser	1, 7
Equatorial X-mode	U N	$10^3 - 10^4$	Cyclotron-maser	8, 3
Bursts, QP[b]/ LF[b]/ ...	E J (U) (N)	$10^3 - 10^4$	Cyclotron-maser ?	9, 10
nKOM[b]	J	10^3	Instabilities near f_{pe} or f_{uh} from Io torus inhomogeneities ?	1, 11
NTC[b]	E J S U N	10^4	Instabilities near f_{pe} or f_{uh} at the magnetopause ?	12

Refs: 1, *Carr et al.* [1983]; 2, *Leblanc* [this book]; 3, *Zarka et al.* [1995]; 4, *Farrell* [this book]; 5, *Treumann* [this book]; 6, *Ladreiter* [this book]; 7, *Zarka* [1998]; 8, *Desch et al.* [1991]; 9, *MacDowall et al.* [1993]; 10, *Steinberg et al.* [1988]; 11, *Reiner et al.* [1993]; 12, *Kurth* [1992].

[a] E=Earth, J=Jupiter, S=Saturn, U=Uranus, N=Neptune.

[b] QP=Quasi-Periodic Jovian bursts, LF=Low-Frequency Terrestrial bursts, nKOM=narrowband Kilometer radio emission from Io's torus, NTC=Non-thermal continuum.

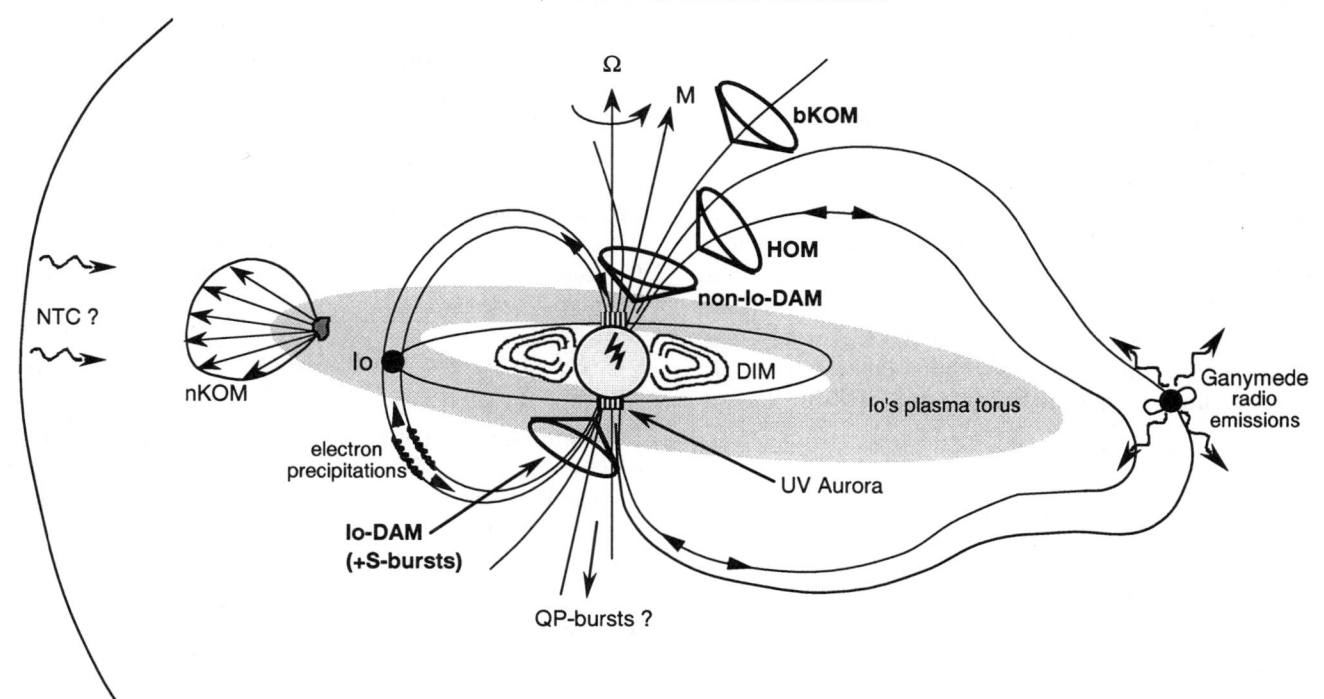

Figure 3. Sketch of radio source locations in the Jovian magnetosphere. Boldface lines emphasize high-latitude emission sources, which actually exist in both hemispheres. bKOM, HOM and DAM are generated at $f \simeq f_{ce}$ and beamed in hollow cones aligned on magnetic field lines with parameter $7 \leq L \leq 15$ [*Ladreiter et al.*, 1994]. Non-thermal continuum (NTC) may originate from density gradients near the magnetopause (not to scale) [*Kurth*, 1992] or be related to the quasi-periodic (QP) bursts thought to originate —at least partly— from southern auroral latitudes [*MacDowall et al.*, 1993]. nKOM is emitted in wide beams by unidentified torus inhomogeneities [*Reiner et al.*, 1993].

Table 2. Radiotelescopes and Spacecraft having made Observations of High-latitude Magnetospheric Radio Emissions

Radiotelescope or Spacecraft	Planet[a]	Measurement Capabilities[b]
Ground-based radiotelescopes ($\geq$ 1955) (Boulder, Nançay, Florida, Kharkov...)	J	I, Q, U, V
RAE (Radio Astronomy Explorers) 1–2	E J	I, 2D-DF
Geos 1–2, Hawkeye, Imp 6–8, ISEE 1–2 ISIS 1–2, Viking, AMPTE	E	I, 1D-DF
DE (Dynamic Explorer) A	E	I, V, 1D-DF
ISEE 3	E	I, Q, U, V, 2D-DF
Voyager 1–2	E J S U N	I, V
Ulysses	J (S)	I, Q, U, V, 2D-DF
Wind (Polar, Geotail)	E J (S)	I, Q, U, V, 2D-DF
Galileo	(V) E J	I

[a] E=Earth, V=Venus, J=Jupiter, S=Saturn, U=Uranus, N=Neptune.
[b] Stokes parameters: I=total intensity, Q,U=linear polarization ratio, V=circular polarization ratio; DF=direction-finding. 1D-DF is obtained through signal modulation on one rotating equatorial dipole and gives the source azimuth, i.e. the projection of $\vec{k}$ on the antenna spin plane. 2D-DF requires $\geq$2 antennas and gives the source azimuth and elevation, i.e. the full $\vec{k}$ vector. (adapted from [Kaiser, 1989]).

"Io-DAM"), whose occurrence and intensity vary as a function of the position of the observer relative to the Io-Jupiter configuration [e.g., Carr et al., 1983].

High-latitude magnetospheric radio emissions consist of slowly varying components and impulsive bursts, superimposed or alternating as a function of time and geometry. We summarize below their main characteristics (spectrum, emitted power, source location, beaming, modulations and frequency-time structures), which are the basic inputs for building and testing theories and models. The data from which these properties have been deduced come from a number of ground-based radiotelescopes and spacecraft radio experiments with various technical capabilities, as listed in Table 2. Synthetic accounts of planetary radio observations can be found in Carr et al., [1983], Genova et al. [1989], and Kaiser [1993] for Jupiter, Kaiser et al. [1984] for Saturn, Desch et al. [1991b] for Uranus, Zarka et al. [1995] for Neptune, and De Féraudy et al. [1988] for Earth. A general review of observations and theories of auroral radio emissions as well as more extensive references can be found in [Zarka, 1998]. Figure 2 of that paper displays typical dynamic spectra (i.e. intensity versus frequency and time) of these emissions for one full rotation of each radio planet.

3.1. Spectrum and Emitted Power

The average spectra of high-latitude magnetospheric radio emissions are given on Figure 2, normalized to a distance of 1 AU. They have been obtained by averaging the total flux received at each frequency, from close range (i.e. allowing to neglect the detection threshold bias), over a few planetary rotations [Zarka, 1992]. As the emissions are highly variable, the accuracy of these average values is not better than a factor 2, and instantaneous flux densities can exceed average values by a factor $\sim 10 - 50$. An alternate way to represent such highly variable spectra is to plot the flux densities that are exceeded $x\%$ of the time, for different values of x, as in Figure 5 of [Kaiser et al., 1984] for SKR.

The bimodal shape of the Jovian HOM+DAM spectrum is attributed to the existence of two DAM components, one of which is of auroral origin and simply the

Table 3. Average Power of High-latitude Magnetospheric Radio Emissions

Planet	Radio Component	Average Power, W
Earth	TKR	10^8
Jupiter	bKOM	10^9
"	HOM	10^{10}
"	DAM	10^{11}
Saturn	SKR	10^9
Uranus	UKR	3×10^7
Neptune	NKR	10^7

See text section 3.1 for details.

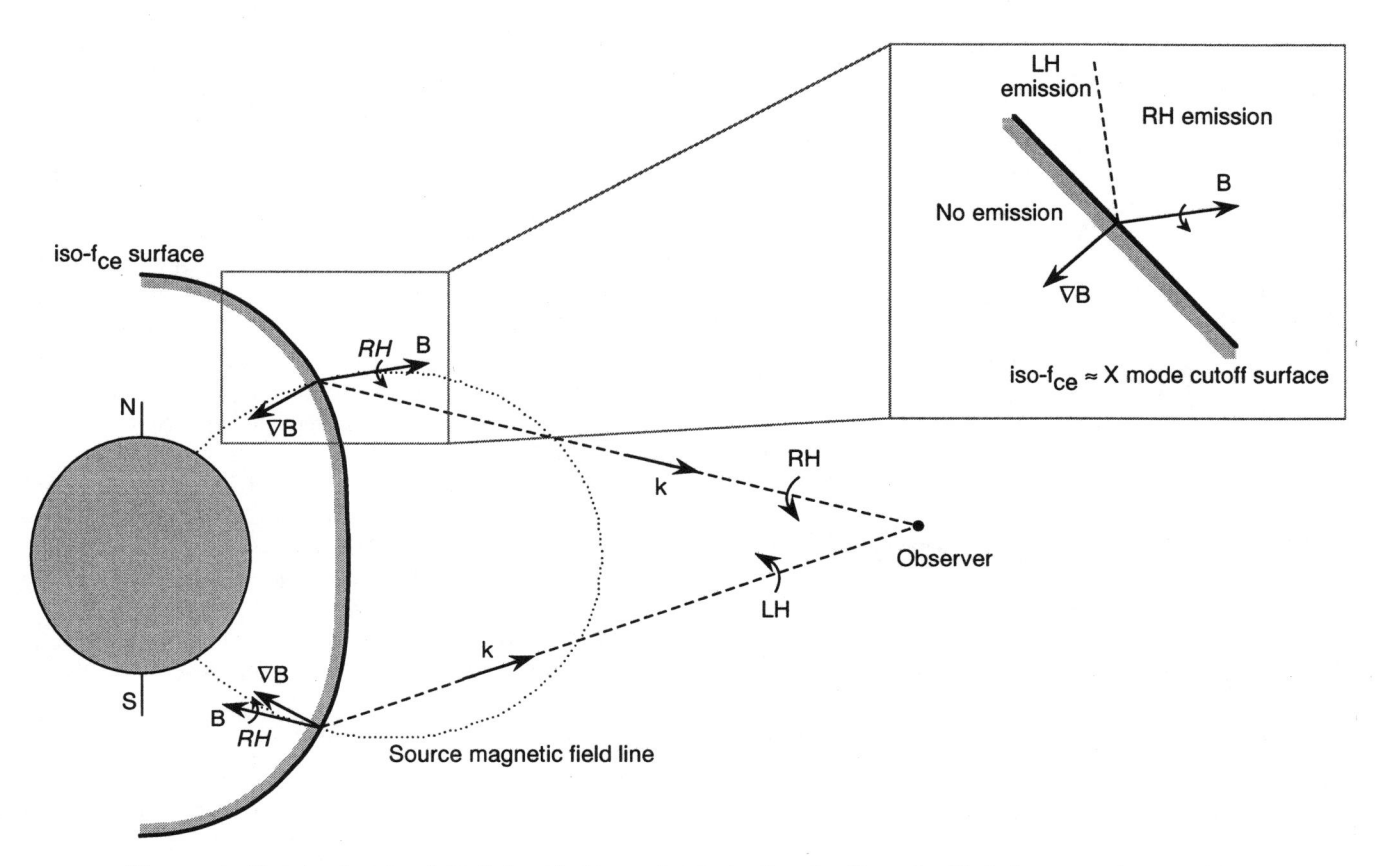

Figure 4. Sketch of emitted (italic) and observed polarization for X mode. Emission cannot propagate at acute angle from ∇B, i.e. "below" the iso$-f_{ce}$ surface (or more exactly, the iso$-f_X$ surface). *Inset:* As $\vec{B}$ is not perpendicular to the X mode cutoff surface, RH-X emission can propagate at obtuse angle from $\vec{B}$ and thus be detected as LH. (adapted from [*Galopeau et al., 1995*]).

high-frequency extent of HOM (non-Io-DAM), and the other is the Io-controlled component (Io-DAM).

From these spectra (flux densities and frequency ranges), and assuming an isotropic emission in 2π sr, one deduces the average powers of Table 3 for the different planets. But the actual beaming pattern of cyclotron-maser generated radio emissions, as deduced from observations and predicted by theory, is —at each frequency— a hollow cone whose axis is aligned on the local magnetic field at the source, with $> 50°$ aperture and a $\sim 1°$ to $2°$ thickness [e.g., *Ladreiter et al., 1994; Queinnec and Zarka, 1998*]. It subtends thus a solid angle about $0.1 - 0.2$ sr, at least 30 times less than an isotropic beam. Instantaneous powers must be reduced accordingly, reminding however that the sporadicity of the emission may occasionally increase them $\sim 10 - 50$ times.

As discussed below, instantaneous sources of these emissions are much smaller than one planetary radius, so that the brightness temperature of all high-latitude radio emissions is larger than $\sim 10^{15}$ K, implying a non-thermal, coherent generation mechanism.

3.2. Polarization and Emission Mode

First, one has to carefully distinguish between the two definitions of wave circular (or elliptical) polarization: the radioastronomical sense is defined with respect to the direction of the wave vector $\vec{k}$, while the plasma physics sense relates to the direction of the magnetic field $\vec{B}$ at the source. As illustrated on Figure 4, the two definitions coincide for acute angles of propagation $(\vec{k},\vec{B})$, but lead to opposite senses of polarization for obtuse angles. RH polarization corresponds to the sense of gyration of electrons around the magnetic field.

The magneto-ionic mode (O or X) of radio waves depends on their polarization sense at the source (in the plasma physics sense). It can be derived from the observed one (radioastronomical) provided some knowledge on magnetic field orientation in the source re-

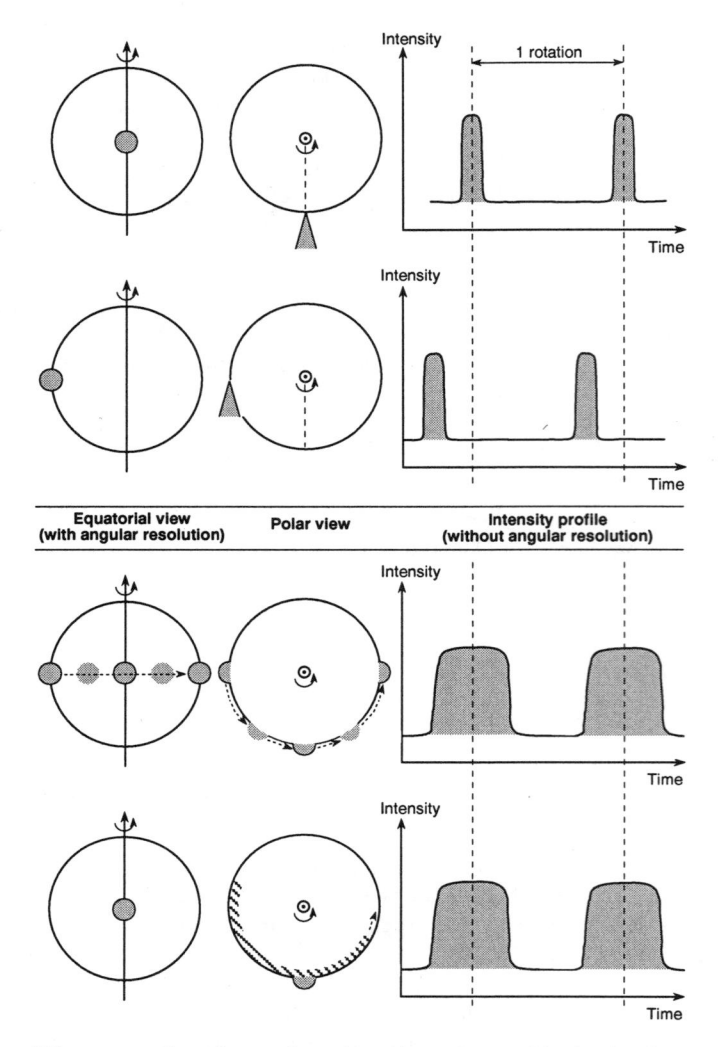

Figure 5. *Top:* Source location / beaming ambiguity in the case of a non-isotropic emission. The observer is downwards in the polar view, and the central meridian is dashed. Identical intensity variations are observed (although with different absolute phases) whatever the direction of the beam relative to the source meridian. *Bottom:* Longitude / Local Time ambiguity for an isotropic emission. Identical intensity profiles are obtained from a rotating source fixed in longitude, and from a source fixed in LT but turned "on" by an "active" sector (hatched). All intermediate situations are possible.

gion. In general, dominant RH polarization is measured for X mode emanating from high northern magnetic latitudes, and LH one for X mode from near a southern magnetic pole. The opposite is true for O mode. Together with the fact that X mode cuts off about the local gyrofrequency (in a strongly magnetized plasma, $f_X \simeq f_{ce}$), while O mode cuts off at the plasma frequency ($f_O = f_{pe}$), it allowed planetary radioastronomers to establish that (i) auroral (and Io-Jupiter) radio emissions are dominantly produced on the RH-X mode, (ii) radiation at any given frequency is produced along the source field line at the altitude where it approximately matches the local X mode LF cutoff f_X, and (iii) all high-latitude radio emissions are 100% circularly polarized [*Lecacheux*, 1988; *Zarka*, 1992].

The only exception to (iii) is the Jovian DAM, which possesses a notable linear component, and is thus 100% elliptically polarized [*Lecacheux et al.*, 1991]. Several explanations have been proposed for this peculiarity, including the existence of extremely low plasma densities at DAM sources, or linear mode coupling in the Jovian magnetosphere along the ray path [e.g., *Shaposhnikov et al.*, 1997].

Note that, as shown by the inset of Figure 4, RH-X emission is able to propagate at obtuse angle relative to the source magnetic field vector, and thus to be detected as LH. This effect can make interpretation of the observed polarization tricky, but alternately it offers a way to put constraints on the source location of a radio emission of known mode [e.g., *Galopeau et al.*, 1995].

3.3. Source Location and Beaming

The determination of these crucial parameters is made extremely difficult by (i) their interdependence, and (ii) the very poor angular resolution at decameter and longer wavelengths ($\frac{wavelength}{antenna\ size} \simeq$ several tens of degrees). Figure 5 illustrates the former problem: with remote observations without angular resolution, it is impossible to distinguish between a rotating searchlight radiating radially from the central meridian (facing the observer) and one radiating perpendicularly to a meridian at $\pm 90°$ of the central one, or between an isotropic source rotating with the planet and a source fixed in local time (or relative to the observer–planet line) but turned "on" during part of the planetary rotation only. The former case corresponds to Jovian DAM sources, and the latter to SKR. Many other complications may arise due to this interdependence.

The only direct information about high latitude radio source locations was obtained (i) at Earth from direct crossings of the TKR sources by the satellite Viking, revealing small-scale, filamentary, depleted source structures [*Hilgers*, 1992], and (ii) at Jupiter by the Ulysses spacecraft using its direction-finding capability during the Jupiter fly-by of 1992 [e.g., *Ladreiter et al.*, 1994]. The latter measurements brought the first direct proof that HOM and bKOM were indeed generated close to the local electron gyrofrequency, and also revealed that these auroral radio sources are instantaneously small-scale and that their position fluctuate with time, as if the auroral "activity" was moving rapidly (along with electron precipitations) and quasi-randomly among high-latitude field lines.

Comparing radio observations to auroral images taken in the UV or visible range is very rewarding because these emissions occur along the same field lines and are due to related —if not the same— electron precipitations, and because UV/visible images offer a good spatial resolution. At Jupiter, Galileo's visible observations revealed a very narrow auroral oval with small, fluctuating intensifications, very similar to the behaviour of HOM and bKOM radio sources located higher along those field lines having their footprints in the auroras (see e.g., Figure 1.9 of [Crary, 1998]). Direct correlations have revealed simultaneous variations of the UV and radio outputs during a strong auroral event at Jupiter [Prangé et al., 1993], while intense UV emissions from Saturn have recently been detected from the footprints of the SKR source locations derived from Voyager radio observations (see Figure 5 of Zarka [1998]).

Besides these multi-spectral correlations, many indirect and ingenuous methods have been invented to derive radio source locations and beaming: analysis of visibility and occultations of the radio emissions (by the planet or a satellite), of polarization reversals (between LH and RH, as mentioned above), modelling of the intensity profile and/or frequency-time profile of specific components, ray tracing (using magnetospheric field and plasma models), and stereoscopic observations, simultaneously from two widely separated directions. As an example of the latter method, the stereoscopic study of SKR with Voyager 1 and 2 demonstrated that the SKR beam is instantaneously narrow but flickers with time to fill a much broader beam on the average [Zarka, 1988]. This presents some analogy with radio direction-finding results at Jupiter.

Most of the above indirect methods take advantage of the fact that LF observations mainly consist in dynamic spectra, where angular resolution information is partly restored via the frequency resolution, considering that a given frequency f can originate only from those magnetospheric regions where $f \simeq f_{ce}$. For the Jovian DAM and at Saturn, Uranus, and Neptune, where no direct information on source locations is available, the simple assumption of wave generation close to the local electron gyrofrequency already constrains source locations to high magnetic latitude regions when comparing the observed radio frequency ranges with electron gyrofrequency maps at the planet's surface and on its vicinity. An example is given in Figure 6 for Saturn: the highest observed SKR frequencies ($\sim$ 1.3 MHz RH and $\sim$ 1 MHz LH), can be emitted only from |latitudes| > 50° in the northern and southern hemispheres.

Finally, one can summarize as follows the main results obtained so far about source location and beaming of auroral and satellite-induced radio emissions (see [Zarka, 1998] for details): (i) radio sources are extended along high magnetic latitude field lines, in strongly magnetized, plasma depleted regions ($f_{pe}/f_{ce} \ll 1$), down to the level where f_{pe}/f_{ce} reaches $\simeq 0.3$; (ii) radiation is produced at the fundamental of the local electron gyrofrequency, along the walls (1° to 2° thick) of hollow conical patterns symmetrical about the local $\vec{B}$ vector; (iii) slowly varying components seem to originate from closed field lines and are beamed $\geq 30°$ away from $\vec{B}$, while impulsive bursts seem to come from open field lines and are beamed $\geq 75°$ away from $\vec{B}$; (iv) sources are fixed in local time (LT) in the Terrestrial and Saturnian magnetospheres (resp. about 2200 LT and 0900-1200 LT), and rotate with the planet at Jupiter and Uranus (Neptune may be an intermediate case).

(i) is consistent with theoretical constraints of the cyclotron-maser mechanism, able to amplify X mode waves only where $f_{pe}/f_{ce} < 0.2 - 0.3$ [Le Quéau, 1988]. (iii) may be related to the cyclotron-maser operation in small-scale cavities [Louarn, 1997]. (iv) is probably indicative of the origin of keV electrons responsible for radio emissions in the different cases: from the magnetotail at Earth (TKR), from the dayside/morningside magnetopause at Saturn (SKR), from the outer edge of Io's torus and the magnetodisc at Jupiter (bKOM, HOM, and non-Io-DAM), and from the interaction of Io and the rotating Jovian magnetic field (Io-DAM).

More uncertainty subsists for Uranian and Neptunian radio sources, observed only by Voyager 2, but an interesting peculiarity is worth a remark: the large tilts between the magnetic and rotation axes at these two planets (resp. 46.9° at Uranus and 58.6° at Neptune) prevents the accumulation of plasma in equatorial regions, where f_{pe}/f_{ce} remains thus low and allows X mode "auroral-like" emission to occur (see Table 1).

3.4. Temporal Modulations

The relative influence of internal (rotation, satellites) versus external (solar wind) control on radio emissions, and auroral activity in general, is a very important clue for understanding the "magnetospheric machine".

The influence of planetary rotation can be readily detected in radio observations of the four giant planets. A "pulsar-like" behaviour is sometimes attributed to the Jupiter magnetosphere [Dessler, 1983], where the prominent rotational control is obviously linked to the

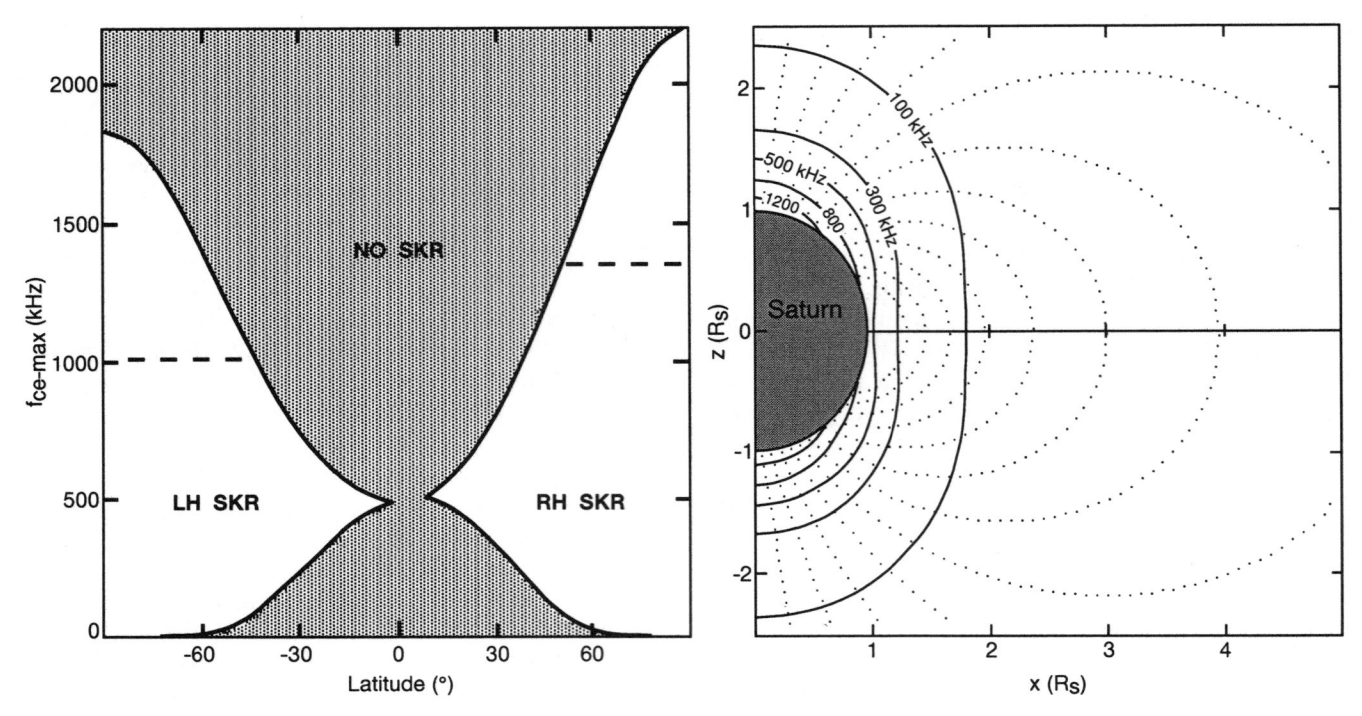

Figure 6. *Left:* The white areas are the ranges of gyrofrequencies reached along a field line with footprint at the latitude in abscissa, between Saturn's surface and the field line apex. The magnetic field model used is the SPV [*Davis and Smith*, 1990], independent of the longitude. The dashed line indicates the highest observed LH and RH SKR frequencies. *Right:* SPV iso−f_{ce} contours (solid lines) and magnetic field lines (dotted) in a meridian plane. (adapted from [*Galopeau et al.*, 1995]).

intense magnetic field and rapid rotation, resulting in a strong corotation electric field. At Uranus and Neptune, the less prominent rotation control is attributed (at least qualitatively) to the large magnetic dipole tilts. At Saturn, the rotational control of SKR is more subtle than a "rotating searchlight" effect: the radio emission turns "on" and "off" with the passage of a certain ("active") longitude range (100° − 130°) about noon LT.

However, it has been realized since a few years that the solar wind "control" of auroral radio emissions is ubiquitous at all radio planets. It has been studied for a long time in the case of TKR (so-called "substorm" activity [e.g., *Gallagher and d'Angelo*, 1981]), and appears to affect also the so-called LF bursts (Table 1). SKR intensity variations are so tightly correlated with solar wind fluctuations that SKR was turned "off" during occasional immersions of Saturn in Jupiter's distant magnetotail, at the time of the Voyager 2 fly-by [*Desch*, 1982, 1983]. UKR and NKR have been shown to be at least partly controlled by solar wind fluctuations [*Desch et al.*, 1989, 1991]. Finally, all Jovian auroral radio components (including the so-called Quasi-Periodic bursts) fluctuate in response to solar wind variations [*Zarka and Genova*, 1983; *Barrow et al.*, 1986; *Genova et al.*, 1987].

Even the nKOM (unpublished) as well as propagation effects attributed to Io's torus show variations with ∼ 26−day periodicity [*Kaiser and MacDowall*, 1998], which suggests that the solar wind influence penetrates into Jupiter's inner magnetosphere. This influence has been exemplified by the so-called "Radio Bode's law", derived by *Desch and Kaiser* [1984], demonstrating the relationship between the auroral radio output of the five radio planets and the incident power of the solar wind flow on each magnetosphere's cross-section (with a proportionality constant of a few $\times 10^{-6}$). As a consequence, it seems largely incorrect to speak of a "pulsar-like" behaviour about Jupiter's radio emissions.

The only Jovian high-latitude radio component independent of the solar wind is the Io-DAM. It is strongly dependent on the rotation, but also modulated by Io's motion in the magnetosphere. This modulation by Io, however, is mainly geometrical: as Io-induced radio emissions are generated along Io's wake field lines, i.e. fixed in Io's frame, and as their beaming is a hollow cone −as for auroral emissions−, the detection of these emissions by a fixed observer is modulated by Io's orbital period, but it is truly the rotation of the Jovian magnetic field and its interaction with Io's ionosphere

(Alfvèn wings...) which powers Io-induced high-latitude radio —and UV / IR— emissions [e.g., *Hill et al.,* 1983].

Finally, recent observations by Galileo's radio experiment suggest a weak control of Jovian HOM/DAM emissions by Ganymede, similar but an order of magnitude weaker than Io's one [*Menietti et al.,* 1998b]. No other satellite appears to control a planetary radio emission, but for the controversial suggestion of a loose control of SKR by Dione [*Kaiser et al.,* 1984].

3.5. Fine Frequency-Time Structures

The very complex structuration of auroral and Io-induced radio emissions in the frequency-time plane is ubiquitous at all radio-planets (Figure 7), and contrasts with the absence of such fine structures in the other radio components of Table 1 (decimeter radiation from Van Allen belts, nKOM, and non-thermal continuum), with the exception of lightning. This should probably be attributed to the fact that cyclotron-maser ensures wave amplification on very short timescales (a few gyroperiods, and thus $\ll 1$ sec), and that it operates in small-scale source regions at high-latitudes, while the other, slowly varying radio components are produced either by non-resonant (incoherent) mechanisms, or in much broader source regions, smoothing out possible fast variations [e.g., *Louarn,* 1992].

It has been suggested that impulsive bursts components could be generated in small-scale filamentary cavities, where a laser-like effect would ensure further amplification and mode selection, and thus generate intense fine structures in dynamic spectra [*Calvert,* 1988]. Bursts beaming characteristics have even be used to infer the existence of auroral cavities at Uranus and Neptune [*Farrell et al.,* 1991], but the predicted effect of such cavities on the radio emission beaming leads to contradictions [*Zarka,* 1998]. The question is thus still wide open.

Finally, it should be noted that proper study of these fine structures is very difficult with the frequency and time resolutions allowed by space-borne radio experiments, and that ground-based measurements (of Jovian short bursts only, detectable above the Earth's ionospheric cutoff) benefit from resolutions orders of magnitude better than spacecraft ones. This is absolutely obvious comparing the center panel of Figure 7 with its neighbours. Such ground-based measurements are thus crucial for the study of radio emissions' fine structures, which represent a unique tool for studying the small-scale source structure and radio generation mechanisms at the outer planets.

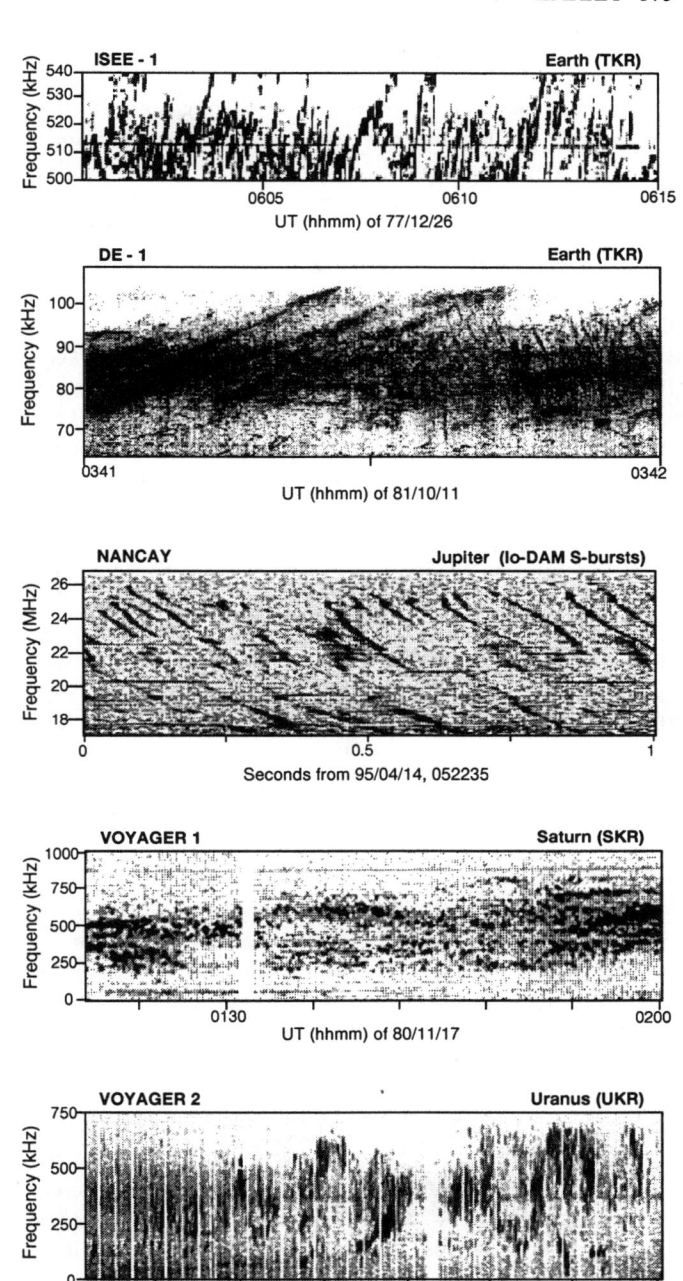

Figure 7. Dynamic spectra of planetary radio emission fine structures at various timescales. *Top to bottom:* Discrete TKR tones, with positive and negative frequency-time drifts; TKR bands and stripes; Jovian DAM "S" (short) bursts recorded in Nançay with 3 ms/spectrum resolution; SKR bursty structure at Voyager's radio astronomy experiment best time resolution (30 ms); Uranus bursts episode. (adapted from [*Zarka,* 1998]).

4. OPEN QUESTIONS

An exhaustive list would be too long and of little interest, so I have selected below a few outstanding questions, distinguishing between "morphological", "exploratory", and "theoretical" issues, which deserve distinct but complementary approaches.

"Morphological" questions include:
(i) What is the relative dependence of high-latitude radio emissions in longitude versus local time ? Local time dependence may be related to substorm-like activity, while the longitude extent of radio sources may be linked to magnetic field anomalies, gradients, etc. The instantaneous longitude extent of Jovian radio sources, and their average position in longitude (if relevant) is poorly known. The nature of the "active" longitude range at Saturn is not understood, and not even well localized (which longitude and which local time are involved in turning "on" and "off" the SKR source ?). The situation at Uranus and Neptune is even less clear, due to the scarce data available.
(ii) Is there a solar wind influence on Io-DAM ? Even if weak, it would be interesting to correlate it with the solar wind influence on nKOM.
(iii) Do satellites —other than Io— control planetary radio emissions ? As written above, it seems to be the case for Ganymede...
(iv) Are all planetary radio bursts unresolved drifting structures (like the Jovian short bursts of Figure 7) ? Answers to these questions request new observations or additional in-depth analyses of existing ones. (i) is obviously very difficult to answer with single-point measurements. Together with (ii) and (iii), it requires long-term series of quasi-continuous measurements. (iv) requires large resources difficult to accomodate on deep-space probes.

The most interesting "exploratory" question concerns the existence of Mercury's radio emissions. As they would probably be trapped in Mercury's tiny magnetosphere, answer to this question requires at least a close fly-by to the planet.

Finally in relation with the above morphological questions, one may raise the following theoretical interrogations:
(i) What is the nature of Saturn's "active" longitude sector, and how does if manifest its effect on SKR ?
(ii) How does the solar wind influence penetrate in Jupiter's inner magnetosphere ?
(iii) How are planetary radio bursts generated ? Is cyclotron-maser able to account for their generation as well as that of slowly varying components [e.g., *Zarka et al.*, 1997b] ?

5. INTEREST

What is the interest of studying in detail planetary auroral (and satellite-induced) radio emissions (or, in other words, "why bother ?"). Beyond interpreting the observations and understanding the generation mechanisms involved, the study of planetary radio emissions presents at least four major "astrophysical" interests:
(i) Non-thermal high-latitude emissions are the only means to determine the rotation rate of the giant planet interiors [e.g., *Lecacheux et al.*, 1993; *Higgins et al.*, 1997].
(ii) Their study is a good testbed for that of solar and stellar radio sources [*Abada-Simon*, 1996].
(iii) They are becoming a powerful remote sensing tool of planetary magneto-ionized environments (magnetic field topology and models [e.g., *Galopeau et al.*, 1991]; plasma density/cavities (see above); electrons characteristic energy [e.g., *Zarka et al.*, 1996]; and in the future parallel electric fields and double-layers...). Multi-spectral observations (especially UV/radio) are very complementary for remote sensing purposes.
(iv) Finally, scaling laws allow to speculate about magnetospheric radio emissions from extra-solar planets [*Farrell et al.*, 1999], the search of which has already begun [*Zarka et al.*, 1997a].

6. FUTURE OBSERVATIONS

The availability and accessibility of large databases (Voyager, Ulysses, Galileo, Nançay, etc.) makes now possible extensive statistical analyses, well adapted to address "morphological" questions. The increase in computing power and the development of new tools (e.g., automated recognition/classification of radio emissions [*De Lassus and Lecacheux*, 1997]) justifies renewed interest for these data sets which are far from exhausted.

In parallel, the Cassini spacecraft will orbit Saturn from mid−2004 to 2008. It carries the most sophisticated radio experiment ever flown to an outer planet, with high resolutions, full polarization, and quasi-instantaneous direction-finding capabilities, and it will probably perform observations during the fly-bys of Venus, Jupiter and the Earth. Spacecraft exploration of Mercury's environment, re-exploration of Uranus and Neptune systems, as well as of Jovian polar regions would also be of high interest. Possibly closer to us (in space and time!), long-term global monitoring of Jupiter and Saturn aurorae from the Earth orbit, especially if multi-spectral, could provide at low-cost significant complements to Galileo and Cassini in-situ observations. Such projects are under consideration in several space agencies (NASA, CNES, ESA). More generally, coordinated studies (ground-based/spacecraft, multi-spacecraft, or multi-spectral) appear the next most promising step in the study of magnetospheric high-latitude regions, and of radio sources there.

Acknowledgments. Many thanks to my colleagues from the "radio/plasmas" group in DESPA for their help to make this tutorial clear ... hopefully.

REFERENCES

Abada-Simon, M., Comparison of the observational data on flare stars, solar and planetary radio emissions, *Planet. Space Sci., 44,* 501-507, 1996.

Barrow, C. H., M. D. Desch, and F. Genova, Solar wind control of JupiterÕs decametric radio emission, *Astron. Astrophys., 165,* 244-250, 1986.

Calvert, W., Planetary radio lasing, in *Planetary Radio Emissions II,* edited by H. O. Rucker et al., pp. 407-423, Austrian Acad. Sci. Press, Vienna, 1988.

Carr, T. D., M. D. Desch, and J. K. Alexander, Phenomenology of magnetospheric radio emissions, in *Physics of the Jovian Magnetosphere,* edited by A. J. Dessler, pp. 226-284, Cambridge Univ. Press, New York, 1983.

Crary, F. J., Io's interaction with the Jovian magnetosphere: models of particle acceleration and scattering, Ph. D. thesis, 147 pp., Univ. of Colorado, Boulder, April 1998.

Davis, L., Jr., and E. J. Smith, A model of SaturnÕs magnetic field based on all available data, *J. Geophys. Res., 95,* 15,257-15,261, 1990.

De Féraudy, H., A. Bahnsen, and M. Jespersen, Observations of nightside and dayside auroral kilometric radiation with Viking, in *Planetary Radio Emissions II,* edited by H. O. Rucker et al., pp. 41-60, Austrian Acad. Sci. Press, Vienna, 1988.

De Lassus, H., and A. Lecacheux, Automatic recognition of low-frequency radio planetary signals, in *Planetary Radio Emissions IV,* edited by H. O. Rucker et al., pp. 359-368, Austrian Acad. Sci. Press, Vienna, 1997.

Desch, M. D., Evidence for solar wind control of Saturn radio emission, *J. Geophys. Res., 87,* 4549-4554, 1982.

Desch, M. D., Radio emission signature of Saturn immersions in JupiterÕs magnetic tail, *J. Geophys. Res., 88,* 6904-6910, 1983.

Desch, M. D., W. M. Farrell, M. L. Kaiser, R. P. Lepping, J. T. Steinberg, and L. A. Villanueva, The role of solar wind reconnection in driving the Neptune radio emission, *J. Geophys. Res., 96,* 19,111-19,116, 1991*a*.

Desch, M. D., and M. L. Kaiser, Predictions for Uranus from a radiometric BodeÕs law, *Nature, 310,* 755-757, 1984.

Desch, M. D., M. L. Kaiser, and W. S. Kurth, Impulsive solar wind-driven emission from Uranus, *J. Geophys. Res., 94,* 5255-5263, 1989.

Desch, M. D., M. L. Kaiser, P. Zarka, A. Lecacheux, Y. Leblanc, M. Aubier, and A. Ortega-Molina, Uranus as a radio source, in *Uranus,* edited by J. T. Bergstralh, E. D. Miner, and M. S. Matthews, pp. 894-925, Univ. of Arizona Press, Tucson, 1991*b*.

Dessler, A. J., Preface, in *Physics of the Jovian Magnetosphere,* edited by A. J. Dessler, pp. xiii-xv, Cambridge Univ. Press, New York, 1983.

Farrell, W. M., M. D. Desch, M. L. Kaiser, and W. Calvert, Evidence of auroral plasma cavities at Uranus and Neptune from radio burst observations, *J. Geophys. Res., 96,* 19,049-19,059, 1991.

Farrell, W. M., M. D. Desch, and P. Zarka, On the possibility of coherent cyclotron emission from extrasolar planets, *J. Geophys. Res.,* in press, 1999.

Gallagher, D. L., and N. DÕAngelo, Correlations between solar wind parameters and auroral kilometric radiation, *Geophys. Res. Lett., 8,* 1087-1090, 1981.

Galopeau, P., A. Ortega-Molina, and P. Zarka, Evidence of SaturnÕs magnetic field anomaly from SKR high-frequency limit, *J. Geophys. Res., 96,* 14,129-14,140, 1991.

Galopeau, P., P. Zarka, and D. Le Quéau, Source location of SKR: the Kelvin-Helmholtz instability hypothesis, *J. Geophys. Res., 100,* 26,397-26,410, 1995.

Genova, F., P. Zarka, and C. H. Barrow, Voyager and Nanay observations of the Jovian radio emission at different frequencies: Solar wind effect and source extent, *Astron. Astrophys., 182,* 159-162, 1987.

Genova, F., P. Zarka, and A. Lecacheux, Jupiter decametric radiation, in *Time-Variable Phenomena in the Jovian System,* edited by M. J. S. Belton, R. A. West, and J. Rahe, pp. 156-174, NASA Spec. Publ., 494, 1989.

Higgins, C. A., T. D. Carr, F. Reyes, W. B. Greenman, and G. R. Lebo, A redefinition of JupiterÕs rotation period, *J. Geophys. Res., 102,* 22033-22041, 1997.

Hilgers, A., The auroral radiating plasma cavities, *Geophys. Res. Lett., 19,* 237-240, 1992.

Hill, T. W., A. J. Dessler, and C. K. Goertz, Magnetospheric models, in *Physics of the Jovian Magnetosphere,* edited by A. J. Dessler, pp. 353-394, Cambridge Univ. Press, New York, 1983.

Kaiser, M. L., Observations of nonthermal radiations from planets, in *Plasma Waves and Instabilities at Comets and in Magnetospheres, Geophys. Monogr. Ser., vol. 53,* edited by B. T. Tsurutani and H. Oya, pp. 221-237, AGU, Washington, D. C., 1989.

Kaiser, M. L., Time-variable magnetospheric radio emissions from Jupiter, *J. Geophys. Res., 98,* 18,757-18,765, 1993.

Kaiser, M. L., M. D. Desch, W. S. Kurth, A. Lecacheux, F. Genova, B. M. Pedersen, and D. R. Evans, Saturn as a radio source, in *Saturn,* edited by T. Gehrels and M. S. Matthews, pp. 378-415, Univ. of Ariz. Press, Tucson, 1984.

Kaiser, M. L., and R. J. MacDowall, Jovian radio "bullseyes" observed by Ulysses, *Geophys. Res. Lett., 25,* 3113-3116, 1998.

Kurth, W. S., Continuum radiation in planetary magnetospheres, in *Planetary Radio Emissions III,* edited by H. O. Rucker et al., pp. 329-350, Austrian Acad. Sci. Press, Vienna, 1992.

Kurth, W. S., D. A. Gurnett, A. Roux, and S. J. Bolton, Ganymede: a new radio source, *Geophys. Res. Lett., 24,* 2167-2170, 1997.

Ladreiter, H. P., P. Zarka, and A. Lecacheux, Direction-finding study of Jovian hectometric and broadband kilometric radio emissions: Evidence for their auroral origin, *Planet. Space Sci., 42,* 919-931, 1994.

Lecacheux, A., Polarization aspects from planetary radio emissions, in *Planetary Radio Emissions II*, edited by H. O. Rucker et al., pp. 311-326, Austrian Acad. Sci. Press, Vienna, 1988.

Lecacheux, A., A. Boischot, M. Y. Boudjada, and G. A. Dulk, Spectra and complete polarization state of two, Io-related, radio storms from Jupiter, *Astron. Astrophys.*, *251*, 339-348, 1991.

Lecacheux, A., P. Zarka, M. D. Desch, and D. R. Evans, The sidereal rotation period of Neptune, *Geophys. Res. Lett.*, *20*, 2711-2714, 1993.

Le Quéau, D., Planetary radio emissions from high magnetic latitudes: The ÒCyclotron-MaserÓ theory, in *Planetary Radio Emissions II*, edited by H. O. Rucker et al., pp. 381-398, Austrian Acad. Sci. Press, Vienna, 1988.

Louarn, P., Auroral planetary radio emissions: theoretical aspects, *Adv. Space Res.*, *12*, (8)121-(8)134, 1992.

Louarn, P., Radio emissions from filamentary sources: a simple approach, in *Planetary Radio Emissions IV*, edited by H. O. Rucker et al., pp. 153-165, Austrian Acad. Sci. Press, Vienna, 1997.

MacDowall, R. J., M. L. Kaiser, M. D. Desch, W. M. Farrell, R. A. Hess, and R. G. Stone, Quasiperiodic Jovian radio bursts; observations from the Ulysses Radio and Plasma Wave experiment, *Planet. Space Sci*, *41*, 1059-1072, 1993.

Menietti, J. D., D. A. Gurnett, W. S. Kurth, J. B. Groene, and L. J. Granroth, Radio emissions observed by Galileo near Io, *Geophys. Res. Lett.*, *25*, 25-28, 1998a.

Menietti, J. D., D. A. Gurnett, W. S. Kurth, and J. B. Groene, Control of Jovian radio emission by Ganymede, *Geophys. Res. Lett.*, *25*, 4281-4284, 1998b.

Ness, N. F., The magnetic fields of Mercury, Mars and the Moon, *Ann. Rev. Earth Planet. Sci.*, *7*, 249, 1979.

Prangé, R., P. Zarka, G. E. Ballester, T. A. Livengood, L. Denis, T. D. Carr, F. Reyes, S. J. Bame, and H. W. Moos, Correlated variations of UV and radio emissions during an outstanding jovian auroral event, *J. Geophys. Res.*, *98*, 18779-18791, 1993.

Queinnec, J., and P. Zarka, Io-controlled decameter arcs and Io-Jupiter interaction, *J. Geophys. Res.*, *103*, 26649-26666, 1998.

Reiner, M. J., J. Fainberg, R. G. Stone, M. L. Kaiser, M. D. Desch, R. Manning, P. Zarka, and B. M. Pedersen, Source characteristics of Jovian narrow-band kilometric radio emissions, *J. Geophys. Res.*, *98*, 13,163-13,176, 1993.

Shaposhnikov, V. E., Vl. V. Kocharovsky, V. V. Kocharovsky, H. P. Ladreiter, H. O. Rucker, and V. V. Zaitsez, On elliptical polarization of the decametric radio emission and the linear mode coupling in the Jovian magnetosphere, *Astron. Astrophys.*, *326*, 386-395, 1997.

Steinberg, J.-L., C. Lacombe, and S. Hoang, A new component of terrestrial radio emission observed from ISEE-3 and ISEE-1 in the solar wind, *Geophys. Res. Lett.*, *15*, 176-179, 1988.

Zarka, P., Beaming of planetary radio emissions, in *Planetary Radio Emissions II*, edited by H. O. Rucker et al., pp. 327-342, Austrian Acad. Sci. Press, Vienna, 1988.

Zarka, P., The auroral radio emissions from planetary magnetospheres: What do we know, what donÕt we know, what do we learn from them?, *Adv. Space Res.*, *12*, (8)99-(8)115, 1992.

Zarka, P., Auroral radio emissions at the outer planets: observations and theories, *J. Geophys. Res.*, *103*, 20159-20194, 1998.

Zarka, P., and F. Genova, Low frequency jovian emission and solar wind magnetic sector structure, *Nature*, *306*, 767-768, 1983.

Zarka, P., B. M. Pedersen, A. Lecacheux, M. L. Kaiser, M. D. Desch, W. M. Farrell, and W. S. Kurth, Radio emissions from Neptune, in *Neptune and Triton*, edited by D. Cruikshank and M. S. Matthews, pp. 341-387, Univ. of Ariz. Press, Tucson, 1995.

Zarka, P., T. Farges, B. P. Ryabov, M. Abada-Simon, and L. Denis, A scenario for Jovian S-bursts, *Geophys. Res. Lett.*, *23*, 125-128, 1996.

Zarka, P., et al., Ground-based high sensitivity radio astronomy at decameter wavelengths, in *Planetary Radio Emissions IV*, edited by H. O. Rucker et al., pp. 101-127, Austrian Acad. Sci. Press, Vienna, 1997a.

Zarka, P., B. P. Ryabov, V. B. Ryabov, R. Prangé, M. Abada-Simon, T. Farges, and L. Denis, On the origin of Jovian decameter radio bursts, in *Planetary Radio Emissions IV*, edited by H. O. Rucker et al., pp. 51-63, Austrian Acad. Sci. Press, Vienna, 1997b.

Philippe Zarka, Space Research Department (DESPA), CNRS UMR 8632, Observatoire de Paris, F-92195 Meudon, France. (e-mail: Philippe.Zarka@obspm.fr)

Planetary Radio Emission from Lightning: Discharge and Detectability

William M. Farrell

NASA/Goddard Space Flight Center, Greenbelt, Maryland

A unifying concept is presented that explains the radio frequency signature of lightning at Earth and the outer planets. It is assumed that the source discharge duration primarily accounts for the character of the radio frequency (RF) spectra observed from space. Planetary-specific propagation effects are considered to be secondary. The advantage of this approach is the ease in explaining both the detection of signals in one frequency band and also the near-simultaneous lack of detection in other frequency bands. Prior to this illustration, planet-unique propagation scenarios had to be invoked to explain the nondetection of an event in a given band. However, according to the present picture, a slow lightning discharge at Jupiter can account for the observation of very low frequency (VLF) whistler signals, but also the lack of substantial high frequency (HF) energy in the form of spheric emissions. A fast discharge at Saturn explains the detection of HF Saturnian Electrostatic Discharges, but also explains the lack of VLF whistler signals, since the radiated energy from the fast current couples most easily to the HF band. From this illustration, estimates of the discharge duration at the outer planets are obtained that are consistent with the observed RF spectra from space.

INTRODUCTION

Although most planetary radio and plasma wave science tends to emphasize natural auroral and magnetospheric emissions, radiation from planetary lightning has significant scientific impact. First, the occurrence of atmospheric lighting indicates the presence of atmospheric convection. Such an inference implies the presence of substantial vertical temperature gradients and vertical winds. Second, lightning/electrical discharges may be part of the basic building blocks of life in forming complex organic molecules. As demonstrated by Miller [1953], methane, ammonia, hydrogen and water, when catalyzed via electrical discharge, formed complex ammino-acid molecules.

All these elements currently exist on Jupiter, and are believed to be present following the formation of the Earth. Finally, recently discovered upward luminous events over thunderstorms on Earth (i.e., sprites and jets) [Franz et al., 1990] suggest an atmospheric/ionosphere/magnetospheric coupling that may be strong but short-lived: a transient process. The implication is that the lower atmosphere may be radiatively connected to the magnetosphere, and energy from one region could affect the electrodynamics of the other.

There are two manifestations of lightning-generated radio emission in space. The first are the well-known very low frequency (VLF) whistler emissions [Helliwell, 1965], so named for their downdrifting whistle-like tone when heard in the audio band. Typically, the emission consists of a short-lived (< 1 second), highly dispersed event below 20 kHz. The emission is considered a "plasma wave" : it is a modified VLF emission from lightning that couples to the ionosphere and magnetosphere via the trapped whistler

Radio Astronomy at Long Wavelengths
Geophysical Monograph 119
This paper not subject to U.S. copyright
Published in 2000 by the American Geophysical Union

Table 1. Observations of lightning-generated radio emission from space.

	HF Spheric	VLF Whistler
Earth	X	X
Jupiter		X
Saturn	X	
Uranus	X	
Neptune		X

mode. The emission is thus intimately tied to the medium, consisting partly as an electromagnetic emission, but also having a large electrostatic component ($k \cdot E$ is nonzero) associated with the medium response. The emission also tends to propagate along the connecting magnetic field line from the source to the spacecraft (emission group velocity is quasi-parallel to planetary magnetic field direction) [Helliwell, 1965]. Consequently, a spacecraft must be near/on the active connecting field lines from the atmospheric source to detect the event. Whistlers and plasma waves associated with lightning have been detected at Earth [Storey, 1953; Helliwell et al., 1965], and at the outer planets Jupiter [Gurnett et al., 1979] and Neptune [Gurnett et al., 1990] by the VLF receiver that was part of the Voyager plasma wave (PWS) experiment [Scarf and Gurnett, 1977].

The second manifestation of lightning-generated radio frequency emission in space is the high frequency (HF) "spheric" signals, typically observed at frequencies above 1 MHz. These bursty events are the discharge-generated radio emission observed directly from their source, with little or no ionospheric/magnetospheric interaction. The emission frequency is greater than the plasma frequency of the ionosphere ($f > fp_{iono}$) and much greater than the ambient magnetospheric plasma frequency ($f \gg fp_{mag}$). Consequently, the medium is considered quasi-transparent with the index of refraction, n, progressively moving closer to unity with increasing frequency. Spheric emissions from atmospheric sources have been detected in space around Earth [Herman, Caruso, and Stone, 1969], and at Saturn [Kaiser et al., 1983] and Uranus [Zarka and Pedersen, 1987] by the HF receiver that is part of the Voyager Planetary Radio Astronomy (PRA) experiment [Warwick et al., 1977].

It is interesting to note that only at Earth have both VLF whistlers and HF spheric signals been observed from space. At Jupiter, VLF whistlers were observed [Gurnett et al., 1979] as Voyager passed by magnetic field lines with footprints connected at the lightning source [Menietti and Gurnett, 1980], but no associated HF spheric signals were observed. Zarka [1985] suggested that ionospheric absorption reduces the HF spheric signal strength to levels below detectability. At Saturn, HF noise bursts (Saturnian electrostatic discharges) were identified as freely-propagating spheric signals generated by atmospheric weather systems [Kaiser et al., 1983]. However, there was a lack of associated VLF whistler activity, presumably because Voyager did not become connected to the appropriate active magnetic field lines. At Uranus, HF spherics were observed and VLF whistlers were not, for the same reasons as Saturn. Finally, at Neptune, evidence for HF spherics is marginal [Kaiser et al., 1991], but a highly dispersed VLF plasma wave signature was observed, most likely a whistler [Gurnett et al., 1990, Menietti et al., 1991] or possibly a z-mode pulse [Farrell, 1996]. Thus, treating the outer planets as a group, either an HF spheric or VLF plasma wave emission was observed, but not both, as indicated in Table 1. There appears to be an "either/or" situation associated with the Voyager observations at the outer planets.

Desch [1994] described the reasons cited for the lack of detection of Jovian spherics, Saturnian whistlers, Uranian whistlers, and Neptunian spherics. In each case, a planet-unique propagation scenario was invoked, this constructed just following the discovery (post-encounter data analysis period) to explain the presence of the signal in one band and complete absence in another. However, reexamining the outer planet observations as a complete set, there is the suggestion of a more fundamental explanation for the band-limited observations that applies to all the planets. Rather than considering a constructed propagation argument on a planet-by-planet basis, there is the possibility that another universal process is occurring to limit the band of observations to those shown in Table 1.

To date, there has been no effort to consider the culpability of the discharge current source itself as the reason for the band-limited observation either in the HF or VLF. As such, we will model discharges of various temporal durations and the emission propagation through a model ionosphere to determine the variation in the RF spectra associated with discharge duration. Specifically, "slower" discharges are expected to couple more easily to the ELF and VLF, with little energy in the HF. Conversely, "fast" discharges are expected to couple directly to the MF and HF, and have little energy in the ELF/VLF. Thus, the "either/or" situation at the outer planets may have less to do with planet-specific propagation effects and more to do with the temporal duration of the discharge. As we quantify below, power level differences between the VLF and HF can vary by as much as 200 dB depending upon the temporal duration of the initial discharge; thereby possibly explaining the band-limited observation.

RADIATION AND PROPAGATION

In order to determine the effect of lightning discharge temporal duration on the RF spectra, an "idealized" planet is considered: one with a steady ionosphere, but with lightning strokes of very different temporal duration ranging from 10 nS to 10 mS. A steady, nonvarying ionosphere (as viewed in the RF) is constructed based upon an ensemble of planetary ionospheres. It is desired to obtain a reasonable ionospheric transfer function that can be held constant (i.e., no diurnal variations, no scintillations, etc.). The temporal duration of the lightning current source (i.e., the rise and fall time of the discharge) is then varied over six orders of magnitude, and the associated RF emission and energy transfer through the idealized, steady ionosphere is determined. The main objective of this exercise is determining the ratio of VLF to HF emission levels (i.e., E(3kHz)/E(30MHz)) for various discharge durations.

This modeling should not be considered overly quantitative, but is more of a demonstration to illustrate the effect of discharge duration on the RF spectra. In fact, we now refer to this exercise as an "illustration" rather than a true model. The illustration is general, while not being specific to the details of any single planet.

The discharge applied is the current wave scenario initially presented in Bruce and Golde [1941]. That model assumes that the RF generated by the lightning discharge is associated with the exponential rise and decay of the discharge current wave. As demonstrated by Le Vine and Meneghini [1978a,b], this current wave approach can explain the zeroth-order features of the lightning spectra, including the emission peak and HF f^{-2} rolloff. At Earth, randomness of the discharge path (also called tortuosity) has a secondary effect of adding power between 0.1–10 MHz [Le Vine and Meneghini 1978a,b] but is not a dominant process. The random power is only about 1% of emission current wave peak power.

Bruce and Golde [1941] described the lightning discharge current as an exponential rising and decaying wave in an ionized channel defined as

$$i(t) = i_o \left(e^{-\alpha t} - e^{-\beta t} \right). \qquad (1)$$

For the case of the typical terrestrial cloud-to-ground strokes, $\alpha = 4 \times 10^4$ sec^{-1} and $\beta = 4 \times 10^5$ sec^{-1} [Bruce and Golde, 1941]. The associated electric field is

$$E(t) = \frac{1}{4\pi\epsilon_o c r^2} \frac{dM}{dt} + \frac{1}{4\pi\epsilon_o c^2 r} \frac{d^2 M}{dt^2} \qquad (2)$$

where M is the electric dipole moment defined as 2Qh, 2h being the length of the electric dipole. The first term represents the induction field while the second term represents the radiation field. As defined by Bruce and Golde [1941], the change in dipole moment, $dM/dt = 2i(t)\gamma^{-1}v_o(1-\exp(-\gamma t))$ with discharge channel speed $v(t) = v_o\exp(-\gamma t)$ with $v_o \sim 0.25$ c and $\gamma < \alpha,\beta$.

Obviously, more complicated discharge models have been derived since the early work of Bruce and Golde [1941]. Le Vine and Meneghini, [1978a,b] constructed a similar current wave model as that defined above, but included retardation time along a segmented, discontinuous (tortuous) discharge path. Their zeroth-order solution was defined by the initiation and cessation of the current wave (the turning on and off of the current, like the Bruce and Golde model), giving rise to a spectrum varying as $1/f^2$ at high frequencies. The tortuous discharge path added a gentler $1/f$ rolloff between 0.1–10 MHz, corresponding to the scale size (L= c/f) of the tortuosity. For the sake of demonstration here, our illustration only considers the initiation and cessation of the current wave and ignores tortuous (nonlinear) effects.

There are some subtle but important difference between terrestrial cloud-to-ground and intracloud lightning: First, intracloud lightning possesses a slightly different temporal profile with $\alpha = 1 \times 10^3$ sec^{-1} and $\beta = 2 \times 10^5$ sec^{-1} (it is slightly slower) and, second, the discharge is more horizontal thereby generating vertically-directed wave propagation vectors [Valdivia et al, 1997, 1998]. Further, they have very complicated and dynamic structure (i.e., tortuous discharge paths) [Villanueva et al., 1994; Mazur et al., 1998]. Despite these differences, the zeroth-order spectra is defined by the current wave with radio spectral emission peaking in the VLF.

Consider now an arbitrary discharge of duration, τ, and the propagation of its radiated electric field through the "idealized" ionosphere to a spacecraft located above the ionosphere (near 2000 km altitude). Ionospheric transmission from source to spacecraft from 1 Hz to 100 MHz will be examined. In the presentation, broadband modes are addressed only, and more complicated propagation at frequencies near plasma cutoffs and resonance are purposely neglected. The latter are planet-specfic situations that allow leakage of a small amount of energy near narrow frequency windows, which for this illustration is a second-order consideration. The ULF transfer function was derived by Dejnakarintra and Park [1974] and Greifinger and Greifinger [1976], with attenuation maximizing on the bottomside of the ionosphere and varying as $1/f$ to $1/f^3$. The transmission function considered is shown in Figure 3 of Greifinger and Greifinger [1976]. The ELF/VLF transfer function is well known based upon whistler studies, and that function applied here is illustrated in Figure 3–31 of Helliwell [1965]. Applications here are for an ionospheric plasma frequency

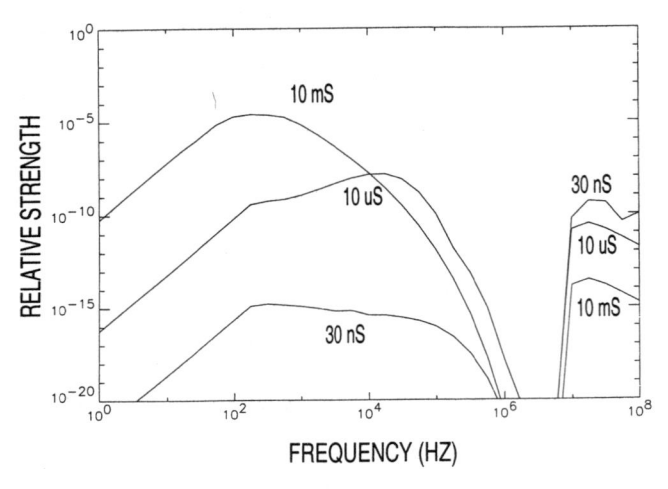

Figure 1. The radio spectra as observed 2000 km above the source over a modeled ionosphere. The Spectra for a discharge duration of 30 nS, 10μS, and 10 mS is shown.

of 7 MHz. The HF transfer function is unity at high frequencies, but displays attenuation as the wave frequency approaches the ionospheric plasma frequency (see Eq. (A7) of Zarka [1985b]). Again, the application is for an ionospheric plasma frequency of 7 MHz.

Figure 1 shows the topside radio spectra for different discharge durations, $\tau = 2\pi/(\alpha\beta)^{1/2} =$ 30 ns, 10μs, and 10 ms. As evident in the figure, given the same ionosphere, there is a wide range of amplitudes possible based upon the variation of discharge duration. For example, near 10 kHz, signal strengths range over 200 dB, based solely upon the variations in discharge duration with slower discharges ($\tau > 1$ μS) tending to couple more efficiently to the VLF/ELF region. In the HF band, a large intensity range is again obtained from the model due to discharge duration, with the faster discharges ($\tau < 1$ μS) tending to couple more efficiently in the HF. Thus, the slow 10 mS discharge has peak radiation energy in the ELF/VLF region, and its strength in the HF is reduce by over 180 dB relative to the maximum. In contrast, the fast 30 nS discharge has peak energy in the HF band, and has greatly diminished strength (by 120 dB) in the ELF/VLF band. It is contended that discharge duration may explain the "either/or" observations in Table 1, with radiation from faster discharges appearing in Voyager's HF radio astronomy instrument and radiation from slower discharges appearing as whistlers in Voyager's VLF plasma wave experiment. The distinct gap in emission near 7 MHz is due to ionospheric reflection/absorption of the radiation near the peak plasma frequency.

Figure 2 shows the ratio of VLF (3 kHz) to HF (30 MHz) emission levels as a function of discharge duration (in seconds) as observed by a spacecraft located at the ionospheric topside. As indicated in the figure, the slower the discharge, the stronger the coupling to lower frequencies (i.e., ELF/VLF). In contrast, the faster the discharge, the stronger the coupling the higher frequencies (i.e., HF). As such, the presentation indicates that discharge duration has a substantial impact on the spectral peak of the radiation and whether it is more likely to be detected in the VLF or HF bands. Discharge scenarios for various planets are indicated in the figure. These are now described more completely below.

APPLICATIONS

The Slow Jovian Discharge. The Lightning and Radio Emission Detector (LRD) onboard the Galileo Probe observed near the 1 Bar level (i.e., below ionospheric layers) a series of slow-varying radio waveforms from atmospheric lightning with peak energies near 500 Hz and an $f^{-1.4}$ high frequency rolloff in the VLF region [Lanzerotti et al., 1996]. A slow discharge with a duration of a few milliseconds could account for such a waveform [Farrell et al., 1999]. Figure 3 shows a model radiated waveform and spectra (as observed below the ionosphere) from a discharge model with a duration of $\tau \sim 3$ mS. The model waveform appears very similar to those observed by the Galileo Probe

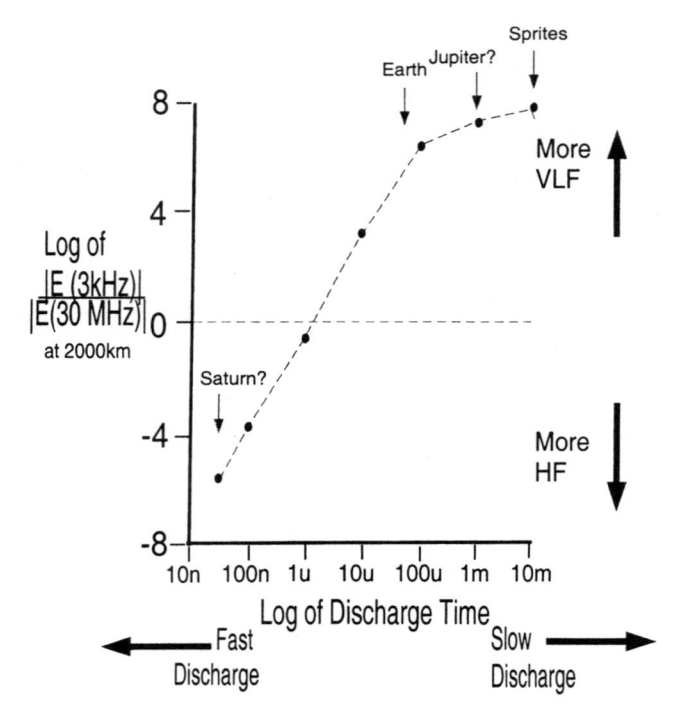

Figure 2. The ratio of VLF-to-HF emission levels as a function of discharge time (in seconds). Note that the faster the discharge, the more coupling to the HF region.

LRD near 1 Bar (see Figure 3 of Lanzerotti et al., [1996]). The spectra in Figure 3 indicate that the power in the HF band (near 30 MHz) from a 3 mS discharge duration is greatly reduced (over 180 dB down) from its peak values near 300 Hz. Figure 2 also indicates that a millisecond discharge will radiate primarily in the ELF/VLF and will only be weakly coupled to the HF.

This slow discharge may explain the lack of detection of Jovian spherics. Zarka [1985a] indicated that ionospheric attenuation by the L7 layer would yield substantial HF attenuation and thus may explain the lack of detection of Jovian spheric signals by the HF receiver onboard Voyager (i.e., the Planetary Radio Astronomy (PRA) Experiment). However, as modeled above, a slow radiation source would only weakly couple to the HF region, generating signals reduced by nearly 180 dB compared to ELF levels. Thus, the dominant effect for the lack of detectable spherics is most likely a source phenomenon, with ionospheric attenuation further reducing the already weak signal strengths. The slow discharge at Jupiter may be analogous to the terrestrial sprite emissions (see below). It should be noted that the signal strength of Jovian VLF whistlers is still relatively large between 1–10 kHz, and will result in an amplitude of 10's of μV/m at 5 R_J [Farrell et al., 1999], a level detectable by the Voyager Plasma Wave PWS receiver. Thus, a slow discharge model explains the Galileo probe results, the Voyager PWS observations of whistlers, but also the lack of detectable spheric signals by the Voyager Radio Astronomy HF receiver.

The Fast Saturnian Discharge. The Voyager HF receiver on the Planetary Radio Astronomy (PRA) experiment detected radio bursts (i.e., spherics) from the Saturnian atmosphere [Kaiser et al., 1983]. However, there was a distinct absence of corresponding lightning-generated whistler emissions in the Voyager VLF plasma wave receiver, presumably because the spacecraft did not intercept connecting magnetic field lines. Since Voyager 1 flew within three planetary radii, the lack of whistlers and direct field line connection to the atmospheric sources implies that the storms are highly localized near the planet equator [Desch, 1994] consistent with the observed periodicity of the overall emission occurrence pattern. A combined spectra of the HF bursts indicates a flat spectra from 10 MHz to the maximum receiver frequency of 40 MHz. Below 10 MHz, the signals become attenuated in association with ionospheric absorption [Zarka, 1985b].

A difficulty in understanding Saturnian spherics has been to explain the flat spectrum. Warwick [1989] presented a discharge model that emphasized random currents in the ionized path and purposely neglected the initiation and cessation (turn on/turn off) of the current wave. The

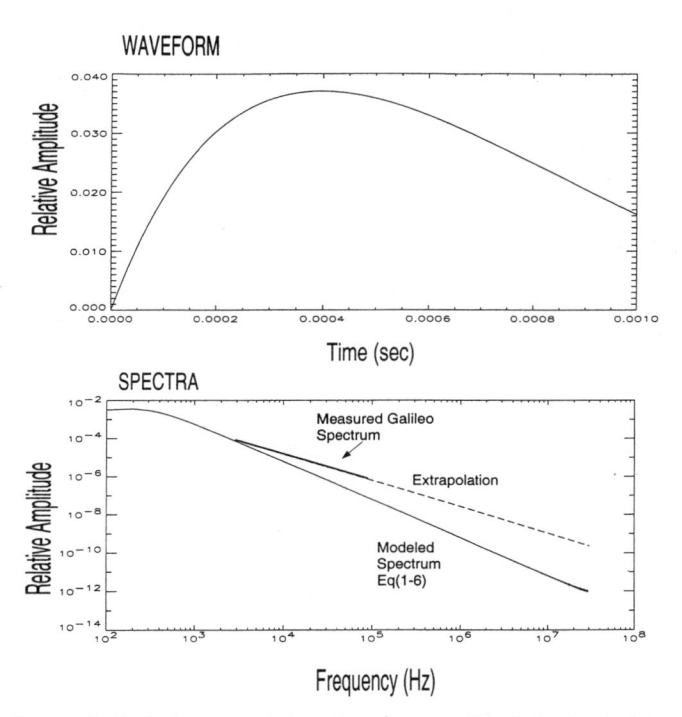

Figure 3. Radiation model from the slow ($\tau \sim 3$ mS) Jovian lightning discharge applicable to observations below the ionosphere. Both the modeled (a) waveform and (b) spectra are consistent with the available observations. See Farrell et al. [1999] for details.

resulting model produced a 1/f spectral variation which appeared similar to the results from the tortuousity models presented by Le Vine and Meneghini [1978 a,b]. However, it did not fit the observations.

Consider now a discharge model that features the initiation and cessation of lightning current, rather than the random discharge approach of Warwick et al., [1989]. In this case, a fast current wave discharge is modeled via Eq (1–3) with a temporal duration of 20 nS. For a spacecraft at the ionospheric topside, the ionospheric absorption [Zarka, 1985b] must also be included in the calculation. The resulting model spectrum is shown in Figure 4. Note that above the ionospheric cutoff, the spectral content is primarily flat. Ionospheric absorption gives rise to the distinct rolloff at the low frequency cutoff, as previously considered by Zarka [1985b]. In essence, the fast discharge accounts for the observed flat spectrum and ionospheric absorption accounts for the rolloff near 7 MHz. Also, the fast discharge model predicts VLF power levels reduced by over 120 dB compared to the HF signal levels (see Figure 1 and 2). These VLF intensities are well below the Voyager VLF plasma wave receiver sensitivity. It is thus concluded that the lack of Saturnian whistlers is their low intrinsic intensity from the fast current source, and is not necessarily related to the

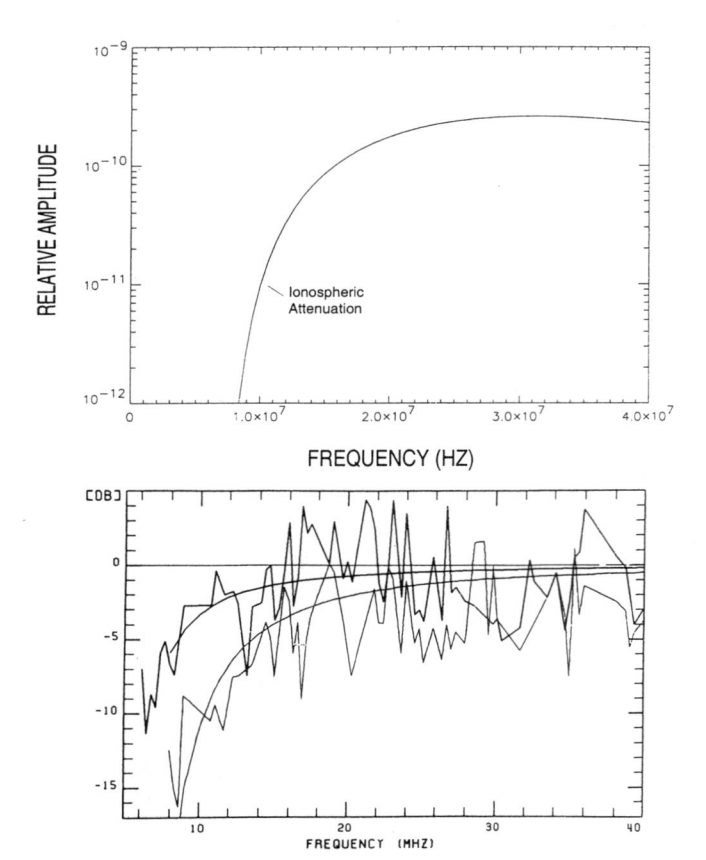

Figure 4. (a) Radiation model from a fast ($\tau \sim 20$ nS) current discharge applicable to the Saturnian spherics (as observed above the ionosphere). Note that the spectrum is nearly flat above about 10 MHz, consistent with the (b) observations (from Zarka, [1985b]).

location of detection. Voyager may have been magnetically-connected to the Saturnian storms, but the fast discharges did not efficiently couple to the VLF regime, and thus the whistler signals were too weak to be detected. This result also relaxes the locational constraint placed on the Saturnian storms, thereby allowing their possible existence at higher latitudes. Consequently, a fast discharge model of Saturnian lightning can explain the observed flat spectra of the spheric emission and the lack of detectable VLF whistler emissions from a storm system at any latitude.

The Slow Terrestrial Sprite Emissions. In 1990, Franz et al. [1990] reported the observation of a large luminous optical emission over thunderstorms, extending upward over 80 km from the cloud tops to the ionosphere. These events were later named "sprites", and were reported to occur following energetic positive cloud-to-ground strokes [Sentman and Wescott, 1993, Sentman et al., 1995]. As reported by Franz et al., the events were slow, having an optical duration of 10–20 milliseconds (i.e., over 1000 times slower than the typical cloud-to-ground lightning event). Also, Franz et al. reported the nondetection of simultaneous VLF band emission.

To explain the lack of VLF emission from these signals, Farrell and Desch [1992] suggested that the events originate from a "slow" current wave extending from the lower atmosphere to the ionosphere. Franz et al's optical waveform steadily rose and descended on time scales of 10's of milliseconds. Farrell and Desch assumed that these optical measurements were driven by a correspondingly slow upward oriented current wave (i.e., peak current near 10 milliseconds), and then determined the emitted radiation via Eq. (1) and (2). An illustration of their model is shown in Figure 5. The resulting modeled radiated pulse from this slow discharge antenna was found to peak at energies below 100 Hz and was determined not to be strongly coupled to the VLF. The results are shown in Figure 5, along with the spectra of the typical cloud-to-ground stroke, for comparison. Subsequent observations [Boccippio et al., 1995; Cummer et al., 1998] confirm the events as having peak spectral energy in the ULF and lower ELF bands, at frequencies well below that of the typical terrestrial cloud-to-ground stroke.

Recently, Pasko [1999] suggested that a "slow" current over a cloud is created after a lightning stroke, as the ionosphere responds to the charge left in the thundercloud. Because the ionospheric conductivity exponentially-increases with height, the response to the underlying cloud charge occurs first at high altitudes near 90 km, and then moves progressively downward as a function of time. Greifinger and Greifinger [1976] have suggested that this ionospheric response is similar to moving a capacitor plate downward toward the thundercloud, with electrical energy density increasing between the plate and the ground, and electric flux being drawn upward toward the plate as it moves close to the cloud charge. The sprite emissions are then associated with the vertical currents that develop to electrically feed this downward moving capacitor plate [Pasko, 1999]. It should be noted that the Pasko model explains the overall sprite structure, but some higher frequency activity [Valdivia et al. 1997, 1998] must be occurring to explain the presence of "dancing sprites" and other faster phenomena.

Uranus and Neptune At Uranus, lightning emissions were only detected in the Voyager HF radio astronomy instrument [Zarka and Pedersen, 1986]. Like at Saturn, the individual spectra were combined by Zarka and Pedersen [1985] to create an overall emission spectrum that appeared to rolloff as $1/f^2$ at frequencies above 5 MHz. This observation implies that the emission peak is at least below 5 MHz, and this can only occur for a discharge duration slower than 0.2 μS ($\tau > 0.2$ μS).

In contrast, at Neptune, potential lightning events were observed in the Voyager VLF plasma wave receiver as a very slow drifting tone between 3–10 kHz propagating in the whistler [Gurnett et al., 1990; Menietti et al., 1991] or z-mode [Farrell, 1996]. The best estimate for an event observed in the VLF is a discharge with a duration $\tau \gg 1\mu S$. Some HF bursts were detected by the Voyager 2 PRA experiment [Kaiser et al., 1991], but they were not uniquely classified as lightning.

CONCLUSIONS

A unifying concept is presented that explains the space-observed radio frequency signatures of lightning at Earth and the outer planets. In this "illustration", the previously-presented planetary-specific propagation effects are assumed

Table 2. A summary of observations and inferred lightning discharge duration.

	HF Spherics	VLF Whistler	Discharge Time	Peak Radiation Frequency
Jupiter	No	Yes	1-2 mS	500 Hz[3]
Saturn	Yes	No	10-30 nS	> 10 MHz
Uranus	Yes	No	> 0.2 μS	<5 MHz
Neptune	No [1]	Yes[2]	$\gg 1\ \mu$S	$\ll 1$ MHz
Earth CG	Yes	Yes	10-100 μS[3]	10 kHz
Sprites	—	—	10-20 mS[3]	10-100 Hz

1. A few candidate events [Kaiser et al., 1991];
2. VLF whistler or Z-mode [Farrell et al., 1996];
3. Actual observation made below ionosphere

to be secondary, and instead the source discharge duration is assumed to primarily account for the character of the RF spectra. The result is that the presentation can explain the band-limited nature of the observations at the outer planets. For example, a slow lightning discharge at Jupiter can account for both the Galileo Probe observations of a slow radiated waveform [Lanzerotti et al. 1996] and VLF whistler signals [Gurnett et al., 1979], but also the lack of substantial HF energy in the form of spheric emissions. A fast discharge at Saturn explains the detection of HF spherics, the relatively flat spectrum at frequencies well above the ionospheric cutoff, the lack of substantial VLF whistler signal levels and also relaxes the location constraint on the atmospheric storm systems.

In essence, the presentation illustrates that discharge time could be a primary factor determining the spectral character of the lightning RF signature in space. Based upon the observed spectra at the Earth and outer planets, the discharge duration can be estimated, and these are presented in Table 2. In the case of Earth and Jupiter, observations have been made below the ionosphere, and thus the discharge character is actually sensed locally. In the case of Saturn, Uranus and Neptune, the discharge duration is inferred from the measured spectra obtained by the Voyager HF radio astronomy receiver (PRA) and Voyager VLF plasma wave (PWS) receiver. It should be duly noted that each planetary

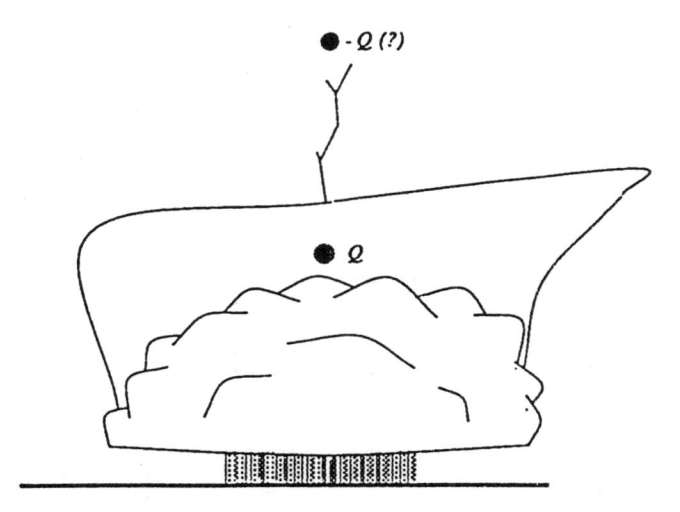

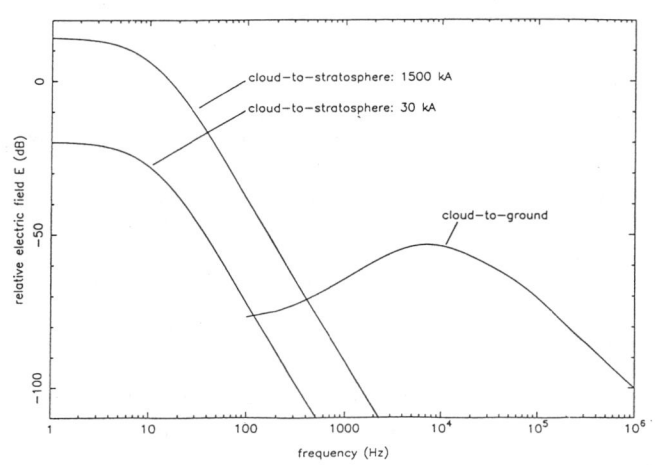

Figure 5. A model of the sprite discharge and associated resulting spectra.

ionosphere is different, and some care must be made in applying the present illustration in detail to specific planets. As such, the illustration may not be a unique explanation for the different radio spectra observed topside, and other explanations may exists. However, previous work did not consider the discharge current source as a possible explanation for the spectral character of the emission. As now demonstrated, the discharge radiation source may in fact be the defining characteristic in determining the spaced-based RF spectra from lightning.

REFERENCES

Bruce C.E.R. and R.H. Golde, The lightning discharge, *J. Inst. Elec Eng,* 88(11), 487, 505, 1941.

Boccippio, D.J., E.R. Williams, S.J. Heckman, W.A. Lyons, I.T. Baker, R. Boldi, Sprites, ELF Transients, and positive ground stokes, *Science,* 269, 1088, 1995.

Cummer, S.A. and U.S. Inan, ELF radiation produced by electrical currents in sprites, *Geophys. Res. Lett.,* 25, 1281, 1998.

Desch, M.D., Lightning at planets in the outer solar system, in *Planetary Radio Emissions III,* ed. H. O. Rucker, SJ Bauer, and ML Kaiser, Austrian Academy of Sceince Press, Vienna, 1994.

Dejnakarintra M., and C.G. Park, Lightning-induced electric fields in the ionosphere, *J. Geophys. Res.,* 79, 1903, 1974.

Farrell, W.M. and M.D. Desch, Cloud-to-stratospheric lightning discharges: A radio emission model, *Geophys. Res. Lett.,* 19, 665, 1992.

Farrell, W.M., Are Neptunes highly dispersed whistler's really z-mode radiation?, *Geophys. Res. Lett.,* 23, 587, 1996.

Farrell, W.M., M. L. Kaiser, and M. D. Desch, A model of the lightning discharge at Jupiter, *Geophys. Res. Lett.,* 26, 2601, 1999.

Franz R.C., R.J. Nemzek, and J.R. Winckler, Television image of a large upward electrical discharge above a thunderstorm system, *Science,* 249, 48, 1990.

Greifinger, C., and P. Greifinger, Transient ULF electric and magnetic fields following a lightning discharge, *J. Geophys. Res.,* 81, 2237, 1976.

Gurnett, D.A., R.R. Shaw, R.R. Anderson, W.S. Kurth, and F.L. Scarf, Whistlers observed by Voyager 1: Detection of lightning on Jupiter, *Geophys. Res. Lett,* 6, 511, 1979.

Gurnett, D.A., W.S. Kurth, I.H. Cairns, and L.J. Granroth, Whistlers in the Neptune's magnetosphere: Evidence of atmospheric lightning, *J. Geophys. Res.,* 95, 20967, 1990.

Helliwell, R.A. *Whistlers and related ionospheric phenomena,* Stanford Univ. Press, Stanford, CA, 1965.

Herman J.R., J.A. Caruso, and R.G. Stone, Radio Astronomy Explorer(RAE)-1 observations of terrestrial radio noise, *Planet. Space Sci.,* 21, 443, 1973.

Kaiser, M.L., J.E.P. Connerney, and M.D. Desch, Atmospheric storm explanation of Saturnian electrostatic discharges, *Nature,* 303, 50, 1983.

Kaiser, M.L., P. Zarka, M.D. Desch, and W.M. Farrell, Restrictions on the characteristics of Neptunian lightning, *J. Geophys. Res.,* 96, 19043, 1991.

Lanzerotti, L.J., K.. Rinnert, G. Dehmel, F.O. Gliem, E.P. Krider, M.A. Uman, J. Bach, Radio frequency signals in Jupiter's atmosphere, *Science,* 272, 858, 1996.

LeVine D.M. and R. Meneghini, Simulation of radiation from lightning return stroke: The effects of tortuosity, *Radio Science,* 13, 801, 1978a.

LeVine D.M. and R. Meneghini, Electromagnetic fields radiated from a lighting return stroke: Applications of an exact solution to Maxwell's equations, *J. Geophys. Res.* , 83, 2377, 1978b.

Mazur, V., X.-M. Shao, and P.R. Krehbiel, "Spider" lighting in intracloud and positive cloud-to-ground flashes, *J. Geophys. Res.,* 103, 19811, 1998.

Menietti, J.D., and D.A. Gurnett, Whistler propagation in the Jovian magnetosphere, *Geophys. Res. Lett.,* 7, 49, 1980.

Miller, S. L. , A production of amino acids under possible primitive earth conditions, *Science,* 117, 528, 1953

Pasko, V.P., U.S. Inan,, T.F. Bell, and S.C. Reising, Mechanism of ELF radiation from sprites, *Geophys. Res. Lett.,* 25, 3493, 1998.

Scarf, F.L. and D.A. Gurnett, A plasma wave investigation for the Voyager mission, *Space Sci. Rev.,* 21, 289, 1977.

Sentman, D.D. and E.M. Wescott, Observations of upper atmospheric optical flashes recorded from an aircraft, Geophys. Res. Lett., 20, 2857, 1993.

Sentman D.D. et al., Preliminary results from Sprites94 campaign: 1) Red sprites, *Geophys. Res. Lett.,* 22, 1205, 1995.

Storey, L.R.O., An investigation of whistling atmospherics, *Phil. Trans. Royal Soc. London, A,* 246, 113, 1953.

Valdivia, J.A., G. Milikh, and K. Papadopoulos, Red Sprites: Lightning as a fractal antenna, *Geophys. Res. Lett.,* 24, 3169, 1997.

Valdivia, J.A., G. Milikh, and K. Papadopoulos, Model of red sprites due to intracloud fractal lightning discharges, *Radio Science,* 33, 1998.

Villanueva, Y., V.A. Rakov, and M.A. Uman, Microsecond-scale electric field pulses in cloud lightning discharges, *J. Geophys. Res.,* 99, 14353, 1994.

Warwick, J.W., J.B. Pearce, R.G. Peltzer, and A.C. Riddle, Planetary radio astronomy experiment for the Voyager missions, Space Sci. Rev., 21, 309, 1977.

Warwick J.W., Power Spectrum of electrical discharges seen on earth and at Saturn, J. Geophys. Res., 94, 8757, 1989.

Zarka, P., On the detection of radio bursts associated with Jovian and Saturnian lightnings, Astron. and Astrophys., 146, L15, 1985a.

Zarka, P. Directivity of Saturn electrostatic discharges and ionospheric implications, Icarus, 61, 506, 1985b.

Zarka, P. and B. M. Pedersen, Radio detection of Uranian lightning by Voyager 2, *Nature,* 323, 605, 1986.

W. Farrell, Code 695, NASA/GSFC Greenbelt, MD 20771

Terrestrial Continuum Radiation in the Magnetotail: Geotail Observations

H. Matsumoto

Radio Atmospheric Science Center, Kyoto University, Uji, Kyoto, Japan.

I. Nagano

Department of Electrical and Computer Engineering, Kanazawa University, Kanazawa, Japan.

Y. Kasaba

Institute of Space and Astronautical Science, Sagamihara, Kanagawa, Japan.

Geotail observations have revealed that the terrestrial continuum radiation observed in the magnetotail consists of several components that are generated at different locations through different processes. We have classified "the continuum-like radiation" into the following components: classical nonthermal continuum radiation generated in the dawnside and dayside plasmapause, short-lived enhancement of the nonthermal continuum radiation generated at the nightside plasmapause, auroral myriametric radiation generated at the auroral plasma cavity, lobe-trapped continuum radiation generated at the plasma sheet boundary layer (PSBL), and narrowband $2f_p$ radiation generated through beam-plasma instability at the PSBL. We summarize their generation and propagation nature. We also discuss possibilities of using their characteristics for the remote sensing studies of the terrestrial magnetosphere.

1. INTRODUCTION

Geotail has extensively observed plasma waves in the terrestrial magnetotail [cf. *Matsumoto et al.*, 1998]. In this region nonthermal continuum radiation is the most common component in the frequency range above the local electron plasma frequency f_p and below several tens kHz. Geotail observations have revealed that the nonthermal continuum radiation observed in the magnetotail is not always the classical one but consists of several components which have different sources and generation processes. The low-frequency radiation provides real-time and global information of local plasma parameters around their source regions or along the propagation paths. Therefore the observation of these radio waves could provide an opportunity for studying the global and fast phenomena in the magnetotail.

In this paper we present a new perspective of 'nonthermal continuum radiation' in the magnetotail based

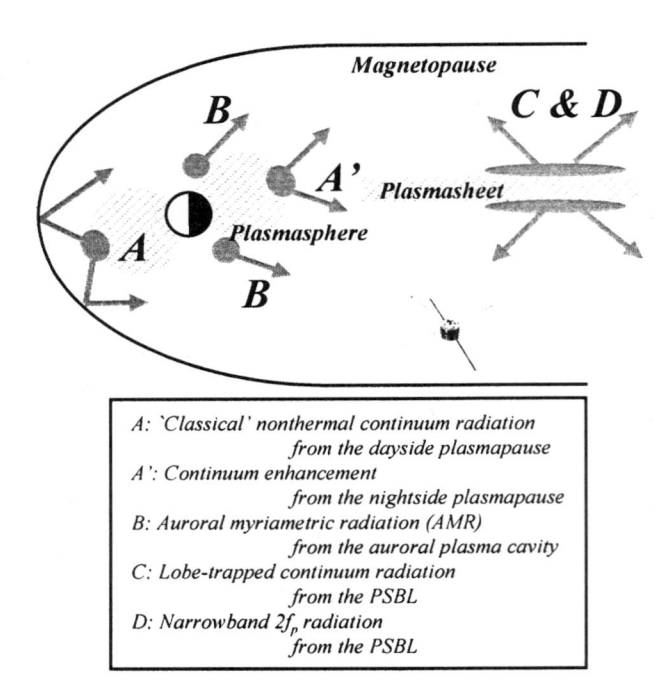

Figure 1. Source locations of "nonthermal continuum radiations" observed in the magnetotail: A) Classical continuum radiation from the dayside plasmapause, A') Continuum enhancement from the nightside plasmapause, B) Auroral myriametric radiation (AMR) from the auroral plasma cavity, C) Lobe-trapped continuum radiation from the plasma sheet boundary layer (PSBL), and D) $2f_p$ radiation from the PSBL.

on the Geotail observations. We have identified several radiation components as listed below (Figure 1): Classical nonthermal continuum radiation (Type A) is generated from the dawnside to dayside plasmapause. In the magnetotail the trapped continuum radiation reflected at the magnetopause is mainly observed. Therefore Type A radiation provides information of the density of the magnetosheath. Short-lived enhancement of the nonthermal continuum radiation (Type A') is generated at the nightside plasmapause by energetic electrons injected from the magnetotail. Type A' radiation is associated with the onset of substorms and carries information of the motion of injected electrons and the location of the plasmapause. Auroral myriametric radiation (Type B) is generated in the auroral region. It is associated with the large enhancement of the auroral kilometric radiation (AKR) and might provide the information in and around the auroral plasma cavity. Lobe-trapped continuum radiation (Type C) and narrowband $2f_p$ radiation (Type D) are generated in the plasma sheet boundary layer (PSBL). These two could be associated with plasma activities during substorms

and can provide the fast and global information of the reconnection processes in the magnetotail. We introduce the nature of these different types of radiation and discuss possibilities of using their characteristics for the remote sensing studies of the magnetotail plasma activities.

In this study we use the data sets of the plasma wave instrument (PWI) [*Matsumoto et al.*, 1994] aboard the Geotail spacecraft. The PWI is composed of three distinct receivers with different frequency and time resolutions: the sweep frequency analyzer (SFA), the multichannel analyzer (MCA), and the waveform capture (WFC). The SFA provides spectral information of plasma waves over the frequency range from 24 Hz to 800 kHz for electric field and 24 Hz to 12.5 kHz for magnetic field with five and three logarithmically ordered linear bands, respectively. Each band has frequency resolution of 1/128 of its frequency bandwidth and time resolution of 8 s above 1.6 kHz or 64 s below 1.6 kHz. The effective dynamic range is approximately 90 dB. The MCA has higher time resolution and coarser frequency resolution for complement of the SFA. The MCA covers from 5.6 Hz to 311 kHz for electric field and from 5.6 Hz to 10 kHz for magnetic field with fixed and logarithmically ordered 20 and 14 channels, respectively. The bandwidth of filters are ±15% below 10 kHz or ±7.5% above 10 kHz of the channel center frequency. The signals are sampled once every 0.25 or 0.5 s. The effective dynamic range is approximately 110 dB for electric field and approximately 100 dB for magnetic field. The WFC has two electric and three magnetic receivers. Waveforms of the analog signal are sampled for a period of 8.7 s with an interval of 5 min and a sampling frequency of 12 kHz, through a filter with an upper frequency cutoff of 4 kHz and a lower frequency cutoff of 10 Hz. The dynamic range of the WFC is 66 dB. These receivers are connected to two sets of electric dipole antennas and two sets of tri-axial search coils. The highest sensitivities of the electric and magnetic sensors are 5×10^{-9} V/$\sqrt{\text{Hz}}$ and 1.5×10^{-5} nT/$\sqrt{\text{Hz}}$, respectively.

The data observed in a period from September 1992 to December 1997 are used in the present analyses. For the first two year period Geotail was injected into the orbits to cover the distant geomagnetic tail region. Then it was maneuvered into much lower apogee of about 30 R_E on November 1994. The orbits of the Geotail for "the distant-tail phase" and "the near-Earth phase" are plotted in Figures 2a and 2b, respectively. Note that we use the modified geocentric solar-magnetospheric coordinates (GSM') which include the effect of the solar wind aberration of 4°, the hinging distance of 10 R_E, and the tilt angle of the geomagnetic dipole axis. The orbits in the distant tail phase cover a wide area in the distant magnetotail. These orbits provide a good opportunity for surveying various "continuum-like" radiations in the middle to distant tail region. On the contrary, the orbits in the near Earth phase are suitable for the studies of the near tail region.

2. NONTHERMAL CONTINUUM RADIATION FROM THE PLASMAPAUSE

First, we show the analyses on the conventional nonthermal continuum radiation generated at the plasmapause. The "classical" nonthermal continuum radiation is mainly observed from the dawnside to dayside zone and lasts for 2~12 hours with a smooth variation of both intensity and frequency. Its frequency range is from the local electron plasma frequency and upwards sometimes up to 100–500 kHz.

This radiation is generated at the geomagnetic equator of the plasmapause from 4h to 14h MLT zone (MLT: magnetic local time) [*Gurnett and Shaw*, 1973; *Brown*, 1973; *Gurnett*, 1975]. In the source region the strong

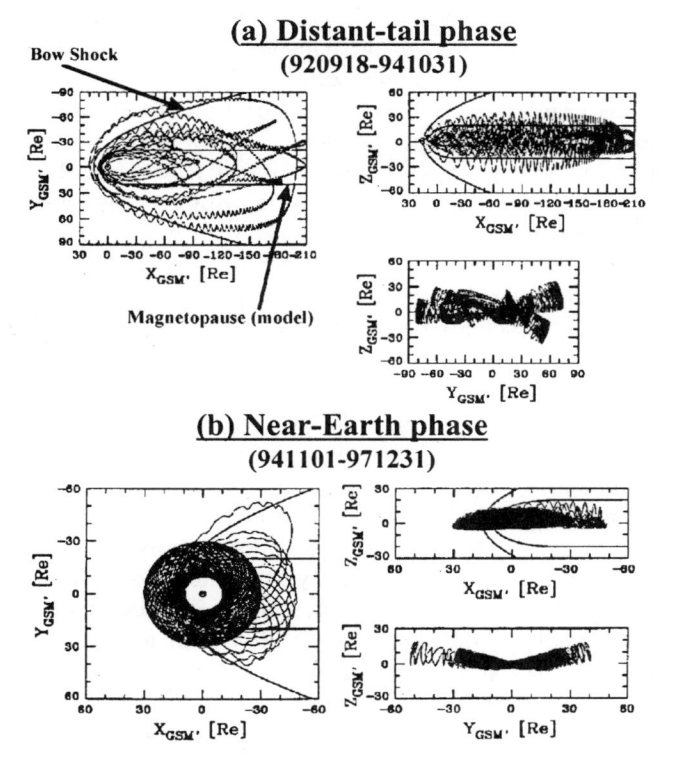

Figure 2. Geotail orbits plotted in the modified GSM coordinate system. (a) Distant-tail phase and (b) near-Earth phase [cf. *Matsumoto et al.*, 1998].

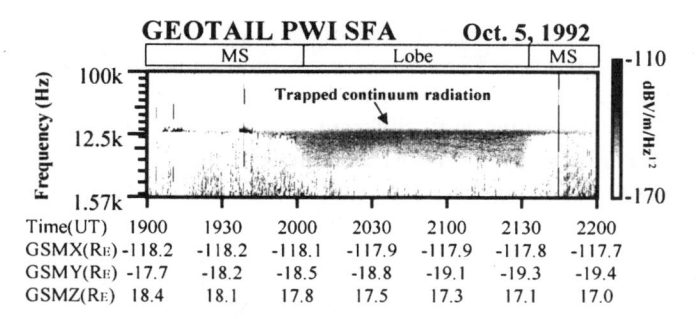

Figure 3. Trapped nonthermal continuum radiation on October 5, 1992 associated with moving of the spacecraft from the magnetosheath to the tail lobe and back to the magnetosheath [cf. *Matsumoto et al.*, 1998].

electrostatic upper hybrid waves are converted to the L-O electromagnetic waves through the conversion window by the linear wave conversion mechanism [*Jones*, 1980; 1981a, b; 1982]. The resultant electromagnetic waves propagate from sharp density gradient at the plasmapause overcoming Landau and cyclotron damping. The upper hybrid resonance by electron temperature anisotropy [cf. *Jones*, 1981a, b] and/or electron beam is most effective at $(n + 1/2)f_{ce}$, where f_{ce} is the electron cyclotron frequency. Therefore generated radiation has harmonic structure whose frequency range and spacing agree with f_p and f_{ce} at the plasmapause, respectively.

2.1. Trapped Continuum

The classical nonthermal continuum radiation observed in the region distant from the plasmapause is further classified into "escaping" and "trapped" components that are observed above and below the electron plasma frequency f_p in the magnetosheath. The former can penetrate the magnetosheath and propagate to the outside of the magnetopause. It keeps its banded frequency structure and large modulation index [cf. *Kurth et al.*, 1975]. The latter is confined to the magnetosphere by reflection at the magnetopause. It shows conglomerate broad band spectrum with low modulation index [cf. *Kurth et al.*, 1981].

In the magnetotail the observed continuum radiation is usually reflected and trapped one by the magnetopause. Figure 3 shows an example of the spectra of the trapped continuum radiation in the magnetotail observed by the SFA on December 5, 1992. In this plot Geotail traverses the magnetopause from the magnetosheath to the tail lobe at about 2000 UT and is back to the magnetosheath at 2130 UT. The trapped continuum radiation is seen in the frequency range below

30 kHz with blurred upper cutoff and clear lower cutoff frequencies. The causes of its wide band nature are thought to be the mixture of the radiation from wide source locations with different magnetic field strength and from a result of the frequency smoothing process via Fermi-Compton scattering in the multiple reflection at the magnetopause [cf. *Kurth*, 1991].

The lower cutoff frequency of the trapped continuum radiation in Figure 3 is generally correlated to the local electron plasma density. Though the lower frequency cutoff cannot be read clearly due to possible enhancement of the density on the path of the propagation, we normally can read out the local electron density. It is noted that the lower cutoff frequency is sometimes modulated by the spin of the antennas. Figure 4 shows a dynamic spectrum of the lower cutoff of the continuum radiation observed by the WFC in the vicinity of the plasma sheet boundary layer observed on November 15, 1992 [*Nagano et al.*, 1994]. Geotail was in the distant tail, $(X_{GSM}, Y_{GSM}, Z_{GSM}) = (-120, -4, 5\ R_E)$. The arrows below the horizontal scale show the timings when the dipole antenna is aligned along the geomagnetic field. The lower cutoff frequency at about 2.5 kHz is modulated with the rotation of receiving dipole antenna. The two alternating lower cutoff frequencies turn out to be f_{R-X} and f_{L-O} [cf. *Gurnett and Shaw*, 1973]. Actually, the cyclotron frequency computed from the difference of $f_{R-X} - f_{L-O}$ is the same as the measured electron cyclotron frequency which is 250 Hz. The polarizations of observed waves with frequency between the f_{R-X} and f_{L-O} is left-handed. Therefore the lower frequency of the modulated cutoff frequency is f_{L-O} which is equal to the electron plasma frequency.

On the other hand, in Figure 3, the upper cutoff frequency of the trapped continuum radiation near 20 kHz is close to the electron plasma frequency of the magnetosheath. Below this frequency the radiation is confined to the magnetosphere so that its modulation index becomes low. It means that the angular intensity distribution of the continuum radiation becomes from isotropic to anisotropic. This nature can be useful to measure the real-time magnetosheath density in global scale. Plate 1 shows two examples of the variation of the spectrum and modulation index of the continuum radiation in the magnetotail on (a) April 12, 1994 and (b) October 16, 1993. On both days Geotail was in the distant tail, $X_{GSM} = -195$ and -172, respectively. In the panels of modulation index, the black line represents a frequency f_T which shows transition between the large and small modulation indices, and the white line is the electron plasma frequency in the solar wind de-

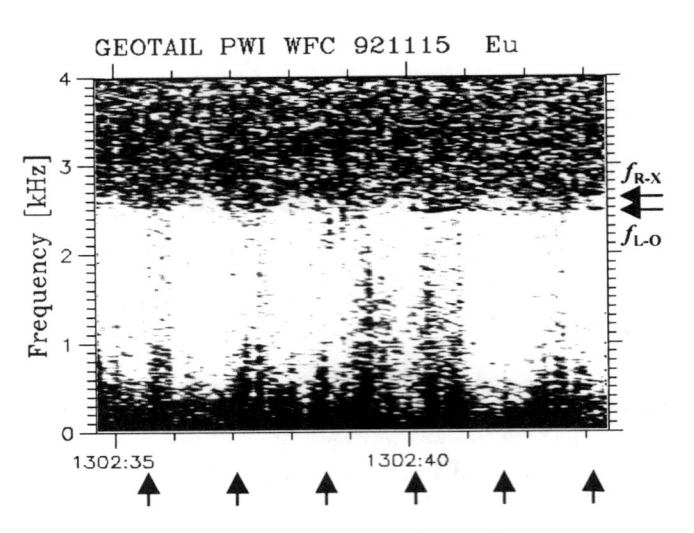

Figure 4. Dynamic spectrum of trapped continuum radiation observed by the WFC for 8.7 s in the tail lobe at the spacecraft location of $(X_{GSM}, Y_{GSM}, Z_{GSM}) = (-120, -4, 5\ R_E)$. The lower cutoff frequency around 2.6 kHz is spin-modulated. The arrows along the time axis show the timings when the dipole antenna is aligned to the local magnetic field [cf. *Nagano et al.*, 1994].

rived from the proton density data of IMP-8. *Nagano et al.* [2000] defined "transit frequency" f_T in the modulation index as the frequency with the deepest modulation index gradient, and studies the variation of the magnetosheath density in global scale. The upper panel of Figure 3a shows that the wave intensity does not show any sudden change around 25 kHz. However the lower modulation panel clearly shows a sudden change around 25 kHz. On the other hand, in Plate 1b, the upstream solar wind density in IMP-8 data is enhanced from 3h to 6h UT associated with the passage of the interplanetary shock. Although the wave intensity profile does not show a good correlation with the solar wind density, the black line which indicates the transition frequency in the modulation index shows much better correlation with a delay. The delay of the variation in the modulation index could indicate the typical reaction time of the magnetosheath density to the variation of the solar wind conditions. The delay is about 3 hours, much longer than the value expected from convection time. Such a long delay may imply that the magnetosheath plasma flows at an Alfven speed of order of several tens of km/s on the average over the global scale. The delay time seems to be enhanced when the large density plasma cloud reaches the Earth. This may suggest a nonlinear magnetosheath response (see *Nagano et al.* [2000]).

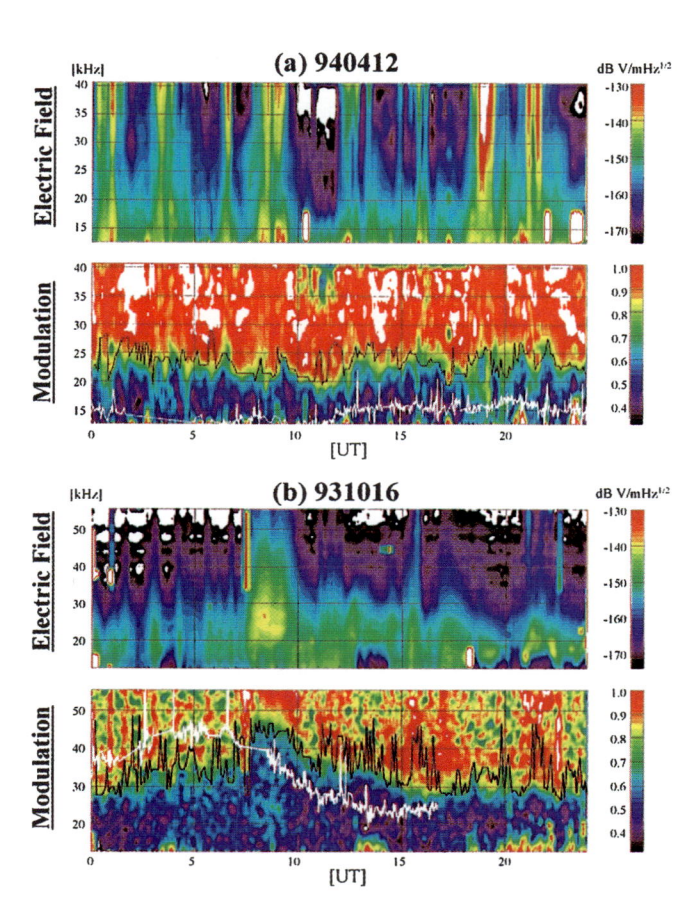

Plate 1. The spectrum and modulation index of the nonthermal continuum radiation observed in the magnetotail on (a) April 12, 1994 and (b) October 16, 1993. Black line represents the modulation transit frequency f_T. White line is the solar wind plasma frequency derived from proton density data of IMP-8 [cf. *Nagano et al.*, 2000].

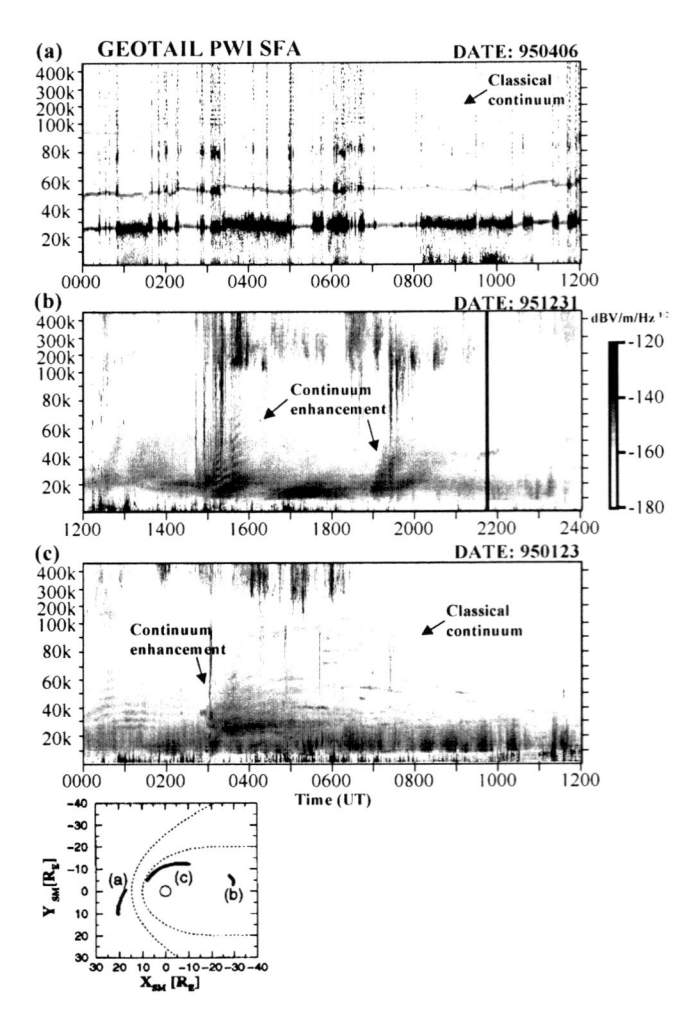

Figure 5. Dynamic spectra observed by the SFA of (a) classical continuum (0h–2h and 5h–10h UT) (b) continuum enhancement (13h–14h, 15h–16h, 19h–20h and 20h–21h UT), and (c) continuum enhancement (3h–4h UT) followed by classical continuum (4h–8h UT). The respective locations and orbits of Geotail corresponding to each case are shown by (a), (b) and (c) in the bottom panel [cf. *Kasaba et al.*, 1998].

2.2. Escaping Continuum : Continuum Enhancement

In contrast to the diffuse and continuous spectrum of the trapped continuum radiation, the escaping continuum radiation is composed of a number of discrete emission lines with a highly structured spectrum. The classical escaping continuum is mainly observed in a region from the dawnside to the dayside sector, and characterized by smooth variation in its intensity and frequency. They normally last for several hours. The escaping continuum radiation observed in the magnetotail, however, shows different features. *Gough* [1982] and *Filbert and Kellogg* [1989] pointed out that the escaping nonthermal continuum radiation observed in the magnetotail had shorter duration time less than a few hours. This radiation is generated at the nightside plasmapause by energetic electrons injected from the tail at the onset of substorms, and has a rising banded frequency spectrum from the local electron plasma frequency up to about 100 kHz and short life only for 1-3 hours. Such shortlived escaping continuum radiation is called as "continuum enhancement."

Figure 5 shows examples of the classical continuum radiation and the continuum enhancement observed by the SFA [*Kasaba et al.*, 1998]. In the top panel when Geotail is located in the dayside zone (the orbit (a) in the bottom panel), the classical continuum is observed at two intervals from 0h to 2h UT and from 5h to 10h UT. In the second panel when Geotail is located in the nightside zone (the orbit (b)), the continuum enhancement is observed at three time intervals from 13h to 14h UT, from 15h to 16h UT and from 19h to 21h UT. The onsets of the continuum enhancement are synchronized with those of the auroral kilometric radiation. Simultaneous observation of Geotail and Wind revealed that the continuum enhancement observed by the nightside spacecraft was occasionally followed by the classical continuum observed by the dayside spacecraft [*Kasaba et al.*, 1998]. The same relation is suggested by the observations of spacecraft in the dawnside. In the third panel when Geotail is located on the dawnside zone (the orbit (c)), the continuum enhancement appearing from 0300 to 0350 UT is followed by the classical continuum which lasts until 8h UT. These results suggest that both types of the continuum radiation are the time sequential phenomena caused by the same onset of substorms.

In the near Earth phase the Geotail orbits cover all MLT zone quite uniformly as shown in Figure 2b. This uniform orbital coverage has given a good opportunity of studying the statistical distribution of the classical continuum and the continuum enhancement. Figure 6 shows (a) the locations and (b) the occurrence histogram of the classical continuum and the continuum enhancement observed by Geotail from November 1994 to December 1995 [*Kasaba et al.*, 1998]. It is clearly seen that the classical continuum is observed mainly on the dawnside and on the dayside, while the continuum enhancement is observed mainly on the night side or the early dawnside.

The direction finding study based on the spin modulation of the electric field intensity with standard Fourier technique [cf. *Manning and Fainberg*, 1980] suggests that the motion of the source of the continuum enhancement agrees with the expected dawnward drift motion of injected electrons [cf. *Filbert and Kellogg*, 1989]. In Figure 7 the continuum enhancement is observed from 13h to 15h UT on September 2, 1995. The top and bottom panels are the frequency-time spectrogram and the pseudocolor plot of the azimuth angle (ϕ_{ce}) from

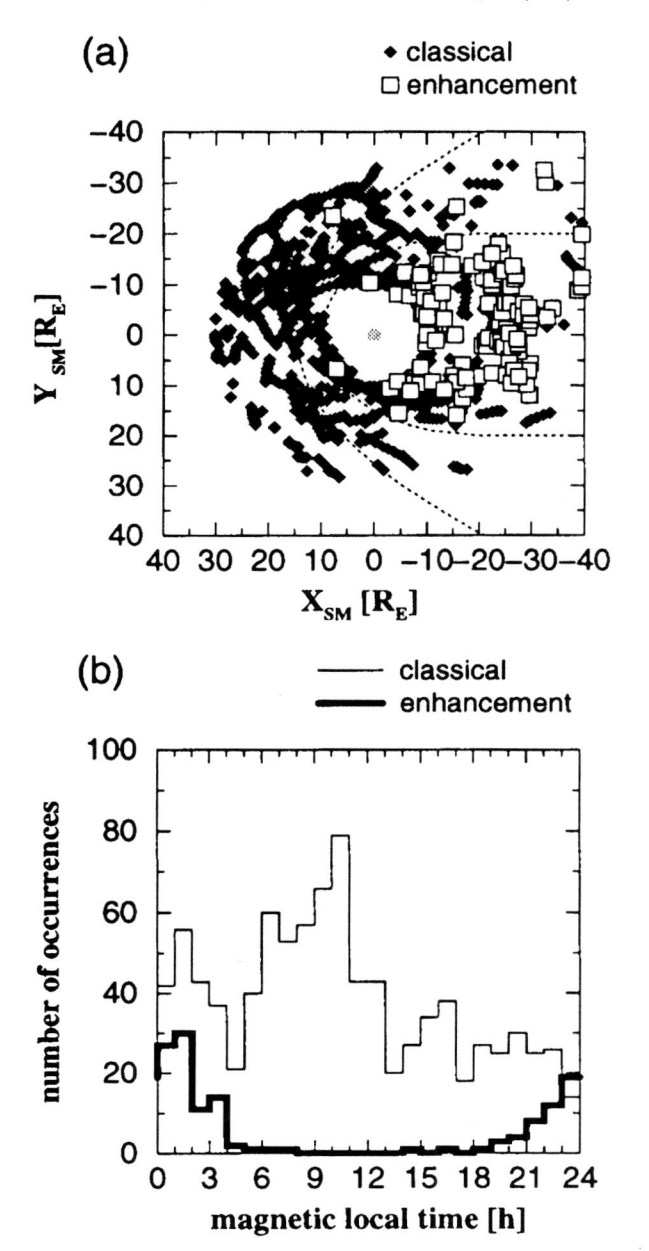

Figure 6. (a) Locations of the observation of the classical continuum (solid diamond) and the continuum enhancement (open square). (b) Occurrence histograms of the two types of continuum versus the magnetic local time [cf. *Kasaba et al.*, 1998].

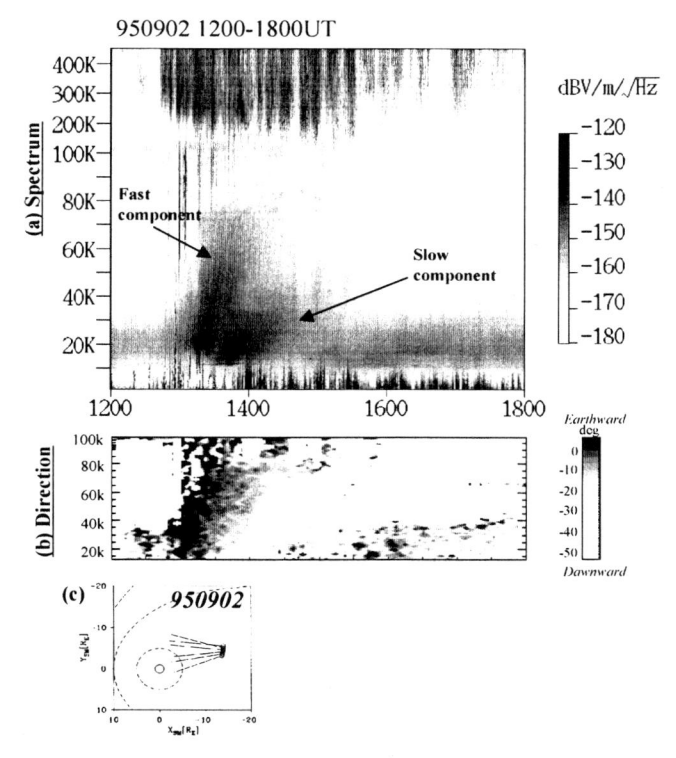

Figure 7. The continuum enhancement from 13h to 16h UT on September 2, 1995. (a) The frequency-time spectrogram of the SFA. (b) The pseudocolor plot of the azimuth angle (ϕ_{ce}) from 12.5 to 100 kHz in 4-min intervals when the flux exceeds −175 dB. Origin of ϕ_{ce} is the direction of the Earth. (c) the observed locations and the azimuth angle ϕ_{ce} at 40 kHz from 13h to 15h UT in SM coordinates [cf. *Kasaba et al.*, 1998].

12.5 to 100 kHz in 4-min intervals when the electric field intensity is above −175 dB. The origin of ϕ_{ce} is the direction of the Earth. The upper right panel also shows the observed locations and ϕ_{ce} at 40 kHz from 13h to 15h UT in SM coordinates. These figures naturally show a physical picture of the electrons injected into the plasmasphere during substorms. At the onset of substorm, energetic electrons are injected into the midnight sector associated with the magnetic field reconnection process in the magnetotail. These injected electrons suffer from $\boldsymbol{E} \times \boldsymbol{B}$ drift and curvature/gradient drift. The former and the latter result in the inward and azimuthal drift motion, respectively. As the curvature and gradient drifts are proportional to the kinetic energy of the injected electrons, the lower energy electrons which have slower dawnward drift velocity penetrate into the higher density region of the inner plasmasphere across the steep density gradient at the plasmapause, and produce the discrete emissions with fast fre-

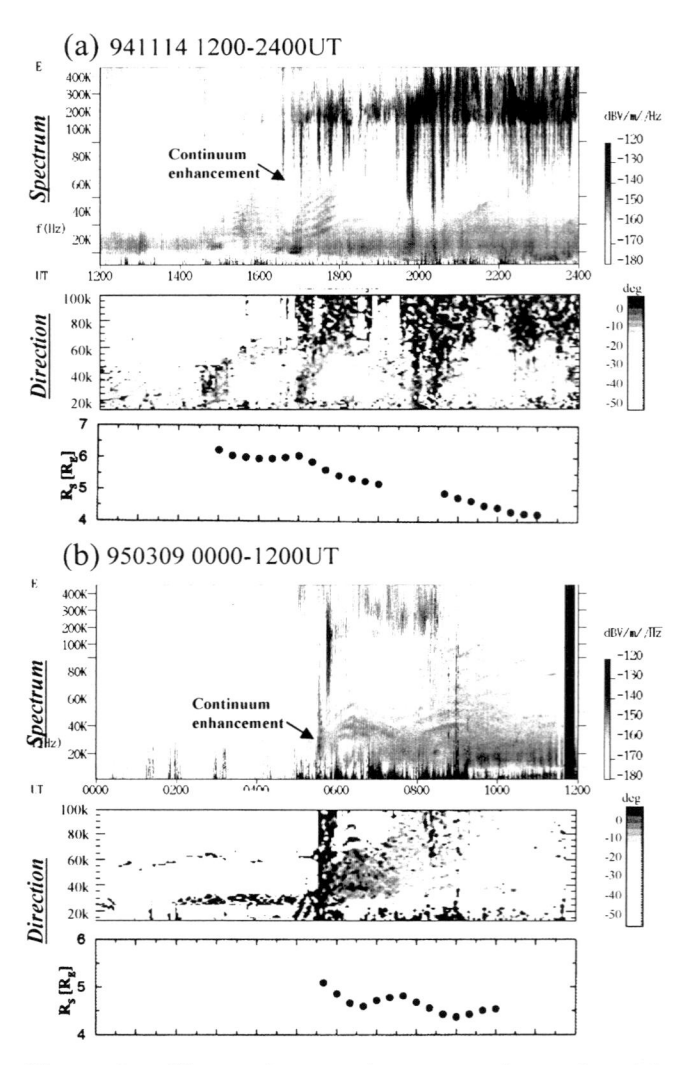

Figure 8. The continuum enhancement observed on (a) November 14, 1994 and (b) March 9, 1995. From top to bottom, the frequency-time spectrogram of the SFA, the pseudocolor plot of the azimuth angle (ϕ_{ce}) from 12.5 to 100 kHz in 4-min intervals when the radiation flux is larger than −175 dB, and the radial distance of the source (R_s) [cf. *Kasaba et al.*, 1998].

quency rising and slow dawnward drift motion. On the other hand, higher energy ($\geq$ 10 keV) electrons suffer faster azimuthal drift and reach the plasmapause after a large dawnward drift motion. They produce the discrete emissions with a slow rate of frequency rise and a fast dawnward drift motion. Actually, in Figure 7, the fast frequency rising part with slow dawnward drift motion is normally followed by the main part with slower rate of the frequency increase and fast dawnward drift motion. Such fast and slow frequency rising components are also seen in the example in the second panel of Figure 5 around 1530 UT. The average drifting time of the injected electrons from the midnight to dawnside is about 1 hour [*Filbert and Kellogg*, 1989]. This value is in the same order of the time interval between the onset of the continuum enhancement and that of the classical continuum.

The spectral nature of the continuum enhancement can provide the fast and global information at the plasmapause associated with substorms. As already described, the continuum radiation is generated at the plasmapause through the upper hybrid resonance. Therefore the frequency range and the spacing of harmonic structure describe the electron plasma frequency and the electron cyclotron frequency in its source region. Since the magnetic field strength correlates to the radial distance of the source from the Earth, we can also derive the radius of the plasmapause from the estimated electron cyclotron frequency in the source when the continuum enhancement is bright enough [*Gough*, 1982]. The continuum enhancement is characterized by the rising of the averaged frequency and the spacing of harmonic structure. Such variations are interpreted as the increase of the plasma density in and the decrease of the radial distance of the source, respectively. These features have been interpreted as the inward motion of the source associated with steepening of the plasmapause [*Gough*, 1982] or the dawnward motion of the source [*Filbert and Kellogg*, 1989] which is due to smaller radius of the dawnside plasmapause [cf. *Carpenter*, 1966].

We show two examples of the spectral variation of the continuum enhancement for two magnetospheric disturbances, large and small. Figure 8a is an example of the continuum enhancement on November 14, 1994 when the magnetospheric activity is high. Figure 8a indicates frequency-time spectrograms, the azimuth angle (ϕ_{ce}), and the radial distance of the source (R_s), determined from the spacing of the harmonic structure (f_{sp}). We assume that f_{sp} agrees with the cyclotron frequency f_{ce} in the radio source at the equatorial plasmapause [cf. *Gough*, 1982]. On the basis of an assumption of a dipole-like magnetic field, R_s is defined as

$$R_s[R_E] = (f_{ce0}/f_{sp})^{1/3} \qquad (1)$$

where $f_{ce0} = 870$ kHz is f_{ce} on the equatorial surface of the Earth. In this plot Geotail moved from (X_{GSM}, Y_{GSM}, Z_{GSM}) = (−38, −8, −1 R_E) to (−41, −12, −4 R_E). After 1600 UT, the AKR activity is much enhanced over 7 hours. Associated with large and continuous magnetospheric activity, series of continuum enhancements were observed from 15h to 16h UT, 17h to 18h UT, and 21h to 22h UT. In this disturbed case

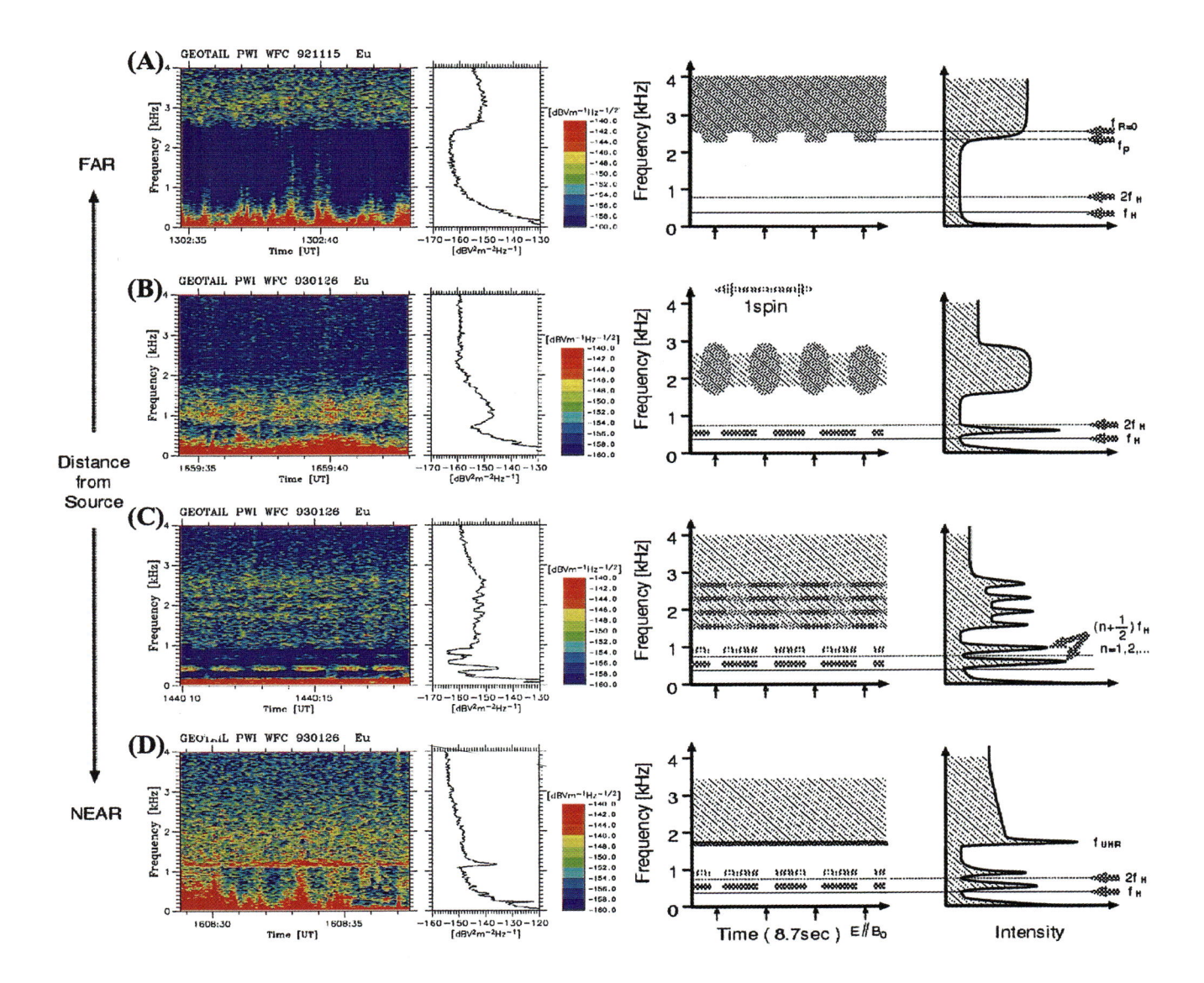

Plate 2. Four different dynamic and frequency spectra of the lobe trapped continuum radiation averaged over 8.7 s. The left two panels are data observed by the WFC. The right two panels are their schematic illustration. The top is the typical lobe trapped continuum radiation with diffused spectrum above the cutoff frequency, while the rest three are the continuum radiation with more or less discrete frequency components (see text) [cf. *Matsumoto et al.*, 1998].

R_s continuously decreased over the entire period. However, the individual event following onset of substorms shows interesting time history of the R_s reduction. In each event, R_s decreases rapidly during the first short period of time of the order of 1 hour. Then the rate of R_s decrease slows down. Figure 8a also shows that the source moves dawnward from the local midnight after each onset of substorms. The dawnward motion of the source itself can contribute to the decrease of R_s [Filbert and Kellogg, 1989]. However, the continuum enhancements generally come from a region close to the local midnight at their onsets. This indicates that the radius of the plasmapause certainly reduces during each subsequent substorm.

On the other hand, Figure 8b is another example of the continuum enhancement observed on March 9, 1995 when the magnetospheric activity is relatively low. In this plot Geotail moved from $(X_{GSM}, Y_{GSM}, Z_{GSM})$ = $(-17, 0, 0\ R_E)$ to $(-3, -10, -1\ R_E)$. Associated with some relatively weak and isolated AKR enhancements, the continuum enhancements are found from 6h to 7h UT and from 7h to 10h UT. The fast and short reduction of R_s is also found for the first 1 hour after each onset in the same manner as shown in Figure 8a. In addition, in the quiet magnetospheric conditions the inverse expansion of R_s is sometimes found at the rate of $+0.1 \sim +0.5\ R_E/h$ for the next 1 hour. Although such reversal expansion is smaller than the scale of the initial reduction, this value is larger than the expected one from the refilling rate by upwelling from the ionosphere.

The 14-month data set of the Geotail plasma wave observations provides 138 events of continuum enhancements. Figure 9 summarizes the variation of R_s in 52 samples that had clear harmonic structure. These data are classified according to Kp index: (a) $Kp < 2$, (b) $2 \le Kp < 3$, (c) $3 \le Kp < 4$, and (d) $4 \le Kp$. The X axis is the time from the onset of the continuum enhancement. The Y axis is the radial distance of the source R_s defined from f_{sp}. Figure 9 demonstrates that R_s in all panels generally decreases in the first 1 hour. The scale of the reduction is not clearly correlated with those of Kp index and R_s at the substorm onset. And after that, some of them with larger radii continue calmed reduction, while others with smaller radii show reversal expansion. Typical variation of R_s is $-1.0 \sim -0.5\ R_E/h$ in the first 1 hour and $-0.5 \sim +0.5\ R_E/h$ in the next 1 hour. The former value might indicate the typical scales of the peeling off of thermal plasma by the magnetospheric convection. On the other hand, inversely expansion after the initial reduc-

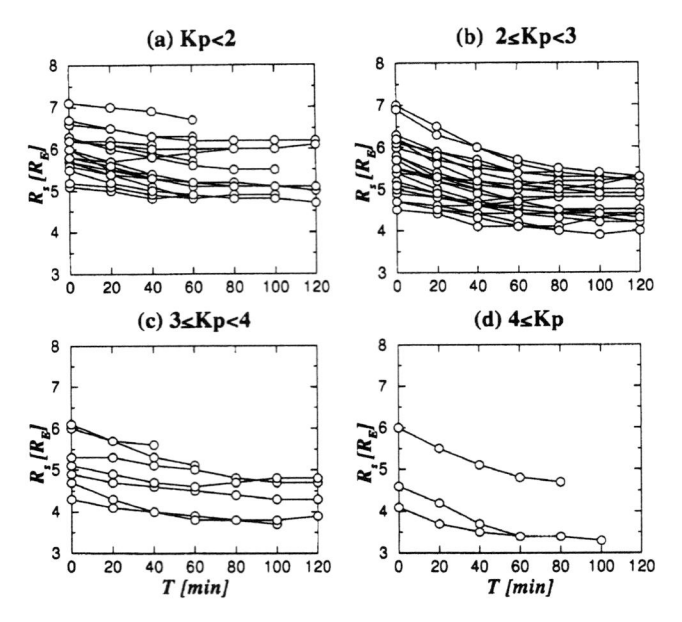

Figure 9. Variation of the radial distance of the source from the Earth (R_s) classified according to Kp index: (a) $Kp < 2$, (b) $2 \le Kp < 3$, (c) $3 \le Kp < 4$, and (d) $4 \le Kp$. The X axis is the time from the onset of the continuum enhancement. The Y axis is R_s defined from f_{sp} [Kasaba et al., 1998].

tion might be explained not only by the slow refilling process but also by the some recovery process of the plasmasphere. After 2 hours from the onset R_s settles around specific values for different levels of geomagnetic activity: from 5 to 6 R_E for $Kp < 2$, from 4.5 to 5.5 R_E for $2 \le Kp < 3$, from 4 to 5 R_E for $3 \le Kp < 4$, and from 3.5 to 4.5 R_E for $4 \le Kp$.

We have showed that the spectral variations of the continuum enhancement indicate fast variation around the plasmapause associated with substorms. Real-time and global information of the plasmapause provided by the continuum enhancement is valuable because there is no other established methods for such information until the launches of future plasmaspheric imaging missions. We expect the detailed comparison with ground-based and other spacecraft observations, including the recent development of the extra ultraviolet imaging.

3. AURORAL MYRIAMETRIC RADIATION (AMR) FROM THE AURORAL PLASMA CAVITY

In the magnetotail we occasionally find an enhancement of electromagnetic waves below 30 kHz associated with the enhancement of the AKR (Figure 10). Although the trapped continuum radiation is also ob-

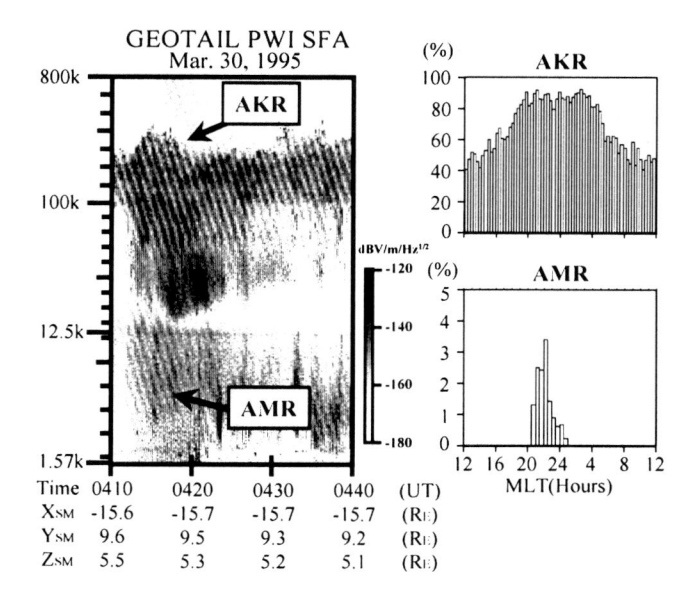

Figure 10. (left) Typical example of dynamic spectrum of AMR which appears below 20 kHz. (right) The magnetic local time dependence of the occurrence rates of the AKR and the direct AMR [cf. *Hashimoto et al.*, 1998].

served in the same frequency range, we have interpreted a part of such enhancement as a radiation from the auroral region. We call such new class of radiation as "auroral myriametric radiation" (AMR). Although the AKR is not always associated with the AMR, the AMR is observed coincidentally with the AKR and has stronger intensity and shorter duration than the trapped continuum radiation. The generation mechanism of the AMR has not been well studied. However, the cyclotron maser instability is not a candidate because the AMR frequency is much lower than the electron cyclotron frequency in the auroral cavity. We think that intense electrostatic waves at a frequency close to electron plasma frequency in the auroral plasma cavity may well be converted to L-O electromagnetic mode through the linear mode conversion process like the nonthermal continuum radiation from the plasmapause. Here we briefly introduce the basic nature of this radiation. Detailed analyses on this radiation are described in *Hashimoto et al.* [1994, 1998].

In order to identify the source region of the AMR, the MLT dependence of the observed location of the AMR is examined [*Hashimoto et al.*, 1998]. The distribution of the occurrence rate of the AMR should be disturbed by the multiple reflection at the magnetopause and not directly correlated to the source location. So *Hashimoto et al.* [1998] only used "direct AMR" with large spin-modulation which was considered to be directly propagated from the Earth without reflections at the magnetopause. The left panel of Figure 10 is an example of the spectrum of the direct AMR observed on March 30, 1995. Clear spin-modulation is seen in frequencies less than 30 kHz. For the statistical analysis the direct AMR is identified by the following criteria: (1) good correlation with AKR enhancement, (2) a frequency range of 2-30 kHz, and (3) the modulation index grater than 0.5. The amount of selected events is 111 from March 1995 to February 1996. The right panels of Figure 10 show the occurrence rates of the AKR and the direct AMR [*Matsumoto et al.*, 1998]. In the latter, the data observed when the Geotail stayed inside the magnetosheath are only used. In Figure 10 the AMR is preferentially observed between 20 and 24 MLT. It is much narrower MLT distribution than AKR, though the locations of peak occurrence rate are same. Such confined distribution supports the idea that the AMR source is located in the auroral plasma cavity around 22 MLT. A possible origin of the less extension of the AMR distribution is its lower frequency, because it is hard for the lower-frequency radiation to propagate into the wide MLT zone. The ray tracing study supports this expectation [*Hashimoto et al.*, 1998].

Statistical studies provided two other suggestions. One is that on the magnetic equator the AMR is more frequently observed in the condition with large Kp index. This might be because in the disturbed phase the shrink of the plasmasphere makes a preferable condition for the propagation to lower latitude region. The other is that the AMR activity is more enhanced in the winter-side hemisphere. The enhancement of activities in the winterside auroral plasma cavity is already suggested in the AKR activity [*Kasaba et al.*, 1997a; *Kumamoto and Oya*, 1998], so that it is another evidence which supports the connection of the AMR to the auroral zone.

4. CONTINUUM RADIATION FROM THE PLASMA SHEET BOUNDARY LAYER

In the magnetotail the spectrum of the trapped continuum is sometimes divided into two parts at about 4–8 kHz which is close to the electron plasma frequency of the plasma sheet. Figure 11 shows a typical example of such spectrum observed by the SFA on September 26, 1992. Geotail was in the distant tail, $(X_{\rm GSM}, Y_{\rm GSM}, Z_{\rm GSM}) = (-135, 8, 14\ R_E)$. The bottom panel is electric field spectrum from 0.5 to 30 kHz at 04:58:30 UT. In the higher frequency part above 8 kHz, the intensity of the radiation is usually steady for several hours. This

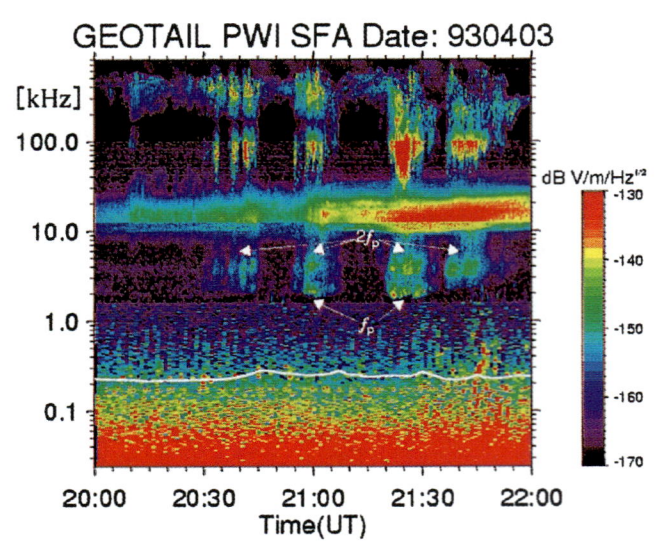

Plate 3. Candidate of the magnetotail $2f_p$ radiation observed by the SFA around 2040, 2100, 2130, and 2145 UT on April 3, 1993.

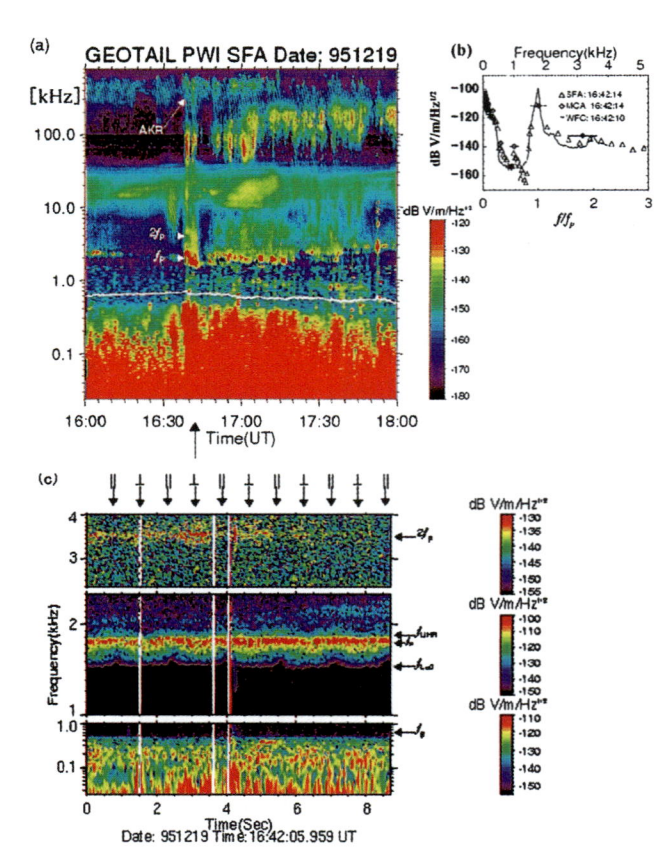

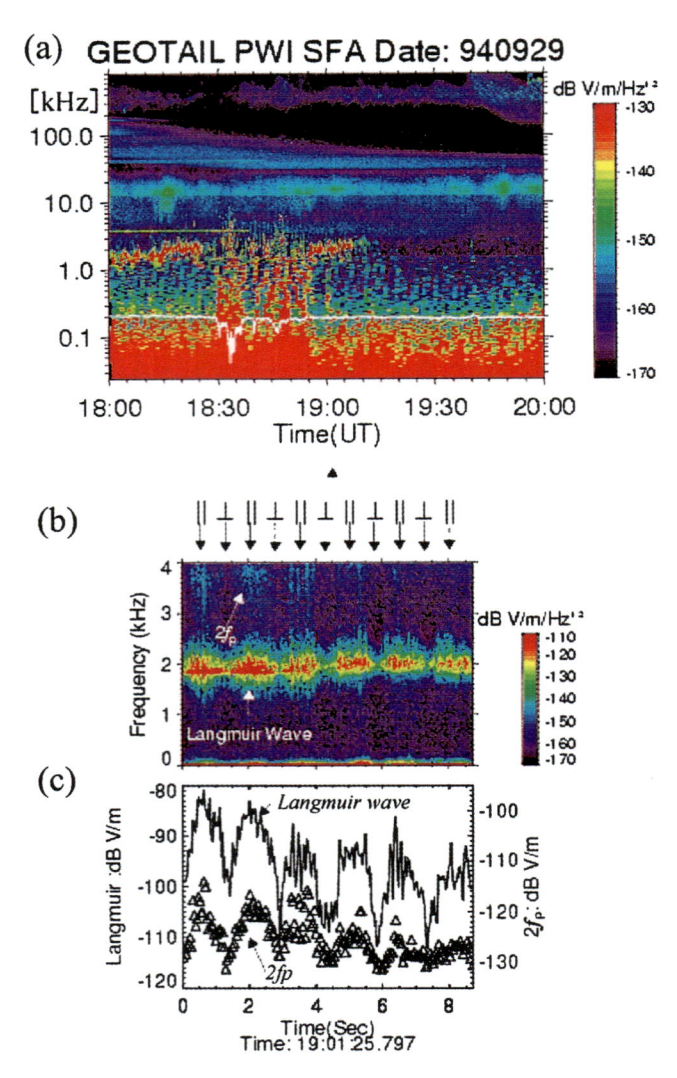

Plate 5. A candidate of electrostatic $2f_p$ waves observed in the magnetotail region by the SFA around 1901 UT on September 29, 1994. (a) Dynamic spectra from 18h to 20h UT observed by the SFA. (b) Dynamic spectra at 19:01:25-33 observed by the WFC. (c) Amplitude at f_p and $2f_p$ in (b).

Plate 4. Spectra of the magnetotail $2f_p$ radiation from 1638 to 1643 UT on December 19, 1995: (a) Dynamic spectra from 16h to 18h UT observed by the SFA. (b) Electric field spectrum at 16:42:14. (c) Dynamic spectra at 16:42:05-13 observed by the WFC.

component is likely propagated from the plasmapause directly and indirectly. In the lower frequency part, however, the amplitude of the radiation is not steady and clearly enhanced associated with substorm activities. The feature of the association with substorms is also found in the AMR (section 3), but the AMR is still brighter above 8 kHz. This radiation component is confined between the electron plasma frequency at the plasma sheet and that in the lobe which is lower than the lowest f_p at the plasmapause. Therefore it could not always be generated at and propagated from the plasmapause.

Figure 12 shows a schematic illustration of the production and propagation of classical "magnetosheath-trapped" continuum radiation and of new "lobe-trapped" continuum radiation. The source region of the former is near the plasmapause, while the source of the latter should be near the PSBL [Nagano et al., 1994]. The physical conditions of the PSBL are similar to those at the plasmapause. Namely, the PSBL also has large density gradient quasi-perpendicular to the magnetic field line and intense electrostatic plasma wave activities. We

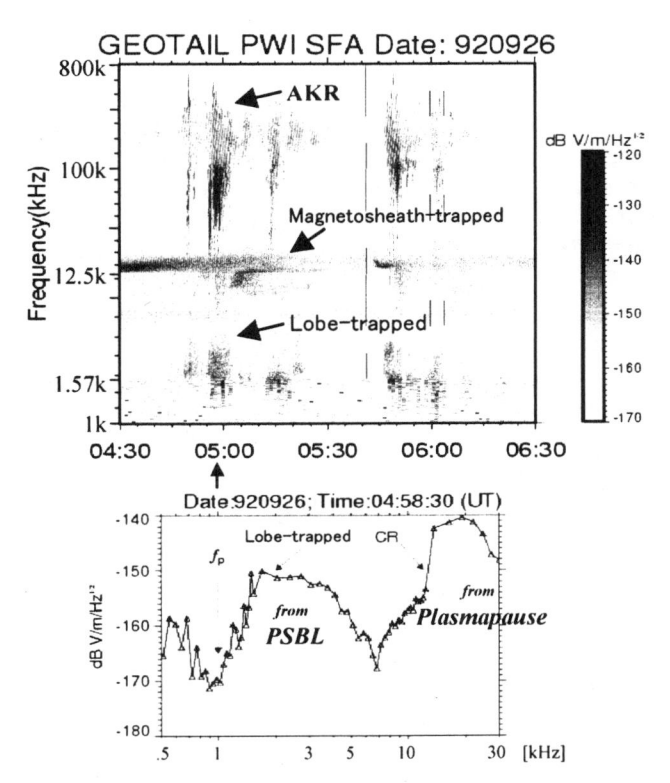

Figure 11. Example of the lobe-trapped continuum radiation observed by the SFA on September 26, 1992. The bottom panel is electric field spectrum from 0.5 to 30 kHz at 04:58:30 UT.

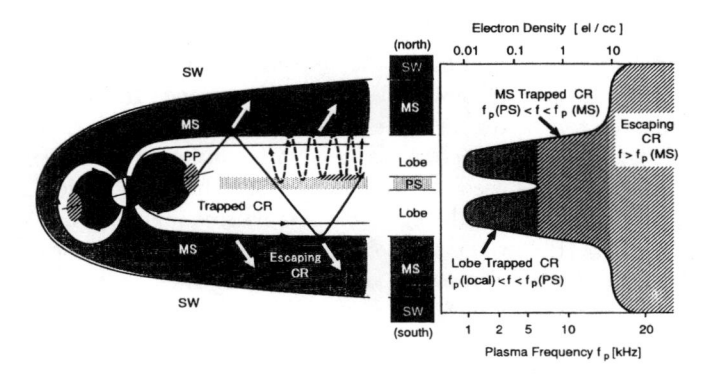

Figure 12. A schematic illustration of the propagation of the magnetosheath-trapped and lobe-trapped continuum radiation. The source region of the former is at the plasmapause, while the source of the latter is in the PSBL [cf. Nagano et al., 1994].

already described in section 2 that in such region electrostatic waves at a frequency close to electron plasma frequency could well be converted to L-O electromagnetic mode near the wall of the steep density gradient through the upper hybrid resonance. And it is known that intense electrostatic waves are frequently enhanced in the PSBL during substorms. Therefore it is expected that such electrostatic waves could produce the lobe-trapped continuum radiation in the PSBL through the same generation process of the classical nonthermal continuum radiation at the plasmapause.

Plate 2 shows four different types of the trapped continuum radiation observed in the tail lobe by the WFC [Matsumoto et al., 1998]. All cases were observed in the distant tail, $(X_{GSM}, Y_{GSM}, Z_{GSM}) = (-120, -4, 5\ R_E)$ in the top one and $(-90, -2, -6\ R_E)$ in the others. The left two panels show the dynamic and frequency spectrum averaged over the 8.7 sec, respectively. The right two panels show the corresponding schematic illustrations. The small arrows on the horizontal scale in the schematic illustrations indicate the timing when the dipole antenna is parallel to the local magnetic field.

The uppermost type in Plate 2, called "Type-A" here, is most often observed. Its dynamic spectrum shows continuous feature above the lower cutoff frequency with a smoothly varying intensity between f_p (f_{L-O}) and f_{R-X} depending on the k-vector of the observed waves (cf. section 2). This feature suggests that the observed radiation component is propagated one from the source distant from the spacecraft. The second type, called "Type-B", has a banded structure with weak discrete lines below the electron plasma frequency (about 800 Hz). These lines are local electron cyclotron harmonic (ECH) waves but do not contribute the radiation component above f_p. This type also suffers from the spin modulation and has the spin-modulated lower cutoff frequency. Therefore Type-B also comes from the distant source region. The third type, called "Type-C", shows clearer discrete emissions with a harmonic structure superimposed on the weak continuous radiation component above the electron plasma frequency (about 900 Hz). Below the local electron plasma frequency local ECH waves are also seen as strong line emissions near $1.5 f_{ce}$ and $2.5 f_{ce}$. The fourth type, "Type-D", has clear waves at the upper hybrid frequency accompanied by ECH waves below the electron plasma frequency. The ECH waves in Type-C and D are of electrostatic nature generated near the PSBL and may well be the source of the lobe-trapped electromagnetic continuum radiation through the linear mode conversion process. It is supported by the fact that the gap between the discrete lines in the lobe-trapped continuum is usually below 1 kHz. This value is corresponding to typical value of f_{ce} in the PSBL and much smaller than the value at the equatorial plasmapause at the distance below 10 R_E from the Earth. This fact is a direct evidence for the hypothesis that the source of the lobe-trapped continuum below 8 kHz is located not at the plasmapause but in the magnetotail.

We suppose that the origin of the difference between Type-A, B, C, and D is the distance of the spacecraft from the radio source. When the spacecraft is too distant from the source (Type-A), the radiation from the PSBL is weak. It does not have clear banded structure because of the mixture of the radiations from the different locations and the reflected ones from the magnetopause. Associated with the decrease of the distance (Type-B), their discrete nature gradually becomes clear. When the spacecraft is very close to the source (Type-C), electromagnetic radio components above f_p include some discrete lines in which the spacing between them agrees with the local electron cyclotron frequency. Just inside the source (Type-D), intense waves at the upper hybrid frequency are observed so that the conversion from the electrostatic waves can occur through the upper hybrid resonance. The distribution of these types and the comparison with the ray tracing studies are now underway with more Geotail data [cf. *Wu et al.*, 1997].

5. $2f_p$ RADIATION FROM THE PLASMA SHEET BOUNDARY LAYER

The $2f_p$ electromagnetic radiation is frequently observed in the upstream region of the Earth's bow shock. It is a narrow band emission at twice the electron

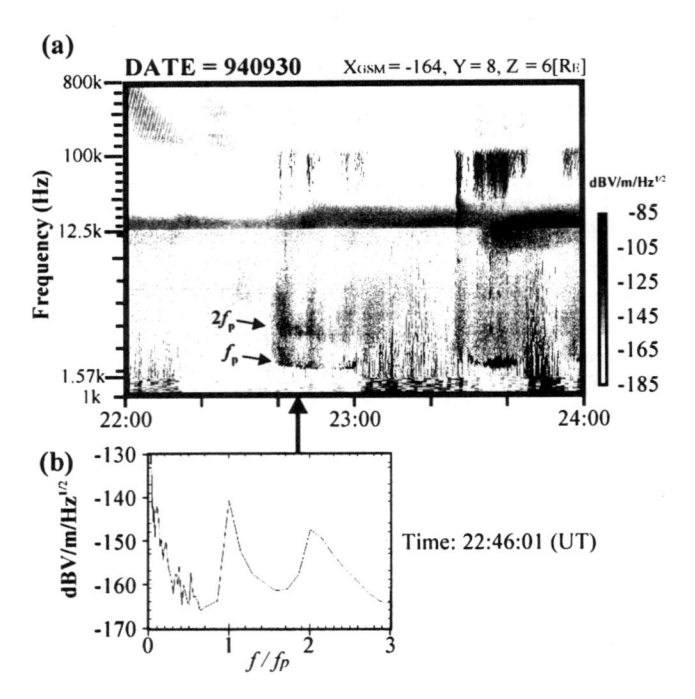

Figure 13. (a) Dynamic spectra of the magnetotail $2f_p$ radiation observed by the SFA from 2240 to 2250 UT on September 30, 1994. (b) Electric field spectrum at 22:46:04.

plasma frequency f_p in the solar wind. The source of this radiation is the electron foreshock on the interplanetary magnetic field (IMF) line tangent to the bow shock [cf. *Hoang et al.*, 1981; *Lacombe et al.*, 1988; *Reiner et al.*, 1996; *Kasaba*, 1997; *Kasaba et al.*, 1997b, 2000]. In this region intense Langmuir waves are excited by energetic electrons backstreaming from the bow shock. The $2f_p$ radiation is thought to be produced through a nonlinear coupling process involving these Langmuir waves. It should be in the same manner as the generation of the type III solar radio burst.

Because Geotail has frequently observed intense Langmuir waves in the plasma sheet boundary layer, we have expected that same excitation process might also be possible in the magnetotail. These Langmuir waves are excited by electron beam accelerated in magnetic field reconnection during substorms. Fortunately, Geotail could find some examples of the enhancement of the $2f_p$ radiation in the magnetotail region associated with substorm activities. Figure 13 shows an example of "magnetotail $2f_p$ radiation" observed in the distant tail lobe on September 30, 1994. In Figure 13 weak enhancement of the AKR is beginning at 2240 UT. The low-frequency radio activity at twice the f_p also starts associated with the one above f_p (about 2 kHz). Figure 13b is the electric field spectrum at 22:46:04 UT.

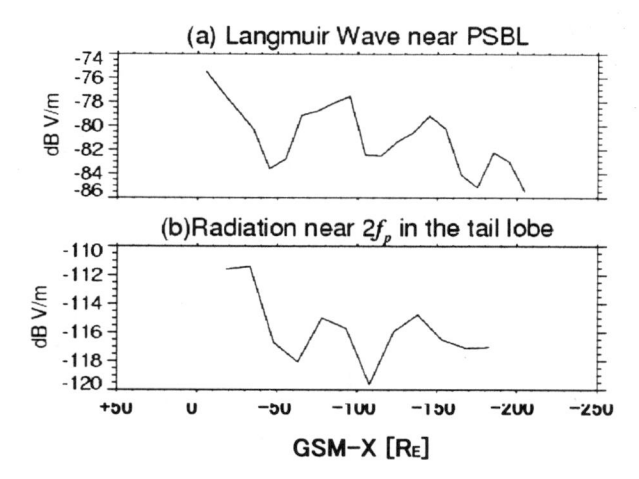

Figure 14. The distribution of the amplitude of (a) Langmuir waves and (b) the $2f_p$ radiation observed in the magnetotail region. The X axis is GSM-X $[R_E]$.

We note that when the SFA is saturated by intense Langmuir waves instrumental noise components are enhanced at the harmonics of f_p. In this case, however, the amplitude at f_p is only about -140 dB V/m/$\sqrt{\text{Hz}}$ which is much lower than the saturation level, so that the activity at $2f_p$ should be real.

Plate 3 shows four candidates of the $2f_p$ radiation observed at 2040, 2100, 2130, and 2145 UT on April 3, 1993. Geotail was in the distant tail, $(X_{\text{GSM}}, Y_{\text{GSM}}, Z_{\text{GSM}}) = (-101, -3, -5\ R_E)$. All wave activities around $2f_p$ are synchronized with the AKR activities. The enhancement of AKR activities is well correlated with substorm activities, so that we can think that plasma activities in the PSBL should be enhanced during the all enhancements of $2f_p$ radio activities. However, the wave activities near f_p do not always appear. It would be difficult for the radiation close to f_p to propagate to the distant region without large density gradient. Therefore the $2f_p$ radiation without the wave activities near f_p is not generated in the region close to the spacecraft but propagated from the distant source. The bandwidth of the magnetotail $2f_p$ radiation is generally wider than the foreshock one. It might be because the electron density in the PSBL is more disturbed than in the upstream region.

In Plate 4 we show polarization of the $2f_p$ radiation observed in the magnetotail. Plate 4 shows dynamic spectra in the magnetotail region on December 19, 1995. Geotail was in the near tail, $(X_{\text{GSM}}, Y_{\text{GSM}}, Z_{\text{GSM}}) = (-26, 7, -4\ R_E)$. The AKR and $2f_p$ radio activities are most enhanced from 1638 to 1643 UT (Plate 4a). In this case the amplitude at f_p is about -100 dB V/m/$\sqrt{\text{Hz}}$ (Plate 4b) which is lower than the saturation level of the SFA. Plate 4c is the dynamic spectrum of the $2f_p$ radiation observed by the WFC at 16:42:05-13. The amplitude of the $2f_p$ component is enhanced when the antenna is perpendicular to the local magnetic field. Because of the instrumental sensitivity limitation we could not identify magnetic field component of the radiation. So we do not neglect the possibility of its electrostatic nature. However, some previous studies suggested that the polarization of electrostatic $2f_p$ waves should be parallel to the mother Langmuir waves [cf. *Klimas*, 1983; *Kasaba*, 1997]. In Plate 4c the weak f_p component without clear spin modulation should not be locally-excited Langmuir waves which might generate the electrostatic $2f_p$. Therefore, we interpret the $2f_p$ component in Plate 4c as a radio wave propagated from the distant source that locates in sunward or tailward. In the latter part of this section we will show a candidate of electrostatic $2f_p$ waves.

We think that the magnetotail $2f_p$ radio activities could give us an opportunity to measure the amount of the global plasma activities in the whole magnetotail region. In order to investigate this hypothesis, we compare the distribution of the amplitude of Langmuir waves observed at the PSBL crossings and $2f_p$ radiation in the magnetotail. The result is in Figure 14. We used data sets from September 18, 1992 to February 25, 1995. In Figure 14a the plane is divided into 20 R_E along the X axis overlapping one another on a 10 R_E grid. On the other hand, in Figure 14b the plane is divided into 15 R_E along the X axis. The samples of the magnetotail $2f_p$ radiation in Figure 14b is 76 events. It is interesting that the Langmuir wave activity in Figures 14a is enhanced where X_{GSM} is below $-40\ R_E$, from -60 to $-100\ R_E$, and from -120 to $-160\ R_E$. This distribution might not be reliable because sampling points are biased by the opportunity of PSBL crossings of the Geotail spacecraft. But we also point out that the region with large $2f_p$ radio activity in Figure 14b is generally overlapped onto the region with large Langmuir wave activity. The sampling method of the $2f_p$ radio flux should not be strongly biased so that such an agreement supports a possibility of the biased distribution of the energetic electron productions in the magnetotail associated with magnetic reconnection during substorms. There are still large ambiguities of this conclusion because the difference of the flux of the $2f_p$ radiation in Figure 14b is small, only 10 dB. We also note that the precise measurement of the amplitude of the $2f_p$ radiation is difficult because the continuum radiation from the plasmapause and from the PSBL is also activated in the same frequency range during substorms. However, we could shown that the magnetotail $2f_p$ radiation had a possibility as a real-time measure of global magnetospheric activities. We will be continue further studies on this point of view.

On the other hand, Geotail also finds some candidates of the electrostatic $2f_p$ component excited by intense Langmuir waves at the vicinity of the spacecraft. Plate 5a is dynamic spectra observed by the SFA from 18h to 20h UT on September 29, 1994. Geotail was in the distant tail, $(X_{GSM}, Y_{GSM}, Z_{GSM}) = (-165, -12, 5\ R_E)$. Before and after the plasma sheet crossings at 1835 and 1847 UT, Geotail observes intense Langmuir wave activities in the PSBL. Plate 5b shows dynamic spectra at 19:01:25-33 observed by the WFC. In this graph Langmuir waves are enhanced at f_p (about 1.8 kHz) when the antenna is parallel to the local magnetic field. And we can also find the enhancement of the $2f_p$ components at about 3.8 kHz which is synchronized with the Langmuir wave activity. Time variation of the amplitude of the f_p and $2f_p$ components for 8.7 s is summarized in Plate 5c. The amplitude of the $2f_p$ component is enhanced when the antenna is parallel to the local magnetic field, and positively correlated with the amplitude of Langmuir waves. The difference of the amplitudes between both waves is about $20 \sim 30$ dB. This might be artificial because the amplitude of Langmuir waves in Plate 5c is about $-80 \sim -90$ dB V/m/$\sqrt{Hz}$ which is close to the saturation level of the PWI preamplifier. However, some proposed theories have predicted the presence of such electrostatic $2f_p$ component associated with intense Langmuir waves [cf. *Klimas*, 1983; *Yoon et al.*, 1994]. And some numerical simulations have generated such electrostatic $2f_p$ waves in the computer space [cf. *Kasaba*, 1997]. However, since the candidate waves are too weak, it is hard to identify the electrostatic nature by magnetic field measurement. Search of the electrostatic $2f_p$ waves is still an open question in our research and will be continued.

In Plate 5b, we also note that the enhancement below 0.1 kHz is also found associated with intense Langmuir waves. Such low frequency component was also observed in the electron foreshock [cf. *Kasaba et al.*, 2000], and might contribute to the excitation of electromagnetic and/or electrostatic $2f_p$ waves through a three-wave process. On the view point of the generation mechanisms of $2f_p$ waves, further progress can be expected by the wave form observation.

6. SUMMARY

In this article we have given an overview of the study of nonthermal continuum-like radiations observed in the magnetotail by the Geotail spacecraft. "Continuum-like radiations" dealt there are the trapped continuum radiation generated from the dayside plasmapause and reflected at the magnetopause, the continuum enhancement from the nightside plasmapause associated with electron injections during substorms, the AMR from the auroral plasma cavity associated with auroral activities, the lobe trapped continuum and the magnetotail $2f_p$ radiation from the plasma sheet boundary layer associated with substorm activities. We have also demonstrated that the continuum-like radiations can bring the information of plasma conditions in their source region and along their propagation paths. Our examples are the density in the magnetosheath, the motion of injected electrons, the location of the plasmapause, and the global plasma activities in the magnetotail during substorms.

The applications of the continuum-like radiations for remote sensing studies are not so easy because these radiations are usually weak and not clear. However, the observation of low-frequency radiations is essential for the study of the magnetosphere of other planets even in the limitation of some technological difficulties (ex. deployment of long antenna) because the simultaneous multi-spacecraft observation by the ISTP-like fleet is still difficult in the orbits around distant planets. Low-frequency radio wave observation could provide an opportunity of expanding the ability of single spacecraft for the studies of global and fast phenomena around planetary objects. Further careful investigations around the Earth is important for the development of the observational technique for such applications.

Acknowledgments. We are grateful to K. Hashimoto, R. R. Anderson, H. Kojima, S. Yagitani, T. Takano, and S. Kudo for doing the analyses of the terrestrial electromagnetic radiations by the Geotail PWI. We express our thanks to T. Okada, M. Tsutsui, I. Kimura, H. Usui, Y-L. Zhang, T. Murata, T. Miyake, K. Ishisaka, and T. Imachi for the discussions on the topics including in this article. We acknowledge S. Kokubun, T. Yamamoto and the Geotail MGF team for the information on the magnetic field. We would also like to thank all the Geotail team members for their collaborations and successful operations.

REFERENCES

Brown, L. W., The galactic radio spectrum between 130 kHz and 2600 kHz, *Astrophys. J., 180,* 359, 1973.

Carpenter, D. L., Whistler studies of the plasmapause in the magnetosphere, 1, Temporal variations in the position of the knee and some evidence on plasma motions near the knee, *J. Geophys. Res., 71,* 693, 1966.

Filbert, P. C., and P. J. Kellogg, Observations of low frequency radio emissions in the Earth's magnetosphere, *J. Geophys. Res., 94,* 8867, 1989.

Gough, M. P., Nonthermal continuum emissions associated with electron injections: Remote plasmapause sounding, *Plant. Space Sci.*, 30, 657, 1982.

Gurnett, D. A., and R. R. Shaw, Electromagnetic radiation trapped in the magnetosphere above the plasma frequency, *J. Geophys. Res.*, 78, 8136, 1973.

Gurnett, D. A., The Earth as a radio source: The nonthermal continuum, *J. Geophys. Res.*, 80, 2751, 1975.

Gurnett, D. A., and L. A. Frank, Continuum radiation associated with low-energy electrons in the outer radiation zone, *J. Geophys. Res.*, 81, 3875, 1976.

Hashimoto, K., H. Matsumoto, H. Kojima, T. Murata, I. Nagano, T. Okada, K. Tsuruda, and T. Iyemori, Auroral myriametric radiation observed by Geotail, *Geophys. Res. Lett.*, 21, 2927, 1994.

Hashimoto, K., S. Kudo, and H. Matsumoto, Source of auroral myriametric radiation observed with Geotail, *J. Geophys. Res.*, 103, 23,475, 1998.

Hoang, S., J. Fainberg, J.-L. Steinberg, R. G. Stone, and R. H. Zwickl, The $2f_p$ circumterrestrial radio radiation as seen from ISEE 3, *J. Geophys. Res.*, 86, 4531, 1981.

Jones, D., Latitudinal beaming of planetary radio emissions, *Nature*, 288, 225, 1980.

Jones, D., Beaming of terrestrial myriametric radiation, *Adv. Space. Res.*, 1, 373, 1981a.

Jones, D., First remote sensing of the plasmapause by terrestrial myriametric radiation, *Nature*, 294, 728, 1981b.

Jones, D., Terrestrial myriametric radiation from the Earth's plasmapause, *Planet. Space. Sci.*, 30, 399, 1982.

Kasaba, Y., Study of radio waves in geospace via spacecraft observations and numerical simulations, *Ph.D. thesis*, Kyoto University, 1997.

Kasaba, Y., H. Matsumoto, K. Hashimoto, and R. R. Anderson, The angular distribution of auroral kilometric radiation observed by the Geotail spacecraft, *Geophys. Res. Lett.*, 24, 2483, 1997a.

Kasaba, Y., H. Matsumoto, and R. R. Anderson, Geotail observation of $2f_p$ emission around the terrestrial foreshock region, *Adv. Space Res.*, 20, 699, 1997b.

Kasaba, Y., H. Matsumoto, K. Hashimoto, R. R. Anderson, J.-L. Bougeret, M. L. Kaiser, X. Y. Wu, and I. Nagano, Remote sensing of the plasmapause during substorm: Geotail observation of the continuum enhancement, *J. Geophys. Res.*, 103, 20,389, 1998.

Kasaba, Y., H. Matsumoto, R. R. Anderson, T. Mukai, Y. Saito, T. Yamamoto, and S. Kokubun, Statistical studies of plasma waves and backstreaming electrons in the terrestrial electron foreshock observed by Geotail, *J. Geophys. Res.*, in printing, 2000.

Klimas, A. J., A mechanism for plasma waves at the harmonics of the plasma frequency in the electron foreshock boundary, *J. Geophys. Res.*, 88, 9081, 1983.

Kumamoto, A., and H. Oya, Asymmetry of occurrence-frequency and intensity of AKR between summer polar region and winter polar region sources, *Geophys. Res. Lett.*, 25, 2369, 1998.

Kurth, W. S., M. M. Baumback, and D. A. Gurnett, Direction-finding measurements of auroral kilometric radiation, *J. Geophys. Res.*, 80, 2764, 1975.

Kurth, W. S., D. A. Gurnett, and R. R. Anderson, Escaping nonthermal continuum radiation, *J. Geophys. Res.*, 86, 5519, 1981.

Kurth, W. S., Continuum radiation in planetary magnetospheres, *Proceedings of the 3rd International Workshop on Radio Emissions from Planetary Magnetospheres*, in Graz, Austria, 1991.

Lacombe, C., C. C. Harvey, S. Hoang, A. Mangeney, J.-L. Steinberg, and D. Burgess, ISEE observations of emission at twice the solar wind plasma frequency, *Ann. Geophys.*, 6, 113, 1988.

Manning, R., and J. Fainberg, A new method of measuring radio source parameters of a partially polarized distributed source from spacecraft observations, *Space Sci. Instrum.*, 5, 161, 1980.

Matsumoto, H., I. Nagano, R. R. Anderson, H. Kojima, K. Hashimoto, M. Tsutsui, T. Okada, I. Kimura, Y. Omura, and M. Okada, Plasma wave observations with Geotail spacecraft, *J. Geomagn. Geoelectr.*, 46, 59, 1994.

Matsumoto, H., H. Kojima, and Y. Omura, Plasma waves in Geospace: Geotail observations, *Geophysical Monograph*, 105, "New perspective on the Earth's magnetotail", 259, 1998.

Nagano, I., S. Yagitani, H. Kojima, Y. Kakehi, T. Shiozaki, H. Matsumoto, K. Hashimoto, T. Okada, S. Kokubun and T. Yamamoto, Wave form analysis of the continuum radiation observed by Geotail, *Geophys. Res. Lett.*, 21, 2911, 1994.

Nagano, I., X.-Y. Wu, S. Yagitani, H. Takano, and H. Matsumoto, Remote sensing the magnetosheath by the spin modulation of terrestrial continuum radiation, *Geophys. Res. Lett.*, submitted, 2000.

Okuda, H., M. Ashour-Abdalla, M. S. Chance, and W. S. Kurth, Generation of nonthermal continuum radiation in the magnetosphere, *J. Geophys. Res.*, 87, 10,457, 1982.

Reiner, M. J., M. L. Kaiser, J. Fainberg, M. D. Desch, and R. G. Stone, $2f_p$ radio emission from the vicinity of the Earth's foreshock: Wind observations, *Geophys. Res. Lett.*, 23, 1247, 1996.

Wu, X. Y., I. Nagano, S. Yagitani, T. Imachi, T. Ogino, T. Murata, and H. Matsumoto, Propagation of continuum radiation in the magnetosphere, in *Proc. 5th International School/Sympo. Space Simulations*, Uji, Kyoto, 291, 1997.

Yoon, P. H., C. S. Wu, A. F. Vinus, M. J. Reiner, J. Fainberg, and R. G. Stone, Theory of $2\omega_{pe}$ radiation induced by the bow shock, *J. Geophys. Res.*, 99, 23,481, 1994.

Y. Kasaba, Center for Planning and Information Systems, Institute of Space and Astronautical Science, Sagamihara, Kanagawa 229-8510, Japan. (kasaba@stp.isas.ac.jp)

H. Matsumoto, Radio Atmospheric Science Center, Kyoto University, Uji, Kyoto 611-0011, Japan. (matsumot@kurasc.kyoto-u.ac.jp)

I. Nagano, Department of Electrical and Computer Engineering, Kanazawa University, Kanazawa 920-8667, Japan. (nagano@ec.t.kanazawa-u.ac.jp)

Terrestrial LF Bursts: Escape Paths and Wave Intensification

Michael D. Desch and William M. Farrell

NASA/Goddard Space Flight Center, Greenbelt, MD

The WAVES instrument on Wind not infrequently observes relatively brief, low-frequency drifting radio bursts (LF bursts), notable for their apparent propagation through the electron plasma frequency cutoff in the Earth's magnetosheath and sometimes extending down nearly to the local plasma frequency of the solar wind itself. We identify for the first time that the characteristic group-velocity dispersion of LF bursts occurs in two separate frequency bands, along with considerable wave intensification. We use these observations to help identify separate frequency-dependent paths of escape of the waves from the magnetosphere. Waves near the solar wind plasma frequency must escape from the very deep tail; waves near the magnetosheath plasma frequency must escape from near the nose. The intensifications are a natural consequence of wave focussing into the boundary normal direction. Finally, an unusual LF event is identified that independently confirms a deep-tail escape of the lowest frequency portion of the waves.

1. INTRODUCTION

At Earth, commonly occurring low frequency magnetospheric emission phenomena include the auroral kilometric radiation, or *AKR* [Gurnett, 1974], nonthermal trapped continuum [Gurnett, 1975], nonthermal drifting continuum [Gough, 1982], and auroral myriametric radiation, or *AMR* [Louarn, et al., 1994; Hashimoto, et al., 1998]. A spacecraft such as Wind, making observations outside Earth's magnetosphere, can easily detect AKR and drifting continuum because these emissions are usually above the 30–50 kHz electron plasma frequency, f_{ms}, of the frontside magnetosheath which acts as an effective shield to lower frequency emissions that might otherwise escape into the solar wind. Trapped continuum (by definition) and AMR are only observable inside the magnetosphere as they are always below the magnetosheath cutoff.

Between these two extremes of trapped and freely-escaping emissions lies a relatively new class of emissions: *LF bursts* [Steinberg et al., 1988]. When observed from outside the magnetosphere, the low-frequency envelope of AKR is cutoff sharply at f_{ms}, as one would expect. However, the distinguishing feature of LF bursts is that they appear to penetrate this stop band, and are often clearly observed nearly to the local (spacecraft) solar wind plasma frequency, a factor of two lower than f_{ms}.

LF bursts as observed by Wind/WAVES in the solar wind are described by Kaiser et al. [1996]. They are usually, but not always, associated with AKR [Steinberg et al., 1990; Desch, 1996]. Direction finding studies with ISEE-3 indicate that the burst source appears large, that is, lacks significant spacecraft spin modulation, and for this reason was referred to by Steinberg et al. [1988] as isotropic terrestrial kilometric radiation, or *ITKR*. Importantly, the apparent direction of the LF burst source does not coincide with the AKR source direction [Steinberg et al., 1990], despite being generally associated in time. The bursts always occur in association with magnetospheric substorm expansion phase onsets [Steinberg, 1990; Anderson et al., 1996] and in a statistical sense are triggered by southward

Radio Astronomy at Long Wavelengths
Geophysical Monograph 119
This paper not subject to U.S. copyright
Published in 2000 by the American Geophysical Union

turnings of the interplanetary magnetic field (IMF) [Desch, 1996]. If one interprets the frequency drift onset of the bursts as due to group velocity dispersion of the waves through the magnetosheath [Desch, 1994], then the source of the emission must be at least 100 – 200 R_e down the tail [Desch, 1996]. Scattering of the waves to account for source size leads to a source ~2000 R_e down the tail [Steinberg et al., 1988, 1990], while more recent modelling of the downstream bow shock leads to source locations nearer to ~230 R_e [Steinberg et al., 1998], the downstream location where the sheath plasma density (i.e., frequency) is low enough to permit LF emission to escape to free space. This location is in closer agreement with the dispersion result.

LF bursts are not periodic on short time scales, unlike QP bursts which were first observed at Jupiter [Kurth et al., 1989; Kaiser et al., 1992]. The Jovian QP (quasi-periodic) bursts appear very similar to LF bursts in that they also have a frequency-drifting burst onset and are driven by the solar wind [MacDowall et al., 1993]; however, the QP bursts also manifest a quasi-periodic tendency to recur about every 15 minutes (Jupiter dayside) and every 40 minutes (Jupiter evening side). The frequency-drifting burst onset of QP bursts is due to group velocity dispersion in the Jovian magnetosheath [Desch, 1994]. The periodic versus aperiodic nature of the QP and LF bursts, respectively, almost certainly has to do with an important difference, not yet identified, between the way in which the terrestrial and Jovian magnetospheres drive radio emissions.

2. OBSERVATIONS

An example of LF burst emission on a particularly active day is shown in Plate 1. There are at least 5 separate and distinct bursts over a 16-hour span. It is clear that the bursts resemble solar type III bursts and Jovian QP bursts in their tendency to drift negatively with time. Although generated within the magnetosphere, they extend well below the plasma frequency of the magnetosheath (identified in the figure as $\sim f_{ms}$). There is no short-term recurrence pattern, only a tendency for LF bursts to recur during periods of enhanced solar wind activity, following sector boundary passages [Desch et al., 1996].

2.1 Two-Band Dispersion and Intensification

Plate 2 shows high-resolution spectrograms recorded on 10 March, 1998 (left) and 28 January, 1997 (right). The drifting onset that is characteristic of LF bursts is clearly evident. This drifting onset is presumably due to group velocity dispersion caused by propagation of the wave through a region close to the wave cutoff condition where the wave frequency, ω, is close to the local plasma frequency, ω_p, so that the index of refraction, n, is close to 0. Since the wave group velocity

$$v_g = c\left(1 - (\omega_p/\omega)^2\right)^{1/2} = cn \qquad (1)$$

then as $\omega \to \omega_p$, $v_g \to 0$ in a highly frequency-dependent fashion and significant retardation of the arrival time of the wave is observed at the lowest frequencies. The dispersion always begins at a frequency just above f_{ms} as is evident from the fact that the frequency drifting appears just above where the burst cuts through the '$2fp$' line (i.e., $\sim f_{ms}$). This point is labelled 'dispersive onset' in Plate 2.

Closer examination of Plate 2 shows a newly-discovered feature of LF bursts. Below f_{ms}, the dispersion stops. Here, centered at a frequency of about 50 kHz (left panel) and 25 kHz (right), the burst onset is frequency independent. Then as the frequency of the burst decreases still further, dispersion begins again as the wave frequency approaches the solar wind plasma frequency, f_{sw}, (labelled 'SW Dispersion'). Thus there appear to be two frequency bands within which dispersion takes place: the first in a band corresponding to propagation and dispersion within the magnetosheath, and the second in a band close to the in situ solar wind plasma frequency.

Also apparent in Plate 2 is what appears to be significant intensification of the LF burst emission in these same two bands where the dispersion is occurring. These enhancements are readily apparent in Figure 1 where the wave intensity as a function of frequency is shown at 1816 GMT for the March 10 event. For comparison, the background noise level (dark shading) at 1800 GMT just prior to the start of the burst shows the f_{sw} and f_{ms} line peaks. It is clear that there are strong intensity enhancements that occur in a band just above the two cutoff frequencies.

2.2 Direct Evidence of LF Burst Escape Point

As has been noted, the free escape of LF burst emission to space at frequencies below f_{ms} is problematic. Steinberg et al. [1998] show that wave escape from a distance about 250 R_E down the tail is consistent with sheath densities at that distance which would permit escape at the lowest frequencies at which ISEE observed LF bursts. As it happens, we have completely independent confirmation of the deep-tail escape of LF bursts owing to an unusual circumstance in the solar wind observed by WIND on 27 June, 1997 (Plate 3). Here we show an LF burst with an unusually high low-frequency cutoff (65 kHz) relative to the solar wind plasma cutoff (45 kHz) at Wind at the time of the burst. At a minimum, one might expect a cutoff at least 70% of the way between f_{ms} and f_{sw} [e.g. Stein-

Terrestrial LF Bursts: Escape Paths and Wave Intensification

Michael D. Desch and William M. Farrell

NASA/Goddard Space Flight Center, Greenbelt, MD

The WAVES instrument on Wind not infrequently observes relatively brief, low-frequency drifting radio bursts (LF bursts), notable for their apparent propagation through the electron plasma frequency cutoff in the Earth's magnetosheath and sometimes extending down nearly to the local plasma frequency of the solar wind itself. We identify for the first time that the characteristic group-velocity dispersion of LF bursts occurs in two separate frequency bands, along with considerable wave intensification. We use these observations to help identify separate frequency-dependent paths of escape of the waves from the magnetosphere. Waves near the solar wind plasma frequency must escape from the very deep tail; waves near the magnetosheath plasma frequency must escape from near the nose. The intensifications are a natural consequence of wave focussing into the boundary normal direction. Finally, an unusual LF event is identified that independently confirms a deep-tail escape of the lowest frequency portion of the waves.

1. INTRODUCTION

At Earth, commonly occurring low frequency magnetospheric emission phenomena include the auroral kilometric radiation, or *AKR* [Gurnett, 1974], nonthermal trapped continuum [Gurnett, 1975], nonthermal drifting continuum [Gough, 1982], and auroral myriametric radiation, or *AMR* [Louarn, et al., 1994; Hashimoto, et al., 1998]. A spacecraft such as Wind, making observations outside Earth's magnetosphere, can easily detect AKR and drifting continuum because these emissions are usually above the 30–50 kHz electron plasma frequency, f_{ms}, of the frontside magnetosheath which acts as an effective shield to lower frequency emissions that might otherwise escape into the solar wind. Trapped continuum (by definition) and AMR are only observable inside the magnetosphere as they are always below the magnetosheath cutoff.

Radio Astronomy at Long Wavelengths
Geophysical Monograph 119
This paper not subject to U.S. copyright
Published in 2000 by the American Geophysical Union

Between these two extremes of trapped and freely-escaping emissions lies a relatively new class of emissions: *LF bursts* [Steinberg et al., 1988]. When observed from outside the magnetosphere, the low-frequency envelope of AKR is cutoff sharply at f_{ms}, as one would expect. However, the distinguishing feature of LF bursts is that they appear to penetrate this stop band, and are often clearly observed nearly to the local (spacecraft) solar wind plasma frequency, a factor of two lower than f_{ms}.

LF bursts as observed by Wind/WAVES in the solar wind are described by Kaiser et al. [1996]. They are usually, but not always, associated with AKR [Steinberg et al., 1990; Desch, 1996]. Direction finding studies with ISEE-3 indicate that the burst source appears large, that is, lacks significant spacecraft spin modulation, and for this reason was referred to by Steinberg et al. [1988] as isotropic terrestrial kilometric radiation, or *ITKR*. Importantly, the apparent direction of the LF burst source does not coincide with the AKR source direction [Steinberg et al., 1990], despite being generally associated in time. The bursts always occur in association with magnetospheric substorm expansion phase onsets [Steinberg, 1990; Anderson et al., 1996] and in a statistical sense are triggered by southward turnings of the interplanetary magnetic field (IMF) [Desch, 1996]. If one interprets the frequency drift onset of the bursts as due to group velocity dispersion of the waves through the magnetosheath [Desch, 1994], then the source of the emission must be at least 100 – 200 R_e down the tail [Desch, 1996]. Scattering of the waves to account for source size leads to a source ~2000 R_e down the tail [Steinberg et al., 1988, 1990], while more recent modelling of the downstream bow shock leads to source locations nearer to ~230 R_e [Steinberg et al., 1998], the downstream location where the sheath plasma density (i.e., frequency) is low enough to permit LF emission to escape to free space. This location is in closer agreement with the dispersion result.

LF bursts are not periodic on short time scales, unlike QP bursts which were first observed at Jupiter [Kurth et al., 1989; Kaiser et al., 1992]. The Jovian QP (quasi-periodic) bursts appear very similar to LF bursts in that they also have a frequency-drifting burst onset and are driven by the solar wind [MacDowall et al., 1993]; however, the QP bursts also manifest a quasi-periodic tendency to recur about every 15 minutes (Jupiter dayside) and every 40 minutes (Jupiter evening side). The frequency-drifting burst onset of QP bursts is due to group velocity dispersion in the Jovian magnetosheath [Desch, 1994]. The periodic versus aperiodic nature of the QP and LF bursts, respectively, almost certainly has to do with an important difference, not yet identified, between the way in which the terrestrial and Jovian magnetospheres drive radio emissions.

2. OBSERVATIONS

An example of LF burst emission on a particularly active day is shown in Plate 1. There are at least 5 separate and distinct bursts over a 16-hour span. It is clear that the bursts resemble solar type III bursts and Jovian QP bursts in their tendency to drift negatively with time. Although generated within the magnetosphere, they extend well below the plasma frequency of the magnetosheath (identified in the figure as ~f_{ms}). There is no short-term recurrence pattern, only a tendency for LF bursts to recur during periods of enhanced solar wind activity, following sector boundary passages [Desch et al., 1996].

2.1 Two-Band Dispersion and Intensification

Plate 2 shows high-resolution spectrograms recorded on 10 March, 1998 (left) and 28 January, 1997 (right). The drifting onset that is characteristic of LF bursts is clearly evident. This drifting onset is presumably due to group velocity dispersion caused by propagation of the wave through a region close to the wave cutoff condition where the wave frequency, ω, is close to the local plasma frequency, ω_p, so that the index of refraction, n, is close to 0. Since the wave group velocity

$$v_g = c\left(1 - (\omega_p/\omega)^2\right)^{1/2} = cn \quad (1)$$

then as $\omega \to \omega_p$, $v_g \to 0$ in a highly frequency-dependent fashion and significant retardation of the arrival time of the wave is observed at the lowest frequencies. The dispersion always begins at a frequency just above f_{ms} as is evident from the fact that the frequency drifting appears just above where the burst cuts through the '$2fp$' line (i.e., ~f_{ms}). This point is labelled 'dispersive onset' in Plate 2.

Closer examination of Plate 2 shows a newly-discovered feature of LF bursts. Below f_{ms}, the dispersion stops. Here, centered at a frequency of about 50 kHz (left panel) and 25 kHz (right), the burst onset is frequency independent. Then as the frequency of the burst decreases still further, dispersion begins again as the wave frequency approaches the solar wind plasma frequency, f_{sw}, (labelled 'SW Dispersion'). Thus there appear to be two frequency bands within which dispersion takes place: the first in a band corresponding to propagation and dispersion within the magnetosheath, and the second in a band close to the in situ solar wind plasma frequency.

Also apparent in Plate 2 is what appears to be significant intensification of the LF burst emission in these same two bands where the dispersion is occurring. These enhancements are readily apparent in Figure 1 where the wave intensity as a function of frequency is shown at 1816 GMT for the March 10 event. For comparison, the background noise level (dark shading) at 1800 GMT just prior to the start of the burst shows the f_{sw} and f_{ms} line peaks. It is clear that there are strong intensity enhancements that occur in a band just above the two cutoff frequencies.

2.2 Direct Evidence of LF Burst Escape Point

As has been noted, the free escape of LF burst emission to space at frequencies below f_{ms} is problematic. Steinberg et al. [1998] show that wave escape from a distance about 250 R_E down the tail is consistent with sheath densities at that distance which would permit escape at the lowest frequencies at which ISEE observed LF bursts. As it happens, we have completely independent confirmation of the deep-tail escape of LF bursts owing to an unusual circumstance in the solar wind observed by WIND on 27 June, 1997 (Plate 3). Here we show an LF burst with an unusually high low-frequency cutoff (65 kHz) relative to the solar wind plasma cutoff (45 kHz) at Wind at the time of the burst. At a minimum, one might expect a cutoff at least 70% of the way between f_{ms} and f_{sw} [e.g. Stein-

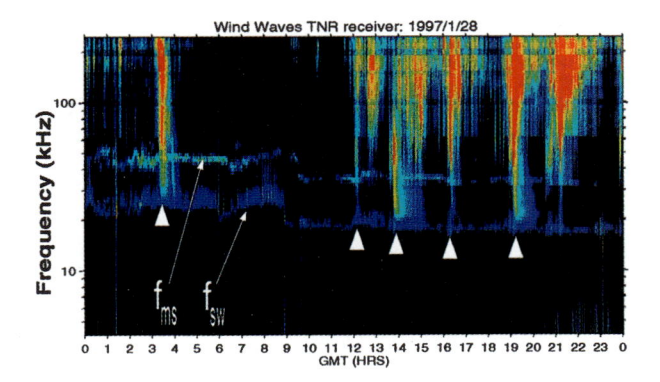

Plate 1. WAVES 24–hour dynamic spectrum from the TNR receiver on Wind, 28 January 1997. Five clear examples of LF bursts are indicated by filled triangles. The plasma frequency (f_{sw}) line from the solar wind and the magnetosheath plasma frequency ($f_{ms} \sim 2 \times f_{sw}$) line from the electron foreshock region are labelled. The LF bursts are seen to extend well below f_{ms}, nearly to f_{sw}.

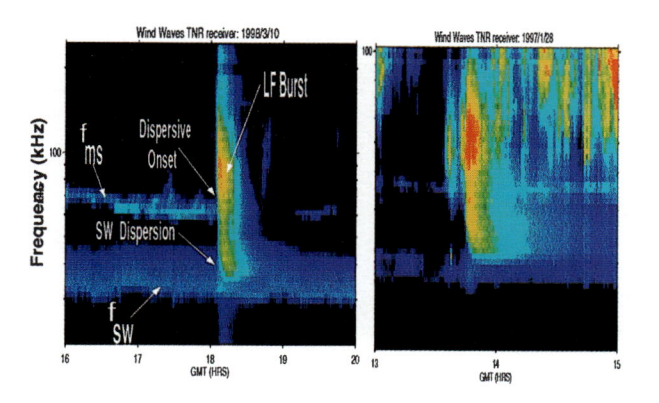

Plate 2. Two WAVES dynamic spectra showing LF-burst dispersion effects. The left panel is from 16 to 20 hours on 10 March 1998, the right panel from 13 – 15 hours on 28 January 1997. Both LF bursts manifest dual-frequency dispersion: first at the magnetosheath plasma frequency, f_{ms} ('Dispersive Onset'), and then near the solar wind plasma frequency ('SW Dispersion'), f_{sw}. Between these two frequency bands there is virtually no dispersion.

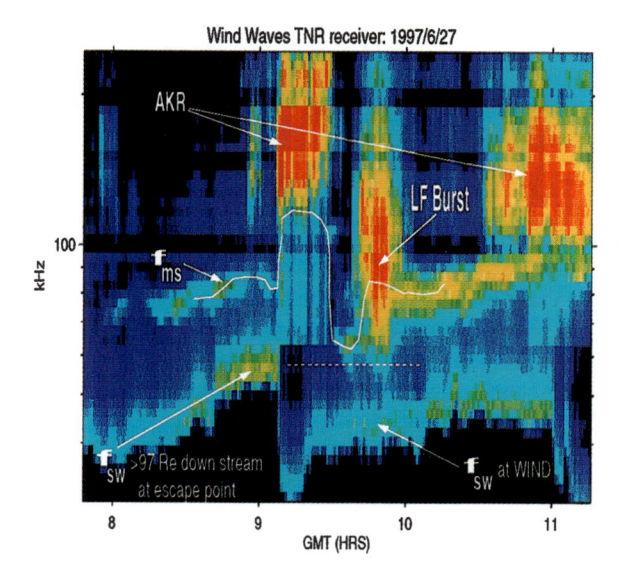

Plate 3. LF burst with unusually high low-frequency cutoff caused by plasma enhancement observed at WIND between 0843 and 0908 GMT. WIND was 80 R_e upstream at this time. By the time of the burst, propagation of the enhancement past Earth places it between 97 and 191 R_e downstream of Earth, effectively locating the escape path of the lowest frequency portion of the waves at $X < -97\ R_e$.

berg et al., 1998], which would place the cutoff at about 55 kHz. (In fact we often observe much lower cutoffs with WAVES, approaching 80–90%.) We will show that this cutoff is due to the absorption of the lowest frequency portion of the burst by an intervening solar wind plasma density enhancement whose location places a definitive limit on the possible escape point of the burst.

In the analysis surrounding Plate 3, we make no assumptions about the falloff of the sheath density away from the nose of the bow shock, and instead use the simple geometry of the density enhancement location relative to Earth's magnetosphere to localize the LF burst escape path. The leading edge of the solar wind density spike is observed at 0843 GMT at Wind (which was located 80 R_e upstream of Earth). The LF burst ends at 0954 GMT, by which time the density enhancement has travelled, at the measured solar wind speed of 400 km/s, $\sim$271 R_e, placing the leading edge of the enhancement $\sim$191 R_e downstream of Earth. Similarly, the trailing edge of the density spike observed at 0908 GMT is $\sim$97 R_e downstream of Earth by 0954 GMT. It is this density enhancement that cuts off the low frequency edge of the LF burst (see dotted line in Plate 3); therefore the escape point must be at least 97 R_e downstream of Earth. If the escape point were closer to the nose than this, the lowest frequency portion of the LF burst would not have encountered the plasma enhancement and the cutoff would be much lower, as is normally observed outside of the unusual circumstances that prevail in this particular case.

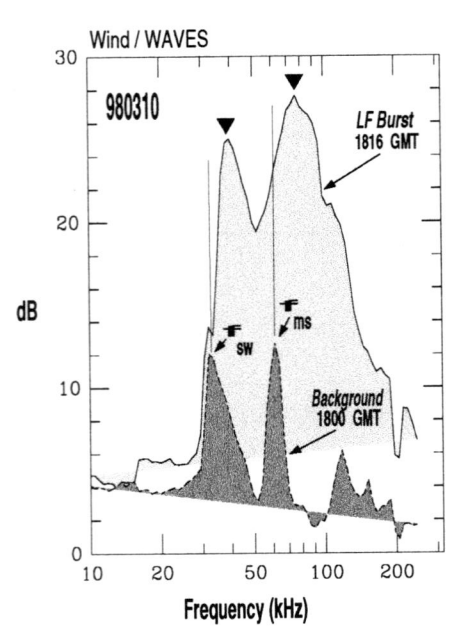

Figure 1. Spectral cuts (intensity in dB vs. frequency) at two separate times (1800 and 1816 GMT) through the 10 March dynamic spectrum of Plate 2. The 1800 GMT spectrum (dark shading) shows the background values just before the LF burst onset. The two significant peaks are the solar wind line at f_{sw} and twice this frequency corresponding to f_{ms}. The second spectral cut (light shading) is through the LF burst at 1816 GMT. This spectrum shows two major intensifications of the LF burst emission (filled triangles) at frequencies just at and above f_{sw} and f_{ms} where most of the dispersion takes place.

3. LF BURST ESCAPE PATHS

The observation of dispersion and wave intensifications at frequencies close to f_{sw} and f_{ms} shows conclusively that the LF burst emission profile, both with regard to its drifting nature (Plate 2) and its intensity spectrum (Figure 1) are determined largely, if not entirely, by propagation effects and not by the burst emission process itself. The burst source, wherever it lies in the magnetosphere, cannot have information about the solar wind plasma frequency and the sheath plasma frequency at the bow shock. One might object that if the source is generated in the magnetosheath close to the nose, then the emission process would reflect plasma conditions there; however, this is not likely since LF burst emission often extends up to hundreds of kHz and so is almost certainly not generated anywhere within the sheath.

Having established with some certainty that the burst characteristics are a consequence of propagation effects, there remains a problem in understanding the precise nature of the escape path of the emission out of the magnetosphere. Leaving aside the question of the source location, suppose the emission escapes from down the tail ($X_{GSE} \sim -250 R_e$) at location A (see Figure 2) where the burst wave frequency $f \sim f_{sw}$ [e.g. Steinberg et al., 1998] and consistent with the data shown in Plate 3. This scenario adequately accounts in general for the escape of the emission from the magnetosphere to 'free space'. In addition, a deep tail escape route can account for the LF dispersion observed near f_{sw} since the ray path through the solar wind to Wind could sometimes be $\gg$500 R_e plus whatever additional dispersion takes place within the sheath itself.

By itself, however, the deep-tail escape route (A in Figure 2) does not explain the quite separate and distinct dispersion observed near f_{ms}, that is, near twice f_{sw}. Dispersion at this frequency can only occur for waves with $f \sim f_{ms}$, that is for rays that propagate through the magnetosheath and bow shock near $Y_{GSE} \sim 0$, corresponding to location B in Figure 2, directly upstream of Earth.

Thus we are forced to admit to a somewhat ad hoc mechanism involving two separate propagation paths, as depicted in Figure 2. The first path deep down the tail is needed to allow for escape of waves with $f \sim f_{sw}$, and to explain the high-frequency cutoff observations shown in Plate 3. The second path near the nose is needed to account for the separate and distinct dispersion observed near f_{ms}. This dual-path escape route, besides being somewhat contrived, also begs the question of uniqueness. Location A is not unique because while the sheath plasma density there permits wave escape at the lowest frequency that can possibly propagate through the solar wind to the spacecraft, it is also quite possible that the actual point of escape is much deeper down the tail, perhaps at a point where the magnetosheath effectively no longer exists and merges indistinguishably with the solar wind. There is also nothing special about location B. On the face of it, B is no different from any other region of the magnetosheath between A and B. In other words, there is no apparent reason why there should not be escape of LF emission at every point between A and B where the wave frequency is greater than the local plasma frequency at that point in the sheath. That this does not seem to occur is apparent from the observations which fail to manifest either dispersion or wave intensification over a broad range of frequencies between f_{sw} and f_{ms} (e.g., Plate 2).

3.1 Explanation of Dual-Band Dispersion

The solution to this apparent paradox is suggested in Figure 3 where we show two hypothetical wave paths from

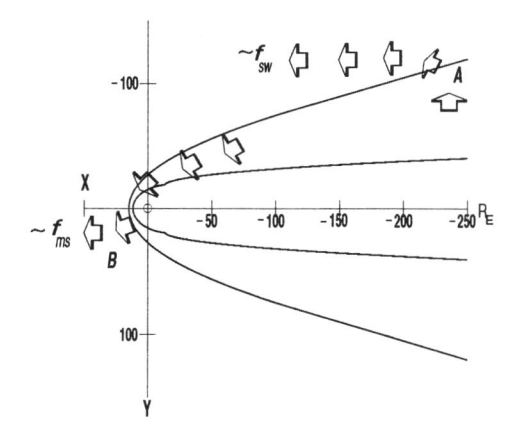

Figure 2. Meridian plane (xy) plot of terrestrial magnetosphere showing possible LF emission 'escape routes' into the solar wind in order to explain observed dual-frequency dispersion. At A, lowest frequency portion of burst escapes at point where magnetosheath plasma frequency is below $\sim$30 kHz, resulting in dispersion through a portion of sheath and also along a long path through solar wind. At B, the nose of the magnetosheath, the wave escapes after ducting down the sheath and finally encountering plasma gradient quasi-parallel to the wave vector direction.

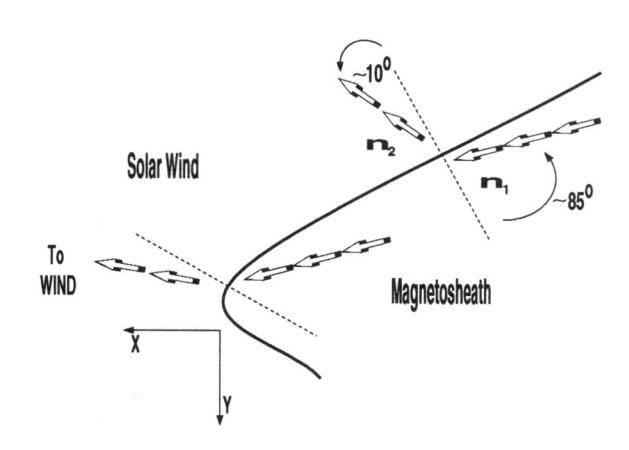

Figure 3. Sketch in the xy plane of the path of LF burst emission across the bow shock based on a simple application of Snell's law across a boundary. Away from the nose of the bow shock, waves near f_{ms} incident at large angles (here 85°) to the boundary normal (dashed line) will exit at a very small angle to the boundary normal, leading to waves that are only visible to spacecraft far from the Earth-Sun line (Y>X). Nearer to the nose, such waves also escape close to the normal direction, but this direction now corresponds to the Wind spacecraft direction throughout most of its orbit (X>Y).

the magnetosheath, across the bow shock boundary to free space. Refraction across the boundary is determined by Snell's law, $n_1 \sin\theta_1 = n_2 \sin\theta_2$, where n_1 and n_2 are the indices of refraction of the sheath and solar wind, respectively, and θ_1 and θ_2 are the wave angles relative to the boundary normals. For significant dispersion to occur (at least 30 sec over a path length of 250 R_e), the wave frequency, f, must be within about 1% of the plasma frequency in the sheath, $f/f_{ms} \sim 1.01$, which leads to the condition, $n_1/n_2 \ll 1$. With θ_1 close to 90° (taken here as 85°), the *group-velocity-dispersed* LF waves will escape at very small angles ($\leq 10°$) to the boundary normal. Along the flanks of the magnetosheath, the normal direction is such that only spacecraft far from the Earth-Sun line would observe such emissions. However, near the nose, the dispersed waves will escape in the general direction of Wind because that is also approximately the boundary normal direction. LF waves can and will escape from far along the flanks and still be observed by Wind; however, only *undispersed* waves will be so observed because undispersed waves, with $f/f_{ms} \gtrsim 1.2$, will have $\theta_2 \gtrsim 45°$. Thus the hyperboloid shape of the bow shock, with boundary normal in the direction of Wind only near the nose, explains the observation of dispersion near f_{ms} distinct from that near f_{sw}.

In this scenario, the escape point for waves near f_{sw} is still far back ($X_{GSE} \leq -250$ R$_E$) along the flanks of the magnetosheath, as discussed earlier, where the shock is very weak [Greenstadt et al., 1990], and $f_{ms} \sim f_{sw}$. Quantitatively, we assume that the sheath density is within about 1% of the solar wind density, and that $f/f_{ms} \sim f/f_{sw} \sim 1.10$. This agrees generally with the more extreme Wind observation of LF burst drift down to within about 10% of f_{sw} (the 30% figure of the ISEE observations [Steinberg et al., 1998] is easier to fit). Under these circumstances, the wave angle θ_2 relative to the boundary normal is about 55°, leading to wave escape within about 9° of the Earth-Sun line given the geometry of the bow shock. Relaxing the requirement that the bursts drift to within 10% of f_{sw} leads to wave propagation even closer to the Earth-Sun line. In this context we note that one of the most extreme cases seen by Wind of LF drift down to within a few percent of f_{sw} occurred on 7 August 1995 when Wind was more than 14° off the Earth-Sun line in the xy plane.

3.2 Explanation of Wave Intensification

There is a natural explanation for the primary wave intensification observed near f_{ms} (e.g., Figure 1), the dominant intensification observed. Above, we assumed that waves were incident on the boundary at an angle of 85° from the normal direction. (This assumption does not affect the results derived above but was done for simplicity and clarity.) In fact, after propagating from the source and scattering through the magnetosheath, LF waves are probably incident at all angles onto the boundary. Under these circumstances, it is easy to show that the emergent beam is narrowly focussed. From Snell's law:

$$sin\theta_2 = n_1/n_2 \ sin\theta_1 \qquad (2)$$

where θ_1 and θ_2 are the incident and emergent wave angles, respectively. Near the nose of the magnetosheath, where the shock is strongest, $n_1/n_2 < 0.16$ for waves within a percent of f_{ms}. By equation (2), θ_2 is thus restricted to a small angular range: the emergent waves are focussed to within less than 10° of the boundary normal for $0° \leq \theta_1 \leq 90°$. Well away from the nose, $n_1/n_2 \to 1.0$, and the angular spectrum broadens considerably. Thus, near the nose, where dispersion is strongest, considerable wave intensification will occur due to angular focussing of the emergent beam. Depending on f/f_{ms}, amplification factors of 10–20 dB are possible, over and above what one would expect from typical AKR levels. (This is illustrated in Figure 4 where we compare LF burst intensity with AKR measured only a few minutes earlier.) Well away from the nose, this focussing

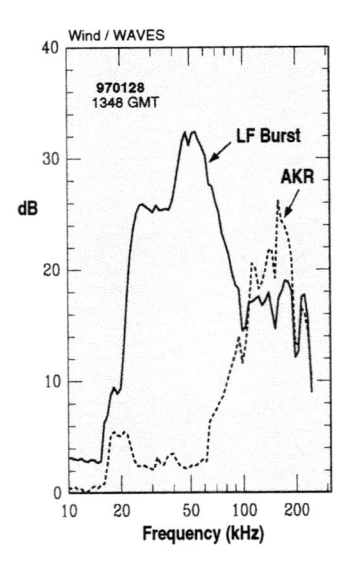

Figure 4. Spectral cuts through radio spectrogram on 28 January 1997 at two time near 1348 GMT (see Plate 1). Heavy line is LF burst, dashed line is from AKR a few minutes earlier. AKR intensity is already falling off toward lower frequencies before LF burst intensification occurs.

falls off smoothly so that wave intensification decreases. This behavior is clear in both Figure 1 and Figure 4.

Amplification by the same mechanism also works to explain the secondary wave intensification, at f_{sw}, which must also occur at the point of escape deep down the tail. However, since the shock is significantly weakened here, there is only a slight boundary gradient ($n_1/n_2 \to 1.0$) and the degree of amplification is less, as is observed. This analysis ignores any possible amplification due to WKB propagation effects which is beyond the present scope.

4. SUMMARY

The following summarizes the LF burst results of this study:

1. The spectral behavior and intensity spectrum are determined largely, if not exclusively, by propagation effects.
2. Observed group velocity dispersion is confined to two bands, one near the solar wind plasma frequency and the other near the magnetosheath plasma frequency appropriate to the nose of the bow shock.
3. There is often significant wave intensification in the same bands where dispersion is observed. Wave intensification is due to angular focussing of the emergent

beam as it crosses the boundary between the magnetosheath and solar wind.

4. Observation of an LF burst with an unusually high low-frequency cutoff has been interpreted in terms of a pronounced solar wind density enhancement leading to independent confirmation of LF burst wave escape from $X \leq -100\ R_e$ down the tail.

5. The LF burst emission is probably escaping to free space all along the magnetosheath boundary with the solar wind, but dispersed (and enhanced) emission is only beamed in the direction of the Earth-Sun line from near the nose ($f \sim f_{ms}$) and from the very deep tail ($f \sim f_{sw}$) where the shock is weak and $f_{ms} \sim f_{sw}$.

We conclude by noting that the relationship between the terrestrial LF bursts and the Jovian QP bursts has received very little attention in the community but is of great interest because of the similarities in form but differences in their short-term repeat behavior. We plan to pursue what significance this has for the magnetospheres of the two planets. Finally, we have left aside the still nagging question of the source of the LF bursts.

REFERENCES

Anderson, R. R., D. A. Gurnett, H. Matsumoto, et al., Observations of low frequency terrestrial type III bursts by Geotail and Wind and their association with isolated geomagnetic disturbances detected by ground and spaceborne instruments, Planetary Radio Emissions IV, (Eds. Rucker, H. O., Bauer, S. J., and Lecacheux, A.), Proceedings of the Fourth International Workshop, *Graz, Austria*, pg 241, 1996.

Desch, M. D., Jupiter radio bursts and particle acceleration, *Astrophys. J.*, 90, 541, 1994.

Desch, M. D., Terrestrial LF bursts: Source and solar wind connection, Planetary Radio Emissions IV, (Eds. Rucker H. O., Bauer, S. J., and Lecacheux, A.), Proceedings of the Fourth International Workshop, *Graz, Austria*, pg 251, 1996.

Desch, M. D., M. L. Kaiser, W. M. Farrell, Control of terrestrial low frequency bursts by the solar wind speed, *Geophys. Res. Lett.*, 23, 1251, 1996.

Gough, M. P., Non-thermal continuum emissions associated with electron injections: Remote plasmapause sounding, *Planet. Space Sci.*, 30, 657, 1982.

Greenstadt, E. W., et al., Observations of the flank of Earth's bow shock to − 110 R_e by ISEE 3/ICE, *Geophys. Res. Lett.*, 17, 753, 1990.

Gurnett, D. A., The Earth as a radio source: Terrestrial kilometric radiation, *J. Geophys. Res.*, 79, 4227, 1974.

Gurnett, D. A., The Earth as a radio source: The non-thermal continuum, *J. Geophys. Res.*, 80, 2751, 1975.

Hashimoto, K., S. Kudo, and H. Matsumoto, Source of auroral myriametric radiation observed with Geotail, *J. Geosphys. Res.*, 103, 23,475, 1998.

Kaiser M. L., M. D. Desch, W. M. Farrell, et al., Ulysses observations of escaping VLF emissions from Jupiter, *Geophys. Res. Lett.*, 19, 649, 1992.

Kaiser, M. L., M. D. Desch, W. M. Farrell, et al., LF band terrestrial radio bursts observed by Wind/WAVES, *Geophys, Res. Lett.*, 23, 1283, 1996.

Kurth, W. S., D. A.. Gurnett, and F. L. Scarf, Jovian type III radio bursts, *J. Geophys. Res.*, 94, 6917, 1989.

Louarn, P., A. Hilgers, A. Roux, et al., Correlation between terrestrial myriametric and kilometric radio bursts observed with Galileo, *J. Geophys. Res.*, 99, 23541, 1994.

MacDowall, R. J., M. L. Kaiser, M. D. Desch, et al., Quasiperiodic Jovian radio bursts: Observations from Ulysses radio and plasma wave experiment, *Planet. Space Sci.,41*, 1059, 1993.

Steinberg, J.-L., C. Laccombe, and S. Hoang, A new component of terrestrial radio emission observed from ISEE-3 and ISEE-1 in the solar wind, *Geophys. Res. Lett.* 15, 176, 1988.

Steinberg, J.-L., C. Lacombe, and S. Hoang, Sounding the flanks of the Earth's bow shock to −230 R_E: ISEE-3 observations of terrestrial radio sources down to 1.3 times the solar wind plasma frequency, *J. Geophys. Res.*, 103, 23565, 1998.

Steinberg, J.-L., S. Hoang and J. M. Bosqued, Isotropic kilometric radiation: A new component of the Earth's radio emission, *Ann Geophys.*, 8, 671, 1990.

M. Desch and W. Farrell, Code 695, NASA/GSFC Greenbelt, MD, 20771

The Influence of the Galilean Satellites on Radio Emissions From the Jovian System

W. S. Kurth, D. A. Gurnett, and J. D. Menietti

Dept. of Physics and Astronomy, The University of Iowa, Iowa City, Iowa

The Galilean satellites influence radio emissions from the Jovian system in a variety of ways. The best and most familiar example of these is the Io control of decametric radiation discovered in 1964 by *Bigg*. Voyager observations of broadband kilometric radiation revealed a low-latitude shadow zone cast by the Io torus at frequencies between a few tens of kHz and about 1 MHz. Voyager also discovered narrowband kilometric radio emissions emanating from the outer edge of the torus. In this paper we will discuss expansions in the suite of satellite influences based on new observations by Galileo. These include the discovery of Ganymede's magnetosphere and evidence of radio emissions generated via mode conversion from upper hybrid waves in the frequency range of about 20 - 100 kHz. There is evidence that Ganymede may control some of the hectometric or low-frequency decametric radio emissions based on occultation measurements and statistical studies of radio emission occurrence as a function of Ganymede phase. Direction-finding measurements in the vicinity of Io suggest that a portion of the hectometric emissions may be generated near the Io L-shell. A rotationally modulated attenuation band in the hectometric emission appears to be the result of scattering at or near the Io L-shell where the waves propagate nearly parallel to the magnetic field. There is even a tantalizing hint of a Europa connection to the source of narrowband kilometric radiation.

1. INTRODUCTION

From early in the study of Jovian radio emissions, it was clear that Io had an important influence on the intensity of the observed decametric signal [*Bigg*, 1964]. In fact, it could be argued that a considerable amount of theoretical work which preceded any missions to Jupiter dwelt on the mechanism by which Io might control these radio emissions. These studies included dynamos, the generation of electric potentials, and current systems which could tie Io to Jupiter via Jupiter's magnetic field [cf. *Goldreich and Lynden-Bell*, 1969]. Voyager caused a revolution not only in the way radio astronomers and magnetospheric physicists viewed the Jovian system by the discovery of numerous new low-frequency radio emissions and their phenomenology [*Science*, Vol. 204, 1 June 1979; *J. Geophys. Res.*, Vol. 86, 30 Sept. 1981], but also by the discovery of the Io torus fueled at the rate of a ton per second of material originating in Io's volcanos [*Morabito et al.*, 1979; *Broadfoot et al.*, 1979]. The new radio emissions included a number of manifestations of Io control. For example, the newly-discovered broadband kilometric radiation exhibited an equatorial shadow zone in its occurrence which was attributed to the existence of the Io plasma torus [*Kurth et al.*, 1980]. And, the outer edge of the Io torus was

Radio Astronomy at Long Wavelengths
Geophysical Monograph 119
Copyright 2000 by the American Geophysical Union

determined to be the source of the narrowband kilometric radiation [*Kaiser and Desch*, 1980]. However, even after the Pioneer, Voyager, and Ulysses flybys, only Io seemed to have any credible influence on Jovian radio emissions.

The Galileo mission has provided two opportunities for further investigation of the influence of satellites on Jovian radio emissions. First, the nature of an orbiting mission is to provide a platform from which maps and statistical studies can be carried out at relatively close range to the planet, as opposed to a flyby mission. Second, Galileo's mission includes multiple flybys of each of the Galilean satellites at close range, allowing direct in situ observations of the interaction of each of those satellites with the Jovian magnetosphere. During its first nearly three years in orbit, Galileo has confirmed a number of the Io influences above and has identified new examples of satellite influences, some of which involve Ganymede.

This review will briefly summarize the satellite - magnetosphere interactions of the four Galilean satellites from the point of view of a wave receiver in Section 2, demonstrate the confirmation or extension by Galileo measurements of previously-known satellite influences in Section 3, and describe new satellite influences discovered by Galileo to date in Section 4. The observations used will be from the Galileo plasma wave investigation described fully by *Gurnett et al.* [1992]. This instrument covers electric fields in the frequency range of 5.6 Hz to 5.6 MHz using an electric dipole antenna with a tip-to-tip length of 6.6 m and magnetic fields in the range of 5.6 Hz to 160 kHz using a pair of magnetic search coil antennas tuned for low and high frequencies.

2. SUMMARY OF PLASMA WAVE OBSERVATIONS OF GALILEAN SATELLITE INTERACTIONS WITH THE JOVIAN MAGNETOSPHERE

Galileo has made a close flyby of Io and several close flybys of Europa, Ganymede, and Callisto since it arrived at Jupiter in December 1995. These encounters have provided our first look at the interactions between the satellites and the Jovian magnetosphere. Before we can discuss how these satellites may influence Jovian radio emissions, it is important to briefly describe these in situ observations. Table 1 summarizes some of the features of the satellite interactions including the peak electron density observed, types of wave emissions observed, and a very subjective classification of the importance or strength of the magnetospheric interaction based primarily on the wave observations thus far.

2.1. Io

Galileo used a flyby of Io to help slow the spacecraft so that it could be captured into orbit on 7 December 1995. Fields and particles observations were made for about a 3 hour interval spanning the radial distance range of about 7.6 to 5.4 R_J (Jovian radii) and culminating in a passage through Io's plasma wake at an altitude of about 900 km. Plate 1a shows electric and magnetic frequency-time spectrograms detailing the plasma wave observations during this interval [*Gurnett et al.*, 1996a]. The display indicates the amplitudes of waves as a function of frequency (ordinate) and time (abscissa) using color to indicate the relative intensity of the waves. Blue areas indicate the weakest amplitudes and red the most intense. The upper panel shows the electric component of the wave spectrum and the lower panel shows the magnetic component. The most prominent emissions at lower frequencies in these spectrograms are electromagnetic whistler-mode emissions. There are indications that kinetic Alfvén waves may also be present, at least in the intervals of enhanced wave intensities such as near 1710 spacecraft event time (SCET) [*A. Roux, personal communication*, 1996]. At the highest frequencies are the hectometric radio emissions. The narrowband emission at a few to several hundred kHz running over the entire duration of the interval is the upper hybrid resonance band that can be used to ascertain the electron density using

$$n_e = \frac{(f_{UH}^2 - f_{ce}^2)}{8980^2} \quad (1)$$

Here n_e is in cm^{-3}, and f_{UH} and f_{ce} are the upper hybrid resonance frequency and electron cyclotron frequency, respectively, in Hz and $f_{UH}^2 = f_{pe}^2 + f_{ce}^2$ where f_{pe} is the electron plasma frequency. The electron cyclotron frequency can be determined from the measured magnetic field, however, it is of order 50 kHz and is negligible in Equation 1 over the time period in Plate 1a. As compared to the densities expected over the Galileo trajectory from the *Bagenal* [1994] Io torus density model based on Voyager measurements, the measured densities are about a factor of two greater [*Gurnett et al.*, 1996a; *Bagenal et al.*, 1997]. Perhaps the most extraordinary result, however, is the density peak of about 40,000 cm^{-3} observed at the closest approach to Io. Based on high plasma densities, low temperatures, and low flow velocities (relative to Io) *Frank et al.* [1996] suggest that Galileo actually passed through Io's ionosphere. Subsequent studies [*Warnecke et al.*, 1997; *Chust et al.*, 1999; *Frank and Paterson*, 1999] report the pickup of heavy ions and protons in the vicinity of Io.

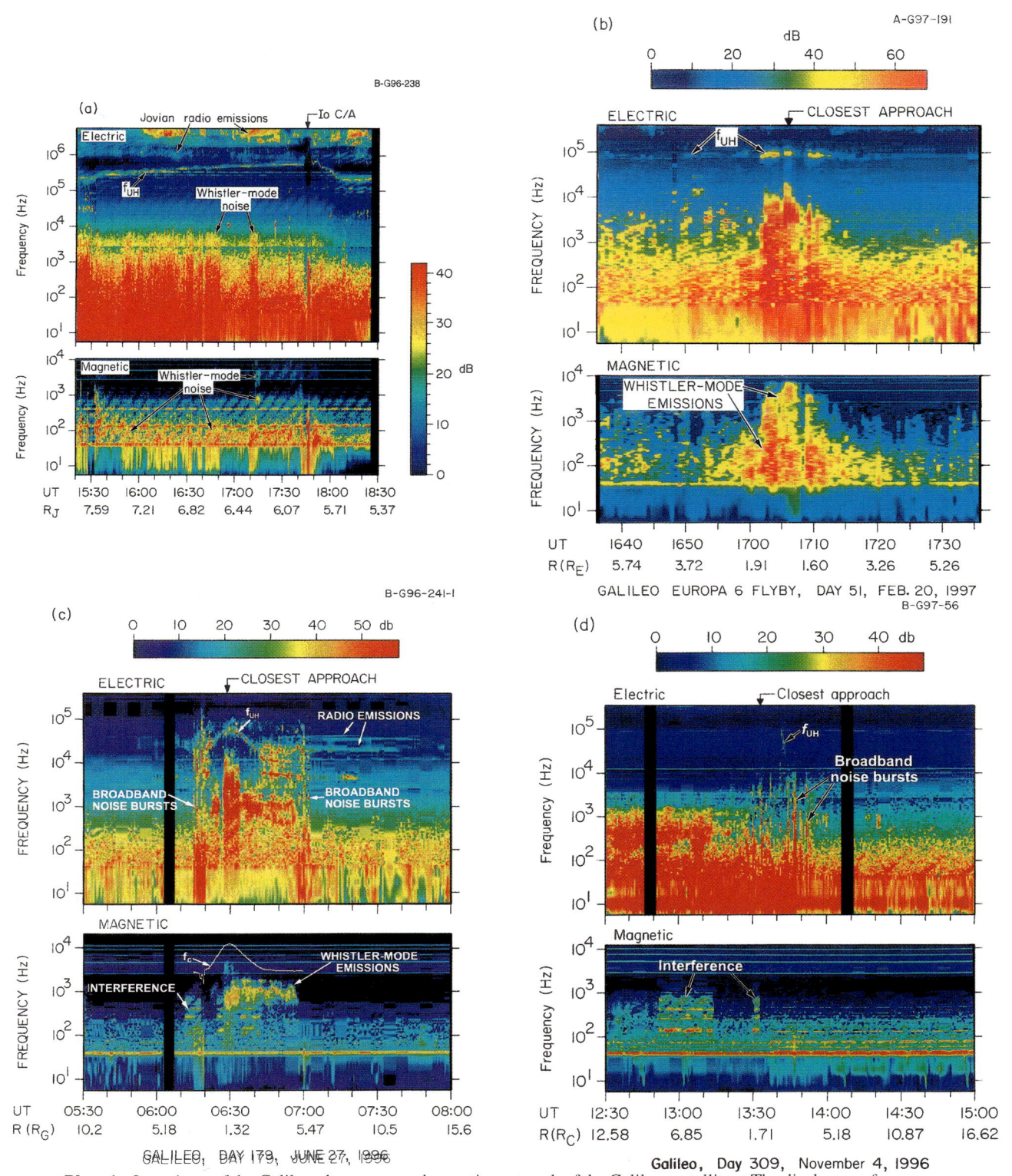

Plate 1. Overviews of the Galileo plasma wave observations at each of the Galilean satellites. The displays are frequency-time spectrograms showing the amplitude of waves as a function of frequency and time. The color bars show that the most intense waves are shown in red and the least intense waves in blue. (a) Io from *Gurnett et al.* [1996a]. [Reprinted with permission from *Science, 274*, 391, 1996; copyright 1996 American Association for the Advancement of Science]. (b) Europa from *Gurnett et al.* [1998a]. [Copyright by the American Geophysical Union]. (c) Ganymede from *Gurnett et al.* [1996b]. [Reprinted with permission from *Nature*, copyright 1996 Macmillan Magazines Limited]. (d) Callisto from *Gurnett et al.* [1997]. [Reprinted with permission from *Nature*, copyright 1997 Macmillan Magazines Limited].

Table 1. A Simple Comparison of the Galilean Satellite Interactions with the Jovian Magnetosphere

Satellite	Distance R_J	Peak n_e (cm^{-3})	Interaction Strength	Waves	Distant Effects
Io	6	41,000	Strongest	Ion cyclotron waves Whistler modes Kinetic Alfvén waves Upper hybrid band Radio emissions	Plasma torus Jovian aurora Radio control, propagation
Europa	9.5	180	Intermediate	Whistler modes Electron cyclotron harmonic bands Upper hybrid band Broadband electrostatic noise	Unknown
Ganymede	15	100	Intermediate	Whistler-mode hiss Chorus Electron cyclotron harmonic bands Upper hybrid band Broadband electrostatic noise Radio emissions	Wake or plume Radio control?
Callisto	26	100	Weakest	Kinetic Alfvén waves Upper hybrid band Broadband electrostatic noise	None known

2.2. Europa

Plate 1b shows the wave observations from the Europa flyby during the E6 orbit on 20 February 1997. The most clear signatures of Europa in this spectrogram are enhanced whistler-mode emissions near 1705 SCET accompanied by enhanced upper hybrid emissions near 100 kHz. During this flyby Galileo passed within 586 km of the surface on the upstream side of the moon. The upper hybrid signature yields density estimates of as high as about 180 cm^{-3} (during the E4 flyby). Given that the ambient density at the orbit of Europa is ~ 80 cm^{-3}, this represents an enhancement of up to about 100 cm^{-3} [*Gurnett et al.*, 1998a]. Plasma observations [*Paterson et al.*, 1999] suggest a source strength of about 2 kg/s generated by Europa's interaction with Jupiter's magnetosphere. The magnetometer observations show an enhanced magnetic field near the moon, but there is no clear evidence that this is due to an intrinsic field. In some sense, it is difficult to characterize the nature of Europa's interaction with Jupiter, probably because the moon is bathed in the outer reaches of the Io torus. This could mask what might be more significant interactions than those apparent in Plate 1b; it is also possible that the position of Europa relative to the plasma torus (elevation with respect to the centrifugal equator) at the time of a Galileo flyby strongly affects the signatures of the interaction [*Kivelson et al.*, 1999].

2.3. Ganymede

A surprise of the Galileo mission is the discovery of an intrinsic magnetic field at Ganymede and the existence of a Ganymede magnetosphere embedded within the Jovian magnetosphere [*Gurnett et al.*, 1996b; *Kivelson et al.*, 1996]. Plate 1c shows wave observations from the Ganymede flyby during the G1 orbit on 27 June 1996. This flyby had a closest approach altitude of 835 km downstream of the moon. Virtually all of the information needed to identify this magnetosphere and to ascertain the existence of an intrinsic field is provided by the data represented by Plate 1c. First, the search coil observations prove that the intense emissions below a few kHz centered near 0630 SCET are electromagnetic and, in particular, whistler-mode emissions. This mode exists only below both f_{pe} and f_{ce}, hence, *Gurnett et al.* [1996b] concluded that the magnetic field intensity was ~ 400 nT near closest approach as compared to the ambient field intensity at 15 R_J of about 100 nT. Second, there are very clear signatures of a magnetopause inbound at about 0615 and outbound near 0700 SCET as signified by broadband electrostatic emissions at these times. Third, the narrowband emission which peaks near 0630 is electrostatic and almost certainly the upper hybrid resonance (UHR) band indicating a relatively dense region at closest approach of about 45 cm^{-3} as compared to the ambient density of order 1 cm^{-3} in this region of the Jovian magnetosphere.

Furthermore, the upper spectrogram in Plate 1c shows evidence for banded electron cyclotron harmonic (ECH) emissions between harmonics of the electron cyclotron frequency. All of these emissions are typical of planetary magnetospheres. Finally, the weak banded emissions seen most prominently after Galileo has exited the magnetosphere (after 0700 SCET) appear to be radio emissions; these will be discussed in detail in Section 4 below. Of course, the Galileo magnetometer investigation [*Kivelson et al.*, 1996] and other instruments confirmed the conclusions listed above either directly or indirectly and the Ganymede magnetosphere has proven to exhibit a number of features common to planetary magnetospheres including evidence of reconnection (albeit with the Jovian magnetic field as opposed to the interplanetary field), a wealth of effects on thermal and energetic plasmas [*Frank et al.*, 1997a,b; *Williams and Mauk,* 1997; *Williams et al.,* 1998; *Paranicas et al.,* 1999], and even evidence for an extended plasma plume extending in the downstream direction [*Khurana et al.,* 1999].

2.4. Callisto

Observations from the Callisto flyby on the C3 orbit are shown in Plate 1d [*Gurnett et al.,* 1997]. This flyby occurred at a distance of 1129 km downstream of the moon on 4 November 1996. The signature of this interaction in the wave data is the least impressive of the four Galilean satellites. The broadband bursty signatures are possibly similar to broadband electrostatic noise signatures seen at various locations at Earth [*Gurnett and Frank,* 1977] and also at Jupiter [*Barbosa et al.,* 1981] but could also be the result of kinetic Alfvén waves. There is a brief burst of electrostatic emission near 90 kHz at about 1342 SCET, near closest approach. This emission is almost certainly an upper hybrid resonance emission and suggests a density of as much as 100 cm^{-3}. The ambient density at 26 R_J in the Jovian magnetosphere is < 1 cm^{-3}. The magnetometer observations do not suggest an intrinsic field at this moon [*Khurana et al.,* 1997].

3. PRE-GALILEO ASSESSMENT OF SATELLITE INFLUENCES ON JOVIAN RADIO EMISSIONS

It is not within the scope of this review to thoroughly cover the pre-Galileo state of understanding of Jovian radio emissions and the influences of the Galilean satellites on their generation and propagation. A number of papers already do this in great detail [c.f. *Boischot et al.,* 1981; *Alexander et al.,* 1981; *Ladreiter and Leblanc,* 1990; *Lecacheux et al.,* 1992]. However, it is fair to summarize that virtually all of the generally-accepted satellite influences are related to Io and the Io torus.

3.1. Io Control of Decametric Emissions

The oldest and most obvious of the Galilean satellite influences are the Io-related decametric radio emissions [*Bigg,* 1964; *Carr and Desch,* 1976]. Galileo's frequency range extends to only 5.6 MHz, just above the peak in the hectometric spectrum, however, *Menietti et al.* [1998c] have shown that even in the frequency range 3.2 to 5.6 MHz enhancements in the radio spectrum can be found when Io is near 90° and 240° in phase. This result raises further questions about the relationship between the decametric and hectometric emissions and whether the hectometric spectrum is just a lower frequency extension of the decametric emissions. The existence of an Io control of the lower-frequency emissions could either suggest the extension of the decametric emissions to lower frequencies, or a similar control of the hectometric emissions. Both classes of radio emissions are presumed to be generated by the cyclotron maser instability; the lower frequency of the hectometric emissions clearly indicates that they are generated at higher altitudes than the decametric radiation where the local electron cyclotron frequency is lower.

3.2. Broadband Kilometric Radiation Equatorial Shadow Zone Cast by the Io Torus

Kaiser and Desch [1980] differentiated the kilometric radiation into two clearly different types of radio emissions and suggested different source locations and mechanisms for them. *Kurth et al.* [1980] showed that there was a magnetic equatorial shadow zone with half-width about 10° within which the broadband kilometric radiation is usually not observed. They attributed this shadow zone to the blocking effect of the Io plasma torus on the relatively low frequency radio emissions. Galileo has confirmed the existence of this equatorial shadow zone [*Kurth et al.,* 1997a] and, in fact, Galileo seldom observed broadband kilometric radiation after the G2 Ganymede flyby. The G2 flyby was used to reduce the few-degree inclination of the initial Galileo orbit to nearly 0°. The 9.6° wobble of the magnetic field as Jupiter rotates is not sufficient to carry Galileo out of the shadow zone.

3.3. Narrowband Kilometric Radiation Source Region in the Outer Io Torus

The narrowband kilometric radiation was isolated from the broadband kilometric radiation by *Warwick et al.* [1979] and *Kaiser and Desch* [1980]. These emissions near 100 kHz are typically about 40 kHz in bandwidth and show an entirely different temporal character than the broadband kilometric radiation. *Kaiser and Desch* first deduced that the narrowband emissions were generated in the outer edge of the Io torus in the radial distance range of about 8 - 9 R_J. Furthermore, the narrowband emissions did not appear completely synchronized with Jupiter's rotation; instead the emissions appeared to slip with respect to the system III rotation period by 3 - 5% [*Kaiser and Desch*, 1980]. This further supported a source in the outer regions of the torus since mass loading will slow the plasma there to below the full corotation speed. Ulysses direction-finding results confirmed the Io torus source region for these emissions and also concluded that they were primarily generated in the extraordinary mode [*Reiner et al.*, 1993].

Galileo routinely observes the narrowband kilometric radiation [*Kurth et al.*, 1997a] appearing episodically in such a way as to suggest the birth of a new source region which lasts for several Jovian rotations before fading away. *Louarn et al.* [1998] have shown that the narrowband kilometric episodes appear to have onsets which coincide with brightening in the hectometric emissions and changes in the trapped continuum radiation spectrum. These onsets, in turn, correlate well with quasi-periodic changes in the energetic particle spectrum in the outer magnetosphere reported by *Krupp et al.* [1998] and *Woch et al.* [1998]. These reports suggest that these quasi-periodic changes are symptoms of magnetospheric dynamics of uncertain origin.

Galileo repeatedly passes through the equatorial magnetosphere into distances of 9 R_J and smaller, hence, there would appear to be ample opportunities to fly through a narrowband kilometric radiation source region. Based on prior studies, we would expect to find intense electrostatic upper hybrid bands at about 100 kHz in the region of 8 - 9 R_J which mode-couple into electromagnetic waves at the same frequency through either a linear mechanism [c.f. *Jones*, 1987] or a nonlinear mechanism [c.f. *Melrose*, 1981] or perhaps to up-converted waves [*Fung and Papadopoulous*, 1987]. While there are a few instances which may prove to be evidence of such source regions in the data acquired to date, none of them are particularly compelling. The quasi-periodicity of the onset of these emissions, however, would seem to reduce the probability of Galileo being at the right place at the right time to actually traverse a source region. Perhaps the most tantalizing aspect of the Galilean observations in this region of the magnetosphere is that Europa appears to be the site of enhanced upper hybrid emissions near 100 kHz [*Gurnett et al.*, 1998a]. It is conceivable that Europa plays a role in the generation of at least some of these emissions.

4. GALILEO CONTRIBUTIONS

Galileo has found evidence for several ways in which the Galilean satellites influence radio waves from the Jovian system which were not previously recognized. In this section we summarize these new Galileo findings.

4.1. Radio Emissions from Ganymede's Magnetosphere

By far the most exciting result from the Galileo mission from the point of view of planetary radio emissions is the discovery of Ganymede's magnetosphere and radio emissions generated at that magnetosphere [*Gurnett et al.*, 1996a; *Kivelson et al.*, 1996; *Kurth et al.*, 1997c]. Plates 1c and 2 show the first and second flybys of Ganymede via measurements from the Galileo plasma wave receiver up to a frequency of 5.6 MHz. The signatures of the familiar planetary plasma wave phenomena pointed out in section 2 and in Plate 1c are also evident in Plate 2. One of these common features is the existence of weak, narrowband emissions most evident following each of these encounters, that is, on the Jupiterward side of Ganymede. There is, however, evidence for these emissions just prior to both encounters as well, although these are not as extensive. These emissions are found between approximately 20 and 50 kHz, although there is some evidence for weaker emissions inside the magnetosphere near 100 kHz in the first encounter (Plate 1c). Wideband observations from the second encounter show that the integrated bandwidth of these emissions is of order 10 kHz and they extend no more than about 18 Ganymede radii from the moon, limited almost certainly by the sensitivity of the Galileo receiver.

Kurth et al. [1997c] argued that these radio emissions have their source in the upper hybrid emissions near the Ganymede magnetopause. The frequency-time character and frequency range of these emissions is unlike other known Jovian radio emissions and their presence only in the near vicinity of Ganymede is a strong indication of a source at Ganymede. The clearest evidence for a source associated with Ganymede's magnetosphere, however, is summarized in Figure 1. Here, the intensity of waves at 20.1 kHz is plotted as a function of the distance from Ganymede as Galileo left the magnetosphere of Ganymede. The larger intensity, bursty emissions early in this time interval are the electrostatic upper hybrid waves at and just inside the magnetopause. Later, the wave intensity decreases more smoothly as the spacecraft leaves the vicinity of the magnetopause, although the amplitude uncertainty arising from the use of a lossy compression algorithm is responsible for the peculiar patterns which appear in this region.

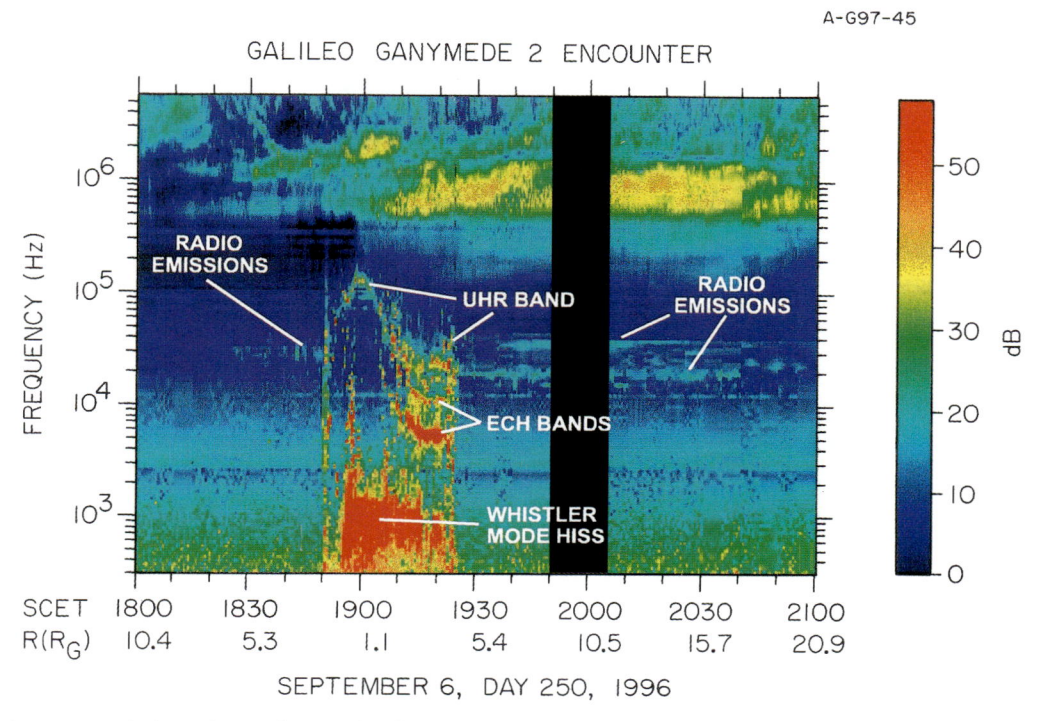

Plate 2. Radio emissions observed emanating from Ganymede's magnetosphere [*Kurth et al.*, 1997c]. [Copyright by the American Geophysical Union].

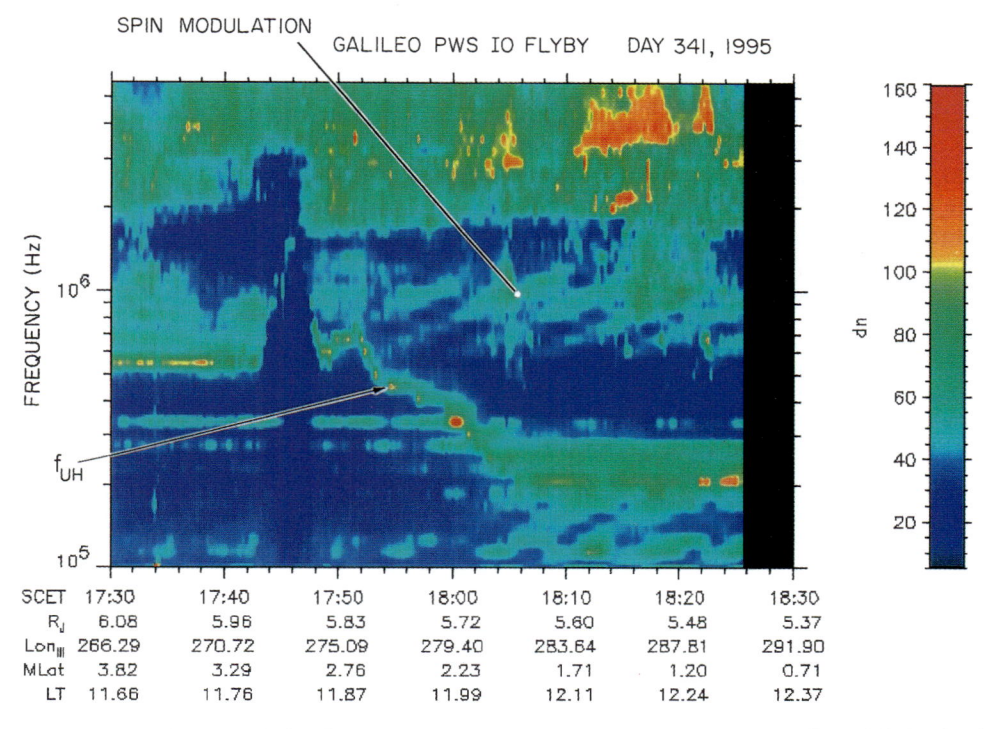

Plate 3. A spin-modulated radio emission feature found close to Io [*Menietti et al.*, 1998b]. [Copyright by the American Geophysical Union].

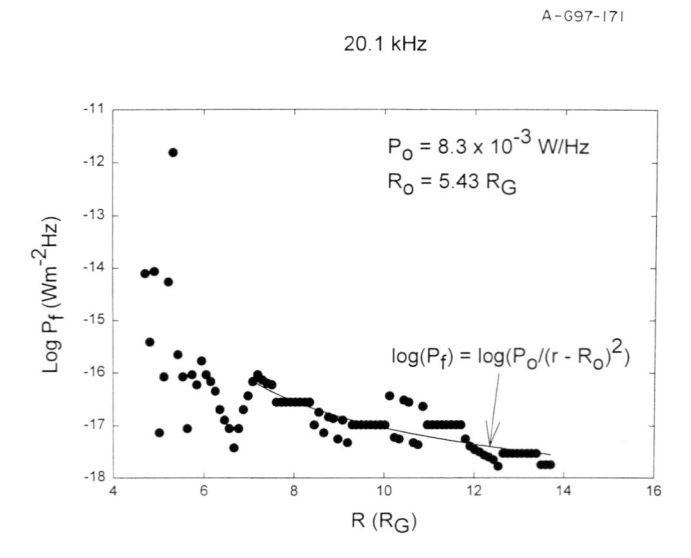

Figure 1. The intensity of waves as a function of distance from Ganymede showing the radio emissions decrease in intensity as r^{-2} from a source at or near the magnetopause [*Kurth et al.*, 1997c]. [Copyright by the American Geophysical Union].

Hypothesizing a source at the magnetopause between Galileo and Ganymede, *Kurth et al.* fit a curve with an r^{-2} dependence to the data showing that the source was near 5.4 R_G (Ganymede radii), consistent with a magnetopause source.

Kurth et al. [1997c] also suggested that the evidence in Plates 1c and 2 strongly supports generation by mode conversion from electrostatic upper hybrid waves, similar to the mechanism commonly accepted for the nonthermal continuum radiation at Earth, Jupiter, and the other outer planets. Particularly in Plate 2, an intense upper hybrid band is evident near the magnetopause at nearly the same frequency (~ 20 kHz) as one of the bands of radio waves. The conditions are just what one would expect for the mode conversion theory to apply: there is evidence of electrostatic upper hybrid resonance emissions at a sharp density gradient and the density gradient is more-or-less perpendicular to the magnetic field (based on the magnetic field configuration near the magnetopause [*Kivelson et al.*, 1996]). Furthermore, the second fit parameter P_0 given in Figure 1 allows one to calculate a total power of approximately 80 W from the source, assuming radiation into a large solid angle and using 10 kHz for the emission bandwidth. This is similar to the total power in the terrestrial continuum radiation [*Gurnett*, 1975].

The importance of the Ganymede radio emissions is not so much their intensity as the mere fact that we now have yet another "planetary" magnetosphere which is the source of one of the two nearly ubiquitous types of radio emissions (nonthermal continuum-like radio emissions and the stronger auroral cyclotron maser emissions). This is even more important given that Ganymede is now the smallest magnetosphere known to support radio emissions and also that this magnetosphere is embedded not in the solar wind but the Jovian magnetosphere. Of course, this begs the question of why there are no obvious cyclotron maser emissions from this magnetosphere. We believe the answer to this is that at least near Ganymede, the plasma frequency as determined by the upper hybrid resonance band seen in Plates 1c and 2 is always greater than f_{ce}. Hence, the usual cyclotron maser requirement that $f_{pe}/f_{ce} < 0.3$ is clearly not met.

4.2. An Io Source for Hectometric Radiation

Menietti et al. [1998a] have developed a direction-finding analysis methodology for the Galileo plasma wave receiver measurements based on the rotating dipole technique [*Fainberg et al.*, 1972; *Kurth et al.*, 1975]. *Menietti et al.* [1998b] have applied this technique to a specific interval just after the Io encounter on 7 December 1995 between ~ 600 kHz and ~ 1 MHz. Plate 3 is taken from the *Menietti et al.* [1998b] study and shows the strong spin modulation of the feature in the hectometric frequency range after the Io flyby.

The rotating dipole technique determines only the plane in which a radio source exists. This plane is defined by the spin axis of the spacecraft and the centroid of the direction of arrival of the radio waves (the direction to the source). Hence, it is not very specific. Additionally, the technique is most useful for sources near the plane perpendicular to the spacecraft spin axis.

During the Io encounter, Galileo was at a local time (with respect to Jupiter) of ~12 hours (local noon) and based on the geometry shown in Figure 2 [*Menietti et al.*, 1998b], it is clear that Jupiter was very close to Galileo's spin axis. This geometry, then, more or less excludes any of the normal hectometric sources at low altitudes in Jupiter's auroral zone since such sources would exhibit little or no spin modulation. Instead, *Menietti et al.* point out that the source direction is consistent with the approximate location of the Io L-shell and perhaps that associated with the high-density plume extending downstream of Io.

This is the first evidence for radio waves originating from the vicinity of Io (as opposed to the foot of Io's flux tube or L-shell near Jupiter) and represents a possible detection of radio emissions in the Io-magnetospheric interaction region. *Menietti et al.* [1998b] note that density gradients, electron beams, and upper-hybrid emissions are all present near Io. The upper hybrid emission may mode-convert into electromagnetic waves [e.g. *Jones*, 1986; *Melrose*, 1981] or

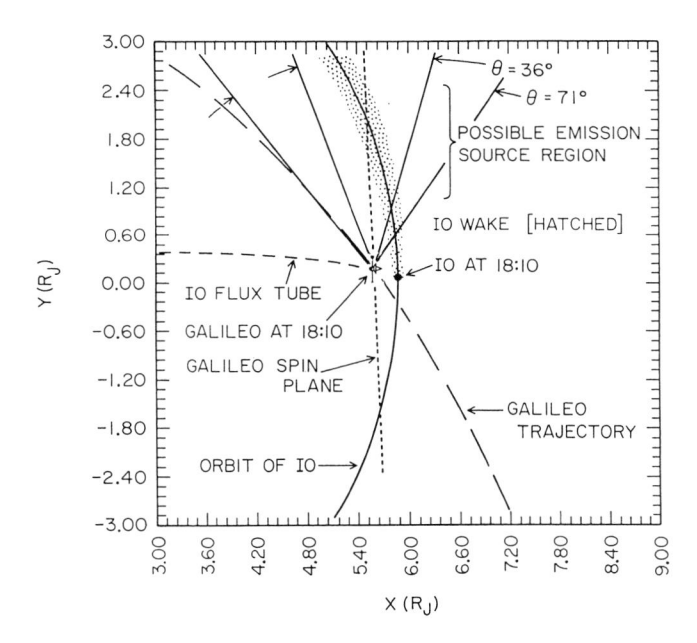

Figure 2. A schematic of the geometry of the direction-finding results of the feature shown in Plate 3 indicating that an extended plume from Io or L-shells associated with the plume are consistent with the source of the waves [*Menietti et al.*, 1998b]. [Copyright by the American Geophysical Union].

the temperature anisotropy beam instability could result in the waves [e.g. *Wong and Goldstein*, 1990]. An alternative to a source near Io, however, has been suggested by *Farrell et al.* [1999] who argue that the waves may simply be the usual auroral hectometric emissions refracted or reflected at the density gradients in the Io L-shell.

4.3. Ganymede Control of Low-Frequency Decametric or Hectometric Radiation

Given the strong interaction now known to exist between the magnetospheres of Ganymede and Jupiter, it is natural to suspect that currents flowing from Ganymede to Jupiter's ionosphere may result in the generation of radio emissions in a manner similar to the Io-related decametric emissions. The existence of such emission is consistent with occultation measurements of hectometric radiation [*Kurth et al.*, 1997b] which suggest that at least some hectometric emission may be generated on or near Ganymede's flux tube near Jupiter.

We have analyzed most of the prime mission radio observations for such a Ganymede influence using the familiar satellite phase vs. system III longitude displays used to highlight Io-related sources. Plate 4 is an example of such an analysis after *Menietti et al.* [1998c] using radio measurements between 3.2 and 5.6 MHz and showing the occurrence of emissions above a threshold of 4×10^{-18} W m^{-2} Hz^{-1} scaled to 100 R_J as a function of Ganymede phase (superior conjunction relative to Jupiter as viewed by Galileo is 0° phase) and system III longitude λ_{III} (of Galileo). Peaks can be seen between 100 - 160° λ_{III} near 80° and 240° Ganymede phase. In this analysis, periods when the phase of Io is between 85° and 100° or 220° and 225° have been filtered out. While the effect shown is not strong, there is an apparent Ganymede emission, consistent with expectations and the occultation measurements of *Kurth et al.* [1997b]. It is almost certainly the case, though, that hectometric emissions are generated by magnetospheric sources situated over a wide range of λ_{III} and local time, hence, one would not expect a Ganymede source to necessarily dominate the hectometric spectrum.

Kaiser and MacDowall [1998] have reported a small number of observations by Ulysses of radio "bullseyes" in the frequency range of 20 to 50 kHz. The bullseye nomenclature arises from the appearance of these features in a frequency-time display. While *Kaiser and MacDowell* show a good correlation between these events and periods of enhanced solar wind ram pressure, they also note that sometimes there is 10.5-hour periodicity between successive events. Since this is the period between successive passages of a given system III longitude past Ganymede, these authors suggest that there may be some Ganymede influence on the radio emissions. To date, there have been no observations of these radio emissions identified in the Galileo data set.

4.4. The Hectometric Attenuation Band

Gurnett et al. [1998b] have reported an attenuation band observed repeatedly in the hectometric emission spectrum which they attribute to a propagation effect at or near the Io L-shell. Plate 5 shows a 24-hour interval of hectometric observations which display parabolic-shaped attenuation features with two features per rotation. These attenuation bands are related to the "lanes" observed by the Voyager spacecraft [*Lecacheux et al.*, 1980; *Higgins et al.*, 1995]. *Gurnett et al.* noted that there are clear indications that the attenuation bands are tied to the rotation of Jupiter. Each of the parabolic features is near a maximum excursion in magnetic latitude by the spacecraft, either north or south of the equator (near λ_{III} = 50° or 185°) and the peak frequency of the feature varies as the radial distance of the spacecraft.

Gurnett et al. [1998b] modeled the frequency of maximum attenuation by assuming scattering or shallow-angle reflection for ray paths nearly parallel to an L-shell near Io. A schematic of the geometry and the model are given in Figure 3. Essentially, for any given geometry, one can locate a hectometric radio source on an auroral field line and at the

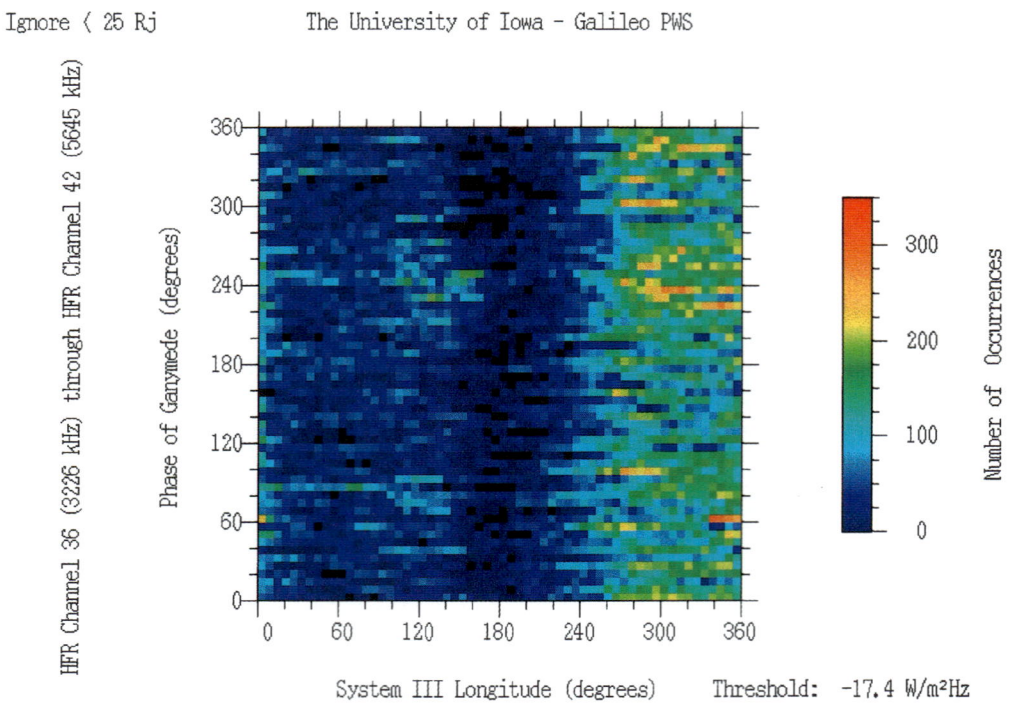

222 SATELLITE INFLUENCES ON JOVIAN RADIO EMISSIONS

Plate 4. An analysis of the occurrence of hectometric radiation as a function of Galileo's system III longitude and the phase of Ganymede relative to Jupiter as viewed by Galileo. Weak peaks in the occurrence rate occur when Ganymede is near 90° and 270° phase [*Menietti et al.*, 1998c]. [Copyright by the American Geophysical Union].

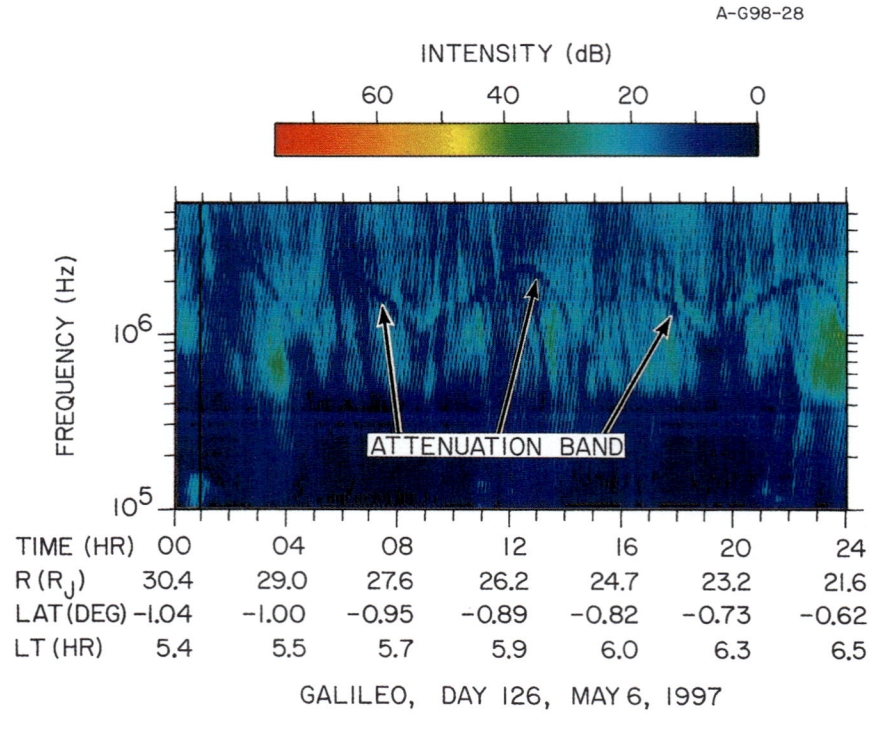

Plate 5. A frequency-time spectrogram showing an attenuation band in the hectometric radio spectrum which varies in frequency at twice the rotation rate of Jupiter [*Gurnett et al.*, 1998b]. [Copyright by the American Geophysical Union].

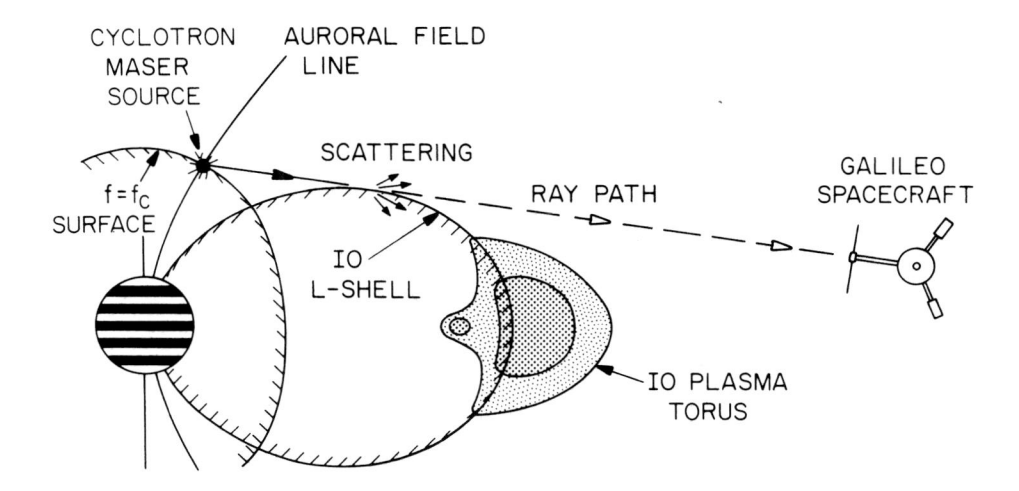

Figure 3. A model by *Gurnett et al.* [1998b] for the attenuation mechanism which accounts for most of the features observed in the spectrogram in Plate 5. Hectometric radio waves which propagate nearly parallel to an L-shell near Io suffer either scattering or shallow-angle reflection; observed wave amplitudes observed in this direction are decreased in amplitude from those at higher and lower frequencies. [Copyright by the American Geophysical Union].

source electron cyclotron frequency (assuming the cyclotron maser instability as a source mechanism) which must propagate nearly tangentially along an L-shell near that of Io. If one assumes scattering by density fluctuations along that L-shell or alternately, shallow-angle reflection at the surface of the L-shell due to field-aligned density variations (gradients perpendicular to the magnetic field), one can reproduce the attenuation pattern observed by Galileo quite accurately. *Higgins et al.* [1999] have carried out additional statistical studies of this feature and have clearly related the effect to the Voyager "lanes." *Menietti et al.* [1998d] have also shown evidence of two nested attenuation bands which provides the first clear evidence of second harmonic cyclotron maser emission in Jovian hectometric radiation. *Menietti et al.* [1999] conclude that shallow-angle reflection can account for the observations, but coherent scattering cannot be ruled out at present.

5. SUMMARY

The Galilean satellites interact with the magnetosphere of Jupiter in a variety of ways and across a spectrum of relative importance. Io certainly has the most important influence on the system, but Galileo has shown that Ganymede, with its intrinsic magnetic field, has a much more interesting interaction with Jupiter than previously thought. Callisto is evidently the least important of the Galilean satellites in terms of its interaction with the magnetosphere. Europa's interaction is largely masked by the Io torus, hence, it is difficult to understand the true nature of its interaction at this time.

The most dramatic contribution of Galileo to our understanding of the satellites' influence on Jovian radio emissions is the discovery of Ganymede's magnetic field. The existence of radio emissions from Ganymede enforces the hypothesis that remotely-sensed radio emissions are a reliable indication of a body with an intrinsic magnetic field, hence, magnetosphere.

Galileo has also provided evidence that Ganymede may control some of the Jovian radio emissions in a manner similar to, but to a lesser extent than, Io.

Galileo observations have provided new insight into effects of the Io torus on the propagation of hectometric radiation in the form of rotationally-modulated attenuation bands. There is also some evidence for radio emissions in the kilometric/hectometric radiation range having a source on the Io L-shell, but in the vicinity of the moon as opposed to the high latitude, low altitude location of most of the Jovian hectometric and kilometric emissions.

Acknowledgments. The research at the University of Iowa was supported by NASA through contract 959779 with the Jet Propulsion Laboratory.

REFERENCES

Alexander, J. K., T. D. Carr, J. R. Thieman, J. J. Schauble, and A. C. Riddle, Synoptic observations of Jupiter's radio emissions:

Average statistical properties observed by Voyager, *J. Geophys. Res., 86*, 8529-8545, 1981.

Bagenal, F., Empirical model of the Io plasma torus: Voyager measurements, *J. Geophys. Res., 99*, 11,043-11,062, 1994.

Bagenal, F., F. J. Crary, A. I. F. Stewart, N. M. Schneider, D. A. Gurnett, W. S. Kurth, L. A. Frank, and W. R. Paterson, Galileo measurements of plasma density in the Io torus, *Geophys. Res. Lett., 24*, 2119-2122, 1997.

Barbosa, D. D., F. L. Scarf, W. S. Kurth, and D. A. Gurnett, Broadband electrostatic noise and field-aligned currents in Jupiter's middle magnetosphere, *J. Geophys. Res., 86*, 8357-8369, 1981.

Bigg, E. K., Influence of the satellite Io on Jupiter's decametric emission, *Nature, 203*, 1008-1010, 1964.

Boischot, A., A. Lecacheux, M. L. Kaiser, M. D. Desch, and J. K. Alexander, Radio Jupiter after Voyager: An overview of the planetary radio astronomy observations, *J. Geophys. Res., 86*, 8213-8226, 1981.

Broadfoot, A. L., M. J. S. Belton, P. Z. Takacs, B. R. Sandel, D. E. Shemansky, J. B. Holberg, J. M. Ajello, S. K. Atreya, T. M. Donahue, H. W. Moos, J. L. Bertaux, J. E. Blamont, D. F. Strobel, J. C. McConnell, A. Dalgarno, R. Goody, and M. B. McElroy, Extreme ultraviolet observations from Voyager 1 encounter with Jupiter, *Science, 204*, 979-982, 1979.

Carr, T. D., and M. D. Desch, Recent decametric and hectometric observations of Jupiter, in *Jupiter*, edited by T. Gehrels, University of Arizona Press, Tucson, pp. 693-737, 1976.

Chust, T., A. Roux, S. Perraut, P. Louarn, W. S. Kurth, and D. A. Gurnett, Galileo plasma wave observations of Iogenic hydrogen, *Planet. Space Sci.*, in press, 1999.

Fainberg, L., L. G. Evans, and R. G. Stone, Radio tracking of solar energetic particles through interplanetary space, *Science, 178*, 743-745, 1972.

Farrell, W. M., R. A. Hess, and R. J. MacDowall, O-mode emission at the Io torus: A real or virtual source? *Geophys. Res. Lett., 26*, 1-4, 1999.

Frank, L. A., and W. R. Paterson, Production of hydrogen ions at Io, *J. Geophys. Res., 104*, 10,345-10,354, 1999.

Frank, L. A., W. R. Paterson, K. L. Ackerson, V. M. Vasyliunas, F. V. Coroniti, and S. J. Bolton, Plasma observations at Io with the Galileo spacecraft, *Science, 274*, 394-395, 1996.

Frank, L. A., W. R. Paterson, K. L. Ackerson, and S. J. Bolton, Outflow of hydrogen ions from Ganymede, *Geophys. Res. Lett., 24*, 2151-2154, 1997a.

Frank, L. A., W. R. Paterson, K. L. Ackerson, and S. J. Bolton, Low energy electron measurements at Ganymede with the Galileo spacecraft: Probes of the magnetic topology, *Geophys. Res. Lett., 24*, 2159-2162, 1997b.

Fung, S. F., and K. Papadopoulos, The emission of narrow-band Jovian kilometric radiation, *J. Geophys. Res., 92*, 8579-8593, 1987.

Goldreich, P., and D. Lynden-Bell, Io, a Jovian unipolar inductor, *Astrophys. J., 156*, 59-78, 1969.

Gurnett, D. A., The Earth as a radio source: The non-thermal continuum, *J. Geophys. Res., 80*, 2751-2763, 1975.

Gurnett, D. A., and L. A. Frank, A region of intense plasma wave turbulence on auroral field lines, *J. Geophys. Res., 82*, 1031-1050, 1977.

Gurnett, D. A., W. S. Kurth, R. R. Shaw, A. Roux, R. Gendrin, C. F. Kennel, F. L. Scarf, and S. D. Shawhan, The Galileo plasma wave investigation, *Space Sci. Rev., 60*, 341-355, 1992.

Gurnett, D. A., W. S. Kurth, A. Roux, S. J. Bolton, and C. F. Kennel, Galileo plasma wave observations in the Io plasma torus and near Io, *Science, 274*, 391-392, 1996a.

Gurnett, D. A., W. S. Kurth, A. Roux, S. J. Bolton, and C. F. Kennel, Evidence for a magnetosphere at Ganymede from plasma-wave observations by the Galileo spacecraft, *Nature, 384*, 535-537, 1996b.

Gurnett, D. A., W. S. Kurth, A. Roux, S. J. Bolton, and C. F. Kennel, Absence of a magnetic-field signature in plasma-wave observations at Callisto, *Nature, 387*, 261-262, 1997.

Gurnett, D. A., W. S. Kurth, A. Roux, S. J. Bolton, E. A. Thomsen, and J. B. Groene, Galileo plasma wave observations near Europa, *Geophys. Res. Lett., 25*, 237-240, 1998a.

Gurnett, D. A., J. D. Menietti, W. S. Kurth, and A. M. Persoon, An unusual rotationally modulated attenuation band in the Jovian hectometric radio emission spectrum, *Geophys. Res. Lett., 25*, 1841-1844, 1998b.

Higgins, C. A., J. L. Green, J. R. Thieman, S. F. Fung, and R. M. Candey, Structure within Jovian hectometric radiation, *J. Geophys. Res., 100*, 19,478-19,496, 1995.

Higgins, C. A., J. R. Thieman, S. F. Fung, J. L. Green, and R. M. Candey, Jovian dual-sinusoidal HOM lane features observed by Galileo, *Geophys. Res. Lett., 26*, 389-392, 1999.

Jones, D., Io plasma torus and the source of Jovian kilometric radiation (bKOM), *Nature, 324*, 40-42, 1986.

Jones, D., Io plasma torus and the source of Jovian narrow-band kilometric radiation, *Nature, 327*, 492-495, 1987.

Kaiser, M. L., and M. D. Desch, Narrow-band Jovian kilometric radiation: A new radio component, *Geophys. Res. Lett., 7*, 389-392, 1980.

Kaiser, M. L., and R. J. MacDowall, Jovian radio "bullseyes" observed by Ulysses, *Geophys. Res. Lett., 25*, 3113-3116, 1998.

Khurana, K. K., M. G. Kivelson, C. T. Russell, R. J. Walker, and D. J. Southwood, Absence of an internal magnetic field at Callisto, *Nature, 387*, 262-264, 1997.

Khurana, K. K., J. Warnecke, M. G. Kivelson, W. S. Kurth, D. A. Gurnett, and D. J. Williams, Ganymede's distant wake, *J. Geophys. Res.*, in preparation, 1999.

Kivelson, M. G., K. K. Khurana, C. T. Russell, R. J. Walker, J. Warnecke, F. V. Coroniti, C. Polanskey, D. J. Southwood, and G. Schubert, Discovery of Ganymede's magnetic field by the Galileo spacecraft, *Nature, 384*, 537-541, 1996.

Kivelson, M. G., K. K. Khurana, D. J. Stevenson, L. Bennett, S. Joy, C. T. Russell, R. J. Walker, C. Zimmer, and C. Polanskey, Europa and Callisto: Induced or intrinsic fields in a periodically varying plasma environment, *J. Geophys. Res., 104*, 4609-4625, 1999.

Krupp, N., J. Woch, A. Lagg, B. Wilken, S. Livi, and D. J. Williams, Energetic particle bursts in the predawn Jovian magnetotail, *Geophys. Res. Lett., 24*, 1249-1252, 1998

Kurth, W. S., M. M. Baumback, and D. A. Gurnett, Direction finding measurements of auroral kilometric radiation, *J. Geophys. Res., 80*, 2764-2770, 1975.

Kurth, W. S., D. A. Gurnett, and F. L. Scarf, Spatial and temporal studies of Jovian kilometric radiation, *Geophys. Res. Lett., 7*, 61-64, 1980.

Kurth, W. S., D. A. Gurnett, S. J. Bolton, A. Roux, and S. M. Levin, Jovian radio emissions: An early overview of Galileo observations, in *Planetary Radio Emissions IV*, edited by H. O. Rucker, S. J. Bauer, and A. Lecacheux, Austrian Academy of Sciences Press, Vienna, pp. 1-13, 1997a.

Kurth, W. S., S. J. Bolton, D. A. Gurnett, and S. Levin, A determination of the source of Jovian hectometric radiation via occultation by Ganymede, *Geophys. Res. Lett.*, 24, 1171-1174, 1997b.

Kurth, W. S., D. A. Gurnett, A. Roux, and S. J. Bolton, Ganymede: A new radio source, *Geophys. Res. Lett.*, 24, 2167-2170, 1997c.

Ladreiter, H. P., and Y. Leblanc, Source location of the Jovian hectometric radiation via ray-tracing technique, *J. Geophys. Res.*, 95, 6423-6435, 1990.

Lecacheux, A., B. Moller-Pedersen, A. C. Riddle, J. B. Pearce, A. Boischot, and J. W. Warwick, Some special characteristics of the hectometric Jovian emission, *J. Geophys. Res.*, 85, 6877-6882, 1980.

Lecacheux, A., B. M. Pedersen, P. Zarka, M. G. Aubier, M. D. Desch, W. M. Farrell, M. L. Kaiser, R. J. MacDowall, and R. G. Stone, In ecliptic observations of Jovian radio emissions by Ulysses comparison with Voyager results, *Geophys. Res. Lett.*, 19, 1307-1310, 1992.

Louarn, P., A. Roux, S. Perraut, W. S. Kurth, and D. A. Gurnett, A study of the large-scale dynamics of the Jovian magnetosphere using the Galileo plasma wave experiment, *Geophys. Res. Lett.*, 25, 2905-2908, 1998.

Melrose, D. B., A theory for the nonthermal radio continua in the terrestrial and Jovian magnetospheres, *J. Geophys. Res.*, 86, 30-36, 1981.

Menietti, J. D., D. A. Gurnett, W. S. Kurth, J. B. Groene, and L. J. Granroth, Galileo direction finding of Jovian radio emissions, *J. Geophys. Res.-Planets*, 103, 20,001-20,010, 1998a.

Menietti, J. D., D. A. Gurnett, W. S. Kurth, J. B. Groene, and L. J. Granroth, Radio emissions observed by Galileo near Io, *Geophys. Res. Lett.*, 25, 25-28, 1998b.

Menietti, J. D., D. A. Gurnett, W. S. Kurth, and J. B. Groene, Control of Jovian radio emission by Ganymede, *Geophys. Res. Lett.*, 25, 4281-4284, 1998c.

Menietti, J. D., D. A. Gurnett, and J. B. Groene, Second harmonic hectometric radio emission at Jupiter, *Geophys. Res. Lett.*, 25, 4425-4428, 1998d.

Menietti, J. D., D. A. Gurnett, W. S. Kurth, and J. B. Groene, Effectiveness of near-grazing incidence reflection in creating the rotationally modulated lanes in the Jovian hectometric radio emission spectrum, *Radio Sci.*, 34, 1005-1012, 1999.

Morabito, L. A., S. P. Synnott, P. N. Kupferman, and S. A. Collins, Discovery of currently active extraterrestrial volcanism, *Science*, 204, 972, 1979.

Paranicas, C., W. R. Paterson, A. F. Cheng, B. H. Mauk, R. W. McEntire, L. A. Frank, and D. J. Williams, Energetic particle observations near Ganymede, *J. Geophys. Res.*, 104, 17,459-17,469, 1999.

Paterson, W. R., L. A. Frank, and K. L. Ackerson, Galileo plasma observations at Europa: Ion energy spectra and moments, *J. Geophys. Res.*, 104, 22,779-22,791, 1999.

Reiner, M. J., J. Fainberg, R. G. Stone, R. Manning, M. L. Kaiser, M. D. Desch, B.-M. Pedersen, and P. Zarka, Source characteristics of Jovian narrow-band kilometric radio emissions, *J. Geophys. Res.*, 98, 13,163-13,176, 1993.

Warnecke, J., M. G. Kivelson, K. K. Khurana, D. E. Huddleston, and C. T. Russell, Ion cyclotron waves observed at Galileo's Io encounter: Implications for neutral cloud distribution and plasma composition, *Geophys. Res. Lett.*, 24, 2139-2142, 1997.

Warwick, J. W., J. B. Pearce, A. C. Riddle, J. K. Alexander, M. D. Desch, M. L. Kaiser, J. R. Thieman, T. D. Carr, S. Gulkis, A. Boischot, Y. Leblanc, B. M. Pedersen, and D. H. Staelin, Planetary radio astronomy observations from Voyager 2 near Jupiter, *Science*, 206, 991-995, 1979.

Williams, D. J., and B. Mauk, Pitch angle diffusion at Jupiter's moon Ganymede, *J. Geophys. Res.*, 102, 24,283-24,287, 1997.

Williams, D. J., B. Mauk, and R. W. McEntire, Properties of Ganymede's magnetosphere as revealed by energetic particle observations, *J. Geophys. Res.*, 103, 17,523-17,534, 1998.

Woch, J., N. Krupp, J. A. Lagg, B. Wilken, S. Livi, and D. J. Williams, Quasi-periodic modulations of the Jovian magnetotail, *Geophys. Res. Lett.*, 24, 1253-1256, 1998.

Wong, H. K., and M. L. Goldstein, A mechanism for bursty radio emission in planetary magnetospheres, *Geophys. Res. Lett.*, 17, 2229-2232, 1990.

D. A. Gurnett, W. S. Kurth, and J. D. Menietti, Department of Physics and Astronomy, University of Iowa, Iowa City, IA 52242 USA.

SL-9: The Impact of Comet Shoemaker-Levy 9 at Jupiter

Yolande Leblanc

Departement de Recherche Spatiale, Observatoire de Paris, 92195 Meudon, France

The collision of Comet Shoemaker-Levy 9 at Jupiter was monitored by a worldwide network of ground-based and spaced-based radiotelescopes, from millimetric to kilometric wavelengths. Some results are: (1) One or two days after the first impact, an increase of flux density by 10-40% was detected at wavelengths from 3 to 90 cm, and the radio spectrum hardened. The increase concerned the synchrotron emission from the Jovian radiation belts. (2) During the week of cometary impacts, the beaming curve was flattened and distorted, and there were long-lasting enhancements of flux density. (3) Images in 2-D show that the radiation belts became brighter than before the impacts, and they remained bright at least one week after the first impact. The increase of brightness at the magnetic equator was confined to limited ranges of longitude, with the principal enhancement being between ≈ 150 to $\approx 240°$. (4) The radial distance of the belts from the planet decreased. (5) Images in 3-D reveal distinct brightenings confined in longitude that are attributed to the longitudes of individual impacts or groups of impacts. (6) Following the week of impacts, the flux density decreased at all wavelengths. A year later, the belts had relaxed to their normal, pre-impact state. (7) The changes produced by the impacts imply the existence of a new or a newly-accelerated population of electrons, a probable increase of the radial diffusion toward small radii, and a possible scattering of the pitch angle of some electrons near the magnetic equator to enhance the brightness at high latitudes. (8) No abnormal emission was detected at decameter or kilometer wavelengths from radio sources located in the auroral regions, although several auroral effects were reported in the UV, IR and X-rays.

1. INTRODUCTION

The crash of Comet Shoemaker-Levy 9 (SL9) into Jupiter, from 16 to 22 July 1994, was an exceptional event in many ways. Astronomers were able to witness the collision of two bodies of the solar system, involving energies similar to some solar flares. The comet was discovered more than one year before the collision, its trajectory, the impact sites and the timing of the individual impacts were known with an exceptional accuracy a long time before the collision. As a consequence, astronomers were able to put many telescopes into operation before, during and after the week of impacts; there were more than 20 individual impacts during six days, which gave many opportunities of comparison.

In this review only the radio observations will be described. Jupiter emits from millimetric to kilometric wavelengths, and distinct types of radiation are responsible for this very wide radio spectrum. (i) Thermal emission emanates from the atmosphere of the planet in the short centimeter and millimeter range; this emission corresponds to the brightness temperature of the disk, nearly constant with wavelength at $\approx 300°$ K; it is not polarized. (ii) Synchrotron emission originating from high energy electrons trapped in the radiation belts at 1.3-4 R_J; this emission is highly polarized (mostly linearly) and its brightness is approximately proportional to the square of the wavelength; as a consequence, this component dominates in the decimetric and metric range. (iii) Cyclotron maser emission originating in the auroral regions of Jupiter dominates in the decameter, hectometer and kilometer wavelength range. (iv) Plasma wave emission originating from sources located in the Io torus; this component is generated by anisotropic, fast electrons within the dense plasma of the torus.

Actually, only the synchrotron emission from the radiation belts was changed by the collision. The radio sources in the auroral regions did not show any abnormal emission during the impacts, nor did the centimetric/millimetric radiation from the atmosphere.

Following, in Section 2 we contrast the present knowledge of the Jovian radiation belts before the impacts, with the changes in flux density and brightness which occurred during the impacts. In Section 3, we comment on the recovery of the radiation belts to their normal pre-impact state, in the week following impacts, and a year later. In Section 4, we mention the physical processes needed to explain the observed changes in the radiation belts. In Section 5 we review the observations at decametric to kilometric wavelengths, and give concluding remarks.

2. CHANGES OF THE SYNCHROTRON RADIATION BELTS DURING THE SL-9 IMPACTS

Predictions of the effects of comet SL-9 on Jupiter's synchrotron radiation concentrated on the diminution of flux that could be produced by the dust and other debris accompanying the major fragments of the comet. It was expected that the relativistic electrons of the radiation belts would be de-energized, and the synchrotron flux would decrease by a little or a lot, depending on the amount of debris and the diameter of the dust particles.

Some 13 major radio telescopes observed Jupiter in the range of 2.8 to 90 cm, before, during and after the collision. Contrary to predictions, the flux density increased, and that increase remained for days to months afterward. The increase was not observed the day of the first impact, but one or two days later. (In IAU telegrams, the first report of a flux increase was on July 18th.) The brightness of the belts was strongly affected by the impacts and the beaming curves were distorted.

2.1. The Jovian Synchrotron Emission

What did we know on the radiation belt emission before the collision? Jupiter's decimeter emission was discovered in 1958 [*Sloanaker*, 1959], the same year as the discovery of the counterpart on Earth, the Van Allen belts. Soon after, the radiation was attributed to synchrotron emission by high energy electrons trapped in the Jovian magnetic field.

There are two different populations of energetic electrons trapped in the Jovian radiation belts: one with pitch angles α near $90°$, giving radiation concentrated at the magnetic equator, with the intensity being largest near $R \approx 1.5$ R_J [*Berge and Gulkis*, 1976]. The other population has a wider pitch angle distribution and produces radiation at a large range of magnetic latitudes, with a secondary peak of intensity at high latitudes [*de Pater and Jaffe*, 1984; *Leblanc et al.*, 1996].

Plate 1a shows a rotation-averaged image of Jupiter at 13 cm made with the Australian Telescope Compact Array (ATCA). In total intensity, the thermal radiation of Jupiter's disk is about as bright as the equatorial synchrotron emission. In linearly polarized intensity (Plate 1b) the thermal radiation is absent. The images show clearly the radiation from the two populations of electrons, one concentrated at the magnetic equator, and the other producing radiation up to high latitudes. The high latitude peaks have been associated with preferred mirroring regions at locations close to the planet where the magnetic field is intense and where the electrons spend most of their time. *Dulk et al.*, [1997] derived the properties of the electrons producing the peaks, and found that they are concentrated near $L \approx 2.4$ and have an average pitch angle of $27°$.

Typical time scales for large pitch angle, 10 MeV electrons spiraling at $R \approx 1.5$ R_J are: a few microseconds for the gyroperiod, a few seconds for the mirror period, 5 days for the drift period around the planet, and 40 days for diffusion toward the planet.

Synchrotron emission depends on the strength and geometry of the magnetic field. Each electron emits radiation with a power proportional to B^2, over a well-defined frequency range. The main characteristics of the Jovian magnetic field have been derived from the decimetric and decametric radio emissions: its magnetic dipole component is inclined by $10°$ from the rotational axis. It has important non-dipolar components, with the highest intensity at the surface being 14 G. Plate 1e is a 3-D image of the synchrotron radiation: it shows the radiation concentrated at the magnetic equator, and the extension toward high latitudes. The surface of the magnetic equator is not plane but distorted by the non-

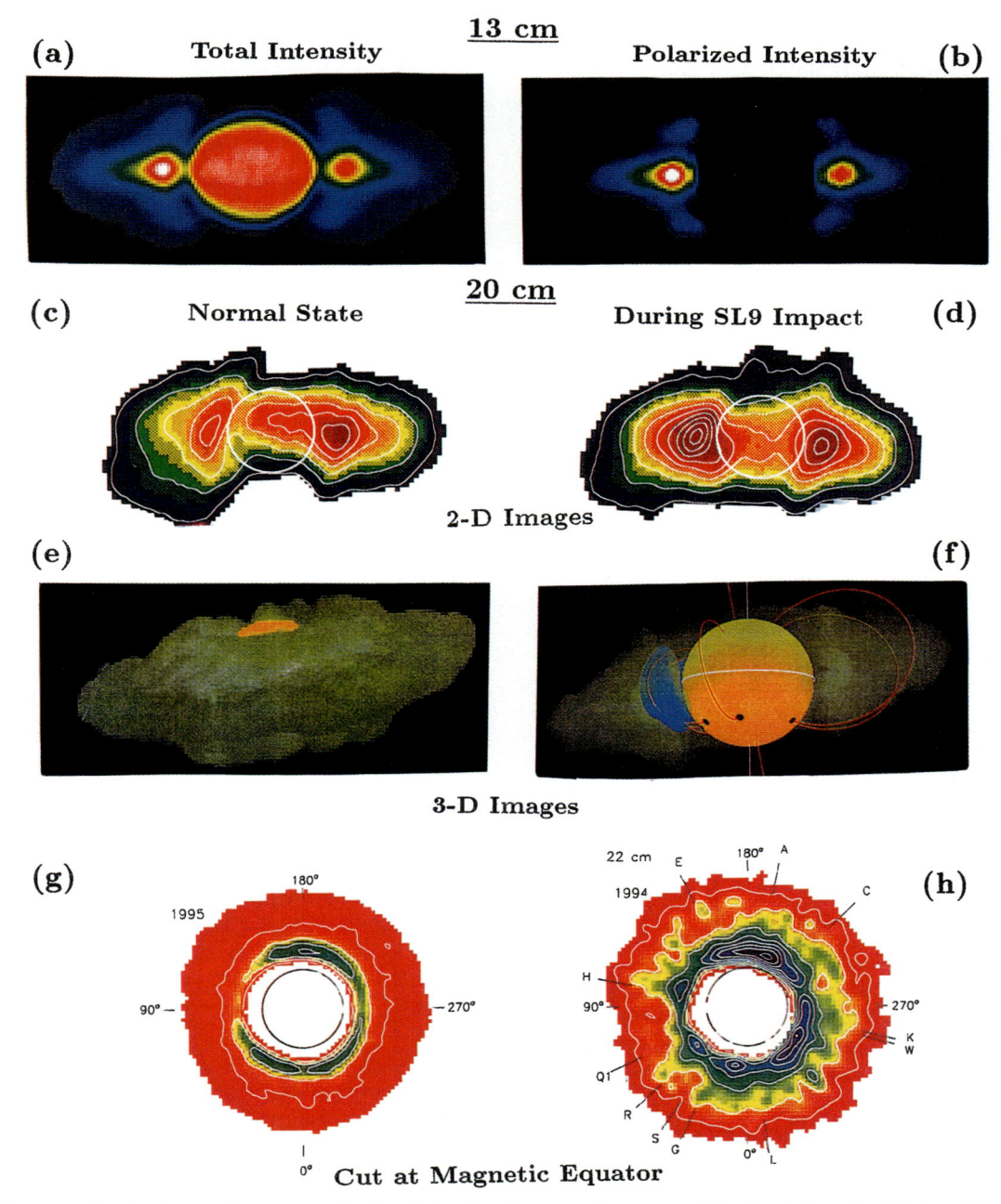

Plate 1. Jupiter's synchrotron radiation belts in the normal state and during/after the SL-9 collision from observations made with the ATCA. (a) Normal state, rotation-averaged image in total intensity at 13 cm, showing the thermal radiation from the disk and the synchrotron radiation on each side of the disk. (b) The same in linearly polarized intensity, where the unpolarized thermal radiation is absent. (c) At 20 cm, a 2-D image of Jupiter in its normal state. Note the East-West asymmetry. (d) The same, 3 days after the first impact: the eastern side (left hand) has almost doubled in brightness. (e) 3-D image of Jupiter in its normal state. (f) Normal state plus effects of the SL-9 collision: the normal state is in green, the principal enhancement due to impacts A-C-E is in blue, the thermal disk is in orange, and the black dots show sites of some impacts with the field lines emanating from them. (g) Cut near the magnetic equator from the 3-D image of the normal state, showing the radial distribution of brightness vs. longitude. The black circle is Jupiter's disc. (h) Average image from observations during/after collision (1994 July 18 to 28). The major localized brightening is at the longitude of impacts A, C, and E, and fainter brightenings are near the sites of other impacts.

dipolar components of the field. Magnetic field models have been derived from direct measurements made by magnetometers aboard the Pioneer and Voyager spacecraft, and from the footprint of the Io flux tube [*Connerney*, 1993; *Connerney et al.*, 1998], but these models are not perfect, particularly at small distances ($R \lesssim 2$ R_J) and low latitudes. High resolution radio observations have been used to suggest changes to these models [*Dulk et al.*, 1996; 1999].

2.2. The Flux Density Increased

The flux density has been monitored at 13 cm since 1971. After regular changes are removed (due mainly to Jupiter's rotation and viewing geometry), there remain no short term variations with time scales of days or weeks [*Klein et al.*, 1997; 1999]. Evidence of long-term variations of flux density with time scales of a few years has been shown by *Klein et al.*, [1989]. The abrupt increase observed during the SL-9 collision is unprecedented.

The flux density increased at all wavelengths, one or two days after the first impact [*de Pater et al.*, 1995; *Leblanc et al.*, 1997]. The flux increase was wavelength-dependent, $\approx 45\%$ at 6 cm, $\approx 30\%$ at 11-13 cm, $\approx 25\%$ at 18-22 cm, $\approx 40\%$ at 36 cm, and 15% at 70-90 cm. Fig. 1 shows the flux density increases and the declines observed at wavelengths from 6 to 90 cm, during and after the collision. These different increases show that the radio spectrum hardened between 3 and 36 cm; at longer wavelengths the situation is not clear due to the large error bars.

2.3. The Beaming Curve Became Distorted and Flattened

The flux density of Jupiter varies as the planet rotates, a consequence of the beamed synchrotron emission at the magnetic equator, and of the tilt of the magnetic axis with respect to the rotation axis. The variation of the flux density with central meridian longitude CML (the Jovian longitude facing the observer, as distinct from λ_{III}, the longitude of a particular feature on Jupiter) shows 2 peaks near the times when the magnetic equatorial plane intersects the observer. These two peaks are of unequal intensity due to the non-dipolar character of the magnetic field, and their intensity changes with D_E, the Jovian declination of the Earth [*Klein et al.*, 1989].

During the week of impacts, the beaming curve was strongly distorted and flattened [*Klein et al.*, 1995; *Bolton et al.*, 1995; *Galopeau et al.*, 1996; *Wong et al.*, 1996]. Localized enhancements of flux density at about the same CML during consecutive days were reported by *Bolton et al.* [1995], *Bird et al.*, [1996] and *Klein et al.*, [1995]. As shown below, the largest of these enhancements can be related to impact A or to the group of impacts A, C, E.

2.4. The Changes of Brightness in the Radiation Belts

The east-west asymmetry of the radiation belts (Plate 1c) was first noticed by *Branson*, [1968] and very soon was attributed to the non-dipolar character of the magnetic field [*Conway and Stannard*, 1972; *Gerard*, 1976]. High resolution 2-D maps reveal that the brightness on each side of the equatorial belt changes as Jupiter rotates [*de Pater and Dames*, 1979; *Dulk et al.*, 1996]. A cut at the magnetic equator from a 3-D image shows that it is the result of the presence of a maximum of brightness at longitude $\lambda_{III} \approx 200°$, and of a minimum brightness at $\lambda_{III} \approx 110°$ (Plate 1g).

During the SL-9 impact, the belts became brighter than before. The principal increase of brightness was confined in longitude, as shown by two dimensional images with good angular resolution produced by aperture synthesis telescopes: at 20 cm with the Very Large Array, and at 13 and 22 cm with the ATCA. The ATCA images of 12 days observation between 9 and 28 of July 1994 [*Leblanc and Dulk*, 1995] show how the belts became brighter, and demonstrated that they remained brighter at least until the end of the observations, 12 days after the first impact.

Plate 1c shows a pre-impact, 2-D image from the ATCA when the CML was 70°, and Plate 1d shows the same thing 3 days after the first impact. The change is impressive, mainly on the eastern side, which has almost doubled in brightness from about 450 K to 800 K. From the set of observations, it was derived that the increase in brightness was not the same at all longitudes; 2-D cuts at the magnetic equator of images at 13 and 22 cm showed that the enhancement was most evident between λ_{III} 150° and 240° [*Leblanc and Dulk*, 1995]. In addition, the VLA images revealed that the high latitude regions increased in intensity up to 30% relative to the equatorial radiation [*de Pater et al.*, 1997].

2.5. Radio Brightenings Localized in Longitude near the Magnetic Equatorial Plane

Images in 3-D constructed from ATCA observations reveal distinct brightenings near the magnetic equator that are confined in longitude. The method of constructing the brightness distribution is based on tomographic techniques and relies on the rotation of Jupiter; it is detailed by *Sault et al.*, [1997a].

Plate 1e shows a 3-D image of Jupiter in its normal state: the magnetic equatorial radiation is most evident, and the magnetic equator is not planar, but is warped. The extentions to high latitudes are due to the

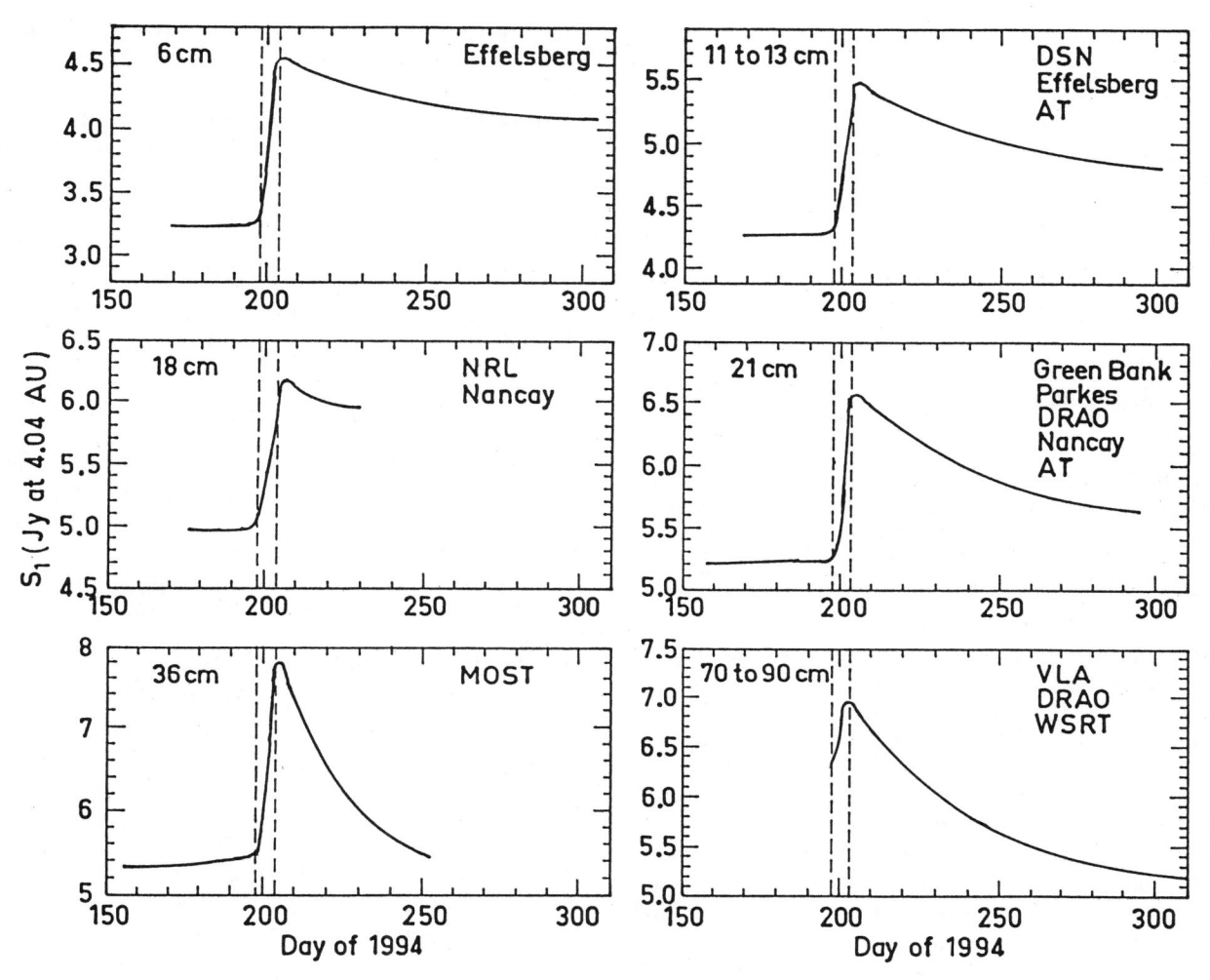

Figure 1. Flux density increases and declines during and after the collision. The week of cometary impacts is enclosed by the two dashed lines. Adapted from *de Pater et al.*, [1995].

population of electrons with smaller pitch angles.

Plate 1f combines an image of the normal state (in green) with the principal enhancement due to the impacts (in blue), the thermal disk (in orange), and sites of some major impacts with the field lines which emanate from them. The strong, localized brightening is on L-shell ≈ 1.5, and is related to field lines emanating from impacts A, C and E, in the longitude range 150-240°. On the other hand, the impacts occurring at longitudes 0-100° were at the footpoints of L-shell ≈ 2.5, near the location of the small pitch angle electron population [*Dulk and Leblanc*, 1995; *Bolton*, 1997].

Plate 1g is a cut near the magnetic equator of a 3-D image at 22 cm, derived from 1995 observations when the belts had recovered their normal state. This image shows a smooth variation of brightness with radius and longitude, with a brightness minimum near 110° and a brightness maximum near 200°. This is to be compared with Plate 1h, which was derived from 8 days of observations, 1994 July 18 to 28. Although it is an average of data taken during and just after impacts, it shows distinct brightenings confined in longitude. The major brightening at $\lambda_{III} \approx 200°$ is attributed to the impacts A, C and E. It is responsible for the long-lasting, localized enhancement observed in the distorted beaming curves by *Bolton et al.*, [1995] and by *Bird et al.*, [1996]. In this averaged image, other localized enhancements can be linked to other impacts.

Sault, Leblanc and Dulk, [1997b] investigated the temporal sequence of changes using 3-D reconstructions of each day's data referenced to earlier data. These "difference images" show that most of the brightenings were located near the longitudes of major impacts or groups of impact. The brightenings were not visible on

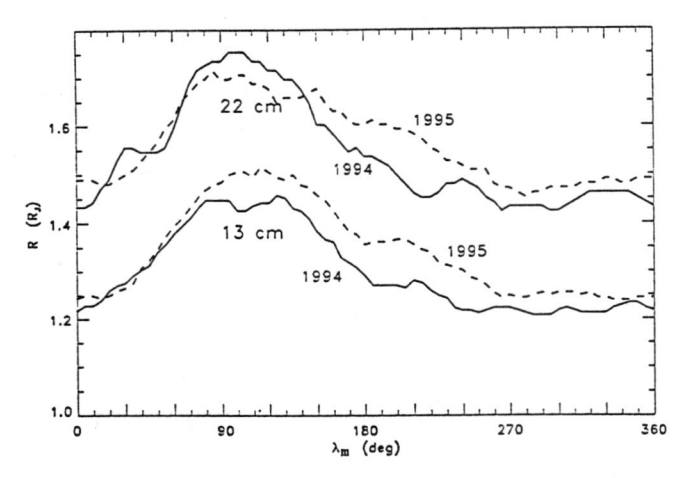

Figure 2. Radial distance of the brightness maximum in the magnetic equator vs. longitude λ_{III}, comparing the normal state (1995) with during/after impacts (1994). (See also Plate 1g and 1h.) The curves at 13 cm have been lowered by 0.2 R_J for clarity. After *Dulk et al.*, [1997].

the days of the impacts but built up over one or two days. They remained visible for several days, slowly fading and becoming more and more diffuse, with little or no drift in longitude. By 28 July, a diffuse brightening was observed all around the planet. The radial distance of the brightness enhancements corresponded to the L-shells of the impacts sites, i.e. $L \approx 1.5$ to 2.5.

The lack of detectable longitude drift is not understood. The drift period of electrons around the planet is about 4-5 days, depending on the electron energy and the L-shell. Why the individual enhancements did not drift at this expected rate of some 70° per day is a mystery.

2.6. Decrease of the Radial Distance of the Belts at the Magnetic Equator

The radial distance of the peak brightness at the magnetic equator has been extracted from 3-D reconstruction, during the impacts, and one year later when the belts had recovered their normal state [*Dulk, Sault and Leblanc*, 1997]. The data of 1995 show the normal variation of the peak distance with longitude: it is the locus of points on the magnetic equator where B=constant ≈ 1.2 G. It varies from 1.4 to 1.75 R_o, the larger distance being where the magnetic field is the highest, close to the major magnetic anomaly. This variation is evident in Plate 1g where it appears that the circle representing Jupiter's disk is offset from the center of the radiation belts. Fig. 2 shows a plot of the radius of maximum brightness vs. longitude.

During the impacts the radial distance of peak brightness decreased by 0.05 to 0.1 R_J. This can be seen by comparing Plate 1h with Plate 1g, and in Fig. 2. The same trend exists at 13 and 22 cm. A possible way that the comet impacts could cause the radial distance of peak brightness to move inward is enhanced radial diffusion of the relativistic electrons due to turbulence created by the explosive events.

3. CHANGES AFTER IMPACTS AND RECOVERY OF THE RADIATION BELTS

After the week of impacts the flux density began to decrease at all wavelengths, and the brightness of the belts diminished progressively.

3.1. Flux Density Decrease

The decrease of the radio emission was observed for more than one year. It was not a simple exponential process. The time scale was more than 100 days at 6, 11, 13 and 20 cm [*Bird et al.*, 1996; *Klein et al.*, 1995; *Wong et al.*, 1996], but less than 30 days at longer wavelengths, 36-70-90 cm [*Dulk, Leblanc and Hunstead*, 1995; *Sukamar*, 1995]. The flux density increases and decreases are shown in Fig. 1. However, even one year after the SL-9 collision, the radio flux had not returned to its pre-impact level at 6 cm [*Bird et al.*, 1996].

3.2. Changes in Polarization

When integrated over the source, the synchrotron emission is significantly linearly polarized, $\approx 25\%$, and very little circularly polarized, $\lesssim 2\%$. Changes in the polarization reflect changes in the radial and latitudinal distribution of the energetic electrons. A small increase ($\approx 1\%$) of the linearly polarized flux density was reported at 11 cm by *Bird et al.*, [1996] and at 20 cm by *Wong et al.*, [1996]. A slight decrease of the circularly flux density was also reported at 11 cm. These global changes, however, are difficult to interpret in terms of a new distribution of the electrons during the impacts.

When observed with high spatial resolution, much of the brightness distribution is strongly linearly polarized, $\approx 70\%$. However, no analysis of cometary effects on the polarized brightness have so far been reported.

3.3. Brightness Decrease of the Belts

The 2-D image made by the ATCA on July 28, twelve days after the first impact, (3.5 days after the last impact, shows that the belts remained brighter than before the collision [*Leblanc and Dulk*, 1995]. Similarly, the high latitude emissions continued changing for weeks after the last impact [*de Pater et al.*, 1997].

One year later, in 1995, the images at 13 and 22 cm were very similar to those obtained before the impacts [*Dulk et al.*, 1997].

4. PHYSICAL PROCESSES INVOLVED

The proposed physical mechanisms described below account only for some of the observed changes in the radiation belts: the increase in intensity of the synchrotron radiation, the inward displacement of the intensity peaks, the flattening of the beaming curve, the broadening of the magnetic equatorial radiation. None of them accounts for all the effects.

The increases of flux density and brightness, with a hardening of the spectrum, have been interpreted as the existence of a new or a newly accelerated population of electrons confined in a portion of the magnetic field of the planet. One process of acceleration, suggested by *Brecht et al.*, [1995], is the formation of collisionless shock waves generated by the impacts. In this mechanism, the shock, in less than one second, could accelerate electrons from low energies to very high energies, up to tens of Mev. However, a shock would have completely dissipated before reaching the electron population near the magnetic equator, and could not have produced delayed, long-lasting brightenings at the individual impact sites.

Another process which leads also to an increase of the energy of the electrons is radial diffusion. This mechanism explains, in general, how the radiation belts become populated with relativistic electrons. The electrons move inward, adiabatically, from regions of weak magnetic field to regions of strong magnetic field. In the process they conserve adiabatic invariants and gain energy. Therefore an increase of radial diffusion could explain the increase of intensity, and also the inward displacement of the peak intensities. An increase in the rate of the radial diffusion can be generated by a transient turbulence in the neutral winds of Jupiter atmosphere [*Ip*, 1995]. However, an increase of radial diffusion by itself would concentrate the electrons near the magnetic equator, which is not the only thing that occurred.

The flattening of the beaming curve and the latitudinal broadening of the magnetic equatorial radiation have been attributed to an increase of the pitch angle distribution of some relativistic electrons. Waves triggered by the impacts may have enhanced pitch angle scattering of the electrons, moving the mirror points of some of them from near the magnetic equator to higher latitudes [*Bolton and Thorne*, 1995]. This mechanism may explain the overall increase of synchrotron radiation and the latitudinal broadening, but it does not explain the increased intensity of the magnetic equatorial radiation, the inward displacement of the equatorial peaks, or the localization of the brightenings.

It is clear from the observations that the longitudes of brightness enhancements at the magnetic equator were close to the longitudes of the individual impacts, and the radial distance of the brightness enhancements corresponded to the L-shells of the impacts sites, 1.5 to 2.5. The most puzzling problem is how the localized brightenings were fixed in longitude for nearly a week. Recently Hill and Dessler, [1999] proposed a mechanism involving enhanced radial diffusion linked to atmospheric winds that localized at the impact sites and are coupled to the magnetosphere by field-aligned currents.

Bolton, [1997] discussed the consistency of proposed mechanisms, and concluded that further theory and modelling is required to understand the changes. In general, the predicted effects were not observed; in particular, the plasma-dust interaction during the traversal by comet debris of the inner magnetosphere. The estimated amount of dust was about 10^8 tons which should have reduced the flux even before the first impact. No decrease of synchrotron emission occurred; rather, an increase of flux density commenced one or 2 days after the first impact.

5. SL-9 EFFECTS AT OTHER RADIO WAVELENGTHS

No change was observed in the emission from the auroral radio sources (at decameter to kilometer wavelengths), although some authors predicted that they would be affected by the dust or the SL-9 fragments.

5.1. Decametric Radio Sources

Several large, decametric, radio telescopes were observing during the impacts, e.g. Nançay in France, Maipu in Chile, Florida in the USA. At these observatories no significant change from the usual level of Jovian decametric activity was observed that could confidently be attributed to the comet impacts [*Carr et al.* 1995]. These authors mention two bursts that might be related to the Q1 and Q2 impacts; however, the bursts were not confirmed by other radio telescopes.

Many other radio telescopes were observing Jupiter at decameter wavelengths for the first time. Some bursts were attributed to individual impacts but a careful inspection of the records does not substantiate a relationship with the comet, and none of the suggested associations have been confirmed by independent observations.

5.2. Kilometric Radio Sources

During the week of impacts, the radio experiment aboard the Ulysses spacecraft was continuously recording radio activity originating from the Sun and Jupiter in the range of 1 to 900 kHz. No systematic intensity changes before, during or after the impact times were recorded [*Desch et al.*, 1995].

6. CONCLUSION

The interest generated by the SL-9 collision at Jupiter has stimulated efforts to understand the synchrotron emission of the radiation belts. In particular, high resolution observations of Jupiter providing 2-D and 3-D images have shown how the warp of the magnetic equator surface affects the observed brightness. In addition, synchrotron radiation belts provide information on the magnetic field at small radii and low latitudes where few direct observations are available. These observations have been used to investigate the fidelity of magnetic field models and to add new constraints to improve the models.

Acknowledgments. All of the images of this paper were acquired with the Australia Telescope Compact Array, a facility of the Australia Telescope National Facility, CSIRO. I thank Prof. George Dulk and Dr. Robert Sault for their much-valued participation in the observations and analysis.

REFERENCES

Berge, G.L., and S. Gulkis, Earth-based radio observations of Jupiter at millimeter to meter wavelengths, In *Jupiter*, Ed. T. Gehrels, Tucson: Univ. Ariz. Press, p. 621, 1976.

Bird, M.K., O. Funke, J. Neidhöfer and I. de Pater, Multi-frequency radio observations of Jupiter at Effelsburg during the SL9 impact, *Icarus*, *121*, 450, 1996.

Brecht, S.H., M.E. Presses, J.C. Lyon, N.T. Gladd and S.W. Mc Donald, An explanation of synchrotron radiation emhancement following the impact of SL-9 with Jupiter, *Geophys. Res. Lett.*, *22*, 1805, 1995.

Bolton, S.J., Interpretation of the observed changes in Jupiter synchrotron radiation during and after the impacts from comet SL-9, *Planet. Space Sci.*, *45*, 1359, 1997.

Bolton, S.J., and R.M. Thorne, Assessment of mechanisms for Jovian synchrotron variability associated with Comet SL9, *Geophys. Res. Lett.*, *22*, 1813, 1995.

Bolton, S.J., R.S. Foster, and W.B. Waltman, Observations of Jupiter's synchrotron radiation at 18 cm during the Comet Shoemaker-Levy/9 impacts, *Geophys. Res. Lett.*, *22*, 1801, 1995.

Branson, N.J.B.A., High-resolution radio observations of the planet Jupiter, *Mon. Not. R. Astron. Soc.*, *139*, 162, 155, 1968.

Carr, T.D. and 14 others, Results of decametric monitoring of the comet collision with Jupiter, *Geophys. Res. Lett.*, *22*, 1785, 1995.

Connerney J.E.P., Magnetic field of the outer planets, *J. Geophys. Res.*, *98*, 18659, 1993.

Connerney, J.E.P., M.H. Acuna, N.F. Ness, and T. Satoh, New models of Jupiter's magnetic field constrained by the Io flux tube footprint, *J. Geophys. Res.*, *103*, 11929, 1998.

Conway, R.G., and D. Stannard, Non dipole terms in the magnetic field of Jupiter and the Earth, *Nature Phys. Sci.*, *239*, 142, 1972.

de Pater, I., and H.A.C. Dames, Jupiter's radiation and atmosphere, *Astron. Astrophys.*, *72*, 148, 1979.

de Pater, I., and W.J. Jaffe, VLA observations of Jupiter's non-thermal radiation, *Astrophys. J. Suppl.*, *54*, 405, 1984.

de Pater, I., and 25 others, Outburst of Jupiter's synchrotron radiation after the impact of comet Shoemaker-Levy-9, *Science*, *268*, 1879, 1995.

de Pater, I., F. van der Tak, R.G. Strom, and S.H. Brecht, Time evolution of the east-west asymmetry in Jupiter's radiation belts after the impact of Comet P/Shoemaker-Levy 9, *Icarus*, *129*, 21, 1997.

Desch, M.D., M.L. Kaiser, W.M. Farrell, R.D. MacDowall, and R.G. Stone, Traversal of Comet SL-9 through the Jovian magnetosphere and impact with Jupiter: radio upper limits, *Geophys. Res. Lett.*, *22*, 1781, 1995.

Dulk, G.A., and Y. Leblanc, Changes in Jupiter's Synchrotron Radiation Belts During and After SL-9 Impacts, *Proc. European SL-9/Jupiter Workshop, ESO Conf. Proc. 52*, Ed. R.M. West and H. Boehnhardt, European Southern Observatory, Garching, Germany, pp. 381–386, 1995.

Dulk, G.A., Y. Leblanc, and R.W. Hunstead, Flux and images of Jupiter at 13, 22 and 36 cm before, during and after SL-9 impacts, *Geophys. Res. Lett.*, *22*, 1789, 1995.

Dulk, G.A., Y. Leblanc, Y., R.J. Sault, H.P. Ladreiter, and J.E. Connerney, The radiation belts of Jupiter at 13 and 22 cm: II. The asymmetries and the magnetic field, *Astron. Astrophys.*, *319*, 282, 1997.

Dulk, G.A., Y. Leblanc, R.J. Sault, Jupiter's radiation belts: at the time of Comet SL-9 and a year later, *Planet. Space Sci.*, *45*, 1231, 1997.

Dulk, G.A., Y. Leblanc, R.J. Sault, S.J. Bolton, J.H. Waite, and J.E.P. Connerny, Jupiter's magnetic field as revealed by the synchrotron radiation belts. I. Comparison of a 3-D reconstruction with models of the field, *Astron. Astrophys.*, *347*, 1999.

Galopeau, P.H.M., E. Gerard, and A. Lecacheux, Long term monitoring of Jupiter's synchrotron radiation with the Nançay radiotelescope including the collision with Comet P/Shoemaker-Levy 9, *Icarus*, *121*, 469, 1996.

Gerard, E., Variation of the radio emission of Jupiter at 21.3 and 6.2 cm wavelength, *Astron. Astrophys.*, *50*, 353, 1976.

Ip, W.H., Time variations of the Jovian synchrotron radiation following the collisional impacts of comet SL-9. Flux enhancemetn induced by neutral atmospheric turbulence, *Planet. Space Sci.*, *43*, 221, 1995.

Klein, M.J., T.J. Thompson, and S.J. Bolton, Systematic observations and correlation studies of variations in the synchrotron radio emission from Jupiter, In: *Time Variable Phenomena in the Jovian System*, NASA SP-494, Eds. M.J.S. Belton, R.A. West, and J. Rahe, p. 151, 1989.

Hill, T.W., and A.J. Dessler, Stirring of the Jovian radiation belt by comet SL-9 impacts, abstract in *Proc. Magnetospheres of the Outer Planets*, Paris, p. 59, 1999.

Klein, M.J., S. Gulkis, and S.J. Bolton, Changes in Jupiter's 13 cm synchrotron radio emission following the impacts of Comet SL-9, *Geophys. Res. Lett.*, *22*, 1797, 1995.

Klein, M.J., S. Gulkis, and S.J. Bolton, Jupiter's synchrotron radiation: observed variations before, during and after the impacts of Comet SL-9, In *Planetary Radio Emissions IV*, Eds. H.O. Rucker, S.J. Bauer and A. Lecacheux, Vienna: Austrian Academy of Science, pp. 217-224, 1997.

Klein, M.J., S. Gulkis, S.J. Bolton, and S.M. Levin, A study of short-term variations in Jupiter's synchrotron emission, abstract in *Proc. Magnetospheres of the Outer Planets*, Paris, p. 58, 1999.

Leblanc, Y., and G.A. Dulk, Changes in brightness during SL-9 impacts, *Geophys. Res. Lett.*, *22*, 1793, 1995.

Leblanc, Y., G.A. Dulk, R.J. Sault and R.W. Hunstead, The radiation belts of Jupiter at 13 and 22 cm. I. Observations and 3-D reconstruction, *Astron. Astrophys.*, *319*, 274, 1996.

Leblanc, Y., R.J. Sault, and G.A. Dulk, Synthesis of magnetospheric radio emissions during and after Jupiter/SL-9 collision, *Planet. Space Sci.*, *45*, 1213, 1997.

Sault, R.J., T. Oosterloo, G.A. Dulk, and Y. Leblanc, The first three-dimensional reconstruction of a celestial object at radio wavelengths: Jupiter's radiation belts, *Astron. Astrophys.*, *324*, 1190, 1997a.

Sault, R.J., Y. Leblanc, and G.A. Dulk, G.A., Localized brightenings in Jupiter's radiation belts resulting from SL-9 impacts, *Geophys. Res. Lett.*, *94*, 2395, 1997b.

Sloanaker, R.M., Apparent temperature of Jupiter at a wavelength of 10 cm, *Astron. J.*, *64*, 346, 1959.

Sukumar, S., Changes in Jupiter's synchrotron radiation due to the impact of Comet Shoemaker-Levy 9, *Astron. J.*, *110*, 1397, 1995.

Wong, M.H., I. de Pater, C. Heiles, R. Millan, R.J. Maddalena, M. Kesteven, R.M. Price, and M. Calabretta, Observations of Jupiter's 20 cm synchrotron emission during the impacts of Comet P/Shoemaker-Levy 9, *Icarus*, *121*, 457, 1996.

Y. Leblanc DESPA, Observatoire de Paris, 92195 Meudon, France. (e-mail: leblanc@obspm.fr)

Long Wavelength Astrophysics

W.C. Erickson

School of Mathematics and Physics, University of Tasmania, Hobart, Tasmania, Australia.

Some aspects of long wavelength astrophysics that have not been the subjects of other more detailed tutorials will be discussed. These include synchrotron emission, the production of steep low frequency spectra, propagation effects in plasmas, spectral turn-overs, and coherent emission. The emphasis will be on simple explanations of the phenomena along with a few of the most relevant equations. Some of the rather unusual consequences of these phenomena will be discussed.

SYNCHROTRON EMISSION

Synchrotron emission is the dominant radiation mechanism at long wavelengths. Synchrotron emission dominates thermal emission because of the $1/\lambda^2$ in the Rayleigh-Jeans approximation to the Planck Law. The Rayleigh-Jeans approximation applies to the radiation from thermal plasmas throughout this wavelength range

$$B = 2kT/\lambda^2 \qquad (1)$$

where B is the surface brightness of an opaque plasma, T is its physical temperature, and λ is the wavelength.

As a single highly relativistic electron spirals in an interstellar magnetic field it radiates at high harmonics of its gyrofrequency as first described by *Scott* [1912] and reviewed by *Ginzburg and Syrovatski* [1965] (see corrections in *Ginzberg, Sazonov and Strovatski* [1968], see also *Landau and Lifshitz* [1951] or *Jackson* [1962]). This emission occurs because the electron travels at nearly the phase velocity of light, causing its radiation to be relativistically beamed into a narrow cone in the instantaneous direction of its motion. Since the fundamental gyrofrequency is only a Hz or so, its harmonics form a virtual continuum in the radio frequency range. The emission peaks near a frequency of ν_c,

$$\nu_c = 16.08 \times B_\perp E^2 \qquad (2)$$

With the magnetic field, $B_\perp$, perpendicular to the electron's motion in μG, and the electron's energy, E, in GeV, ν_c is in MHz. For typical values of $1\mu G$ and 1 GeV, $\nu_c = 16$ MHz. The spectrum radiated by a single electron is illustrated in Fig.1. It is given by a fairly complicated expression involving an integral of a modified Bessel function. For frequencies $\nu \ll \nu_c$ the emission increases as $\nu^{1/3}$ while for $\nu \gg \nu_c$ it falls as e^ν.

Of course, the radiation that we observe is generated by a large ensemble of electrons. It is easily shown that if the radiating electrons have an energy spectrum given by

$$n(E) = n_o E^{-\gamma} \qquad (3)$$

the power per unit frequency radiated by the ensemble is

$$P_\nu = \nu^\alpha \qquad (4)$$

where $\alpha = -(\gamma - 1)/2$ is the spectral index of the emission.

Most radio sources have $\alpha \approx -0.75$ which implies $\gamma \approx 2.5$. This value also agrees with the measured energy

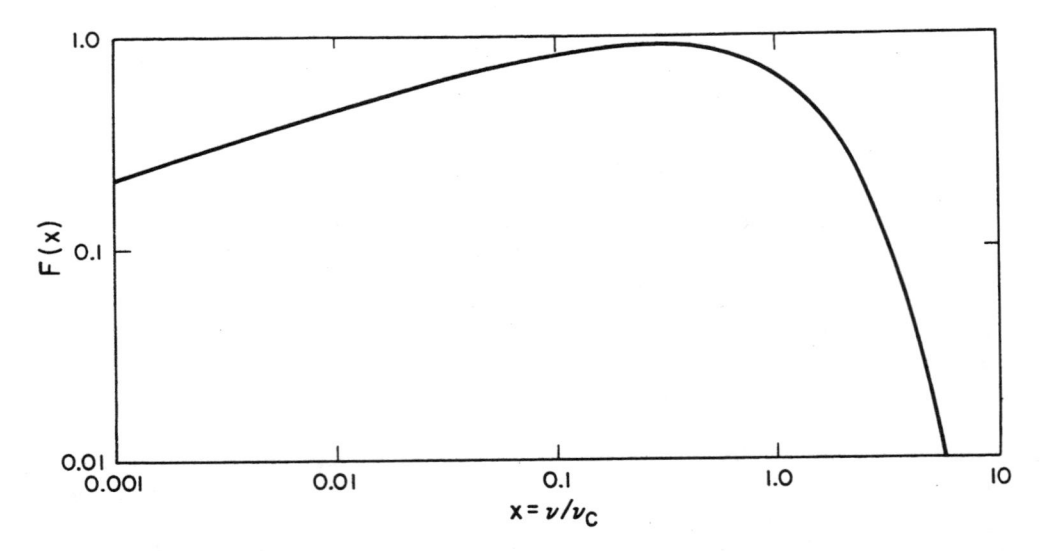

Figure 1. The single-electron synchrotron emission spectrum.

spectrum of cosmic ray electrons in the vicinity of the Earth. However, various sources have spectral indices that range from -0.5 to -2.0, implying $-2.0 > \gamma > -5.0$.

STEEP SPECTRA

Components of sources that have particularly steep spectra are of special interest at long wavelengths where they become most prominent. These components are generally of much larger angular size than flat spectrum ones. Most of the steep spectrum components are bridges between double radio sources, halos surrounding sources, or halos in clusters of galaxies. They are assumed to be produced by source regions containing electrons having particularly steep energy spectra. The large $|\gamma|$ could be produced intrinsically by the electron acceleration mechanism or by various energy loss mechanisms.

Electrons can lose energy through their synchrotron radiation or through inverse-Compton scattering by, primarily, microwave background photons. These losses go as E^2. They may also lose energy through bremsstrahlung with atomic material or because of the work expended if the source region expands. Bremsstrahlung and expansion losses go as E. Energy may also be lost through ionization of ambient matter. Ionization losses are nearly energy independent in the relevant energy ranges. Assuming that the energy loss rate can be expressed by a polynomial (*see Kardashev* [1962])

$$-dE/dt = a + bE + cE^2 \qquad (5)$$

and electrons are being continuously injected with a power-law spectrum

$$q(E) = KE^{-\gamma_o} \qquad (6)$$

then an equilibrium electron energy spectrum is eventually established which, over restricted energy ranges, can be characterized by power-laws with indices of $-(\gamma_o - 1)$, $-\gamma_o$, and $-(\gamma_o + 1)$. The radio emission spectrum from such an ensemble will possess three power-law segments with $\alpha = -(\gamma_o - 2)/2$ at low frequencies, $\alpha = -(\gamma_o - 1)/2$ at intermediate frequencies, and $\alpha = -\gamma_o/2$ at high frequencies. Thus the spectrum will have changes of slope of -1/2 at two frequencies ν_1 and ν_2. This mechanism is more likely to produce a flat low frequency spectrum than a steep one. Only if the energy losses are so severe that ν_2 is below the observing frequency will a steepening of the low frequency spectrum result.

Many source spectra are curved and exhibit steepening at high frequencies (*Kuhr et al*, 1981). This behavior is probably caused by electron energy losses.

Red shifts do not change spectral indices but shift all spectral features by a factor of $1/(1+Z)$ and thus may shift steep higher frequency portions of a radio source's spectrum to lower frequencies. This effect has been used to identify distant, highly red-shifted radio galaxies (*Rawlings et al*, 1996).

However, if the electron acceleration is not continuous, synchrotron losses will produce a radio spectrum that evolves with time (*Kardashev*, 1962). After an initial injection of electrons with an energy spectrum index of γ_o, the radio spectral index changes from $-(\gamma_o-1)/2$ below a break frequency, ν_b, to $-(2\gamma_o+1)/3$ above ν_b. The break frequency decreases with time as:

$$\nu_b \approx 340 B^{-3} t^{-2} \qquad (7)$$

where ν_b is in MHz, B is in μG, and t is in 10^9 years. If $B = 1\mu G$ and $t = 10^9$ years, then $\nu_b = 340$ MHz. In a much stronger field of $10\mu G$, ν_B would be only 0.3 MHz. Therefore, this effect could be very important in interpreting long wavelength observations.

A class of red, dusty, ultra-steep spectrum quasars has recently been identified (*de Breuck et al*, 1998). These are potentially very important objects for study at long wavelengths. It is not yet clear just where they fit into a picture of quasar or cosmological evolution.

ABSORPTION, REFRACTION, SCATTERING, SCINTILLATION, AND FARADAY ROTATION

Long wavelength radiation is particularly susceptible to absorptive and refractive effects. This is both a blessing and a curse since it allows us to perform many interesting studies of the media that intervene between the source and the observer while these same propagation effects often mask interesting source properties.

As shown by many authors (*e.g. Ratcliffe*, 1959) the propagation of radio waves in an ionized medium is controlled by three parameters:

$$X = \nu_P^2/\nu^2 \qquad (8)$$

$$Y = \nu_H/\nu \qquad (9)$$

$$Z = \nu_c/2\pi\nu \qquad (10)$$

where ν is the wave frequency in Hz, $\nu_P = 8.984\sqrt{n}$ is the plasma frequency with n equaling the electron density in m^{-3}, $\nu_H = 2.8 \times 10^6 H_{gauss}$ is the gyrofrequency and ν_c is the collision frequency. Normally $1 \gg X \gg Y$ or Z and in the absence of a magnetic field the index of refraction is $\mu = \sqrt{1-X}$ and the coefficient of absorption is $K = X\nu_c/c\mu$.

Since the absorption coefficient of an ionized medium is proportional to X which is, in turn, proportional to λ^2, thermal (or free-free) absorption is often a very important phenomenon at low frequencies. In dense media, such as the ionosphere and solar corona, absorption is often strong but also over the long path lengths through the interstellar medium or through HII regions almost complete absorption can occur.

Absorption by the tenuous ionized component of the interstellar medium provides data concerning the density and distribution of this component. Absorption in HII regions is particularly important because these regions act as opaque walls at varying distances and permit estimates of the Galactic synchrotron emissivity along the paths to these regions.

Net refraction is usually unimportant (except in the ionosphere and solar corona) because μ is so close to unity but multiple, small angular deviations caused by small-scale structures in the interplanetary and interstellar medium generate scattering and scintillation effects. These effects have been adequately discussed during this meeting by *Rickett* and by *Cordes*.

Faraday rotation will frustrate most attempts to measure linear polarization at long wavelengths. The problem is the large magnitude of the rotation. A common value for the interstellar rotation measure near the Galactic plane would be $RM = 100$ rad/m^2, the total Faraday rotation along such a line-of-sight would be about 1000 radians at 100 MHz and 10,000 radians at 30 MHz. Emission distributed at various distances along the line of sight will be scrambled in polarization. Discrete sources normally have huge amounts of rotation which also scrambles their polarization. At $RM = 100$ rad/m^2, the frequency band over which depolarization occurs is also small, about 100 kHz at 100 MHz and 2 kHz at 30 MHz. Linear polarization is to be expected only for solar system sources and nearby pulsars. Early reports of linear polarization in solar bursts have been refuted, so linear polarization measurements at long wavelengths will probably be confined to Jupiter, Saturn, Earth, and those nearby pulsars that have low rotation measures.

Intrinsic polarization angles probably cannot be measured from the ground in any case because of the large and variable ionospheric rotation at long wavelengths. With hundreds of rotations in the ionosphere, it seems unlikely that ionospheric corrections can be made with sufficient accuracy to recover the intrinsic polarization of a source.

SPECTRAL TURN-OVERS

Some sources display spectral turn-overs or flattening at long wavelengths. These phenomena are of considerable interest because of the important information they provide concerning the physical conditions in the sources. Several mechanisms can produce these turn-

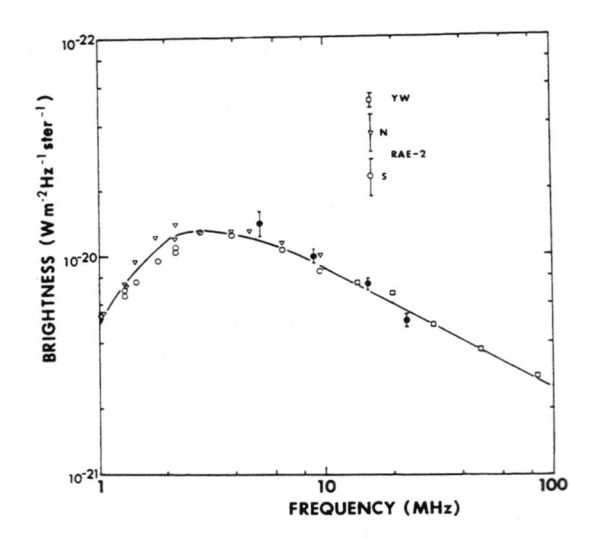

Figure 2. The surface brightness spectrum of the Galactic polar regions presented by *Cane*, [1979].

overs or flattenings. The first one is thermal absorption as discussed above. Observations of spectral turn-overs in supernova remnants were used by *Dulk and Slee* [1975] and by *Kassim* [1989] to estimate ionized gas densities in extended HII region envelopes. Thermal absorption in normal spiral galaxies has been used to provide evidence for cool ionized gas in these objects (*Israel and Mahoney*, 1990).

Spectral turn-overs in highly compact sources have been attributed to synchrotron self-absorption as discussed by many authors (*e.g. Ginzburg and Syrovatski*, 1962). In this case the absorption of photons by the synchrotron-radiating electrons is comparable with their emission. Such absorption occurs when the surface brightness temperature of the source is similar to the kinetic temperature of the relativistic electrons, about 10^{12} K. Because the electrons have a power-law energy distribution, the spectral index predicted for an opaque, synchrotron self-absorbed source is $\alpha = +2.5$ rather than the $\alpha = +2.0$ produced by a source with a Maxwell-Boltzmann distribution of electron energies.

To my knowledge, spectral indices of $+2.5$ have not been observed but this can probably be attributed to the fact that different small portions of the source region become self-absorbed at various frequencies. Clear measurement of an $\alpha = +2.5$ spectrum at long wavelengths would provide firm evidence for this mechanism, but any such long wavelength measurements present a challenge. Consider a synchrotron self-absorbed source component that peaks at 100 MHz with a flux density of 1 Jy and an angular size of about 0.01 arcsec. With an $\alpha = +2.5$ spectrum this component would have a flux density of 49 mJy at 30 MHz and 3 mJy at 10 MHz, making it difficult to discern among other larger and, probably, much stronger source components.

Another effect which could suppress long wavelength emission is the *Tsytovitch* [1951] or *Razin-Tsytovitch* effect (*Razin* 1960). This is the suppression of synchrotron emission caused by the lowering of the index of refraction by a plasma and the subsequent increase of the phase velocity of light. Thus relativistic electrons radiating synchrotron radiation from within a plasma do not approach the phase velocity of light as closely as they would in a vacuum, and less relativistic beaming occurs. It is not clear that this effect has yet been observed in astrophysics. Thermal absorption is generally expected to occur before the *Razin-Tsytovitch* effect becomes important except in sources of very small sizes and with weak magnetic fields (*Moffet*, 19??).

The majority of spectral flattenings and turn-overs at low frequencies are probably caused by flattenings in the electron energy spectra within the source region. All electron energy spectra must flatten at some point or the low energy electrons would possess infinite energy. Note that, since the low energy tail of the single electron synchrotron spectrum has a slope of $\alpha = +1/3$, the spectrum of an ensemble can rise no faster than this. However, most flattenings or turn-overs are modest and are in agreement with this limit.

COHERENT EMISSION

From the Larmor Law we know that the total power radiated by an accelerated charge is proportional to $q^2 a^2$ where q is the total charge and a is its acceler-

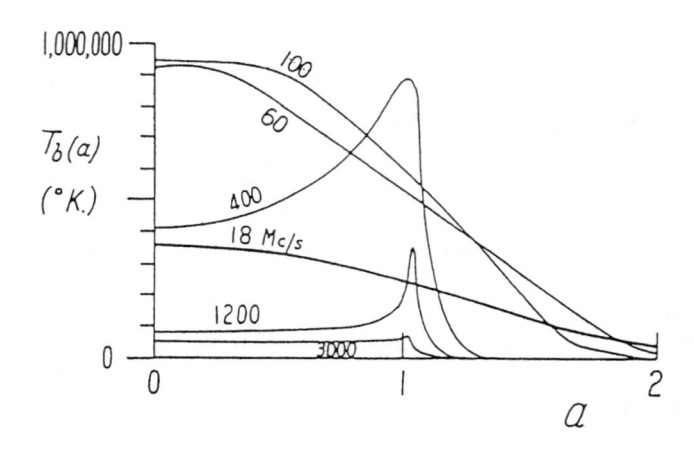

Figure 3. 18 MHz ray paths in the solar corona calculated by *Bracewell and Preston*, [1956].

ation. If N electrons are accelerated coherently, their total radiation goes as N^2. In astronomical sources one might expect very large values of N so if coherent motions occur the radiation produced is likely to be many orders of magnitude stronger than incoherent, single-electron emission. Coherent emission occurs primarily at long wavelengths for, if the volume containing the coherently moving charges has dimensions larger than the wavelength, retardation effects between the emissions from different parts of the volume will cause interference and cancelation of the coherent emission. Therefore, the magnitude of the volume that can produce coherent emission scales approximately as λ^3 and, since the total radiation goes as N^2 and N may be proportional to volume, one might expect the total emission to scale something like λ^6.

Plasma oscillations in the solar corona and in Jupiter's and Earth's magnetospheres generate coherent accelerations and emissions. Coherent accelerations of charge in the polar caps of pulsars also produce these emissions. In all cases, the emissions are characterized by very steep low frequency spectra. Coherent emissions are also characterized by rapid time variations of intensity and spectra. When coherent emission occurs, one obtains a wealth of information concerning the physical conditions in the plasma that generates the radiation.

It seems almost certain that coherent plasma motions must occur in other classes of radio sources such as active galactic nuclei, quasars, and various transient sources. One may expect that coherent emissions will eventually be found from various other classes of sources.

LONG WAVELENGTH PHENOMENA

There are a number of somewhat unusual phenomena that distinguish the long wavelength sky. Most of them were well known before about 1960 and are familiar to "old-timers" but some of them are unknown to younger workers in the field, so it seems worthwhile to mention them.

It is well known that the brightness of the Galactic background rises with wavelength. At $\nu \sim 50$ MHz the background surface brightness rises above that of HII regions and these regions begin to appear as cool areas against the hot background. A cool absorption trough develops along the Galactic plane which deepens and expands as the wavelength increases and lower density ionized gas becomes opaque. At $\nu \sim 3$ MHz the Galactic poles become brighter than lower latitude regions.

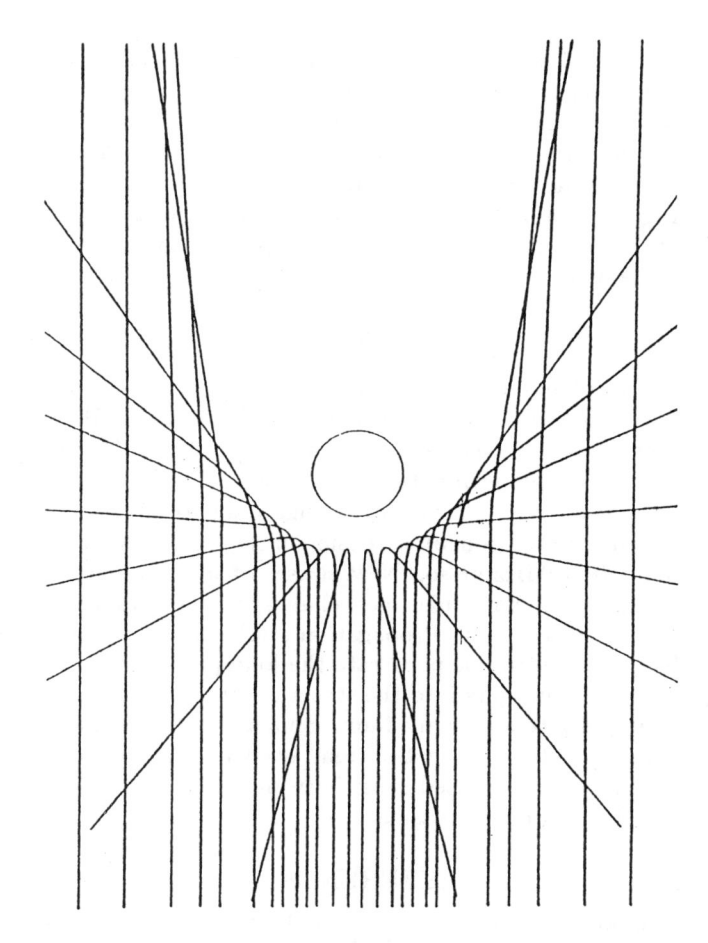

Figure 4. The solar brightness temperature across the disk at various frequencies (*Bracewell and Preston*, 1956).

Finally, by $\lambda \sim 1$ MHz, strong absorption has developed even at the poles and Galactic background brightnesses fall in all directions and our view is limited to only local features. These phenomena are indicated in Fig. 2.

The quiet Sun also displays some novel features at long wavelengths. At $\nu > 100$ MHz the corona is semi-transparent and one observes a mixture of 10^6K radiation from the corona and lower temperature emission from cooler, lower layers. At $\nu < 100$ MHz the corona becomes opaque and the full 10^6K emission from the corona is found. This persists to $\nu \sim 60$ MHz, but at lower frequencies sufficient refraction occurs that some ray paths are reflected back into interplanetary space without encountering strong absorption. The brightness temperature and total flux density of the corona fall (*Erickson*, 1977) as the corona becomes more and more reflective. At $\nu \approx 15$ MHz the brightness of the corona is less than that of the Galactic background. At still lower frequencies the corona would then appear to be a cool, reflective disk, reflecting the Galactic background radiation. It will occult a large solid angle but will be almost indistinguishable against the background. Figures 3 and 4 are taken from a paper by *Bracewell and Preston* [1956] in which these phenomena are well explained.

SUMMARY

The phenomena discussed in this paper have been mostly investigated with ground-based telescopes. However, these studies point the way for future work from space. The determination of long wavelength features in radio source spectra, steepenings and flattenings, will benefit greatly from an extension of the frequency range down to 1 MHz or so. New classes of sources may be found and, in particular, new classes of coherent sources may be discovered. Propagation studies at long wavelengths that do not penetrate the ionosphere will yield many new insights into the structure of the interplanetary and interstellar medium. Similar studies of intergalactic radio propagation may even become possible utilizing the increased sensitivity of the longest wavelengths to tenuous plasmas.

REFERENCES

Bracewell, R.N. and G.W. Preston, Radio reflection and refraction phenomena in the high solar corona, *Astrophys. J. 123*, 14-29, 1956.

Cane, H.V., Spectra of the non-thermal radio radiation from the galactic polar regions, *Mon. Not. R. ast. Soc. 189*, 465-478, 1979.

de Breuck, C., M.S. Brotherton, H.D. Tran, W. van Breugel, and H.J.A. Rottgering, Discovery of an ultra-steep spectrum, highly polarized red quasar at z = 1.462, *Astron. J. 116*, 13-19, 1998.

Dulk, G.A. and O.B. Slee, Spectral turnovers of galactic supernova remnants, *Astrophys. J. 199*, 61-68, 1975.

Erickson, W.C., T.E. Gergely, M.R. Kundu, and M.J. Mahoney, Determination of the decameter wavelength spectrum of the quiet sun, *Solar Physics 54*, 57-63, 1977.

Ginzberg, V.L., V.N. Sasonov, and S.I. Syrovatsky, On the magnetobremsstrahlung (synchrotron radiation) and its reabsorption, *Usp. Fiz. Nauk 94*, 63, 1968.

Ginzberg, V.L. and S.I. Syrovatsky, Cosmic magnetobremsstrahlung (synchrotron radiation), *Ann. Rev. Astron. Astrophys. 3*, 297-350, 1965.

Israel, F.P. and M.J. Mahoney, Low-frequency radio continuum evidence for cool ionized gas in normal spiral galaxies, *Astrophys. J. 352*, 30-43, 1990.

Jackson, J.D., *Classical Electrodynamics* pp. 481-488, John Wiley & Sons, New York 1962.

Kassim, N.E., Low-frequency observations of galactic supernova remnants and the distribution of low-density ionized gas in the interstellar medium, *Astrophys. J. 347*, 915-924, 1989.

Kuhr, H., A. Witzel, I.I.K. Pauliny-Toth, and U. Nauber, A catalogue of extragalactic radio sources having flux densities greater than 1 Jy at 5 GHz, *Astron. Astrophys. Suppl. 45*, 367-430, 1981.

Landau, L. and E. Lifshitz, *The Classical Theory of Fields*, pp. 213-218, Addison-Westley Press, Cambridge, 1951.

Moffett, A.T., Strong nonthermal radio emission from galaxies, *Stars and Stellar Systems, Vol. 9, Galaxies and the Universe*, Chap. 7, pp. 211-281, University of Chicago Press, Chicago, 1975.

Ratcliffe, J.A., *Magnetoionic Theory and its Application to the Ionosphere*, Cambridge University Press, Cambridge, 1959.

Rawlings, S., M. Lacey, K.M. Blundell, S.A. Eales, A.J. Bunker, and S.T. Garrington, A radio galaxy at redshift 4.41, *Nature 383*, 502-505, 1996.

Razin, V., *Radiofizika 3*, 584-594, 1960.

Scott, G.A., *Electromagnetic Radiation*, Cambridge University Press, Cambridge, 1912.

Tsytovitch, V.N., *Vestnik. Mosk. Univ., Ser. Phys. 4*, 27, 1951.

W.C. Erickson, School of Mathematics and Physics, University of Tasmania, Hobart, Tasmania, Australia. (email:wce@astro.umd.edu)

The Promise of Long Wavelength Radio Astronomy

K. W. Weiler

Naval Research Laboratory, Code 7213, Washington, DC

The past five decades have seen many of the most exciting discoveries in astronomy made at low frequencies. Now, with new ground-based instrument plans and space-based and lunar-based initiatives, the field promises a bright future. There is a challenging list of LF astrophysics problems to be solved and a series of exciting instrumental concepts to investigate them. Solar astronomy, planetary science, the thermal interstellar medium, supernova remnants, pulsars, interstellar-plasma refractive and diffractive scattering, cosmic rays, old "fossil" electron populations, quasars, radio galaxies, galactic background and halo studies, coherent emission mechanisms, and possible serendipitous discoveries all promise valuable insights at low frequencies. Also, the well understood technology and relatively low cost of low-frequency instruments will allow workers to plan very cost effective programs of ground-based arrays, ground-to-space VLBI, Earth-orbit and deep-space synthesis telescopes, and large mapping arrays on both the near and far side of the Moon.

1. INTRODUCTION

Even though radio astronomy was originated by Jansky at decametric wavelengths, the urgent quest for ever higher angular resolution and the fact that ionospheric structure has, in the past, limited interferometric imaging at long wavelengths (LW, taken to be $\lambda \gtrsim 2$ m or $\nu \lesssim 150$ MHz) to short (<5 km) baselines has left the LW range among the most poorly explored in the entire radio spectrum. This is despite the many important astrophysical questions which can only be addressed by LW observations, plus those problems which can be addressed much more accurately in the LW regime. The recent demonstration with the 74 MHz VLA[1] system [*Kassim, et al.,* 1993] that phase correction techniques and self-calibration can effectively and accurately remove ionospheric distortions over long baselines now offers the exciting opportunity to efficiently and economically open a new high-resolution, high-sensitivity window on the electromagnetic spectrum from the ground [*Kassim and Erickson,* 1998]. Also, refined Low Frequency Space Array (LFSA) concepts [*Basart, et al.,* 1997a,b and links from *http://rsd-www.nrl.navy.mil/7214/weiler/*] promise to extend this mapping capability from the ionospheric cutoff down to the lowest frequencies which can penetrate the interstellar and interplanetary media.

For the low frequency range from ~15 MHz to ~150 MHz, which can still be observed from the ground with

Radio Astronomy at Long Wavelengths
Geophysical Monograph 119
This paper not subject to U.S. copyright
Published in 2000 by the American Geophysical Union

---

[1] The VLA is operated by the National Radio Astronomy Observatory of the Associated Universities, Inc. under a cooperative agreement with the National Science Foundation.

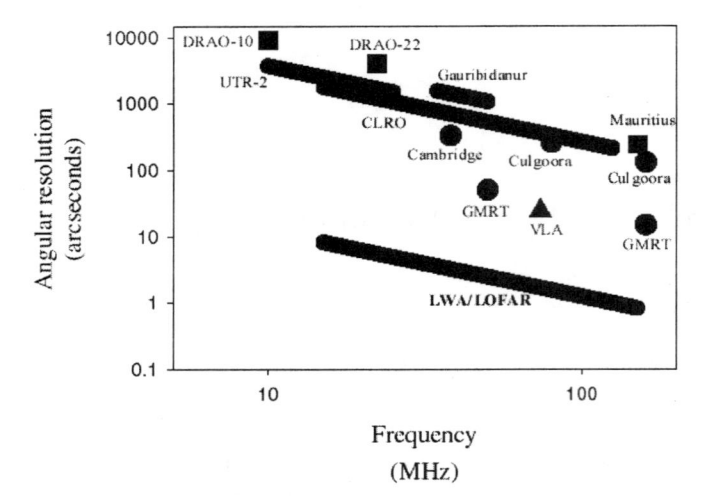

Figure 1. Angular resolution vs. Frequency for a number of past and present LW ground-based telescopes and the planned LWA/LOFAR. Note that the LWA/LOFAR will provide at least an order-of-magnitude improvement in resolution, as well as frequency coverage, over the entire band from ~15 - ~150 MHz.

high resolution and sensitivity, a new, large, ground-based Long Wavelength Array (LWA; also known as the LOw Frequency ARray or LOFAR) has been proposed [see *Kassim and Erickson*, 1998; *Erickson and Kassim*, 2000a and http://rsd-www.nrl.navy.mil/7213/lazio/decade_web/index.html with associated links]. LWA/LOFAR resolution is shown at different frequencies in Figure 1. Note that it provides at least an order-of-magnitude improvement over past LW instruments.

At very low frequencies ($\nu \lesssim 30$ MHz), high resolution ($\lesssim 1'$) observations are difficult to impossible from the ground. The ionosphere is a severe restriction and ionospheric phase and amplitude fluctuations are so large that even the strongest calibration sources become decorrelated and selfcal techniques begin to break down. At $\nu \lesssim 15$ MHz on the day side of the Earth near sunspot maximum and $\nu \lesssim 10$ MHz on the night side near sunspot minimum, observations are below the critical frequency of the ionospheric plasma, making the layer opaque. Even at preferred sites near the magnetic poles, such as Canada and Tasmania, at night near sunspot minimum when ground-based observations can be taken as low as ~2 MHz, the available resolution is extremely poor (several degrees). This situation is illustrated in Figure 2 which shows the resolution of a Low Frequency Space Array (LFSA) concept and a number of modern, ground-based low frequency telescopes. Note that all ground-based instruments are above the ~10 MHz ionospheric cutoff except for the Llanherne array in Tasmania [*Whitham*, 1975; *Ferris, et al.*, 1980], which is near the magnetic pole in the South. Note also that the RAE (Radio Astronomy Explorer) satellites [*Alexander and Novaco*, 1974; *Alexander, et al.*, 1975] from the late 1960s (RAE-1) and early 1970s (RAE-2), even though they were in Earth and lunar orbit, respectively, avoiding the ionospheric limitation, had extremely poor resolution because they were single element antennas and not interferometric arrays. An LFSA would give several orders-of-magnitude improvement in resolution over past or existing capability at all frequencies except the very highest which are accessible from the ground. The presently most promising concept to implement a LFSA is the Astronomical Low Frequency Array (ALFA; [*Jones, et al.*, 2000; see also http://sgra.jpl.nasa.gov/html_dj/ALFA.html].

Additional references for description of the astrophysical need and possible instrumental designs to open the LW window on the universe can be found in the proceedings from Workshops held in Green Bank, West Virginia in September 1986 [*Weiler*, 1987] and in Crystal City, Virginia in January 1990 [*Kassim and Weiler*, 1990] as well as in a number of scientific papers (see http://rsd-www.nrl.navy.mil/7214/weiler/kwhtml/lfrabiblio.html), a popular article [*Weiler, et al.*,1988], in *Erickson and Kassim*, [2000b], on the web site http://rsd-www.nrl.navy.mil/7214/weiler/, and elsewhere in this volume.

As an example of the need for improved LW observing capability, Figure 3 shows the currently best available

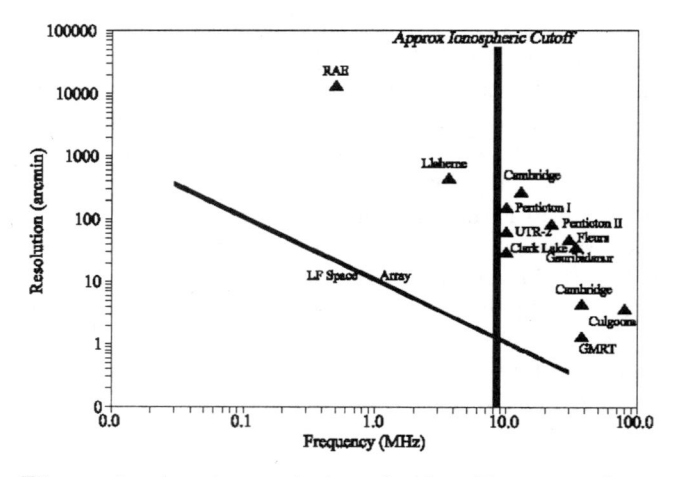

Figure 2. Angular resolution of a Low Frequency Space Array (LFSA) as a function of frequency. The capability of a number of ground-based telescopes, as well as the Radio Astronomy Explorer (RAE) satellites, is shown along with the approximate ionospheric cutoff.

map of the sky at 10 MHz while Figure 4 shows the vast improvement which can be obtained even with relatively low sensitivity and minute of arc resolution by a test map made with the prototype 74 MHz system on the VLA. Clearly, there is a great deal of new astrophysical information to be obtained at low frequencies with a combination of ground-based and space-based instruments to open this last, broad, unexplored window on the Universe.

2. SCIENCE

Since the articles by *Kaiser and Weiler* [2000], *Jones, et al.*, [2000], *Erickson and Kassim* [2000b], and others in this volume all make a very strong science case for new LW capability, I will not repeat the arguments in detail here. However, let me enumerate some of the many areas where low frequency measurements will have an impact.

- Astrophysics
  - Cosmology and large scale structure
  - Formation and evolution of galaxies
    * Cluster halos and intergalactic magnetic fields
    * Evolutionary studies of radio galaxies
      · Very high redshift radio galaxies
      · Emission mechanisms and jet physics
      · Star forming galaxies
    * Galaxy halo emission
      · Cosmic ray diffusion times away from galactic disks
    * Emission and absorption mechanisms in galaxies
    * "Fossil" radio galaxies
  - Interstellar processes
    * The origin, acceleration, distribution, and energetics of cosmic rays
    * The structure and distribution of diffuse ionized hydrogen
    * The origin and transport of turbulence in the ISM
    * Interstellar refractive and diffractive scattering
  - Structure and properties of HII regions
    * Temperature, density, and clumpiness

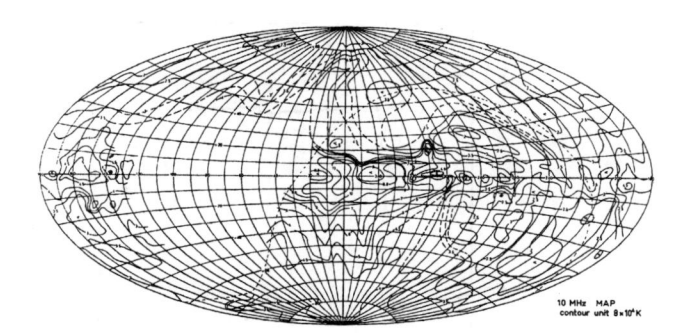

Figure 3. A 10 MHz map (resolution ∼5°) of the southern sky in galactic coordinates [*Cane and Erickson*, 2000]

    * Separation of thermal and non-thermal components
  - Structure, properties, and statistics of supernova remnants (SNRs)
    * Relation to cosmic ray acceleration sites
    * Location of "missing" SNRs
    * Improvement of stellar birth/stellar death statistics
    * Test of diffusive shock acceleration models
  - Discovery of new phenomena
    * Searches for millisecond pulsars
    * Searches for coherent emitters
    * Searches for bursting extra-solar planets

- Solar Physics
  - Solar variability
    * Type II and Type III burst studies
  - Solar-terrestrial interaction
    * Solar wind-Earth's magnetosphere interaction
  - Space weather forecasting
    * Coronal mass ejection (CME) tracking to 1 AU
  - Solar radar

Interferometric arrays, ground-based (above the ionospheric cutoff) and space-based (below the ionospheric cutoff) will permit high resolution, high sensitivity surveys and study of individual objects anywhere on the sky to open up entirely new areas of astrophysics.

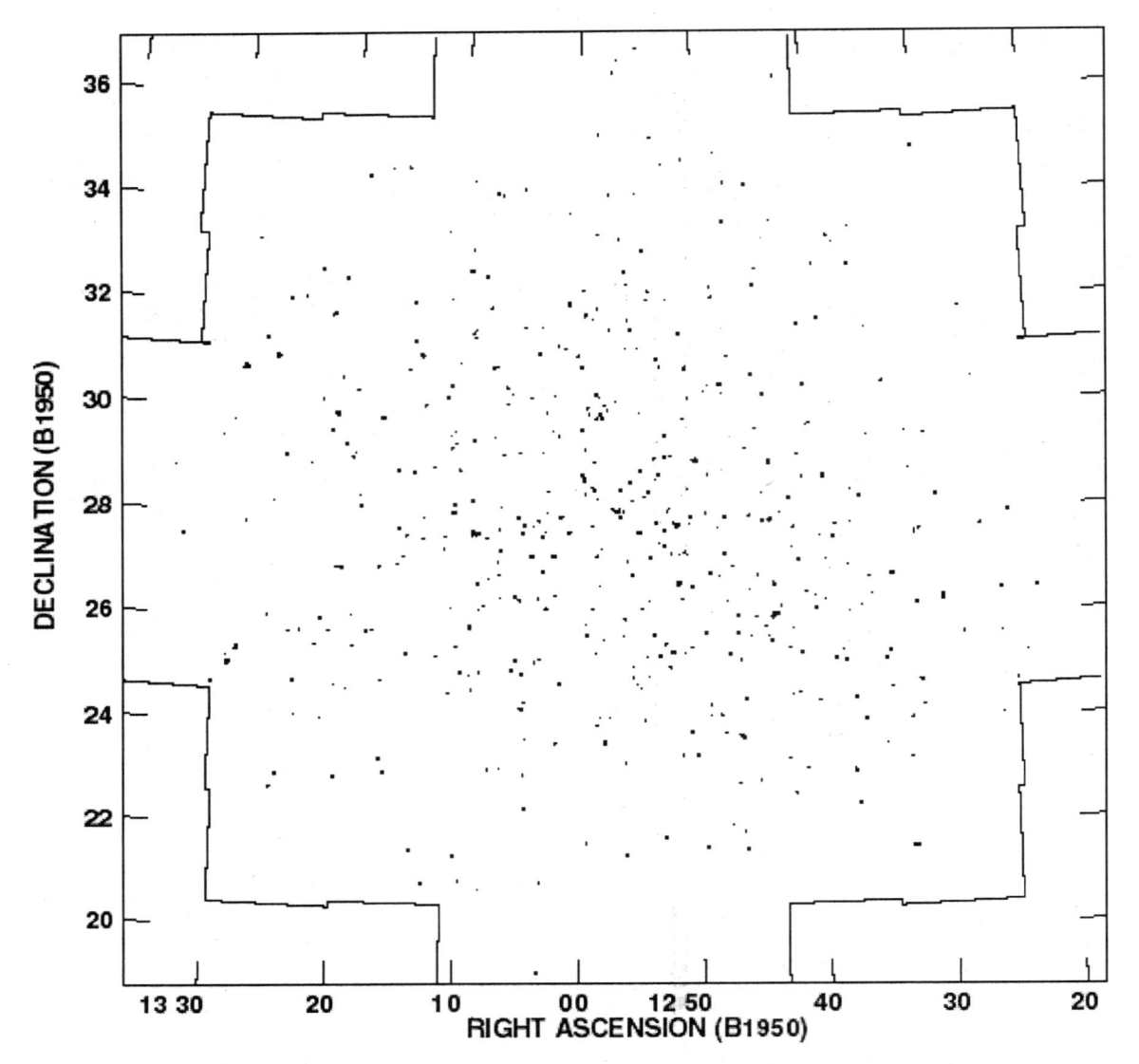

Figure 4. 74 MHz VLA image. Note the myriad sources which appear even with the relatively low resolution (~1′) and low sensitivity (100s of mJy) on this approximately 16° x 16° section of the sky centered on the Coma Cluster [Enßlin, et al., 2000]. Contour levels are 7, 10, 20, 40, 80, 160, and 320 x 30 mJy beam^{-1}.

3. LOW FREQUENCY INSTRUMENTS – GROUND-BASED

There have been many instruments built for LW ground-based observations over the years. However, all have been relatively low resolution and most are no longer operating. Of those listed in Figure 1, only the UTR-2 ([Braude, et al., 1978], the Mauritius telescope [Golap, et al., 1995a,b], and the VLA at 74 MHz [Kassim, et al., 1993; Erickson and Kassim, 2000a] are currently producing data at low frequencies. The GMRT [Ananthakrishnan, 1995] is now operating at higher frequencies and is scheduled to add the frequencies of 38 and 153 MHz in the near future.

The newest proposed instrument, the LWA/LOFAR, will have the highest sensitivity and resolution of any past or present LW telescope in the 15 to 150 MHz frequency range. It is presently envisioned to consist of large banks of crossed-dipole, log-periodic antenna arrays. The individual receiving elements will be pointed vertically and be mechanically fixed. Each bank will be about 100 m square and contain 16 to 256 receiving elements. The beam of a bank can be steered to any point in the sky by adjusting the phases of the individual elements within it and beam steering will be entirely electronic with no mechanical devices used. A possible LWA/LOFAR layout in illustrated in Figure 5.

Deciding on the high and low frequency limits of a new telescope is somewhat subjective and there is always pressure to make an instrument as broad band as possible. However, for the LWA/LOFAR reasonably solid upper and lower frequency limits can be established.

The high-frequency limit of the LWA/LOFAR is determined by the scientific work that can be accomplished with the higher resolution available at higher frequencies and by the necessity of obtaining instrumental and ionospheric phase correction information at the higher frequencies where the ionospheric effects are smaller. On the other hand, the effective collecting area of the system is proportional to the wavelengths squared (λ^2) and at too high a frequency the antennas do not collect sufficient signal power for self calibration. From these considerations, the LWA/LOFAR high frequency limit is taken to be ~150 MHz, although an innovative plan to utilize active dipole elements might expand the upper end of the frequency range to ~300 MHz.

The low frequency limit of the instrument determines the physical size of the individual antenna elements and, assuming that dipoles are used in the log-periodic elements, the dipoles must be approximately one-half wavelength long. Since a major cost of the system is in the antenna elements, larger elements are much more costly. Also, the antenna elements must be separated by $> \lambda/2$ at the low frequency limit, and such wide spacings can lead to spurious grating responses at higher frequencies.

Another low frequency limitation is ionospheric amplitude scintillation. Ionospheric phase variations scale as λ, provided that no amplitude scintillation is present, so that high frequency phase measurements can be used for low frequency phase corrections. However, below some limit, angular refraction in ionospheric irregularities is sufficiently large that rays passing through different irregularities can intersect, causing constructive and destructive interference or amplitude scintillation. At that point, the ionospheric phase wanders in an almost random fashion, making phase correction virtually impossible. At most mid-latitude sites amplitude scintillation will occur ~30% of the time at frequencies <30 MHz so that, with continually diminishing efficiency, observations down to ~15 - 20 MHz should be possible at some times of particularly quiet ionospheric condi-

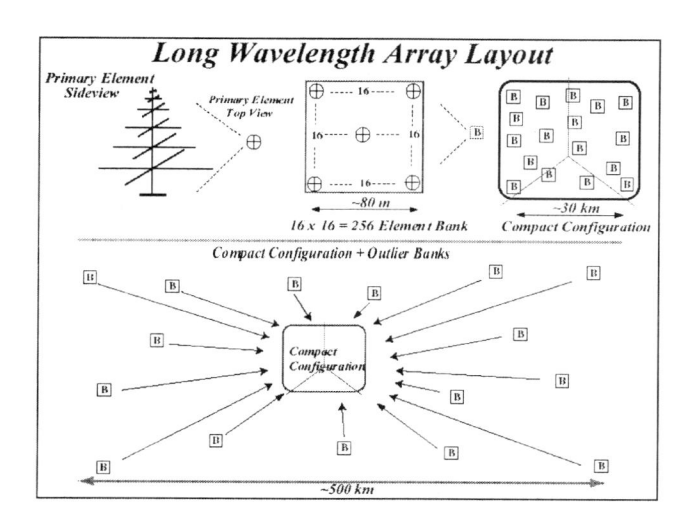

Figure 5. Possible parameters and layout of the LWA/LOFAR.

tions. Thus, the LWA/LOFAR is proposed to operate in the frequency range 15 - 150 MHz, with lower efficiency in the range 15 - 30 MHz.

The log-periodic elements will be arranged into banks with each bank equivalent to a single antenna in a multiple dish array like the VLA or the WSRT. Each bank must provide enough collecting area for self-calibration at 150 MHz, which will require ~256 elements per bank, or ~350 m^2 effective collecting area at 150 MHz, giving ~35,000 m^2 effective collecting area per bank at 15 MHz.

The LWA/LOFAR will consist of ~30 banks with 15 banks in an area ~30 km in diameter to form a "compact" array with angular resolutions of ~2' and ~14" at 15 MHz and at 150 MHz, respectively. The additional 15 banks will be in a Y-shaped configuration up to ~500 km in diameter to form an "extended" array with angular resolutions of ~8" at 15 MHz and ~1" at 150 MHz (see Figure 5). If the compact array is sited at the VLA, the extended array could co-locate with the VLA Expansion Project whose "A+" configuration will be similarly sized. More detail can be found at *http://rsd-www.nrl.navy.mil/7213/lazio/decade_web/index.html*.

4. VERY LOW FREQUENCY INSTRUMENTS – SPACE-BASED

As can be seen from Figure 2, because of ionospheric limitations the only telescopes which have been successful in mapping at frequencies $\lesssim$10 MHz are the Llanherne array [*Whitham*, 1975; *Ferris, et al.*, 1980]

Plate 1. Artists conception of the ALFA spacecraft in a 16 element spherical array with a blowup of one spacecraft illustrated (middle panel). The top panel illustrates a CME ejection from the Sun and the bottom panel illustrates the image of a bright, double-lobed radio galaxy.

in Tasmania and the two Radio Astronomy Explorer (RAE) [*Alexander and Novaco*, 1974; *Alexander, et al.*, 1975] satellites launched in the late 1960s and early 1970s. These instruments had extremely low resolution, at least two orders-of-magnitude lower than the concept for a Low Frequency Space Array (LFSA). There has never been a very low frequency, space-based, imaging array. Although solar system probes of radio emissions from the Sun and planets have been highly successful and are included on most planetary missions [see the review by *Kaiser and Weiler*, 2000 and other papers in this volume], they are generally a single dipole with limited sensitivity and no mapping capability.

The most promising concept for a LFSA at the present time is that of the Astronomical Low Frequency Array (ALFA). ALFA will consist of an array of sixteen identical satellites operating together as a synthesis interferometer. Very low frequency radio radiation will be sampled by a pair of orthogonal dipole antennas on each of the satellites, and each dipole will feed signals to a simple but flexible high dynamic range receiver. Observing frequency, bandwidth, sample rate, and phase switching are controlled by a central spacecraft processor, and can be changed at will.

The receivers cover 0.03-30.0 MHz, with Nyquist sampled bandwidths up to 125 kHz and an ability to handle a very wide range of input levels. They can be constructed from commercially available components and require no new development.

ALFA will have a maximum baseline length of $\sim$100 km, which provides a good overall match to interstellar and interplanetary angular broadening, and will be placed in a nearly circular distant retrograde orbit (DRO) about the Earth-Moon barycenter, with a typical distance from Earth of $\sim 10^6$ km. A DRO has the advantage of being sufficiently far from Earth to minimize terrestrial interference but close enough for each satellite to communicate directly with relatively small (11-meter) and affordable ground stations. This approach involves no reliance on a "mother ship" for data relay, or any other mission critical function, giving an extremely robust (16-way redundancy) array data path all the way to the ground. Similarly robust is the technique for continuously monitoring the relative positions of the satellites by measuring the separations between all pairs of satellites. This provides far more constraints than are needed to solve for all of the relative positions; should one or more of the array satellites fail, observing by the rest of the array can continue unhampered.

Each satellite will receive an uplink X-band carrier (to which the local oscillators are phase locked) and

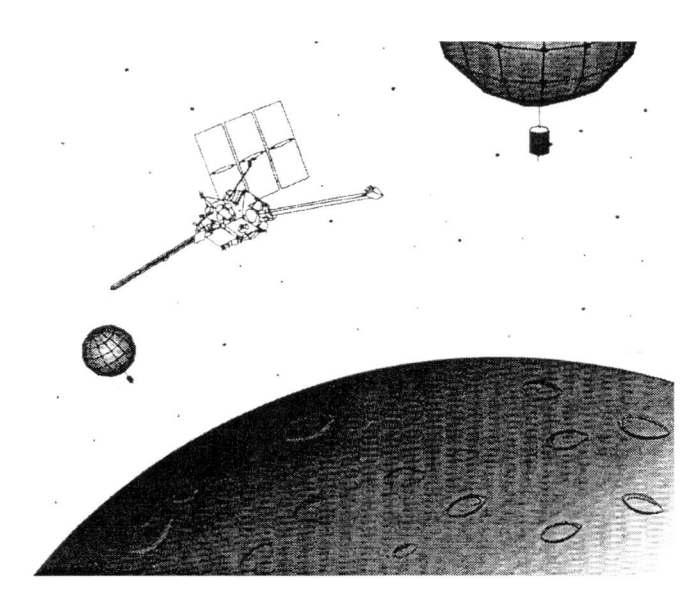

Figure 6. Lunar Orbiting Radio Astronomy Experiment (LORAE) [*Burns*, 1990]

low-rate command telemetry, and will send an X-band data downlink to the ground continuously at 0.5 Mb/s per satellite. The distance of the DRO and its location in the ecliptic plane allows continuous coverage of the array by three ground stations. At the ground station, telemetry headers are removed and the remaining interferometry data from each satellite are recorded on tapes for transport to the correlation computer. Small subsets of the data for rapid solar snapshot imaging can be stored on disk and retrieved from the stations via the Internet.

The ALFA concept is illustrated in Plate 1 and described in more detail by *Jones, et al.*, [2000] and on http://sgra.jpl.nasa.gov/html_dj/ALFA.html and associated links.

5. VERY LOW FREQUENCY INSTRUMENTS – MOON-BASED, ORBITING

The "holy grail" of very low frequency radio astronomy is to build an array on the shielded backside of the Moon. Since no lunar concepts are discussed elsewhere in this volume, let me describe some of the ideas for establishing a lunar-based array, either in orbit around or on the surface of the Moon.

LORAE – The Lunar Orbiting Radio Astronomy Experiment (LORAE) concepts, one of which is illustrated in Figure 6, have been proposed by *Burns* [1990] and collaborators. The LORAE in Figure 6 would consist of two small satellites, and a mother ship, forming a

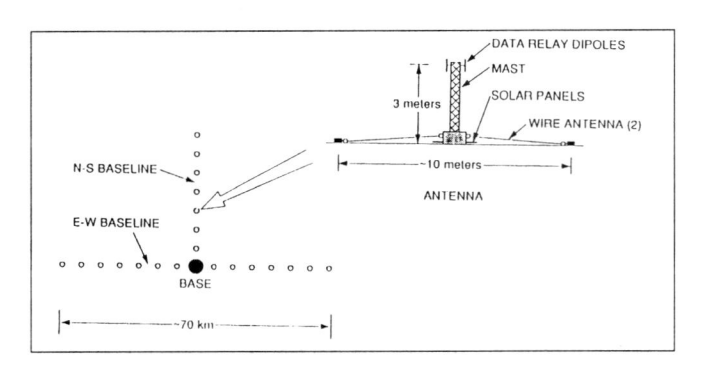

Figure 7. Possible configuration for a Lunar Near-Side Array (LNSA) [*Kuiper, et al.,* 1990].

single baseline interferometer operating at 13 and 26 MHz and collecting data when on the lunar backside. Over the course of a year, a two element, orbiting interferometer would assume sufficient baseline projections and orientations to produce good u,v plane coverage and permit mapping to several arcsecond resolution.

The artist's conception pictured in Figure 6 has each of the two array elements carrying an antenna consisting of a novel, phased dipole array imprinted on a 7 – 10 m inflatable structure [*Basart, et al.,* 1997a,b]. Such a balloon antenna would permit some antenna gain and directivity for the individual array elements.

LORI – The Lunar Orbiting Radio Interferometer (LORI) concept [*Allen,* 1996] consists of a large number (~50) of very simple microsats, each carrying 3 mutually perpendicular dipole antennas. Orbiting the Moon with many different orbital inclinations, to establish a large grid of satellites within view of each other at any given time, the array could form numerous instantaneous interferometers. Those on the shielded lunar backside would operate as a coherent interferometric array for LW mapping. A mother ship would collect information on the various baseline lengths and orientations, maintain phase stability of the array, and assemble the IFs from the microsats into a combined data stream for local correlation or transmission to Earth. The receivers would be tunable and cover the band from 0.5 to 20 MHz.

LOREX – The Lunar Orbiting Radio astronomy EXplorer (LOREX) concept [*Desch,* 1996] proposes to use the lunar occultation technique to image radio sources in the 1 to 10 MHz band while on the backside of the moon in a lunar orbit from 3,000 to 10,000 km in radius. While the proposed satellite would have only a single, half-wave dipole ~90 - 120 m long, the concept of using a single "long wire" antenna several km in length to provide some antenna gain is also suggested for more detailed investigation. (Note: A single wire which is many wavelengths long has antenna gain along the axis of the wire.)

While there are certainly many more lunar orbit concepts which are possible, the above give some of the plans which have been investigated to the point of proposals to funding agencies. On the solid lunar surface, there are many other interesting possibilities for LW arrays.

6. VERY LOW FREQUENCY INSTRUMENTS – MOON-BASED, SURFACE

LNSA - Astronomers at the JPL [*Kuiper, et al.,* 1990] have developed a Lunar Near-Side Array (LNSA) concept for a LW telescope on the lunar surface to cover the band from 0.15 to 30 MHz and provide resolutions of 18' at 1 MHz, 6' at 3 MHz, and 2' at 10 MHz.

The LNSA proposes a very simple system with multiple crossed dipole receiving elements in a "T" shaped array illustrated in Figure 7. Each station has two, 10-m long orthogonal dipoles which feed very simple, commercially available receivers like pocket digital shortwave radios. The digitization is done with the circuitry used in digital audio tape recorders. A 3-m spring-loaded mast at each station has a small UHF antenna at the top to receive timing signals and tuning commands from a central "mother station" and sends out the digitized data streams using commercial transceiver technology. Power is provided by solar panels during the lunar day and lithium batteries during the lunar night.

Because the Moon is small, with a large curvature, each station relays signals from other stations. With each station visible to two other stations in both the inward and outward directions, relay redundancy is provided in case of single station failures.

Data processing is done by the mother station at the array center, which contains a correlator and a narrow band data link back to Earth. Alternatively, the individual station IFs can be combined by the mother station and downlinked on a wide bandwidth link for correlation on the Earth.

VLFA - Consideration of placing an array on the more inaccessible far-side of the moon has been undertaken by *Basart and Burns* [1994] as the Very Low Frequency Array (VLFA). In their study, they consider a phased array deployment. The first phase consists of 169 elements within a 17 km diameter circle, giving a resolution of 2' at 30 MHz. The second phase consists of 361 dipoles covering baselines out to 1000 km and provides 1.5'' resolution at 30 MHz.

The VLFA concept is developed more fully in a report by the European Space Agency (ESA) which has undertaken a study of projects for the lunar surface, including LW observatories [*Lunar Study Group,* 1992]. The ESA concept would consist of 300 dipole antennas within a 17 km diameter circle and have a mother station which would perform the command, control, and correlation for the array. The individual dipole receiving elements would be deployed by lunar rovers.

7. TECHNICAL CHALLENGES

While the hardware and required precision for LW arrays is generally simple, well tested, and inexpensive, there remain many technical challenges for both ground-based and space-based instruments.

- Interference (RFI) suppression
  - Multichannel/high dynamic range frontend
  - High array gain
  - High array element gain
  - High frequency resolution backend
  - High time resolution backend
  - Interference rejection algorithms
- Mapping
  - On-line calibration and mapping
  - Wide-field, 3-D mapping
  - Long integration and good u,v coverage
  - Suppression and/or removal of moving and variable sources in the field
  - Suppression of main beam and sidelobe confusion
- Antennas
  - Large effective area and high gain antennas
  - Wide bandwidth antenna arrays
  - Simple, low cost construction and deployment
- Receivers
  - Wide bandwidth with high dynamic range
  - Inexpensive and stable
- Digital backend
  - Light weight and low power
  - Thousands of spectral channels
- Data Handling
  - Real-time correction and correlation
  - Full data archiving
- Calibration
  - Isolation of calibrators in a bright and crowded sky
  - Correction for ionospheric and IPM isoplanatic patches
  - Suppression of confusion
  - Compensation for ISM scintillation

7.1. Interference (RFI) Suppression

For ground-based LW arrays, the most obvious source of shielding from interference is the horizon for an isolated observing site. Beyond that, interfering signals are generally at low enough levels that they do not overload the highly linear electronics modules that are available on today's market at low prices. Experience has shown that sufficiently clear gaps exist between the interfering signals to allow sensitive observations. In fact, the 74 MHz band at the VLA is one of the cleanest bands available to that instrument, and the radio band below 100 MHz is one of the few portions of the spectrum where RFI levels actually seem to be decreasing as many communication services move to higher frequencies or to optical fibers. Nevertheless, it will almost certainly be necessary to incorporate some RFI avoidance and excision procedures into any LW telescope system. With modern digital technology, this can be done relatively easily even for a correlator with many thousands of channels.

In Earth orbit, or even as far as the near side of the Moon, interference from terrestrial transmitters could be a problem and requires suppression techniques. This is best illustrated with the measurements made by RAE-2 in lunar orbit (Figure 8). It is easy to see how much the interference background decreased while the satellite was shielded from terrestrial interference behind the moon and how strong, even at the distance of the Moon, interference was while the satellite was visible from the Earth.

While there are various techniques for suppressing interference, including primary antenna and array beams, array delay beam, multiple frequency channels, and location on the shielded lunar backside, the problem can also be minimized by distance from the Earth. Figure 9 shows that at a distance of 160 Earth radii (160

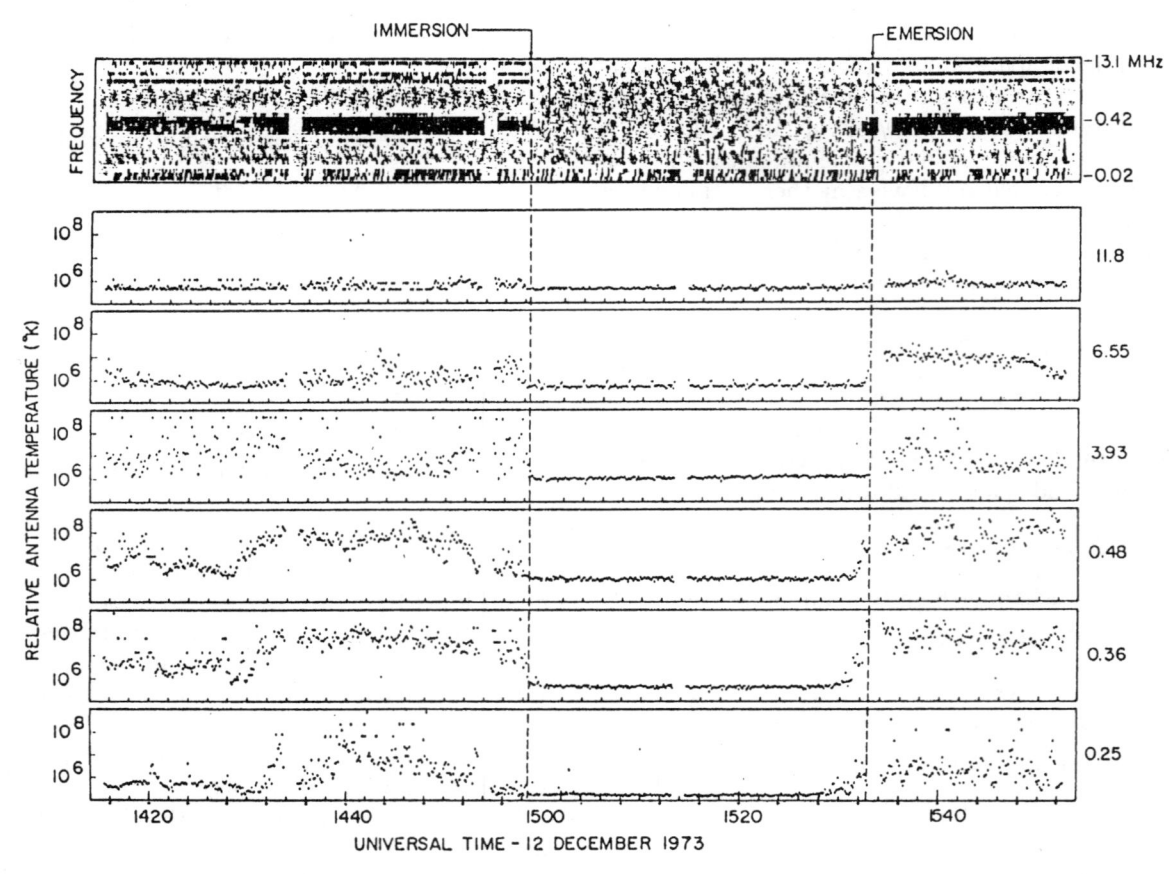

Figure 8. Data from RAE-2 in lunar orbit showing the dramatic disappearance and reappearance of interference from the Earth [*Alexander, et al.*, 1975].

Re or $\sim 10^6$ km), except for a few narrow frequencies with strong terrestrial interference, most observing bands have noise levels limited only by the galactic background emission.

7.2. Mapping

As with RFI, a ground-based (or lunar-based) array has similar mapping problems to a space-based array, but less severe. A ground- or lunar-based array can have large, coherent beds of antennas to delimit the field-of-view and the FoV can, at most, be $\sim 2\pi$ sterradians in size. Nevertheless, a conventional two-dimensional inversion of the three-dimensional visibility function measured by a non-coplanar array introduces phase errors which can become severe at long wavelengths [*Cornwell and Perley*, 1992]. The phase error goes as $w\theta^2$, where w is the third dimension in the visibility data, perpendicular to the standard u,v plane, and θ is the distance from the phase center. The "3-D problem" arises when the entire primary beam (i.e., θ large), which may contain hundreds of discrete sources at the wavelengths under consideration, must be properly deconvolved. Numerical solutions to this problem have now been successfully implemented within two widely used astronomical data reduction packages (SDE and AIPS), with their only limitation being the high computational expense. For example, SDE implements a polyhedron algorithm in which the 3-D "image volume" is split into many 2-D "facets," and is currently used to generate thermal-noise-limited images from 330 MHz VLA observations on a routine basis. Examples of successful wide-field, non-coplanar image deconvolutions are shown in this volume (Figure 3 in *Kaiser and Weiler* [2000], and in *Kassim, et al.*, [2000] for the Galactic center at 330 MHz), and in Figure 4 for the Coma cluster field at 74 MHz.

Extending the polyhedron algorithm to lower frequencies is limited by the number of facets (N_{facet}) that are required to divide up the surface of the image sphere into quasi-planar regions and $N_{facet} \propto \lambda B/D^2$

where B is the baseline length, λ is the observing wavelength, and D is the aperture size of the individual antenna elements [Cornwell and Perley, 1992]. For the 74 MHz VLA $N_{facet} \sim 225$, so that for the LWA/LOFAR with $B \gg 100$ km, thousands of facets may be required. However, current scaled-processing platforms (e.g., SGI Origin 2000) have already demonstrated the capability of reducing data with hundreds of facets and, in light of the continuing rapid advance in available computational power, practical solutions to the wide-field imaging problem should soon be readily available for ground-based arrays.

For a purely space-based array, the challenges of imaging the sky at very low radio frequencies are even more severe. There is no horizon present, so emission from the full 4π sterradians of the sky is visible at all times. Also, because individual radio antennas of reasonable size have very low directivity at these frequencies, there is very little limitation of the interferometer field-of-view and very strong radio sources will create sidelobes in directions far from their positions. High dynamic range imaging will require removal of the effects of strong sources from all sky directions, not just from the region immediately adjacent to the sources of interest. Even more complex, some of the interfering sources such as the Sun, Moon, and Earth will be moving rapidly across the sky and refracting and occulting background radio sources as well as producing their own emission. While conceptually closely related to the 3-D imaging problems for ground-based arrays discussed above, actual implementation of a mapping package for a space-based array has not yet been carried out.

7.3. Antennas

At these long wavelengths, antennas are a difficulty. Focusing elements such as parabolic dishes which are at least several wavelengths across become prohibitively expensive for both ground and space. Thus, one returns to the basic element of crossed dipole antennas used either singly (for space) or in sub-array beds (for ground or lunar surface) to give some primary beam field delimitation and gain. Unfortunately, unlike frequency insensitive parabolic reflectors, essentially all receiving element based arrays have intrinsically narrow bandwidths compared to the several decade wide frequency bands being considered. Conventional solutions, such as wide bandwidth log-periodic antennas for the ground or electrically short dipoles for space, appear likely to yield satisfactory results. Additionally, more exotic technologies such as active dipoles or magnetic dipoles are being considered.

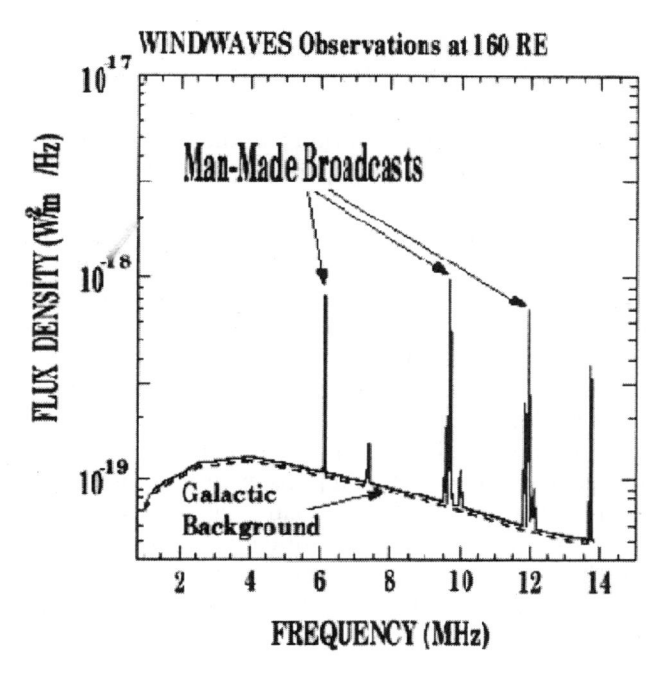

Figure 9. Terrestrial interference observed by the WAVES instrument on the WIND spacecraft $\sim 10^6$ km from Earth [Desch, 1998]. At this distance 93% of the measurements above 6 MHz are within 3% of the galactic background. Interference from Earth will not be a serious problem for an array such as ALFA in a Distant Retrograde orbit (DRO).

7.4. Receivers

Receivers at these wavelengths are, fortunately, not a difficulty. Small, light, wide-bandwidth, high dynamic range receivers are readily available commercially at low cost. Additionally, since receiver noise temperature is always dominated by the bright galactic background emission at these frequencies, no expensive and complex cooling systems are needed.

7.5. Digital Backends and Data Handling

As with receivers, the relatively limited bandwidths and numbers of channels needed for LW arrays can easily be handled by commercial systems to perform the digitization, correlation, integration, and storage of the instrument data streams.

7.6. Calibration

A ground-based LW array can maintain its phase coherence through the conventional technique of distributing a timing signal to all array elements from a central frequency standard through coaxial cables or, more likely these days, an optical fiber or radio link. Instrument calibration can be performed at the higher end of the instrument's frequency range where the sky is simpler, the ionospheric effects are more tractable, and the calibration can be performed in a more-or-less standard fashion. These calibration parameters can then be extrapolated to the lower operating frequencies of the array.

For a space-based array, life is somewhat more complicated. For example, ALFA will maintain phase calibration by an X-band uplink carrier, to which all satellite oscillators are locked. Amplitude calibration is provided by : 1) periodically injecting a known calibration signal into the signal path between the antennas and very low frequency receivers, 2) comparison with know astronomical sources at the high end of ALFA's frequency range, and 3) comparison with ground-based observations of solar bursts using antennas of known gain. Because LW systems are generally quite stable, and the deep space environment is unchanging, calibration parameters should be only very slowly changing, if at all.

8. SUMMARY

It is clear that new ground- and space-based LW arrays can open a new window on the Universe with high resolution mapping at the relatively unexplored frequencies from the IPM cutoff at ~0.03 MHz up to the relatively well exploited frequencies above ~150 MHz. New parameter space will be explored in sensitivity, resolution and frequency; a new sky will appear because of the great differences in absorption and emission parameters from those seen at higher frequencies; and new astrophysical sources and processes will become available for study.

While the hardware is generally well tested, easily available, and relatively inexpensive, challenges remain in the areas of data handling – calibration, correction, mapping, and analysis. However, no insurmountable difficulties appear to exist and planned new instruments promise to finally open this last frontier of astrophysics.

Acknowledgments. I wish to thank the Office of Naval Research (ONR) for the 6.1 funding supporting this work.

REFERENCES

Alexander, J. K. and Novaco, J. C., Survey of the galactic background radiation at 3.93 and 6.55 MHz, *Astron. J.*, *79*, 777-785, 1974.

Alexander, J. K., Kaiser, M. L., Novaco, J. C., Grena, F. R., and Weber, R. R., Scientific instrumentation of the Radio-Astronomy-Explorer-2 satellite, *Astron. Astrophys.*, *40*, 365-371, 1975.

Allen, R. J., *private communication,* 1996.

Ananthakrishnan, S., The Giant Meterwave Radio Telescope/GMRT, *J. Astrophys. Astron.*, *16*, 427, 1995.

Basart, J. P. and Burns, J. O., *private communication,* 1994.

Basart, J. P., Burns, J. O., Dennison, B. K., Weiler, K. W., Kassim, N. E., Castillo, S. P., and McCune, B. M., Directions for space-based low frequency radio astronomy, Part I: System considerations, *Radio Sci.*, *32*, 251-264, 1997a.

Basart, J. P., Burns, J. O., Dennison, B. K., Weiler, K. W., Kassim, N. E., Castillo, S. P., and McCune, B. M., Directions for space-based low frequency radio astronomy, Part II: Telescopes, *Radio Sci.*, *32*, 265-276, 1997b.

Braude, S. I., Men, A. V., and Sodin, L. G., The UTR-2 decametric-wave radio telescope, *Antenny*, *26*, 3-15, 1978.

Burns, J. O., The Lunar Observer Radio Astronomy Experiment (LORAE), *Proc. Conf., Low Frequency Astrophysics from Space,* edited by N. Kassim and K. Weiler, Springer-Verlag, Berlin, 1990.

Cane, H. V. and Erickson, W. C., A 10 MHz map of the Galaxy, *Radio Sci., in press,* 2000.

Cornwell, T. J. and Perley, R. A., Radio-interferometric imaging of very large fields - The problem of non-coplanar arrays, *Astron. Astrophys.*, *261*, 353-364, 1992.

Desch, M., *private communication,* 1996.

Desch, M., *private communication,* 1998.

Enßlin, T. A., Kronberg, P. P., Perley, R. A., and Kassim, N. E., 74 MHz VLA observations of Coma Bernices with subarcminute resolution - observation, data reduction, and preliminary results, *Proc. Conf., Diffuse Thermal and Relativistic Plasma in Galaxy Clusters,* edited by H. Boehringer, L. Feretti, and P. Schuecker, MPE Report No. 271, MPE Garching, in press, 2000.

Erickson, W. C. and Kassim, N. E., The VLA at 74 MHz and future plans for a long wavelength array, in *Space-Based Radio Observations at Long Wavelengths,* edited by R. Stone, K. Weiler, M. Goldstein, and J-L Bougeret, AGU, in press, 2000a.

Erickson, W. C. and Kassim, N. E., Long wavelength astrophysics, in *Space-Based Radio Observations at Long Wavelengths,* edited by R. Stone, K. Weiler, M. Goldstein, and J-L Bougeret, AGU, in press, 2000b.

Ferris, R. H., Turner, P. J., Hamilton, P. A., and McCulloch, P. M., The Llanherne VHF radio telescope, *Astron. Soc. Australia, Proc.*, *4*, 26-28, 1980.

Golap, K., Issur, N. H., Somanah, R., Dodson, R. G., Modgekar, M., Sachdev, S., Udayashankar, N., and Sastry, C. V., The Mauritius radiotelescope, *J. Astrophys. Astron.*, *16*, 447, 1995a.

Golap, K., Issur, N. H., Somanah, R., Dodson, R., Modgekar, M., Sachdev, S., Udayashankar, N. U., and Sastry, Ch. V., The Mauritius radiotelescope, *Astrophys. and Space Sci.*, *228*, 373-377, 1995b.

Jones, D., et al., The Astronomical Low Frequency Array: A proposed Explorer mission for radio astronomy, in *Space-Based Radio Observations at Long Wavelengths,* edited by R. Stone, K. Weiler, M. Goldstein, and J-L Bougeret, AGU, in press, 2000.

Kaiser, M. L. and Weiler, K. W., The current status of low frequency radio astronomy from space, in *Space-*

Based Radio Observations at Long Wavelengths, edited by R. Stone, K. Weiler, M. Goldstein, and J-L Bougeret, AGU, in press, 2000.

Kassim, N. E. and Weiler, K. W. (eds.), *Proc. Conf., Low Frequency Astrophysics from Space,* Springer-Verlag, Berlin, 1990.

Kassim, N. E. and Erickson, W. C., Meter/decameter wavelength array for astrophysics and solar radar, *Proc. Conf., Advanced Technology MMW, Radio, and Terahertz Telescopes,* SPIE, 1998.

Kassim, N. E., Perley, R. A., Erickson, W. C., and Dwarakanath, K. S., Subarcminute resolution imaging of radio sources at 74 MHz with the Very Large Array, *Astron. J., 106,* 2218-2228, 1993.

Kassim, N. E., LaRosa, T. N., Lazio, T. J. W., and Hyman, S. D., Wide field radio imaging at 74 MHz., *Proc. Conf., The Central Parsecs – Galactic Center Workshop,* edited by A. Cotera and H. Falcke, *Publ. A. S. P.,* in press, 2000.

Kuiper, T. B. H., Jones, D. L., Mahoney, M. J., and Preston, R. L., A simple low-cost array on the lunar nearside for the early lunar expeditions, *Proc. Conf., Low Frequency Astrophysics from Space,* edited by N. Kassim and K. Weiler, Springer-Verlag, Berlin, 46-51, 1990.

Lunar Study Group, Mission to the Moon: Europe's priorities for the scientific research exploration and utilization of the moon, *ESA SP-1150,* 1992.

Weiler, K. W. (ed.), *Proc. Conf., Radio Astronomy from Space,* Workshop No. 18, NRAO, 1987.

Weiler, K. W., Dennison, B. K., and Johnston, K. J., Radio astronomy looks to space, *Astronomy, 16,* 18-25, 1988.

Whitham, P. S., Hardware and software for the Llanherne low frequency sky surveys, *Astron. Soc. Australia, Proc. 2,* 338-340, 1975.

K. W. Weiler, Naval Research Laboratory, Code 7213, Washington, DC 20375-5320, USA
(e-mail: weiler@rsd.nrl.navy.mil)

What Would the Sky Look Like at Long Radio Wavelengths?

K. S. Dwarakanath

Raman Research Institute, Bangalore, INDIA

At wavelengths of about 10 meters the appearance of the sky is dominated by the intense synchrotron background radiation of the Galaxy. Several ionized-Hydrogen regions in the Galaxy make their appearance as discrete absorption features seen against this intense background radiation. Some of the Galactic supernova remnants are resolved even at moderate resolutions ($\sim 0.^{\circ}5$) of most of the existing long-wavelength sky surveys. Away from the Galactic plane a large number of extragalactic sources are seen, most of which are unresolved at this resolution and are identified with radio galaxies and quasars while a very small fraction of them are due to clusters of galaxies. Only a couple of these sources are millisecond pulsars seen as continuum sources in the confusion-limited long wavelength surveys. At the resolution ($\sim 20''$) and sensitivity (~ 1 mJy/beam) that the Low Frequency Space Array is expected to achieve at wavelengths of about 10 meters, most of the supernova remnants, ionized-Hydrogen regions, halos in near-by clusters and halos around galaxies should be resolved. An all-sky survey carried out by the Space Array is expected to be at least two orders of magnitude more sensitive compared to the existing large scale sky surveys. A large population of millisecond pulsars might be detectable as continuum sources in such a survey. At wavelengths much longer than 10 meters, the Galaxy is no longer optically thin and at about 100 meters wavelength even the Warm Ionized Medium becomes optically thick for a path length of about 2 kpc. Hence, at such long wavelengths, the Galactic radio emission from only the solar neighborhood would be detectable. Observations in the wavelength range of 10 to 100 meters can thus lead to a 'tomographic' study of the Galaxy.

1. INTRODUCTION

The phrase 'Long Wavelengths' in the present context of the Low Frequency Space Array is meant to indicate wavelengths longer than 10 m or frequencies smaller than 30 MHz. The 10 m wavelength marks the low end of the electro magnetic spectrum up to which the sky has been explored in some detail from ground-based observations. I will adopt 30 MHz and 3 MHz as two canonical frequencies at which to explore the appearance of the sky.

There are three factors which play important role in changing the appearance of the sky as one tunes down the observing frequency (ν). First, the free-free optical depth which changes as $\nu^{-2.1}$ results in the turn-over of the spectra (marked 'HII') as shown in Fig. 1. At low frequencies such optically thick HII regions absorb the non-thermal background radiation and the Galaxy

Radio Astronomy at Long Wavelengths
Geophysical Monograph 119
Copyright 2000 by the American Geophysical Union

258 SKY AT LONG RADIO WAVELENGTHS

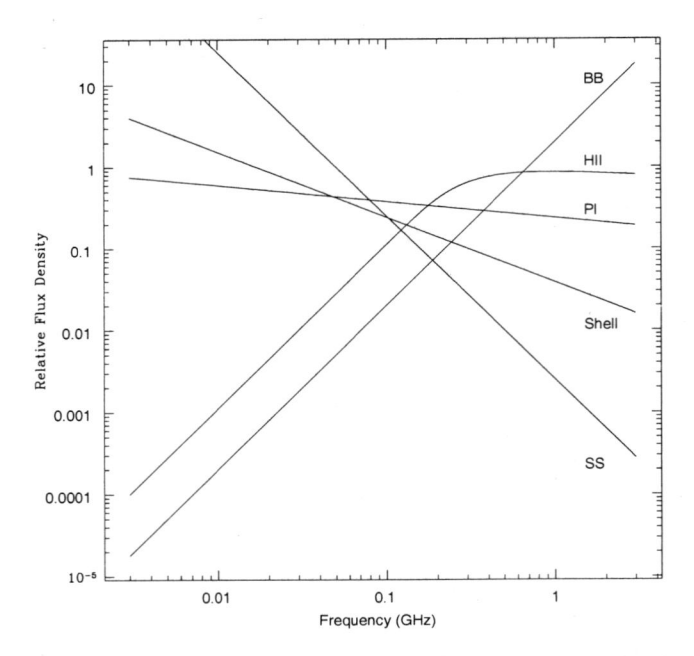

Figure 1. Typical spectra of different kinds of sources. 'BB' refers to Black-Body with an $\alpha = +2$, where flux density $\propto$ (frequency)$^\alpha$. The spectrum of radiation from ionized gas (HII) is optically thin at higher frequencies where $\alpha = -0.1$, while at lower frequencies it is optically thick with $\alpha = +2$. The spectra of plerions ('Pl') is flat with an $\alpha \sim -0.15$, while the shell-type supernova remnants ('Shell') have an $\alpha \sim -0.5$. Also shown in the figure is steep-spectrum source (SS) with an $\alpha = -2$.

no longer remains transparent to radio waves as it does at much higher frequencies ~ 1 GHz. Second, the relative prominence of the sources change due to the differences in their spectra. For e.g., a shell-type supernova remnant ('Shell' in Fig. 1) appears brighter and more dominant compared to a plerion ('Pl' in Fig. 1) as the former has a steeper spectral index compared to the latter. Similarly, steep-spectrum sources ('SS' in Fig. 1) like the millisecond pulsars or the halos in clusters of galaxies appear relatively bright at low frequencies. Third, improved angular resolution and sensitivity lead to the detection of new sources and reveal a rich variety of structures in known sources. This is particularly important at low frequencies as most of the ground-based telescopes have only a moderate resolution due to their limited aperture sizes and large wavelengths. At frequencies like 30 MHz all-sky images are available at $\sim 0^\circ.5$ resolution (Williams, Kenderdine, & Baldwin 1966, Jones & Finlay 1974, Dwarakanath & Udaya Shankar 1990). These surveys are confusion-limited with point source sensitivities ~ 10 Jy. Only a limited portion of the sky has been imaged at $12'$ (Kassim 1988) and $4'$ resolution (Rees 1990) at frequencies ~ 30 MHz. On the other hand, at frequencies ~ 1 GHz, synthesis instruments routinely achieve arc second resolutions and milli-Jansky sensitivities.

The proposed Astronomical Low Frequency Array (ALFA) is expected to improve upon the existing res-

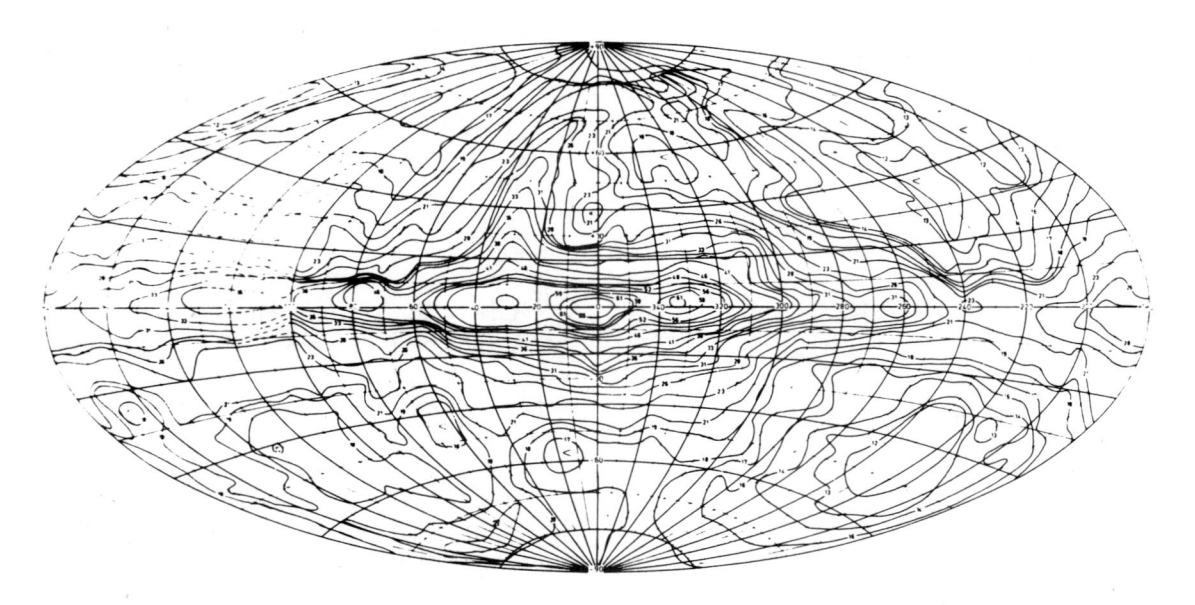

Figure 2. All-sky 30 MHz image at 10^0 resolution from Cane (1978). The contour unit is 1000 K of equivalent brightness temperature. The brightness temperature varies from $\sim 10^4$ to 6.6×10^4 K.

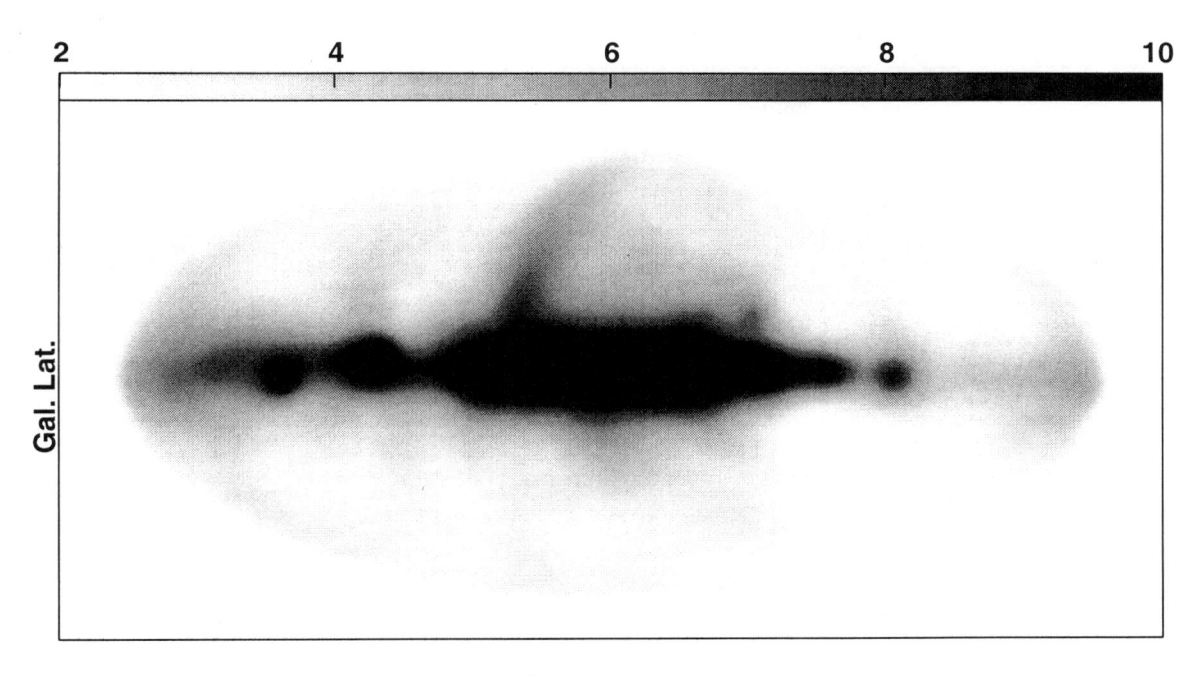

Figure 3. All-sky image at 408 MHz from Haslam et al. (1982) in equal area projection and in Galactic Coordinates. The image presented here is at 10^0 resolution. The brightness temperature in this image varies from a minimum of $\sim$ 15 K to a maximum of $\sim$ 260 K.

olutions and sensitivities by several orders of magnitude. Around 30 MHz the expected resolution is $\sim$ 20″ with a confusion limit $\sim$ 1 mJy/beam. Around 3 MHz, the resolution and confusion limits are $\sim$ 3′ and $\sim$ 75 mJy/beam respectively. Needless to emphasize that ALFA will detect new sources and reveal detailed structures of known sources.

2. LARGE SCALE FEATURES

An image of the sky at 30 MHz highlighting the large scale features is shown in Fig. 2. This is from Cane (1978) who combined the single dish observations from Jodrell Bank (Milogradov-Turin & Smith 1973) and Parkes (Mathewson et al. 1965) to produce this image. It is instructive to compare this with the 408 MHz all-sky image shown in Fig. 3. The original 408 MHz all-sky image, at $0.^085$ resolution (Haslam et al. 1982), was convolved to 10^0 to produce this image with a resolution similar to that of the 30 MHz image in Fig. 2. A rich variety of loops and spurs of non-thermal emission can be seen at 408 MHz, the most prominent of which is the North Polar Spur starting from the Galactic plane at l $\sim$ 30^0 and going toward the North Galactic Pole. A comparison of figures 2 & 3 reveals that the low frequency sky is similar in appearance to the high frequency sky with the various loops, spurs, minima and maxima identifiable in both the images. However, the ratio of maximum to minimum intensity is about a factor of 3 smaller at 30 MHz compared to that at 408 MHz. This is indicative of free-free absorption of the non-thermal background by the foreground thermal gas in the Galactic plane. Further discussions will elaborate on this point.

A 2^0 smoothed image from the 408 MHz all-sky survey in the range $340^0 < l < 100^0$ and $-20^0 < b < +20^0$ is shown in Fig. 4. The ridge of non-thermal emission at b$\sim$ 0^0 for l$<$ 50^0 is evident here. In addition, the unresolved extragalactic source Cygnus A and the resolved Galactic source Cygnus X are clearly seen at l$\sim$ 75^0. The equivalent brightness temperature, T_B of the non-thermal radiation at 408 MHz ranges from $\sim$ 20 K away from the Galactic Plane to $\sim$ 1000 K towards the Galactic Center. It is interesting to compare this image with that made at 34.5 MHz from the GEETEE Survey (Dwarakanath & Udaya Shankar 1990). The same region of the Galaxy as in Fig. 4, but from the GEETEE Survey smoothed to 2^0 resolution, is shown in Fig. 5. This figure looks remarkably different from Fig. 4. To begin with, the Cygnus X complex seen clearly in Fig.

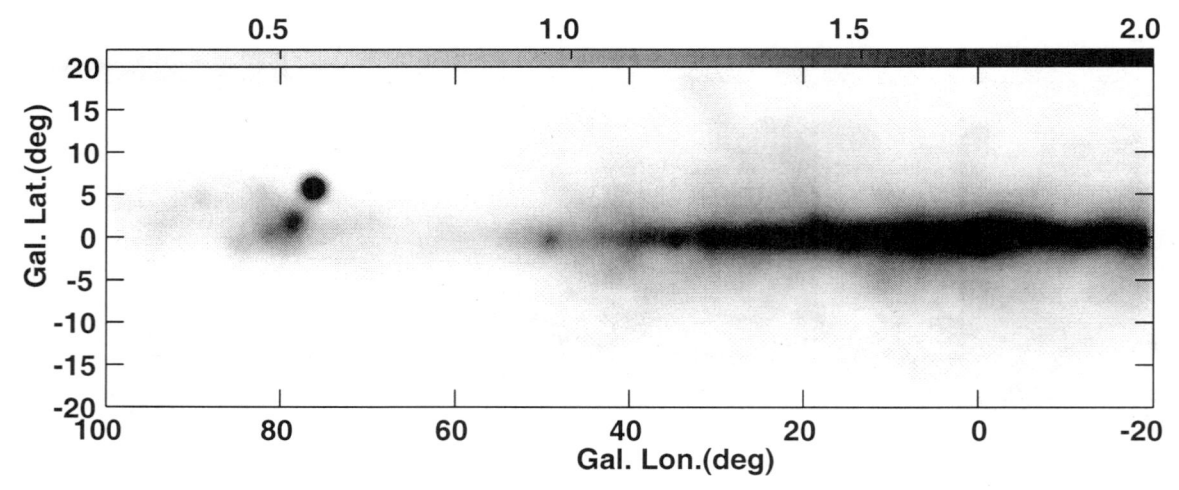

Figure 4. Inner Galaxy at 408 MHz from Haslam et al. (1982). The resolution of this image is 2^0. The brightness temperature in this image varies from 20 to 1000 K.

4 is almost absent in Fig. 5 indicating that most of its radio emission is of thermal origin. In addition, we see a flat top emission from the plane at 34.5 MHz as opposed to the ridge of emission seen at 408 MHz. The profiles of the brightness temperature across the plane at 34.5 and 408 MHz (Fig. 6) make this even more evident. The flat-top emission at 34.5 MHz is clearly due to the absorption of the non-thermal background emission by the foreground thermal gas in the Galactic plane. The thermal gas responsible for the absorption at 34.5 MHz is seen clearly in Fig. 7 which is at the full resolution of the GEETEE Survey of $\sim 0.^05$. Several discrete absorptions can be seen in the Galactic plane leading to a patchy appearance of the non-thermal emission. Most of these absorption features can be identified with ionized Hydrogen regions seen in emission at other wavelengths.

What further changes do we expect in the appearance of the sky as one tunes the observing frequency down to ~ 3 MHz ? This is best answered by starting with the existing images at frequencies close to this. The lowest frequency ground-based observations were carried out by Reber (1968) at 2 MHz taking good advantage of an ionospheric hole above Tasmania. An image of the sky from his observations is shown in Fig. 8. Although this image has only 8^0 resolution, it has valuable information. In this image the South Galactic Pole (SGP) has a $T_B \sim 3.5 \times 10^6$ K. The effective spectral index for

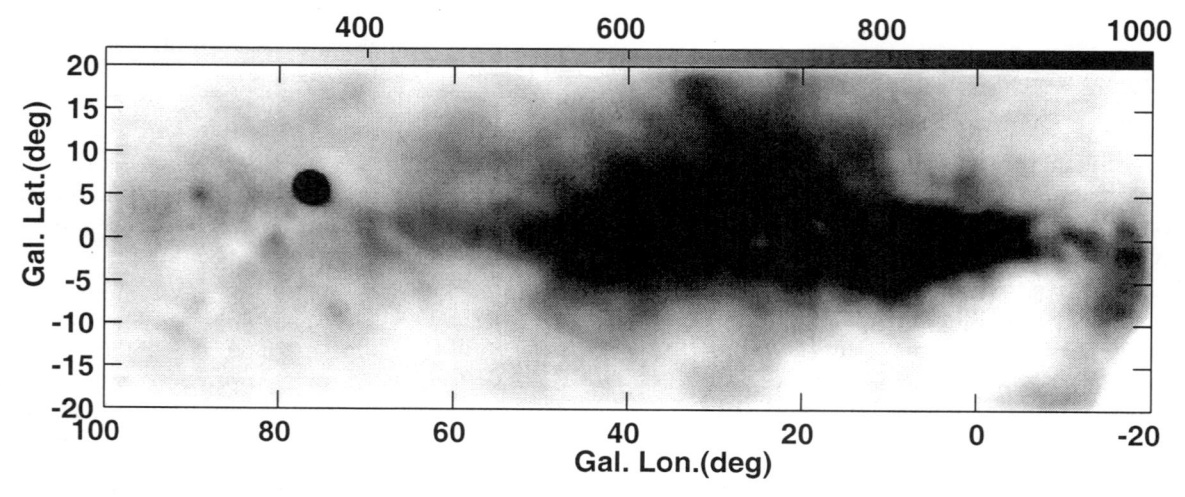

Figure 5. Inner Galaxy at 34.5 MHz from the GEETEE Survey (Dwarakanath & Udaya Shankar 1990). The resolution of this image is 2^0. The brightness temperature in this image varies from 10^4 to 1.8×10^5 K.

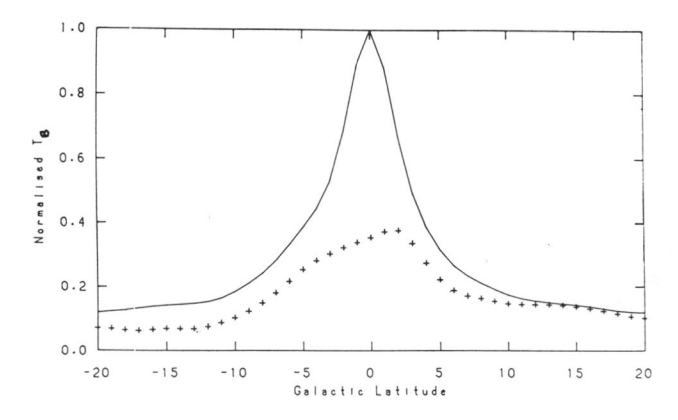

Figure 6. The solid curve represents the normalized brightness temperature profile at 408 MHz at l= 10°. The crosses represent the same at 34.5 MHz. The 34.5 MHz profile is aligned with that at 408 MHz at b~ 20°. Both profiles were obtained from images at 2° resolution.

the temperature, β, between 2 and 34.5 MHz is -2.0 ($T_B \propto \nu^\beta$). Between 34.5 and 408 MHz the β value for the South Galactic Pole is -2.55. A detailed analysis of the intensity of the non-thermal radiation between 1 and 100 MHz (Cane 1979) leads to the conclusion that there is a spectral turn over below $\sim$ 6 MHz, leading to the flatter spectral index between 2 and 34.5 MHz estimated here. One of the components required in fitting the spectrum between 1 and 100 MHz is free-free absorption due to a warm ionized gas in the Galaxy. This diffuse gas becomes optically thick around frequencies $\sim$ 2 MHz. An interesting consequence of this is that in the 2 MHz image of Reber the intensity of the non-thermal background *decreases* to $\sim 10^5$ K towards the Galactic plane from its larger value towards the South Galactic Pole ! This is one of the first indications of the existence of a new phase of the Interstellar Medium, viz. the Warm Ionized Medium (WIM). In the absence of such an absorption, the Galactic plane at 2 MHz would have been much brighter with a maximum $T_B \sim 2 \times 10^8$ K !! The WIM has been inferred from other observations also. It is now believed that the WIM is responsible for producing the diffuse H_α emission, the dispersion of pulsar signals, and the Faraday Rotation observed in every direction in the Galaxy.

The implications of the WIM to low-frequency ($\leq$ 3 MHz) observations are dramatic. One of the advantages radio astronomy boasts off, compared to optical astronomy, is the transparency of the Galaxy to radio waves. However, the observable region of the Galactic plane is once again limited to the Solar neighborhood at low frequencies. Instead of feeling limited by this, we can turn it to our advantage. If one can observe the Galactic plane over a range of frequencies, like 1 to 10 MHz, it is possible to solve for non-thermal emissivity and free-free absorption of the Galaxy. The non-thermal emissivity at a frequency ν of a length L of the Galaxy is $\propto LB_\perp^{(\gamma+1)/2}\nu^{-(\gamma-1)/2}$, where $B_\perp$ is the perpendicular component of the magnetic field of the Galaxy and γ is the power-law index in the distribution of the number density (N (E)) of relativistic particles of energy (E), such that $N(E) \propto E^{-\gamma}$. The free-free opti-

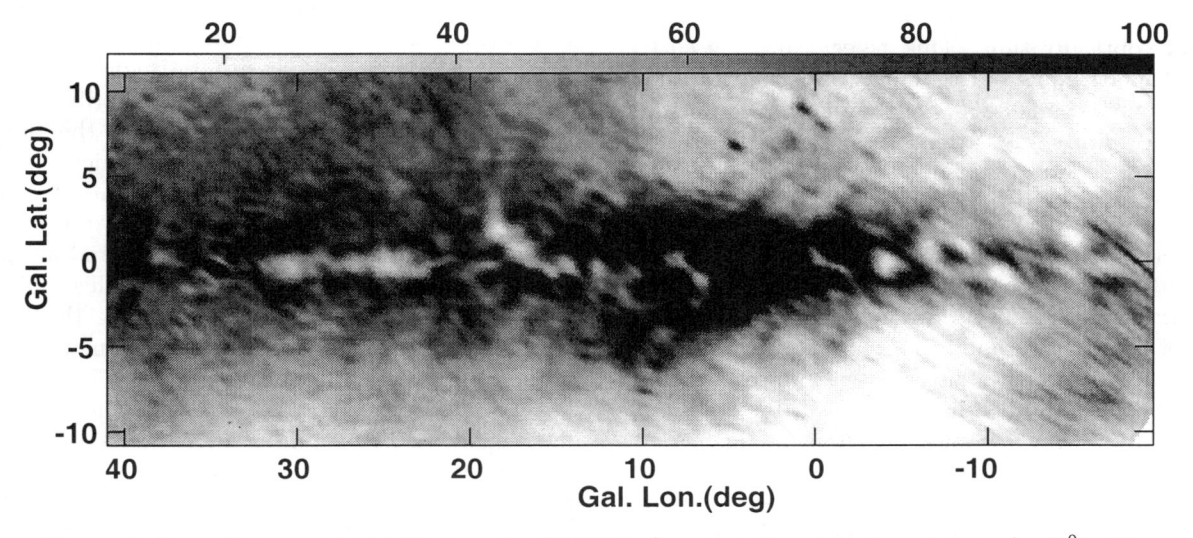

Figure 7. Inner Galaxy at 34.5 MHz from the GEETEE Survey at the original resolution of $\sim 0.°5$. The maximum brightness temperature here is $\sim 2.5 \times 10^5$ K. The white patches close to Galactic Latitude $\sim 0°$ are due to absorption of the non-thermal background by the foreground optically thick thermal gas (HII regions).

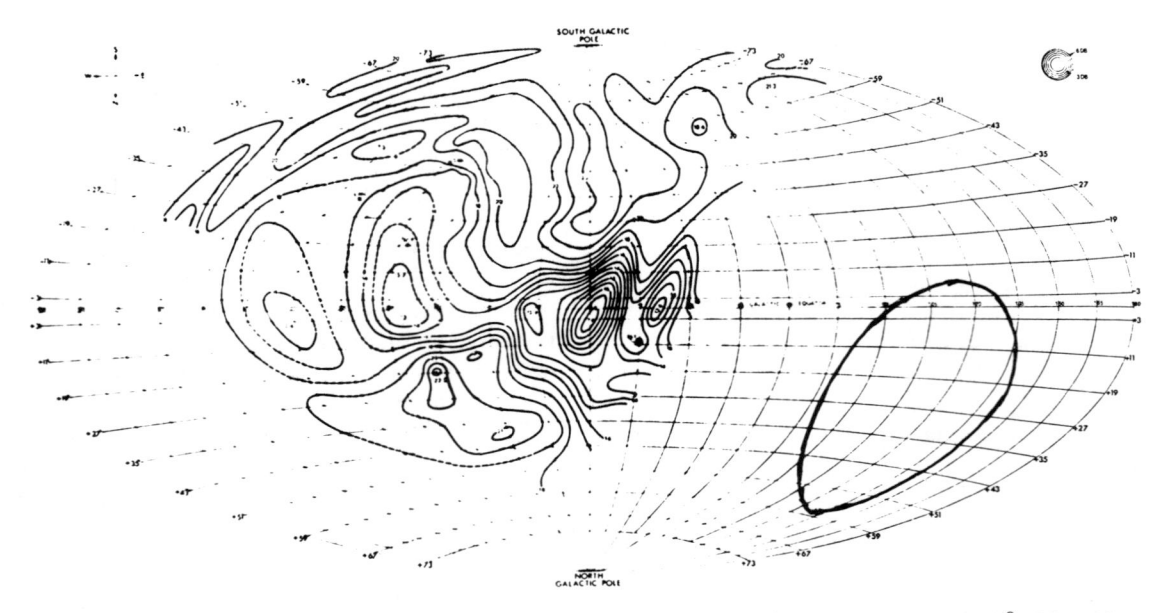

Figure 8. Sky at 144 m (~ 2.1 MHz) as seen by Grote Reber (1968). The resolution is 8^0. Note that the South Galactic Pole is at the top – as seen by a southern observer !. The numbers are in db's. The intensity at the SGP is 21.3 db above 1 μV and is equal to 3.5×10^6 K. Note that the intensity *decreases* toward the Galactic plane. The thick, dark line encloses ~ a steradian of the sky which has been imaged at 38 MHz (see Fig. 16).

cal depth (τ) is related to the electron temperature (T_e) and thermal electron density (n_e) as $\tau \propto T_e^{-1.35} \nu^{-2.1} n_e^2 L$. The free-free optical depth becomes unity for L ~ 2 kpc at frequencies ~ 3 MHz for T_e ~ 10^4 K and $<n_e> \sim 0.03$ cm^{-3}. At frequencies ~ 1 MHz, most of the non-thermal emissivity comes from the local neighborhood, while at frequencies ~ 10 MHz, the Galaxy is essentially optically thin. This aspect can be used to attempt 'tomography' of the Galaxy and solve for both the non-thermal emissivity and the distribution of the thermal gas in a self-consistent way.

3. SMALL SCALE FEATURES

In order to get a feeling for the small-scale features of the sky at long wavelengths as will be viewed by ALFA, it is instructive to look at the existing low-frequency images of the sky. I would like to start with the sky survey at 34.5 MHz and illustrate what we see at moderate resolutions and sensitivities. I will further show some examples, drawn from high frequency images, to demonstrate what we would expect to see with the vastly improved resolution and sensitivity at these low frequencies. The existing high frequency images have resolutions and sensitivities comparable to what ALFA is expected to achieve at low frequencies and hence are appropriate examples.

3.1. The GEETEE Survey

A synthesis image of the sky at 34.5 MHz was made using the Gauribidanur T-array (GEETEE). This survey (Dwarakanath & Udaya Shankar 1990) covers the declination rage of -50^0 to $+70^0$ and the full 24 hours of right ascension. The resolution of the survey is ~ $0.^05$ with a sensitivity for point source detection (3σ) of ~ 15 Jy/beam. This sensitivity is confusion limited. About 5000 sources were detected away ($|b| > 5^0$) from the Galactic plane in this survey above this detection limit. There are a variety of sources which can be expected to be detected in a survey like this. (a) Galaxy Clusters – where the radio emission is expected to be due to powerful cDs, head-tail sources, and halos. The spectral index α (flux density $\propto \nu^\alpha$) of this radio emission is known to be steep and is ~ -1.5. (b) Globular Clusters in the Galaxy which are known to contain a large number of millisecond pulsars. The ms-pulsars have very steep spectra ($\alpha \sim -2.5$) and the average flux density of all the pulsars in a Globular Cluster can can add up to a bright radio continuum source. (c) field millisecond pulsars with steep spectra. (d) radio galaxies and quasars with a mean spectral index $\alpha \sim -0.8$. Most of the sources in this survey are in the category (d) with ~ 1 to 2 % of the sources being in the category (a). There were 5 marginal detections of (b) types and

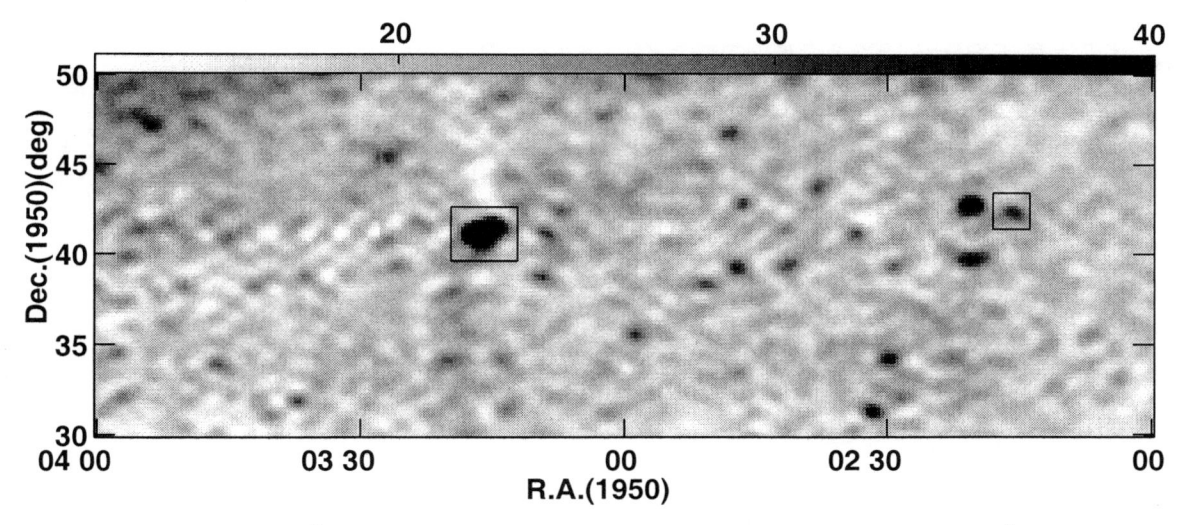

Figure 9. A 2 Hour × 20° region of the sky at 34.5 MHz from the GEETEE Survey at ~ 0.°5 resolution. There are ~ 50 sources in this field above a detection limit of ~ 15 Jy. The source at $\alpha \sim 03^h16^m$ (within a box) is due to radio emission from the Perseus Cluster of galaxies. The integrated flux density of this source at 34.5 MHz is 440 Jy. The flux density of the source at $\alpha \sim 02^h15^m$ (within a box) is 50 Jy, about half of which is due to the millisecond pulsar J0218+4232.

2 to 3 source in the (c) category. So, at this resolution and sensitivity the low-frequency sky is dominated by the radio galaxies and quasars. As an illustration, I will show in Fig. 9 a small portion of the sky at 34.5 MHz made with the GEETEE.

3.2. An Example of Spectral Effects

Even at coarse resolutions of ~ 0°.5 currently available at frequencies ~ 30 MHz, it is possible to discern the nature of sources by comparing the low frequency images with those at higher frequencies (~ 408 MHz) and estimating their spectra. One such illustrative example is the Vela-Puppis Complex in the southern skies. I show in Figs. 10 and 11 a 2°-smoothened image of this complex at 34.5 and 408 MHz respectively. It can be seen from these two images that Vela Y and Vela Z are more prominent compared to Vela X in the low frequency images. This is borne out by their spectra : Vela X has a flatter spectral index ($\alpha \sim -0.16$) compared to Vela YZ ($\alpha \sim -0.53$). In fact, Vela YZ is a shell-type supernova remnant which owes its radio emission to a blast wave while Vela X is a plerion being maintained by the Vela Pulsar (Dwarakanath 1991). The source seen to the south-east of the Vela supernova remnant in the 408 MHz image is completely absent in the 34.5 MHz image. Not surprisingly, this turns out to be an HII region.

3.3. Cluster Halos

There are only a limited number of sources that are amenable to the kind of treatment mentioned in the previous subsection. In most cases the limited resolution and sensitivity of ground-based observations mask the

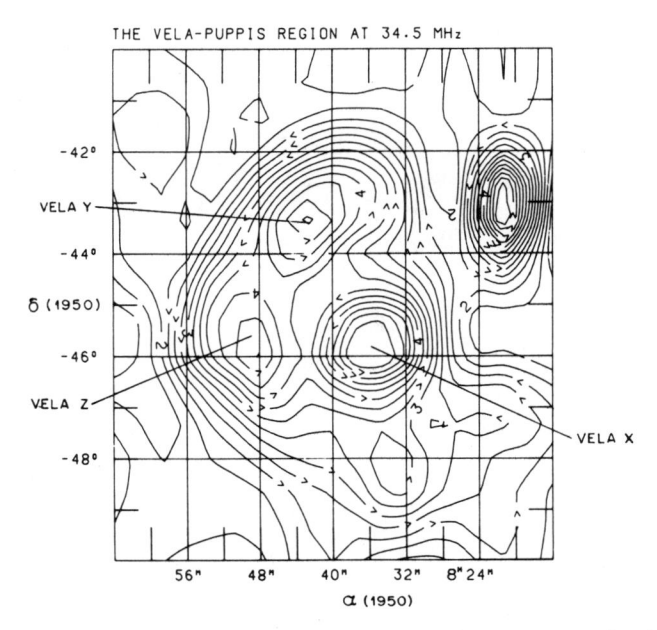

Figure 10. The Vela-Puppis region at 34.5 MHz convolved to a resolution of $1° \times 1.°5$. The contours are labeled in units of 10^4 K.

264 SKY AT LONG RADIO WAVELENGTHS

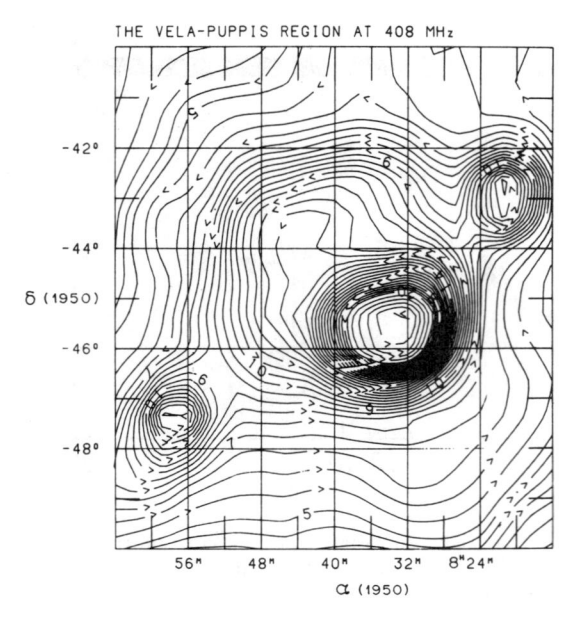

Figure 11. The Vela-Puppis region at 408 MHz convolved to a resolution of $1°\times 1.°5$. The original data at a resolution of $51'\times 51'$ is from Haslam et al. (1982). The contours are labeled in units of 10 K.

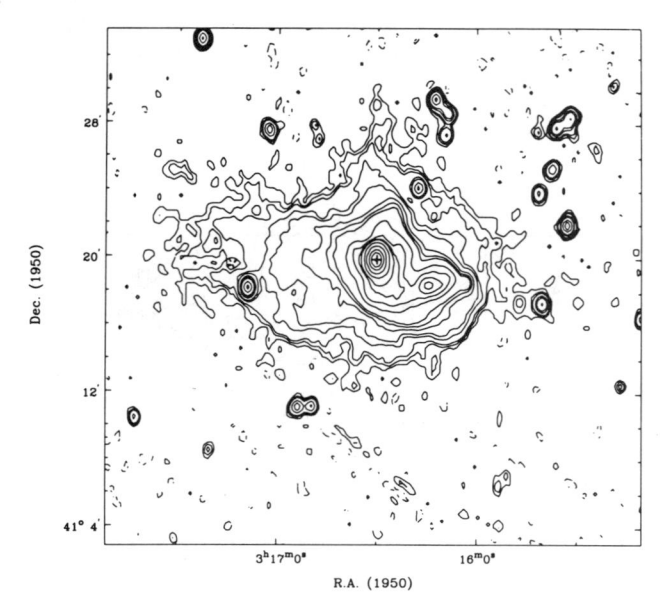

Figure 12. A 49 cm (612 MHz) Westerbork Synthesis Radio Telescope image of the central region of the Perseus cluster of galaxies (the source enclosed in the left box in Fig. 9). This image at a resolution of $29''\times 44''$ is from Sijbring (1993). The weakest contour in this image is at 0.8 mJy/beam. The central source (3C 84) as well as many others surrounding it are members of the Perseus cluster.

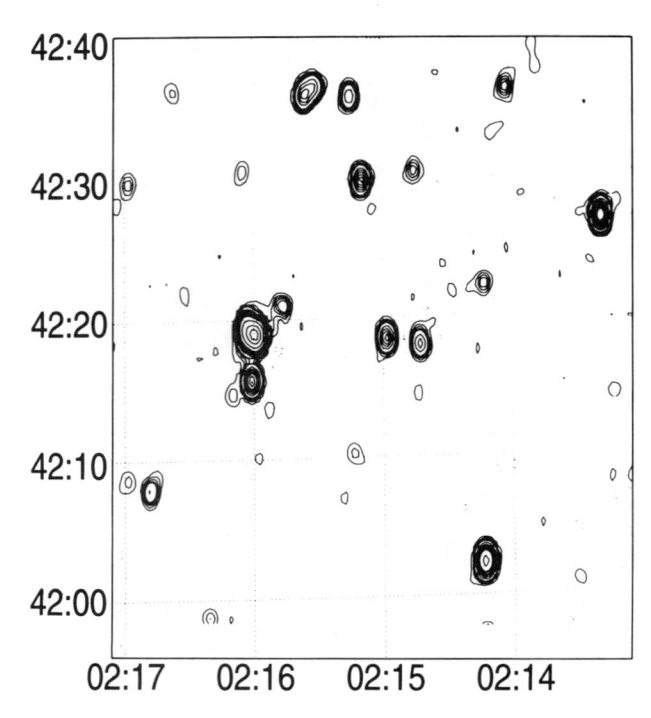

Figure 13. A 325 MHz WSRT image of the source in the right hand side box in Fig. 9. This image at a resolution $\sim 1'$ is from Navarro et al. (1995). The source at 02:15 has an $\alpha \sim -2.3$ and contributes about half the flux density of the low frequency source seen in Fig. 9. This source is a ms pulsar.

interesting details. Consider, for example, the bright source at 0316+41.3 in Fig. 9. How might this appear when seen by ALFA ? This can be assessed from Fig. 12. This is an image at 49 cm at a resolution $\sim 30''$ made with the WSRT (Sijbring 1993). The weakest feature in this image, ~ 0.8 mJy/beam, will be ~ 30 mJy/beam at frequencies ~ 30 MHz for a spectral index of -1.2 estimated for the halo emission from high frequency measurements. This will be easily detected by ALFA. At lower frequencies ~ 3 MHz, the halo is expected to be even more extended. At these low frequencies the lifetime of the radiating electrons $\sim 3\times 10^9$ yr - comparable to the cluster age. We could very well be looking at the fossil cosmic ray electrons in the low frequency images of this halo.

3.4. Short-period pulsars

The 'source' at 0215+42.3 (Fig. 9) has been imaged at $1'$ resolution using the WSRT at 325 MHz (Navarro et al. 1995). This is shown in Fig. 13 – an image similar to what can be expected from ALFA at low frequencies. The source at $\sim 0215+4220$ in this figure turns out to be a compact, steep-spectrum ($\alpha \sim -2.3$), continuum source. It turns out to be a 2.3 ms, 2-day binary pulsar in the field. This is very reminiscent of the very first ms-pulsar (B1937+21) that was detected in a similar way at low frequencies (Erickson 1983) and subsequently B1957+20 also. It is believed that a large population ($\sim 10^4$) of such field binary ms-pulsars in our Galaxy are yet to be discovered (Lorimer 1995). Low-frequency surveys like the ones by ALFA can reap a rich harvest of these objects as candidates for targeted pulsar searches at higher frequencies.

One other class of objects which are interesting to observe with ALFA will be Globular Clusters (in our Galaxy). The enhanced stellar density in the cores of Globular Clusters is considered conducive to the formation of Low Mass Binary Systems which lead to millisecond pulsars. An estimate seems to indicate that the Globular Cluster Terzan 5 probably has 50 pulsars beamed toward the Earth. However, only one (eclipsing) pulsar PSR 1744-24A has been detected so far in this cluster (Fig. 14). There are two other radio continuum sources in the direction of this cluster - one with a 20 cm flux density of 1 mJy and the other with a 20 cm flux density of 2 mJy (Fruchter & Goss 1994). Both

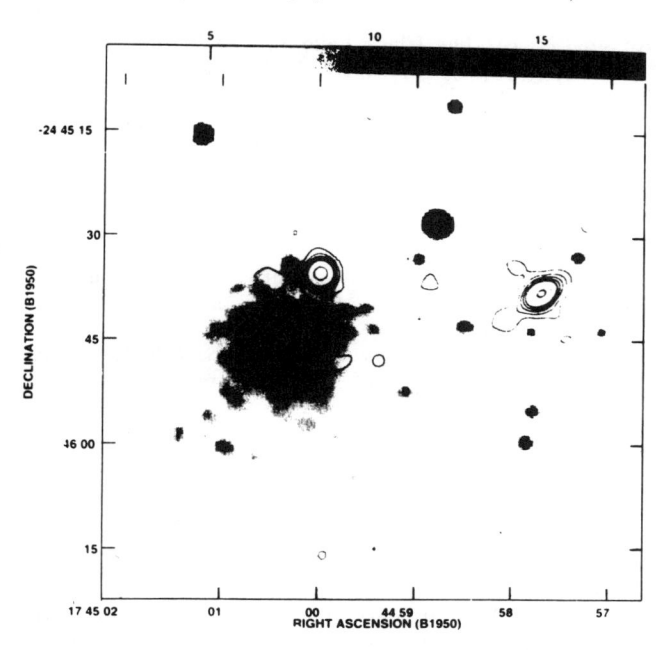

Figure 14. A 20 cm Very Large Array image of the Globular Cluster Terzan 5 from Fruchter & Goss (1994). The 20 cm radio image in contours is overlaid on an I-band image in grey scale. The strength of the source to the west of the center is 1 mJy at 20 cm and turns out to be an eclipsing pulsar, PSR 1744-24A. The source to the east of the center is 1 mJy at 20 cm with an $\alpha \sim -2$. The extended dark area below this source has a diffuse radio emission of 2 mJy at 20 cm and is expected to have an $\alpha < -1.4$.

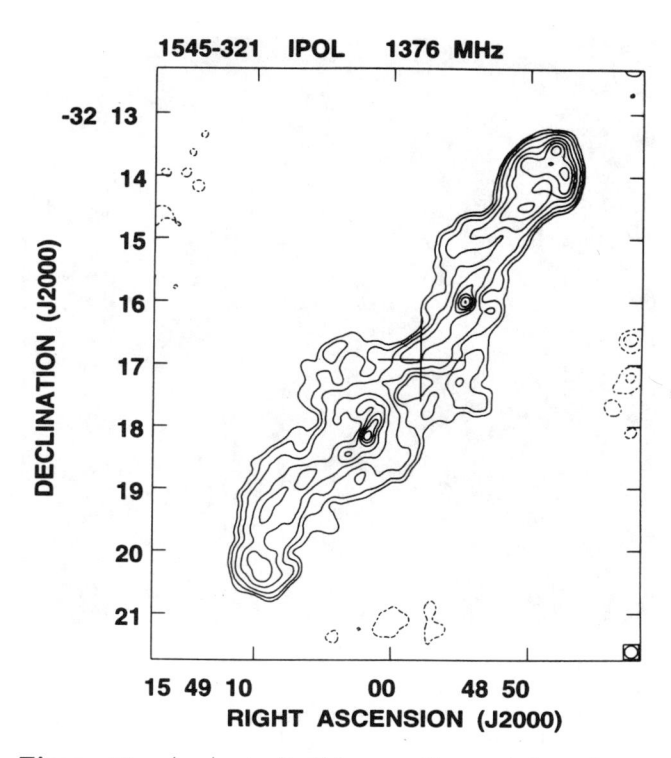

Figure 15. An Australia Telescope Compact Array image of a Giant Radio Galaxy (z=0.1085) at $12''$ resolution by Subrahmanyan et al. (1996). This image is at 1376 MHz. The lowest contour is at 1 mJy/beam. The '+' marks the position of the galaxy identification.

these sources are believed to have spectra steeper than $\alpha \sim -1.5$. At frequencies ~ 30 MHz the cluster core will be a 6 Jy continuum source if the steep spectra continue up to low frequencies. Detecting radio continuum emission from Globular Clusters is a useful first step in detecting short-period pulsars. So far, only $\sim 10\%$ of the Globular Clusters have been observed in radio continuum. A low-frequency survey like the one proposed by ALFA is bound to pick up many Globular Clusters as bright continuum sources like the one displayed in Fig. 14.

3.5. Giant Radio Galaxies

These are some of the largest FR-II radio sources with sizes > 1 Mpc. One such example is shown in Fig. 15 (Subrahmanyan et al. 1996). There is considerable interest in understanding the evolution of these sources and their surroundings, viz. the Intergalactic Medium. The important ingredients in such an understanding are the production and acceleration of particles, and the strength of magnetic field. A detailed spectral index distribution across the source is vital in understanding these issues. The source shown in this figure appears

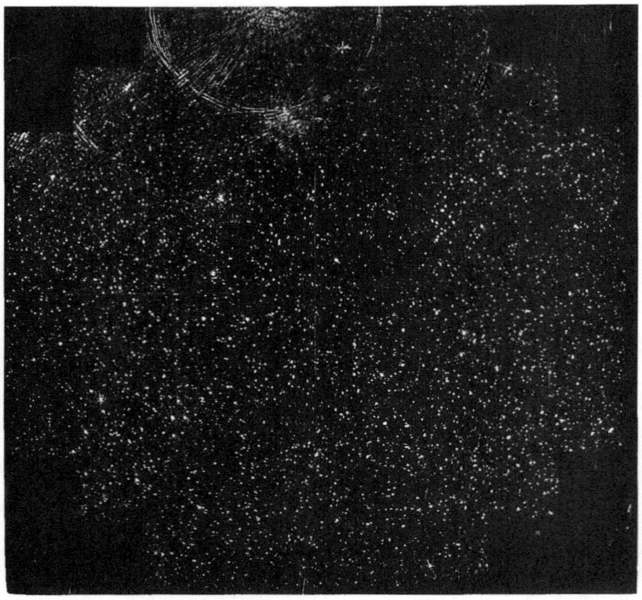

Figure 16. A steradian centered on the North Celestial Pole imaged at 38 MHz by Nick Rees (1990). This image has 4′ resolution, 40,000:1 dynamic range, and sub-Jy sensitivity. The approximate size of this image is shown as an oval in Fig. 8

as an unresolved 75 Jy source in the GEETEE survey. However, due to its large angular extent, future space missions with their improved resolution and sensitivity will be able to obtain detailed images of such sources at frequencies as low as 3 MHz. The range of frequencies available for analyses of spectral indices will then be an unprecedented 3 orders of magnitude !

4. A GIANT LEAP

As was mentioned in the first section, the currently available best image of the sky at frequencies of a few MHz is that by Reber (1968). The space mission ALFA is expected to achieve a resolution of $\sim 3'$ and a corresponding confusion limit ~ 75 mJy/beam at these frequencies. An impression of the sky at these resolutions can be obtained by looking at the 38 MHz image produced by Rees (1990) using the Cambridge 5-km array. This image (Fig. 16) is at a resolution of 4′ and has sub-Jy sensitivity for point sources. This image is a steradian of the sky centered around the North Celestial Pole. The approximate size of this image is marked in Fig. 8. The dynamic range on this image is 40,000:1. Needless to emphasize the wealth of details that will be brought about by images like these at low frequencies.

Acknowledgments. I would like to thank the National Science Foundation, and the National Aeronautics and Space Administration for supporting my travel to the Chapman Conference on Space Based Radio Observations at Long Wavelengths held in Paris (October 19-23, 1998). I would also like to thank A. A. Deshpande & N. Udaya Shankar for useful discussions and comments on the manuscript.

REFERENCES

Cane, H. V., A 30 MHz map of the whole sky, *Aust. J. Phys.*, *31*, 561-565, 1978.

Cane, H. V., Spectra of the non-thermal radio radiation from the galactic polar regions, *Mon. Not. R. Astr. Soc.*, *189*, 465-478, 1979.

Dwarakanath, K. S., and N. Udaya Shankar, A synthesis map of the sky at 34.5 MHz, *J. Astrophys. Astr.*, *11*, 323-410, 1990.

Dwarakanath, K. S., Low-frequency observations of the Vela supernova remnant and their implications, *J. Astrophys. Astr.*, *12*, 199-211, 1991.

Erickson, W. C., What is 4C 21.53 ?, *Astrophys. J.*, *264*, L13-L17, 1983.

Fruchter, A. S., and W. M. Goss, Deep radio observations of the rich globular clusters Terzan 5 and Liller 1, *Proc. of the Diamond Jubilee Symposium of the Indian Academy of Sciences*, 245-253, 1994.

Haslam, C. G. T., C. J. Salter, H. Stoffel, and W. E. Wilson, A 408 MHz all-sky continuum survey II - The atlas of contour maps, *Astr. Astrophys. Suppl.*, *47*, 1-142, 1982.

Jones, B. B., and E. A. Finlay, An aperture synthesis survey of the Galactic plane, *Aust. J. Phys.*, *27*, 687-711, 1974.

Kassim, N. E., The Clark Lake 30.9 MHz Galactic plane survey, *Astrophys. J. Suppl.*, *68*, 715-733, 1988.

Lorimer, D. R., The birth rate of low-mass binary pulsars in the Galactic disk, in *The Lives of the Neutron Stars*, edited by M. A. Alpar, U. Kiziloglu, and J. van Paradijs, pp. 477-492, Kluwer, Dordrecht, 1995.

Mathewson, D. S., N. W. Broten, and D. J. Cole, A survey of the southern sky at 30 Mc/s, *Aust. J. Phys.*, *18*, 665, 1965.

Milogradov-Turin, J., and F. G. Smith, A survey of the radio background at 38 MHz, *Mon. Not. R. Astr. Soc.*, *161*, 269-279, 1973.

Navarro, J., A. G. de Bruyn, D. A. Frail, S. R. Kulkarni, and A. G. Lyne, A very luminous binary millisecond pulsar, *Astrophys. J.*, *455*, L55-L58, 1995.

Reber, G., Cosmic static at 144 meters wavelength, *J. Franklin Inst.*, *285*, 1-12, 1968.

Rees, N., The new Cambridge 38 MHz radio survey - 1 steradian at 4 arcmin resolution and sub-Jansky sensitivity, in *Low Frequency Astrophysics from Space*, edited by N. E. Kassim, and K. W. Weiler, pp. 204-213, Springer-Verlag, Berlin, 1990.

Sijbring, D., A radio continuum and HI line study of the Perseus cluster, Ph. D. Thesis, 177 pp., University of Groningen, October 1993.

Subrahmanyan, R., L. Saripalli, and R. W. Hunstead, Morphologies in megaparsec-size powerful radio galaxies, *Mon. Not. R. Astr. Soc.*, *279*, 257-274, 1996.

Williams, P. J. S., S. Kenderdine, and J. E. Baldwin, A Survey of radio sources and background radiation at 38MHz, *Mem. R. Astr. Soc.*, *70*, 53-110, 1966.

K. S. Dwarakanath, Raman Research Institute, C. V. Raman Avenue, Sadashivanagar, Bangalore 560 080, INDIA

Capabilities And Limitations Of Long Wavelength Observations From Space

Graham Woan

Department of Physics and Astronomy, University of Glasgow, Glasgow, UK

There are compelling reasons why we should leave the Earth to carry out radio astronomy at the long wavelength limit. The terrestrial ionosphere cannot influence the performance of a space based telescope, and we can escape much of the man-made (and natural) radio interference emitted from the Earth. The solution is not complete however. The limitations on angular resolving power, temporal resolution and sensitivity set by the solar wind, interstellar medium and galactic radio emission still apply. Also, the effects of the Sun, by far the brightest source in the sky, on observations of other sources are potentially more damaging for a free-flyer than for an instrument that can hide behind a planet. Here we review how considerations such as these influence the design of a space or Moon-based radio telescope, concentrating on the fundamental issues of sensitivity and resolving power.

1. INTRODUCTION

The effects of the Earth's ionosphere and of strong terrestrial interference have largely prevented effective, ground-based, imaging radio astronomy below 30 MHz. Although pioneering work has been done at frequencies as low as 2 MHz by *Cane* [1977] and others, it is now accepted that good angular resolution cannot be attained from below the ionosphere, and a space-based solution is required.

With the exception of the RAE 1 and RAE 2 satellites launched in 1968 and 1973, no broad-band low-frequency measurements were made from space until the launch of the WAVES experiment on the Wind spacecraft in 1994. The lack of attention this band has received is remarkable, considering it remains the only part of the accessible electromagnetic spectrum yet to be well investigated by the astronomical community.

We expect the performance of space-based very low-frequency radio telescopes to be limited by the effects of the interplanetary medium (IPM) and the interstellar medium (ISM). Density fluctuations in these plasmas lead to variations in their refractive indices that scale as the square of the observing wavelength, and their effects are therefore most severe at the lowest frequencies. This is of particular importance in the design of space-based radio telescopes as they largely define the parameters of any practical instrument. Indeed, the combined effect of the ISM and IPM greatly restrict the available angular and temporal resolutions. Furthermore, the warm component of the interstellar medium is absorptive at low frequencies, descending to a general galactic 'fog' at about 1 MHz [*Reynolds*, 1990]. These thresholds are shown graphically in Figure 1, together with the limiting operating frequencies of some important ground-based instruments.

In addition, Wind spacecraft observations (Figure 2) show that terrestrial interference dominates this portion of the spectrum out to several tens of Earth radii so there is a need not only to get away from the Earth's ionosphere but also its man-made signals. The notion

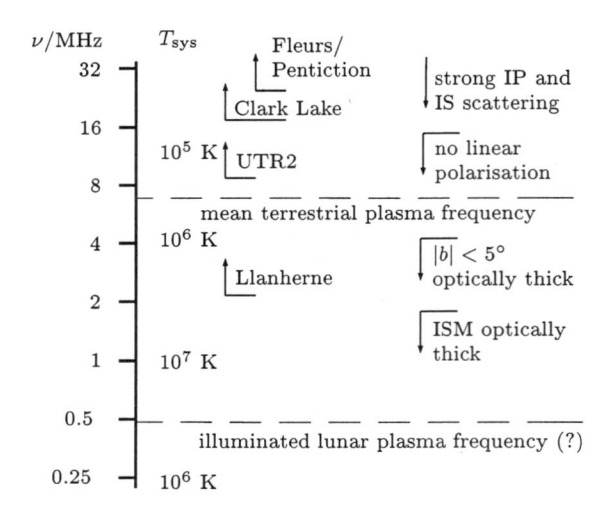

Figure 1. Threshold frequencies for important propagation effects, and lower-limiting observing frequencies for some ground-based low-frequency telescopes. The sky-dominated system temperature (T_{sys}) is also shown.

of achieving this by putting a decametric/hectometric radio telescope in distant Earth orbit or on the far side of the Moon is becoming increasingly popular, not least due to the renewed interest of both NASA and ESA in the Moon as a target for future missions, and the idea of detecting exoplanets from their decametric emission [*Zarka et al.*, 1997].

ESA recently commissioned a design study for a low-frequency radio telescope on the Moon [*Bély et al.*, 1997], (see also *Kuiper and Jones*, in this volume), and several consortia have proposed viable space missions aimed at putting radio telescopes in free-flying constellations (e.g., *Weiler et al.*, 1988; *Basart et al.*, 1997; *Jones et al.*, in this volume). Although these instruments would largely escape the effects of the Earth's ionosphere, magnetosphere and terrestrial interference, there are still many challenges to be overcome before they can perform effectively as imaging interferometers. The electrically short dipole antennas used for very low frequency observations are each sensitive to radiation from most of the sky, giving an unusually large primary beam pattern and making source subtraction and Fourier imaging particularly problematic. If the antennas are flying freely, there is the additional complication of imaging with a three-dimensional dynamic constellation. Furthermore, the performance will be affected by strong solar bursts and variable density structures within the solar wind.

There are clearly many factors that influence the design of a space or Moon-based long wavelength radio telescope. Here we review in detail several of the most important and show how they affect both the instrument and its location.

2. SCATTERING

As already highlighted, interplanetary and interstellar scattering greatly limit long wavelength astronomy from space. It is therefore helpful to consider the basic processes that underpin scattering in these media.

2.1. Angular Broadening

At frequencies well above the plasma frequency of the medium, a region of space in which the electron density deviates by Δn_e from the mean deviates in refractive index by

$$\Delta \eta \simeq -\frac{r_e}{2\pi} \lambda^2 \Delta n_e, \quad (1)$$

where r_e is the classical radius of an electron (2.82×10^{-15} m) and λ the wavelength of the radiation.

It is often instructive to model the density fluctuations from an extended region of space as being compressed to a thin screen of thickness L at some distance z along the line of sight to the radio source under observation. We may further simplify the model by assigning a scale-size of a for the clouds of excess density within the screen. This is the so-called 'Gaussian thin-screen model' and, although conceptually simple, demonstrates most of the major scattering properties of real turbulent astrophysical plasmas.

There is an excess change of phase, $\Delta\phi$, of $2\pi\Delta\eta a/\lambda = r_e \lambda a \Delta n_e$ through each cloud relative to the mean path, and each part of the wavefront incident on the thin screen encounters about L/a clouds. If the clouds are arranged randomly this introduces phase fluctuations of

$$\Delta\phi_{\rm rms} = r_e \lambda a \, \Delta n_e \left(\frac{L}{a}\right)^{1/2}$$
$$= r_e \lambda \, \Delta n_e (La)^{1/2} \quad (2)$$

over the screen by the time the wavefront has emerged. At frequencies below 10 MHz we expect the magnitude of these phase fluctuations to be large ($\Delta\phi \gg 1$) in both the ISM and the IPM so the effect is one of a random assembly of small prisms, each deflecting the radiation over a characteristic angle

$$\theta_s = \frac{\lambda \Delta\phi}{2\pi a}. \quad (3)$$

Strong scattering therefore spreads the radiation from a point source over a solid angle of characteristic width

$$\theta_{\rm s} = \frac{r_{\rm e}}{2\pi}\lambda^2 \Delta n_{\rm e}\left(\frac{L}{a}\right)^{1/2}, \quad (4)$$

[*Uscinski*, 1968; *Thompson et al.*, 1986; *Lyne & Graham-Smith*, 1990]. An alternative approach is to use a Kolmogorov power law [$P(k) \propto k^{-11/3}$, where k is the spatial wave number] for the spatial spectrum of the density fluctuations rather than a fixed scale-size, a [*Rickett*, 1977]. This gives a similar relation but with $\theta_{\rm s} \propto \lambda^{2.2}$ rather than λ^2. The result can be usefully expressed in terms of the integral of the 'turbulence strength' of the medium, C_N^2, along the line-of-sight to the source, resulting in a scattering angle of

$$\theta_{\rm s} \simeq 4.1 \times 10^{-13} \left[\int_0^L C_N^2(x)(x/L)^{5/3}\,{\rm d}x\right]^{3/5} \lambda^{11/5} \quad (5)$$

arcsec, for a point source a distance L away at $x=0$ [*Cordes et al.*, 1985; *Thompson et al.*, 1986]. However, there is good evidence from scintillation studies for a true inner scale size to interplanetary turbulence of about 200 km [*Cohen et al.*, 1967; *Readhead et al.* 1978; *Harmon*, 1991] making a Gaussian spectrum more appropriate for the IPM.

Both the IPM and ISM have been observed extensively at decimetre wavelengths [*Cordes et al.*, 1986; *Erickson*, 1964; *Rickett*, 1977], allowing us to calibrate the above expressions for $\theta_{\rm s}$. Under the strong scattering conditions that we expect below 30 MHz we have:

$$\text{IPM}: \quad \theta_{\rm s} \sim 100/(P\nu_{\rm MHz})^2 \quad (6)$$
$$\text{ISM}: \quad \theta_{\rm s} \sim 22/\nu_{\rm MHz}^2 \quad (7)$$

arcmin [*Erickson*, 1964; *Duffett-Smith and Readhead*, 1976], where P is the 'impact parameter' of the line of sight to the source relative to the Sun in AU, and the ISM value is for high galactic latitudes. Here we use the Gaussian model for both the IPM and ISM, but the two models (Gaussian and Kolmogorov) give roughly similar results over the frequency range considered. Indeed, $\theta_{\rm s}$ is proportional to λ^2 in both models if the angular size is determined by a single baseline interferometric measurement and fitted to a Gaussian sky brightness distribution [*Rickett*, 1977]. We see from this that interplanetary scattering exceeds interstellar scattering by a factor of about four. Although interplanetary scattering changes with the solar elongation (see *Spangler & Armstrong* [1990] for a careful analysis), most of the material that contributes to the scattering process is within about 1 AU of the Earth and can therefore be thought of as 'local'. The result is that interplanetary scattering does not vary greatly for elongations of more than about 90°.

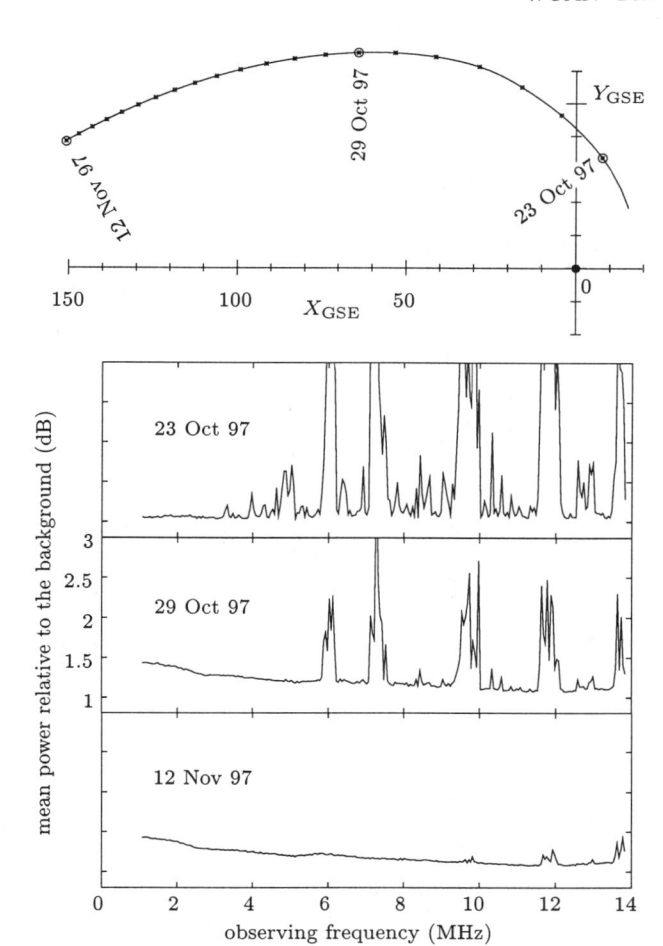

Figure 2. Typical man-made interference received by the WAVES instrument on Wind. The orbital dimensions (top panel) are in Earth radii ($R_{\rm E}$), and the bottom panels show the signal strengths at 40, 93 and 157$R_{\rm E}$, averaged over 24 hours. Note the prominent short-wave broadcast bands.

2.2. Temporal Broadening

The scattered rays that broaden the angular size of a point source also introduce temporal broadening. This arises from the range of propagation paths, and hence times, available between the source and the receiver, and will smear transient signals such as pulses from pulsars [*Cronyn*, 1970; *Cordes*, 1990].

Taking small-angle approximations, the range in differential propagation times from a source at infinity, broadened to an angular extent of $\theta_{\rm s}$ radians is approximately

$$\Delta\tau_{\rm b} = \frac{z\theta_{\rm s}^2}{2c}, \quad (8)$$

where z is the distance between the screen and the observer and c is the speed of light. The analysis for a thick screen has also been carried out [*Williamson*, 1972; *Williamson*, 1973; *Williamson*, 1974; *Cordes et al.*, 1985], showing that, for a suitable choice of z, the broadening times are similar.

Again, this expression can be given an approximate calibration using observations of pulsars and angular broadening at frequencies of a few tens of megahertz [*Cordes*, 1990] to give

$$\text{IPM}: \quad \Delta \tau_b \sim 0.1 \, \nu_{\text{MHz}}^{-4} \quad (9)$$
$$\text{ISM}: \quad \Delta \tau_b \sim 2 \times 10^8 \, \nu_{\text{MHz}}^{-4} \quad (10)$$

seconds. We see immediately that the comparatively vast distances involved in interstellar propagation make the effects of interstellar temporal broadening severe. At 1 MHz the temporal broadening is approximately 5 years. Clearly little, if any, temporal work can be done with extra-solar system objects at frequencies below 10 MHz.

Interplanetary temporal broadening is less severe, amounting to about 0.1 sec at 1 MHz and 10 sec at 300 kHz, although it must be stressed again that the effects of the IPM can be highly variable. At still lower frequencies we cannot justify the small angle approximation involved in deriving Equation (8): at 100 kHz the true temporal broadening is about a factor of three less than that predicted by the equation. The limiting situation arises when the source is so severely scattered that all directionality in the signal is lost. However, signals scattered with delays much greater than the light travel time to the source will not contribute significantly to the overall power due to inverse-square losses. This limits the expected temporal broadening of a signal from Jupiter received near Earth to less than about 40 min [*Barrow et al.*, 1999].

3. POLARISATION

A magnetic field component, $B_{\|}$, parallel to the direction of propagation, will cause the Faraday rotation of a linearly polarised signal by an angle

$$\Phi = \frac{2.36 \times 10^4}{\nu^2} \int n_e B_{\|} \, dl \quad \text{radians}, \quad (11)$$

where ν is in Hz, n_e in m^{-3}, $B_{\|}$ in T and l in m. Within the IPM both n_e and $B_{\|}$ decrease as $\sim 1/r^2$, where r is the radial distance from the Sun, so that the Faraday rotation is always dominated by the near-Earth environment, giving

$$\Phi \simeq \frac{46}{\nu_{\text{MHz}}^2} \quad \text{radians}, \quad (12)$$

in the antisolar direction, assuming $B_{\|} \simeq 6 \times 10^{-9}$ T and $n_e \simeq 6.5 \times 10^6$ m^{-3} at 1 AU. However, both n_e and $B_{\|}$ are highly variable quantities within the solar system, each changing by factors of up to 10 on timescales of minutes to hours. As a result, polarisation angle is poorly defined at these frequencies, although dual-frequency measurements could conceivably go some way to 'unwind' the rotation introduced at higher frequency ranges.

In contrast, circular polarisation remains well defined, the only effect of the intervening plasma being the introduction of a phase-lag between the two polarisation components. Unfortunately, the intrinsic low luminosity of cyclotron processes in comparison to synchrotron processes means that strong circular polarisation is rarely seen outside the solar system. Within the solar system circularly polarised cyclotron emission from the Sun and planets could be detected in Earth orbit, notably from Jupiter and Saturn. Indeed, there is some hope that characteristic decametric planetary emission may be useful in identifying planets around other stars [*Zarka et al.*, 1997]. Furthermore, under certain circumstances there is the possibility of observing an appreciable circular component to synchrotron radiation at low frequencies, as the relevant electrons will be less strongly relativistic than those predominant at higher frequencies [*Sciama & Rees*, 1967]. The additional requirement is for a highly anisotropic electron velocity distribution [*Ginzburg*, 1979]. Those pulsars with emission that is known to contain a uniformly handed circular component would be of particular interest.

The radiation from a source can travel to the observer over a multiplicity of paths, each with its own, possibly different, Faraday rotation. This gives rise to (linear) Faraday depolarisation and represents a fundamental limit to linear polarisation measurements as, unlike simple rotation, it cannot be compensated for by making dual-frequency observations.

The depolarisation within the solar system can be estimated using the Gaussian thin screen-model. Equation (11) shows that each cloud (scale-size a) in the screen (thickness L) will introduce a differential rotation of

$$\Delta \psi = \alpha \Delta n_e \lambda^2 a B_{\|}, \quad (13)$$

where $\alpha = 2.6 \times 10^{-13}$ (assuming SI units). The analysis proceeds in a manner similar to that for angular broadening. The wavefront encounters about L/a clouds during its passage through the screen, giving the emerging wavefront a random polarisation structure with an rms polarisation angle of

$$\Delta\psi_{\rm rms} = \alpha\Delta n_{\rm e}\lambda^2 aB_{\parallel}\left(\frac{L}{a}\right)^{1/2}. \qquad (14)$$

This can be combined with Equation (4) to give an expression in terms of scattering angle:

$$\Delta\psi_{\rm rms} = \frac{2\pi\alpha}{r_{\rm e}}aB_{\parallel}\theta_{\rm s}. \qquad (15)$$

Taking a to be about 200 km for the IPM gives

$$\Delta\psi_{\rm rms} \simeq \frac{10^{-2}}{\nu_{\rm MHz}^2}. \qquad (16)$$

Although not significant at frequencies greater than about 1 MHz, this could be a limiting factor for linear polarisation measurements at hectometric wavelengths. Again, only the *phase* of circular polarisation is affected by this process.

On the larger scale, we do not expect to see any linear polarisation from galactic or extragalactic signals. Both the amount of rotation and the degree of depolarisation scale as λ^2 at low frequencies [*Tribble*, 1991]. If we take the galactic electron number density to be about 0.3×10^5 m^{-3} and the magnetic field about 5×10^{-10} T, a rotation of 1 radian would be seen over a plasma cloud of size $\sim 10^{-3}$ pc at 1 MHz. As we can expect many such clouds within the telescope beam for all but the closest sources, the radiation can be expected to be linearly unpolarised. A fuller analysis of this can be found in *Linfield* [1996]. Furthermore, the total Faraday rotation over interstellar distances is large, amounting to about 10^4 radians pc^{-1} at 1 MHz and scaling as λ^2. The *differential* rotation over the observing band is therefore correspondingly great. At 3 MHz, the differential rotation amounts to 1 radian over a 1 kHz bandwidth after only 1.4 pc.

Any circularly polarised galactic signals would preserve their intrinsic degree of polarisation, but, as has been mentioned, such sources are rare and dim. Unpolarised galactic background emission would add in to reduce the overall degree of polarisation and make detection difficult.

4. ABSORPTION EFFECTS

We expect free-free (thermal bremsstrahlung) absorption to render the ionised interstellar medium optically thick (optical depth, τ, of 1) at some turnover fre-

Table 1. Expected antenna temperatures for a low-frequency radio telescope in space. At frequencies above about 5 MHz the temperature is dominated by galactic synchrotron emission, with a turnover at lower frequencies due to free-free absorption in the interstellar medium (adapted from *Novaco & Brown*, 1978.)

$T_{\rm A}$ (K)	ν (MHz)
3.3×10^5	10
2.6×10^6	5
2.0×10^7	1
2.6×10^7	0.5
5.2×10^6	0.25

quency $\nu_{\rm T}$. This frequency depends on both the emission measure of the medium and its electron temperature. Specifically,

$$\nu_{\rm T,GHz} = 5.21 \times 10^{-7}\, T_{\rm e}^{-0.64}\left(\int n_{\rm e}^2\, {\rm d}x\right)^{0.48}, \qquad (17)$$

[*Gordon*, 1988] where x is in parsecs along the line-of-sight. Note that this expression assumes the temperature to be constant in the region of interest. The diffuse, warm, interstellar medium has $T_{\rm e} \simeq 10^4$ K and $n_{\rm e} \simeq 3 \times 10^4$ m^{-3} [*Weisberg et al.*, 1980], giving an approximate measure of the distance (in parsecs) one can see through the medium before $\tau = 1$ of

$$l_{\rm pc} \simeq (34\,\nu_{\rm MHz})^{2.1}. \qquad (18)$$

This result is of particular importance in both low-frequency extra-solar system planet searches and extragalactic observations. Because the galactic disc has a thickness of ~ 1 kpc, the sky appears uniformly 'foggy' at frequencies below 1–2 MHz [*Reynolds*, 1990]. At higher frequencies however, we can see out of the plane, and the plane itself will appear mottled due to the presence of discrete regions of particularly high electron density (H II regions) absorbing the background synchrotron emission.

At very low frequencies the free-free process absorbs much of the bright background synchrotron emission from the galaxy before it reaches us, causing the telescope system temperature to *decrease*. Observations by the RAE2 spacecraft [*Novaco & Brown*, 1978] predict the antenna temperatures, $T_{\rm A}$, as a function of observing frequency, ν, shown in Table 1.

Of course, any absorption of this sort will limit the usefulness of an instrument for extra-solar system observations, both because the sky fogs over at some distance, and because the extinction will vary across the sky. This variation will give rise to uncertainties in intrinsic flux density similar to interstellar reddening seen at optical wavelengths. However, it does allow us to study in some detail the distribution of both ionised material and synchrotron emission throughout the galaxy, which is an important subject in itself.

5. INVESTIGATING THE INTERSTELLAR MEDIUM

A good understanding of the interstellar medium is central to the planning and interpretation of radio observations below about 10 MHz. The effects of interstellar scattering have already been considered, but the ISM is itself one of the major science targets for very low-frequency astronomy.

The ionised hydrogen seen in the galaxy at low frequencies can be divided into H II regions (often seen as discrete photo-ionised clouds, mostly in the galactic plane) and warm, diffuse gas permeating the plane and extending into the halo. The denser H II regions become optically thick at about 30 MHz, whereas the diffuse emission becomes thick at 1–5 MHz. The background synchrotron emission, and the radiation from discrete background sources, is therefore seen against a complex absorption pattern generated by the ISM. This pattern is overlaid with further foreground synchrotron emission to build up an even more complex picture of absorption and emission over the sky.

To disentangle the various components contributing to the observed temperature distribution we must exploit both their frequency dependence and additional ground-based observations. The procedure we be rather involved, but it is instructive to consider the major contributing effects, and determine how they might be identified and measured.

5.1. Galactic Synchrotron Emission

As the dominant radiation mechanism in the galaxy at low frequencies, most of the signal a radio telescope will detect will be galactic synchrotron emission modulated by interstellar absorption. At about 50 MHz the sky brightness of this emission scales as $\nu^{-2.4}$, slightly flattened from the generic index of -2.8 found at frequencies greater than about 400 MHz [Webster, 1974]. This intrinsic spectral index shows some further flattening at lower frequencies before free-free absorption turns the spectrum over at about 4 MHz. The variation of the synchrotron intensity and spectral index over the sky is an important driver for low-frequency space astronomy.

5.2. Extragalactic Radio Sources

Free-free emission modulates our measurements of intrinsic sky brightness and the flux densities of discrete extragalactic radio sources. The optical depth [cf. Equation (17)] is

$$\tau \simeq 6.5 \times 10^5 \, \nu^{-2.1} \int T_e^{-1.35} n_e^2 \, dx, \qquad (19)$$

using SI units throughout. The apparent spectral index of a source will therefore depend on its intrinsic index and the emission measure and the electron temperature along the line-of-sight. Each source will also show a spectral turnover at a frequency $\nu_T \propto S_T^{2/5} B^{1/5}/\theta^{4/5}$ due to synchrotron self-absorption [Slish, 1963] and the Razin-Tsytovič effect ($\nu_T \propto n_e/B$), where B and n_e are the magnetic field and electron density local to the source, S_T its flux density at turnover and θ its effective angular size. There are clearly many contributions to the apparent flux density of any single source, and much of the interpretation from discrete sources will therefore be statistical in nature.

5.3. Pulsars

We know that temporal broadening prevents us from measuring pulsar dispersion at very low frequencies, but ground-based observations at frequencies > 20 MHz are routine [Phillips & Wolszczan, 1989]. The dispersion,

$$\frac{\partial \tau}{\partial \nu} \propto \nu^{-3} \int n_e \, dx, \qquad (20)$$

determines the dispersion measure (the integral in the above equation) to the pulsar. This can be combined with the emission measure determined from free-free absorption observations to give some indication of the nature of the spatial distribution of n_e along the line-of-sight.

5.4. Optical Line Emission

Balmer recombination (Hα) emission can help to disentangle the various contributing factors to sky brightness. In regions of the sky that are optically thin to Hα (galactic latitudes $\gtrsim 10°$), the emission intensity is proportional to $\int T_e^{-0.92} n_e^2 \, dx$ [Martin, 1988]. Hα emission therefore also depends on emission measure but with a slightly different weighting from the electron temperature as compared to free-free absorption. The two can therefore be used for mutual verification in region of the sky optically thin to Hα.

5.5. Angular Broadening

There is some benefit from measuring the degree of angular broadening of sources over the sky. *Hajivassiliou* [1992] showed that if a sufficient number of background sources exist in a region of sky, the degree of broadening for each can be used to map the local ISM density. This is similar to the approach used successfully to map density structures in the heliosphere via interplanetary scintillation [*Woan*, 1995]. In principle, such observations are a sensitive way of measuring weak density structures, as the angular broadening for a given turbulence strength scales as λ^2. In practice, however, it may prove difficult to detect enough background sources in the field of interest and the small variations in angle of scatter may be swamped by the more powerful general structure.

6. TELESCOPE LOCATION

The two most popular locations proposed for a space-based low-frequency radio interferometer are the surface of the Moon and in distant Earth orbit. A lunar telescope would be shielded from the Sun for some of the time and, if on the far-side, from the Earth. On the other hand a free-flying interferometer would be considerably cheaper and could have much greater baseline flexibility. Some of the points for and against various locations are summarised in Figure 3. The table should be regarded as a guide only, and this is reflected in its suitably vague notation. Also the list is not exhaustive, for example a telescope in lunar orbit also has some attractions. Here we will consider the relative merits of free-flying and Moon-based instruments.

6.1. Free-flying Instruments

As has already been pointed out, an effective method of reducing terrestrial interference is to place the instrument at sufficient distance from the Earth that inverse-square losses attenuate the signals to an acceptable level. Figure 2 shows that the terrestrial interference drops to the background noise level at a distance of about 150 Earth radii ($\sim 10^6$ km). Despite this however, the signals will still show strongly in the correlated output after very little integration, as they have a high degree of temporal and spatial coherence. The instrument would therefore need high spectral resolution, allowing these narrow-band signals to be removed, and a high dynamic range so that the front-end remains linear throughout periods of interference. This is the strategy employed for the ALFA mission (*Jones et al.*, in this volume). One must also keep in mind that some forms of 'interference', such as the Earth's auroral kilometric radiation, are subjects of study in themselves.

A free-flying interferometer array of short dipole antennas has a nearly-isotropic primary beam. This imposes serious demands on the mapping procedure and, for uniqueness, requires that the array elements be distributed in three dimensions rather than the usual two. Nearly any practical design of low frequency interferometer will suffer from similar problems, but for a free-flyer, they are particularly acute. In particular the Sun and Jupiter, both strong variable sources, will be present in the beam continuously.

There are, however, very definite advantages to a free-flying geometry. The surface brightness sensitivity of an interferometer depends on its filling factor – the fraction of the synthesised aperture filled by real collecting area – and for a fixed geometry this is a strong function of observing frequency. In contrast, the relative locations of free-flying elements can be freely adjusted to optimise both angular resolution and filling factor for a particular observation. In addition, there is a greater degree of built-in redundancy in a free-flyer, as the individual elements are identical in all respects. The ALFA design exploits this further by giving each element an independent telemetry link to Earth, so that the loss of a small number of elements would not seriously affect overall performance. Finally, a free-flying mission would be considerably cheaper and easier to implement than one based on the lunar surface.

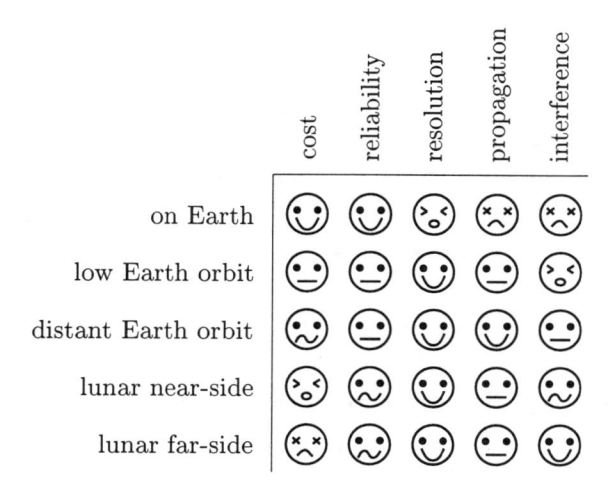

Figure 3. Performance guide for popular telescope locations. The cost column progresses smoothly from good/low on Earth to bad/high on the Moon.

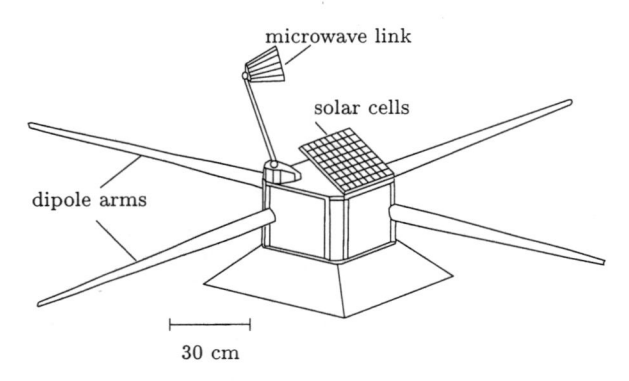

Figure 4. Proposed lunar element for the ESA Very Low Frequency Array [*Bély et al.*, 1997]. 300 such elements would be deployed on the far-side of the Moon.

6.2. Moon-based Instruments

There are good reasons for believing that the ultimate location for a low frequency instrument is the far-side of the Moon. Despite the formidable engineering challenges that it presents, the lunar far-side is the only location in the solar system permanently facing away from the Earth and therefore shielded from terrestrial interference. Furthermore, the Sun is below the horizon for about half the time, giving additional shielding from the strong solar transients that could limit the performance of a free-flyer. *Bély et al.*, [1997] carried out a detailed investigation into a practical lunar instrument and showed that, as part of a larger Moon programme, a 300-element instrument could be deployed from a single Ariane 5 launch. Each element (Figure 4) is solar powered and communicates with a central station over a microwave link. The central station is linked to Earth via a relay satellite in a halo orbit at the Earth-Moon L2 position. The elements must have a line-of-sight to the central station, and a suitable location for this appears to be the Tsiolkovsky crater (129°E, 21°S) which has a large, unusually flat, floor. The elements are arranged with a power-law spacing along three spiral arms (Figure 5) to maintain a reasonable filling factor at a fixed resolution of 0.5° over the frequency range.

Although very tenuous, the lunar atmosphere could prove to be an important factor for any lunar instrument. Dual-frequency measurements made in the 1970s by the Luna 19 and Luna 22 spacecraft gave good evidence that an ionised layer builds up on the side of the Moon facing the Sun [*Vyshlov*, 1976]. The inferred variation of electron density with height on different days is shown in Figure 6. The maximum lunar plasma frequency consistent with these data is about 0.46 MHz (corresponding to number densities of 2 100 cm^{-3}) and could therefore severely affect performance at the lower reaches of the observing band. However, measurements taken above the dark side of the Moon showed no detectable electron concentration. This, together with need to shield the telescope from solar radio bursts, ensures that observations must be carried out during the night. The apparent drop in electron content close to the lunar surface is something of a puzzle. *Bauer* [1996] has pointed out that this is unexpected, as the lunar exosphere is optically thin to ionising radiation, however the magnitude of the drop is just within the quoted errors.

The issue of radio propagation close to the lunar surface is an important one. The regolith has a very low electrical conductivity so we can place the antennas horizontally, and receive the component of the electric field parallel to its surface. Because the loss tangent of the regolith is quite low (0.001 to 0.1) we can expect waves to penetrate to between 1 and 100 λ, opening up the possibility of sub-surface reflections. Clearly, a careful assessment of factors such as this is vital before any lunar mission in undertaken.

7. SUMMARY

One of the driving forces behind space-based radio telescopes is the need to get above the Earth's own ionosphere, which is largely opaque to radio waves at frequencies below 10 MHz. What we are left to contend with is the interplanetary medium and the interstellar medium. Their main effect is to broaden the apparent angular sizes of sources to discs of around 1° at 1 MHz, so limiting the effective angular resolution of the

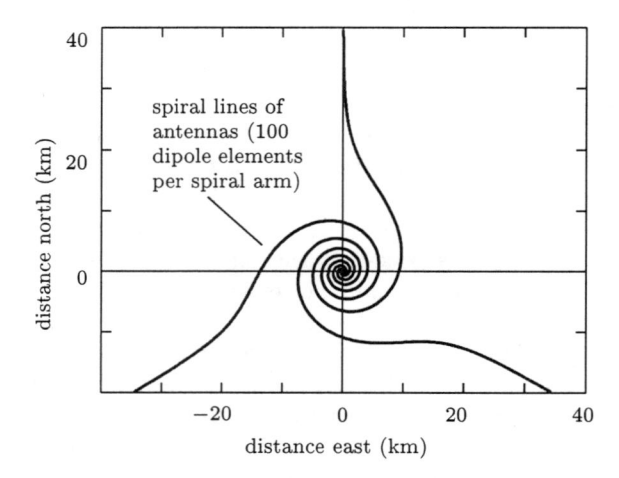

Figure 5. A possible configuration for a lunar far-side array [*Bély et al.*, 1997]. The spiral design maintains a good filling factor over a broad range of frequencies.

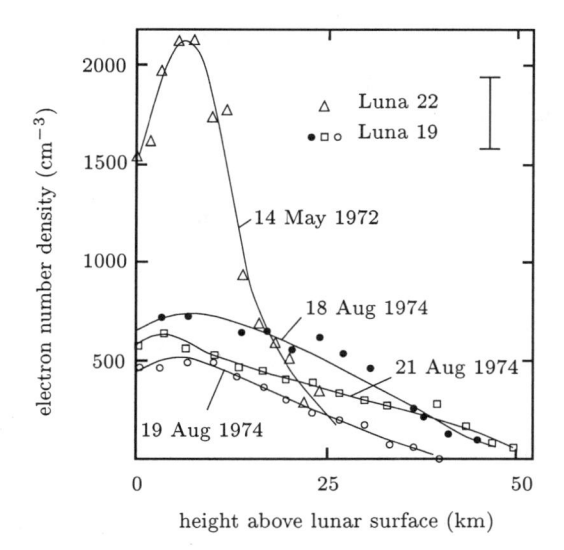

Figure 6. Day-side lunar ionosphere, as inferred from Luna 19 and 22 spacecraft measurements [*Vyshlov*, 1976]. The error bar is ± 200 cm^{-3}.

instrument. The scattering would also limit the temporal resolution of any instrument. This is strong enough for it to be impossible to detect pulsed emission from nearly all known pulsars and to reduce the instruments sensitivity to transient signals ($\lesssim 1$ s) in the solar system at frequencies below about 600 kHz.

Although the severe effects of Faraday rotation and depolarisation preclude the measurement of the linear polarisation state of signals from extra-solar system (and probably solar system), sources below about 5 MHz, circular polarisation may still be readily detected. At frequencies just above about 4 MHz, when the ISM becomes optically thick, we may expect to see an increased amount of circular polarisation from low energy electrons involved in synchrotron emission.

One of the clearest signatures we can expect to see is that of free-free absorption in the ISM. This has the effect of dimming both the synchrotron background radiation and signals from discrete sources. The amount of extinction will show strong point-to-point variations as well as a strong dependence on galactic latitude. At frequencies below 1–2 MHz we can expect the sky to appear almost uniformly foggy. At higher frequencies the absorption will give the background emission a more 'blotchy' appearance. The study of these effects gives important information on both the ISM and the Galaxy's synchrotron background radiation and is a primary goal of any mission.

Both distant Earth orbit and the Moon are attractive locations for a telescope. Both have some disadvantages, but at present the free-flyer option is more attractive on grounds of cost, flexibility and deployment. The ultimate site – the Mauna Kea of VLF radio astronomy – is probably the Tsiolkovsky crater on the lunar far-side. This is a flat, low-latitude site offering good isolation from both solar and terrestrial interference. This site would become accessible following the development of a broader lunar programme.

REFERENCES

Barrow, C.H., G. Woan, and R.J. MacDowall, Interplanetary scattering effects in the jovian bKOM radio emission observed by Ulysses, *Astron. Astrophys.*, *344*, 1001, 1999.

Basart, J.P., J.O. Burns, B.K. Dennison, K.W. Weiler, N.E. Kassim, S.P. Castillo, and B.M. McCune, Directions for space-based low frequency radio astronomy. 1. System considerations, *Radio Sci.*, *32*, 251, 1997.

Bauer, S.J., Limits to a Lunar Ionosphere, *Anzeiger Abt II*, *133*, 17, 1996.

Bély P.Y., R.J. Laurance, S. Volonte, R.R. Ambrosini, A. Ardenne, C.H. Barrow, J-.L. Bougeret, J-.M. Marcaide and G. Woan, *Very low frequency array on the lunar far side*, ESA report SCI(97)2, European Space Agency, 1997.

Cane, H.V., and P.S. Whitham, Observations of the southern sky at five frequencies in the range 2–20 MHz, *Mon. Not. R. Ast. Soc.*, *17*, 21, 1977.

Cohen M.H., E.J. Gundermann, H.E. Hardebeck, and L.E. Sharp, Interplanetary Scintillations II. Observations, *Ap. J.*, *147*, 449, 1967.

Cordes, J.M., J.M. Weisberg, and V. Boriakoff, Small-scale electron density turbulence in the interstellar medium, *Ap. J.*, *288*, 221, 1985.

Cordes, J.M., A. Pidwerbetsky, and R.V. Lovelace, Refractive and diffractive scattering in the interstellar medium, *Ap. J.*, *310*, 737, 1986.

Cordes, J.M., Low frequency interstellar scattering and pulsar observations, in *Low Frequency Astrophysics from Space*, edited by N.E. Kassim, and K.W. Weiler, Springer-Verlag Lecture Notes in Physics 362, pp. 165, 1990.

Cronyn, W.M., Interstellar Scattering of Pulsar Radiation and Its Effect on the Sprctrum of NP0532, *Science*, *168*, 1453, 1970.

Duffett-Smith, P.J., and A.C.S. Readhead, The angular broadening of radio sources by scattering in the interstellar medium, *Mon. Not. R. Ast. Soc.*, *174*, 7, 1976.

Erickson, W.C.,The Radio-Wave Scattering Properties of the Solar Corona, *Ap. J.*, *139*, 1290, 1964.

Ginzburg, V.L., *Theoretical Physics and Astrophysics*, Pergamon Press, 1979.

Gordon, M.A., in *Galactic and Extragalactic Radio Astronomy* (Second Edition), edited by G.L. Verschuur, and K.I. Kellerman, Springer-Verlag, 1988.

Hajivassiliou, C.A., Distribution of plasma turbulence in our Galaxy derived from radio scintillation maps, *Nature*, *355*, 232, 1992.

Harmon J.K., and W.A. Coles, The density fluctuation power spectrum of the inner solar wind, in *Solar Wind Seven*, Proceedings of the 3rd COSPAR Colloquium, edited by E. Marsch, and R. Schwenn, Goslar, Germany, pp. 461, 1991.

Linfield, R., IPM and ISM Coherence and Polarization Effects on Observations With Low-Frequency Space Arrays, *Astron. J.*, *111*, 2465, 1996.

Lyne, A.G., and F. Graham-Smith, *Pulsar astronomy*, Cambridge University Press, Cambridge, 1990.

Martin, P.G., Hydrogenic radiative recombination at low temperature and density, *Ap. J. Supp.*, *66*, 125, 1988.

Novaco, J.C., and L.W. Brown, Nonthermal galactic emission below 10 MHz, *Ap. J.*, *221*, 114, 1978.

Phillips, J.A., and A. Wolszczan, Interpulse emission from pulsars at 25 MHz, *Ap. J. lett.*, *344*, L69, 1989.

Readhead A.C.S., M.C. Kemp, and A. Hewish, The spectrum of small-scale density fluctuations in the solar wind, *Mon. Not. R. Ast. Soc.*, *185*, 207, 1978.

Reynolds, R.J., The low density ionised component of the interstellar medium and free-free absorption at high galactic latitudes, in *Low Frequency Astrophysics from Space*, edited by N.E. Kassim, and K.W. Weiler, Springer-Verlag Lecture Notes in Physics 362, pp. 121, 1990.

Rickett, B.J., Interstellar scattering and scintillation of radio waves, *Ann. Rev. Astr. Astrophys.*, *15*, 147, 1977.

Sciama, D.W., and M.J. Rees, Possible Circular Polarization of Compact Quasars, *Nature*, *216*, 147, 1967.

Slish, V.I., Angular Size of Radio Stars, *Nature*, *199*, 682, 1963.

Spangler, S.R., and J.W. Armstrong, Low frequency angular broadening and diffuse interstellar plasma turbulence, in *Low Frequency Astrophysics from Space*, edited by N.E. Kassim, and K.W. Weiler, Springer-Verlag Lecture Notes in Physics 362, pp. 155, 1990.

Thompson, A.R., J.M. Moran, and G.W. Swenson, *Interferometry and Synthesis in Radio Astronomy*, Wiley-Interscience, 1986.

Tribble, P.C., Depolarization of extended radio sources by a foreground Faraday screen, *Mon. Not. R. Ast. Soc.*, *250*, 726, 1991.

Uscinski, B.J., The multiple scattering of waves in irregular media, *Phil. Trans. R. Soc. A*, *262*, 609, 1968.

Vyshlov, A.S., Preliminary results of circumlunar plasma research by the Luna 22 spacecraft, *Space Res. XVI*, Proc. of Open Meetings of Workshop Groups on Physical Sciences, pp. 945, Akademie-Verlag, 1974.

Webster, A.S., The spectrum of the galactic non-thermal background radiational Observations at 408, 610 and 1407 MHz, *Mon. Not. R. Ast. Soc.*, *166*, 355, 1974.

Weisberg, J.M., J. Rankin, and V. Boriakoff, H I absorption measurements of seven low-latitude pulsars, *Astron. Astrophys.*, *88*, 84, 1980.

Weiler, K.W., K.J. Johnston, R.S. Simon, B.K. Dennison, W.C. Erickson, M.L. Kaiser, H.V. Cane, M.D. Desch, L.M. Hammarstrom, A low frequency radio array for space, *Astron. Astrophys.*, *195*, 372, 1988.

Williamson, I.P., Pulse broadening due to multiple scattering in the interstellar medium, *Mon. Not. R. Ast. Soc.*, *157*, 55, 1972.

Williamson, I.P., Pulse broadening due to multiple scattering in the interstellar medium-II, *Mon. Not. R. Ast. Soc.*, *163*, 345, 1973.

Williamson, I.P., Pulse broadening due to multiple scattering in the interstellar medium-III, *Mon. Not. R. Ast. Soc.*, *166*, 499, 1974.

Woan, G.,Observations of long-lived solar wind streams during 1990-1993, *Annales Geophysicae*, *13*, 227, 1995.

Zarka, P., J. Queinnec, B.P. Ryabov, V.B. Ryabov, V.A. Shevchenko, A.V. Arkhipov, H.O. Rucker, L. Denis, A. Gerbault, P. Dierich, and C. Rosolen, Ground-based high sensitivity radio astronomy at decameter wavelengths, in *Planetary Radio Emissions IV*, editors H.O. Rucker et al., Austrian Acad. Sci. Press, Vienna, 1997.

G. Woan, Department of Physics and Astronomy, University of Glasgow, Glasgow, G12 8QQ, United Kingdom (email: graham@astro.gla.ac.uk)

Low-Frequency Radio Astronomy and the Origin of Cosmic Rays

Nebojsa Duric,[1]

Department of Physics and Astronomy, University of New Mexico

One of the longest standing mysteries in physics and astronomy is the origin of cosmic rays. Space-based low-frequency radio astronomy holds the promise of providing answers to key pieces of this important puzzle. There are a number of important reasons why the study of Galactic low frequency emission will offer new insights into this 87 year-old problem. (i) The comparison of gamma-ray and radio data may lead to a better understanding of how low energy "seed" electrons are distributed in the galaxy. (ii) It may be possible to build a 3-dimensional model of the cosmic ray electron distribution in the Galaxy. Such a model would offer new insights into the geometry of the Galactic cosmic ray disk and/or possible halo. (iii) A combination of gamma-ray and low-frequency radio observations also affords the possibility of indirectly measuring the ratio of nuclei to electrons as a function of energy and position in the Galaxy, a critical parameter in understanding cosmic ray acceleration and particle energetics. (iv) Observations of point sources "blurred" by interstellar scintillation offer the unique possibility of making maps of plasma turbulence in the vicinity of SNR shocks. Characterization of shock turbulence is of profound importance in constraining theoretical models of particle acceleration.

1. OVERVIEW

Some 87 years after their discovery (Hess, 1912), cosmic rays continue to confound physicists and astronomers alike and their origin remains an unsolved mystery. In this paper, we make a case for using low frequency radio astronomy to address the issue of cosmic ray (CR) origin. Yet, it is interesting to note that the connection between cosmic rays and galactic radio emission was not accepted until some 50 years after Hess's discovery and some 20 years after the birth of radio astronomy. Viable cosmic ray origin theories did not appear for another 15 years after that. Since then, our understanding of cosmic ray properties has grown enormously. We now know that cosmic rays are mainly atomic nuclei and electrons that have been accelerated to energies greater than their rest mass energies. They follow a nonthermal power-law distribution with a differential energy index of ≈ 2.5. Their energy density in space is ≈ 1 eV cm^{-3} which translates into a number density of $\approx 10^{-9}$ cm^{-3}. Thus, the cosmic rays can be collectively characterized as a *collisionless, nonthermal gas*. It is instructive to compare the properties of this nonthermal gas with that of the various components of the interstellar medium (ISM). The comparison is summarized in Table 1.

Column 1 lists the various phases of the ISM. Approximate filling factors are shown in column 2. The number density is shown in column 3. The kinetic en-

Radio Astronomy at Long Wavelengths
Geophysical Monograph 119
Copyright 2000 by the American Geophysical Union

Table 1: Components of the ISM

Phase	f	n(cm^{-3})	kT(eV)	nkT(ev cm^{-3})
Molecular Clouds	10^{-3}	> 100	< 10^{-2}	-
Cold Neutral Medium	0.025	40	$\approx 10^{-2}$	≈ 0.4
Warm Neutral Medium	≈ 0.5	≈ 0.5	≈ 1	≈ 0.5
Warm Ionized Medium	≈ 0.25	≈ 0.2	≈ 1	≈ 0.2
Hot Ionized Medium	≈ 0.2	$\approx 3 \times 10^{-3}$	$\approx 10^2$	≈ 0.3
Cosmic Rays	≈ 1	$\approx 10^{-9}$	$\approx 10^9$	≈ 1

ergy per particle from each phase is shown in column 4 in electron volts. Column 5 lists the energy density (obtained by multiplying column 3 by column 4) of each phase.

Cosmic rays are energetically important, *containing at least as much energy as the other phases of the ISM*. They therefore play an important role in the evolution of the ISM and should be considered as a legitimate phase of the ISM worthy of detailed study.

2. MEASURING GALACTIC COSMIC RAYS

Being charged particles, cosmic rays are easily deflected by Galactic, interplanetary, and geophysical magnetic fields. It is therefore generally impossible to deduce the origin and complete spectrum of the cosmic ray particles from direct measurements. There is, however, a wide array of measurement tools available for studying Galactic cosmic rays indirectly. The most commonly used tracers of the global distribution of cosmic rays are summarized in Table 2.

Column 1 lists the cosmic ray component that is detectable. The mechanism by which the component is detected is shown in column 2. Column 3 shows the component of the ISM with which the mechanism is associated. The range of energies of the cosmic ray particles contributing to the observed emission is shown in column 4. The quantity actually represented by the observed emission is shown in the last column.

It is clear that multiple tools exist for tracing cosmic rays in the Galaxy. However, it is also obvious that all global tracers of cosmic rays are tied to other components of the ISM and that no global tracer measures the cosmic rays directly. Furthermore, examination of Table 2 shows that there is little overlap in electron energy coverage among the three electron tracers. In fact, there is *no* overlap between cm-wave radio emission and the two γ−ray tracers. This absence of overlap, as it applies to CR electrons, is illustrated in figure 1. (based on a similar figure in Longair (1990)).

However, at the lower radio frequencies the gap, shown in figure 1, is bridged for frequencies $\approx$10 MHz - 30 MHz, because relativistic electrons with E $\approx$ 200 - 300 MeV are traced under typical interstellar conditions (this is a region of the energy spectrum that is not possible to study through direct detections because of solar modulation). These same electrons also radiate relativistic bremsstrahlung at the lower γ−ray photon energies (Table 2). Consequently the equations in rows 1 and 2 of Table 2 can be combined (in conjunction with the well established data base of Galactic HI emission) to uniquely determine the distribution of cosmic ray electrons in the Galaxy, *separately from the magnetic field distribution*.

It is interesting to note that although low frequency radio astronomy paved the way for opening up the field of cosmic ray astrophysics it has had a relatively minor role to play since the pioneering efforts of Jansky (1935) and Reber (1940). The major reasons for the irrelevancy of low frequency radio astronomy to cos-

Table 2: Global Measurements of Cosmic Rays

Components	Tracer	Related ISM Component	CR Energy Range	Quantity Measured
Electrons	Radio Synchrotron	$\vec{B}$ Field	200 MeV - 10 GeV	$\int N_{CRe} B^{1.8} dr$
Electrons	Relativistic Bremsstrahlung	Thermal ISM	100 MeV - 300 MeV	$\int N_H N_{CRe} dr$
Electrons	Inverse Compton	Photons	< 100 MeV	$\int N_* N_{CRe} dr$
Protons	Neutral Pion Decay	Thermal ISM	300 MeV - 10 GeV	$\int N_H N_{CR} dr$

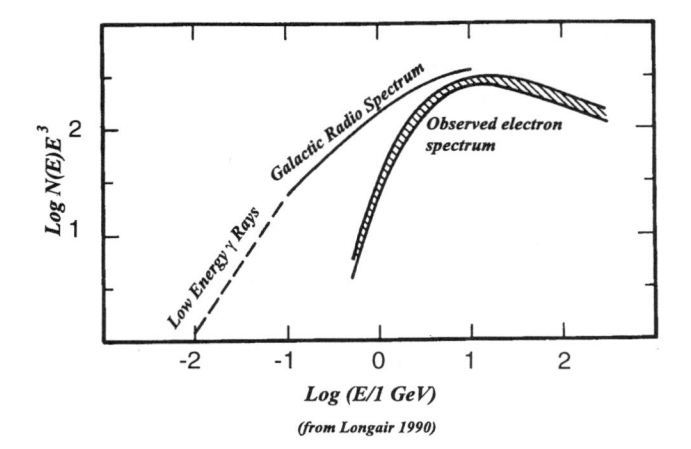

Figure 1. The spectrum of relativistic electrons in the local ISM (from [Longair 1990]). The directly observed electron energy spectrum is shown in the hatched area. Modulation by the solar wind prevents direct measurement of the intrinsic spectrum below roughly 10 GeV. The electron spectrum derived from the spectra of the Galactic low frequency radio emission (solid line) and low energy gamma ray emission (dashed line) are also shown. Note that absence of overlap between the radio observations and the γ-ray observations.

mic ray studies has been the limitations imposed by the Earth's ionosphere. A space-based array would obviate these limitations and open a powerful new window on the universe.

The ability to trace CR electrons directly is the foundation for making the case that space-based low-frequency radio astronomy has the potential to answer the question of CR origin. The "big" question of CR origin can be decomposed into smaller more focussed questions. We have chosen the following three questions.

How are cosmic rays distributed in space?
How are cosmic rays accelerated?
What are the energetics of Galactic cosmic rays?

We now discuss how a space-based array can answer these questions.

3. DISTRIBUTION OF COSMIC RAYS IN THE GALAXY

3.1. Radio-γ-Ray Comparisons

All sky surveys at (cm-wave) radio and γ-ray wavelengths show similar emission morphologies. A detailed comparison of γ-ray and radio emission was first made by Haslam et al (1981) using the Haslam radio survey and COS-B data. Their figure 1 shows latitude-averaged longitudinal profiles of radio and γ-ray emission. Although the profiles differ substantially in directions of specific sources such as Cas A, Vela and the Galactic center, the overall distributions, reflecting the diffuse emission, are remarkably similar even showing the enhancement along the spiral arm tangents. The correlation is even more striking in two dimensions as shown in figure 2.

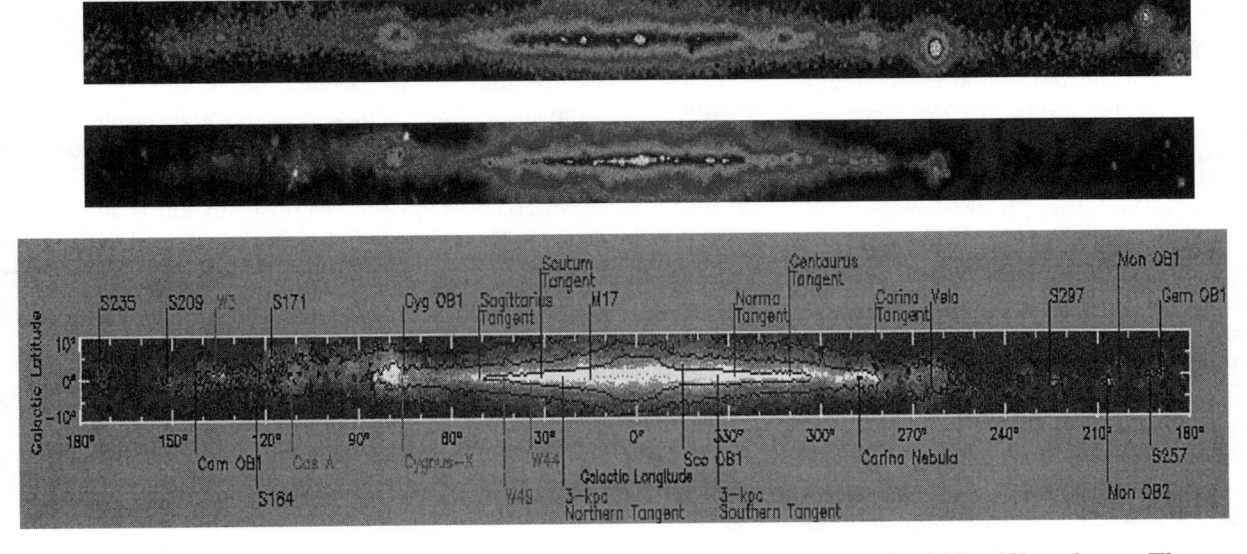

Figure 2. A comparison of γ-ray (top) and radio (middle) maps of the Milky Way plane. The bottom panel can be used to identify features in the top two panels. Note the overall similarity in the distribution of radio and γ-ray data. These images were obtained from the web site: adc.gsfc.nasa.gov/mw/milkyway.html

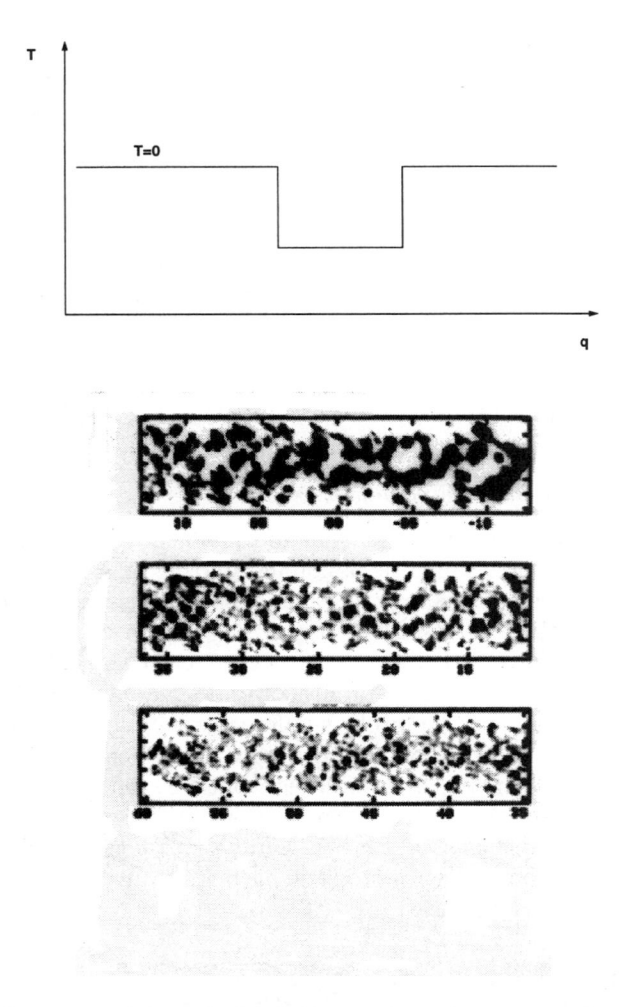

Figure 3. The top sketch shows the 1-d response of an interferometer along the spatial coordinate q. When seen against an extended background of emission, the absorbed sources show up as "negative" features. The panels on the bottom show actual low frequency observations (from Kassim, 1988) of the Milky Way plane. The negative sources show up as white features in these maps.

The γ-ray data measure the integral product of cosmic rays and thermal gas along a given line of sight (column 5, Table 2) while the radio data represent the same for cosmic ray electrons and magnetic fields. Thus, the above spatial correlations could be caused by the Galactic cosmic rays but they could equally well represent a correlation of gas and magnetic fields. A further complicating factor is that cm-wave radio emission traces >GeV electrons while the γ-ray (relativistic Bremsstrahlung) emission traces sub-GeV electrons (figure 1).

As discussed earlier, low frequency radio emission from the ISM traces the same sub-GeV electrons (for frequencies below about 100 MHz). Consequently, a comparison of low frequency radio emission with mid-energy γ-ray emission, in conjunction with HI emission data, should yield a much improved estimate of the cosmic ray electron distribution. A favorable consequence of estimating the cosmic ray distribution is that the line-of-sight averaged magnetic field can then be estimated by inverting the synchrotron equation (row 1, Table 2).

It needs to be stressed that any estimates of the CR electron distribution, produced in this manner, are subject to two major limitations. First, the absence of "depth" information means that only column densities of CR electrons can be obtained. In fact, both the γ-ray emission and the radio continuum emission are optically thin so that the column densities represents integrals along path lengths through the whole Galaxy. Secondly, knowledge of the CR electron column density is further limited by the poor resolution of present and foreseeable future γ- instruments which limits the quality of γ-ray data against which the comparison with radio data can be made. In section 3.2 (below) we describe a novel method of estimating CR electron densities that circumvents the "depth" problem. This method is unique to low-frequency radio astronomy.

3.2. 3-D Distribution of Cosmic Rays

At low radio frequencies (10 - 100 MHz), HII regions become optically thick. Observations of low frequency synchrotron emission in the direction of such remnants offers a unique opportunity to measure synchrotron emissivity along columns whose lengths are given by the distances to the HII regions. It may therefore be possible to build a 3-dimensional model of the electron distribution in the Galaxy. Such a model would offer new insights into the geometry of the Galactic cosmic ray disk and/or possible halo.

An HII region becomes optically thick when the frequency dependent optical depth approaches unity, that is when $\tau_\nu \approx 1$. The corresponding critical frequency is given by

$$\nu_0 \approx \left[RT^{-3/2}N_e^2\right]^{-2},$$

where R is the characteristic size of the emitting region while T and N_e are the temperature and number density of the ionized region in *cgs* units. Inverting the above equation to solve for R yields an expression for the minimum size an HII region needs to have in order to become optically thick. Thus,

$$R = \frac{10^6}{\nu_0^{1/2} N_e^2} \text{pc}.$$

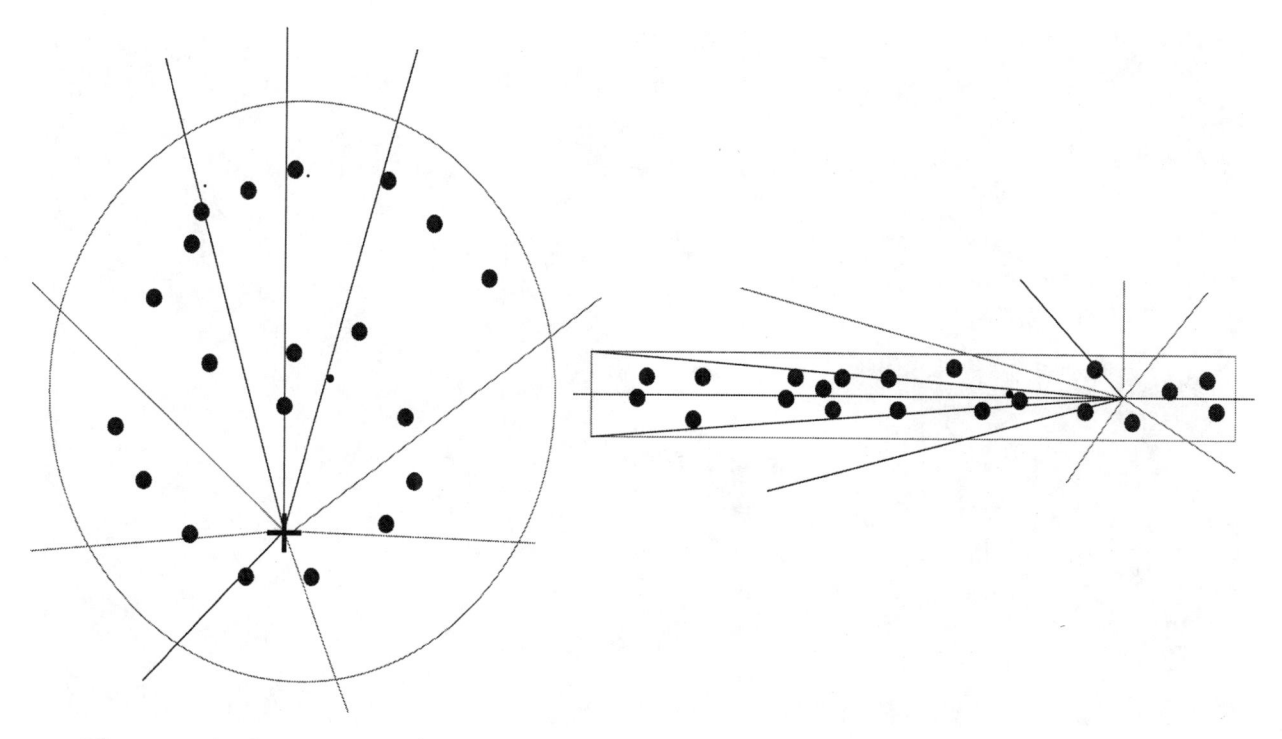

Figure 4. A schematic view of our location relative to HII regions in the Galaxy. The left figure shows the face-on view while the edge-on view is shown on the right. Line segments indicate lines of sight relative to the observer situated at the convergence point of the line segments.

At $\nu_0 = 30$ MHz,

$$\rightarrow R \approx \frac{200}{N_e^2} \text{pc}.$$

The Milky Way is characterized by an average density, $\langle N_e \rangle \approx 0.1$ cm^{-3} which yields $R \approx R_{MW}$. In other words, the Milky way itself is, on average, marginally opaque at 30 MHz. Thus any region with a substantially greater column density of ionized gas is opaque. Consequently, just about all known HII regions are opaque at 30 MHz. For a typical HII region with $\langle N_e \rangle \approx 100$ cm$^{-3} \rightarrow R \ll R_{HII}$. The HII regions have a very thin "skin depth" and are therefore seen in absorption against the Galactic background.

Such opaque HII regions manifest themselves in a unique manner when observed with an interferometer. Since an interferometer cannot measure DC signal levels (extended background emission) any such level appears as a zero-level of emission. Consequently, any absorption against such a DC level will appear negative. Such holes have been observed at low frequencies (e.g. Kassim, 1988). They have the effect of giving the Milky Way plane a "swiss cheese" appearance (figure 3). The absorption holes are not completely dark because of foreground synchrotron emission produced by CR electrons between the observer and the HII region and the emission from the "surface" of the HII region. Consequently the emission actually measured by an interferometer, in the direction of an absorbed HII region, contains information on the column density of CR electron. The emission contributed by the HII region can be estimated from data obtained at higher frequencies and thereby removed so that an estimate of the foreground synchrotron emissivity can be obtained (averaged along the line of sight). In principle, an estimate of the average number density of CR electrons is then possible.

Since a large number of HII regions have known distances they can be used to determine, collectively, the synchrotron emissivities along many different lines of sight. An ensemble of such measurements can then be used to construct a 3-dimensional distribution of synchrotron emissivity, something that is not possible with conventional measurement techniques (figure 4).

In order for the above technique to be applicable it is necessary for the instrument to be sensitive to the levels of foreground emission expected at very low frequencies. We have made such an estimate using a recently made 22 MHz map of the northern sky made by Roger et al

Figure 5. A 22 MHz map of the northern sky as obtained by Roger et al. using the DRAO synthesis telescope.

(1999). It is shown in figure 5. At 30 MHz,

$$j_\nu \approx 10^{-38} \to 10^{-34} \text{ erg s}^{-1} \text{ cm}^{-3} \text{ Hz}^{-1}$$

which corresponds to a column surface brightness of

$$= 0.01 \to 100 \text{ mJy / arcsecond}^2 \text{ / kpc}.$$

Such emissivities provide a working constraint on the sensitivity of a low-frequency interferometer. To make the above technique practical a minimum sensitivity of 0.1 mJy/arcsecond^2 is required.

The CR electron distribution follows from the emissivity measurements via the inversion of

$$S_\nu \propto \int N_{CRe} B^{1+q/2} \, d\Omega dr,$$

where Ω represents solid angle, r is the distance to the HII region and q is the spectral index of the CR electron energy distribution. The essential unknown in the inversion process is the B field. However, as noted above, comparison of radio emission with that of medium energy γ-ray emission can yield line-of-sight averaged values of the B field making the inversion of the above equation possible. However, the γ-ray emission is not affected by the presence of the HII regions and therefore represents a longer path length compared to the radio emission. Thus, in order to extract the B field it is necessary to do the radio- γ-ray comparison along lines-of-sight that are adjacent to the HII region. The column averaged B field is then assumed to be representative of the field in the foreground of the HII region. Such an estimate is not ideal since we don't know how strongly the B fields changes along a radial line of sight. However, the resulting estimates should lead to a substantially better understanding of the 3-D distribution of CR electron number densities, a key unknown in the CR origin puzzle.

The above technique only measures the distribution of CR electrons. What about the CR protons? Although it may well be that the protons and electrons are distributed in a similar manner it may also be possible to check that assumption by comparing low-frequency radio and high energy γ-ray emission. As shown in table 2, this is possible because the high energy γ-ray emission is dominated by the interaction of protons with the ISM while the low-frequency radio emission is the result of CR electrons interacting with magnetic fields.

Since the magnetic field can be extracted independently (see the above discussion) and since good HI data are available to describe the proton interaction it is possible to compare (at least the projected) distributions of protons and electrons across the sky. This comparison could act as an important check on the similarity of the two distributions.

4. THE ACCELERATION OF COSMIC RAYS

The current paradigm holds that high energy phenomena, related to supernovae and/or active galactic nuclei (AGNs), are the sources of cosmic rays. However, no direct connection between the particles that we observe locally and any identified cosmic sources has been made, leaving their origin uncertain.

It is interesting to note, that as early as the 1930's, Zwicky (1939) suggested that supernovae were somehow responsible for the acceleration of cosmic rays. This idea was proposed even before the connection between cosmic rays and galactic radio emission was made. Although the supernova idea was much closer to the currently accepted view of cosmic ray acceleration it was recognized early on that strong adiabatic losses would prevent significant particle acceleration during the supernova expansion. Instead, it was proposed that it is the interaction of the supernova remnants with the ISM that leads to the acceleration of cosmic rays. The idea was first developed by Bell in 1978 using the concept of first order Fermi acceleration. Subsequent work has evolved this idea into what is currently known as diffusive shock acceleration (DSA). DSA is fully reviewed by Ellison et al (1994) and, to a lesser extent but more recently, by Jones et al (1998).

The DSA mechanism in conjunction with SNRs appears to be able to account for the vast majority of cosmic rays (those with energies between $\approx 10^9$ and 10^{15} eV). However, the classic problem of the injection mechanism (the ability to accelerate seed particles from the thermal pool) remains unsolved (see, for example, Ellison et al, 1994; Jones et al, 1998). A separate mechanism is required to energize particles to sub-GeV energies in order to inject the necessary seed particles into the DSA environment. Fortunately, low frequency radio astronomy is ideally placed for addressing this problem. In supernova remnants (SNRs), where magnetic fields range from 10 microGauss to 1 milliGauss the synchrotron emitting electrons have significantly lower energies at a given critical frequency. At 10 - 30 MHz, the electrons span the energy range of 15 MeV to 750 MeV, exactly the range needed to study the nature of the injection mechanism. The shape of the electron spectrum and the spatial correlation of hot-spots with shock features, at these energies, would provide robust constraints on the mechanism that accelerates seed particles in SNRs.

Table 3: Scattering Disks

ν (MHz)	θ''
300	0.01
30	3
10	30

Perhaps even more intriguing is the possibility of measuring plasma turbulence in and around SNRs. Interstellar scintillation has the effect of "blurring" radio sources in much the same way that terrestrial atmospheric turbulence affects optical astronomical images (e.g Cordes and Rickett, 1998). Such blurring sets a finite angular size to a point radio source. The apparent angular size of a point source depends on the degree of turbulence and the wavelength of the emission and is therefore a function of the Galactic coordinates and the observing wavelength. The wavelength dependence $\propto \lambda^2$ and is illustrated in table 3 for a typical source in the Galactic plane. Using table 3 as a guide it is evident that in order to measure the diameters of the scattering disks it is necessary for the array to have a resolving power of $\approx 1''$ at 30 MHz thereby ensuring that at lower frequencies the disks are resolved even better since the ratio of source size to beam size grows as λ. Thus, low frequency imaging of background sources (extragalactic point sources) along lines of sight passing near SNR shocks has the potential to produce "plasma turbulence maps" of the shock regions (at cm wavelengths this kind of work requires VLBI techniques, e.g. Spangler and Cordes, 1998).

One of the predictions of DSA theory is that turbulence should be enhanced just upstream and just downstream of SNR shock fronts, thereby allowing the particles to be accelerated many times. Detailed mapping of turbulence in SNRs would provide a direct test of the theory. In a sense, it may be possible to catch SNRs *in the act* of accelerating cosmic rays. Such observations, applied to SNRs of varying age and in differing environments, would pin down the role of SNRs in producing CRs by determining the duration and timing of such acceleration over an SNRs lifetime and by assessing the role of the environment in determining the CR yield. Identification of these processes would fill another major hole in our knowledge of CR origin by determining whether SNRs are major source of cosmic rays in the Galaxy.

5. ENERGETICS OF COSMIC RAYS

A very important constraint on cosmic ray origin is the energy density of cosmic rays in the Galaxy. Knowing the distribution of cosmic ray energy densities in the ISM makes it possible to estimate the energy budget of Galactic cosmic rays. Any plausible sources of cosmic rays must be consistent with such an energy budget.

Direct measurements of cosmic ray fluxes at the top of the Earth's atmosphere suggest that the energy density of cosmic rays is about 0.5 eV cm^{-3}. However, this number is a lower limit because the flux is reduced by the outflowing solar wind which strongly modulates the cosmic ray flux, particularly below about 1 GeV. Pioneer and Voyager spacecraft have been making *in situ* measurements of cosmic ray fluxes as they proceed out of the solar system (e.g. Webber, 1991). Their measurements indicate that the energy density of cosmic rays is slowly increasing with distance from the Sun. At 70 AU, the value is about 1.5 eV cm^{-3}. The spacecraft have not yet reached the heliospheric boundary. When they do, they will be making measurements in the true ISM, local to the Sun.

Unfortunately, these direct measurements are very local and do not tell us about the energy density of cosmic rays in other regions of the Galaxy. The energy densities have to be inferred from the synchrotron emission of the relativistic electrons. Interpretation of the synchrotron all sky maps by Beuermann et al (1985) suggest that the energy density of cosmic rays increases from about 1 eV cm^{-3} at the solar circle to about 6 eV cm^{-3}, 4 kpc from the Galactic center. The interpretation of γ-ray data suggests similar numbers (e.g. Strong et al, 1996). However, these studies are severely hampered by the projection of emission along the line-of-sight. Low frequency radio astronomy offers the advantage of being able to measure actual path lengths thereby greatly improving estimates of energy densities in the Galaxy. This is achieved by spatially resolving optically thick HII regions (very common at low frequencies), of known distance, against which the distributed synchrotron emissivity can be accurately determined ([Kassim 1990]). As discussed above, the power of this approach is the availability of relatively well determined path lengths to the HII regions, allowing the true three-dimensional space distribution of cosmic ray induced synchrotron emissivities to be determined. A comparison of the local synchrotron emissivity with the known local energy density of cosmic rays would form the calibration needed to convert synchrotron emissivities to cosmic ray energy densities.

Once the energy budget of cosmic rays is determined, then it is possible to critically compare the energetics of proposed cosmic ray sources. For example, the supernova rate for our Galaxy is estimated to be about 1/30 years. If a supernova produces 10^{51} ergs of kinetic energy, the average energy input rate is $\approx 10^{42}$ erg s^{-1}. This number is comparable to the energy rate needed to maintain the Galactic cosmic rays (making specific assumptions about the poorly known energy density of cosmic rays in other parts of the Galaxy, e.g. Jones et al, 1998).

The above approach can be carried to external galaxies (which by virtue of their better perspective minimize the "forest for the trees" problem associated with our galaxy). Studies of SNRs in external galaxies, at cm-wavelengths, have led to calculations of energy budgets showing that SNRs may in fact provide the needed energy (e.g. Duric et al, 1993; Duric et al, 1995; Gordon et al, 1998; Lacey, 1998). However, these studies could be greatly improved at lower frequencies. The above studies have found a strong spatial correlation of SNRs with HII regions. Consequently, the thermal emission from the HII regions strongly contaminates the radio emission from the SNRs. A low-frequency array capable of 1" imaging and 0.1 mJy sensitivity at frequencies below 100 MHz should allow for the detection of SNRs even in the most confused regions of galaxies. The improved statistics would prove invaluable in gauging the role of SNRs in regulating the cosmic ray production in external galaxies.

SUMMARY

Long wavelength space-based radio astronomy has the potential to revolutionize our knowledge of Galactic cosmic rays by addressing the subject on a number of different fronts. However, in order to do this effectively any space-based array must meet some challenging specifications (referred here to an observing frequency of 30 MHz).

- A resolution of $\approx 1''$ is needed to properly resolve SNRs and self-absorbed HII regions.

- A sensitivity of at least 0.1 mJy/arcsecond^2 is required to record emission in front of absorbing HII regions.

- The array should be able to survey the whole sky in order to map out the 3-D distribution of emission and to fully characterize the scintillation characteristics of the ISM.

- In order to make meaningful comparisons with $\gamma-$ray observations and to probe injection-level energies in SNRs the array should operate down to 10 MHz. This would allow probing of cosmic ray electron energies below 300 MeV in the general ISM and below 50 MeV in supernova remnants.

- The challenge: build an imaging array to perform a very sensitive, high-resolution survey of the sky at wavelengths of 3 - 30 meters.

Acknowledgments. I thank Dr. Namir Kassim for the many useful discussions on low-frequency radio astronomy. I also thank the AGU for a travel grant that made my trip to the Chapman conference possible.

REFERENCES

Bell, A.R. MNRAS, *182*, 147, 1978.
Beuermann, K., G.O. Kanbach, and E.M. Berkhuijsen A&A, *153*, 17, 1985.
Cordes, J. M. and B. J. Rickett ApJ, *507*, 846, 1998.
Duric, N., F. Viallefond, W.M. Goss, J.M. van der Hulst A&AS, *99*, 217, 1993.
Duric, N., S.M. Gordon, W.M. Goss, F. Viallefond, and C. Lacey ApJ, *445*, 173, 1995.
Gordon, S. M., R. P. Kirshner, K. S. Long, W. P. Blair, N. Duric, R. C. Smith ApJ, *117*, 89, 1998.
Ellison, D.C., et al. PASP, *106*, 780, 1994.
Haslam, C.G.T., C.J. Salter, H. Stoffel, and W.E. Wilson A&AS, *47*, 1, 1981.
Hess, V. Phys. Z., *13*, 1084, 1912.
Jansky, K. H. Proc. I.R.E., *23*, 1158, 1935.
Jones, T.W. et al. PASP, *110*, 125, 1998.
Kassim, N. E. ApJS, *68*, 715, 1988.
Kassim, N. E. 1990, In: Low frequency astrophysics from space; Proceedings of an International Workshop, Crystal City, VA, Jan. 8, 9, (A91-57026 24-89). Berlin and New York, Springer-Verlag, 1990, p. 144-151, 1990.
Lacey, C.K. PhD Thesis, University of New Mexico, 1998.
Longair, M.S. 1990, In: Low frequency astrophysics from space; Proceedings of an International Workshop, Crystal City, VA, Jan. 8, 9, 1990 (A91-57026 24-89). Berlin and New York, Springer-Verlag, p. 227-236, 1990.
Reber, G. ApJ, *91*, 621, 1940.
Roger, R. S., C. H. Costain, T. L. Landecker, C. M. Swerdlyk, A&AS, *137*, 7, 1999.
Strong, A.W. et al. A&AS, *120*, 381, 1996.
Spangler, S. R. and J.M. Cordes ApJ, *505*, 766, 1998.
Webber, W.R. in ASP Conf. Ser. 16, The Interpretation of Modern Synthesis Observations of Spiral Galaxies, ed. N. Duric and P.C. Crane (San Francisco: ASP), 37, 1991.
Zwicky, F. Phys. Rev., *55*, 986, 1939.

N. Duric, Department of Physics and Astronomy, University of New Mexico, Albuquerque, NM, 87131, USA.
(e-mail: duric@tesla.phys.unm.edu)

Long Wavelength Observations of Supernova Remnants

Namir .E. Kassim

Code 7213, Naval Research Laboratory, Washington, DC, USA.

Farhad Yusef-Zadeh

Dearborn Observatory, Northwestern University, Evanston, IL, USA.

Low frequency or long wavelength ($LW \equiv \lambda \geq\sim 1\ m$) observations offer unique insights into supernova remnants (SNRs) and their interaction with the interstellar medium (ISM). The relatively high surface brightness sensitivity and large field of view are well matched to the nonthermal radiation and extended morphology of SNRs. They address the incompleteness in SNR catalogs by uncovering new remnants, with discovery of the barrel-shaped SNR G7.06-0.12 presented here as an example. Continuum spectra constrain particle acceleration and cosmic ray theories, since the radio spectrum reflects the energy distribution of the synchrotron emitting relativistic electrons. Furthermore, LW observations compliment higher frequency images because spectral measurement errors are reduced due to the wide frequency coverage. They also can detect the unshocked ejecta inside young shell-type SNRs through absorption, and can also delineate other forms of ionized material inside remnants such as filaments. LW SNR observations can also constrain the distribution of ionized gas in the ISM by measuring the line of sight free-free absorption. Measurements of co-distant samples of SNRs in nearby galaxies are especially useful for studying the energetics, star formation history, ISM, and cosmic ray gas properties of external systems. These studies are enhanced by extending them to longer wavelengths if good angular resolution can be maintained, and new and planned ground-based systems such as the 74 MHz VLA, GMRT, and the Low Frequency Array (LOFAR) are beginning to achieve this. Extending such studies to space-based observations at even longer wavelengths will be even more powerful.

1. WHY STUDY SNRS AT LOW RADIO FREQUENCIES?

The nonthermal radio emission from SNRs makes them intrinsically brightest at long wavelengths (LWs). Furthermore because they are typically extended the relatively reduced angular resolution is not a great disadvantage and is usually more than made up for by the increased surface brightness sensitivity. Similarly the generally larger field of view makes it convenient for studying large individual Galactic remnants or for efficiently surveying whole remnant populations in nearby galaxies. These reasons often make it convenient to study known SNRs or find new ones at low radio frequencies. Furthermore existing continuum spectrum studies may be significantly enhanced by sensitive low frequency measurements because of the increased lever arm achieved in frequency space, while thermal absorption (intrinsic and extrinsic) effects can be measured only at low frequencies. Scattering effects related to SNR shocks may also be best studied at long wavelengths.

2. FINDING NEW SNRS

2.1. Galactic

Galactic SNR catalogs are severely limited by selection effects and notoriously incomplete. We know we are missing many low surface brightness, extended SNRs, with current catalogs estimated to be complete only to surface brigthness levels of $\approx 8 \times 10^{-21} W\ m^{-2}\ Hz\ sr^{-1}$ (*Green 1991*). Also known to be undersampled in current catologs is the population of bright, compact, "young" SNRs. Complete samples are required for understanding SN/SNR birthrates, statistics, and their energy input to the interstellar medim (ISM). Furthermore their distribution is important for comparison with that of the presumed progenitor population. A good SNR census is also important for understanding the statistical implications of pulsar/SNR associations, such as the radio lifetime of SNRs and the distribution of pulsars at birth (*Frail, Goss, and Whiteoak 1994*).

As noted earlier, LW systems provide high surface brightness sensitivity, and this makes them ideal for finding older, extended SNRs, or other extended nonthermal sources such as are found at the Galactic center. For example G0.3+0.0, the closest known SNR to the Galactic center after G0.0+0.0 (Sgr A East, whose identification as an SNR is quite controversial, see for example *Metzger et al,* [1989] or *Khokhlov and Melia* [1996]. G0.3+0.0 was first identified as a discrete nonthermal source from relatively low angular resolution (5') observations at 80 MHz (*LaRosa and Kassim* 1985, *Kassim, Erickson, and LaRosa* 1986) and was later confirmed as a remnant by *Kassim and Frail* [1996] from 330 MHz VLA observations with 1 resolution. The full image used to identify this new SNR is shown in Plate 1, and has been used to identify a number of previously unrecognized Galactic center nonthermal sources (*Kassim et al*, 1999, *LaRosa et al*, 2000), including the new nonthermal filament named the Pelican (*Lang et al*, 1999). Similarly *Frail, Goss, and Whiteoak* [1994] found several new low surface brightness SNRs near young pulsars with the 330 MHz VLA, significantly increasing the number of known SNR/PSR associations, an important result pertaining to the radio lifetime of SNRs remnants and the distribution of pulsar velocities at birth.

As an example of how readily new SNRs emerge from sensitive meter wavelength interferometer images, we catalog here a new Galactic SNR, G7.1-0.1 (or more accurately G7.06-0.12), first recognized in an ongoing study of this region by *Yusef-Zadeh et al*, [2000]. This SNR shows itself readily on an arc-minute resolution wide-field 330 MHz VLA image centered on the W28 SNR and shown in Figure 1. Protruding north of the W28 SNR near RA 17h 58.5 m is actually the western half of G7.1-0.1, whose eastern half is smothered by the bright HII region M20 centered near RA 17h 59.3 m. The higher resolution 20 cm image shown in Figure 2 reveals that G7.1-0.1 is a classic barrel shaped SNR aligned roughly with its major axis parallel to the Galactic equator as is common (*Gaensler* 1998). This SNR is also revealed on resinspection of the higher resolution (20") wide-field 330 MHz image of the SNR G5.4-0.1 (*Frail, Kassim, and Weiler* 1994), thanks to the large primary beam of the VLA antennas at 330 MHz (FWHM $\sim$ 156'). A rough estimate of the spectrum comes from comparing the integrated flux of the well defined western half of the barrel, giving $\sim$ 3.9 Jy at 330 MHz and $\sim$ 1.9 Jy at 1452 MHz. This yields the canonical $\alpha \sim -0.5 \pm 0.1$ nonthermal spectrum of a typical shell type SNR, as expected. A comprehensive analysis of the new data revealing this source is currently under way (*Yusef-Zadeh et al*, 1999). Previously unidentified, low surface brightness features on 330 MHz maps such as Figure 2 abound, and this simple example shows that a sensitive ($< 10\ mJy\ beam^{-1}$) meter-wavelength Galactic plane survey (say with the VLA or GMRT) with an interferometer of moderate resolution (1') would undoubtedly reveal dozens of such previously unidentified low surface brightness remnants; like shooting fish in a barrel (*Frail* 1995).

Low frequency radio observations should also be able to address the problem of the missing, compact SNRs since they are also nonthermal. But here the ionospheric-limited angular resolution of past low frequency instruments has resulted in very poor confusion-limited point source sensitivity despite large collecting areas. This makes identification of moderate to weaker compact nonthermal sources, particularly in the Galactic plane, difficult. This situation should soon change since the new 74 MHz VLA system has demonstrated that self-calibration can remove ionospheric phase effects (*Kassim et al*, 1993, *Erickson, Kassim, and Perley* in these proceedings), thus opening the door for future low frequency instruments with much greater angular resolution and point source sensitivity.

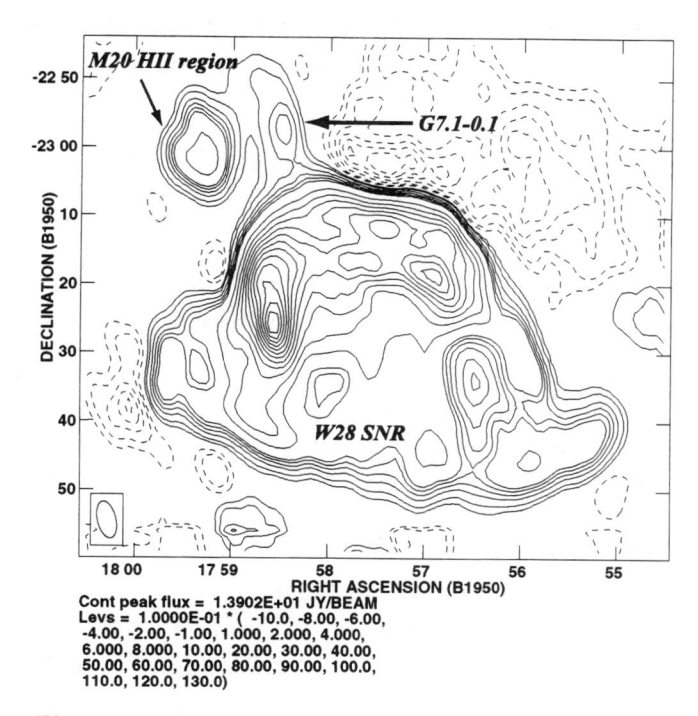

Figure 1. Contour image of 330 MHz D array observations W28 supernova remnant. The angular resolution is approximately 3', and the sensitivity is about 50 mJy/beam. The western arm of the newly discovered barrel shaped SNR G7.1-0.1 can be seen protruding from the north of the larger W28 SNR. The eastern half of G7.1-0.1 is obscured by the M20 HII region. Contour levels are as shown at the bottom.

2.2. Extragalactic

Extragalactic SNR radio, optical, and X-ray studies are passing through a renaissance of sorts, driven by the power of satistical studies of co-distant samples of SNRs in nearby normal galaxies (*Jones et al*, 1998). High angular resolution LW observations are ideal for finding and identifying new SNRs since they distinguish between thermal and nonthermal discrete sources, and they can anchor spectral identifications made at higher frequencies. An unexpected result has been that many remnants show LW turnovers in their radio continuum spectra at relatively higher frequencies ($\geq 100 MHz$) than those seen towards Galactic SNRs (*Kassim* 1989a,b). In M82 this has been measured at 408 MHz and is attributed to free-free absorption by ionized gas in their immediate surroundings with emission measures $\sim 10^6 \ cm^{-6}$, comparable to properties of Galactic giant HII regions (*Wills et al*, 1997). These measurements place the discrete sources relative to the ionized component of the host galaxy ISM. The powerful statistical studies which follow from co-distant SNR samples can measure particle acceleration efficiency (*Duric et al*, 1995), relate cosmic ray origin to specific SNe types (*Lacey* 1997), and measure SN rates, star formation, and ISM properties of the host galaxies (*Jones et al*, 1998). Observations below 100 MHz have not had sufficient angular resolution to conduct studies of discrete sources, but Clark Lake studies of the integrated flux of normal galaxies resulted in an inclination/optical depth relationship (*Israel and Mahoney* 1990), *Israel et al*, 1992) which is consistent with thermal absorption from the ionized ISM of those galaxies. More sensitive studies at higher resolution should soon follow with the advent of systems like the 74 MHz VLA (*Erickson, Kassim, and Perley* in these proceedings).

3. STUDYING KNOWN SNRS

LW observations of SNRs offer both unique information (e.g. from absorption studies) as well as data

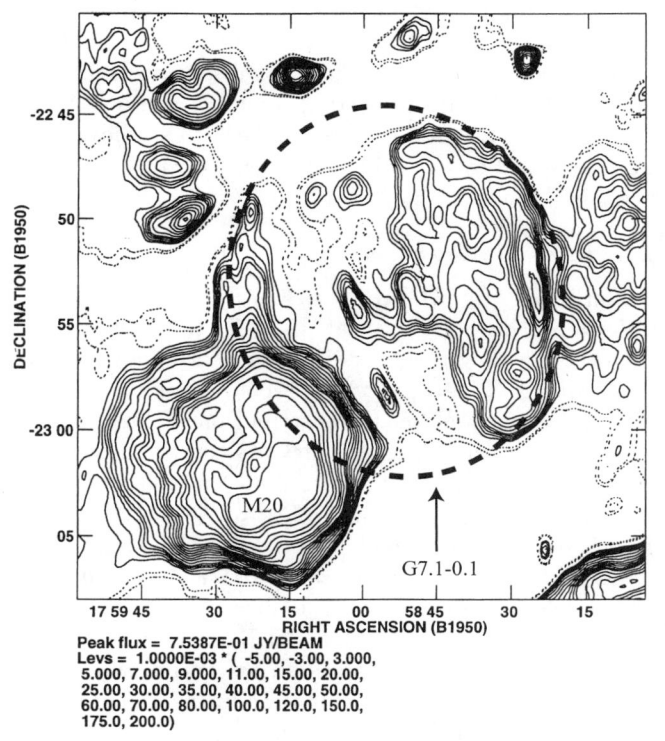

Figure 2. 1450 MHz VLA image of M20 field delineating the barrel shaped morphology of the newly discovered SNR G7.1-0.1. The major axis of G7.1-0.1 is oriented along the Galactic plane as is common for such remnants (*Gaensler* 1998). On this image with approximately 20" resolution the eastern half of the new SNR can now be seen superimposed on the M20 HII region. Contour levels are again as shown at the bottom of the plot.

which especially compliments higher frequency measurements (e.g. spectral studies). LW measurements can be useful for distinquishing between various SNR morphological types, as demonstrated by the Ooty 34 MHz observations of Vela (*Dwarakanath* 1991). In this case the relatively flat spectrum of the Vela X plerion ($\alpha \sim -0.16$) was easily distinguished form the relatively steeper spectrum Vela YZ shell ($\alpha \sim -0.53$). Similarly the flattening of the shell spectrum of the SNR G5.4-1.2 anchored at 330 MHz were used to infer the influence of the pulsar wind from PSR 1757-24 on the SNR shell emission (*Frail, Kassim, and Weiler* 1994). Furthermore LW measurements are ideal for disentangling the relative superposition of HII regions and SNRs along complex lines of sight towards the inner Galaxy (*Kassim and Weiler* 1990). LW observations may also place useful limits on shell-like emission around "naked" plerions, such as the Crab. The Crab pulsar tells us that there was an explosion of a massive star, yet the failure to detect a shell remains a mystery. The present best radio limit on the existence of a shell around the Crab comes from LW observations (*Frail et al*, 1995). Finally the integrated and spatially resolved continuum spectra of SNRs are important for all theories of SNR emission, and LW observations make such studies much more powerful by extending them over a wider range in frequency space.

3.1. External Absorption

Clark Lake (at 31 MHz) and Culgoora (at 80 MHz) LW studies have shown that many SNRs show *extrinsic* low frequency turnovers in their continuum spectra (*Dulk and Slee* 1975, *Kassim* 1989a,b). This offers a powerful constraint on the distribution of low density ionized gas in the ISM. The *lack* of absorption to some distant SNRs places an upper limit on the density of the Warm Ionized Medium (WIM) of $\leq 0.26\ cm^{-3}$ (for $T \sim 8000$ K), a limit which would be significantly improved by lower frequency (< 30 MHz) measurements. The patchy absorption which is measured is consistent with absorption by Extended HII Region Envelopes or EHEs. EHEs were first suggested by centimeter wavelength Galactic ridge recombination lines (RRLs) which appeared to come from regions free of discrete sources, and were directly inferred from 325 MHz *stimulated* radio RRLs (*Anantharamaiah* 1986). EHE physical properties include $T \sim 3000$-8000 K, $n \sim 0.5 - 10\ cm^{-3}$, and sizes ~ 50-200 pc. Alternative interpretations are that GRRLs may originate in old, evolved HII regions, since not as much absorption was seen at 34 MHz (*Dwarakanath* 1989) as might be expected if all HII regions had EHEs. This work is in relatively primitive state, since the highly confused Clark Lake and Culgoora SNR spectrum studies were severely limited in frequency, angular resolution, and sensitivity. The advent of sub-arcminute resolution imaging systems at frequencies below 100 MHz such as the 74 MHz VLA (*Erickson, Kassim, and Perley* in these proceedings) and the GMRT should be able to make great strides in such ISM studies which use shell-type SNRs as well calibrated beacons with instrinsically straight power-law spectra.

As noted earlier, LW observations may constrain the relative position of thermal/nonthermal emitters in complex regions, with among the most well known examples being the 330 MHz observations which place the thermal absorbing Sgr-A West in front of the nonthermal Sgr-A East (*Pedlar et al*, 1989).

3.2. Internal Absorption

The discovery of internal thermal absorption towards the first two SNRs observed with the 74 MHz VLA came as somewhat of a surprise, and suggests similar LW observations should provide a powerful constraint on the thermal material inside SNRs. The Cas-A internal absorption at 74 MHz is only the second case for direct detection of unshocked ejecta inside a young SNR, and only the first in the radio (*Kassim et al*, 1995); the first case came from uv observations of iron absorption lines towards SN1006, which only by chance happened to have a suitable background source of uv radiation located behind it (a white dwarf, *Hamilton and Fesen* 1988). This suggests LW observations may be the best way of approaching this problem. The Cas-A result is actually expected from theory and provides information on the state of the ionized but unshocked ejecta interior to the reverse shock in a young SNR. This material has proven elusive to study and provides an important constraint for the evolutionary theory of young SNRs. The 74-330 MHz spectral index map showing the Cas-A absorption is shown in Figure 3.

The second SNR observed with the 74 MHz VLA was the Crab nebula, which showed internal absorption from its thermal filaments. Though not particularly surprising, this measurement is nonetheless important because it helps constrains the location of the thermal filaments relative to the pulsar powered nonthermal emitting material (*Bietenholz et al*, 1997).

These early examples of LW internal thermal absorption at the relatively high frequency (for thermal absorption) of 74 MHz in both a shell-type and plerionic

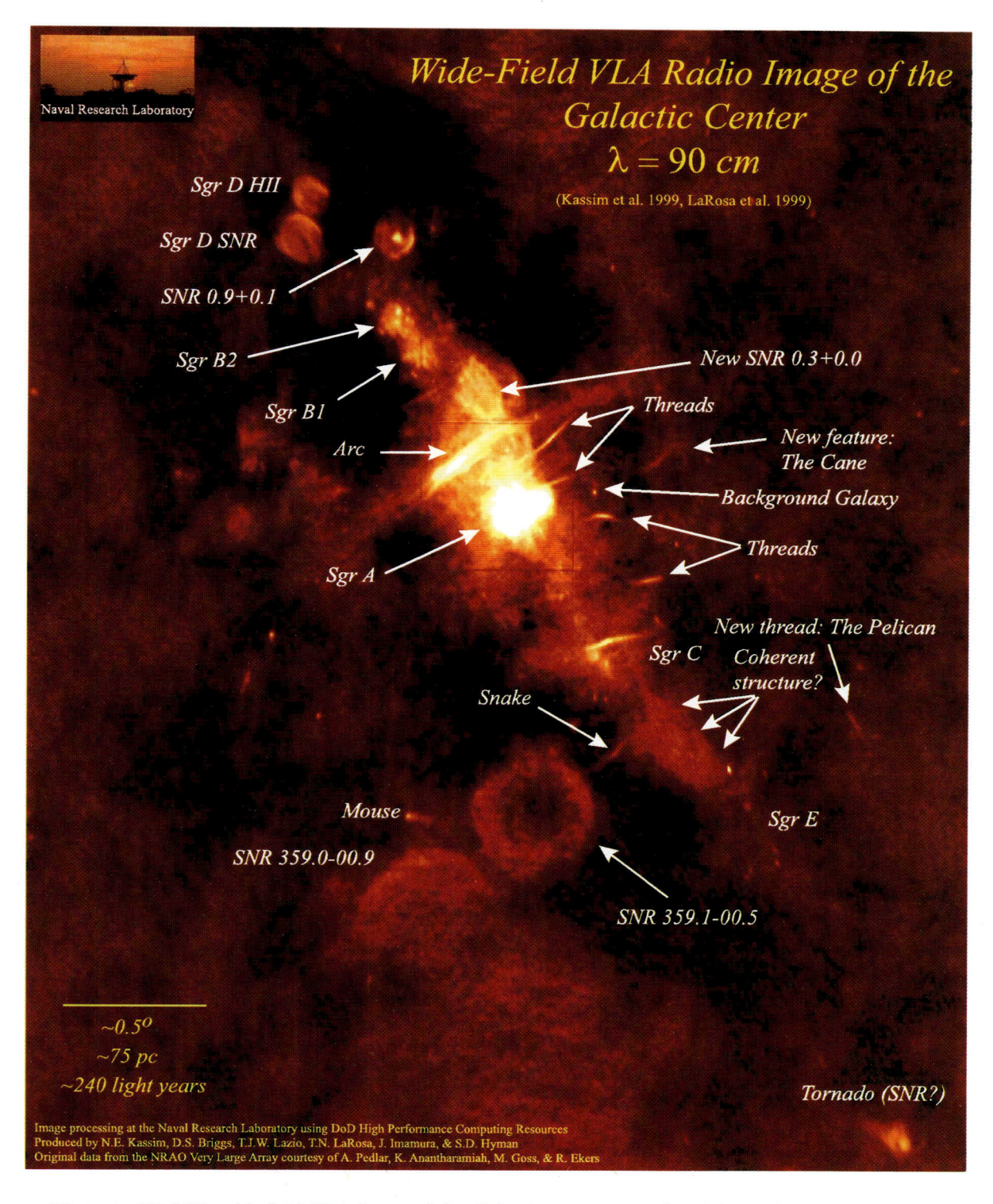

Plate 1. 330 MHz wide-field VLA image of the Galactic center, reproduced from *Kassim et al,* 1999). The angular resolution is approximately 45", and the rms noise is about 5 mJy/beam. Among the new sources discovered on this image were the SNR G0.3+0.0 (*Kassim and Frail* 1996), and two new nonthermal filaments designated the Cane and the Pelican (*Lang et al,* 1999).

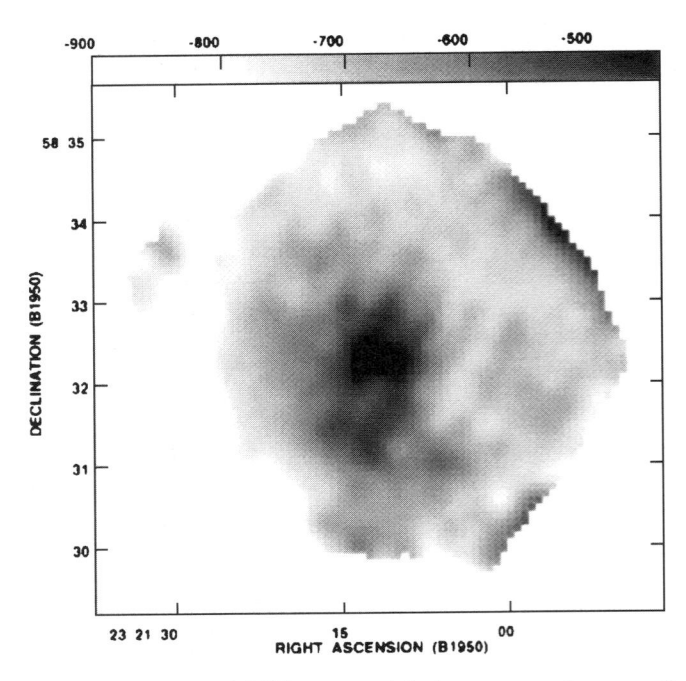

Figure 3. 74-440 MHz spectral index grey-scale map of Cas A made with the VLA. Dark corresponds to a flatter spectrum, and this region of absorption is believed to be caused by free-free absorption from unshocked ejecta interior to the reverse shock (*Kassim et al*, 1995).

SNR certainly wets the appetite for future LW investigations of this phenomena.

3.3. Continuum Radio Spectrum

3.3.1. Integrated Spectrum. Low frequency measurements are key to determining integrated spectra, as these data are the weakest link in presently known measurments. Integrated spectra provide significant constraints on predictions of diffusive shock acceleration theory, particularly in young SNRs. *Reynolds and Ellison* [1992] claim concave integrated spectra in Tycho and Kepler, consistent with the simplest Fermi test particle predictions, and this can constrain the mean magnetic field. The integrated spectrum is also important for understanding van der Laan emission in older SNRs, where spectral breaks can be linked to compression of cosmic ray gas with a break near 200 MHz, ($\alpha \sim -0.4$ below, $\alpha \sim -0.9$ above) (*Bridle* 1967). Such breaks have been claimed in the Cygnus Loop (*Sastry et al*, 1981), but the lowest frequency (< 40 MHz) data is controversial (see *Green* 1990) and new low frequency systems with even modest ($\sim 1'$) resolution (such as the Giant Meter Radiowavelength Telescope) can make important contributions.

3.3.2. Spatial Variations in Spectrum. Understanding spatial variations in continuum spectral index across SNRs is important to all forms SNR emission, whether by van der laan compression or shock acceleration. Variations have been claimed over the years in the literature, but equally as often they have been challenged. Again the past poor quality of the lowest frequency images is often the major source of error (see for example *Moffett and Reynolds [1994a,b]*). Explaining such variations, whether real or imaginary, is quite complicated, with various aspects of theory seemingly able to produce any perceived observational result.

Examples of explained variations include intrinsic manifestations of details in particle acceleration parameters. For example the compression ratio r is related to the radio spectral index alpha through $\alpha = [2(r-1)] - 1$. Hence the relatively steep spectrum ($\alpha \sim -0.75$) of the CTB-1 breakout region is interpreted in terms of a weaker shock (*Pineault et al*, 1997, 1998). On the other hand an underlying curved electron energy spectrum together with magnetic field variations can produce spectral spatial variations which may nothing to do with the details of ongoing or past acceleration processes. Such curved spectra may be produced by compression of the cosmic ray gas or by particle acceleration with diffusion, and magnetic fields commonly change strength across SNRs depending on whether one is in regions dominated by radiative or non-radiative shocks. Green's discussion of the reality of spatial variations across the (older) Cygnus Loop touches on these issues. Cosmic-ray mediated shocks and mixed thermal gas components add further levels of complexity.

In their 1993 review *Anderson and Rudnick* [1993] reported two trends. In older SNRs they found high emissivity shells having flatter indices than diffuse regions, while the opposite trend, with significant exceptions, seemed to exist in younger remnants. Their main conclusion, echoing the view which emerged from the Centenial Minnesota SNR workshop in 1996 (*Jones et al*, 1998) was that the agent most likely to be responsible for spatial variations in radio spectral index was the environment into which the SNR evolved. This in turn is established by initial conditions, such as the location of the SNR relative to associated molecular clouds and the pre-supernova mass loss (winds) history of the progenitor star.

The situation for Cas-A is worth noting, since this was a source whose bright knots of emission were taken as strong circumstantial evidence for second order (stochastic) Fermi acceration. Yet Anderson and Rudnick's important (and ongoing) study concluded exactly the opposite. They found evidence for a radial trend in knot spectral index, with flatter spectrum knots residing preferentially in the radio ring, while a spectrally steeper population of knots resides beyond this. Their conclusion is that the bright knots are not the sites of present acceleration, but simply light up and reveal the history of past particle acceleration processes in the enviroment through which they now move. A similar trend in the extended emission around Cas-A emerges from 74 MHz VLA observations (*Kassim et al*, 1995). However an alternative interpretation, in which the bright knots are the acceleration sites, has also emerged (*Atoyan et al*, 1999).

New techniques are also revolutionizing techniques of spectral index analysis, such as the utility of T-T (temperature-temperature) plots (*Anderson and Rudnick* 1993) and convolution differential spectral index techniques (*Zhang et al*, 1997). Such measurements claim to be robust against offset errors and large-scale background variations. For example a consistent trend of variations for G78.2+2.1 at 232 MHz, 408 MHz, 1420 MHz, 2695 MHz, and 4800 MHz would appear to be solid (*Zhang et al*, 1997). The trend here was for higher spectral index near the bright periphery, and a lower index towards the interior. The argument is made against this being due to compression of cosmic ray gas, since the basic emission requires ongoing particle acceleration. Hence spectral differences must be related to its details, though why and how are not cleas. The powerful technique of spectral tomography, applied in an ongoing spectral index study of Tycho's SNR (*Katz-Stone et al*, 1999), is joining this suite of new techniques.

We are likely on the verge of a renaissance of sorts in radio spectral index studies of remnants. This is because instruments (for example VLA, DRAO, MOST, and GMRT) which can deliver the more sensitive low frequency images required for comparison with existing centimeter wavelength maps, together with the sophisticated analysis techniques required to properly interpret them, are finally at hand. From an observer's point of view, this quiet revolution in the quality of observational results which we are presently witnessing, should finally provide theorists with many good cases of believable spectral index variations in wide varieties of remants, thereby reeling in what have been very poorly constrained theoretical interpretations.

3.4. Scattering Near or Through SNRs

Scattering measurements, often convenient to measure at long wavelengths, can also test predictions of shock acceleration theory. A key element of diffusive shock acceleration theory says that their must be turbulence (for example in the form of tangled magnetic field lines), both upstream and downstream of the shock, to allow particles to scatter back and forth across the main shock and gain energy in a statistical sense. Evidence from scattering for such turbulence would provide theory with a badly needed smoking gun. However the poor angular resolution and sensitivity of past low frequency observations have precluded such findings, although low frequency VLBI techniques have been used to explore for the scattering of compact extragalactic sources by SNRs. For example the third heaviest scattered line of sight known in the Galaxy is towards 1849+005, perhaps a result of its being 10' from the line of site towards SNR G33.6+0.1 (*Spangler et al*, 1986), though this not at all certain. Plate 2 shows a 330 MHz wide-field VLA image around the W44 supernova remnant field. The λ^2 dependence of scattering may well bring such measurements into the realm of VLA measurements at 330 MHz (angular resolution 6") and 74 MHz (angular resolution 20"), where as such measurements were previously considered in the VLBI regime. In any case it is a possible important line of study best attacked at low frequencies, and perhaps the outer Galaxy would be an easier place to start such a study, away from the myriad sources of scattering from which exist along lines of sight towards inner Galaxy.

4. SUMMARY

SNR observations clearly play to the strength of low frequency radio measurements, since these sources are nonthermal and often large. Low frequencies are ideal for finding new Galactic and extragalactic SNRs, by making differentiation with thermal sources relatively straight forward, and providing high surface brightness sensivity for finding weaker, larger, and presumably older SNRs. The discovery of SNR G7.06-0.12 presented in this paper is an example of this. Planned sensitive high resolution low frequency instruments should also be useful in finding missing bright remnants. Low frequency maps are helpful for morphological and spectral classification of plerions, shells, and composites, and for better delineating the important class of SNR-PSR associations. At the lowest frequencies (<408 MHz), images may also allow seperation of relative positions of SNRs and HII regions.

Radio spectrum studies, both integrated and spectral mapping, are crucial to theory, but only recently have low frequency images of sufficient quality for meaningful comparisons with centimeter wavelength maps come

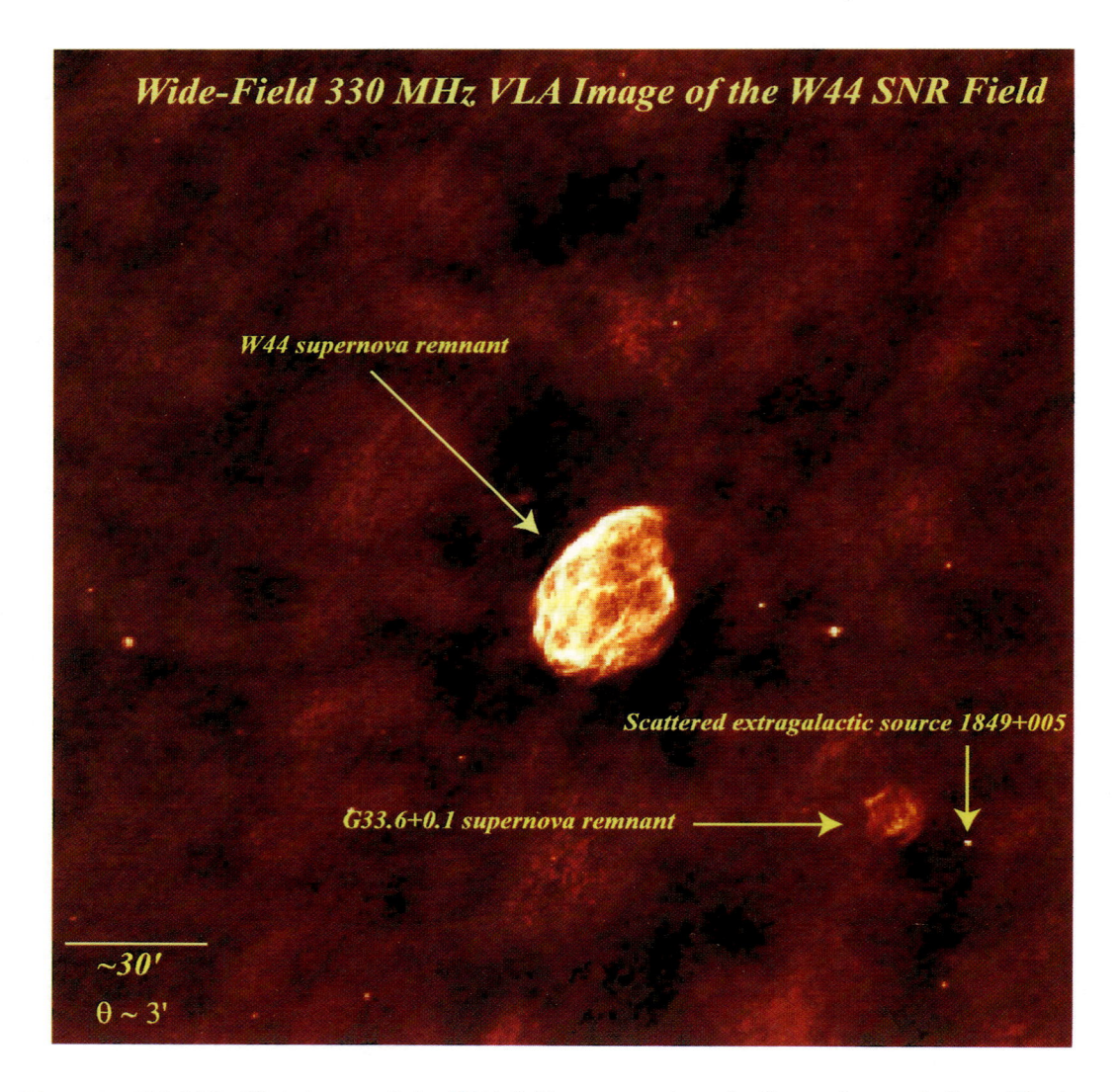

Plate 2. 330 MHz VLA image of the W44 field at approximately 3' angular resolution. The image shows the W44 SNR as well as the nearby SNR G33.6+0.1 and the heavily scattered extragalactic source 1849+005.

on the scene. As probes of external absorption, they are a powerful means of studying the properties and distribution of ionized gas, and for delineating the relative suerposition SNRs, HII regions, and the general ISM, both in our Galaxy and in nearby ones. The discovery of internal absorption is important to reverse shock physics and early blast wave evolution theory, and for constraining the ionized gas in SNRs in general. Finally, scattering measurements at low frequencies may provide the crucial evidence for the turbulence predicted and required by shock acceleration theory.

All these rich areas of study have been hindered by the limited angular resolution and sensitivity of previous low frequency observing systems. But the shattering of the ionospheric barrier by the 74 MHz VLA has inspired the ~5-150 MHz Low Frequency Array (LOFAR) which will have 2-3 orders of magnitude improved imaging power over previous low frequency instruments in its range. Taken together with the sensitive meter wavelength maps now being generated at the Very Large Array, Molonglo Synthesis Telescope, Westerbork Synthesis Telescope, Dominion Radio Astronomy Observatory Synthesis Telescope, and soon at the Giant Meter Radiowavelength Telescope, the situation is rapidly improving. Therefore we conclude optimistically that we are on the verge of a significant step forward in supernova remnant research, and one in which new low frequency radio measurements will play a key roll.

Acknowledgments. Basic research in radio astronomy at the Naval Research Laboratory is supported by the Office of Naval Research. The VLA is a facility of the National Radio Astronomy Observatory operated by Associated Universities, Inc., under a cooperative agreement with the National Science Foundation.

REFERENCES

Anantharamaiah, K. R., On the origin of the Galactic ridge recombination lines, *Indian Journal of Astrophys. & Astron. 7*, 131, 1986

Anderson, M. C. and Rudnick, L., Spatial spectral index variations in Galactic shell supernova remnants G39.2-0.3 and G41.1-0.3, *Astrophys. J. 408*, 514, 1993.

Atoyan, A. M., Tuffs, R. J. Aharonian, F. A. and Völk, H. J., On energy-dependent propogation effects and acceleration sites of relativistic electrons in Cassiopeia A, *Astrophys. J. 408*, 514, 1993.

Bietenholz, M. F., Kassim, N. E., Frail, D. A., Perley, R. A., Erickson, W. C., and Hajian,A. R., The Radio Spectral Index of the Crab Nebula, *Astrophys. J. 490*, 291, 1997.

Bridle, A., The spectrum of the radio background between 13 and 404 MHz, *Mon. Not. Roy. Astron. Soc. 136*, 219, 1967.

Dulk, G. A. and Slee, O. B., Spectral turnovers of galactic supernova remnants, *Astrophys. J. 199*, 61, 1975.

Duric, N., Gordon, S. M., Goss, W. M., Viallefond, F., and Lacey, C., The relativistic ISM in M33: Role of the supernova remnants, *Astrophys. J. 445*, 173, 1995.

Dwarakanath, K. S., A synthesis study of the radio sky at decametre wavelengths, PhD Thesis, Raman Research Institute, Bangalore, India, 1989.

Dwarakanath, K. S., Low-frequency observations of the Vela supernova remnant and their implications, *Indian Journal of Astrophysics and Astronomy 12*, 199, 1991.

Erickson, W. C., Kassim, N. E., and Perley, R. A., The VLA at 74 MHz and plans for a long wavelength array, in these proceedings.

Frail, D. A., Kassim, N. E., Cornwell, T. J., and Goss, W. M., Does the Crab Have a Shell?, *Astrophys. J. 454*, L129, 1995.

Frail, D. A., Kassim, N. E., and Weiler, K. W., Radio imaging of two supernova remnants containing pulsars, *Astrophys. J. 107*, 1120, 1994).

Frail, D. A., private communication (1995).

Frail, D. A., Goss, W. M., and Whiteoak, J. B. Z., The radio lifetime of supernova remnants and the distribution of pulsar velocities at birth, *Astrophys. J. 437*, 781, 1994).

Gaensler, B. M. The Nature of Bilateral Supernova Remnants, *Astrophys. J. 493*, 781, 1998.

Green, D. A., The Cygnus Loop at 408 MHz; Spectral variations, and a better all around view, *Astron. J. 100*, 1928, 1990.

Green, D. A., Limitations imposed on statistical studies of Galactic supernova remnants by observational selection effects, *Pub. Astr. Soc. Pac. 103*, 209, 1991.

Hamilton, A. J. S. and Fesen, R. A., The reionization of unshocked ejecta in SN 1006, *Astrophys. J. 327*, 178, 1988.

Israel, F. P., Mahoney, M. J., and Howarth, N., The integrated radio continuum spectrum of M33 - Evidence for free-free absorption by cool ionized gas, *Astron. & Astrophys 261*, 47, 1992.

Israel, F. P. and Mahoney, M. J., Low-frequency radio continuum evidence for cool ionized gas in normal spiral galaxies, *Astrophys. J. 352*, 301, 1990.

Jones, T. W., Rudnick, L., Jun, B., Borkowski, K., Dubner, G., Frail, D. A., Kang, H., Kassim, N. E., and McCray, R., 10^{51} Ergs: The Evolution of Shell Supernova Remnants, *Pub. Astron. Soc. Pac. 110*, 125, 1998.

Kassim, N. E., Radio spectrum studies of 32 first quadrant Galactic supernova remnants, *Astrophys. J. Suppl. 71*, 799, 1989a.

Kassim, N. E., Low-frequency observations of Galactic supernova remnants and the distribution of low-density ionized gas in the interstellar medium, *Astrophys. J. 347*, 915, 1989b.

Kassim, N. E., Perley, R. A., Erickson, W. C., and Dwarakanath, K. S., Subarcminute Resolution Imaging of Radio Sources at 74 MHz with the Very Large Array, *Astron. J. 106*, 2218, 1993.

Kassim, N. E., Perley, R. A., Dwarakanath, K. S., and Erickson, W. C., Evidence for thermal absorption inside Cassiopeia A, *Astrophys. J. 455*, L59, 1995.

Kassim, N. E., Erickson, W. C., and LaRosa, T. N., Aperture synthesis observations at 80 MHz of the Galactic center region - Possible evidence for Seyfert-like activity, *Nature 322*, 522, 1986.

Kassim, N. E., and Frail, D. A., A new supernova remnant over the Galactic Centre, *MNRAS 283*, L51, 1996.

Kassim, N. E. and Weiler, K. W., W30 revealed - Separation and analysis of thermal and nonthermal emission in a Galactic complex , *Astrophys. J. 360*, 184, 1990.

Kassim, N. E., LaRosa, T. N., Lazio, T. J. W., and Hyman, S. D., Wide field radio imaging of the Galactic center, *Proc. Conf., The Central Parsecs of the Galaxy,*, edited by H. Falcke, A. Cotera, W. Duschl, F. Melia, and M. Reicke), ASP COnference Series, *186*, 403-414, 1999.

Katz-Stone, D. M., Kassim, N. E., and Lazio, T. J. W., and O'Donnell, R., Spatial Variations of the Synchrotron Spectrum with Tycho's Supernova Remnant (3C10): A Spectral Tomography Anlysis of Radio Observations at 20 and 90 cm, *Astrophys. J. 529*, 2000 (in press).

Khokhlov, A. and Melia, F., Powerful ejection of matter from tidally disrupted stars near massive black holes and a possible application to Sagittarius A East, *Astrophys. J. 457*, L61, 1996.

Lacey, C. K., The origin of Cosmic Rays: Role of Type II SNe, PhD Thesis, University of New Mexico, Albuquerque, New Mexico, 1997.

Lang, C., Anantharamaiah, K. R., Kassim, N. E., and Lazio, T. J. W., Discovery of a nonthermal Galactic Center filament (G358.85+0.47) parallel to the Galactic plane, *Astrophys. J. Letters 521*, L41-44, 1999.

Larosa, T. N. and Kassim, N. E., Aperture synthesis observations at 80 MHz of the Galactic center region - Possible evidence for Seyfert-like activity, *Astrophys. J. 299*, L13, 1985.

Larosa, T., N., Kassim, N. E., Lazio, T., J., W., Hyman, S. D., A Wide-Field 90 Centimeter Image of the Galactic Center Region, *Astron. J. 119*, 207, 2000.

Mezger, P. G., Zylka, R., Salter, C. J., Wink, J. E., Chini, R., Kreysa, E., and Tuffs, R., Continuum observations of Sgr A at mm/submm wavelengths,*Astron & Astrophys. 209*, 337, 1989.

Moffett, D. A., Reynolds, S. P., Multifrequency studies of bright radio supernova remnants. 2: 3C391, *Astrophys. J. 425*, 668, 1994a.

Moffett, D. A. and Reynolds, S. P., Multifrequency studies of bright radio supernova remnants. 2: W49B, *Astrophys. J. 437*, 705, 1994b.

Pedlar, A., Anantharamaiah, K. R., Ekers, R. D., Goss, W. M., Van Gorkom, J. H., Schwarz, U. J., and ZHAO, J-H., Radio studies of the Galactic center. I - The Sagittarius A complex, *Astrophys. J. 342* 769, 1989.

Perley, R. A. and Erickson, W. C., A proposal for a large, low frequency array located at the VLA Site, *VLA Scientific Memorandum 146*, 1984.

Pineault, S., Landecker, T. L., Swedlyk, C., and Reich, W., The supernova remnant CTA-1 - (G119.5+10.3): a study of the breakout phenomenon, *Astron. & Astrophys. 324*, 1152, 1997.

Pineault, S., Landecker, T. L., Swedlyk, C., and Reich, W., The supernova remnant CTA-1 - Putting particle acceleration theories to test, *J. R. Astron. Soc. Can. 92*, 31, 1998.

Reynolds, S. P. and Ellison, D. C., Electron acceleration in Tycho's and Kepler's supernova remnants - Spectral evidence of Fermi shock acceleration, *Astrophys. J. 399*, L75, 1992.

Sastry, C. V., Dwarakanath, K. S., and Shevgaonkar, R. K., The structure of the Cygnus Loop at 34.5 MHz, *Indian Journal of Astrophysics and Astronomy, 2*, 339, 1981.

Spangler, S. R., Mutel, R. L., Benson, J. M., Cordes, J. M., Interstellar scattering of compact radio sources near supernova remnants, *Astrophys. J. 301*, 312, 1986.

Wills, K. A., Pedlar, A. Muxlow, T. W. B., and Wilkinson, P. N., Low-frequency observations of supernova remnants in M82, *M.N.R.A.S. 291*, 517, 1997.

Yusef-Zadeh, F., Shure, M., Wardle, M., and Kassim, N., Radio Continuum Emission from the Central stars of M20 and the Detection of A New Supernova Remnant Near M20, *Astrophys. J.*, 2000 (in press).

Zhang, X., Zheng, Y., Landecker, T. and Higgs, L. A., Multifrequency radio spectral studies of the supernova remnant G 78.2+2.1, *Astron. & Astrophys. 324*, 641, 1997.

N.E. Kassim, Code 7213, Naval Research Laboratory, Washington, DC 20375-5351, USA. (email: kassim@rsd.nrl.navy.mil)

F. Yusef-Zadeh, Dearborn Observatory, Northwestern University, 2131 Sheridan Road, Evanston, IL 60208-2900, USA. (email: zadeh@nwu.edu)

The Giant Metrewave Radio Telescope

G. Swarup

National Centre for Radio Astrophysics, TIFR, Post Box No.3, Pune 411 007, India

The Giant Metrewave Radio Telescope consists of 30 parabolic dishes of 45-m diameter each. It has been set up in a relatively radio quiet zone about 80 km north of Pune in Western India. Twelve of the dishes are located in a central array of about 1 km×1 km in size and other eighteen along three Y-shaped arms of about 14 km length each. GMRT has been designed for operation in the frequency range of about 38 to 1700 MHz. Receivers in 40 MHz bands near 153, 233, 325, 611 and 1000-1430 MHz have been installed in all the 30 antennas. Some of the scientific programmes, which have been initiated, particularly with reference to the scope of the Chapman Conference are briefly described.

INTRODUCTION

There are many outstanding astrophysical problems which are best studied at metre wavelengths. However, a radio telescope operating at long wavelengths is required to have a large collecting area since the galactic background increases as $\lambda^{2.7}$ where λ is the wavelength. Further a metrewave radio telescope is required to be extensive in size as the resolving power is inversely proportional to the size of the aperture. Taking advantage of the lower level of radio frequency interference prevailing in India compared to that observed in highly industrialized countries, a Giant Metrewave Radio Telescope has been recently constructed in India for operation in the frequency range of about 38 MHz to 1700 MHz [*Swarup et al.* 1991]. In Sections 2 and 3 are given brief description of the Giant Metrewave Radio Telescope and of some of the scientific programmes which have been initiated, particularly with reference to the scope of the Chapman Conference.

Radio Astronomy at Long Wavelengths
Geophysical Monograph 119
Copyright 2000 by the American Geophysical Union

THE GIANT METREWAVE RADIO TELESCOPE (GMRT)

The Giant Metrewave Radio Telescope consists of thirty fully steerable parabolic dishes of 45-m diameter. It has been set up recently by NCRA-TIFR near Khodad (19°06′ N, 74° 03′ E, altitude 650 m), which is about 80 km North of Pune in Western India [*Swarup* 1990; *Swarup et al.* 1991]. Twelve of the thirty antennas are located in a Central Array of about 1 km × 1 km in size and the other eighteen are distributed along three 14 km long arms of a Y-shaped array shown in Figure 1. An economical design has been chosen for the 45-m diameter parabolic dishes. The reflecting surface consists of 0.55 mm diameter stainless steel wire mesh, which has a size of 10×10, 15×15 and 20×20 mm in the central, middle and outer one third area of the dish. The wire mesh has a width of about 1.2 m and is stretched along its length by attaching it to a series of parallel wire rope trusses made of 4 mm diameter stainless steel ropes, fixed to sixteen parabolic frames forming the back-up structure of the 45-m dish. This novel concept has been nicknamed as SMART (Stretched Mesh Attached to Rope Trusses). The 45-m dish, weighing only 34 tonnes, is mounted on a cradle with a weight of

LOCATIONS OF GMRT ANTENNAS (30 dishes)

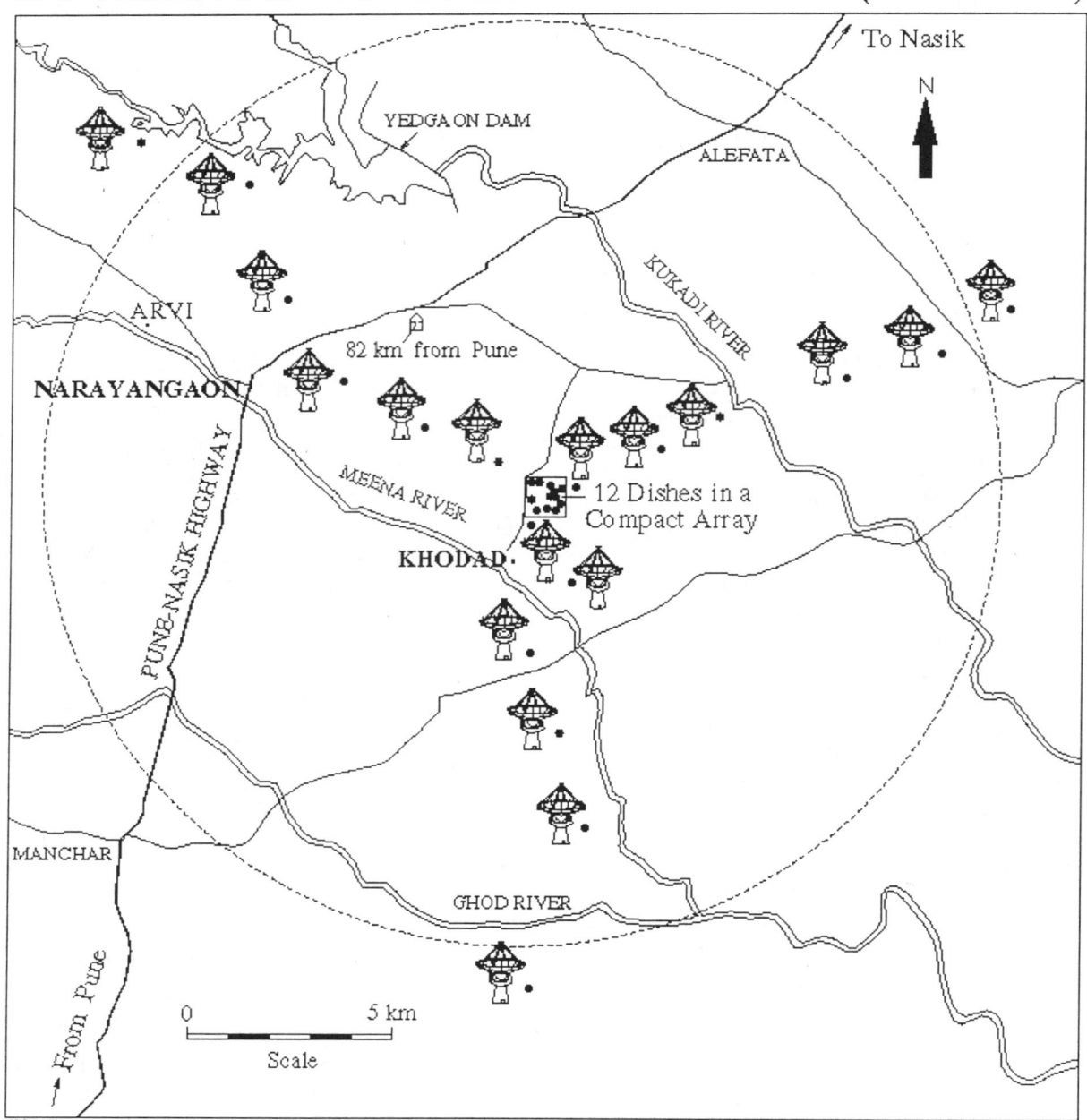

Figure 1. Y-array of GMRT

14 tonnes. The cradle also supports a counter weight of about 35 tonnes. The cradle is mounted on two elevation bearings placed on the top of a Yoke, with a weight of about 34 tonnes. The 8-m high Yoke is mounted on a slew ring bearing of 3.6 m diameter which is placed on a 12-m high concrete tower. A counter-torque servo system consisting of two 6 kVA motors controls the slewing and tracking of the antennas in elevation and azimuth, with maximum slewing rates of 20°/min and 30°/min respectively.

A rotating turret is placed near the focus of the parabolic dish for supporting antenna feeds operating in 8

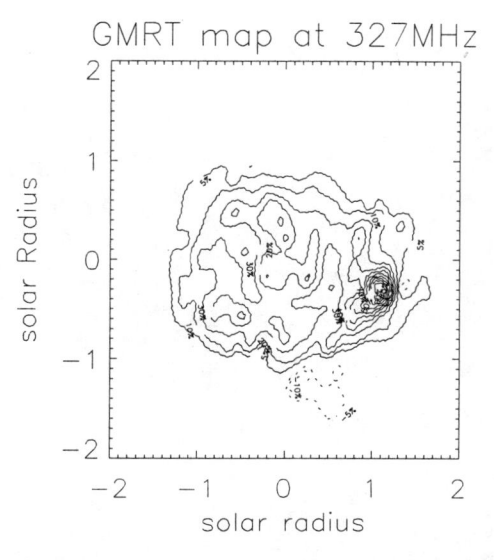

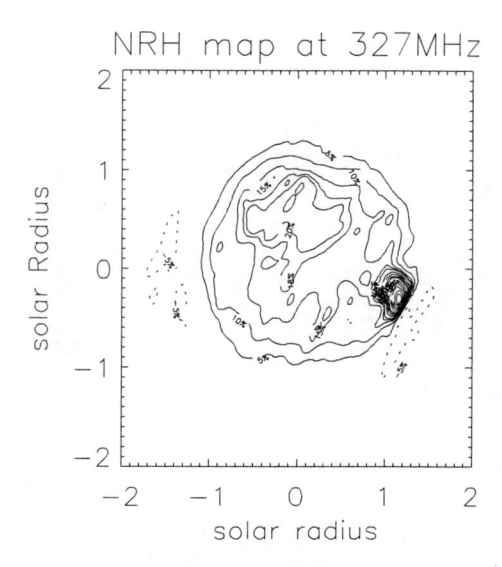

Figure 2. Cleaned map of the radio intensity during Sep. 18 at 327 MHz. From GMRT on the left, and NRH on the right. The two maps are comparable.

frequency bands from 38-1670 MHz. The antenna feeds and front-end RF electronics for operation near 150, 233, 325, 610 and 1420 MHz have already been installed in all the 30 antennas but the feeds for operation in the bands 40-60 MHz, 1610-1613.8 and 1660-1670 MHz are under development. A single mode optical fibre system has been installed for bringing the received signals from each of the 30 antennas to a central electronics building and also for distributing telemetry and local-oscillator phase-reference signals to all the 30 antennas. The RF system has a bandwidth of 40 MHz in the bands 150, 233, 325 and 610 MHz but the 21-cm feeds allow observations in the band of about 1000 to 1430 MHz. The 21-cm feed and amplifier system has been built by the Raman Research Institute. Using selectable filters, the bandwidth of the received signals can be varied in 9 steps from 64 kHz to 32 MHz. A 30-station F-X (Fourier-transform and multiplier) correlator system is now operational in which the base-band signals are applied to the samplers, delay lines and FFT cards producing 128 channels over 16 MHz for each of the antennas. The output of each of the 128 channel outputs of a given antenna is cross-multiplied and integrated with itself (self) and with outputs from all the other 29 antennas (giving correlated-signal outputs). A second side-band system is under fabrication. GMRT uses the principle of Earth's rotation synthesis for mapping celestial radio sources [*Thompson et al.* 1986]. The estimated system parameters for GMRT are summarized in Table 1.

ASTRONOMY WITH GMRT

Some of the design features of GMRT were selected considering two out-standing scientific objectives, viz. (i) search for highly redshifted 21-cm radiation from massive neutral hydrogen clouds from proto-clusters or groups of proto-galaxies and (ii) studies of pulsars including searches for new short-period pulsars. Because of its wide frequency coverage, high sensitivity and high angular resolution, GMRT is a very versatile instrument for studying a wide range of astrophysical problems at metre and decimeter wavelengths. It provides a quantum jump in the available facilities at metre wavelengths.

The Solar Radio Emission

The Sun exhibits a wide variety of transient events associated with active regions which give rise to solar radio bursts, such as Type-I, II, III and IV. Some of these phenomena give rise to interplanetary disturbances [*Ananthakrishnan et al.* 1998]. With its high time resolution and mapping capabilities, GMRT will provide valuable information about the physics of these events and also their association with a variety of phenomena seen at optical and X-ray wavelengths. Another important problem is to understand the relationship between solar flares, coronal holes and interplanetary disturbances.

In Figure 2(a) is shown a map of the Sun made at 327 MHz using only 8 of the 12 antennas of the Central

GMRT Maps of Type III Solar Radio Bursts on 21/09/1997

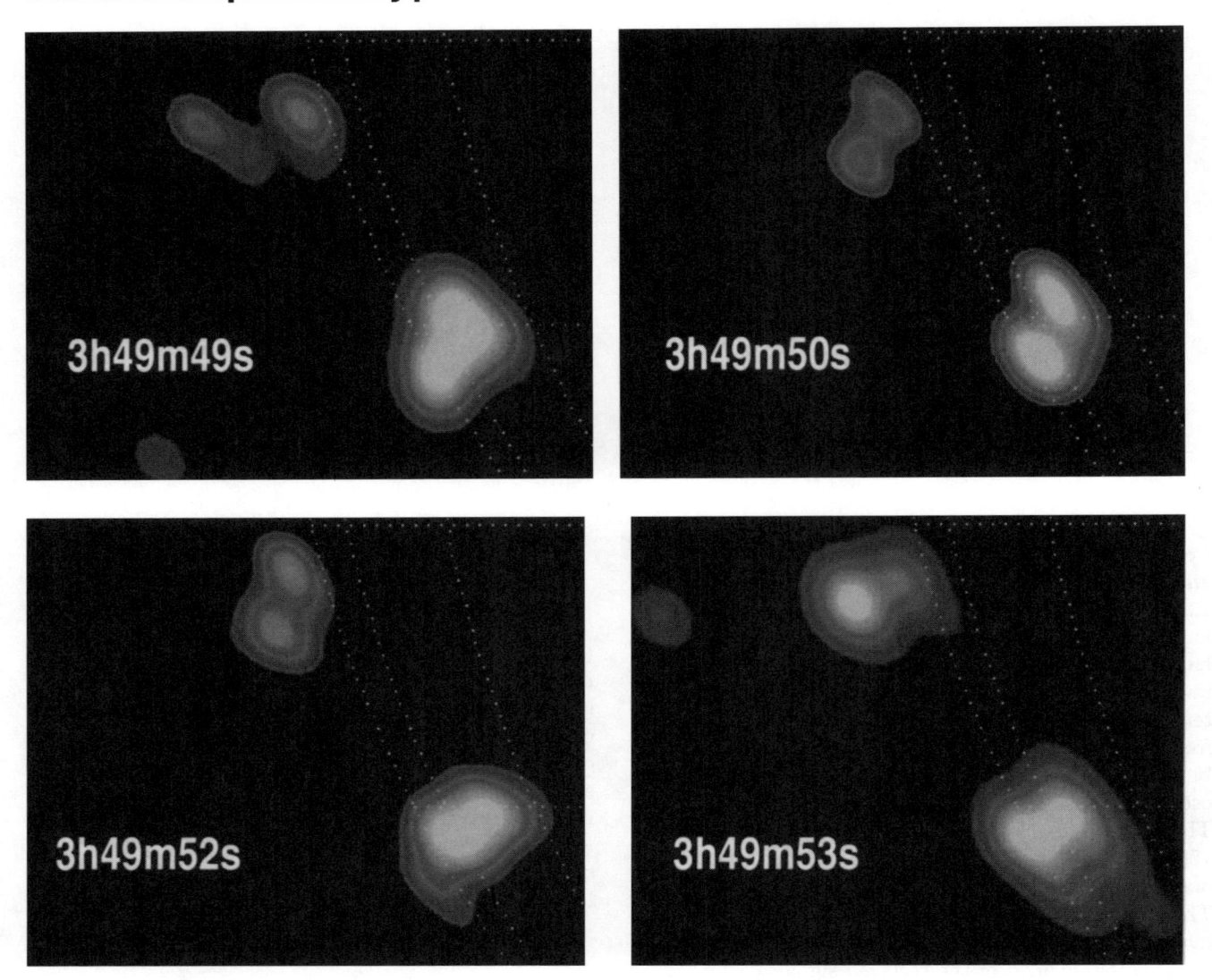

Figure 3. GMRT Maps of Type III Solar Radio Bursts on Sep 21, 1997

Array of GMRT [*Delouis*, 1998]. A comparison with the map shown in Figure 2(b) made at 327 MHz using the Nancay Radio Heliograph (NRH) shows a good correlation, even though the 8 antennas of the GMRT provided relatively poor coverage of the low spatial visibilities compared to the well filled-up coverage of the Nancay Heliograph. With a much higher resolution and sensitivity of the full array of GMRT, it should be possible to obtain valuable data particularly at the longer wavelengths for which GMRT provides a good low spatial frequency coverage. An important problem is to understand the relationship between solar flares, coronal holes and traveling interplanetary disturbances for which combined observations with GMRT, Gauribidanur [*Ramesh* 1998] and Nancay radio heliographs will be invaluable. Coordinated observations are also planned using the Ooty Radio Telescope which allow measurements of the enhanced electron density and velocity of interplanetary disturbances in the range of 0.2 to 1 A.U. [*Manoharan* 1997; *Ananthakrishnan et al.* 1998].

In Figure 3 is shown 4 out of 12 snapshot maps of two pairs of Type-III bursts which were made every

0.5 second on 21st September 1997 using 8 antennas of GMRT [*Delouis et al.* 1999, in prep.]. Each of the two pairs show occurrence of Type-III bursts at two nearby locations which are likely to be fundamental and harmonic components. A detailed analysis seems to rule out their origin due to the refraction in the solar corona [*Delouis et al.* 1999, in prep.].

In Figure 4 are shown the measured values of the interferometric visibilities of the GMRT for a Type-III burst observed using the 8 antennas of GMRT on 6th September 1997, with baselines lying upto about 15,000 wavelengths at 327 MHz. These observations indicate that there exists considerable fine spatial structure in the solar Type-III burst. From the observed visibilities, [*Delouis et al.* 1999, in prep.] it is estimated that there are present roughly 600 fibres in a region of about 120 arcsec × 25 arcsec in an active region of the Sun giving rise to the Type-III burst. The details of the fine structure support the fibrous jet model for the Type-III bursts [*Vlahos and Raoult* 1995]

Pulsars

The sensitivity of GMRT is similar to that of the Arecibo Radio Telescope but it has much wider declination coverage. Hence GMRT will be a valuable instrument for studies of the origin and evolution of pulsars, their emission mechanisms and studies of short period pulsars for detecting their gravitational radiation. The GMRT group and the Raman Research Institute have built several complex digital back-ends for using GMRT as a phased array, de-dispersion of the pulsar signals, pulsar search and polarization studies. Observational programmes have been initiated for detailed studies of some of the pulsar characteristics.

Galactic Studies

Radio maps have been made with the GMRT for HII regions, supernova remnants and centre of our Galaxy.

Table 1. Some estimated system parameters for GMRT

	Frequency (MHz)					
	38	150	233	327	610	1429
Primary beam (deg)	13	3.1	2.0	1.4	0.8	0.32
Synthesized beam Total array (arcsec)	80	20	13	9	5	2
Central compact array (arcmin)	28	7	4.5	3.2	1.7	0.7
Total System Temperature (K)	10280	580	250	110	100	70
RMS noise in image (Jy)*	1420	55	24	11	10	12

* For assumed bandwidth of 16 MHz, integration time of 10 h, and natural weighting and 2 polarizations.

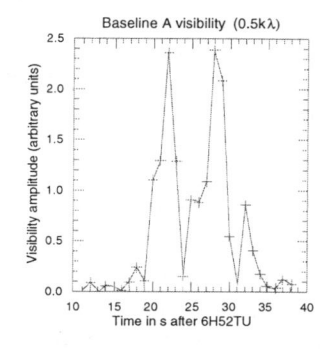

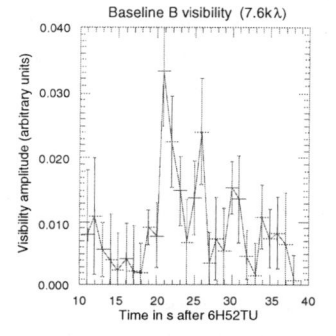

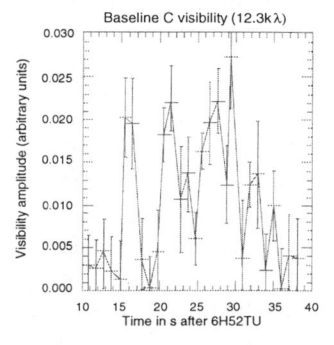

Figure 4. Visibility amplitude during Sep 6, 97 event at 327 MHz for 3 different baselines versus time. Even at 12kl the visibility amplitude varied during the event. It shows that the visibility amplitude are linked to the type III burst.

Comparison of these maps with those obtained with the VLA and elsewhere at centimeter wavelengths will determine spectral-index variation across the sources giving information about their physical characteristics. Observations of HII regions at metre wavelengths provide direct measurement of their electron temperature. A detailed mapping of the Southern Sky is likely to identify some unknown supernova remnants. Their detailed study is also valuable for pulsar-supernova association. Detailed studies of the central regions of our Galaxy at metre wavelengths and comparison with the data at other wavelengths including IR maps will provide information about the thermal and non-thermal components and about the star-formation rate near the Centre of our Galaxy. Detailed studies are also planned for studying the inter-stellar medium from studies of the recombination lines and HI absorption of distant radio sources in selected directions.

Extra-Galactic Radio Sources

Metre wavelengths are also appropriate to investigate synchrotron emission from the oldest population of relativistic electrons in radio galaxies allowing the ages of the emission regions to be derived from their spectral forms. Studies have been initiated or are planned for studying edge-on galaxies, cluster of galaxies, relic sources, giant radio galaxies and other extended extragalactic radio sources.

Cosmological Studies

Highly sensitive surveys of extragalactic radio sources in the direction of clusters and IR, optical and X-ray deep-fields should provide valuable information about the evolution of radio sources with cosmic epoch. Using GMRT, HI absorption lines have been measured for two damped Ly-α system at redshifts of 0.2211 and 0.0912 towards OI 363 [*Chengalur and Kanekar* 1999]. The spin temperature derived from the GMRT spectra for the low redshift damped Ly-α systems are about 1000 K which are similar to the damped systems at high redshifts but are much higher than those measured in nearby spiral galaxy disks (100-200 K)

An important objective of GMRT is to search for the highly redshifted 21-cm radiation from primordial hydrogen condensates which gave rise to the formation of galaxies and proto-clusters [*Swarup* 1996]. It is proposed to make extensive searches for HI condensates using GMRT at 325 MHz in a few selected regions for which multi-wavelength data at optical and IR wavelength is available. A foreground rich cluster may increase the probability of detecting HI condensates due to gravitational lensing. Searches will also be made near 150, 235 and 610 MHz.

CONCLUSION

GMRT is a powerful facility for studying universe at low frequencies. It is a complimentary facility to several large synthesis radio telescopes designed primarily for observations at centimeter wavelengths, viz. the Very Large Array, Westerbork Radio Telescope and Australia Telescope.

Acknowledgment. GMRT has been built by a team of astronomers, engineers and other staff of TIFR. The Raman Research Institute (RRI) has designed and fabricated the 1000-1430 MHz antenna feeds and front-end electronics and also part of the pulsar backend equipment.

REFERENCES

Ananthakrishnan, S., Tokumaru, M., Kojima, M., Balasubramanian, V., Janardhan, P., Manoharan, P.K. & Dryer, M., Study of Solar Wind Transients Using IPS, *AIP Conference Proceedings 471*, Solar Wind Nine, Nantucket, Massachusetts, 321, 1998.

Chengalur, J.N., and Kanekar, N., GMRT Observations of Low-z Damped Lyα Absorbers, *Mon. Not. R. Astron. Soc. (Let)*, **302**, 29, 1999.

Delouis, J.M., *Ph.D thesis*, University of Paris, 1998.

Manoharan, P.K., Solar activity dependence of interplanetary disturbances, *Astrophysics and Space Science*, **243**, 221, 1997.

Manoharan, P.K., The solar cause of interplanetary disturbances observed in the distance range 0.25-1 AU, *Geophysical Research Letter*, **24**, 2623, 1997.

Ramesh, R., Multi-frequency observations of the outer solar corona with the Gauribidanur radioheliograph, *Ph.D. Thesis*, Indian Institute of Astrophysics, 1998.

Swarup, G., Giant Metrewave Radio Telescope (GMRT) — Scientific Objectives and Design Aspects, *Ind. J. Radio & Space Physics, SK Mitra Memorial Issue 19*, 493, 1990.

Swarup, G., Ananthakrishnan, S., Kapahi, V.K., Rao, A.P., Subrahmanya, C.R., and Kulkarni, V.K., Giant Metrewave Radio Telescope, *Current Science*, **60**, 95, 1991.

Swarup, G., Searches for HI Emission From Protoclusters using the Giant Metrewave Radio Telescope, in *Cold Gas at High Redshift*, edited by M.N. Bremer, P.p. van der Werg, H.J.A. Rottgering and C.. Carilli, Kluwer Academic Publishers, Dordrecht, Netherlands, 457, 1996.

Thompson, A.R., Moran, J.M., and Swenson, G.W., Interferometry and Synthesis in *Radio Interferometry*, Wiley-interscience publication, New York, 1986.

Vlahos L., and Raoult, A., Beam fragmentation and type III bursts, *Astron. Astrophys.* **296**, 844, 1995.

G. Swarup, National Centre for Radio Astrophysics, Tata Institute of Fundamental Research, Post Bag 3, Ganeshkhind, Pune University Campus, Pune 411007, India. (e-mail: gswarup@ncra.tifr.res.in)

The VLA at 74 MHz and Plans for a Long Wavelength Array

William C. Erickson

School of Mathematics and Physics, University of Tasmania, Hobart, Tasmania, Australia.

Namir E. Kassim

Code 7213, Naval Research Laboratory, Washington, DC, USA.

Richard A. Perley

National Radio Astronomy Observatory, Socorro, NM, USA.

The new 74-MHz observing system at the VLA is described and samples of the data that it has produced are presented. Proposals are also described for new and powerful long wavelength arrays that have been generated at the NFRA in Dwingeloo and at the NRL in Washington.

1. THE VLA AT 74 MHZ

Progress is slow in the field of long wavelength radio astronomy, largely because there are few workers and instruments in the field, and little competition. The development of a 74 MHz system at the VLA illustrates this fact very well. In 1984 *Perley and Erickson* [1984] proposed that a large, low frequency array be located at the VLA site. It was proposed that banks of low cost, low frequency elements, probably crossed log-periodic dipole elements, be located near each of the A-array stations at the VLA site. The system would utilize unused bandwidth on the VLA waveguide to transport the signals to the central laboratory for processing in a special, narrow-bandwidth correlator. The system would be a stand-alone array which would not interfere with

VLA operation at higher frequencies and it could be built very cheaply because the major cost of any such instrument would normally be the signal transport system which was already in place at the site.

However, it was recognized that calibration and correction of ionospheric phase fluctuations would be a major problem. The problem seemed to be tractable, but *Perley and Erickson* felt that tests using the 25-m dishes would be judicious before embarking on a project to build the stand-alone array. In order to conduct such tests a simple, crossed-dipole, 74-MHz feed for the 25-m dishes was developed. Blockage of higher frequencies was minimized by placing the dipoles within the shadow of the dish's feed support legs. The crossed dipoles were suspended between the feed support legs and the subreflector was used as a crude backplane. Plate 1 shows pictures of the 74 MHz VLA dipoles. A simple 74-MHz preamplifier/receiver was developed, and its output was added to the 330 MHz signal for transport to the control building.

Because of budgetary constrains only a few of these

Radio Astronomy at Long Wavelengths
Geophysical Monograph 119
Copyright 2000 by the American Geophysical Union

Plate 1. Photos of the 74 MHz VLA dipoles before and after mounting near the primary focus of a VLA 25-m antenna dish.

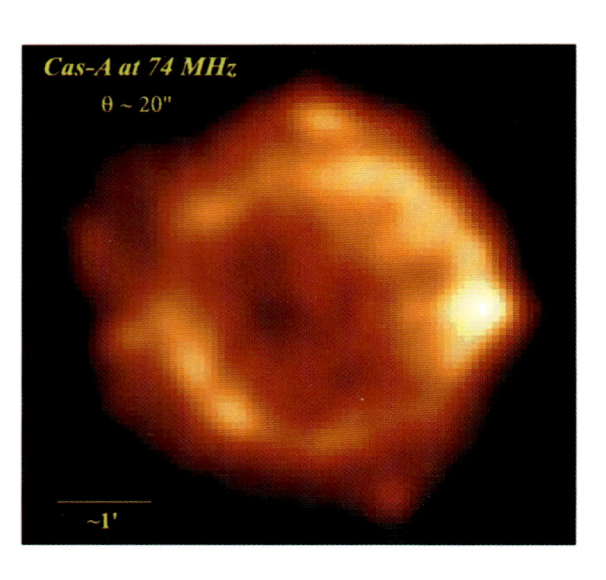

Plate 2. 74 MHz VLA image of the shell-type Galactic supernova remnant Cas A. The angular resolution is approximately 20". This is one of the few sources which was bright enough to image with the initial 8 antenna protoype system.

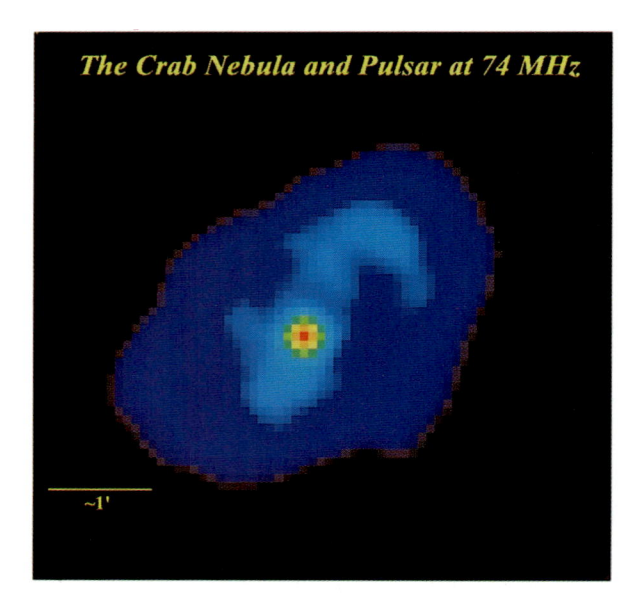

Plate 3. 74 MHz VLA image of the plerionic Galactic supernova remnant the Crab. The angular resolution is approximately 20", and the steep spectrum pulsar can be seen as a point source near the center of the source.

Plate 4. 74 MHz VLA image of the Coma Cluster field. This image was made from data obtained from the B and C configurations of the VLA, and has an angular resolution of approaximately 1' with an rms noise in the image of about 25 mJy. These data, in which the halo of the Coma cluster has been detected, is part of an ongoing study of this region by P. Kronberg and collaborators.

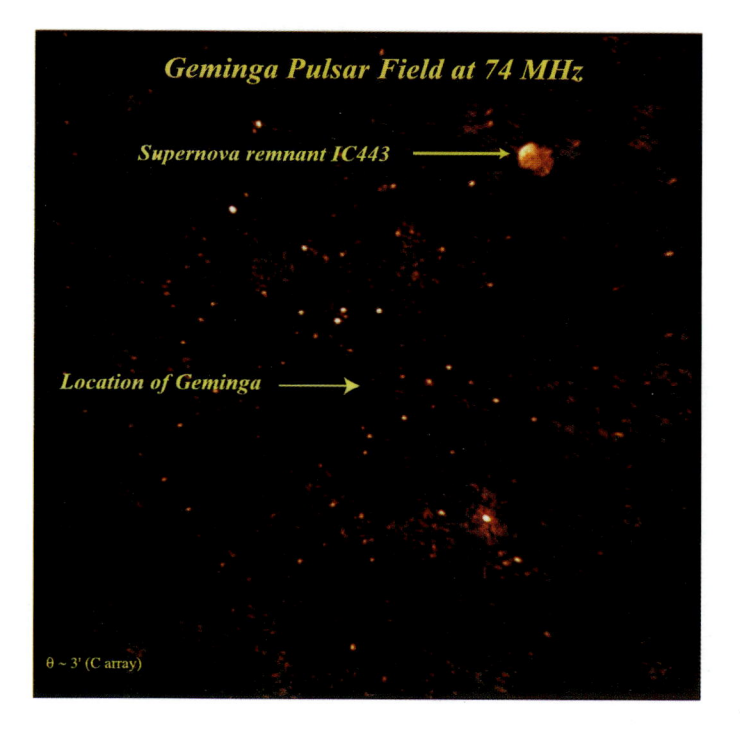

Plate 5. 74 MHz VLA image of the Geminga pulsar field made in the C configuration. A deep A array search for continuum emission at 74 and 326 MHz from this gamma-ray pulsar have failed to detect any radio emission, placing strong constraints on previous claimed yet somewhat controversial detections ((*Kassim and Lazio* 1999). This C array image has an angular resolution of approximately 3' and is sensitive to extended emission, such as the supernova remnant IC443 visible in the northwest of the image.

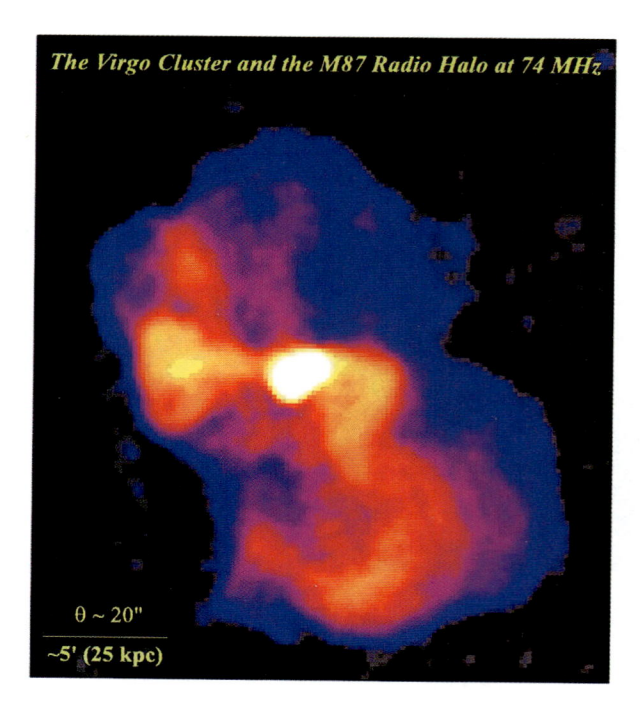

Plate 6. 74 MHz VLA image of the radio galaxy Virgo A made with approximately 20" resolution. The close correspondence of details in the steep spectrum halo across a wide variety of wavelengths suggests that this structure is relatively young and has been generated in response to recent activity by the jet-black hole system near the center of M87.

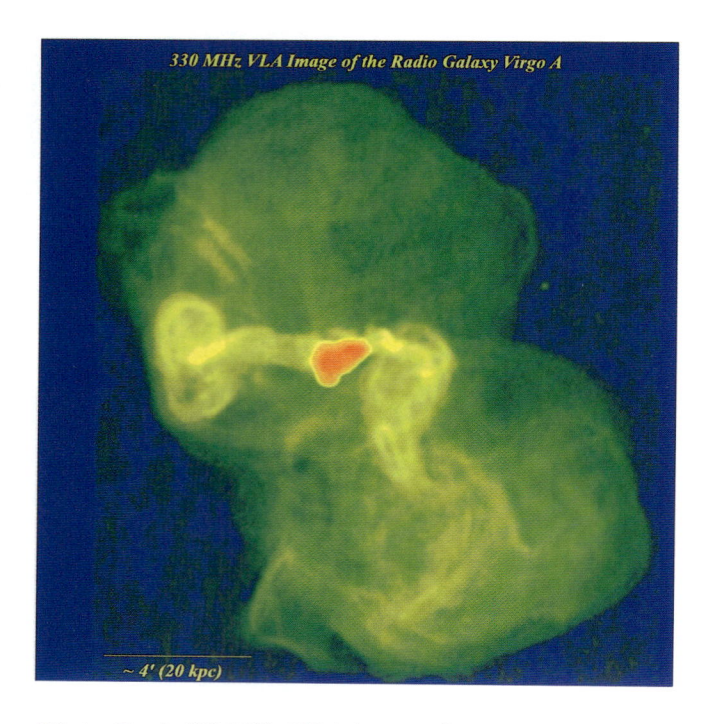

Plate 7. A 330 MHz VLA image of Virgo A at approximately the same angular resolution as the 74 MHz image in Plate 6. The close correspondence with the 74 MHz image is obvious.

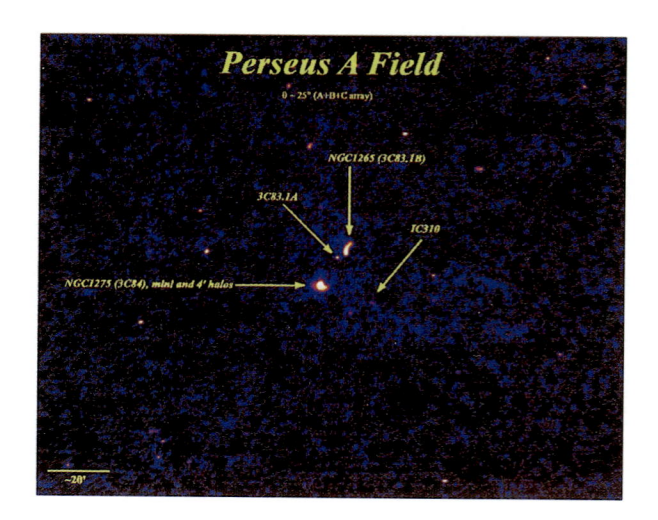

Plate 8. 74 MHz VLA image of the Perseus A field. The various radio components from higher frequency measurements are all present in the 74 MHz image.

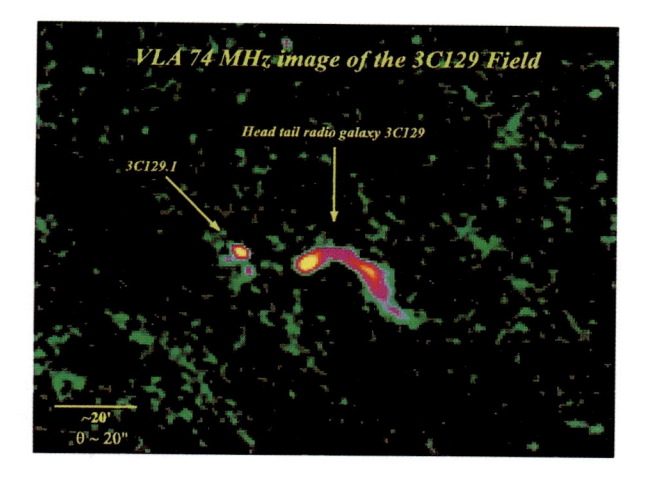

Plate 9. 74 MHz VLA image of the 3C129 field made from data in the B and C configurations. This prominent head tail galaxy has an integrated flux of only 25 Jy at 74 MHz. It had note been possible to image sources such as this with the early 8 antenna prototype.

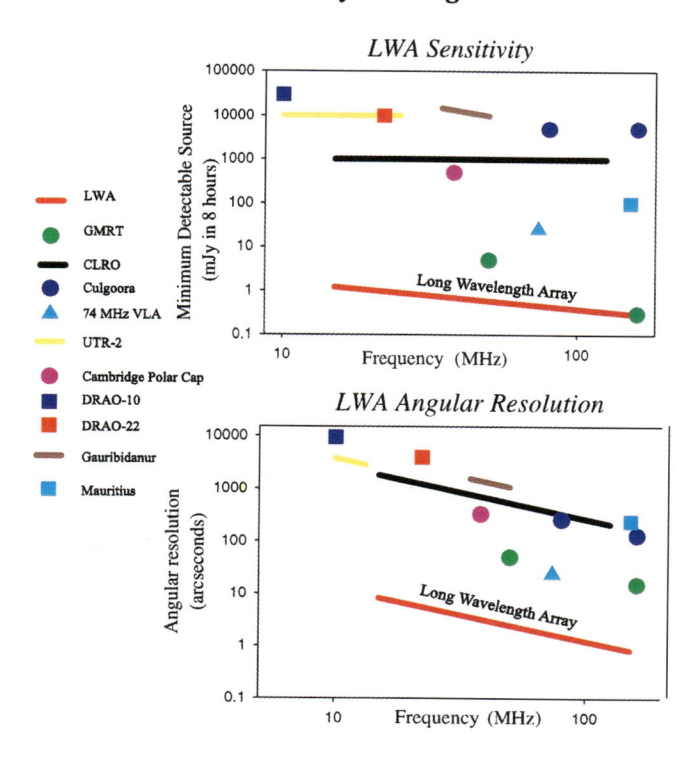

Plate 10. Sensitivity and angular resolution versus frequency of the proposed Long Wavelength Array as compared to existing and planned low frequency systems. The LWA will exceed by 2-3 orders of magnitude the performance of other instruments within its main 15-150 MHz operating range.

systems were built from time to time over the years. By 1991 eight dishes were equipped for 74 MHz; using them we found that we could successfully calibrate and map fairly strong sources with the 20 arcsec beamwidth that the system provided (*Kassim et al*, 1993). By the time that we had demonstrated that successful calibration could be accomplished at this frequency, much of the impetus and support for the original, stand-alone array had evaporated.

Having the 330 MHz frequency available simultaneously with the 74 MHz signal allowed us to develop a dual-frequency phase referencing procedure that was applicable to weaker sources. Self-calibration is performed at 330 MHz to determine the ionospheric phase shifts from the antenna-based solutions. These phases are then applied to the 74 MHz signals to largely eliminate the ionospheric phase fluctuations. However, radio source mapping at 74 MHz was tedious because, to obtain sufficient u-v coverage for useful mapping, it was necessary to observe each object over several array configuration cycles with the 74 MHz equipped dishes at placed at different locations along each arm of the array.

In 1997 a plan was initiated whereby new 74 MHz receivers would be built at the Naval Research Laboratory and provided to the VLA in order to equip all of the VLA dishes. This was carried out and the first A-array observations with the full VLA were scheduled for early 1998. These observations turned out to be highly successful; with 27 antennas calibration was robust and use of the phase referencing scheme was unnecessary. In every field there are sufficient sources for calibration. A sensitivity of ~ 20 mJy was achieved fairly easily and, with further software development to remove sidelobes from sources far from the field center, a sensitivity of 10 mJy should be achievable. Note: The primary beam of the 25 m dishes is large, $\sim 13°$, and contains many sources that cannot be included in a single self-calibration solution because they lie outside of the central isoplanatic patch ($\sim 4°$ in diameter) over which the ionospheric phase shift is relatively constant. Some examples of fields mapped with the 27-dish system are shown in Plates 2-9.

It is interesting to note that the 74 MHz system is troubled by interference, but all of the observed interference is generated by the VLA itself. The greatest effort in developing the system was involved with reducing the VLA-generated interference to manageable levels. We have yet to document a single instance of externally-generated interference. The system operates in the 73.8 MHz radio astronomy band that appears to be completely clear.

2. A NEW LONG WAVELENGTH ARRAY

Plans are being developed both at the NFRA in Dwingeloo and the NRL in Washington for new, powerful long wavelength arrays. The NFRA effort is tied closely with development of instrumentation for the Square Kilometer Array (SKA). The NFRA project, called LOFAR (Low Frequency Array), would be a test bed for many SKA concepts which can be much more easily implemented at decameter wavelengths than at the decimeter wavelengths proposed for the SKA. The proposed LOFAR system is highly sophisticated, employing thousands of short, active dipoles designed to operate from 15 to 300 MHz with complex beam-forming networks that would adaptively place nulls in the directions of interfering signals. Its goal is to make sky surveys at a number of frequencies using E-W Earth-rotation synthesis. For this reason, LOFAR would be built, primarily, on an E-W baseline.

The NRL effort has many similarities and differences with LOFAR. This project, called LWA (Long Wavelength Array), would be a sensitive array with good snap-shot capability for solar radar studies and passive solar observations. It would operate in the 15 to 150 MHz range and would employ Earth-rotation synthesis to obtain excellent u-v coverage and low sidelobe levels for sidereal observations. During daytime periods the system would be used as an imaging receiver for bi-static solar-radar/space-weather studies in conjunction with a powerful radar transmitter. The transmitter would probably be at Arecibo while the VLA site is the most likely location for the LWA. In Plate 10 the angular resolution and sensitivity of the proposed system are compared with existing instruments. A "Y" configuration is envisioned for a fairly compact array utilizing areas along the arms of the VLA. Some banks of elements would be placed at outlying stations over ~ 300 km baselines for high angular resolution studies.

There is great commonality in the LOFAR and the LWA plans. The two projects are in the process of merging.

Acknowledgments. Basic research in radio astronomy at the Naval Research Laboratory is supported by the Office of Naval Research. The VLA is a facility of the National Radio Astronomy Observatory operated by Associated Universities, Inc., under a cooperative agreement with the National Science Foundation.

REFERENCES

Kassim, N.E. and Lazio, T.J.W., Upper limits on the Continuum Emission from the Geminga at 74 and 326 MHz, *Astrophys. J. Letters*, 1999 (submitted).

Kassim, N.E., R.A. Perley, W.C. Erickson, and K.S. Dwarakanath, Subarcminute Resolution Imaging of Radio Sources at 74 MHz with the Very Large Array, *Astron. J. 106*, 2218, 1993.

Perley, R.A. and W.C. Erickson, A Proposal for a Large, Low Frequency Array Located at the VLA Site, *VLA Scientific Memorandum 146*, 1984.

W.C. Erickson, School of Mathematics and Physics, University of Tasmania, Hobart, Tasmania, Australia. (email: wce@astro.umd.edu)

N.E. Kassim, Code 7213, Naval Research Laboratory, Washington, DC 20375-5351, USA. (email: nkassim@shimmer.nrl.navy.mil)

R.A. Perley, National Radio Astronomy Observatory - Very Large Array, P.O. Box 0, Socorro, NM 87801, USA. (email: rperley@nrao.edu)

Ukraine Decameter Wave Radio Astronomy Systems and Their Perspectives

A. A. Konovalenko

Institute of Radio Astronomy, Kharkov, Ukraine

Perhaps the main way of low frequency radio astronomy progress in the next century will be space-based. But ground-based decameter radio astronomy certainly has not reached its limits. Successful operation of a number of effective instruments proves this. The world biggest decameter wave radio telescopes UTR-2 and VLBI system URAN are among them. They have yielded a considerable amount of astrophysical information and could be very useful for generation of new ideas and investigation methods. This will help effective development of a new generation of very low frequency ground-based antenna arrays and space-based projects.

1. INTRODUCTION

It is known that radio astronomy began with Karl Jansky's cosmic radio emission detection at just decameter wavelengths of 15 MHz more than 60 years ago. But further progress was based mainly on observations with shorter waves up to centimeter and millimeter ranges. The cause of this is simple and clear - when aperture size or interferometer base is fixed in order to reach maximum angular resolution (one of the principal goals of observational astronomy), you have to observe with the shortest possible wave. Until the end of 1950's decameter wave instruments had limited capabilities. They had low angular resolution, narrow operating range, and lacked two coordinate steering. In spite of some important achievements, such as the detection of Jovian sporadic emission [*Burke and Franklin*, 1955], decameter waves were not properly mastered at that time. But astrophysical investigations in this range are of great interest. At such low frequencies, a great number of astrophysical processes and phenomena are displayed very uniquely. Among those worth mentioning include: effective interaction of matter with radio emission (such as free-free absorption in plasma, dispersion, and refraction), formation of synchrotron radio emission spectra (which are most intense at low frequencies), sharp variation of spectral indices in this range, intense impulse and sporadic radio emission generation due to waves and particle beams moving through magnetically active plasma, coherent emission, fine atomic effects, and others. Some features of low frequency radio astronomy are summarized in the paper of *Weiler et al.* [1988]. For example, radio emission of decameter wavelength near 15m differs from centimeter range emission by three orders in frequency. Such a huge frequency distance always leads to a considerable distinction in corresponding radio images, which is confirmed by observations. So we can confidently say that observations at low frequencies yield principally unique information not duplicating the other radio astronomical branches.

2. DECAMETER WAVES INSTRUMENTS UTR-2 AND URAN

At the beginning of 1960's, professor S.Ya. Braude pointed out good perspectives of astrophysical studies of cosmic radio emission with the longest radio waves, which could reach the Earth's surface.

Figure 1. North-South arm of the UTR-2 radio telescope.

Decametric ranges are extremely difficult for radio astronomical observation. Natural and artificial sources of interference of various kind (quasi-monochromatic, wide band, impulse, and others) are very numerous and often much more intensive than space radio emission. Moreover, the influence of the medium (ionosphere, interplanetary and interstellar space) is very strong. This leads to absorption, refraction and scattering of radio waves. Also in decameter range the brightness temperature of non-thermal galactic radio emission (which determines the noise temperature of receiving systems), is very high.

The above mentioned circumstances and some other problems prompted the Radio Astronomy Institute team of the National Academy of sciences of Ukraine (NASU) to search for principally new ways of effective decameter wave instrument design. The project leaders were Prof. S. Ya. Braude and Prof. A. V. Megn, and the principal investigators were Prof. L. G. Sodin and Dr. Yu. M. Bruck. It became clear that radio telescopes must have a size not less than several kilometers and effective area more than 100,000 m². The operating range should be from 10 to 30 MHz with simultaneous receiving in several spatially scattered beams; the field of view had to be at least more than the refraction angle; quick directing of the beam to every direction in the main hemisphere should be provided also. Telescopes must have wide dynamic range, high reliability, and stable parameters. The chosen solution was a big wide band T-shaped array with electrical beam steering (correlation telescope) which provides the optimum ratio of sensitivity and angular resolution. Thus, at the beginning of the 1970's the Ukrainian T-shaped Radio telescope, second modification (UTR-2) was built near Kharkov, [*Braude et al.*, 1978a; *Braude et al.*, 1978b]. It has the following parameters:

- Operating range: 8 – 40 MHz.
- Beam width at 25 MHz for zenith direction: 25'.
- Maximum effective area: 150,000 m².
- Number of array elements: 2040.
- Sector of beam steering: ±70° from zenith for both coordinates.
- The time of electrical beam directing to every point of the sector: less than 0.1 s.
- Step of beam moving: about 4'.

- The whole number of beam positions in the sector: about $2*10^6$.
- The number of simultaneously working beams: 5–8.
- The side lobe level: regulated in range from −13 dB to −30 dB.
- System temperature corresponding to brightness temperature of galactic background: about 30,000 K at 25 MHz.
- Sensitivity (3-σ level, band width of 10 kHz, integration time 60 s): about 10 Jy at 25 MHz.

It is possible to work simultaneously at every frequency of the operating range and independently using the North-South (1800m×60m) and West-East (900m×60m) antennas.

The outlook of North-South UTR-2 antenna is shown on Fig. 1.

Radio telescope structure is flexible and provides easy changing between configuration and working modes. With outputs of eight sections of the North-South antenna, four sections of the West-East antenna, and corresponding equipment, it is possible to get different beam patterns and different UV plane filling (Fig. 2).

The radio telescope is equipped with various analog and digital devices for receiving and registration. They are used correspondingly to performed scientific programs and to kind of received radio emission. The telescope equipment includes multi-channel receivers of continuous emission, correlation spectral analyzers, an acoustic-optical analyzer, dynamic spectrographs, polarization measurement devices, a heliograph, magnetic tape recorders and computer systems for data collecting. The telescope is controlled using personal computers.

Observation with UTR-2 for almost 30 years has displayed its good characteristics. It turns out that the influence of the above mentioned negative factors can be decreased considerably. For example, high dynamic range of front-end preamplifiers, ferrite devices, and electronic and relay commutators limit the intermodulation effects that arise due to intense interference at the very low level. Simultaneous multi-beam receiving provides a way to observe under ionosphere refraction and scattering. The large effective area of UTR-2, which is probably more than the total effective area of all existing radio astronomy antennas, provides good sensitivity even under high galactic background temperature. This sensitivity (about 10 Jy) is obtained with a very narrow analysis band (only 10 kHz) and rather little integration time (about 1 min). For a given beam size of UTR-2, the sensitivity at the pointed out level is limited by confusion effects. Coarse comparison with sensitivity that could be achieved at higher frequency radio astronomy gives the following. For a compact radio source of synchrotron radio emission with spectral index of about unity and decameter wave flux density of about 10 Jy, flux density would be 10

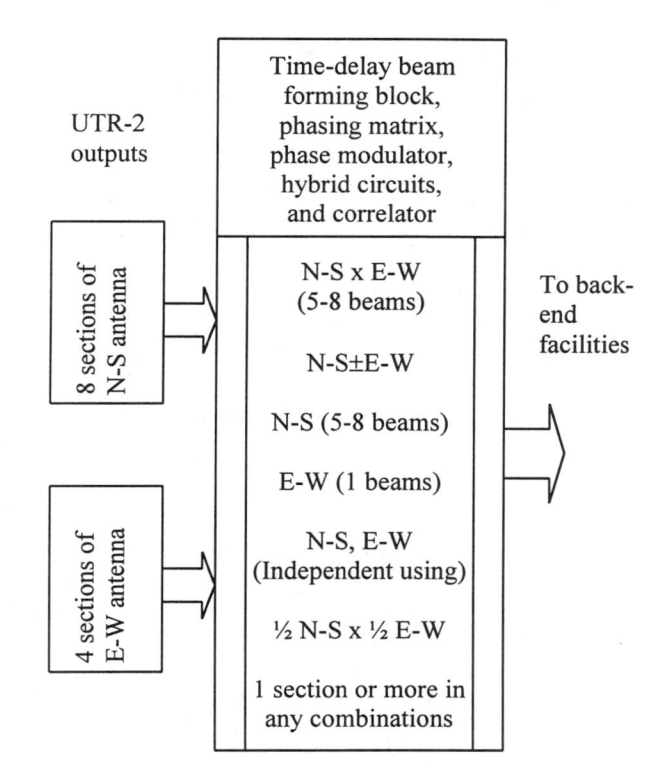

Figure 2. Possible UTR-2 configurations.

mJy in the centimeter range. This is not a bad sensitivity level for high frequency radio astronomy. Moreover, at low frequencies, radio sources could have steeper spectra. So with a sensitivity of 10 Jy at decameter waves, it is possible to detect sources not observed at high frequencies. This is confirmed by the all sky survey on UTR-2 [*Braude et al.,* 1994], in which a large percentage of the sources were detected for the first time. When investigating not only spatial and energetic characteristics (as for continuum radio source studies), but frequency and time features (as for spectral lines and pulsars studies) the limitations due to confusion effects are not so important. Thus, in some cases, it is possible to improve sensitivity very considerably by increasing integration time and/or analyses band. As UTR-2 is wide band and capable of durable source tracking, this method is frequently used. In radio spectroscopy experiments, the measured brightness temperature ratio limit was up to 0.0001 at integration times of several hundreds of hours and frequency resolution of 1 kHz. This corresponds to an absolute sensitivity of about several tens of mJy per beam.

Radio telescope UTR-2 yields flux density and brightness temperature measurement accuracy of about 20 % for every beam position. This is provided by the thorough manufacturing of all radio telescope elements, taking into account systematic and random errors of the time phasing system, accurate measurements and analytic calculation of

Figure 3. UTR-2 and URAN systems on the territory of Ukraine.

system parameters, including the influence of ground parameters.

On the base of radio telescope UTR-2, the decametric VLBI system URAN was built [*Megn et al., 1997*]. Besides the UTR-2, it includes another four radio telescopes of smaller size. URAN-1 and URAN-4 belong to Institute of Radio Astronomy. URAN-2 is owned by Gravimetric Observatory of NASU (Poltava) and URAN-3 belongs to Institute of Physics and Mechanics of NASU (Lviv). Fig. 3 shows the distribution of URAN elements on Ukraine territory. The system has bases from 40 km to 900 km. The angular resolution reaches 1 angular second (which corresponds to the fundamental limit imposed by scattering in the interstellar medium at these frequencies). Its ideology is analogous to that of high frequency VLBI networks. Comparatively narrow analysis band (up to several tens of kHz) and long wavelengths give the possibility to use rubidium frequency standards instead of hydrogen ones. Besides, quality commercial video recorders and personal computers for correlation data processing are employed. The main problem at these low frequencies is the strong influence of the medium through which radio waves propagate. In order to smooth the dependence on this factor, all URAN antennas can simultaneously receive two orthogonal polarizations. Now, the measurements with URAN are limited to determination of visibility function module for different bases and hour angles [*Megn et al., 1998*].

3. SCIENTIFIC PROGRAMS AND SOME RESULTS

During the years of operation of UTR-2 and the URAN system, a great amount of new astrophysical information was obtained [*Braude, 1992; Konovalenko, 1996*]. Most of the Universe's objects, including Earth's near space, the Solar system, the Galaxy, as well as the farthest objects such as radio galaxies and quasars, became accessible for investigation. It was proved that decameter wave radio astronomy could be very informative. Precise measurements of widely varying energetic, spatial, spectral, timing and polarization characteristics of space radio emission convey information about processes of the Universe scale, as well as fine atomic effects.

Among the main observational programs of decameter radio astronomy are the following:

- Survey of extra galactic sources and composition of corresponding catalogue.
- Investigation of radio emission parameters and spectra of quasars, galaxies, and galaxy clusters.

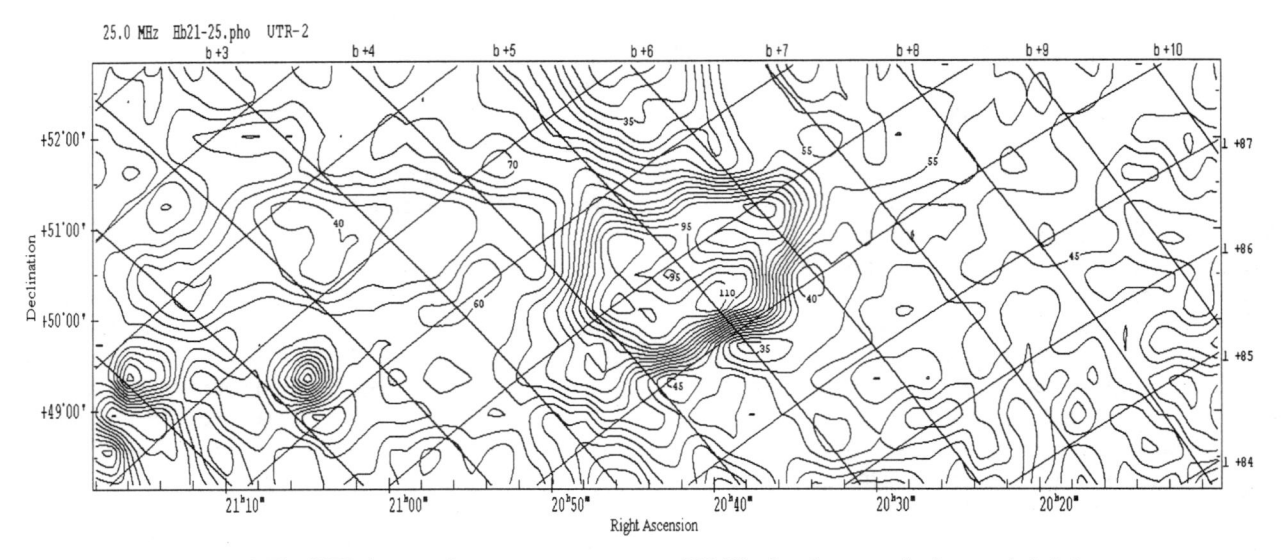

Figure 4. The UTR-2 map of supernova remnants at 25 MHz. Isophotes are in thousands kelvins.

- Investigation of distributed non-thermal radio emission of Galaxy.
- Investigation of emission nebulas and ionized gas regions.
- Investigation of supernova remnants and their interaction with interstellar medium.
- Radio spectroscopy of interstellar medium with low frequency spectral lines of different kinds.
- Investigation of impulse and continuum emission of pulsars.
- Investigation of decametric emission of flare stars.
- Investigation of sporadic and quiet emission of the Sun.
- Investigation of emission from Jupiter and other planets of solar system and search for exoplanets.
- Investigation of interplanetary medium and solar wind.
- Decametric radar study of solar system objects.
- Investigation of fine spatial structure of radio sources with VLBI, Moon occultation and scintillations methods.

To illustrate the possibilities of the instrumental base, some results for a number of objects with different kinds of radio emission (continuum, pulsed, monochromatic ones) are presented lower.

Fig. 4 shows the continuum emission map of supernova remnant HB-21, obtained with UTR-2 near 25 MHz by

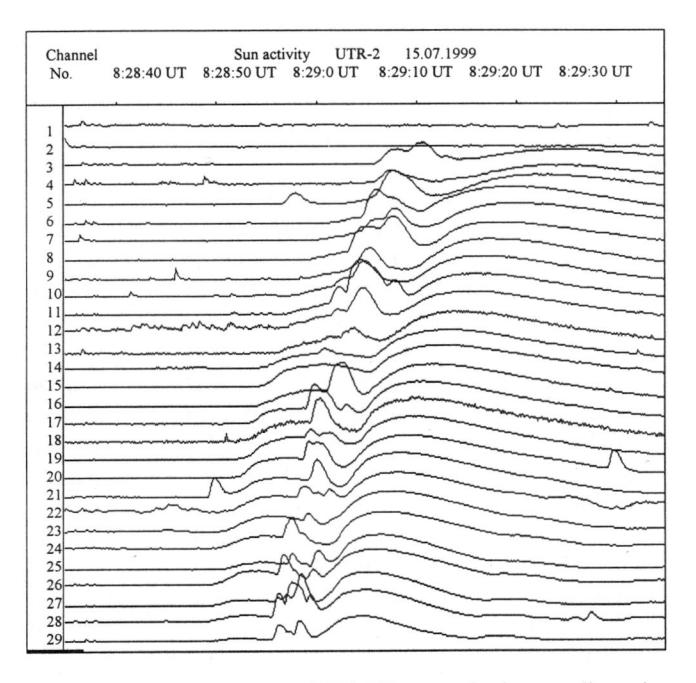

Figure 5. Dynamic spectra of III b-III type solar burst radio emission. Intensity is in relative units.

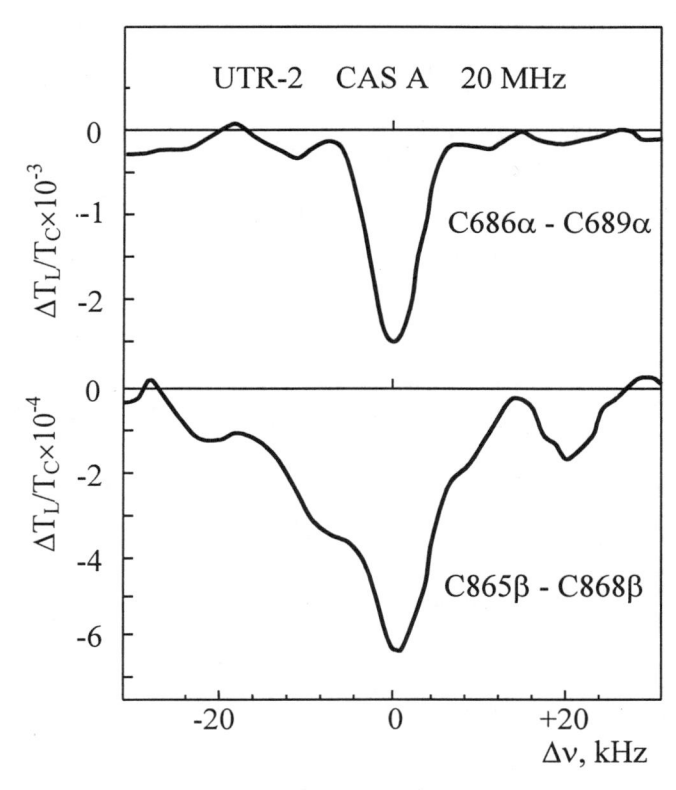

Figure 6. Carbon recombination spectral lines with most high principal quantum numbers detected on UTR-2.

Abramenkov, Krymkin, and Sidorchuk. This data permits the study of spectral index distribution, interaction with ambient interstellar medium, energetic and evolution characteristics and spatial structure of objects and comparing them with higher frequency radio images.

In Fig. 5, the results of drifting solar burst observations of III b-III type are shown, obtained by Abranin, Bazelyan, and Lisachenko. Dynamic spectrum was measured with a 29 channel analyzer. Each channel corresponds to a range of 9 - 30 MHz, with a step of about 700 kHz (the first channel is near 9 MHz), bandwidth of 10 kHz and time resolution of 100 ms. Decametric range is saturated with quickly varying phenomena of different intensity. They could be sporadic and periodic. But the large effective area of UTR-2 provides the detection of even single pulsar impulses.

With UTR-2, an exotic phenomenon of spectral lines from highly excited interstellar carbon atoms has been detected. Corresponding principal atomic numbers are more than 600. These lines turn out to be very effective means of low density cosmic plasma diagnostic, yielding information not available with other astrophysical methods [*Konovalenko*, 1990].

Fig. 6 shows an average spectrum of several such lines (C865β–C868β) detected in the direction of radio source

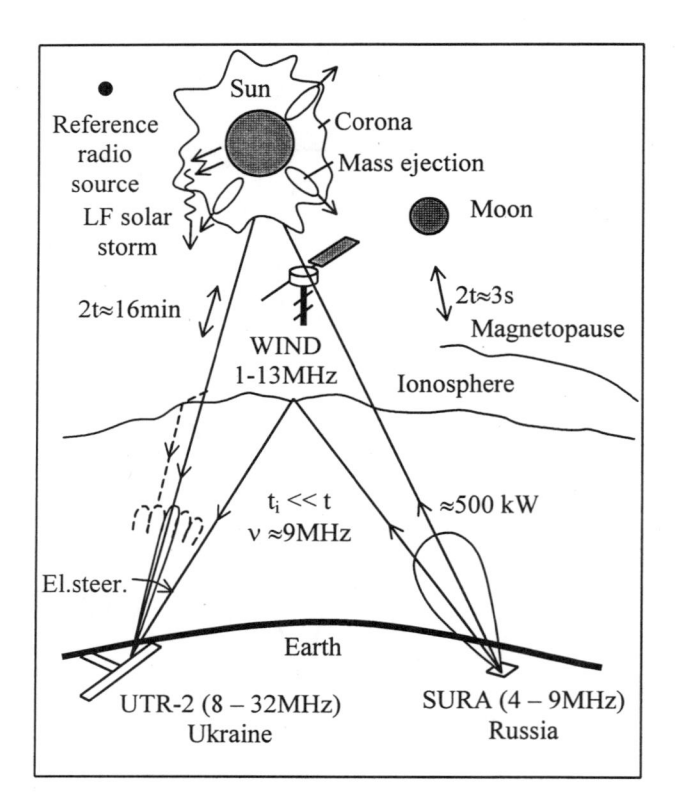

Figure 7. Diagramme of very low frequency Sun-Moon bistatic radar experiments.

Cassiopeia A, at frequencies near 20 MHz by Konovalenko, Stepkin, and Shalunov. These lines correspond to record high levels of space observed atomic excitation (up to ~860). The diameter of such highly excited atoms is huge, reaching 0.1 mm.

One of the new UTR-2 scientific programs is connected with the bistatic radar study of solar system objects such as the Moon, solar corona, coronal mass ejections (CME), and near Earth space plasma turbulence (simultaneously with passive observation at UTR-2 and the NASA spacecraft WIND). Experiments are carried out in cooperation with Russia (NIRFI), the USA (NRL, NASA), France (Meudon observatory) and the Netherlands (ESTEC). The diagram of experiments is shown in Fig. 7. For the transmission of radar signals, the powerful receiving-transmitting antenna SURA of Institute of Radio Physics, N. Novgorod, Russia, [*van't Klooster et al.*, 1995] is used, and UTR-2 is used for receiving. Experiments near 9 MHz give reliable detection of signals reflected from the Moon. The possible detection of CME is of particular interest. It is important for the study of the Sun-Earth relations [*Rodriguez*, 1996].

The detailed description of all results obtained with UTR-2 and URAN is in several hundreds of scientific papers published during last 30 years.

4. PERSPECTIVES OF UTR-2 AND URAN OPERATION

The development of radio astronomy investigations with the UTR-2 and URAN radio telescopes is connected mainly with the upgrade of their equipment sets, and the improvement of corresponding observational methods. These will provide a new qualitative level of the investigations described in chapter 3, and open ways for new problem solving. The improvements will concentrate primarily on the solution of the principal problem – reduction of interference and medium influence. A considerable modernization, jointly with special methodical investigations, was fulfilled on UTR-2 during the last several years. New sets of matching devices of antenna elements and electronic commutators were installed. A wide band high sensitive system of antenna amplification with high dynamic range was developed and put into service. Also installed were the following: a new multi-channel receiving system, new system of automatics and digital data collecting based on high speed multibit AD converters and powerful personal computers, and special registration devices (digital correlometers, VLBI equipment set, and precise time system with GPS receivers). Some of the above mentioned devices and equipment sets are described in the papers [*Abranin et al.*, 1997; *Stepkin*, 1996; *Konovalenko et al.*, 1997]. But the possibilities of UTR-2 and URAN systems have not reached their limits yet, and further modernization is being planned now.

4.1. Continuum Observation Methods

4.1.1. Broadening of receiving band. The fight with interference could be successful. It is necessary to have a front-end with high dynamic range (as with UTR-2), and to build a wide band back-end with high frequency resolution and dynamics. Of course, there must be some "clear" windows between interference in the receiving band, and frequency resolution should be less than interference widths. In this case, signals could be received and analyzed even when interference is numerous, particularly using combinations of spectral analysis and full power measurements. Moreover, it is almost always possible to choose selection criteria to distinguish useful signals from interference. For example, in Fig. 5 monochromatic signals look like horizontal lines and impulse wide band interference (lightning) are vertical lines.

Broadening of the receiving band up to 5 MHz will maintain, or even increase sensitivity of UTR-2 simultaneously with a sharp reduction of integration time (up to 0.1 s). This is useful for studying rapidly variable processes and impulse interference selection. The power pattern of a T-shaped telescope beam has sign alternating lobes, and broadening of the receiving band provides their smoothing.

4.1.2. Increasing scanning speed. One lacking feature of the UTR-2 is a comparatively small field of view. Five or eight spatial beams cover a region of 2–4°, with a step of 0.5°. When mapping a continuum radio source the telescope beam is directed motionlessly in a given position and scanning is provided by the Earth's rotation. When the scan interval is ended, the beam is switched to hour angle. One-shot mapping of a celestial sphere region near the zenith with a size of 10°x10° takes more than three hours. Such a scanning regime with a motionless beam was formed historically. It gives maximum sensitivity because antenna parameters do not change during scan time. In this case, the principal merit of an electrically scanning antenna – high speed beam directing – is not used. But observations with UTR-2 show that the deviation of antenna parameters from the known rule when beam direction is changed is less than 1%. Mapping of the above mentioned area needs about 100 beam shifts. When integration time is less than 1 s, one-shot mapping takes about 100 s. So the mapping process could be accelerated more than by two order. It is a problem of interest to install such a technique on UTR-2.

4.1.3. Aperture synthesis. At the present time, beams of UTR-2 are formed with analogue devices before the inputs of digital registration equipment. Modern means and methods of aperture synthesis provide ways to construct images with the existing UTR-2 configuration. There are outputs of twelve sections (eight for antenna North-South and four for antenna West-East). This gives 32 bases and yields the same UV-plane filling as when beams are formed via analogue. The view field, defined as one section, is of several degrees. The section has a sufficiently large size (180 elements). Such a regime of measurements could smooth the influence of the medium.

The structure of the UTR-2 phasing system has a Christmas tree configuration and provides an easy way to collect signals from smaller antenna parts and obtain a bigger view field. But this demands additional transmission lines or signal multiplexing.

Aperture synthesis for T-shaped radio telescopes was successfully used at low frequency antennas in Clark Lake (USA) [*Erickson et al.*, 1982] and Garibidanur (India) [*Dwarakanath and Udaya Shankar*, 1990].

4.1.4. Increasing of absolute measurement accuracy. The main error of brightness temperature and flux density measurement is connected to the uncertainty of ground parameters (as a rule, the average parameters are used). An effective method of precise and prompt determination of ground parameters has been developed [*Falkovich et al.*, 1994]. Their monitoring and consideration provides absolute measurement error reduction up to 10%.

4.2. Multi-channel Correlometer

The main problem of low frequency radio recombination line investigation is connected with the low intensity of features. But highly excited atoms form series of such lines, and they are much more condensed at lower frequencies. In the range of 20 – 30 MHz, there are 90 lines of α type and several hundreds of β. When they are simultaneously observed, averaged sensitivity could be improved by nearly one order.

Fast multi-channel digital correlometers are very suitable for carrying out such measurements. In cooperation with W.C. Erickson, Maryland University, a 4096 channel correlometer with an analysis band of 30 MHz was developed and manufactured. It can also be used for pulsar observations, radar experiments, interference monitoring and other investigations.

4.3. Real Time High Dynamic Range Digital Spectral Processor

Promising investigation opportunities with UTR-2 opens a new highly effective digital spectrum analyzer, manufactured in terms of cooperation with Meudon observatory, France and Space Research Institute, Austria. The device was developed by these organizations and successfully probed at the Nancay Decameter Array [*Lecacheux et al.*, 1998]. It has an analysis band of ~10 MHz, frequency resolution of ~20 kHz, time resolution up to 1 ms and dynamic range about 70 dB (it has 12-bit ADC at the input).

The device provides a significant improvement of dynamic spectrum studies (Sun, Jupiter, planets, pulsars, exoplanets and flare stars). Its use is also reasonable for continuum source investigations, both in additive mode and in beam multiplication mode, which could be effective in fighting interference (see 4.1.1).

4.4. VLBI Experiments

The problem of the reduction of ionosphere influence is complicated. Special investigations are needed for its study and solution. This is especially important when mapping with URAN system. Perhaps the experience of ionospheric influence at the VLA with 74 MHz will help [*Kassim et al.*, 1993].

In the near future, the number of URAN-2 cross dipoles will be increased up to 512. The whole number of independent wide band dipoles will be 1024. URAN-2's sensitivity will be only two times less than that of UTR-2. So the larger antenna will improve the conditions of VLBI experiments, and will be useful as an instrument for independent investigations. It would be interesting to equip this system with a transmitter and use it for radar studies.

5. FUTURE OF VERY LOW FREQUENCY RADIO ASTRONOMY

The interest in very low frequency radio astronomy (decameter wave range) has risen considerably during recent years. In spite of difficult and laborious means of carrying out of qualitative measurements at such low frequencies, the high astrophysical value of the data yielded by this range justifies the efforts.

Now, proposals to put low frequency radio astronomical instruments outside the Earth's ionosphere (in open space and on the Moon) are being actively developed. One of the first is in the paper [*Weiler et al.*, 1988]. Also impressive is the project of a telescope on the far side of the Moon [*Bély et al.*, 1997]. This way the influence of a great number of hindering factors could be dramatically reduced. But the construction of a big "instant" effective area in space, which is indispensable for so many astrophysical investigations, is questionable. Therefore, development of ground based decameter wave radio astronomy, including the upgrade of the biggest existing instruments, and the building of new telescopes with effective area up to 1,000,000 square meters, remains actual. Such an approach has gained support in the world radio astronomy community. These proposals and conceptions are under discussion and development in a number of countries [*Perley and Erickson*, 1984; *Lecacheux and Rosolen*, 1996; *Kassim and Erickson*, 1998; *van Ardenne et al.*, 1999; *Thide and Boström*, 1994].

A considerable part of the international project INTAS 97-1964 "New Frontiers In Decameter Radio Astronomy" (Ukraine, France, Austria, Russia) is devoted to the development of the new generation of giant ground-based decameter wave instruments. The maximum utilization of existing opportunities, and experience accumulated with the biggest low frequency instruments such as UTR-2, URAN and Nancay Decameter Array, is proposed. The consolidation of efforts of the entire "low frequency" community seems to be natural.

Radio astronomy at extremely low frequencies has a great number of special features both from the astrophysical and the methodological points of view. Particularly, for an array consisting of low directivity elements and fixed effective area, the required number of elements is approximately proportional to $1/\lambda^2$. Therefore it is easiest to reach a giant effective area at the lowest frequencies. Moreover, the newest achievements of modern electronics can be effectively installed, especially in the decameter wave radio astronomy instrumental base. This promises that new low frequency projects would not be very expensive. So we can suppose that radio astronomy of the next century will effectively combine space-borne and giant ground-based instruments.

6. CONCLUSION

The world biggest decameter wave radio telescope, UTR-2, works in the non-stop regime. For VLBI investigations, the telescopes of the URAN system are used. Constant improvement of observational means and methods leads to new astrophysical results.

On 1 April, 1999 the radio telescope UTR-2 and the VLBI system URAN were included in the national list of scientific objects of prime national and world importance by the bill of the Ukrainian government.

Close international collaboration in the field of decameter wave radio astronomy, and the great value of corresponding scientific investigations, ensures good perspectives of its further progress.

Acknowledgments. The author is thankful to his colleagues due to whose activities the radio telescopes UTR-2 and URAN were constructed and have operated. The author gives thanks to all the enthusiasts of low frequency radio astronomy who happily are not rare in the world and with whom the author has been glad to contact. Many thanks to Kurt Weiler for his proposition to do this review and support. Ukrainian radio astronomers are thankful to President of National Academy of sciences of Ukraine B. E. Paton for the thorough support from the moment of the creation of decameter wave instruments until the present time.

Described investigations were carried out with partial financial support of ESO, ISF, INTAS, AAS and Fund of fundamental investigations of Ukraine. Present-day investigations are supported by the grants INTAS 96-0183, INTAS 97-1964 and INTAS-CNES 97-1450.

REFERENCES

Abranin, E. P., Yu. M. Bruk, V. V. Zakharenko, and A. A. Konovalenko, Structure and parameters of new system of antenna amplification of UTR-2 radio telescope (in Ukrainian), *Radio Physics and Radio Astronomy, 2,* 95-102, 1997.

van Ardenne, A., J. D. Bregman, M. P. van Haarlem, A. Kokkeler, G. W. Kant, J. E. Noordam, A. B. Smolders, G. H. Tan, J. G. bij de Vaate, R. de Wild, and E. E. M. Woestenburg, Low frequency option for SKA, in *Pre-feasibility Report,* pp. 1-39, Netherlands Foundation for Research in Astronomy, Dwingeloo, 1999.

Bély, P.-Y., R. J. Laurence, S. Volonte, R. R. Ambrosini, A. van Ardenne, C. H. Barrow, J.-L. Bougeret, J.-M. Marcaide, G. Woan, E. Rey, F. Monjas, F. Lamela, C. M. Pascual, and C. Cagigal, Very low frequency array on the lunar far side, in *Report by the Very Low Frequency Astronomy Study Team,* pp. 1-67, European Space Agency, Paris, 1997.

Braude, S. Ya., A. V. Megn, and L. G. Sodin, Decameter waveband UTR-2 radio telescope (in Russian), *Antennas, 26,* 3-15, 1978a.

Braude, S. Ya., A. V. Megn, B. P. Ryabov, N. K. Sharykin, and I.

N. Zhouck, Decametric survey of discrete sources in the northern sky, I. The UTR-2 radio telescope, *Astrophys. and Space Sci., 54*, 3-36, 1978b.

Braude, S. Ya., K. P. Sokolov, and S. M. Zakharenko, Decametric survey of discrete sources in the northern sky, XI, *Astrophys. and Space Sci.,213*, 1-61, 1994.

Braude, S. Ya., Decametric radio astronomy, in *Astrophysics on the Threshold of 21st Century,* edited by N. S. Kardashev, *7,* pp. 81-102, Gordon and Breach Sci. Publ., 1992.

Burke, B. F., and K. L. Franklin, Observations of a variable radio source associated with the planet Jupiter, *J. Geophys. Res., 60,* 213-217, 1955.

Dwarakanath, K. S., and N. Udaya Shankar, A synthesis map of the sky at 34.5 MHz, *J. Astrophys. Astr., 11,* 323-410, 1990.

Erickson, W. C., M. J. Mahoney, and K. Erb, The Clark Lake Teepee-Tee telescope, *Astrophys. J. Suppl., 50,* 403-419, 1982.

Falkovich, I. S., N. N. Kalinichenko, L. G. Sodin, and A. A. Stanislavskij, An effective technique for measuring the dielectric constant of the ground to determine parameters of antenna above an interface, in *Journess Internationale De Nice,* pp. 370-375, Conference Proceedings, Nice, 1994.

Kassim, N. E., R. A. Perley, W. C. Erickson, and K. S. Dwarakanath, Subarcminute resolution imaging of radio sources at 74 MHz with the Very Large Array, *Astron. J., 106,* 2218-2228, 1993.

Kassim, N. E., and W. C. Erickson, Meter/decameter wavelength array for astrophysics and solar radar, in *http://rsd-www.nrl.navy.mil/7213/nkassim/spie98-3.html,* pp. 1-20, 1998.

van't Klooster, C. G. M., Yu. Belov, Yu. Tokarev, J.-L. Bougeret, B. Manning, and M. Kaiser, Experimental 9 MHz transmission from Vasil'sursk in Russia to the WIND spacecraft of NASA, in *Preparing for the Future, 5,* pp. 1-3, ESA, Publications Division, Noordwijk, the Netherlands, 1995.

Konovalenko, A. A., Review of decameter wave recombination lines: problems and methods, in *Radio Recombination Lines: 25 Years of Investigations,* edited by M. A. Gordon and R. L. Sorochenko, pp. 178-188, Kluwer Acad. Publishers, Dordrecht, 1990.

Konovalenko, A. A., Large broad-band array antennas and their use in interferometer configuration, in *Workshop on Large Antennas in Radio Astronomy,* pp. 139-143, ESTEC, Noordwijk, the Netherlands, 1996.

Konovalenko, A. A., K. P. Sokolov, and S. V. Stepkin, Determination of the optimum operating frequencies for observations with the UTR-2 radio telescope in the sky surveying mode (in Ukrainian), *Radio Physics and Radio Astronomy, 2,* 188-198, 1997.

Lecacheux, A., and C. Rosolen, Un radiotelescope geant (10^6 m^2) pour la radioastronomie sur ondes longues (> 10 m) (in French), in *Contribution a la Prospective en Astronomie,* pp. 1-4, Departement de Radioastronomie ARPEGES, Observatoire de Paris, France, 1996.

Lecacheux, A., C. Rosolen, V. Clerc, P. Kleewein, H. O. Rucker, M. Boudjada, and W. Van Driel, Digital techniques for ground based, low frequency radio astronomy, in *Recherche Instrumentale en Radioastronomie,* pp. 1-10, Observatoire Paris-Meudon, France, 1998.

Megn, A. V., S. Ya. Braude, S. L. Rashkovskij, N. K. Sharykin, V. A. Shepelev, G. A. Inyutin, A. D. Khristenko, V. G. Bulatsen, A. I. Brazhenko, V. V. Koshevoj, Yu. V. Romanchev, V. P. Tsesevich, and V. V. Galanin, Decameter radio interferometer system URAN (in Ukrainian), *Radio Physics and Radio Astronomy, 2,* 385-401, 1997.

Megn, A. V., S. Ya. Braude, S. L. Rashkovskij, N. K. Sharykin, V. A. Shepelev, G. A. Inyutin, A. D. Khristenko, A. I. Brazhenko, and V. G. Bulatsen, Experimental study of quasar 3C154 angular structure at decameter wavelength (in Russian), *Astronomical J., 75,* 818-826, 1998.

Perley, R. A., and W. C. Erickson, A proposal for a large, low frequency array located at the VLA site, in *VLA Scientific Memorandum 146,* pp. 1-65, NRAO, Socorro, 1984.

Rodriguez, P., High frequency radar detection of Coronal Mass Ejections, in *Solar Drivers of Interplanetary and Terrestrial Disturbances,* edited by K. S. Balasubramanian, S. L. Keil, and R. N. Smartt, *95,* pp. 180-188, ASP Conference Series, 1996.

Stepkin, S. V., Digital sign correlometer for radio astronomical spectroscopy (in Ukrainian), *Radio Physics and Radio Astronomy, 1,* 255-258, 1996.

Thide, B., and R. Boström (Eds.), Hi Scat international radio observatory, in *Feasibility Study Report,* pp. 1-66, Swedish Institute of Space Physics, Uppsala, Sweden, 1994.

Weiler, K. W., B. K. Dennison, K. J. Johnston, R. S. Simon, W. C. Erickson, M. L. Kaiser, H. V. Cane, M. D. Desch, and L. M. Hammarstrom, A low frequency radio array for space, *Astron. Astrophys., 195,* 372-379, 1988.

A. A. Konovalenko, Institute of Radio Astronomy, 4 Chervonopraporna Str., Kharkov, 310002, Ukraine.
(e-mail: akonov@ira.kharkov.ua)

The Nançay Decameter Array: A Useful Step Towards Giant, New Generation Radio Telescopes for Long Wavelength Radio Astronomy

Alain Lecacheux

Observatoire de Paris - CNRS UMR 8644, 92195 Meudon, France

The Nançay Decameter Array, operating in the 10-80 MHz frequency range, consists in two phased antenna arrays in opposite senses of circular polarisation - with a 4000 m^2 effective aperture each -, and a series of powerful spectrum analysers allowing for wide band, high resolution and sensitive spectroscopy of Jovian and Solar Corona radio emissions. Coupled with spacecraft observations (*Voyager, Ulysses, Wind, Galileo, SOHO*) or working alone, it has provided some key information on radiation processes in Solar System magnetised objects. The used antenna concept and the newly developed, digital receiver system are described. Both have demonstrated their ability to solve some of the main difficulties in ground-based observing at decametre wavelengths: struggling against man made radio frequency interference, perturbations from terrestrial ionosphere, low signal to noise ratio inherent to the high sky brightness. Some ideas for future giant decameter radio telescope projects (10^6 m^2 effective area at least) are proposed.

1. GUIDELINES IN DESIGNING THE NANÇAY DECAMETER ARRAY

The Nançay Decameter Array was designed and built in the mid seventies by A. Boischot and his team [*Boischot*, 1980] for studying the impulsive, broadband radio emissions originating from Jupiter and the Solar Corona at decametric wavelengths. Because of the nature of the studied radio emissions, an antenna scheme based on twin compact, filled aperture, phased antenna arrays, made of intrinsic wide band antenna elements, was chosen at the expense of angular resolution capability. The privileged capabilities rather were: i) high sensitivity, by maximising the physical area of the telescope (10^4 m^2 were built, taking into account available resources), ii) wide instantaneous bandwidth (one octave) by using log-periodic primary antenna element and an adequate phasing system, iii) long tracking time, by optimising the phasing system, iv) spectroscopy at high time and frequency resolutions (dynamic spectrum) and v) full polarisation capability, by developing specific, high performance spectrometers.

These uncommon choices are the main reasons of the scientific usefulness of the instrument, in spite of its rather moderate size, as compared to other famous decameter radio telescopes, like the Clark Lake TPT array [*Erickson and Fisher*, 1974] or the Kharkov UTR-2 radio telescope [*Braude et al.*, 1978].

Moreover the Nançay Decameter Array was proven to be quite robust to the pollution by RF interference

Figure 1. General view of the Nançay Decameter Array, taken from the south-west corner.

and could contribute to the understanding and to the accurate analysis of several propagation effects caused by the terrestrial ionosphere at decameter wavelengths.

In the next sections, the characteristics of the Nançay Decameter array are described, including the receiver system, and some original results are briefly commented. We conclude in delineating some guidelines for the next generation of ground-based, decameter radiotelescopes, that are directly inspired by the lessons from the Nançay instrument.

2. DESCRIPTION OF THE NANÇAY DECAMETER ARRAY

2.1 Antenna system

The Nançay Decameter Array (Figure 1) is made of 144 conical helices of the type used at Clark Lake Radio Observatory [*Erickson and Fisher*, 1974], filling a square aperture of about 10^4 m^2. The antenna array is divided in two half parts of 72 antennas (6×12 in East-West and North-South directions, respectively), wound in opposite senses, giving two sub-arrays with the same characteristics but sensitive to opposite circular polarisations. The measured effective area of each sub-array are equal and maximise to 4000 m^2 at about 30 MHz.

2.2 Phasing

In order to maximise both the tracking time and the instantaneous bandwidth of the arrays, a two-stage phasing scheme was used. The first stage uses the fact that each helix antenna is made of eight copper-steel wires wound on the surface of a cone and connected to the output coaxial cable by diode switches; only six wires are used at a time to form the antenna; the other two, diametrically opposite, are left disconnected. By changing the connections through the diode switches, the antenna can be electrically rotated around the cone axis, corresponding to a phase change of the antenna by steps of 45°. In each polarised array, the antennas are arranged in nine groups of eight antennas (2×4 in East-West and North-South), inside which the phases are matched by appropriate connection switches. This first process is chromatic and then reduces the instantaneous array bandwidth from 10-120 MHz to about one octave around the chosen centre frequency. The second phasing stage is achieved by using analogue delay lines between the nine antenna groups (for each polarisation), with an elementary delay step of 0.2m.

The entire array is fully steerable within the broad, 90° half power beam width pattern of the elementary antenna, centred on the antenna cone axis. The axis is tilted 20° South in the meridian plane, to enhance the array gain towards ecliptic directions. The resulting achievable tracking time - with nearly constant gain independent on the frequency over one octave - is ± 4 hours around the meridian transit time, within the −20° to +50° declination range (Figure 2).

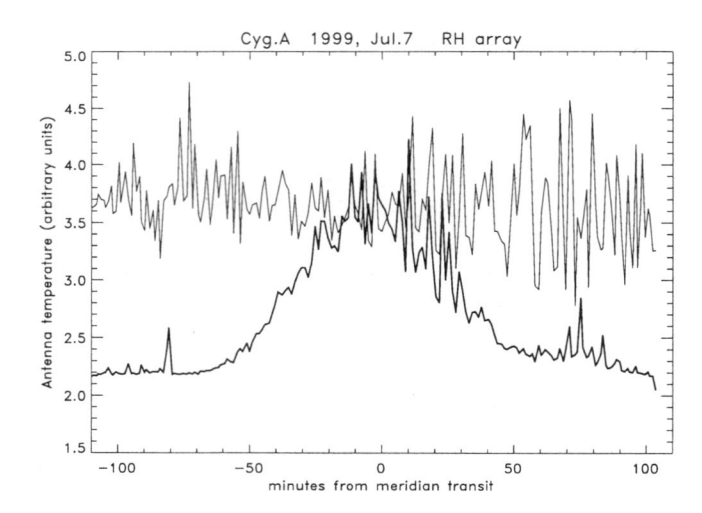

Figure 2. Observation of Cygnus A at 35 MHz. The antenna array was phased with the main beam alternately synthesised in the direction of Cygnus A (source tracking mode, thin curve), then in the meridian plane at the declination of Cygnus A (source drift mode, thick curve). The measurement demonstrates the gain stability of the instrument over the 3 hours tracking time. The broad amplitude oscillations are ionospheric scintillations.

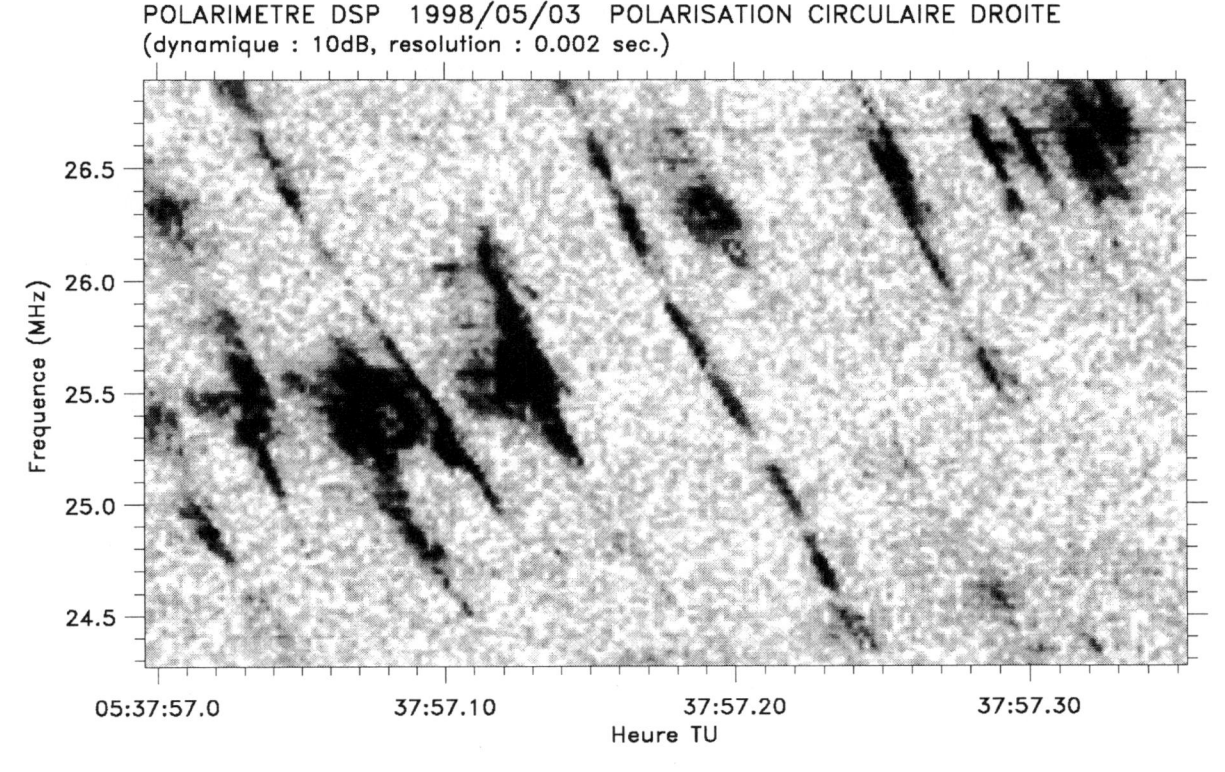

POLARIMETRE DSP 1998/05/03 POLARISATION CIRCULAIRE DROITE
(dynamique : 10dB, resolution : 0.002 sec.)

Figure 3. High resolution, real time spectral analysis of Jovian DAM radiation on 1998, May 3, by using the Nançay Decameter Array and the DSP receiver. The time resolution is 1 ms and the frequency resolution 12 kHz. About 300 ms of recorded data are displayed, showing fast drifting structures with 20 MHz/s slope in the time-frequency plane.

2.3 Signal distribution and back-ends.

The phased antenna signal is further filtered - several pass band filters can be selected in HF, depending on the ionosphere and RFI conditions -, then amplified and distributed to the back-end system.

The backend system is made of several spectrometers providing complementary capabilities of spectral analysis. In the configuration operated since mid 1997, three different spectrometers are available.

The overall decameter activity and observing context are recorded by using a wide band spectrum analyser, - the "survey" swept-frequency spectrometer -, which sweeps 400 frequency channels linearly spaced over the full observed bandwidth (10-40 MHz resp. 20-70 MHz for Jovian resp. solar observations), by alternating the sense of circular polarisation every 0.5 second.

Sensitive, high time resolution spectral analysis is provided by an acousto-optical spectrometer, having a frequency resolution of 35 kHz over a bandwidth of 24 MHz with 2048 channel output and a time resolution down to 3 ms.

Finally, the DSP digital spectro-polarimeter [*Kleewein and Lecacheux*, 1999] provides the ultimate spectral analysis capability of the Nançay instrument. Unlike existing analysers, the new machine performs spectral analysis digitally in real-time using digital signal processing (DSP) techniques. Both theoretical study and practical experience have shown that such an approach offers significant advantages over conventional analysers, namely swept frequency analysers, filterbanks, acousto-optical spectrometers or autocorrelators. None of these systems indeed can simultaneously provide wide bandwidth, high time and frequency resolutions, high dynamic range and no loss in sensitivity.

The device that has been developed can operate down to a time resolution of one millisecond with 1024 frequency channels over an instantaneous bandwidth of 12.5 MHz, including full wave polarisation measurements (Figure 3). The measured linear dynamic range for broadband noise is higher than 60 dB. Thanks to its high dynamic range, its excellent selectivity and accuracy, this receiver immediately appeared as a real

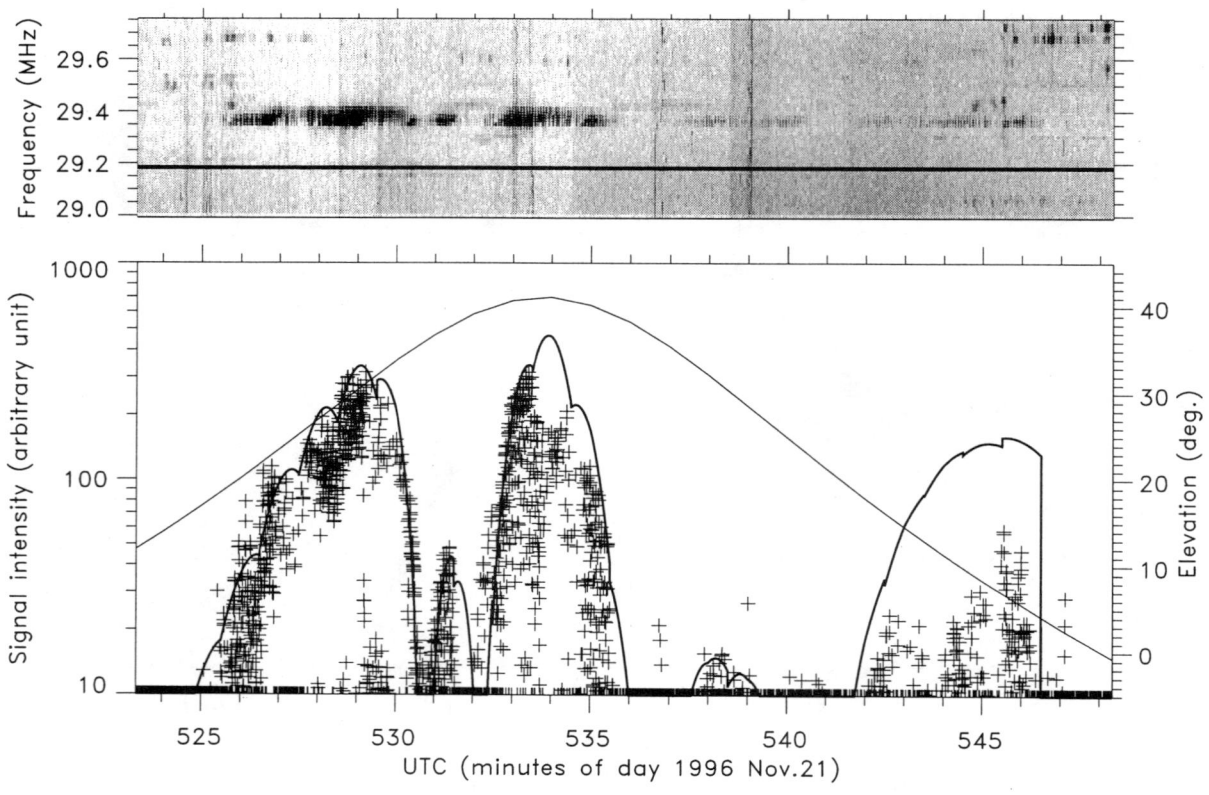

Figure 4. Check of the phasing of the Nançay Decameter Array by using Earth orbiting, radio emitting satellite as reference signal. The used satellite is the radio amateur satellite RS 15, launched in 1994 along an elliptical orbit (1800/2200 km, 2 hours revolution period). The used signal comes from the downlink transponder at 29.357 and 29.397 MHz (5W power) and the typical path duration above Nançay is 25 mn. The displayed example was a check of the backward lobes of the Nançay Decameter Array, by using a low elevation orbit (culmination at 41° elevation towards the north-north-east). The synthesised main beam was pointed along the satellite trajectory and changed every 1 mn. The signal was recorded with a bandwidth of 10 kHz and a time constant of 0.25 s. The top panel (dynamic spectrum) displays the variable intensity and bandwidth of the transponder signal. The bottom panel displays the excellent agreement between the recorded amplitude signal vs. time at 29.36 MHz (crosses) and the computed, expected response of the array (full line). Note that the signal is not corrected for the satellite spin (about 9 s. period) and that the satellite direction went out the available range of delay lines. The attenuation by ionosphere is visible after minute 542, but the signal could be tracked down to the horizon (the RS-15 elevation above horizon is given by the light line, right scale).

improvement, in particular regarding the RF interference problem.

2.4 Calibration

Accurate polarisation measurements require a good knowledge of the antenna response and continuous control of their performances.

The balance of the two polarised antenna outputs can be achieved by replacing antenna signals by a pair of stable noise diodes through a 4 × 10-dB steps programmable attenuator. Absolute amplitude and check of gain linearity are further provided by switching receiver inputs against a reference noise diode calibrator, through a 64 × 1-dB step programmable attenuator. These measurements are periodically carried out and recorded during each observing sessions (the typical period being one hour), in particular those of the survey receiver: a recent examination of all the calibration records available since 1990, has shown an excellent overall stability (within a fraction of 1 dB) of the instrument characteristics over the past ten years.

Check of the array phasing is regularly done by using celestial radio sources. Unfortunately, only a few radio sources (Cygnus A, Cassiopea A, Taurus A) are

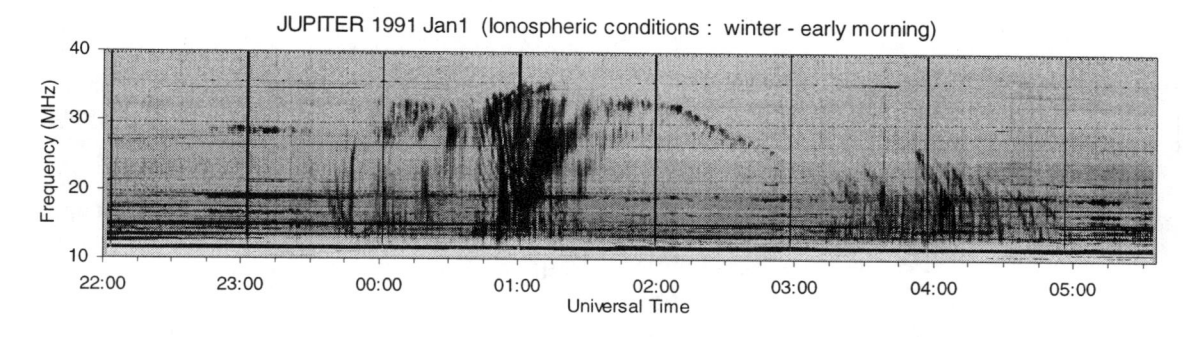

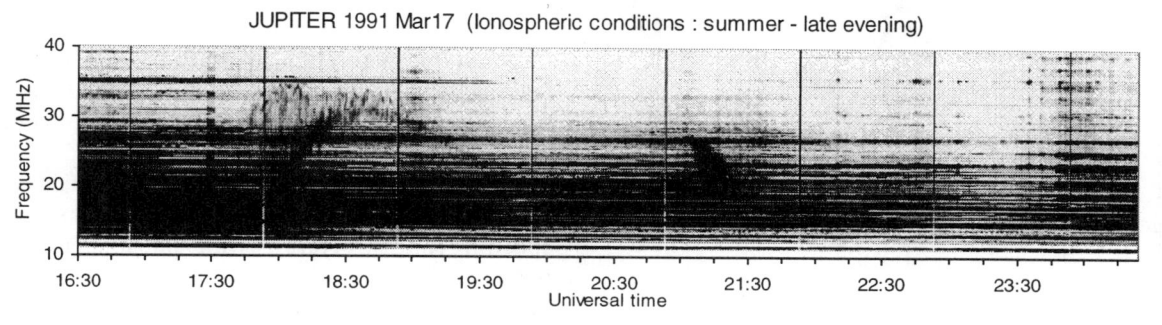

Figure 5. Examples of Nançay wide band records of Jupiter decametric emission illustrating the changing conditions of observations, occurring at decameter wavelengths. The top panel displays typical night time, winter observation in transparent ionospheric conditions : the Jupiter radiation can be followed down to the lowest analysed frequency. The bottom panel shows day time, summer observation in very bad ionospheric conditions and high RFI level : Jupiter is barely visible, especially below 25 MHz, although accurate measurements are still possible after adequate data processing. The one-hour spaced, vertical strikes are calibrations.

available with a signal level high enough to provide accurate calibration. An alternative method was developed by using radio amateur satellites (Figure 4).

2.5 Observing methods

Most of the observations are done in wide band mode, by fast recording of high resolution dynamic spectra (Figure 5). A primary reason is that the decameter radio emissions from the Sun and Jupiter most often occur as broadband, varying continuums. An additional reason is that the dynamic spectrum method is very powerful for disentangling the various RF interference kinds (AM broadcasts and CB, ionospheric sounders, radars, lightnings, etc...) and trying to eliminate them. Powerful algorithms, dealing with signal processing in the time-frequency plane, have been developed and are in use at Nançay for cleaning records from RF interference. An example is given by the successful detection of PSR 0834+06, at the sensitivity level of $10^{-4} \cdot T_{SYS}$, in spite of numerous RF interferences over the 20 to 40 MHz bandwidth (Figure 6).

So far applied in post processing, these algorithms are presently being integrated in the on board software of the DSP receiver, for more efficient usage in real-time.

2.6 Data distribution and archiving

The Nançay Decameter Array is fully automatized and can be operated remotely. Tracking information and low resolution dynamic spectrum display are available in real time (through Internet), as an help for multi site, coordinated observations.

Digital data records are produced since 1990 and can be distributed on request. A low resolution quick look data collection (taken from both Jovian and solar surveys) is freely available from the Web.

3. SOME RESULTS (1978-1999)

The Nançay Decameter Array started working in early 1978 near the time of Jupiter's flybys by *Voyager* spacecraft. The Nançay wide band observations of Jupiter (Figure 5) provided useful comparisons with the

Dynamic spectrum of PSR 0834+06

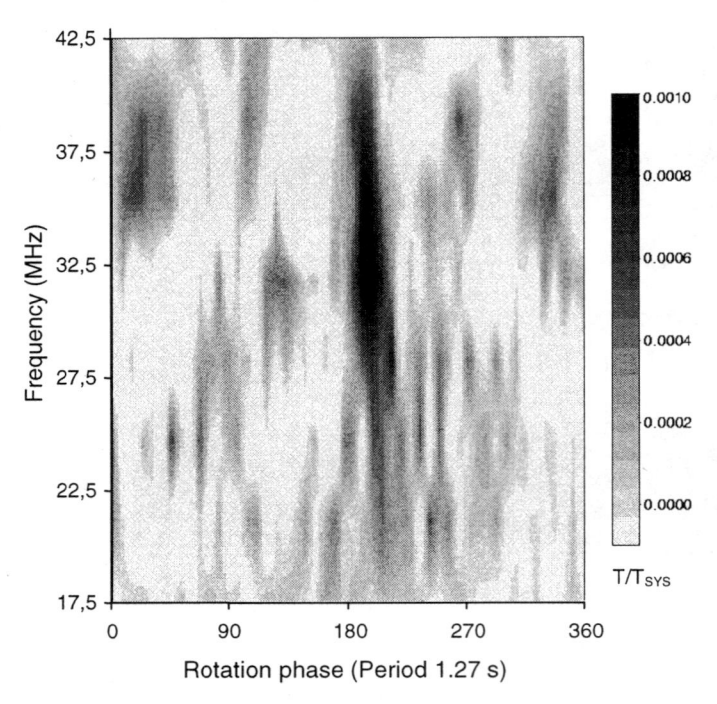

Figure 6. Detection of PSR 0834+06 with the Nançay Decameter Array. The 18-42 MHz band was analysed by the AOS spectrometer with a frequency resolution of 35 kHz and a time resolution of 20 ms. The recorded data were first cleaned from RF interference, then folded and averaged at the pulsar period (1.27 s), and finally corrected for the interstellar dispersion. The integration time is 45 s. The contrast (T/T_{SYS}) of the pulse signal is about 0.001, corresponding to a flux density of 50 Jy. Note the spectral turn over at about 28 MHz In spite of numerous RFI occurring in the wide observed bandwidth, the achieved sensitivity, after data processing, get close of the maximal capability of the instrument.

measurements made by the on board Planetary Radio Astronomy experiment [Boischot et al., 1981], in particular regarding the directivity of the Jovian decametric emission [Barrow et al., 1982] and the understanding of the recurrent features in its dynamic spectrum, linked to the Io-Jupiter electro-dynamic interaction [Leblanc, 1981].

Very high resolution spectral analyses of the Jovian decameter radio emission, - not possible so far with instruments on board interplanetary spacecraft because of telemetry constraints -, have been carried out in Nançay, in wide band mode (> 10 MHz) and with temporal resolution down to 1 millisecond (Figure 3). These observations have allowed to characterise several types of very short spectral structures, some being due to interplanetary scattering [Genova and Boischot, 1981], other originating in Jupiter's magnetosphere, like the intriguing "S-bursts phenomenon", still not really understood [Leblanc et al., 1980a, 1980b; Genova and Calvert, 1988; Zarka et al., 1996].

The capability of measuring the four Stokes parameters of the incident wave has been applied to the Jovian decameter radiation, that was found to be 100% elliptically polarised [Lecacheux et al, 1991], with a well defined polarisation ellipse which does not depend on the changing observing conditions [Dulk et al., 1994]. The electron column density of the plasma Io torus could be directly measured [Dulk et al., 1992], leading to a possible method for remotely monitoring the dynamics of this fundamental component of the Jupiter magnetosphere.

Broadband survey of Solar Corona emissions has been carried out in two main directions: studies of spectral fine structures, with phenomenological and theoretical comparison with those occurring in planetary radiations [e.g. Barrow et al, 1994]; and spectroscopic study of wide band radio emissions (like solar Type II radiation) linked with developing solar flare and matter ejection out of the Corona (CME) [e.g. Bougeret et al, 1998; Dulk et al, 1999]. Ground based decametre radio astronomy appears as a powerful tool for defining CME's start time and shock propagation velocity.

Some propagation effects due to the terrestrial ionosphere could be recognised and well understood thanks, in particular, to the broad band spectroscopic capability of the Nançay Decameter Array: an example is given by the modelling of TID focusing effects on decameter radio waves in terms of discrete refractive scattering event [Meyer-Vernet et al, 1981, Lecacheux et al, 1981]. This kind of scattering effects was later identified on centimetric wavelength pulsar radiation propagating through the interstellar medium [Fiedler et al., 1987].

4. DRIVERS FOR DESIGNING LOW FREQUENCY RADIO TELESCOPES OF NEW GENERATION

From the experience gained in Nançay, we believe that the key characteristics for defining new generation, ground based, low frequency radio telescopes depend primarily on the wanted working frequency range. Above 30 MHz or so, the techniques needed to reach most of the astrophysical objectives remain ordinary radio astronomy techniques and mainly involve imaging capability, high sensitivity and some robustness against RFI. However, below 30 MHz, - which might be the most promising frequency domain for low frequency astrophysics-, difficulties increase dramatically and the primary drivers become the telescope collecting area, the terrestrial ionosphere blockage and the huge amount of RF interference from man made and natural origins.

4.1 Effective area must be $>> 10^6$ m^2

Because of the very high brightness temperature of the sky background ($T_B \approx 10^5$ K at 30 MHz, $\approx 10^7$ K at 3 MHz), which unavoidably limits the system temperature, the only practicable way to achieve acceptable sensitivity is to build a telescope collecting area quite unusual in size: assuming a time-bandwidth product $b \cdot \tau = 10^6$ (e.g. integration time $\tau=100$ s over bandwidth b=10 kHz), the detection of a source of flux density S = 1 Jy with $T_{SYS} = T_B \approx 10^6$ K indeed requires an effective area as large as

$$A_e = 2k \cdot (T_{SYS} / \sqrt{b \cdot \tau}) / S \approx 3 \; 10^6 \; m^2.$$

Effective area this large implies a telescope configuration based on the concept of filled aperture, compact phased antenna array. The elementary antenna can be a simple dipole or a circularly polarised antenna like the Nançay helices, making the main part of the instrument relatively low cost.

The large effective area of the instrument is also an advantage against the RFI since, in principle, higher the antenna directivity is, lower is the RFI signal contribution, mainly entering through the secondary lobes, as compared to that of a radio source seen in the main beam.

Moreover, the whole instrument might be distributed in several distant sites, in order to decorrelate the local interference and the local effects of the ionosphere, and to provide some imaging capability as well.

4.2 RF Interference

With the increase of telecommunications, the RF interference had become a very difficult problem, now faced by all the radio astronomers at any wavelength. Solutions are actively searched, which involve various techniques like i) antenna beam forming (in phased array), ii) ultra linear, high dynamic range electronics (both in front end and back end), iii) real time signal processing (for interference excision), iv) multi-site observing (discrimination of local RFI). Any new generation low frequency telescope must primarily use most of these promising techniques to overcome the RFI challenge. It is worth noticing that RFI level seems no longer very much increasing at decameter wavelengths and, hopefully, might even decrease in the next future.

4.3 Ionosphere blockage / propagation effects

The maximum electron plasma frequency (f0F2) of the terrestrial ionosphere is highly variable in a given location. The main variations are due to seasonal (solar illumination) and diurnal (sunset and sunrise) effects, or due to the solar activity. Higher the magnetic latitude is (but less than auroral), lower is the average f0F2: in many sites in Northern hemisphere (Canada, North Europe) or in Southern hemisphere (Tasmania, South Africa), the f0F2 frequency is as low as 5 MHz a notable percentage of the time.

Because of the modern understanding and measuring of the terrestrial ionosphere dynamics [e.g. *Mannuci et al.,* 1998], several propagation effects due to the ionosphere, like absorption, refractive and diffractive scintillations of radio waves, should be modelable and even correctable. Some progress have been done recently in this matter [*Kassim et al,* 1993]. They could be enhanced by additional techniques, like ionosphere tomography by using GPS satellite and calibration of ionospheric effects by using dedicated transmitting μ-satellites.

5. CONCLUSION

There are great promises of low frequency radio astronomy in Planetary Sciences, Solar Physics and Astrophysics [cf. the reviews and contributions by *Kaiser et al., Zarka et al., Bougeret et al., Dulk et al.,* and *Weiler et al.,* this book]. In the latter domain, very few has been done so far at wavelengths longer than 10 meters, and yet this is a domain of choice for the search of non thermal, radio emissions from known or still unknown astrophysical objects.

In spite of its relatively small collecting area, the Nançay Decameter Array, operating in the 3-30 meter wavelength range, has proven to be quite powerful in studying non thermal radiations from Jupiter and the Solar Corona. Its wide instantaneous bandwidth and high resolution spectral analysis capabilities are of prime interest in the understanding of such plasma radiations, and brought some efficient solutions to the radio spectrum invasion by man made, radio telecommunication signals.

The ultimate radio telescope for exploring the electromagnetic spectrum below 30 MHz, will one day be deployed in space, to avoid radio wave blockage by the terrestrial ionosphere. But giant collecting area ($\gg 10^6$ m^2) is required for Astrophysics: premature use of small antennas on spacecraft might be disappointing for galactic and extra galactic astronomy.

In the meantime, building on the ground a new generation, ground based giant radio telescope makes sense. Cheap solutions do exist, based on the concept of phased antenna arrays. The new instrument must include modern digital techniques (massively parallel computing, etc...) to overcome RFI / ionospheric effects problems.

The new instrument could efficiently serve as a useful, low cost demonstrator and be built during the next decade. Then, if scientific promises are still appealing, a more ambitious telescope could eventually be deployed in space or on the far side of the Moon.

REFERENCES

Barrow, C.H., Lecacheux, A., and Leblanc Y., "Arc structures in the Jovian Decametric Emission observed from the Earth and from Voyager", *Astron.Astrophys.*, **106**, 94-97, 1982.

Barrow, C.H., Zarka, P., and Aubier, M.G., "Fine structures in solar radio emission at decametre wavelengths", *Astron.Astrophys.*, **286**, 597-606, 1994

Boischot, A., Rosolen, C., Aubier, M.G., Daigne, G., Genova, F., Leblanc, Y., Lecacheux, A., de la Noë, J. and Pedersen, B.M., "A New High Gain, Broadband, Steerable Array to Study Jovian Decametric Emission", *Icarus*, **43**, 399-407, 1980.

Boischot, A., Lecacheux, A., Kaiser, M.L., Desch, M.D., and Alexander J.K., "Radio Jupiter after Voyager: An Overview of the Planetary Radio Astronomy Observations", *J.Geophys.Res.*, **86**, 8213-8226, 1981.

Bougeret, J-L, Zarka, P., Caroubalos, C., Karlicky, M., Leblanc, Y., Maroulis, D., Hillaris, A., Moussas, X., Alissandrakis, C.E., Dumas, G., Perche, C., "A shock-associated (SA) radio event and related phenomena observed from the base of the solar corona to 1 AU", *Geophys.Res.Lett.*, **25**, 2513, 1998

Braude, S.Ya., Megn, A.V., Sodin, L.G., "The UTR-2 decametric-wave radio telescope", *Antennas* (in Russian), **26**, 3-15, 1978.

Dulk, G.A., Lecacheux, A., and leblanc, Y., "The complete polarisation state of a storm of millisecond bursts from Jupiter", *Astron.Astrophys.*, **253**, 292-306, 1992.

Dulk, G.A., Leblanc, Y., and Lecacheux, A., "The complete polarisation state of Io-related radio storms from Jupiter: a statistical study", *Astron.Astrophys.*, **286**, 683-700, 1994.

Dulk, G.A., Leblanc, Y., and Bougeret, J-L., "Type II shock and coronal mass ejection from the corona to 1 AU", *Geophys.Res.Lett.*, **26**, 2331, 1999

Erickson, W.C., and Fisher, J.R., "A new wideband, fully steerable decametric array at Clark Lake", *Radio Science*, **9**, 387-401, 1974.

Fiedler, R.L., Dennison, B., Johnston, K., Hewish, A., *Nature*, **326**, 675-678, 1987.

Genova, F., and Boischot, A., "Structure of the source of jovian decametric emission and interplanetary scintillation", *Nature*, **293**, 382-383, 1981.

Genova, F., and Calvert, W., "The source location of Jovian millisecond radio bursts with respect to Jupiter's magnetic field", *J.Geophys.Res.*, **93**, 979-986, 1988.

Kassim, N.E., Perley, R.A., Erickson, W.C., Dwarakanath, K.S., "Subarcminute resolution imaging of radio sources at 74 MHz with the Very Large Array", *Astron.J.*, **106**, 2218-2228, 1993

Kleewein, P. and Lecacheux, A., "A digital spectrum analyser for ground based decameter radio astronomy", submitted to *Astron.Astrophys.*, 1999.

Leblanc, Y., Genova, F., de la Noë, J., "The Jovian S-bursts: occurrence with L-bursts and frequency limit", *Astron.Astrophys.*, **86**, 342-348, 1980a.

Leblanc, Y., Aubier, M.G., Rosolen, C., Genova, F., de la Noë, J., "The Jovian S-bursts: frequency drift measurements at different frequencies throughout several storms", *Astron.Astrophys.*, **86**, 349-354, 1980b.

Leblanc, Y., "On the arc structure of the DAM Jupiter emission", *J.Geophys.Res.*, **86**, 8564-8568, 1981.

Lecacheux, A., Meyer-Vernet, N. and Daigne, G., "Jupiter's Decametric Radio Emission: A Nice Problem of Optics", *Astron.Astrophys.*, **94**, L9-L12, 1981.

Lecacheux, A., Boischot, A., Boudjada, M.Y., and Dulk, G.A., "Spectra and complete polarisation state of two, Io-related, radio storms from Jupiter", *Astron.Astrophys.*, **251**, 339-348, 1991.

Mannuci, A.J., Wilson, B.D., Yuan, D.N., Ho, C.H., Lindqwister, U.J., and Runge, T.F., "A global mapping technique for GPS-derived ionospheric total electron content measurements", *Radio Sci.*, **33**, 565-582, 1998.

Meyer-Vernet, N, Daigne, G., and Lecacheux, A., "Dynamic spectra of some terrestrial ionospheric effects at decametric wavelengths - Applications in other astrophysical contexts", *Astron.Astrophys.*, **96**, 296-301, 1981.

Zarka, Ph., Farges, T., Ryabov, B.P., Abada-Simon, M., and Denis, L., "A scenario for Jovian S-bursts", *Geophys.Res.Lett.*, **23**, 125-128, 1996.

Instrumentation For Space-based Low Frequency Radio Astronomy

Robert Manning

Dt. de Recherche Spatiale, CNRS UMR-8632, Observatoire de Paris, Meudon, France

This paper will provide a general overview of the techniques that have been used to measure low frequency radio waves. It will first cover basic antenna systems and problems associated with their implementation on a spacecraft. Then a discussion of receiver types and measurable parameters will follow. Resolution and precision will be addressed. The limits to sensitivity will be described and illustrated.

1. INTRODUCTION

In the history of space exploration it appears that only two major spacecraft have been built and flown expressly for radio astronomy (RAE-1 and RAE-2) [*Alexander et al*, 1975]. In all other cases it has been necessary to co-exist with other instrumentation, physically and electrically. The types of problems encountered will be addressed in the following chapters.

2. ANTENNA SYSTEMS

There are two main types of antennas that have been used to measure electric fields at low frequencies: the double-sphere probe (Figure 1a) and the thin cylindrical antenna (Figure 1b).

The double-sphere antenna uses the voltage potential collected on the spheres for the measurement. The spheres have often contained preamplifiers. The principle is that these are located away from the spacecraft and are less disturbed by the cloud of photoelectrons surrounding it. Different schemes are also used to reject photoelectrons present on the cable up to the spheres. For a description of a typical double-sphere system see [*Petersen et al*, 1984].

These antennas have been designed for very low frequency receivers and DC measurements and in fact are not well adapted for radio observations. Their frequency response is limited because of the difficulty of driving the cable capacitance and the field measurement is not accurate because of the distortion of the field caused by the cable to the spheres [*Manning*, 1998].

Thin cylindrical antennas are normally implemented with thin conducting wire or with metallic tubes. Because wire has no lateral strength it can only be used on spinning spacecraft and deployed perpendicular to the spin axis. But greater lengths are possible.

There are also antennas which have been developed to measure the magnetic component of electromagnetic waves. These consist of either wire loops with many turns and a large area or else multiple windings around ferrite cores (generally known as search coils). As with the double-sphere probes these are principally used at low frequencies. At frequencies of interest for radio observations they do not have sufficient sensitivity. Even if a coil is built with enough theoretical sensitivity to detect the magnetic field there is a high probability that it will be even more sensitive to the electric field present - remember that the ratio E/B is equal to c, the speed of light, for an electromagnetic wave.

For the rest of this paper we will only consider cylindrical antennas although most of the remarks are also valid for the double-sphere probes as well.

a) double-sphere probe

b) cylindrical antenna

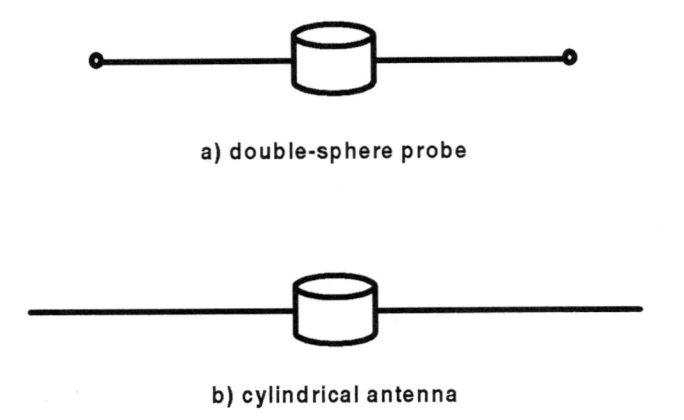

Figure 1. Principal electric field antenna types.

3. DEPLOYMENT MECHANISMS

Very short antennas can sometimes be launched in a deployed state or be released by a simple spring mechanism. But in the general case for antennas of length greater than about five meters the antenna element is coiled up within a box and is extended with the help of a motor or in some cases by its own stored energy.

Wire antennas are almost exclusively used on spinning spacecraft and deployed perpendicular to this axis, preferably in the plane of the center of gravity of the spacecraft. The wire is coiled up on a spool which is driven by a motor. Motor power is needed for the start of the deployment where there is no centrifugal force and then has the feature of regulating the erection speed. A small mass is attached to the tip of the antenna to help the initial deployment. The wire is held straight by centrifugal force along a line going through the exit point of the mechanism and the center of gravity (around the spin axis) of the spacecraft.

Wire used for these antennas is usually very thin (0.4 mm diameter wire has been used on quite a few recent spacecraft) and can be very long - the IMAGE spacecraft has four 250 meter wires. Common material for the wires is beryllium-copper (BeCu) because of its relatively high strength, good conductivity, and lack of magnetism.

Tubular antennas are used when centrifugal force is not available to ensure the deployment - along the spin-axis of spinning spacecraft and on 3-axis stabilized spacecraft. The most widely used tubular antennas are based on the original STEM (Storable Tubular Extendible Member) elements developed by Dehavilland (Canada) in the 1960's. The element is composed of one or two strips of thin metal (again usually BeCu) which have undergone a heat treatment such that when unconstrained a cylinder is formed. The strips are rolled flat on a drum before deployment. When only one strip is used the antenna has little strength. In most present-day cases a double-strip element is used, where the two strips are attached together along the sides by various methods.

A motor is most often used for the deployment. However there are some examples where the stored energy of the rolled-up strip assures the deployment. These have been single-strip units. The range of diameters of the tubes has been from 6 to 30 mm.

Another type of tubular antenna worth considering is the STACER developed by the University of California at Berkeley. It is a self-deploying (motorless) device which results in an element which is slightly conical.

The weight of the deployer units can vary considerably, particularly for the tubular units as a function of diameter. The known range of tubular units is 250g to 7 kg. For wire units the range is 500g to 3 kg (with the exception of the IMAGE deployers - $\simeq 4.7$ kg - where there was a high voltage constraint).

4. ANTENNA CONFIGURATIONS

Spinning spacecraft are much more favorable for long antennas, as explained above. Depending on the scientific goals of the instrument and the resources available there will be from one to three axes of measurement which are practically constrained to be among those shown in Figure 2.

The spin-axis antenna may consist of only a monopole. This might be imposed by the presence of a telemetry antenna or another instrument in the second direction. Monopoles function almost as well as dipoles but dipoles are preferable because 1) there will be less effect of the spacecraft on the antenna pattern and 2) a dipole configuration will reject common-mode signals that are present on both antennas.

It would be of great interest in some cases to be able to use a tilted dipole which would give extra information of source characteristics [*Manning and Fainberg*, 1980].

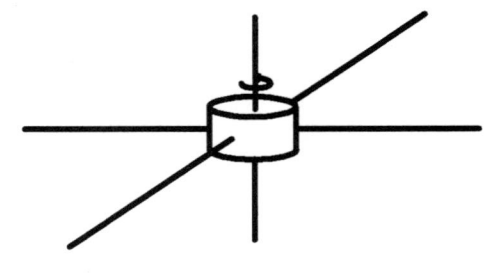

Figure 2. Spinning spacecraft antenna possibilities.

Physically this is not possible because of the centrifugal force effects. But it is possible to perform this electronically, by summing the outputs of the preamplifiers (not necessarily with equal gains). This was first performed on the Ulysses spacecraft [*Stone et al*, 1992] with considerable success.

On non-spinning spacecraft the mounting options are usually more constrained because of incompatibilities with other instruments and with spacecraft equipment. On the other hand there are no imposed directions due to spacecraft rotation. Again one to three axes are used. Sometimes it is not possible to place more than one antenna (monopole or dipole). On the Voyager spacecraft there were two monopoles set in a Vee configuration. On Cassini there are three monopoles in a tetrahedron configuration, as in Figure 3.

It is important to realize that the antenna system cannot be dissociated from the spacecraft, and *vice versa*. On the one hand, the actual physical presence of the spacecraft will have an effect on the receiving pattern of the antennas, and spacecraft movements can cause the antennas to oscillate and bend. On the other hand, the flexible nature of the antennas can cause spacecraft movements and instabilities.

With semi-rigid booms on a non-spinning spacecraft the size of the satellite is often of the same order of magnitude as the boom length. This "mass" will necessarily have an effect on the directional properties of the antenna, especially if a monopole is used. For instance, the null direction of a short monopole, theoretically physically aligned with the antenna, can easily shift 10 or 20° if it is in a very asymmetrical position. This effect can be estimated before launch by several methods, the best one appearing to be the rheographic model measurement [see *Rucker et al*, 1996].

Potentially more severe than the modification of the antenna radiation pattern are the effects the flexible nature of the antennas can cause on the spacecraft, which can be catastrophic. One case is that of a spinning spacecraft with a tubular antenna erected along its spin-axis. The antenna is along a line of astable equilibrium, and as soon as the antenna swings away from the axis centrifugal force will want to bend the antenna even further. If the antenna is stiff enough an equilibrium may be reached and this may be a stable configuration. However, excessive coning may result and the spacecraft may become unstable. This effect is well-known and characterized [*Meirovitch*, 1974] and the maximum allowable antenna length can be determined as a function of antenna characteristics, and the spacecraft's spin rate and moment of inertia.

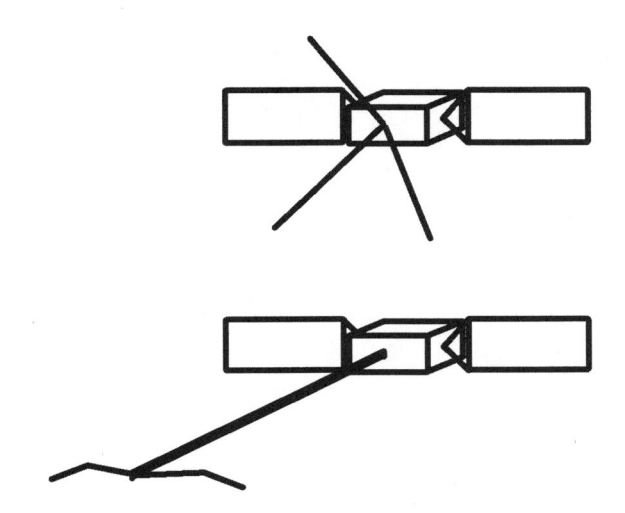

Figure 3. Some non-spinning spacecraft possibile configurations.

Problems associated with non-spinning spacecraft are numerous. A small step change in spacecraft position can cause the antennas to start to oscillate, which in turn will induce a small oscillation of the spacecraft itself (due to conservation of angular momentum). This may be large enough to disturb an optical instrument with pointing requirements of the order of arc-seconds. These oscillations can be very long to damp. The antenna elements should be either very light and flimsy or else as stiff as feasible, depending on the particular circumstances.

A major problem with tubular antennas is thermal flutter. These are oscillations, generally at the boom's cantilever resonant frequency, which are caused by the sun heating one side of the element, causing it to bend away from the sun-direction until the spring force brings it back. The solar flux supplies the energy to keep the system in oscillation. This is typically brought under control by perforating the antenna element such that some sun-light heats up the shaded side of the element.

Finally the field-of-view issue must not be neglected. The antennas are not transparent and it is sometimes very difficult to find locations on the spacecraft where the antennas are not in the field-of-view of another instrument (including indirect stray light from glint).

5. ANTENNA ELECTRICAL CHARACTERISTICS

At low frequencies, where the wavelength is long with respect to the antenna dimensions, the impedance of a dipole antenna is dominated by a series capacitance, which can be expressed (at low frequencies) as:

Table 1. Examples of base capacitance effects

				30pF < C_b < 150pF (monopole) 15pF < C_b < 75pF (dipole)		
	$C_a = \epsilon\pi h/(\ln(2h/d)-1)$					
h(m)	d(cm)	C_a(monopole)	C_a(dipole)	C_b(monopole)	$C_a/(C_a+C_b)$	
2	1	22pF	11pF			
2	0.1	15	7.6			
5	1	47	23.5			
20	1	152	76			
5.3	2.8	60	30	90pF	0.4	== Wind
50	0.04	243	122	40	0.86	== Wind
10	2.8	100	50	150	0.4	== Cassini

$Ca = \epsilon\pi h/(\ln(2h/d)-1)$ [*King*, 1956]

where h is the length of one arm (half-length) and d is the diameter of the wire or cylinder. There is a resistive term: $Ra = 80(\pi fh/c)^2$

which is very small and can be neglected in the equivalent circuit and if one considers the response of the system to an external electric field E. The open-circuit voltage collected by the antenna is simply given by $Va = E \bullet h$ (neglecting the usually small effects due to the physical presence of the spacecraft). If a high-impedance preamplifier is connected to the antenna ports (the general case) it will not be able to measure directly Va because of unavoidable stray capacitance that will be present. This capacitance will originate in the deployment mechanism, cables to the preamplifiers, and in the preamplifiers themselves. It is generally called base capacitance (Cb). The voltage measured will be

$Va' = Va * Ca/(Ca+Cb)$

The value of the base capacitance will of course vary considerably depending on the type of deployer used, the proximity of the preamplifier, and the efforts made to keep it to a minimum. Tubular antennas generally have a much larger base capacitance than wire antennas because of the remainder of the tube element still in the case.

A reasonable bracket for the base capacitance would be 30-150 pF per monopole, which in terms of the dipole gives 15-75 pF (the two monopoles are in series and the resultant is divided by two). Let us compare this to some (theoretical) values of antenna capacitance Ca (Table 1).

The last three cases correspond respectively to the tubular and long wire antennas on Wind and the tubular antennas on Cassini. It is clearly seen that it is important to know well the value of Cb and all efforts should be made to measure it very carefully before launch. This is not an easy task for the contribution from the deployment mechanism in most cases.

At frequencies where the antenna length is no longer short compared to the wavelength Ca will increase up to the half-wave resonance ($h = \lambda/4$) where it will become practically ∞ and then change sign to become inductive. Up until the half-wave resonance this will just cause a change in the capacitive division Ca/(Ca+Cb) but shortly afterwards there will be a point where Ca=-Cb and a resonance will be observed. The resistive component Ra will also increase (and limit the amplitude of the resonance) but will not have as much effect on the system response as will the reactive effects.

6. RECEIVERS

Receivers used for radio reception can be classified into four types:
sweeping
multichannel
correlator
waveform capture

6.1. Sweeping Receivers

A sweeping receiver measures the signal received in a given frequency band and then steps to measure the next band. Frequency resolution and coverage can be very good but a certain time is needed to sweep through all the frequencies and the time resolution suffers.

Often sweeping receivers are of the super-heterodyne type, where the input signal is mixed (multiplied) with a variable drive frequency. The mix creates two frequency components at the sum and difference of the desired input band frequency F_{in} and the drive frequency F_1. One of these components (usually the difference) is selected by a band pass filter centered at F_2 and with a

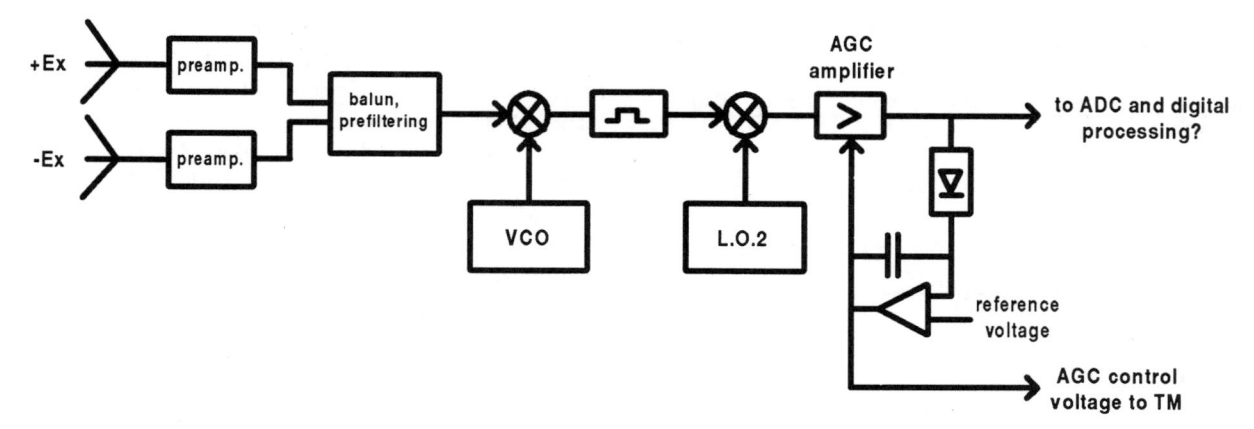

Figure 4. Block diagram for a typical super-heterodyne receiver.

width of ΔF. The output of the filter contains in principle only components of the original signal that were in the band F_1-$F_2 \pm \Delta F/2$ (or F_2-$F_1 \pm \Delta F/2$ if the sum is used). Further treatment of the signal (amplification, detection, extended frequential analysis) is now possible without dependence on the original input frequency.

A block diagram for a typical super-heterodyne receiver is shown in Figure 4. It shows the case of reception on a dipole. The output of the preamplifiers are combined in a differential circuit sometimes called a balun (balanced-unbalanced). Some filtering should be applied here. Frequencies below the desired analysis band should be filtered out and a low-pass filter should be implemented to eliminate the "image" frequencies, i.e. where $F_2 = F_{in}$-F_1, $F_{in} = F_1 + F_2$ over the range of F_1.

The variable frequency drive F_1 is shown in this diagram as a VCO (Voltage Controlled Oscillator) which is generally part of a frequency synthesizer system. In the future it is expected that VCOs will no longer be needed as a new type of synthesizer "on a chip" is being proposed (DDS = Direct Digital Synthesizer).

There are various mixers available or that can be built. They must contain non-linear elements to perform their functions, which are generally transistors or diodes. They will not be discussed here but they are one of the most critical elements of the chain because of their noise contribution and their limitations in dynamic range. On Figure 4 we have not shown intermediate amplifier stages which are generally required to minimize these problems.

The bandpass filter, often called the IF filter, is very important also. It is this component which sets the bandwidth of the receiver. It is important that its insertion loss (usually 3-6 dB) and bandwidth be stable over the temperature range and over time. The rejection of frequencies outside the passband must be very good, > 60 dB. Phase matching is important if the output of two receivers is to be compared for correlation measurements (see below). Clearly the best filters for this usage are crystal filters, or a slight variant made from lithium tantalate, which have a large heritage. Frequency range of these filters is from about 4 to 100 MHz, for bandwidths from 1 to 50 kHz. It is recommended to select when possible 10.7 or 21.4 MHz because there are many filter studies that have been made for commercial use (FM radio). For low frequency receivers there is also the possible of using ceramic filters. There, a frequency of choice is 470 kHz.

At the output of the filter one has a signal in a fixed band, proportional to, and with all the characteristics of the original signal. It will need further amplification and a means of detecting its intensity. The amplification can continue at the same frequency, or as shown in the figure, after a second frequency translation to a lower frequency easier to handle. In general it is not sufficient to have a linear amplification followed by a level detector. The dynamic range of a detection circuit is rarely adequate to cover the range needed for the observations. At least three techniques are used to increase the dynamic range. One is to place several amplifier stages in series, and then select the output of the last one that is not saturated. A limitation of this technique is the time required to change channels when the signal is fluctuating rapidly. Another option is to use an amplifier whose gain decreases with larger signals, as a form of clipping. The better of these amplifiers supply an output amplitude proportional to the log of the input amplitude. Each of these options then require a level detector that can operate over a reduced range.

The third option is the one shown in the figure and is the one employed by the DESPA in its receivers. It consists of a variable amplifier and a feedback loop af-

Table 2. Some typical flight-quality ADC converters

Part type	# of bits	Max. sampling rate (x10^6/sec)	Power (mW)
AD7821	8	1	50
AD9048	8	35	550
AD9002	8	150	750
AD9060	10	75	2800
AD7870	12	0.1	60
AD1671	12	1.25	570

ter level detection to maintain the level detected at a constant level. This is called Automatic Gain Control, or AGC. It has the advantage of supplying signal which has a constant power level for the detector that can eventually be used for further spectral analysis and/or correlation. There are variable amplifiers that respond to the log of the control voltage, which satisfies the required need for dynamic range compression.

In all these cases the level detector should detect rms voltage rather than peak voltage because of the "noisy" nature of radio signals.

6.2. Multichannel Receivers

Multichannel receivers generally don't imply superheterodynage because the hardware needs would be too great. Thus they have been limited in frequency range to less than 1 MHz. They can be all analog, all digital, or a combination of the two. An analog multichannel receiver consists of a bank of filters (usually L-C) and the means to separately amplify, compress, and detect each channel. Thus there is an important weight penalty.

Modern fast analog-to-digital converters (ADC) and signal processing devices allow multichannel spectral analysis to be performed in near real-time. The principal limitations are power requirements and dynamic range. Power drain increases with the number of bits supplied and is proportional to the sampling speed. The number of bits determines the number of quantification values the signal is converted to. Each extra bit adds 6 dB to the dynamic range. This means that an 8-bit converter (the most common and rapid) can permit only about 48 dB of dynamic range. Table 2 shows some typical flight-worthy converters presently available with their maximum frequencies and power consumption (source: online catalog of Analog Devices, March 1999). This material is evolving very rapidly and faster and less power-hungry converters will be available soon.

In many or most cases the ADC converters cannot handle the required dynamic and frequency ranges by themselves. In this case it is necessary to add an analog "front end". This analog section can perform an overall compression of the input signal range. However, although this increases the range where large signals can be observed, it does not allow distinction between two in-band signals with amplitude differences larger than allowed by the number of bits digitized. Also, the compression should not be done with an amplifier that deforms the signal as this will create unwanted frequency artifacts. Cascaded amplifiers can be used but gain switching has to be avoided during the digital analysis. AGC amplifiers as described above work well in this application, with the advantage of supplying a fixed overall level to the ADC converter, thus making the best use of its intrinsic dynamic range.

Depending on the application the analog section could divide the frequency range to be analyzed digitally into smaller bands. The DESPA uses two- to three- octave bands, at a cost of increased hardware. In this manner large signals at low frequencies will not swamp out the typically smaller signals at higher frequencies.

The signal must be sampled at a rate greater than twice the highest frequency in the band of interest, and should be at least 2.5 to 3 times this. This is because an input signal at F_{in} sampled at F_s will give the same output as a signal at F_s-F_{in}. The input signal must be low-pass filtered very heavily so that at F_s-F_{in} for the highest frequency of interest the signal is attenuated by the amount of dynamic range desired. An example: for a ratio of 2.5 and a dynamic range of 40dB the filter must attenuate by a factor of 100 between $.4F_s$ and $.6F_s$. This is difficult to obtain in practice. The greater the dynamic range (number of bits) the higher the sampling rate has to be.

Processing nowadays is performed with specialized signal-processing microcomputers (DSP). The most frequently used are of the TMS320 series from Texas Instruments and the ADSP2100 series from Analog Devices. This are fast RISC processors with instruction sets especially designed for this type of application. Processing is most often performed with FFT analysis for which these devices have been optimized. DESPA has developed its own wavelet-type transform that has the advantage of performing a logarithmically-spaced frequency analysis with constant relative bandwidth.

6.3. Correlators

Correlation is an important measurement when it is desired to determine the arrival direction and/or the polarization of the incoming signals on two or more (non-collinear) antennas, and eventually to sort out only the desired types. For instance, receivers can be designed to be only sensitive to left- or right- hand polarization such as one on the Voyager spacecraft [*Warwick et al*, 1977] In earlier times instruments were launched that measured only the correlation index, sometimes with only one bit sampling. Times have evolved and now correlation measurements are often part of a receiver. A digital receiver makes this function readily obtainable, by multiplication of the output spectra obtained from signals that were sampled simultaneously on each receiver.

A type of analog correlation that has been used successfully is done by simple addition of the signals from two orthogonal antennas on a spinning spacecraft, with and without a 90° phase shift [*Manning and Fainberg*, 1980]. This has been used on the Ulysses and Wind spacecraft [*Stone et al.*, 1992; *Bougeret et al.*, 1995].

6.4. Waveform Capture

By waveform capture is meant a burst of rapidly taken samples that allow a time-domain analysis for a short period of time. Often for radio receivers this creates much more information than can be transmitted in real-time so that the digitized information is sent more slowly to the ground and the resulting duty cycle can be quite low. An improvement over taking a burst of samples at random moments is to use an event-driven trigger such as, for instance, a high count level on a particle instrument (not really a radio measurement in this case) or a detected level above a commandable threshold. A good example of an event-driven waveform sampler can be found on the Wind spacecraft [*Bougeret et al*, 1995,].

The hardware required is the same as for the digital multichannel analyzer described above. The constraint of using a sufficiently high sampling frequency is the same as for the multichannel analysis but here it hurts even more because it directly affects the bit rate.

In some applications a slice of the spectrum is taken. A mixer and an oscillator are used to select a portion of the spectrum and translate it to "baseband". This is not easy to do "cleanly" because of frequency folding. A description of one such system flown on ISEE-1 can be found in [*Gurnett et al.*, 1995]. With modern rapid digitizers it is becoming possible to use the sampler as the final translator to baseband, taking advantage of the fact that $nF_s - F_{in}$ gives the same result as F_s. However the strict pre-filtering is still necessary.

7. RECEIVER CHARACTERISTICS

The primordial characteristic of a radio receiver is its linearity in amplitude in order to eliminate risks of intermodulation and the apparition of false signatures. A typical manifestation of intermodulation would be a small artificial signal perceived at the difference (or sum) frequency of two large signals near the top end of the receiver's dynamic range.

Secondly, the receiver should have adequate sensitivity which implies that its own noise floor is below a certain limit. Generally, this noise is expressed in $nV/\sqrt{Hz}$. An achievable number for this is $10\ nV/\sqrt{Hz}$ which allows practically all standard measurements to be performed without much degradation from the receiver itself.

The precision, or accuracy, of a receiver is particularly important in certain applications such as, for instance, where determination of the source direction of the received emissions is desired.

Adequate integration time is necessary to achieve a degree of precision. All signals received by radio receivers are of the stochastic type (noise) and only an estimate of the amplitude is possible. The precision of the estimate is given as $\Delta S/S \sim 1/\sqrt{(\beta\tau)}$ here $\Delta S/S$ is the relative rms error in the spectral estimate, β is the bandpass of the receiver, and τ is the integration time of the measurement. As an example, let us suppose that $\beta = 1$ kHz, $\tau = 10$ ms. This gives $\Delta S/S = .3$ (30%) which is rarely good enough!

Note that for digital receivers like an FFT the β is that of the output bandwidths (not the input) and the effective τ is usually reduced by windowing schemes.

Multichannel receivers allow better precision in a given measurement time.

8. BACKGROUND LIMITATIONS AND EMC

Besides preamplifier/receiver noise there are basically three sources of natural noise that limit the sensitivity of a space-borne radio receiver system. First is the photoelectric shot-noise, caused by the cumulative effect of small electric pulses generated each time a solar photon causes an electron to be ejected from the surface of an antenna or the spacecraft body, and also when the photoelectron returns to the surface. This is a low-frequency noise of a $1/f$ nature. A second source of background noise is the quasi-thermal noise, caused by Langmuir waves generated as electrons fly close to the antennas. These two noise sources have been comprehensively addressed in [*Meyer-Vernet and Perche*, 1989] and [*Issautier et al*, 1999].

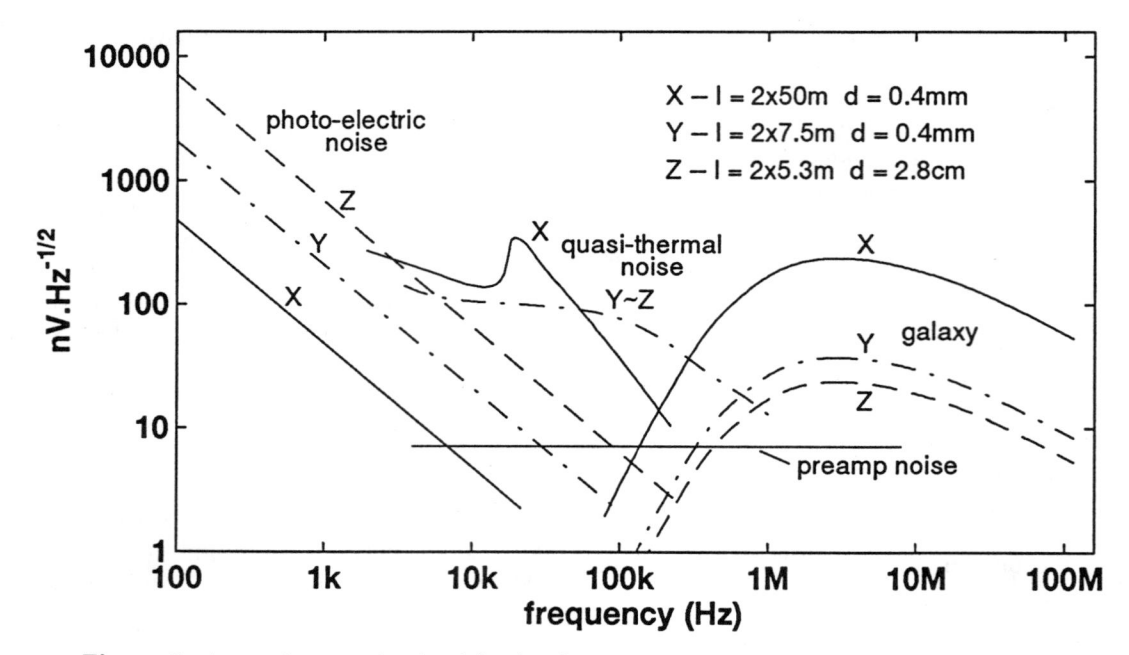

Figure 5. Approximate noise densities for three antenna configurations in the solar wind.

Finally, the third source is the background radiation of the galaxy. This noise starts to be significant at ∼ 200-300 kHz, and remains dominant up into the GHz region. See [*Cane,* 1979] for a model of the galactic noise spectrum.

Figure 5 illustrates the three phenomena for a practical case, the 3-axis antenna system on the Wind spacecraft in the solar wind at 1 a.u. The amplitudes shown are approximate. In this figure three different antenna lengths and diameters are used. The frequency extends quite high, further in fact than the range where these antennas are really usable. The curves ignore the resonant effects of the antennas as discussed previously. Figure 6 shows a real-life measured background spectrum which includes an antenna resonance, for the RAD2 (Y) receiver on the Wind spacecraft, covering 1 to 13.8 MHz. The large bump to the right is the main resonance, where the antenna reactance (now inductive) is equal in magnitude to that of the base capacitance, while the smaller "hook" towards the middle has been identified as an effect due to the electrical resonance of the 12-meter extendible booms holding the magnetometers.

Sensitivity can sometimes be increased by integrating the receiver output over a longer time, with high amplitude resolution, if the reduced time resolution is acceptable. However this is in practice only helpful at higher frequencies where the galactic noise is the limitation because it is a stable source. Photoelectric noise and especially quasi-thermal noise are time-variable depending on local plasma characteristics and spacecraft orientation.

Non-natural noise in a receiver system generally is a result of EMI (Electro-Magnetic Interference). These are perturbations coming from other instruments and sub-systems on the spacecraft (and in some cases from one's own instrument!). They can be caused by a radi-

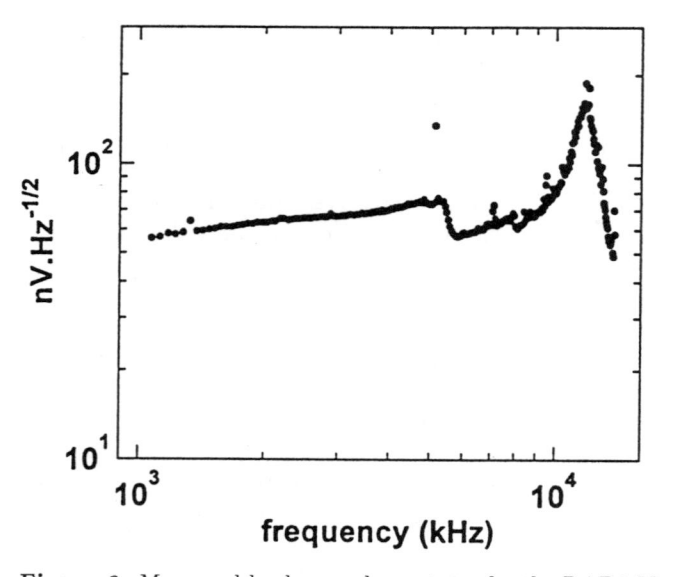

Figure 6. Measured background spectrum for the RAD2-Y receiver on the Wind spacecraft.

ated electric field onto the antennas, or pickup caused by a pulsating magnetic field, by conductive coupling in the harnesses interconnecting the instruments, or by noise generated on the spacecraft structure which couples into the antenna system *via* the base capacitance. The area of expertise in keeping the spacecraft-generated noise down is known as EMC, for Electro-Magnetic Compatibility. Much experience has been gained in the last decade and it is now much easier and less costly to build a EMC "clean" spacecraft.

9. CONCLUSION

The most difficult problem in designing and implementing a radio receiver instrument is the choice and accommodation of the antenna system. The choice will depend on the physical characteristics of the spacecraft and the requirements levied on its dynamics.

In a clean EMC environment it is natural noise sources that limit the sensitivity of a receiver. This paper has limited its discussions to electrically-short antennas. At high frequencies sensitivity can be gained by directive antennas but this is not often an allowable option.

For experiments where polarization measurements and direction finding are important it is necessary that the receivers be well calibrated. It is a wise choice to foresee an internal on-board calibration source. Pseudo-random noise generators are useful for this application.

REFERENCES

Alexander J.K., Kaiser, M.K., Novaco, J.C., Grena, F.R., Weber, R.R., Scientific instrumentation of the Radio-Astronomy-Explorer-2 satellite, *Astron and Astrophys*, 40, 365, 1975.

Bougeret, J.-L., et al, Waves: The radio and plasma wave investigation on the Wind spacecraft, *Space Sci. Rev., 71*, 231-263, 1995.

Cane, H.V., Spectra of the non-thermal radio radiation from the galactic polar regions, *M.N.R.A.S. 189,* 465, 1979

Gurnett, D.A., et al, The Polar plasma wave instrument, *Space Sci. Rev., 71,* 597-622, 1995.

Issautier, K., N. Meyer-Vernet, M. Moncuquet, S. Hoang, and D. J. McComas, Quasi-thermal noise in a drifting plasma: Theory and application to solar wind diagnostic on Ulysses, *J. Geophys. Res., Vol. 104,* 6691, 1999.

King, R.W.P., The theory of linear antennas, Harvard University Press, Cambridge, Mass., 1956.

Manning, R., A simulation of the behavior of a spherical probe antenna in an AC field, AGU Monograph 103: Measurement techniques in space plasmas - Fields, 181-184, 1998.

Manning, R. and Fainberg, J., A new method of measuring radio source parameters of a partially polarized distributed source from spacecraft observations, *Space Sci. Instr., 5,* 161-181., 1980.

Meyer-Vernet, N. and Perche, C., Tool kit for antennae and thermal noise near the plasma frequency, *J. Geophys. Res., Vol. 94, No. A3,* 2405-2415, 1989.

Meirovitch, L., Bounds on the extension of antennas for stable spinning satellites, *J. Spacecraft, Vol. 11, No. 3,* 202-204, 1974.

Pedersen, A., et al, QuasiStatic electric field measurements with spherical double probes on the GEOS and ISEE satellites, *Space Science Reviews, 37,* 269-, 1984.

Rucker, H.O., Macher, W., Manning, R., Ladreiter., H.P., Cassini model rheometry, *Radio Science, 31,* 1299, 1996.

Stone R.G., et al., The unified radio and plasma wave investigation, *Astron. Astrophys., Suppl 92,* 291-316, 1992.

Warwick, J.W., Peace, J.B., Peltzer, R.G., Riddle, A.C., Planetary radio astronomy experiment for Voyager missions, *Space Sci. Rev, 21,* 309-327, 1977.

R. Manning, DESPA, Observatoire de Paris, 92195 Meudon Cedex, France. (e-mail: manning@obspm.fr)

The Astronomical Low Frequency Array: A Proposed Explorer Mission for Radio Astronomy

D. Jones,[1] R. Allen,[2] J. Basart,[3] T. Bastian,[4] W. Blume,[1] J.-L. Bougeret,[5] B. Dennison,[6] M. Desch,[7] K. Dwarakanath,[8] W. Erickson,[9] D. Finley,[4] N. Gopalswamy,[7] R. Howard,[10] M. Kaiser,[7] N. Kassim,[11] T. Kuiper,[1] R. MacDowall,[7] M. Mahoney,[1] R. Perley,[4] R. Preston,[1] M. Reiner,[7] P. Rodriguez,[11] R. Stone,[7] S. Unwin,[1] K. Weiler,[11] G. Woan,[12] and R. Woo[1]

A radio interferometer array in space providing high dynamic range images with unprecedented angular resolution over the broad frequency range from 0.03 – 30 MHz will open new vistas in solar, terrestrial, galactic, and extragalactic astrophysics. The ALFA interferometer will image and track transient disturbances in the solar corona and interplanetary medium - a new capability which is crucial for understanding many aspects of solar-terrestrial interaction and space weather. ALFA will also produce the first sensitive, high-angular-resolution radio surveys of the entire sky at low frequencies. The radio sky will look entirely different below about 30 MHz. As a result, ALFA will provide a fundamentally new view of the universe and an extraordinarily large and varied science return.

---

[1] Jet Propulsion Laboratory, California Institute of Technology, USA
[2] Space Telescope Science Institute, USA
[3] Iowa State University, USA
[4] National Radio Astronomy Observatory, USA
[5] Observatoire de Paris, France
[6] Virginia Polytechnic Institute, USA
[7] Goddard Space Flight Center, USA
[8] Raman Research Institute, India
[9] University of Maryland, USA, and University of Tasmania, Australia
[10] Orbital Sciences Corp., USA
[11] Naval Research Laboratory, USA
[12] University of Glasgow, UK

Radio Astronomy at Long Wavelengths
Geophysical Monograph 119
Copyright 2000 by the American Geophysical Union

1. BACKGROUND

The scientific promise of high resolution radio observations at long wavelengths, and the need to make such observations from space, has been recognized for many years. Mission concept studies and proposals for low frequency radio arrays in space were developed at several institutions during the past two decades. These included arrays in Earth orbit, solar orbit, and on the lunar surface. The mission described in this paper, the Astronomical Low Frequency Array (ALFA), was proposed to NASA as a medium-class Explorer mission in 1998. Although the proposal was not selected for funding by NASA in 1999, the scientific peer reviews in both solar and astrophysics areas were favorable. Sooner or later a low frequency array in space will be created. The scientific goals of such an array, which are largely independent of any specific array design, configuration, or location, are compelling. This paper summarizes the scientific goals of ALFA as an illustration of the case to be made for high resolution radio observations at long wavelengths.

2. SCIENTIFIC GOALS

2.1. Introduction

The Astronomical Low Frequency Array mission will produce the first low frequency, high resolution radio images of the solar corona and interplanetary disturbances such as shocks driven by coronal mass ejections (CMEs). Equally important, for the first time we will be able to image and track these solar disturbances from the vicinity of the Sun all the way to 1 AU, which requires observing frequencies from tens of MHz to tens of kHz. Since Earth's ionosphere severely limits radio interferometry from the ground at frequencies below $\sim$ 30 MHz, these measurements must be made from space. The ALFA imaging interferometer will operate from 0.03 to 30.0 MHz.

One of the major space weather goals of the ALFA mission is accurate prediction, days in advance, of the arrival of CMEs at Earth. CMEs interacting with Earth's magnetosphere can result in geomagnetic storms which are capable of damaging satellite and electric utility systems and disrupting communications and navigation services. Solar disturbances can also pose a threat to astronauts. For this reason, successors to ALFA may become as indispensable for future space weather forecasting as weather satellites are today. In addition, ALFA will image Earth's magnetospheric response to such solar disturbances, providing a unique global view of the magnetosphere from the outside.

The ALFA mission will also produce the first sensitive, high resolution radio images of the entire sky at frequencies below 30 MHz – a region of the spectrum that remains unexplored with high angular resolution. Many physical processes involved in the emission and absorption of radiation are only observable at low radio frequencies. For example, the coherent emission associated with electron cyclotron masers, as seen from the giant planets, Earth, and several nearby stars, is not only expected to occur and be detectable elsewhere in the galaxy but to be ubiquitous. Incoherent synchrotron radiation from fossil radio galaxies will be detectable by ALFA, revealing the frequency and duration of past epochs of galactic nuclear activity. It is also likely that unexpected objects and processes will be discovered by ALFA. Indeed, one of the exciting aspects of the mission is its very high potential for discovery.

Because the solar and extra-solar-system astrophysics programs are managed separately within the NASA Office of Space Science, we discuss their science goals separately in this paper. However, it should be kept in mind that ALFA will simultaneously address key goals in both the Sun-Earth Connection (SEC) and Structure and Evolution of the Universe (SEU) science theme areas with no significant increase in cost or compromise in mission design.

In the SEC area, our science goals address:

- solar variability – physics of solar transient disturbances, the evolution of coronal and solar wind structures, and interactions of plasma and magnetic field topology.

- terrestrial response – interactions with Earth's magnetosphere, geomagnetic storms, and space weather.

- implications for humanity – forecast the arrival of coronal mass ejections.

In the SEU theme area, ALFA will address:

- galaxy evolution – detection of fossil radio galaxies and very-high-redshift radio galaxies, and cosmic ray diffusion times and magnetic field distributions in galaxies.

- life cycles of matter – distribution of diffuse ionized hydrogen in the interstellar medium, energy transport via interstellar plasma turbulence, origin of cosmic ray electrons, and the detection of old galactic supernova and gamma-ray remnants.

- discover new phenomena and test physical theories – new sources of coherent radio emission, pulsar emission regions, shock acceleration, physics of electrically charged dusty plasmas, and new classes of objects not seen at higher frequencies.

The fundamental technique of ALFA is aperture synthesis, in which interferometric data from a large number of baseline lengths and orientations are combined to produce images with an angular resolution comparable to that of a single aperture the size of the entire interferometer array. This is the basis of ground-based arrays such as the VLA and VLBA and the VSOP space VLBI mission, and results in many orders of magnitude improvement in angular resolution. The concept was endorsed by the radio astronomy panel of the *Bahcall* [1991] decade review committee, which recommended "...establishing a program of space radio astrophysics during the next decade leading to the establishment of a Low Frequency Space Array, a free flying hectometer wavelength synthesis array for high resolution imaging, operating below the ionospheric cutoff frequency." The technology now exists to carry out this mission inexpensively.

The ALFA imaging interferometer consists of 16 identical small satellites with dipole antennas and low frequency radio receivers, distributed in a spherical array 100 km in diameter. The array will be placed in a nearly circular retrograde orbit $\sim 10^6$ km from Earth. The size of the array is determined by a fundamental limit to angular resolution created by the scattering of radio waves in the interstellar and interplanetary media. However, this scattering limit is a strong function of direction and observing frequency. To allow for this, it will be possible to vary the size of the array during the mission to increase or decrease the maximum angular resolution.

2.2. The Sun-Earth Connection Science Goals

The study of the nature and evolution of solar transient phenomena is essential to understanding the Sun-Earth connection. Disturbances traveling through interplanetary space generate radio emissions at the characteristic frequencies of the plasma, which range from a few kilohertz to several gigahertz. The higher frequency emissions occur very close to the Sun where the electron density and plasma frequencies are high, while lower frequency emission occurs in the less dense regions far from the sun. Ground-based radio observations are limited by Earth's ionosphere to frequencies above 20 to 30 MHz and therefore to solar emissions that are generated close to the Sun (<2 $R_\odot$). For the vast spatial region between the Sun and Earth, radio observations from space provide a proven way to observe transients in the Sun's extended atmosphere. There have been numerous space-based radio instruments that have made and continue to make low-frequency observations of the Sun. However, without exception, these observations are made with simple dipole antennas from single spacecraft and provide very poor angular resolution. Even the proposed STEREO mission will only use dipole antennas and will consequently only be able to track the centroids of radio bursts.

Meter wavelength Type II bursts have been observed in the solar corona by ground-based observatories since the 1950s, and are associated with coronal shocks [*Wagner and MacQueen*, 1983]. However, kilometric wavelength Type II bursts, which are generated by CME-driven shocks propagating through the interplanetary medium, are observed only by sensitive low frequency spacecraft radio receivers. These interplanetary Type II bursts are associated with the fastest shocks, and the radio emission regions can be very large [*Lengyel-Frey and Stone*, 1989]. As the CME propagates different regions of the shock front become the site of particle acceleration and therefore the site of radio emissions. The Type II radio emission mechanism depends on the local plasma density, so Type II burst observations provide information on plasma density in the vicinity of the shock. *Reiner et al.* [1998a,b] demonstrated that kilometric Type II emissions originate in the upstream regions of CME-driven shocks. Images of kilometric Type II bursts obtained by ALFA, at different frequencies, will therefore provide the first high resolution maps indicating the range of densities in the upstream region of CME-driven shocks, and will provide new information about the sites of particle acceleration along the shock front. Plate 1 shows the expected appearance of a type II radio burst in a $2° \times 2°$ field of view imaged by ALFA.

The temporal and frequency behavior of these type II burst images will enable us to distinguish those CMEs that are directed toward Earth from those that are not. This will permit accurate (to within hours) predictions to be made, days in advance, of their arrival time at Earth.

In addition, ALFA will provide a new, exterior view of Earth's magnetosphere. The ALFA array will be in a near circular orbit about 160 R_E from Earth, providing an ideal opportunity to image Earth's magnetosphere and bow shock from many vantage points over the course of the mission. At the distance of ALFA from Earth, the frontside magnetosphere will subtend $21°$ and the magnetotail will extend over $40°$ [*Alexander et al.*, 1979]. Bow shock and magnetospheric emissions, while occurring simultaneously, are well separated in frequency and therefore will constitute distinct radio images. Imaging is possible not only on a routine basis as a complement of the IMAGE mission (scheduled for launch in 2000 to observe from within the magnetosphere), but especially when reconnection is occurring on the frontside. These will be the first images of Earth's magnetosphere from the outside.

2.3. Galactic and Extragalactic Science Goals

The multi-frequency, all-sky radio images produced by ALFA will extend our knowledge of phenomena in galactic and extragalactic objects into a vast unexplored spectral region, addressing many SEU goals. Unique information will be obtained on how galaxies evolve, on matter in extreme conditions, and on the life cycles of matter in the Universe. In addition, new objects and phenomena unseen at higher frequencies are almost certain to be found, an exciting aspect of the ALFA mission. In addition to its very high potential for discovery, the ALFA mission will address several key issues in NASA's SEU science area, including: 1) understanding the evolution of galaxies, 2) the exchange of matter and energy among stars and the interstellar medium, and 3) testing physical theories and revealing new phenomena. In each case, the key contribution of ALFA will be unprecedented angular resolution and sensitivity in a nearly unexplored frequency range. This gain in resolution and sensitivity will enable ALFA to detect and resolve individual objects anywhere on the sky and determine their low frequency spectra from imaging at multiple frequencies. As a result, ALFA will open up an entirely new regime of astrophysical investigation. Among the specific science objectives of ALFA are:

1) The evolution of galaxies

- Search for "fossil" radio galaxies to obtain information on the frequency and duration of the active phases of galactic nuclei, and on the intergalactic magnetic field.

How large a fraction of all galaxies had active nuclei in the distant past, but are now quiescent? ALFA can answer this fundamental question. The discovery of a significant number of "fossil" radio galaxies would provide important new constraints on galaxy evolution and specifically on the frequency and lifetime of active phases. Information on early epochs of galactic activity and its duration will help constrain models for the evolution of massive black holes in galactic nuclei. ALFA will search for fossil radio components associated with presently radio quiet galaxies [e.g., *Goss et al.*, 1987; *Cordey*, 1987; *Reynolds and Begelman*, 1997] as well as presently active galaxies to determine how often and for how long galaxies were active in the past. The long radiative lifetimes of electrons at low frequencies will preserve evidence of early phases of activity in galaxies which are too faint at higher frequencies to be included in existing radio catalogs. For example, a synchrotron source with an initial spectral index of -0.7 and 10^{-5}G magnetic field will have a spectral index of -2 above 3 MHz after ~ 0.4 billion years due to radiation losses. Such a source could have a flux density of 400 Jy at 3 MHz (easily detectable by ALFA) and yet be < 2 mJy at 1.4 GHz, below the detection limit of even the recent VLA Sky Survey [*Condon et al.*, 1998].

- Search for very high redshift radio galaxies, which typically have steeper spectra than closer radio galaxies.

ALFA will expand the number of known galaxies at the highest redshifts. Extragalactic sources with steep spectra at low frequencies are typically high-z galaxies [e.g., *Krolik and Chen*, 1991]. ALFA is well suited to select the steepest spectrum sources from, for example, the 6C catalog at 151 MHz [*Blundell et al.*, 1998], which are expected to be the most distant galaxies ($z > 4$). This distance limit increases with decreasing survey frequency, and consequently the ALFA sky survey will provide a glimpse of the Universe at a time when galaxies were young, possibly back to the protogalaxy era [*Silk and Rees*, 1998]. AFLA's observations of very distant galaxies will complement FIRST and SIRTF.

2) Life cycles of matter in the universe

- Accurately map diffuse interstellar ionized hydrogen, the last major component of the interstellar medium whose distribution is not currently well determined.

Ionized hydrogen is "the only major component of the interstellar medium that has not yet been surveyed" [*Reynolds*, 1990]. It is important to determine the large-scale distribution of diffuse H II both to improve our understanding of the heating and ionization processes in the interstellar medium and to account for the emission or absorption by this gas in other parts of the spectrum. ALFA will determine the galactic distribution of diffuse H II by measuring the free-free absorption of radiation along the lines of sight to a large number of bright galactic and extragalactic sources [e.g., *Kassim*, 1989]. Free-free absorption due to intervening ionized hydrogen produces a more steeply inverted spectrum below the turnover frequency than synchrotron self-absorption or internal free-free absorption, and thus the processes can be distinguished. These measurements by ALFA will cover high galactic latitudes, which are not well sampled by pulsar dispersion measure observations, and can be combined with recent Hα surveys. A critical question about the diffuse ionized medium is whether its energy comes from absorbing nearly all of the kinetic energy from stellar winds and supernovae or from radiation escaping from dense H II regions surrounding massive stars. The only way to answer this question is by mapping the distribution of diffuse H II with ALFA.

- Study the origin and transport of turbulence in the interstellar medium; constrain models for the dissipation of turbulent energy.

A better understanding of all aspects of interstel-

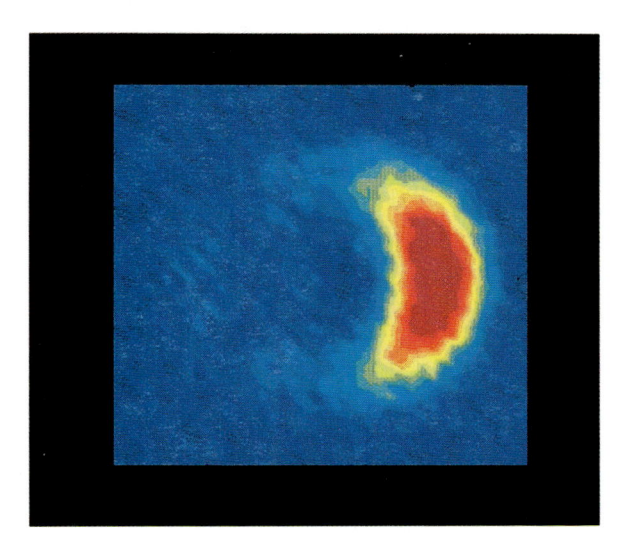

Plate 1. Simulated 2° × 2° ALFA image of a solar radio burst associated with a coronal mass ejection.

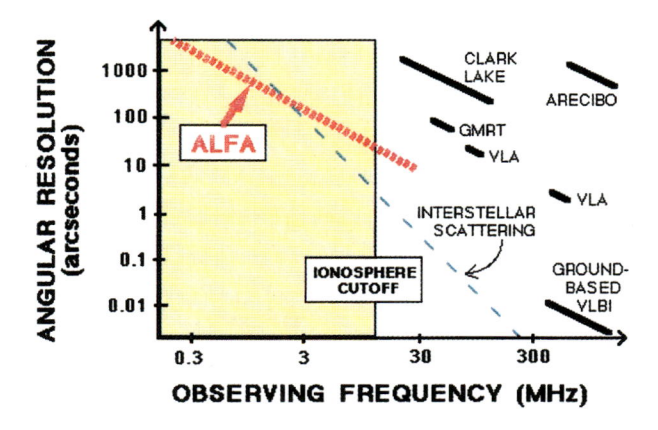

Plate 2. Frequency coverage and angular resolution of the ALFA imaging interferometer.

lar scattering will tell us more about the turbulence properties of the interstellar plasma (energy input and transport) and consequently will shed light on its role in star formation and galactic evolution. ALFA can contribute to this goal in several ways. ALFA will map the distribution of scattering material by measuring the angular broadening of distant sources in all directions [e.g., *Taylor and Cordes,* 1993], including high galactic latitudes. For paths which pass only through the diffuse "Type A" interstellar scattering medium [*Cordes, Weisberg, and Boriakoff,* 1985] it will be possible to directly measure the inner scale of the turbulence by determining the baseline length at which the visibility amplitude curve of a scattered source changes from a Gaussian shape to an exponential. This is not possible with ground-based VLBI arrays operating at higher frequencies because the scattering disk is too small at these frequencies. Only scattering by the clumpy "Type B" component of the interstellar medium is observable with ground-based VLBI on baselines corresponding to the inner scale of turbulence. ALFA's determination of the inner scale for Type A interstellar scattering will help identify the physical properties of the plasma responsible, and thus help understand heating of the diffuse interstellar medium by turbulent energy dissipation.

3) Test theories and reveal new phenomena

- Rigorously test the hypothesis that cosmic ray electrons originate in galactic supernova remnants.

ALFA will test the hypothesis that cosmic ray electrons originate in galactic supernova remnants by directly detecting the presence of low energy electrons needed to initiate shock acceleration mechanisms. In addition, comparison of low and high frequency images of supernova remnants will allow spectral variations to be mapped, providing information on locations and efficiency of shock acceleration.

2.4. Discovery

- Search for expected new sources of coherent radiation and determine the conditions in extreme plasma environments, including emission regions of millisecond pulsars.

ALFA will detect coherent radio emission from a wide range of objects by measuring their low frequency spectra and variability. Coherent emission processes, capable of producing extremely high brightness temperature radio emission, are common at long wavelengths.

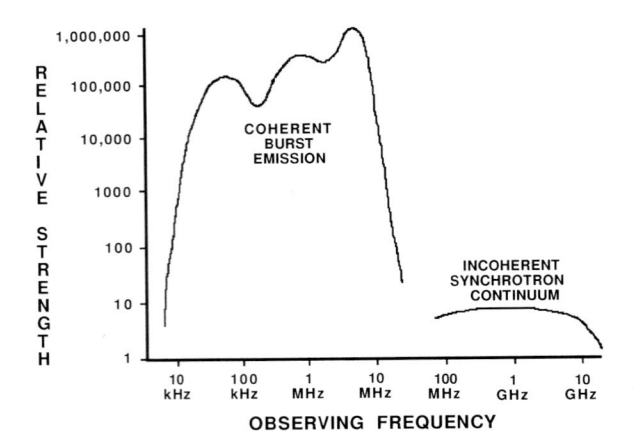

Figure 1. Time-averaged radio emission from Jupiter. Peak burst intensities are two orders of magnitude higher than shown here (figure adapted from *Carr, Desch, and Alexander* [1983]).

The Sun, giant planets and Earth's magnetosphere all display extremely strong coherent emission at low radio frequencies. As an example, below about 10 MHz electron-cyclotron maser emission from Jupiter is many orders of magnitude more intense than the incoherent synchrotron emission at higher frequencies (see Figure 1).

With the exception of transient bursts from the Sun and some active flare stars, this type of emission (and the information it contains on local plasma and gyro frequencies) can only be detected and studied at low frequencies. Basic physics predicts that similar coherent processes will produce strong low frequency radio emission from objects such as supernova remnants, active galaxies, and quasars. In our galaxy, coherent emission is not only expected to occur and be detectable, but to be ubiquitous. The best way to identify coherent emission from objects outside the solar system will be based on their spectra measured by ALFA.

At present, pulsars, circumstellar masers, and a few flare stars are the only objects outside our solar system which are known to radiate coherently. At frequencies of a few MHz, however, many of the most intense discrete galactic sources in the sky (outside of the solar system) will be coherently-emitting sources driven by accretion and outflow. These are objects at the extremes of stellar evolution, with protostellar systems on one end and binary white dwarfs and neutron stars on the other.

Pulsars whose radio spectra have no observed low-frequency turnover, such as the Crab pulsar and some millisecond pulsars, are promising targets for ALFA.

These pulsars will be among the strongest sources in the sky at frequencies below about 10 MHz [e.g., *Erickson and Mahoney*, 1985]. Timing data at higher frequencies provide an upper limit on the size of the coherent emitting region [*Phillips and Wolszczan*, 1990] while measurement of a low-frequency spectral turnover by ALFA will provide a lower limit. In this way ALFA will probe one of the most extreme environments known.

A large increase in low frequency radio brightness is also expected from coherent emission in relativistic jets [e.g., *Baker et al.*, 1988; *Benford*, 1992]. This suggests that a much larger number of extragalactic objects may be detected by ALFA than would otherwise be possible.

- Search for new sources of galactic nonthermal emission in supernova remnants, H I supershells (possible gamma-ray burst remnants), large-scale ionized filaments, the "galactic center fountain", and presently unknown explosive remnants.

ALFA will image galactic supernova remnants and other extended structures to search for new sources of nonthermal emission. For example, the question of whether the Crab nebula is located inside a previously undetected fast shock, as expected from mass and energy considerations, is still unresolved [*Frail et al.*, 1995; *Sankrit and Hester*, 1997; *Jones et al.*, 1998]. Because of its relatively steep spectrum, emission from the blast wave will be most easily seen at low frequencies. For the same reason, nonthermal emission associated with galactic features such as H I supershells (possibly remnants of previous γ-ray bursts; *Loeb and Perna* [1998]; *Efremov et al.* [1998]), very old supernova remnants, the galactic center fountain, and large-scale ionized filaments [*Haffner et al.*, 1998] will likely be found at the low frequencies measured by ALFA. Such emission contains information on the location and strength of shocks, and can allow very extended (old) remnants to be detected. A more complete sample of old explosive remnants provided by ALFA is essential to improve estimates of the frequency of supernovae and γ-ray bursts in our galaxy and the total kinetic energy input to the interstellar medium from explosive stellar events.

- Search for cyclotron emission from giant planets orbiting nearby stars.

The radio emission from giant planets orbiting nearby stars may be detectable with ALFA, because emission by stellar-wind-driven cyclotron masers is a very strong function of the stellar wind energy flux. In some systems this flux will be much greater than in our solar system, leading to radio luminosities orders of magnitude greater than that of Jupiter at low frequencies.

3. MISSION CONCEPT

3.1. Instrument Description

The "science instrument" for ALFA is composed of the entire array of sixteen satellites operating together as an interferometer. Low frequency radio radiation will be sampled by a pair of orthogonal dipole antennas on each of the identical satellites, and each dipole will feed signals to a simple but flexible high dynamic range receiver. The dipoles are 10 m long, determined by the availability of self-deploying, flight-proven antenna elements. Observing frequency, bandwidth, sample rate, and phase switching of the receivers are controlled by the central spacecraft processor, and can be changed at will. The receiver is a straightforward, single channel design based on commercially available components. It covers 0.03–30.0 MHz, with Nyquist sampled bandwidths up to 125 kHz and an ability to handle a wide range of input levels. No new development is required for the low frequency antenna or receiver systems.

ALFA will have a maximum baseline length of ~ 100 km, which provides a good over-all match to interstellar and interplanetary angular broadening (see Plate 2).

The array will be placed in a nearly circular distant retrograde orbit (DRO) about the Earth-Moon barycenter, with a typical distance from Earth of one million km. There are many advantages of a DRO for this mission, including sufficient distance from Earth to minimize terrestrial interference combined with the ability of each satellite to communicate directly with relatively small (11-meter) and affordable ground stations. These ground stations are currently used to support the VSOP space VLBI mission.

Note that this approach involves no reliance on a single spacecraft for data relay or any other mission critical function; the array data path is extremely robust (16-way redundancy) all the way to the ground. Similarly robust is our technique for continuously monitoring the relative positions of the satellites by measuring the separations between all pairs of satellites. This provides far more constraints than are needed to solve for all of the relative positions. Should one or more of the array satellites fail, observing by the rest of the array continues unhampered.

ALFA will observe in all directions continuously, at frequencies determined by solar emission during solar radio bursts, and at one of several sky survey frequencies during periods between solar bursts. Each satellite

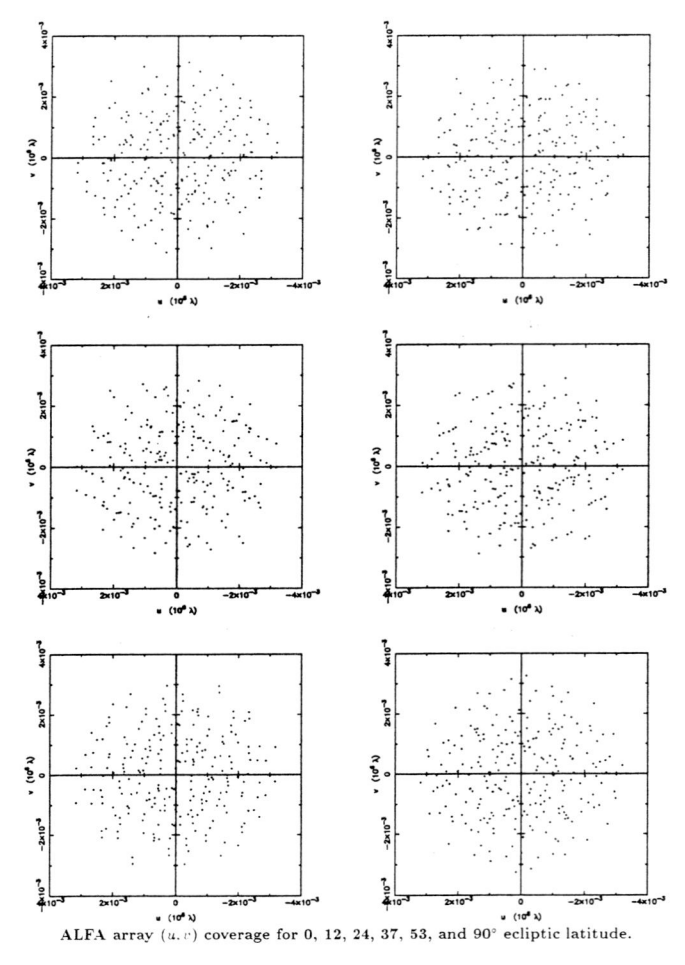

Figure 2. Instantaneous aperture (u,v) plane coverage in six widely different directions.

ALFA array (u,v) coverage for 0, 12, 24, 37, 53, and 90° ecliptic latitude.

receives an X-band carrier (to which the local oscillators are phase locked) and low-rate command telemetry, and transmits X-band data to the ground continuously at 0.5 Mb/s per satellite. The distance of the DRO and its location in the ecliptic plane allows continuous coverage of the array by ground tracking stations located at the three sites of the Deep Space Network. At the ground station telemetry headers are removed and the remaining interferometry data from each satellite are recorded on tapes for transport to the correlation computer. Small subsets of the data for rapid solar snapshot imaging will also be stored on disk and retrieved from the stations via internet. The DSN 11-m ground stations are currently operational, and are designed to be fully automated. Operator intervention will be required only for occasional (once every several days) tape changes or in case of station equipment failure.

3.2. Imaging

Among the challenges of imaging the sky at low radio frequencies is the need to image the entire sky at the same time. This is necessary because individual radio antennas of reasonable size have very low directivity at these frequencies (which is the motivation for using an interferometer array in the first place). Consequently very strong radio sources will create sidelobes in directions far from their positions, and high dynamic range imaging will require that the effects of strong sources be removed from all sky directions, not just from the region immediately adjacent to the sources. This in turn requires an array geometry which produces highly uniform aperture plane coverage in all directions simultaneously, a requirement that no previous interferometer array has had to meet. A quasi-random distribution of antennas on a single spherical surface was found to provide excellent aperture plane coverage in all directions with a minimum number of antennas. This concept was developed by Steve Unwin, who noted the importance of using a minimum separation constraint when computing antenna locations to avoid an excessive number of short projected spacings. Figure 2 shows the instantaneous coverage in six separate directions for a 16-antenna Unwin sphere.

Cross-correlation of the signals will be done by Fourier transforming data from each satellite into a time series of frequency spectra and then cross-multiplying pairs of spectra to obtain the cross-power spectra for each baseline and for each phase center. The computing power required to cross-correlate all data in less than the observing time (3 GFLOPS) can be obtained from a cluster of workstations, for example eight Sun Ultra 60s. This approach offers greater flexibility and less cost than a dedicated hardware correlator, and will directly benefit from future workstation performance improvements.

3.3. Interference and Calibration

Prior to cross-multiplication, all spectra will be multiplied by a combination of Gaussian and cosine functions to filter the frequency response of the array. This greatly reduces the delay beam sidelobes (see Figure 3), and consequently reduces the interferometer response to sources far from the phase center.

The interferometer response is reduced below 1% when the geometric delay exceeds 21 μs. For delays greater than 29 μs the interferometer response falls below 10^{-3}; the maximum geometric delay of the array is about 330 μs.

The delay-beam technique for suppressing interferometer response to emission far from the nominal phase center will fail for narrow-band signals. The most obvious source of narrow signals is terrestrial transmitters. This problem is minimized for ALFA by a combination of observing frequency selection, a high dynamic range receiver, ionospheric shielding, spectral data editing prior to cross-multiplication, and distance from Earth.

Phase calibration of the array is provided by the X-band uplink carrier, to which all satellite oscillators are locked. Amplitude calibration is provided by: 1) periodically injecting a known calibration signal into the signal path between the antennas and low frequency receivers, 2) comparison with known astronomical sources at the high end of ALFA's frequency range, and 3) comparison with ground-based observations of solar bursts using antennas of known gain. The measured total power is iteratively divided between the fields of view, starting with the fields containing the strongest sources.

The array geometry is determined in two steps. First, the relative positions of all satellites are measured by on-board UHF ranging systems with an accuracy of better than ±3 m. This gives the three-dimensional array geometry except for an over-all rotation, to which the inter-satellite ranging data are insensitive. Second, the orientation of the array in inertial space is determined by combining ground-based differential Doppler measurements (made simultaneously for all satellites) and the difference between predicted and measured changes in inter-satellite ranges caused by their slightly different orbital motion between station-keeping maneuvers. The angular orientation of the array will be determined to within half the highest-frequency fringe spacing every few days.

The theoretical array sensitivity at 3 MHz is ~ 10 Jy (1σ) in 5 minutes. The coherence time limits imposed by fluctuations in the solar wind do not prevent useful imaging even at the lowest frequencies [Linfield, 1996]. However, it will be confusion noise and dynamic range rather than the Galactic background which will determine the number of detectable sources. Confusion effects will be minimized by imaging all strong sources on the sky simultaneously so their flux can be taken into account for each field of view. Dynamic range is determined mainly by the number and distribution of visibility samples, the data signal-to-noise ratio, the quality of calibration, and the complexity of the sources being imaged. Based on our imaging simulations, we expect to obtain a dynamic range of $10^2 - 10^3$ for relatively compact sources (<100 beams in size), depending on frequency. For very extended sources or the lowest observing frequencies the dynamic range will still be a

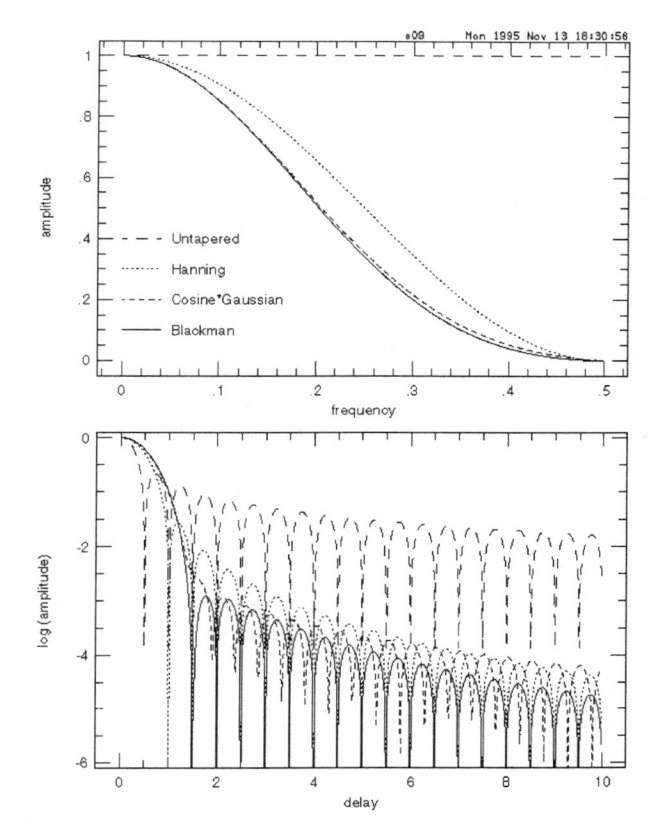

Figure 3. Bandpass weighting (top) and resulting delay beam (bottom). In bottom figure, delay is in units of 1/bandwidth. The dashed line is unweighted and the solid line is ALFA weighting.

few tens, which is entirely adequate for imaging strong, rapidly evolving sources. The use of linearly polarized dipole antennas is not a problem at very low frequencies because radiation will be depolarized by interstellar differential Faraday rotation across any source [Linfield, 1995]. Interplanetary differential Faraday rotation across the array will be negligible for solar elongations >90°, even at 1 MHz, and will also be averaged out for angularly large solar radio sources.

3.4. Computing Resouces

Aperture synthesis imaging of very wide fields requires 3-D Fourier transforms, but regions of limited angular size (over which the effects of sky curvature are small) can be imaged with separate transforms in which one dimension is much smaller than the other two [Cornwell and Perley, 1992]. This approach lends itself naturally to parallel processing. For ALFA the imaging problem is most difficult at the highest frequency (30 MHz) where the synthesized beam is smallest ($\approx 20''$). We plan to make 4096 × 4096 pixel images with 6 arcsecond pixels, so each image will cover an area of 6.8° × 6.8°. Thus, $\approx$ 1000 images are needed to cover the entire sky. Each image will require a 16 pixel Fourier transform in the "radial" direction to allow for sky curvature over the largest scale structure to which the data are sensitive. We will divide each image into $\sim$ 100 smaller areas which will each be deconvolved with the appropriate synthesized or "dirty" beam [e.g., *Frail, Kassim, and Weiler*, 1994]. All clean components are subtracted from the data for each field and each field is transformed again to produce residual images. This process continues until no residual sidelobes remain.

The computing cost for each uncleaned image at 30 MHz is $\sim$ 230 GFLOP, or less than 15 minutes on a Sun Ultra 60 workstation. A cluster of ten such workstations could produce an uncleaned all-sky image consisting of 1000 separate fields in a single day. At lower frequencies the synthesized beam is larger so fewer pixels (and consequently less time) will be needed for each image. Deconvolution of an all-sky image is expected to take 10-30 residual image iterations, which would take an excessive amount of time on a workstation cluster (up to 1 month for the worst case at 30 MHz). A large parallel machine of the sort available at JPL, NRL, GSFC, and other institutions will reduce the time needed for deconvolution by two orders of magnitude. This makes the deconvolution problem tractable.

Solar snapshot imaging will be much faster because 1) a smaller field of view is required, 2) the pixel size is generally larger, and 3) deconvolution may not be needed at all because solar radio bursts will be the strongest sources anywhere on the sky during active periods. Without deconvolution, a 20° × 20° image centered on the sun at 1 MHz could be made in less than 5 minutes on a single workstation.

4. CONCLUSIONS

The progress made during the past years in solar system radio studies from individual spacecraft and in many areas of astrophysics from ground based interferometer arrays has covered large areas of the long-wavelength observational parameter space, whose dimensions are frequency, sensitivity, angular resolution, temporal resolution, and sky coverage. The region in this space least well covered by existing facilities is the combination of low frquencies, high angular resolution, and large sky coverage. This is the domain of space-based interferometer arrays. The first decade of the next century is likely to see the deployment of such instruments. When this happens, solar and planetary radio astronomers will be able to see the morphology of source regions, and galactic and extragalactic astronomers will have their first look at the low frequency sky with high enough resolution to distinguish large numbers of sources.

Acknowledgments. Portions of this research have been carried out at the Jet Propulsion Laboratory, California Institute of Technology, under contract with the National Aeronautics and Space Administration.

REFERENCES

Alexander, J. K., and M. L. Kaiser, Terrestrial kilometric radiation 2. Emission from the magnetospheric cusp and dayside magnetosheath, *J. Geophys. Res.*, *82*, 98-104, 1977.

Bahcall, J. N. (Ed.), *Decade of Discovery in Astronomy and Astrophysics*, 181 pp., National Acad. Sci., Washington, DC, 1991.

Baker, D. N., J. Borovsky, G. Benford, and J. Eilek, The collective emission of electromagnetic waves from astrophysical jets – Luminosity gaps, BL Lacertae objects, and efficient energy transport, *Astrophys. J.*, *326*, 110-124, 1988.

Benford, G., Collective emission from rapidly variable quasars, *Astrophys. J.*, *391*, L59-L62, 1992.

Blundell, K. M., S. Rawlings, S. Eales, G. Taylor, and A. Bradley, A sample of 6C radio sources designed to find objects at redshift z>4 – I. The radio data, *Mon. Not. R. Astron. Soc.*, *295*, 265-279, 1998.

Carr, T. D., M. Desch, and J. Alexander, Phenomenology of magnetospheric radio emissions, in *Physics of the Jovian Magnetosphere*, edited by A. Dessler, pp. 226-284, Cambridge Univ. Press, Cambridge, 1983.

Condon, J. J., W. Cotton, E. Greisen, Q. Yin, R. Perley, G. Taylor, and J. Broderick, The NRAO VLA sky survey, *Astron. J.*, *115*, 1693-1716, 1998.

Cordes, J. M., J. Weisberg, and V. Boriakoff, Small-scale electron density turbulence in the interstellar medium, *Astrophys. J.*, *288*, 221-247, 1985.

Cordey, R. A., IC 2476 – A possible relic radio galaxy, *Mon. Not. R. Astron. Soc.*, *227*, 695-700, 1987.

Cornwell, T. J., and R. Perley, Radio-interferometric imaging of very large fields – The problem of non-coplanar arrays, *Astronomy and Astrophysics*, *261*, 353-364, 1992.

Efremov, Y. N., B. Elmegreen, and P. Hodge, Giant shells and stellar arcs as relics of gamma-ray burst explosions, *Astrophys. J.*, *501*, L163-L165, 1998.

Erickson, W. C., and M. Mahoney, The radio continuum spectrum of PSR 1937+214, *Astrophys. J.*, *299*, L29-L31, 1985.

Frail, D. A., N. Kassim, and K. Weiler, Radio imaging of two supernova remnants containing pulsars, *Astron. J.*, *107*, 1120-1127, 1994.

Frail, D. A., N. Kassim, T. Cornwell, and W. Goss, Does the Crab have a shell?, *Astrophys. J.*, *454*, L129-L132, 1995.

Goss, W. M., W. McAdam, K. Wellington, and R. Ekers,

The very low-brightness relic radio galaxy 1401-33, *Mon. Not. R. Astron. Soc., 226,* 979-988, 1987.

Haffner, L. M., R. Reynolds, and S. Tufte, Faint, large-scale H alpha filaments in the Milky Way, *Astrophys. J., 501,* L83-L87, 1998.

Jones, T. W., L. Rudnick, B.-I. Jun, K. Brodowski, G. Dubner, D. Frail, H. Kang, N. Kassim, and R. McCray, 10^{51} ergs: The evolution of shell supernova remnants, *Publ. A. S. P., 110,* 125-151, 1998.

Kassim, N. E., Low-frequency observations of galactic supernova remnants and the distribution of low-density ionized gas in the interstellar medium, *Astrophys. J., 347,* 915-924, 1989.

Krolik, J. H., and W. Chen, Steep radio spectra in high-redshift radio galaxies, *Astron. J., 102,* 1659-1662, 1991.

Lengyel-Frey, D., and R. Stone, Characteristics of interplanetary type II radio emission and the relationship to shock and plasma properties, *J. Geophys. Res., 94,* 159-167, 1989.

Linfield, R. P., Faraday rotation effects on ALFA observations, *JPL IOM 4-5-95,* 2 pp., Jet Propulsion Laboratory, California Institute of Technology, Pasadena, 1995.

Linfield, R. P., IPM and ISM coherence and polarization effects on observations with low-frequency space arrays, *Astron. J., 111,* 2465-2468, 1996.

Loeb, A., and R. Perna, Are HI supershells the remnants of gamma-ray bursts?, *Astrophys. J., 503,* L35-L37, 1998.

Phillips, J. A., and A. Wolszczan, Pulsar astronomy at meter and decameter wavelengths: Results from Arecibo, in *Low Frequency Astrophysics from Space* (Notes in Physics, vol. 362), edited by N. Kassim and K. Weiler, pp. 175-186, Springer-Verlag, Berlin, 1990.

Reiner, M. J., M. Kaiser, J. Fainberg, J.-L. Bougeret, and R. Stone, On the origin of radio emissions associated with the January 6-11, 1997, CME, *Geophys. Res. Lett., 25,* 2493-2496, 1998a.

Reiner, M. J., M. Kaiser, J. Fainberg, and R. Stone, A new method for studying remote type II radio emissions from coronal mass ejection-driven shocks, *J. Geophys. Res., 103,* 29651-29664, 1998b.

Reynolds, R. J., The low density ionized component of the interstellar medium and free-free absorption at high galactic latitudes, in *Low Frequency Astrophysics from Space* (Lecture Notes in Physics, vol. 362), edited by N. Kassim and K. Weiler, pp. 121-129, Springer-Verlag, Berlin, 1990.

Reynolds, C. S., and M. Begelman, Intermittant radio galaxies and source statistics, *Astrophys. J., 487,* L135-L138, 1997.

Sankrit, R., and J. Hester, The shock and extended remnant around the Crab nebula, *Astrophys. J., 491,* 796-807, 1997.

Silk, J., and M. Rees, Quasars and galaxy formation, *Astronomy and Astrophysics, 331,* L1-L4, 1998.

Taylor, J. H., and J. Cordes, Pulsar distances and the galactic distribution of free electrons, *Astrophys. J., 411,* 674-684, 1993.

Wagner, W. J., and R. MacQueen, R., The excitation of type II radio bursts in the corona, *Astronomy and Astrophysics, 120,* 136-138, 1983.

D. L. Jones, Jet Propulsion Laboratory, Mail Code 238-332, 4800 Oak Grove Drive, Pasadena, CA 91109, USA. (e-mail: dj@bllac.jpl.nasa.gov)

Lunar Surface Arrays

T. B. H. Kuiper and D. Jones

Jet Propulsion Laboratory, California Institute of Technology, Pasadena, California, USA

During the latter half of the 1980's, three concepts for low frequency arrays on the Moon were independently studied for NASA, leading to two workshops in 1990. Perhaps not surprisingly, when one considers the constraints, the concepts were all quite similar. Each consisted of tens to hundreds of dipoles deployed over tens of kilometers. Each element had a superheterodyne receiver, a digitizer, and a data transmitter and antenna mast. Each team envisioned that the array would start small and grow with time. The main technical challenges were those of deploying the array on the Moon, and of correlating the data on the Moon or returning all the raw data to Earth. We review these lunar low-frequency array concepts and note possible alternative approaches to some of the concept features.

1. INTRODUCTION

In 1984, Jim Douglas and Harlan Smith presented a concept for very low frequency radio observatory (VLFRO) on the Moon at a NASA-sponsored conference [*Douglas and Smith*, 1985]. They noted the advantages which the Moon offers:

- It can hold any number of elements in perfectly stable relative positions.
- It can start modestly and be expanded with time.
- The antenna element wires can be laid directly on the regolith.
- Lunar rotation provides full sky coverage.
- The lunar farside is shielded from terrestrial interference.

The VLFRO was envisioned as a 15 × 30 km array operating between 300 kHz and 30 MHz.

Four years later NASA, the University of New Mexico, and BDM Corporation sponsored a workshop to define the science goals and preliminary specifications for a Very Low Frequency Array (VLFA) on the Lunar Far-Side. The strawman design was an initial 17 km circle, with the array growing with time to a 1000 km diameter [*Basart and Burns*, 1989].

In 1989, at the request of NASA's Astrophysics Division, JPL conducted a study of possible lunar astrophysics experiments which could be conducted at a Lunar Outpost to be established as part of the Lunar/Mars Human Exploration Initiative. One of the concepts studied was a 70 × 35 km T-shaped Lunar Low Frequency Imaging Array (LLFIA) [*Kuiper et al.*, 1989], also known as the Lunar Near Side Array (LNSA).

These various concepts [*Smith*, 1990; *Kuiper et al.*, 1990b; *Basart and Burns*, 1990] came together in 1990 at workshops held in Crystal City [*Kassim and Weiler*, 1990] and Annapolis [*Mumma and Smith*, 1990].

In the mid-90's, ESA also conducted a study of a lunar low frequency array [*Bély et al.*, 1997; *Woan*, these proceedings], also called the Very Low Frequency Array. (We will designate it VLFA-2 here.)

Although there was little technical interaction between the studies, the concepts were all quite similar.

Table 1. Comparison of Specifications

Concept	VLFRO	VLFA	LNSA	VLFA-2
Authors	Douglas & Smith	Bassart & Burns	Kuiper et al.	Bély et al.
Frequency (MHz)	0.3-30	1,3,10,30	0.15-30	0.5-30
Aperture (km)	15×30	$17 \to 1000$	35×70	40
Configuration	$T \to$ filled	non-uniform circle	T	spiral-Y
Location	far-side preferred	far-side	near-side	far-side
Antenna type	short dipole	short dipole	crossed dipoles	crossed dipoles
No. of Antennae	$300 \to 10,000$	$169 \to 361$	19	300
Dipole length (m)		1	5	4
Receiver		superheterodyne	superhet.	superhet.
Bandwidth (kHz)	1 kHz	≤ 5 MHz	≤ 22 kHz	100 kHz
Data format	digital	digital	digital	digital
Data transmission	radio or optic fiber	radio or optical	radio	radio
Deployment	manned rover	rover	rover	rover
Correlation	on Moon	on Moon	on Earth	on Moon

In this paper we review the concepts, identify design choices that need to be made, and technical challenges to be overcome.

2. CONCEPT COMPARISON

Table 1 summarizes the specifications for the four VLF array concepts. The considerations which led to the various choices are discussed below.

2.1. Frequency Coverage and Array Size

The lower end of the frequency range is constrained to be > 20 kHz by the local interplanetary plasma frequency, and probably $\gtrsim 90$ kHz by the lunar ionosphere [*Douglas and Smith*, 1985]. On the day-side, the plasma frequency of the lunar ionosphere is estimated to be about 500 kHz [*Woan*, these proceedings], which would prevent observations much below 1 MHz.

Scattering and scintillation in the interplanetary and interstellar media broaden the apparent angular size of sources as roughly the square of the wavelength. Extrapolations from existing data are a bit uncertain, but interstellar scintillation should produce an apparent source size of $0.3° - 2°$ at 1 MHz, depending on Galactic latitude. Interplanetary scintillation should give a somewhat larger size, depending on solar elongation and solar cycle phase. Based on this consideration, Fig. 1 shows that an array size of ~ 100 km gives as much resolution as can be used frequencies < 3 MHz, where Earth-based observations are absolutely ruled out.

The upper end is softly constrained by the possibility of doing high resolution observations from Earth which, while very challenging, have not really been tried with enough determination to establish where the practical frequency limit is. If 30 MHz is that limit, then potentially arcsecond resolution is obtainable, justifying an array size which approaches the lunar radius (1738 km) [*Basart and Burns*, 1990].

2.2. Antenna Elements

All the array concepts used short dipoles. In Table 1 the specified length is that of one dipole arm, or half the total length. All the dipoles are very short ($\ll \lambda$)

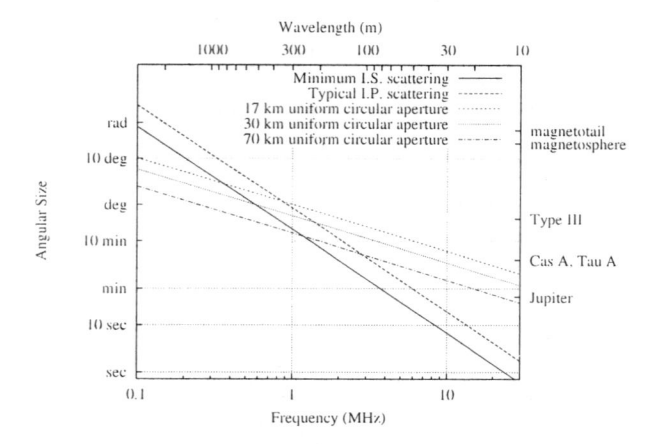

Figure 1. The achievable angular resolution is constrained by plasma inhomogenities which cause interplanetary and interstellar scattering.

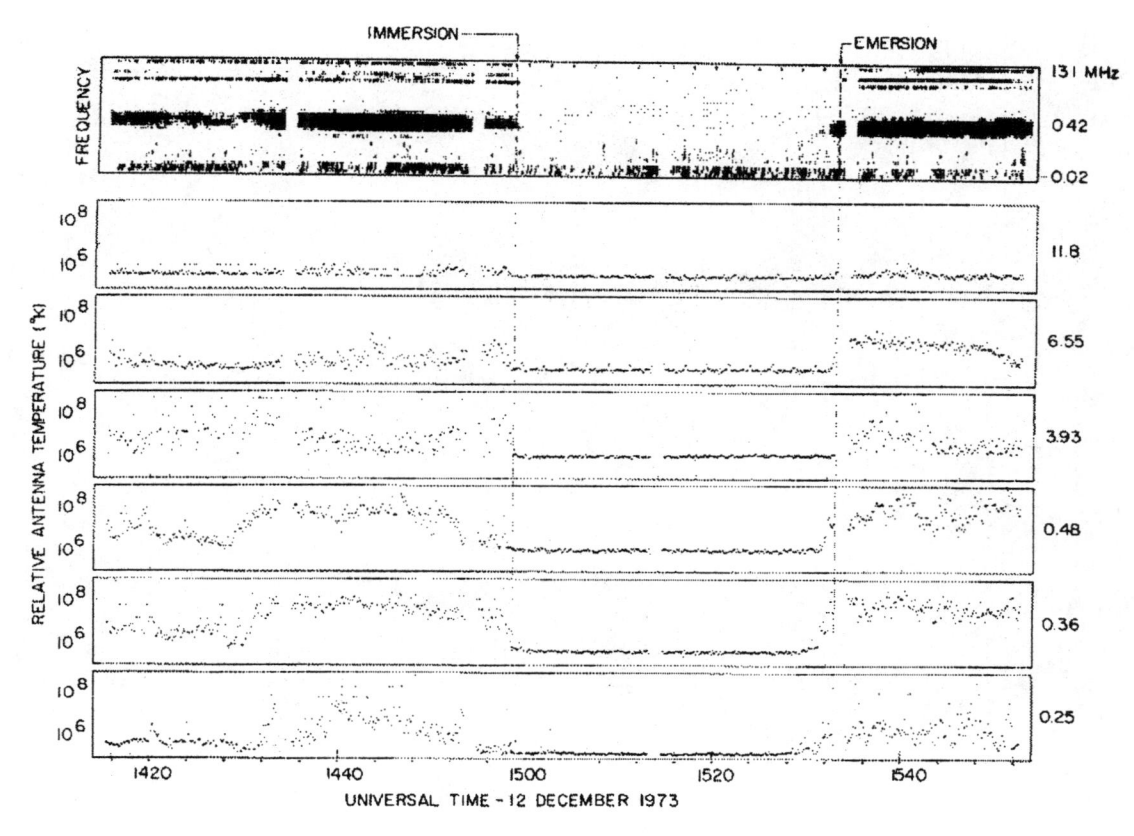

Figure 2. Example of noise levels observed by RAE-2 as it passed behind the Moon [*Alexander et al.*, 1975]. Noise from the Earth was as much as 40 dB above that from the Galactic background.

at most operating wavelengths. Such dipoles have essentially all-sky coverage.

The main advantage of short dipoles over more optimally designed elements is that they are compact and simple to deploy.

2.3. Location

As seen in Fig. 2 the lunar near-side is exposed to a very harsh interference environment [*Alexander et al.*, 1975]. However, one person's interference is another person's research. The Earth's AKR could be one of the main research goals of an initial array on the Lunar Near Side. The experience obtained early in a renewed lunar exploration program will be very valuable for an eventual, much larger array on the Far Side.

The crater Tsiolkovsky (Fig. 3), photographed during two Apollo missions[1], located at 128.5° E 20.5° S, provides a smooth surface 113 km across [*Taylor*, 1989]. The entire crater floor is visible from the central peak, providing an ideal location for the central.

---

[1] //cass.jsc.nasa.gov/expmoon/Apollo15/A15metric0757.gif

2.4. Receivers

The receivers should be fully tunable. This is important because the VLF domain is qualitatively very different from the frequencies studied by most radio astronomers. At cm and mm wavelengths, astronomers generally study static or slowly varying spectra, or time series at one or a few frequencies. The decametric through kilometric wavelength domain however is replete with richly complex dynamic events. Coherent emissions abound, because the scale of relevant structures is often smaller than the wavelength. Receivers will need to be able to record dynamic spectra as a matter of routine. (However, temporal variations from sources at interstellar distances and beyond due to multipathing [*Woan*, these proceedings].)

The requisite receiver electronics exists on VLSI chips. Shortwave radios with the necessary sensitivity, tunablity, and bandwidth are available as hand-held consumer items, with most of the mass being in the speaker, batteries, switches, and display. Likewise, VLSI consumer electronics exists for digitizing the data in, for example, digital audio tape recorders.

Figure 3. Tsiolkovsky is centered near 129 degrees east longitude and 21 degrees south latitude. The flat floor of Tsiolkovsky is much darker than the surrounding lunar surface. The dark material is about 125 kilometers. The central peak, which stands as an "island" within the dark material, is about 40 kilometers long.

2.5. I.F. Data Handling

All the studies concluded that radio communication links would suffice to handle the digital data streams, though optic fiber was mentioned as an alternative for the VLFRO [*Douglas and Smith*, 1985], and photodiode lasers with small telescopes for the VLFA [*Basart and Burns*, 1990]. The demand on the link depended on what was assumed for the instantaneous bandwidth. The VLFA was envisioned as having up to 5 MHz bandwidth for up to 361 elements, which, allowing for guard bands, would put the communication bands into the GHz range. The VLFRO and LNSA concepts, because they assumed narrower bandwidths, could be accomodated with VHF links. All array concepts require a signal from the central site to which to lock the local oscillators at each element, and additionally signals to control the receivers.

Because of the curvature of the Moon, an antenna on top of a 3-m mast has a line-of-sight to a similar antenna up to 12 km away [*Kuiper et al.*, 1990a] (see Fig. 4). Barring a central antenna at great height (e.g. Crater Tsiolkovsky) or a physical link (e.g. optic fiber), most elements in the array will need to serve as relays for antennas farther out. Conceivably the telecommunications transceivers will be more complex than the observing receivers, though not more complex that cellular modems.

If signals are to be relayed, then the system must provide for redundancy, so that the loss of a single element will not disconnect all the elements farther out along that arm. This means that two or more elements must be visible from any element except the last one in an arm.

An array on the far-side will also need a relay satellite. The VLFA-2 concept uses a satellite in a halo L2 orbit to connect the central station (on the central peak of crater Tsiolkovsky) with the Earth.

Another concept that should be studied, especially for a far-side array, is the possibility of having the "central station" on a satellite in an elliptical orbit chosen to be able to have the array in view for a large fraction of its orbit, and also have periodic links to Earth of long enough duration to download the correlated data (see below).

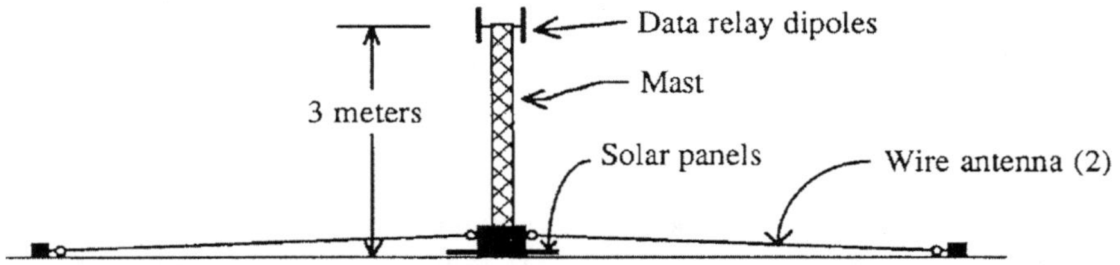

Figure 4. Schematic of a sinle LNSA station, showing the extended communications mast.

2.6. Data Processing and Delivery

The various concepts have quite a range of data processing requirements, as illustrated by Table 2.

The concept with the least demanding data processing requirement (LNSA – within range of an upper end workstation) proposes to return the raw data to Earth. This is because the combination of being on the lunar nearside, plus the modest total data bandwidth, allows all the data to be returned and stored for both quick-look processing and later re-processing.

The upper end of the data processing requirements may be considered one of the "engineering challenges", which we consider next.

3. ENGINEERING CHALLENGES

The lunar low frequency array concept is mature enough for a detailed engineering design study. At the Albuquerque workshop, Stewart Johnson summarized the engineering challenges for the VLFRO [*Johnson*, 1989].

3.1. Environment

Most of the environmental factors – vacuum, radiation, UV degradation, micrometeor impacts – are common to all operations in space and should not present any special challenges.

3.1.1. Temperature Variation. Temperature variation constitutes a serious engineering challenge. During the lunar night, the temperature of the lunar surface drops to $\sim$ 100 K. During the lunar day, the lunar surface temperature rises to $\sim$ 380 K. Experience with Mars Pathfinder and planning for future Mars mission has driven home the difficulty of coping with extreme temperature variations. Electronics are perhaps the least susceptible. Silicon-based chips can easily operate over the 100-380 K temperature range. The very challenging problems are mechanical and chemical. Chemical batteries operate in a typical range of $\pm 40°$C (230-310 K). Without special techniques, solder joints and epoxy bonds will crack when subjected to such extreme temperature cycling. Studying a Mars sample return mission, it was decided that it would be simpler to store the sample container in Mars orbit than on the surface of Mars.

One possible solution is to bury the electronics package in the lunar regolith. At a depth of 10 cm, the diurnal temperature variation is < 35 K [*Keihm et al.*, 1973].

Another potential solution to this problem is the "V-groove radiator". Fig. 5 illustrates the concept, but the details of the design need to be tailored to the application.

In the LNSA concept [*Kuiper et al.*, 1990a], the station power subsystem (solar cells and battery) constituted 90% of the station mass. Providing heating and cooling to regulate the temperature of a package on the surface will further increase the size of the batteries and solar cells. Alternately, burial makes the deployment more complex. Finally, the rated life of lithium batteries is 80 cycles, or $\sim$ 7 years. Some consideration should therefore be given to the possibility of a simple and less expensive exploratory array operating without batter-

Table 2. Data Processing Requirements

Array Concept	Num. of Elements	Num. of Baselines	Bandw. (kHz)	Operations/sec (Mflops)
VLFRO	300 $\to$ 10,000	$4.5 \times 10^4 \to 5 \times 10^7$	1	$90 \to 10^5$
VLFA	169 $\to$ 361	$(1.4 \to 6.5) \times 10^4$	5000	$(1.4 \to 6.5) \times 10^5$
LNSA	19	171	2×22	15
VLFA-2	300	44850	100	3.6×10^4

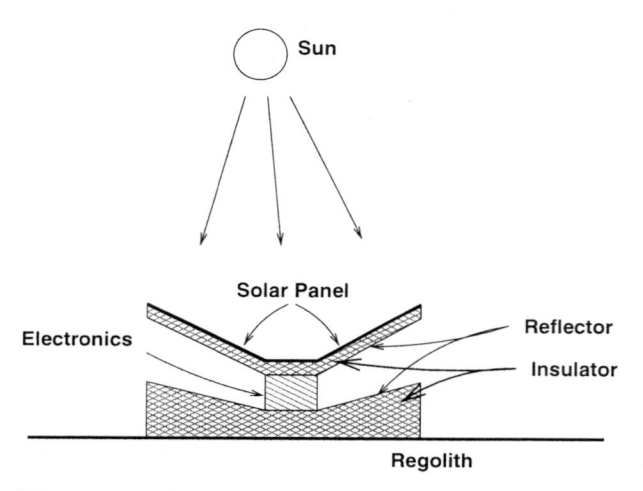

Figure 5. The V-groove radiator is designed so that the electronics package only "sees" cold sky.

ies, in the day-time only. The drawbacks are that the lunar ionosphere probably limits observations to > 1 MHz, and that the solar radio noise would always be present. In any case, the array should be designed to operate in this way after the batteries fail. Sensitive observations could still be done during solar minima.

3.1.2. Dust. Dust may also presents an engineering challenge. Dust adhering to reflective surfaces can increase solar energy absorption by a factor of two or three. Dust on solar panels degrading their performance may also merit further study. Experience with recent operations on Mars suggests that dust is less of a problem than first feared, although this experience may not be relevant to the airless Moon.

The solution to this problem may be as simple as waiting for the dust to settle before opening the solar panels.

3.2. Deployment

The main challenge is to find a cost-effective method of deploying the antenna elements. All the concepts assumed that the deployment vehicle would bear some family resemblance to the Apollo LRV, though operated autonomously or remotely. It would have to contend with craters and boulders, but in view of the experience with the Apollo LRV and the Mars Rover, this should present no insurmountable obstacles.

In view of the success of the Mars Pathfinder mission, some consideration should be given in future studies to having self-deploying antenna elements, using Mars Rover technology to move the elements to their assigned locations. This would be particularly attractive to dilute arrays with small numbers of elements, which could reconfigure to optimize UV coverage for particular needs or to fill in UV coverage. The next Mars rover is being designed to travel up to 10 km or more from the base station. Crucial navigation decisions will still be made by controllers on Earth, but it will have enough intelligence to negotiate simpler obstacles without help. Similar technology could be used for self-deploying array stations. One challenge to be addressed would be position determination. One possible solution would be to use VLBI techniques to solve for the baselines as the stations move out from the central site.

Deployment by impact does not yet seem to be practical yet. The current Mars microprobe design, presumably close to the state-of-the-art for battery-powered impactors, is supposed to survive an impact velocity of 600-700 km/h or $\sim$ 0.2 km/s. Free-fall onto the moon from infinity gives an impact velocity of almost 2.5 km/s. The difference is a factor of $\sim$ 200 in kinetic energy. Thus, current technology appears to be unable to provide a survivable hard lander for a low frequency array. (Smart artillery shells have electronics which survive much higher accelerations. However, they do not have deployable antennas.)

4. CONCLUSION

Lunar low frequency astronomy is an ideal candidate for lunar based science. A modest telescope could be deployed near a temporary or permanent lunar base early in a revived lunar exploration program. The telescope could be designed to grow by adding additional elements during successive missions to that base. (A prototype telescope could even be tested by deploying elements in a remote terrestrial region, such as the Antarctic polar plateau.) Eventually, as human presence on the Moon expands, an array could be started and expanded on the far side of the Moon.

Acknowledgments. The authors thank R. Manning for information on deployment, dust, and temperature issues, and S. Levin, M. Seiffert, T. Velusamy, and H. Yorke for making helpful comments after a trial presentation of this paper.

This work was performed at the Jet Propulsion Laboratory, California Institute of Technolgy, under contract to the National Aeronautics and Space Administration.

REFERENCES

Alexander, J. K., Kaiser, M. L., Novaco, J. C., Grena, F. R., and Weber, R. R., Scientific instrumentation of the radio-astronomy-explorer-2 satellite, *Astr. & Astroph.*, *40*, 365, 1975.

Basart, J. P. and Burns, J. O., Lunar far-side very low frequency array, in Burns, J. O., Duric, N., Johnson, S., and Taylor, G. J., editors, *A Lunar Far-Side Very Low Frequency Array*, NASA Conference Publication 3039, 1989.

Basart, J. P. and Burns, J. O., A very low frequency array for the lunar far-side, in Kassim, N. E. and Weiler, K. W., editors, *Low Frequency Astrophysics from Space*, in Lecture Notes in Physics 362, Springer-Verlag, Berlin, pp. 52–56, 1990.

Bély, P. Y., Laurance, R. J., Volonte, S., Ambrosini, R. R., Ardenne, A. v., Barrow, C. H., Bougeret, J. L., Marcaide, J. M., Woan, G., Very low frequency array on the lunar far side, ESA report SCI(97)2, European Space Agency, 1997.

Douglas, J. N. and Smith, H. J., A very low frequency radio astronomy observatory on the moon, in Mendell, W. W., editor, *Lunar bases and space activities of the 21st century*, National Aeronautics and Space Admininstration, Lunar and Planetary Institute, Houston, pp. 301–306, 1985.

Johnson, S. W., Engineering for a lunar far-side very low frequency array, in Burns, J. O., Duric, N., Johnson, S., and Taylor, G. J., editors, *A Lunar Far-Side Very Low Frequency Array*, in Conference Publication 3039, NASA, 1989.

Kassim, N. E. and Weiler, K. W., editors, *Low Frequency Astrophysics from Space*, Lecture Notes in Physics 362, Springer-Verlag, Berlin, 1990.

Keihm, S. J., Peters, K., and Langseth, M. G., Apollo 15 measurement of lunar surface brightness temperatures: thermal conductivity of the upper $1\frac{1}{2}$ meters of regolith, *Earth & Plan. Sci. Let.*, 19, 337, 1973.

Kuiper, T., Jones, D., Mahoney, M., and Preston, R., Lunar low frequency imaging array, JPL Internal Document D-6703, Rev. A, Vol. 2, Sec. 5C, Jet Propulsion Laboratory, Pasadena, California, 1989.

Kuiper, T. B. H., Jones, D. L., Mahoney, M. J., and Preston, R. L., Lunar low-frequency radio array, in Mumma, M. J. and Smith, H. J., editors, *Astrophysics from the moon*, in AIP Conference Proceedings 207, American Institute of Physics, New York, pp. 522–527, 1990a.

Kuiper, T. B. H., Jones, D. L., Mahoney, M. J., and Preston, R. L., A simple low-cost array on the lunar near-side for the early lunar expeditions, in Kassim, N. E. and Weiler, K. W., editors, *Low Frequency Astrophysics from Space*, in Lecture Notes in Physics 362, Springer-Verlag, Berlin, pp. 46–51, 1990b.

Mumma, M. J. and Smith, H. J., editors, *Astrophysics from the moon*, AIP Conference Proceedings 207, American Institute of Physics, New York, 1990.

Smith, H. J., Very low frequency radio astronomy from the moon, in Kassim, N. E. and Weiler, K. W., editors, *Low Frequency Astrophysics from Space*, in Lecture Notes in Physics 362, Springer-Verlag, Berlin, pp. 29–33, 1990.

Taylor, G. J., Site selection criteria, in Burns, J. O., Duric, N., Johnson, S., and Taylor, G. J., editors, *A Lunar Far-Side Very Low Frequency Array* in NASA.Conference Publication 3039, 1989.

Woan, G. Capabilities and limitations of long wavelength observations from space, these proceedings.

T.B.H. Kuiper, Jet Propulsion Laboratory 169-506, California Institute of Technology, Pasadena, CA 91109. (e-mail: kuiper@jpl.nasa.gov)

Dayton Jones, Jet Propulsion Laboratory 238-600, California Institute of Technology, Pasadena, CA 91109. (e-mail: dj@logos.jpl.nasa.gov)

Radio Sounding in the Earth's Magnetosphere

J. L. Green[1], R. F. Benson[1], S. F. Fung[1], W. W. L. Taylor[2], S. A. Boardsen[2], and B. W. Reinisch[3]

The radio sounding technique has been successfully used for probing and determining the characteristics of remote plasma regions for nearly seventy-five years. Ground-based radio sounders have made extensive measurements of the bottomside of the Earth's ionosphere, while satellites such as ISIS have investigated the topside ionosphere, but there have been no similar measurements made in the Earth's magnetosphere up to this time. Significant advancements in radio sounding have occurred that will enable radio sounding to be performed in the Earth's magnetosphere using the sounding technique. The Radio Plasma Imager (RPI) will be a first-of-its-kind instrument, designed to use radio wave sounding techniques to perform repetitive remote sensing measurements of electron number density (N_e) structures and to study the dynamics of the magnetopause and plasmasphere. RPI will fly on the Imager for Magnetopause-to-Aurora Global Exploration (IMAGE) mission, to be launched in the year 2000. RPI will operate at frequencies from 3 kHz to 3 MHz and will provide quantitative N_e values from 10^{-1} to 10^5 cm^{-3}. Using ray tracing calculations, combined with specific instrument characteristics, simulations of what RPI will measure dramatically show that radio sounding can be used quite successfully to measure a wealth of magnetospheric phenomena. The radio sounding technique will provide a truly exciting opportunity to study global magnetospheric dynamics in a way that was never before possible.

1. INTRODUCTION

In the last thirty years of ionospheric and magnetospheric research, there have only been a few spacecraft instruments that have been able to make remote sensing observations of extended plasma regions. As discussed in the review by *Williams et al.* [1992], many of the techniques reviewed could produce images of various regions of the inner magnetosphere such as the auroral zone, plasmasphere, ring current, geocorona, and plasma sheet in a way never before achieved in space plasma observations, which has been dominated predominantly by *in-situ* measurements. As reported by a task group of the Space Science Board of the National Academy of Sciences [*Scarf*, 1988], imaging of the Earth's magnetosphere will be as important for studying magnetospheric morphology and dynamics as auroral images have become for studying the aurora.

NASA has recently selected a new mission called the Imager for Magnetopause-to-Aurora Global Exploration (IMAGE) as the first mid-sized explorer in magnetospheric physics, to be launched in early 2000. IMAGE will produce forefront science by quantifying the response of the

[1]NASA Goddard Space Flight Center, Greenbelt, Maryland, USA
[2]Raytheon, NASA Goddard Space Flight Center, Greenbelt, Maryland, USA
[3]Center for Atmospheric Research, University of Massachusetts, Lowell, Massachusetts, USA

Radio Astronomy at Long Wavelengths
Geophysical Monograph 119
Copyright 2000 by the American Geophysical Union

magnetosphere to the time variable solar wind. It will acquire, for the first time, a variety of three-dimensional images of magnetospheric boundaries and plasma distributions extending from the magnetopause to the inner plasmasphere. On a time scale of minutes, the IMAGE instruments will produce images which are needed to answer important questions about global solar wind — magnetosphere interactions such as magnetospheric storms and substorms.

The IMAGE mission will address the above objectives in unique ways using both existing and new imaging techniques. IMAGE will make use of two new observing techniques to accomplish its mission. The first technique is radio plasma imaging, which senses plasma densities between 10^{-1} and 10^5 cm^{-3} and determines the distance and motion of key plasma boundaries. The second new technique is neutral atom imaging at energies from 10 eV to 500 keV. Finally, IMAGE will also use optical imagers in the far and in the extreme ultraviolet to image the auroral zone, geocorona, and the plasmasphere (by resonance He$^+$ emission). The auroral observations will enable the observations from the IMAGE instruments to be placed in context with the appropriate substorm phase.

Although originally not designed to measure neutral atoms, there have been a number of neutral atom measurements made by the IMP 7/8, ISEE, and Polar spacecraft [see for example: *Roelof et al.*, 1985; and *Henderson et. al.*, 1997]. The radio imaging technique, however, has not yet been tried in the Earth's magnetosphere. On IMAGE, it will be performed using the Radio Plasma Imager (RPI), which will use the same basic principles of operation that have been used by ground-based and spaceborne radio sounders to investigate the ionosphere for decades.

This paper will provide a brief overview of the field of radio sounding/imaging as used in space science research, with primary emphasis on the expected results from the RPI instrument on IMAGE. In the next sections, we will review the basic principles of radio sounding, discuss background material and results from previous studies, and then illustrate through simulations from ray tracing calculations providing a realistic glimpse of what the RPI instrument will accomplish.

2. BASIC PRINCIPLES OF RADIO WAVE IMAGING

Radio wave sounding in the magnetosphere uses the same principles as ionospheric sounding. A cold magnetized plasma supports two freely propagating electromagnetic waves, the L-O (for left hand - ordinary) and R-X (for right hand - extraordinary) modes with two distinct phase velocities and polarizations [*Budden*, 1985; *Stix*, 1992]. The propagation characteristics of these electromagnetic waves are determined by the electron plasma frequency (f_p) and gyrofrequency (f_g) along the propagation path in the plasma. Wave propagation at frequencies above the cutoff frequencies for these modes is unimpeded by the plasma until the wave frequency approaches or is at the cutoff frequency. The cutoff frequency is where the index of refraction is zero causing complete reflection. The cutoff for the L-O mode is the local f_p,

$$f_p = \frac{1}{2\pi}\sqrt{\frac{N_e e^2}{\varepsilon_0 m}} \approx 9\sqrt{N_e}\, kHz \qquad (1)$$

in which N_e is the electron density (cm^{-3}), ε_0 is the permitivity of free space, e is the electronic charge and m is the electronic mass. For the X mode, the cutoff frequency is:

$$f_x = \frac{1}{2}f_g + \sqrt{f_p^2 + \frac{1}{4}f_g^2} \qquad (2)$$

where

$$f_g = (1/2\pi)eB/m = 2.80\, \mathbf{B} \qquad (3)$$

and **B** is the magnetic field strength in gauss and f_g is in MHz.

A swept-frequency sounder, generally consisting of a radio transmitter and receiver, transmits pulses typically 100 microseconds long (for ionospheric sounding) at sequentially increasing frequencies. The receiver operates at the same frequency for several tens of milliseconds after each pulse. When the freely-propagating transmitted waves enter a density or magnetic field gradient, they will be reflected upon encountering their respective plasma cutoff (*i.e.*, when a plasma cutoff frequency equals the wave frequency) [*Budden*, 1985; *Stix*, 1992]. Thus, the measurements of the time delays and frequencies of the radio echoes by the receiver will produce data records of the echoed signal amplitude as a function of echo delay and frequency. Such records obtained during ionospheric soundings are called ionograms; the records to be obtained by RPI in the magnetosphere will be called plasmagrams.

From the swept-frequency sounder measurements, one can determine the line-of-sight N_e profiles in remote plasma regions. The analysis procedure, known as "true range" analysis, is straightforward and well developed for analyzing ionospheric sounder data. It is an inversion technique that takes into account the density profile dependence on the refractive index between the sounder and the point of reflection at a given frequency. By starting with echoes having the lowest frequencies and shortest time delays, hence the nearest echoes, and extending to signals of higher frequencies and longer delays, one can then recover the plasma profile between the sounder and the remote plasma location [*Jackson*, 1967, 1969a; *Huang and Reinisch*, 1982]. Ground and space-based radio sounders have used this technique quite successfully for decades in the bottomside and topside of the ionosphere, respectively.

3. BACKGROUND AND PREVIOUS STUDIES

The investigation of the ionosphere using radio sounding techniques dates back more than a half century to the experiments performed using selected fixed frequencies by *Breit and Tuve* [1926]. These early experiments indicated the need for swept-frequency sounders which were developed and evolved into a global network of sophisticated instruments that produced high-resolution ionograms (displays of echo delay versus frequency) on 35 mm film [*Brown*, 1959]. The resulting N_e profiles of the bottomside of the ionosphere provided a major input toward achieving the goals of the International Geophysical Year [*Berkner*, 1959]. Plate 1 is an example of the type of ionogram obtained from a ground-based digital ionosonde from the University of Massachusetts, Lowell. These ionosondes are known as Digisondes. In this example, both the O (red) and X (blue) traces coming from the ionosphere directly above the Digisonde, along with the corresponding electron density profile derived from these traces, and echoes from directions other than the zenith (called oblique echoes), are shown. For further information on ionospheric sounding, see the excellent book by *Hunsucker* [1991].

As technology advanced, ionospheric swept-frequency sounders were incorporated into satellites in order to obtain N_e profiles of the topside of the ionosphere, above the density maximum, and thus inaccessible from ground-based sounders. ISIS (International Satellites for Ionospheric Studies), one of the most successful long-lasting international space programs, produced more than 50 satellite-years of ionospheric topside-sounder data. Plate 2 shows three separate ISIS topside ionograms taken as the spacecraft approached the polar cusp (panels A and B) and was within the polar cusp (panel C). Panel A of Plate 2 shows the O and X echoes with the frequency of the ionospheric X reflection directly below the spacecraft. In addition, the labeled cusp echoes are believed to be coming from the enhanced densities in the cusp that the spacecraft encounters along the trajectory of ISIS (panel C). It is important to note that ISIS did not have an echo direction capability but from extensive data analysis and multi-spacecraft comparisons some echo directions can be assumed with reasonable accuracy [*Dyson and Winningham*, 1974]. Using the standard inversion techniques [*Jackson*, 1969a], the density profiles corresponding to the ISIS data in Plate 2 were produced and are shown in Figure 1. The densities from each of the panels in Plate 2 are correspondingly labeled in Figure 1. Figure 1 clearly shows the increase in the cusp density by up to an order of magnitude.

Topside/bottomside density comparisons and multi-spacecraft rendezvous studies have indicated that the uncertainty of the ISIS sounder-derived N_e values (even at the most remote distances) is typically a few percent and no greater than 10% [*Jackson*, 1969b; *Whitteker et al.*, 1976; *Hoegy and Benson*, 1988 and references therein]. In addi-

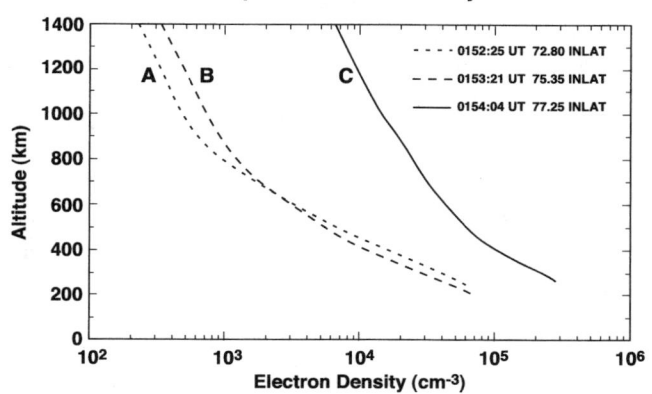

Figure 1. The N_e profiles from the ionograms A, B, and C of Plate 2.

tion, the ISIS satellites also demonstrated that the sounders could operate in a manner compatible with other instruments on the same spacecraft. This compatibility was particularly well illustrated with ISIS 2 which, in addition to producing N_e profiles, produced the first monochromatic auroral images from space for scientific investigations [*Lui and Anger*, 1973; *Shepherd et al.*, 1976].

Later spacecraft-borne sounders, employing digital technology, were developed in Japan for ISS-b (Ionosphere Sounding Satellite) [*Maruyama and Matuura*, 1984], Ohzora (also called EXOS-C) [*Oya et al.*, 1985], and for Akebono (or EXOS-D) [*Oya et al.*, 1990] and in the former USSR for Intercosmos 19 and Cosmos 1809 [*Shuiskaya et al.*, 1990]. Relaxation sounders, which transmit very low power and are designed to stimulate plasma resonances in the ambient medium near the spacecraft for diagnostic purposes, have been flown on a variety of spacecraft over the last 15 years: GEOS 1 & 2 and ISEE 1 [*Etcheto et al.*, 1983], Jikiken (EXOS B) [*Oya and Ono*, 1987], Viking [*Perraut et al.*, 1990], Oedipus A [*James*, 1991], and Ulysses [*Stone et al.*, 1992].

Figure 2 illustrates how the echoes received by ionospheric sounders are related to the vertical N_e distribution in an assumed horizontally-stratified ionosphere. It is composed of two separate complete sounding measurements and their resulting densities combined to illustrate how a complete ionospheric vertical N_e profile is produced from topside and ground-based ionosonde data [*Jackson*, 1969b]. The data were simultaneously collected during a pass of the Alouette 1 satellite over Wallops Island, Virginia. The sounder data are presented in the form of echo time delay as a function of the sounder frequency. The time delay t is often expressed in terms of a virtual (or apparent) range given by $ct/2$, *i.e.*, assuming that the propagation is at the free space speed of light c. The solid lines represent the O-mode traces as they appeared on the ionograms from both

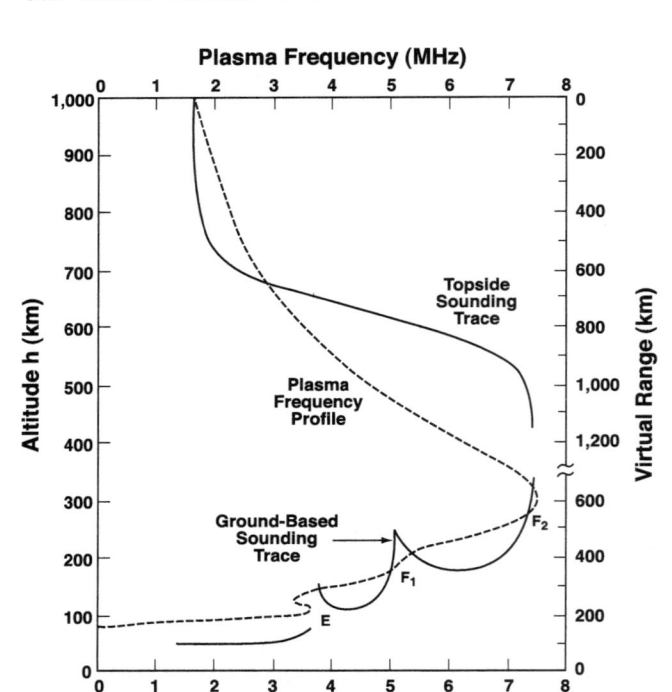

Figure 2. Vertical N_e profiles (dashed line) calculated from topside (Alouette 1) and ground-based (Wallops Island, Virginia) ionosonde O-mode traces (solid curves) presented in the form of virtual range (right scale). The vertical true height (left scale) N_e distribution is expressed in terms of plasma frequency units (top scale) in order to compare with the O-mode ionogram traces, which are based on the bottom scale. In order to keep the topside and ground-based sounding traces from overlapping, they are plotted using a scale (right) compressed by 1/2 of the altitude scale (left). The data correspond to 1651 UT, 10 June 1968.

the ground-based Wallops Island ionosonde and the Alouette 1 topside sounder (1,000 km altitude). The echoes from Wallops Island ionogram (labeled as ground-based sounding trace) were produced by reflections from the ionospheric E and F regions, and are displayed in virtual range (lower right axis) as a function of sounder frequency (bottom axis). In a similar manner, the Alouette 1 ionogram echoes (labeled topside sounding trace) are displayed in virtual range (top right axis) as a function of sounder frequency (bottom axis). The vertical N_e distributions with true altitude (left scale) calculated from both sets of ionosonde data are shown as a dashed curved in Figure 2. In order to aid in the comparison of the N_e profile with the ionogram traces, N_e has been converted to f_p (top scale) using equation 1.

It is instructive to compare the ionogram traces (solid lines) with the corresponding portions of the vertical f_p profile (dashed line). In the ionosphere, the wave energy travels with a speed less than c, thus the apparent range on the right scale is greater than the true range, converted to altitude, on the left scale; the right and left vertical scales have been adjusted for optimum comparison of the ionogram traces with the f_p profile. The smooth topside (above 300 km) f_p profile produces a smooth ionogram trace with steep gradients near the wave cutoff (between 1 and 2 MHz) at 1,000 km and near the f_p peak (between 7 and 8 MHz). The structured E and F regions (dashed line below 300 km), however, produce major signatures on the corresponding (ground-based) ionogram trace. The local minimum at the E layer (called the E valley) produces a clear break in the ionogram trace. The minor inflection at the F_1 layer, produces a dramatic "cusp" feature in the ionogram trace. These signatures have become easy to recognize by ionospheric scientists and they provide instant insight of the reflecting medium since they are characteristic of density enhancements. A similar situation is expected in the case of magnetospheric radio sounding using RPI on IMAGE.

4. RPI ON THE IMAGE MISSION

IMAGE will be launched into a highly elliptical polar orbit with an initial apogee of 8 R_E geocentric radial distance, 1,000 km perigee and 90° inclination. The IMAGE period will be about 13.5 hours. Its prime mission will be of two years duration with one additional year of data analysis. During the prime mission, the IMAGE orbit apogee will precess over the north pole as illustrated in Figure 3. During most of its prime operation, the apogee of IMAGE will be above 45° north geographic latitude. In this region, the spacecraft will be in the magnetospheric N_e cavity extending from the plasmapause to the magnetopause. When in the magnetospheric N_e cavity, RPI will be able to simultaneously receive echoes from the magnetopause and the plasmapause.

One important distinction between RPI and previous topside sounders is that the RPI will be able to determine the direction of the returned echoes using its three-axis antenna system. RPI is called an imager rather than a sounder because, in addition to measuring echo signal strength and delay time as a function of sounding frequency, it will be capable of measuring the echo direction-of-arrival and Doppler spectrum in order to produce "echo maps" [*Reinisch et al.*, 1999]. The 500 m tip-to-tip X and Y (spin plane) dipole antennas and 20 m tip-to-tip Z axis dipole antenna on RPI will be used to measure the direction of arrival of the echoes coming from distances of many R_E [see *Reinisch et al.*, 1999].

In previous ionospheric topside sounder experiments, echoes from different directions were identified based on experience gained from the observed changes in echo characteristics with satellite motion and by comparing the observations with ray tracing calculations. The echo attrib-

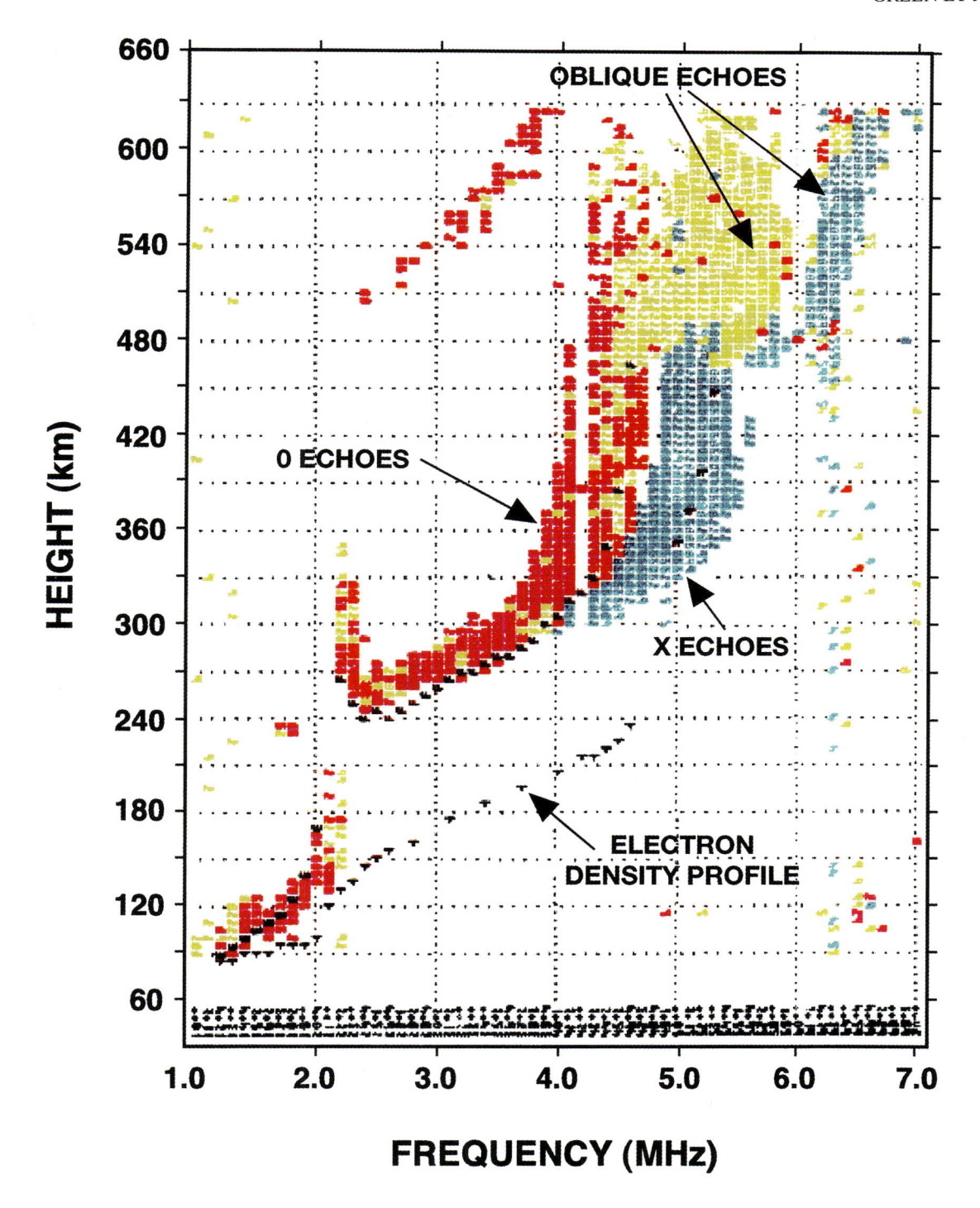

Plate 1. An example of an ionogram obtained from a ground-based digital ionosonde from the University of Massachusetts, Lowell. In this example, both the O and X traces coming from the ionosphere directly above the Digisonde, along with the corresponding electron density profile derived from these traces, and echoes from directions other than the zenith (called oblique echoes), are shown.

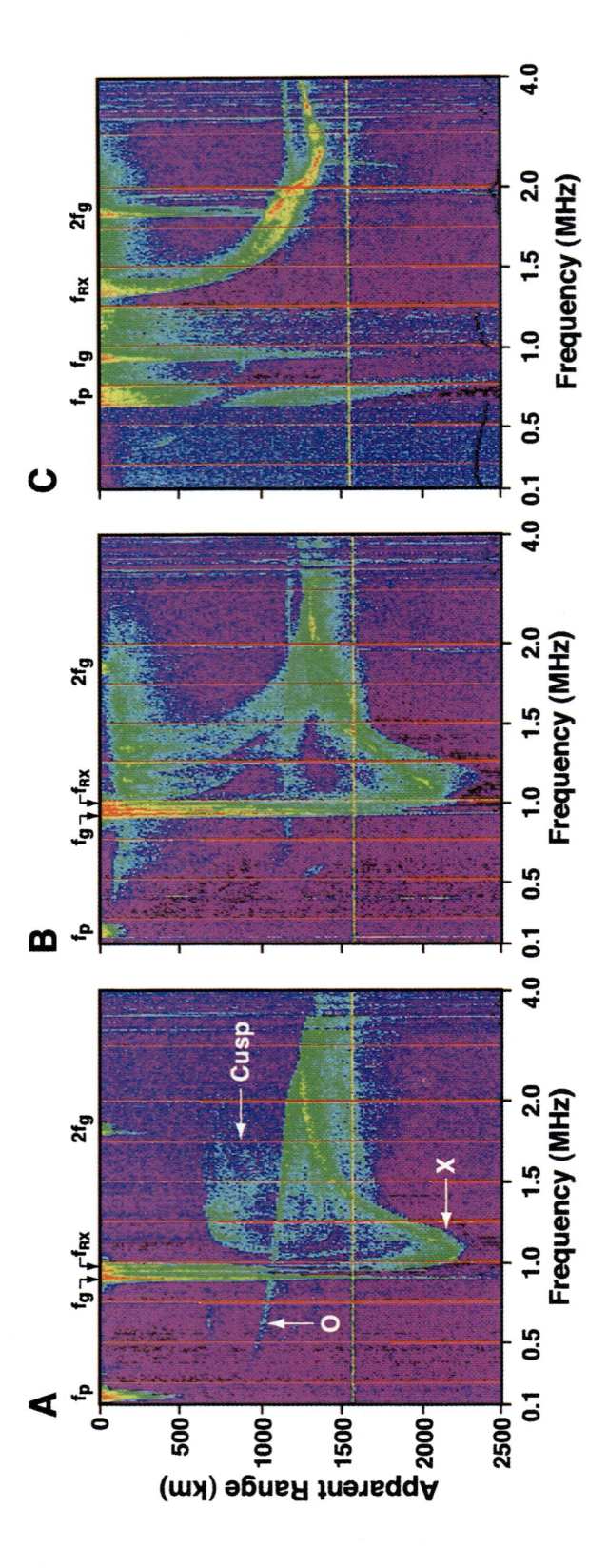

Plate 2. Ionograms from ISIS 2 during an encounter with the polar cusp. Panels A and B are from times just before the cusp is encountered, while panel C is when ISIS 2 is well situated in the cusp. Panel A shows the O and X traces, which are echoes from the ionosphere below the spacecraft and the echoes from the cusp.

RAY TRACING AT 100 kHz

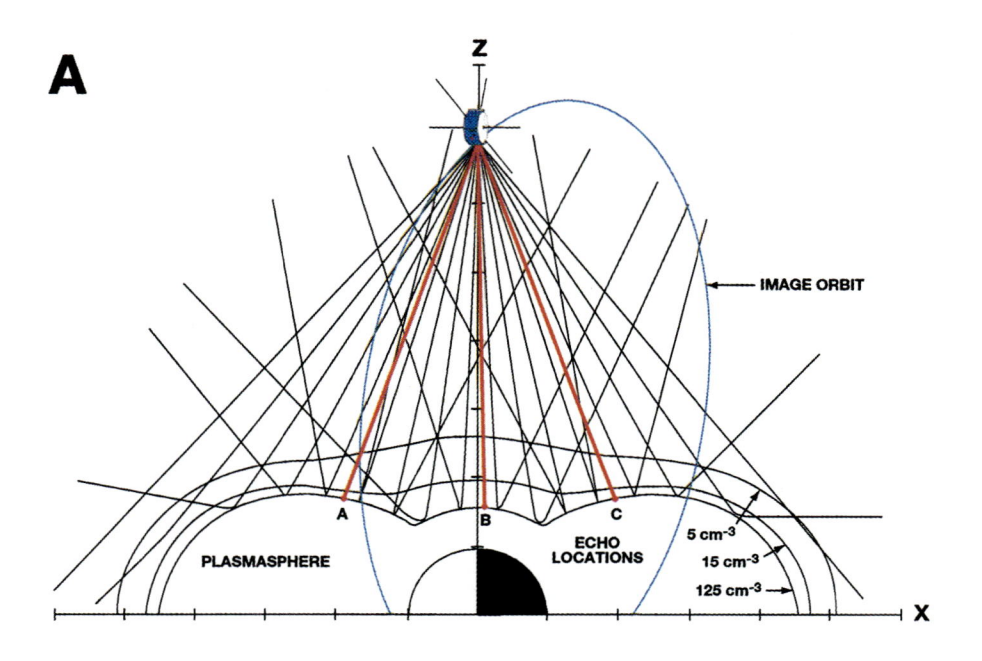

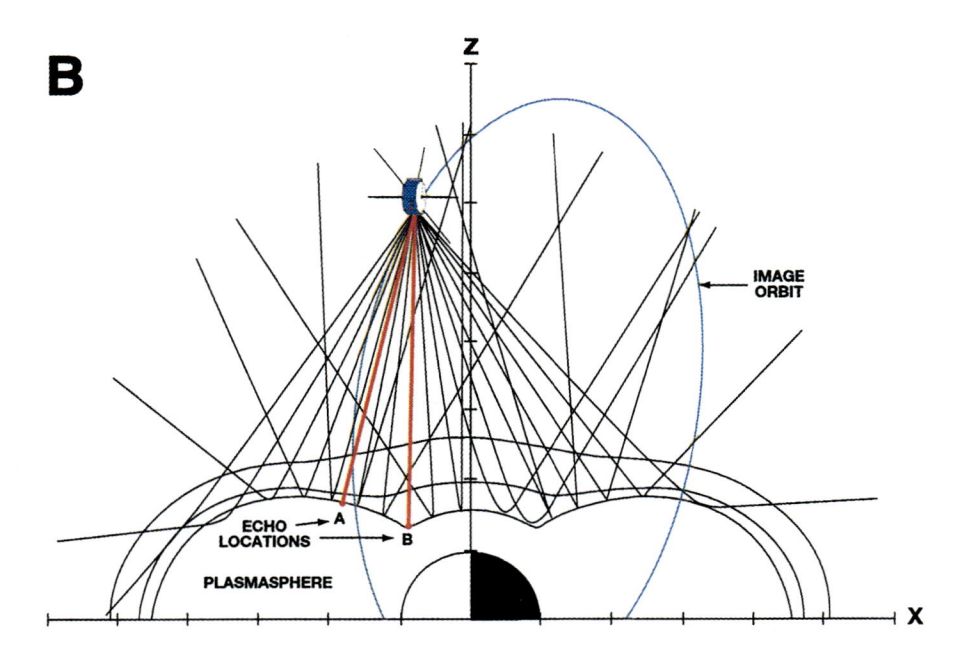

Plate 3. Ray tracing calculations at 100 kHz of O mode waves from two locations in the IMAGE orbit. The rays' paths in red are reflections from density regions that return to the spacecraft and therefore would be measured as echoes. Panel A shows the location of three echoes from the day- and nightside plasmapause and polar ionosphere. Panel B show only two echoes, one from the dayside plasmapause and the other from the trough region.

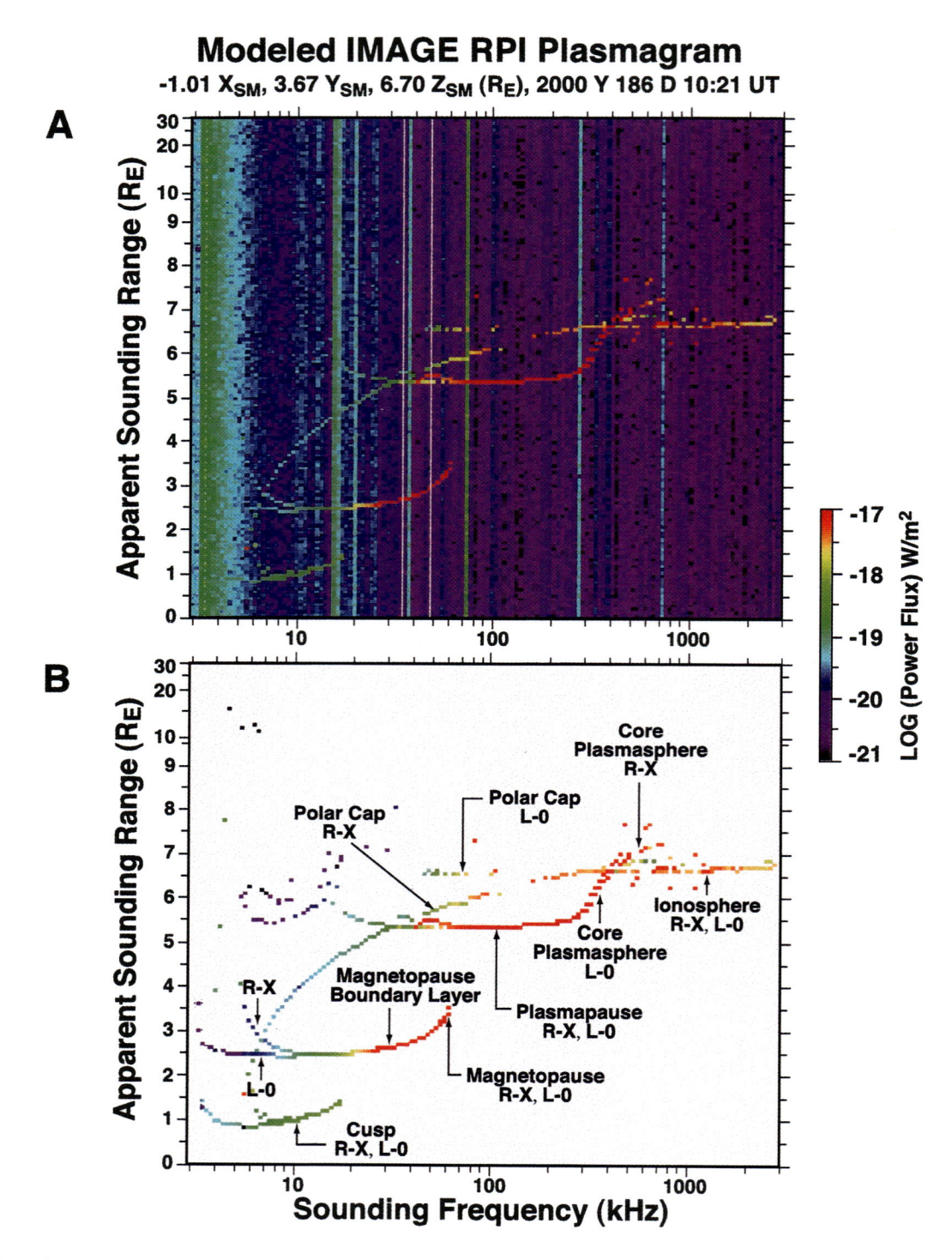

Plate 4. Two simulated plasmagrams when IMAGE is near apogee. Panel A contains the echoes from the ray tracing calculations, the actual receiver noise level measured during calibration, and a model of the local thermal plasma line. In order to better delineate the echoes from the background, panel B contains only the ray tracing calculations. Each characteristic plasma region produces a family of echoes which are labeled, along with the propagation mode, in panel B. This simulation illustrates the structure of the echoes anticipated from a routine sounding by RPI.

uted to the vertical direction was then used to deduce the "true height" electron density profile between the satellite position and the height of the ionospheric peak density, *i.e.*, the "topside" electron density profile.

In the density cavity of the magnetosphere, RPI radio soundings will be generated in nearly a spherical pattern. Plasma density structures will reflect these waves producing echoes from multiple directions for each sounder frequency. Modeling these echoes have confirmed that RPI will obtain many echoes from a variety of directions. The challenge has been to visualize these three-dimensional measurements. The "echo map" techniques will allow us to facilitate the display of RPI echoes over specific frequency ranges from many different directions in a fish-eye-like plot. This visual technique should be particularly effective for specific RPI frequencies in which steep density gradients exist (*e.g.*, plasma boundaries of the plasmapause, magnetopause, and polar cusp). The echo maps, which form the RPI images, can be created from a single frequency sounding or multiple frequency soundings, and therefore range in time from a few seconds to several minutes.

RPI will operate at frequencies between 3 kHz and 3 MHz and will provide quantitative N_e values from 10^{-1} to 10^5 cm^{-3} with high precision. Table 1 presents a list of representative magnetospheric targets for the RPI. Over many years of in-situ measurements, the average range in density of these magnetospheric targets has been determined and is shown in Table 1. The corresponding frequency range for RPI is also shown. All targets are well within the RPI sounder frequency range. An obvious difference between ionospheric and magnetospheric sounding is in the distances, or ranges, involved between the sounder and the reflecting medium. In the ionosphere, the range is on the order of hundreds to thousands of km; in the magnetosphere, the range will be many R_E, *i.e.*, greater by several factors of 10. As shown in the feasibility studies by *Green et al.* [1993] and *Calvert et al.*, [1995] the optimal range to the targets is from 1 to 5 R_E. RPI will detect these targets sufficiently often to define their structures and motions.

5. RAY TRACING TECHNIQUE

Simulations of magnetospheric radio sounding have been studied by a number of authors [see for example; *Ondo et al.*, 1978; *Fung and Green*, 1996; *Reiff et al.*, 1996; and *Green et al.*, 1998]. Using similar techniques as in these previous studies, ray tracing calculations have been performed to simulate the return pulses or echoes from RPI transmissions on IMAGE located in a model magnetosphere. The first formulation of ray-tracing equations that were suitable for integration by standard numerical methods using computers was done in the mid-1950's by Haselgrove. *Haselgrove* [1955] developed six first-order differential equations that describe the motion of the energy

Table 1: Representative RPI Targets

Target	Electron Density Range (cm^{-3})	Plasma Freq. Range (kHz)
Magnetosphere' boundary layer	5-20	20-40
Polar Cusp	7-30	25-50
Magnetopause	10-60	30-70
Plasmapause	30-800	50-250
Plasmasphere	800-8x10^3	250-800
Ionosphere	8x10^3-1x10^5	800-3000

in electromagnetic waves propagating in an anisotropic medium in three dimensions. The ray tracing equations of Haselgrove allowed for the inclusion of one of many formulations for the index of refraction (N). The expression for N used in the RPI simulations is for radiation in a cold plasma that has been developed by *Stix* [1992].

In order to obtain realistic echoes, the plasma and magnetic field models that are used in the ray tracing code must be acceptable representations of the physical environment that influences the radiation. The magnetic field model employed in the ray tracing simulations is a simple dipole model, while the plasma density model is a combination of models of diffusive equilibrium by *Angerami and Thomas* [1964], of the ionosphere and plasmasphere by *Kimura* [1966], of the plasmapause by *Aikyo and Ondoh* [1971], and of the magnetopause by *Roelof and Sibeck* [1993].

5.1. Ray Path Calculations

An example of the ray tracing calculations used to simulate RPI observations is shown in Plate 3. Plate 3 shows selected ray path calculations at 100 kHz for two locations in the IMAGE orbit for O mode rays. A ray path that reflects at a wave cutoff in which the density gradient vector is along the ray path will return to the spacecraft and therefore be measured as an echo. The travel times and ray path directions for all echoes are calculated and are used in creating a simulated plasmagram (next section). In Plate 3, all echoes are shown in red. In Plate 3, panel A, echo locations A, B, and C are located at the dayside and nightside plasmapause (A and C) and at the polar cap ionosphere (B). At a different location in the IMAGE orbit, Plate 3, panel B, shows only two echoes with reflection locations at the dayside plasmapause (A) and in the trough region (B). The ray tracing calculations in Plate 3 clearly illustrate that not all rays are echoes and that the spacecraft position plays an important role in what regions of the magnetosphere will produce measurable echoes.

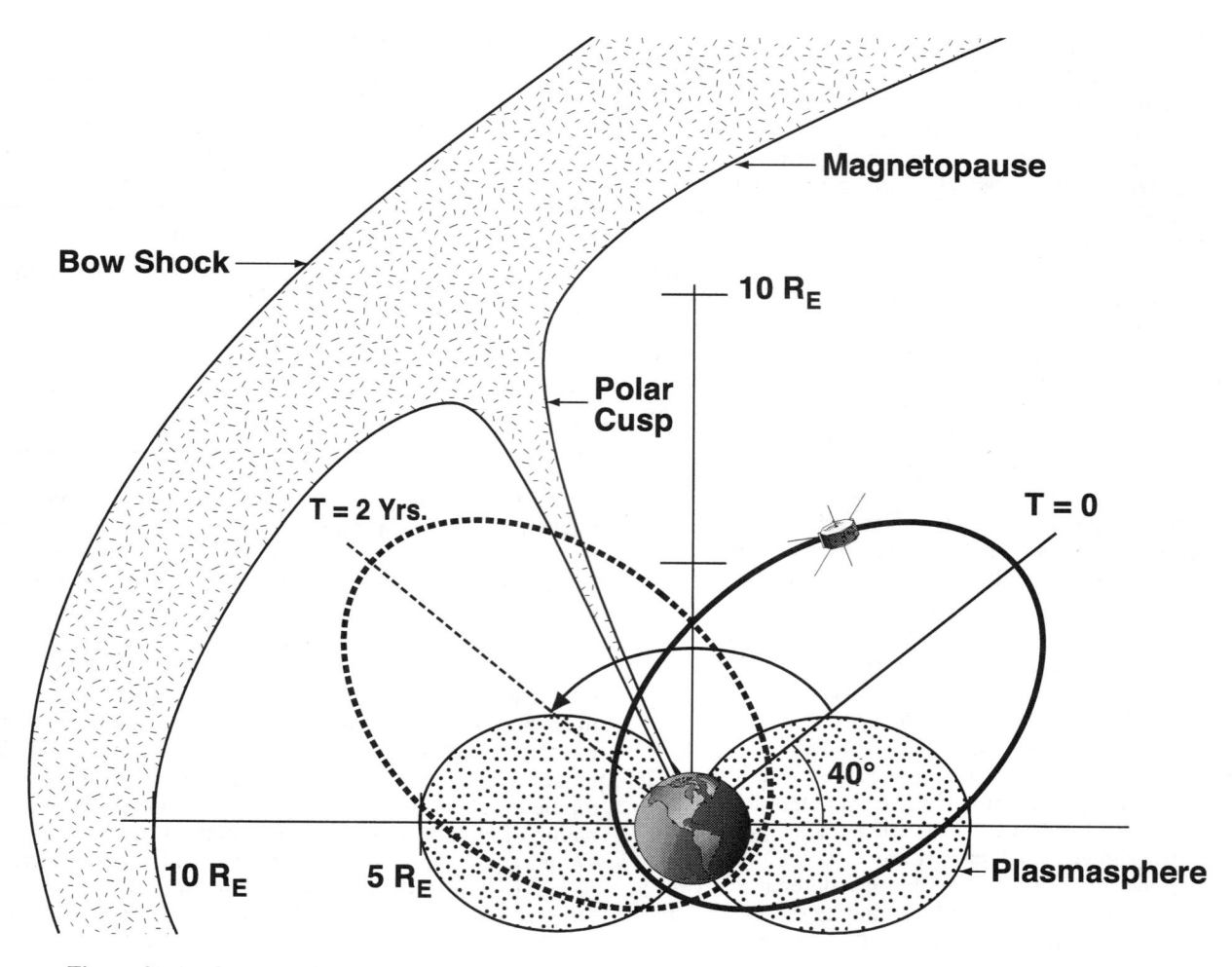

Figure 3. A schematic of the orbit of the IMAGE spacecraft showing the evolution of the orbit over a two-year period. The IMAGE spacecraft spends much of its time in the magnetospheric density cavity between the plasmasphere and the magnetopause.

5.2. Plasmagrams

The primary presentation of RPI data will be in the form of plasmagrams, which are the magnetospheric analogs of ionograms. A plasmagram is a plot of echo power as a function of frequency and echo delay. Simulated magnetospheric X and O-mode echoes, as anticipated to be received by RPI and derived from ray tracing calculations, are presented in Plate 4. Both plasmagrams in Plate 4 are identical except that the plasmagram in Panel A is the best representation of the RPI observations since it contains, in addition to the simulated echoes, the measured RPI instrument noise level and the local thermal plasma emission at f_p. The local thermal plasma emission was modeled after the results of *Meyer-Vernet and Perche*, [1989]. Panel B of Plate 4 shows only the simulated echoes that are above the background noise levels of panel A. The RPI simulated echoes are presented in the form of echo time delay (t), expressed in terms of apparent range (right scale), as a function of the sounder frequency (bottom scale). The apparent range corresponds to ct/2 where c is the speed of light. The intensity is color-coded and has been calculated for each echo in this Plate. The location of the spacecraft for this simulation is at approximately 19 hours magnetic local time near apogee (approximately 8 R_E) at about 60° magnetic latitude.

Moderately intense to intense natural noises, in RPI's frequency range, consist of Type III solar noise bursts and storms, auroral kilometric radiation (AKR), and the non-thermal continuum (escaping and trapped). These natural noises have the potential to effect RPI's ability to clearly distinguish the echoes that will be generated. However, due to several mitigation strategies and a favorable orbit, RPI should be able to operate over nearly the entire frequency range over a significant part of the orbit [see *Reinisch et al.*, 1999].

It was commonly believed that the non-thermal continuum radiation uniformly filled the Earth's magnetosphere and posed the greatest problem for observing the most distant RPI echoes. In a recent study by *Green and Boardsen* [1999], the angular distribution of the non-thermal continuum radiation was studied with observations from the Hawkeye spacecraft and modeled with ray tracing calculations. From these results, it is clear that the trapped continuum radiation does not uniformly illuminate the magnetospheric cavity but is mainly confined to low latitudes. Thus, RPI should rarely encounter continuum radiation since IMAGE will be in a high-inclination orbit.

The simulated echoes in panel B of Plate 4 are labeled with their appropriate propagation modes (R-X, L-O) and in regions of the magnetosphere in which the echoes have occurred (*i.e.:* the plasmasphere, plasmapause, magnetopause, polar cap, and polar cusp). The plasmagram contains the simulated echo measurements from a complete RPI instrument cycle when the spacecraft is near apogee. An RPI instrument cycle (transmission plus echo reception) typically takes several minutes to complete. As can be seen, the resulting plasmagram can be quite complicated with echoes from several regions over an extended frequency range and a wide range of distances.

In order to reduce the complexity of the plasmagram of Plate 4 and to illustrate how these echoes are used in the analysis of magnetospheric density structures and motions, simulated echoes from the magnetopause boundary layer have been isolated and shown in Figure 4. Panels B, D, and F on the right, in Figure 4, show three distinct density profiles which may occur in the boundary layer and magnetopause under different solar wind and interplanetary magnetic field conditions. The spacecraft in panels B, D, and F show the location of the RPI relative to these density structures. The three left-hand panels, A, C, and E in Figure 4, depict the corresponding simulated plasmagrams. For comparison, the echo structure of the magnetopause boundary layer in panel C of Figure 4 can easily be seen in plasmagram of Plate 4.

The plasmagram in Figure 4 panel A shows the start of the simulated echoes occurring at the local f_p and with no delay and monotonically increasing to high frequency and longer delay time. The corresponding N_e used in this simulation is shown in Figure 4, panel B. It is clear that the N_e increases monotonically from the spacecraft to the magnetopause. The simulated plasmagrams of Figure 4 (panels C and E) have echoes that start at a large time delay (apparent range) because the transmitted wave going toward the magnetopause initially propagates into a region of decreasing N_e and must travel some distance before encountering an N_e value capable of causing total reflection. The delay time then decreases with increasing frequency for a short frequency interval, even though the wave is penetrating deeper into the medium, because the wave propagation speed increases as the frequency increases. When the frequency of the sounder exceeds the maximum plasma frequency at the magnetopause, the echo trace stops. The additional density in the boundary layer structure of panel F has the effect of slowing down (increasing delay time) the echoes at the lower frequencies causing a "cusp" like structure in the corresponding plasmagram of panel E. Panels A, C, and E illustrate the significant differences in the echo traces since they are sensitive indicators of the density structures involved in the reflection.

As in the case of ionospheric topside-sounders, the N_e profiles from RPI will be deduced from the reflection traces (Plate 4, panels A, C, and E of Figure 4) using the standard inversion techniques developed by *Jackson* [1969a] and *Huang and Reinisch* [1982]. This technique has been applied to a number of echoes of the simulated plasmagrams and the resulting density profiles are nearly identical to those used in the ray tracing calculations. RPI will be transmitting and measuring echoes in both the X and O modes. Wave propagation in these two plasma modes is governed by two distinctively different dispersion relationships. The inversion process itself that is used for magnetosphere echo analysis will depend on the dispersion relationships and not models of the plasma environment. As long as the electron density profile between the satellite and the reflection point is monotonic, the inversion process is unique. If density minima (valleys) exist along the ray path, the profiles may show errors.

In addition to deriving N_e profiles, it will be possible to construct N_e contours in the orbital plane as was done using ionospheric topside sounding [see for example, *Warren*, 1969; *Benson and Akasofu*, 1984]. Using the RPI measurements, instantaneous locations of magnetospheric boundaries (magnetopause and plasmapause) and density profiles along a given radio echo ray path can be obtained by inverting the plasmagram trace.

The RPI on IMAGE will also be able to deduce magnetospheric properties at extremely large distances. A number of studies based on the Alouette/ISIS data have shown that signal enhancements of 20-40 dB (relative to vertical propagation) can be obtained from echoes resulting from waves guided along N_e field-aligned irregularities [*Lockwood*, 1973, and references therein]. Such ducted echoes have been used to demonstrate the extended field-aligned nature of equatorial ionospheric bubbles and to determine N_e profiles within them [*Dyson and Benson*, 1978]. *Calvert* [1981] has shown that a sounder capable of generating 5 μW within a magnetospheric duct would detect long-range magnetospheric echoes suitable for remote N_e measurements near the cusp and auroral zone and the detection of dayside compressions and nighttime expansions of the geomagnetic field.

6. CONCLUSIONS

The technique of probing remote plasma characteristics by radio waves has been well established for ionospheric studies, but no such U.S. instrument has yet been flown in

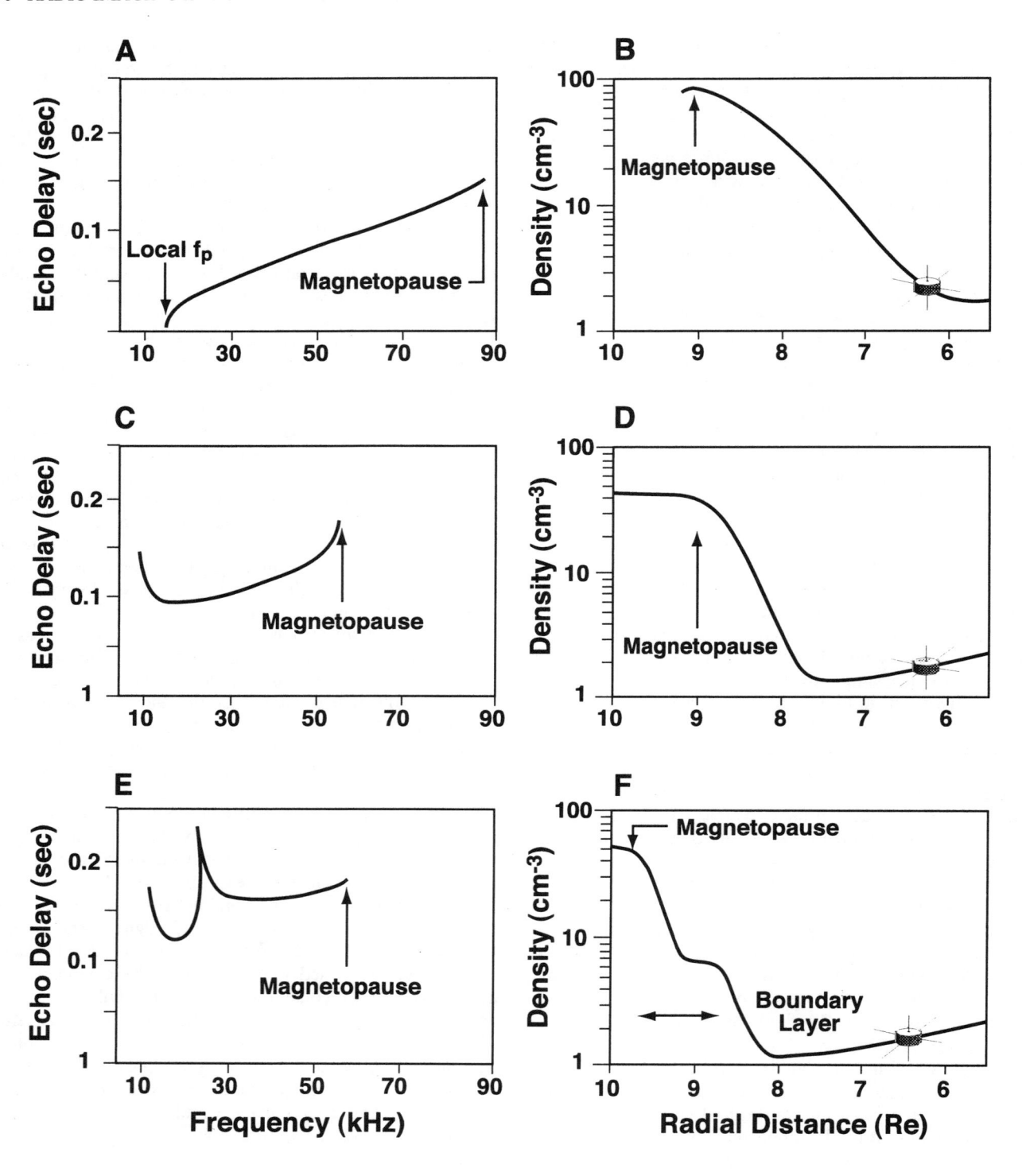

Figure 4. Simulations of three representative magnetopause boundary layers (panels A, C, E) and their corresponding density structures (panels B, D, and E). The echoes are sensitive indicators of the boundary layer structure and location.

the magnetosphere. The successful ground-based ionosondes and the Alouette-ISIS programs have spawned numerous investigations, significantly contributing to our current understanding of the ionosphere [see for example, Hunsucker, 1991]. With the new generation of ground-based digital radio signal instrumentation and processing techniques developed over the last several years, it is now possible to apply the same radio wave sounding techniques to measure distant plasma parameters in the Earth's magnetosphere on the upcoming IMAGE mission.

The overall scientific objective of the IMAGE mission is to understand the global magnetospheric response to changing conditions in the solar wind. This goal will be accomplished by using a variety of remote sensing techniques. The RPI instrument on the IMAGE mission is a swept frequency adaptive sounder, with on-board signal processing that transmits and receives coded electromagnetic pulses over the frequency range from 3 kHz to 3 MHz. The primary mode of operation for RPI is to generate radio pulses that propagate as free-space waves through the magnetosphere and are reflected upon encountering their plasma cutoff frequencies. The RPI will measure the amplitudes, Doppler shift, and direction of arrival as a function of frequency and echo delay. The echo arrival angles will be calculated from the amplitudes and phases of the signals from three orthogonal receiving dipole antennas.

Ray tracing calculations have been used to determine the feasibility and limitations of the RPI and to simulate the frequency time structure of the returning RPI echoes, which will greatly aid in the analysis of the actual data. At times when RPI is situated in the density cavity of the magnetosphere, the ray tracing calculations illustrate how it will be able to simultaneously determine the location and dynamics of remote boundaries such as the plasmapause and magnetopause. In addition, it will be able to provide N_e profiles in different directions on time scales of a few minutes or less. In summary, the RPI should be able to provide unprecedented global magnetospheric observations.

Acknowledgments. The work at Raytheon ITSS under NASA contract NASW-97002.

REFERENCES

Aikyo, K., and T. Ondoh, Propagation of nonducted VLF waves in the vicinity of the plasmapause, *J. Radio Res. Labs.,* 18, 153, 1971.

Angerami, J. J., and J. O. Thomas, The distribution of ions and electrons in the Earth's exosphere, *J. Geophys. Res.,* 69, 4537, 1964.

Benson, R. F., and S.-I. Akasofu, Auroral kilometric radiation/aurora correlation, *Radio Sci.,* 19, 527-541, 1984.

Berkner, L. V., The international geophysical year, *Proc. IRE,* 47, 133-136, 1959.

Breit, G., and M. A. Tuve, A test for the existence of the conducting layer, *Phys. Rev.,* 28, 554-575, 1926.

Brown, J. N., Automatic sweep-frequency ionosphere recorder, Model C-4, *Proc. IRE,* 47, 296-300, 1959.

Budden, K. G., *The propagation of Radio Waves,* Cambridge Univ. Press, New York, 1985.

Calvert, W., The detectability of ducted echoes in the magnetosphere, *J. Geophys. Res.,* 86, 1609-1612, 1981.

Calvert, W., R. F. Benson, D. L. Carpenter, S. F. Fung, D. L. Gallagher, J. L. Green, D. M. Haines, P. H. Reiff, B. W. Reinisch, M. F. Smith and W. W. L. Taylor, The feasibility of radio sounding in the magnetosphere, *Radio Sci.,* 30, 1577-1595, 1995.

Dyson, P. L. and R. F. Benson, Topside sounder observations of equatorial bubbles, *Geophys. Res. Lett.,* 5, 795-798, 1978.

Dyson, P. L. and J. D. Winningham, Topside ionospheric spread F and particle precipitation in the dayside magnetospheric clefts, *J. Geophys. Res.,* 79, 5219-5230, 1974.

Etcheto, J., G. Belmont, P. Canu, and J. G. Trotignon, Active sounder experiments on GEOS and ISEE, in Active Experiments in Space (Proc. Alpbach, Austria 24-28 May 1983 symposium), *ESA SP-195,* pp. 39-46, 1983.

Fung, S. F., and J. L. Green, Global imaging and radio remote densing of the magnetosphere, radiation belts: Models and standards, *AGU Monograph 97,* 285-290, 1996.

Green, J. L., R. F. Benson, W. Calvert, S. F. Fung, P. H. Reiff, B. W. Reinisch, and W. W. L. Taylor, A study of radio plasma imaging for the proposed IMI mission, NSSDC Technical Publication, February 1993.

Green, J. L., W. W. L. Taylor, S. F. Fung, R. F. Benson, W. Calvert, B. Reinisch, D. L. Gallagher, and P. Reiff, Radio remote sensing of magnetospheric plasmas, measurement techniques for space plasmas: Fields, *AGU Monograph 103,* 193-198, 1998.

Green, J. L., and S. A. Boardsen, Confinement of non-thermal continuum radiation to low latitudes, *J. Geophys. Res.,* 104, 10307-10316, 1999.

Haselgrove, J., Ray theory and a new method for ray tracing, London Phys. Soc., Report of Conference on the Physics of the Ionosphere, 355, 1955.

Henderson, M. G., G. D. Reeves, H. E. Spence, R. B. Sheldon, A. M. Jorgensen, J. B. Blake, and J. F. Fennell, First energetic neutral atom images from Polar, *Geophys. Res. Lett.,* 24, 1167-1170, 1997.

Hoegy, W. R., and R. F. Benson, DE/ISIS conjunction comparisons of high-latitude electron density features, *J. Geophys. Res.,* 93, 5947-5954, 1988.

Huang, X., and B. W. Reinisch, Automatic calculation of electron density profiles from digital ionograms. 2. True height inversion of topside ionograms with the profile-fitting method, *Radio Science,* 17, 4, 837-844, 1982.

Hunsucker, R. D., *Radio Techniques for Probing the Terrestrial Ionosphere,* Vol. 22, Phys. Chem. Space, Springerverlage, Berlin, 1991.

Jackson, J. E., Analysis of topside ionograms, NASA/GSFC Report X-615-67-452, Greenbelt, MD, Sept. 1967.

Jackson, J. E., Reduction of topside ionograms to electron-density profiles, *Proc. IEEE*, 57, 960-976, 1969a.

Jackson, J. E., Comparison between topside and ground-based soundings, *Proc. IEEE*, 57, 976-985, 1969b.

James, H. G., Guided Z-mode propagation observed in the OEDIPUS-A tethered rocket experiment, *J. Geophys. Res.*, 96, 17865-17878, 1991.

Kimura, I., Effects of ions on whistler mode ray tracing, *Radio Science, 1 (New Series)*, 3, 269, 1966.

Lockwood, G. E. K., Side band and harmonic radiation from topside sounders, *J. Geophys. Res.*, 78, 2244-2250, 1973.

Lui, A. T. Y., and C. D. Anger, Uniform belt of diffuse auroral emissions seen by the ISIS 2 scanning photometer, *Planet. Space Sci.*, 21, 799-809, 1973.

Maruyama, T., and N. Matuura, Longitudinal variability of annual changes in activity of equatorial Spread F and plasma bubbles, *J. Geophys. Res.*, 89, 10903-10912, 1984.

Meyer-Vernet, N. and C. Perche, Toolkit for antennae and thermal noise near the plasma frequency, *J. Geophys. Res.*, 94, 2405, 1989.

Ondoh, T., Y. Nakamura, and T. Koseki, Feasibility of plasmapause sounding from a geostationary satellite, *Space Sci. Instrum.*, 4, 57-71, 1978.

Oya, H., A. Morioka, and T. Obara, Leaked AKR and terrestrial hectometric radiations discovered by the plasma wave and planetary plasma sounder experiments on board the Ohzora (EXOS-C) satellite-instrumentation and observation results of plasma wave phenomena, *J. Geomagn. Geoelectr.*, 37, 237-262, 1985.

Oya, H., and T. Ono, Stimulation of plasma waves in the magnetosphere using satellite JIKIKEN (EXOS-B), Part II: Plasma density across the plasmapause, *J. Geomagn. Geoelectr.*, 39, 591-607, 1987.

Oya, H., A. Morioka, K. Kobayashi, M. Iizima, T. Ono, H. Miyaoka, T. Okada, and T. Obara, Plasma wave observation and sounder experiments (PWS) using the Akebono (EXOS-D) satellite – Instrumentation and initial results including discovery of the high altitude equatorial plasma turbulence, *J. Geomagn. Geoelectr.*, 42, 411-442, 1990.

Perraut, S., H. de Feraudy, A. Roux, P. M. E. Decreau, J. Paris, and L. Matson, Density measurements in key regions of the Earth's magnetosphere: Cusp and auroral region, *J. Geophys. Res.*, 95, 5997-6014, 1990.

Reiff, P. H., C. B. Boyle, J. L. Green, S. F. Fung, R. F. Benson, W. Calvert, and W. W. L. Taylor, Radio Sounding of Multiscale Plasmas, in *Physics of Space Plasmas, 14*, T. Chang and J. R. Jasperse, Ed., MIT Press, Cambridge, MA, 415-429, 1996.

Reinisch, B. W., D. M. Haines, K. Bibl, G. Cheney, I. A. Galkin, X. Huang, S. H. Myers, G. S. Sales, R. F. Benson, S. F. Fung, J. L. Green, W. W. L. Taylor, J.-L. Bougeret, R. Manning, N. Meyer-Vernet, M. Moncuquet, D. L.Carpenter, D. L. Gallagher, and P. Reiff, The Radio Plasma Imager investigation on the IMAGE spacecraft, *Submitted to Space Sciences Review*, 1999.

Roelof, E. C., D. G. Mitchell, and D. J. Williams, Energetic neutral atoms (E~50 keV) from the ring current: IMP 7/8 and ISEE 1, *J. Geophys. Res.*, 90, 10991-11008. 1985.

Roelof, E. C., and D. G. Sibeck, Magnetopause shape as a bivariate function of interplanetary magnetic field Bz and solar wind dynamic pressure, *J. Geophys. Res.*, 98, 21421-21450, 1993.

Scarf, F. L., *Space Science in the Twenty-First Century—Imperative for the Decades 1995 to 2015, Volume 1 Solar and Space Physics*, National Academy Press, Washington D.C., 1988.

Shepherd, G. G., J. H. Whitteker, J. D. Winningham, J. H. Hoffman, E. J. Maier, L. H. Brace, J. R. Burrows, and L. L. Cogger, The topside magnetospheric cleft ionosphere observed from the ISIS 2 spacecraft, *J. Geophys. Res.*, 81, 6092-6102, 1976.

Shuiskaya, F. K., Yu. I. Galperin, A. A. Serov, N. V. Baranets, Yu. V. Kushnerevsky, G. V. Vasil'ev, S. A. Pulinets, M. D. Fligel, and V. V. Selegey, Resonant heating of the ionospheric plasma by powerful radiopulses aboard the Intercosmos-19 and Cosmos-1809 satellites, *Planet. Space Sci.*, 38, 173-180, 1990.

Stix, T. H., *Waves in Plasmas*, AIP, New York, 1992.

Stone, R. G., J. L Bougeret, J. Caldwell, P. Canu, Y. de Donchy, N. Cornilleau-Wehrlin, M. D. Desch, J. Fainberg, K. Goetz, and M. L. Goldstein, The unified radio and plasma wave investigation, *Astron. Astrophys. Suppl. Ser.*, 92, 291-316, 1992.

Warren, E. S., The topside ionosphere during geomagnetic storms, *Proc. IEEE*, 57, 1029-1036, 1969.

Whitteker, J. H., L. H. Brace, E. J. Maier, J. R. Burrows, W. H. Dodson, and J. D. Winningham, A snapshot of the polar ionosphere, *Planet. Space Sci.*, 24, 25-32, 1976.

Williams, D., E. C. Roelof, and D. G. Mitchell, Global magnetospheric imaging, *Rev. Geophys.*, 30, 183-208, 1992.